BIOFUELS AND BIOENERGY

BIOFUELS AND BIOENERGY

Opportunities and Challenges

Edited by

Baskar Gurunathan

Department of Biotechnology, St. Joseph's College of Engineering, Chennai, Tamil Nadu, India

Renganathan Sahadevan

Department of Biotechnology, Alagappa College of Technology, Anna University, Chennai, Tamil Nadu, India

Zainul Akmar Zakaria

School of Chemical and Energy Engineering, Faculty of Engineering, Universiti Teknologi Malaysia, UTM Johor Bahru, Johor, Malaysia

ELSEVIER

Elsevier
Radarweg 29, PO Box 211, 1000 AE Amsterdam, Netherlands
The Boulevard, Langford Lane, Kidlington, Oxford OX5 1GB, United Kingdom
50 Hampshire Street, 5th Floor, Cambridge, MA 02139, United States

Notices

Knowledge and best practice in this field are constantly changing. As new research and experience broaden our understanding, changes in research methods, professional practices, or medical treatment may become necessary.

Practitioners and researchers must always rely on their own experience and knowledge in evaluating and using any information, methods, compounds, or experiments described herein. In using such information or methods they should be mindful of their own safety and the safety of others, including parties for whom they have a professional responsibility.

To the fullest extent of the law, neither the Publisher nor the authors, contributors, or editors, assume any liability for any injury and/or damage to persons or property as a matter of products liability, negligence or otherwise, or from any use or operation of any methods, products, instructions, or ideas contained in the material herein.

Library of Congress Cataloging-in-Publication Data
A catalog record for this book is available from the Library of Congress

British Library Cataloguing-in-Publication Data
A catalogue record for this book is available from the British Library

ISBN: 978-0-323-85269-2

For information on all Elsevier publications visit our website at https://www.elsevier.com/books-and-journals

Publisher: Candice Janco
Acquisitions Editor: Peter Adamson
Editorial Project Manager: Michelle Fisher
Production Project Manager: Sojan P. Pazhayattil
Cover Designer: Vicky Pearson

Typeset by TNQ Technologies

Contents

III

Bioethanol and biodiesel production

5. Refining lignocellulose of second-generation biomass waste for bioethanol production

E. RAJA SATHENDRA, R. PRAVEENKUMAR, BASKAR GURUNATHAN, S. CHOZHAVENDHAN, AND MANI JAYAKUMAR

6. Various methods of biodiesel production and types of catalysts

V. AMIRTHAVALLI, ANITA R. WARRIER, AND BASKAR GURUNATHAN

7. Biodiesel production using enzymatic catalyst

NURFADHILA NASYA BINTI RAMLEE, HILMAN IBNU MAHDI, AND NUR IZYAN WAN AZELEE

8. Importance of nanocatalyst and its role in biofuel production

ELSA CHERIAN, G. KALAVATHY, T. JAYASREE JOSHI, M. GRACE LYDIA PHOEBE, AND BASKAR GURUNATHAN

9. Potentials and challenges in biodiesel production from algae—technological outlook

S. CHOZHAVENDHAN, G. KARTHIGADEVI, R. PRAVEENKUMAR, M. ANISKUMAR, MANI JAYAKUMAR, AND BASKAR GURUNATHAN

IV

Biogas production

10. Sustainable approaches for biohydrogen and biogas production from corn wastes: prospects and challenges

ACHMAD SYAFIUDDIN, FIFI KHOIRUL FITRIYAH, DIFRAN NOBEL BISTARA, ABDUL HAKIM ZAKKIY FASYA, AND ZAINUL AKMAR ZAKARIA

11. Biogas production potential in India, the latest biogas upgradation techniques and future application in a fuel cell

THIRUSELVI DEVARAJ, YUVARANI MANI, SALMA AATHIKA, SHANMUGAM PALANIYANDI, BASKAR GURUNATHAN, AND SIVANESAN SUBRAMANIAN

12. Anaerobic digestion of agrowastes: end-of-life step of a biorefinery

IPSITA CHAKRAVARTY AND SACHIN A. MANDAVGANE

13. Biomass gasification: a step toward cleaner fuel and chemicals

MAHENDRA RAM AND MONOJ KUMAR MONDAL

V
Bioenergy production using pyrolysis

14. Pyrolysis: an alternative approach for utilization of biomass into bioenergy generation

GOUTAM KISHORE GUPTA AND MONOJ KUMAR MONDAL

VI

Bioenergy production using microbial fuel cell

15. Production of biofuels in a microbial electrochemical reactor

MAMTA PAL, RAJAT GIRI, AND RAKESH KUMAR SHARMA

16. Advances in microbial fuel cell technology for zero carbon emission energy generation from waste

MPUMELELO THOMAS MATSENA AND EVANS MARTIN NKHALAMBAYAUSI CHIRWA

17. Concomitant bioenergy production and wastewater treatment employing microbial electrochemical technologies

M.M. GHANGREKAR, SANTOSH KUMAR, AZHAN AHMAD, AND SOVIK DAS

VII

Microalgae as source of biofuels and bioenergy

18. Microalgae—the ideal source of biofuel

C.V. VIDHYA

19. Principles and technology advances in wastewater treatment through microalgae for bioenergy production

SAPTASWA BISWAS, RAHULKUMAR MAURYA, AND KAUSTUBHA MOHANTY

20. Weeds as a renewable bioresource: prospects for bioconversion to biofuels and biomaterials through a cascade of approaches

RUPAM KATAKI AND MANIKA D. KATAKI

21. Critical parameters affecting large-scale production of microalgal biomass in outdoor open raceway ponds

E. BALDEV, D. MUBARAK ALI, R. SATHYA, AND N. THAJUDDIN

22. Fostering single cell oil synthesis by de novo and ex novo pathway in oleaginous microorganisms for biodiesel production

RADHA PANJANATHAN, SANJANA NARAYANAN, ANGANA CHAUDHURI, SAMEENA ANJUM, AND RAMANI KANDASAMY

23. Production of biofuel from genetically modified microalgal biomass and its effects on environment and public health

PRATYUSH KUMAR DAS, BIDYUT PRAVA DAS, PATITAPABAN DASH, AND BASKAR GURUNATHAN

VIII

Techno-economic analysis and life cycle assessment

Contributors

Salma Aathika Department of Applied Science and Technology, Alagappa College of Technology, Anna University, Chennai, Tamil Nadu, India

Azhan Ahmad Department of Civil Engineering, Indian Institute of Technology Kharagpur, Kharagpur, West Bengal, India

D. Mubarak Ali School of Life Sciences, B.S. Abdur Rahman Crescent Institute of Science and Technology, Chennai, Tamil Nadu, India

V. Amirthavalli Department of Petroleum Engineering, Academy of Maritime Education and Training, Chennai, Tamil Nadu, India

M. Aniskumar V.S.B. Engineering College, Karur, Tamil Nadu, India

Sameena Anjum Biomolecules and Biocatalysis Laboratory, Department of Biotechnology, School of Bioengineering, SRM Institute of Science and Technology, Kattankulathur, Kanchipuram, Tamil Nadu, India

Renu Bala Department of Chemical Engineering and Technology, Indian Institute of Technology, Banaras Hindu University, Varanasi, Uttar Pradesh, India

E. Baldev Sreedhareeyam Farmherbs India Pvt. Ltd., Ernakulam, Kerala, India

Debraj Bhattacharyya Department of Civil Engineering, Indian Institute of Technology Hyderabad, Kandi, Telangana, India

Difran Nobel Bistara Institute of Research and Community Service (LPPM), Universitas Nahdlatul Ulama Surabaya, Surabaya, East Java, Indonesia

Saptaswa Biswas School of Energy Science and Engineering, Indian Institute of Technology Guwahati, Guwahati, India

Gustavo Amaro Bittencourt Federal University of Paraná, Centro Politécnico, Curitiba, Paraná, Brazil

Ipsita Chakravarty Visvesvaraya National Institute of Technology, Nagpur, Maharashtra, India

Angana Chaudhuri Biomolecules and Biocatalysis Laboratory, Department of Biotechnology, School of Bioengineering, SRM Institute of Science and Technology, Kattankulathur, Kanchipuram, Tamil Nadu, India

Elsa Cherian Department of Food Technology, Saintgits College of Engineering, Kottayam, Kerala, India

S. Chozhavendhan Department of Biotechnology, V.S.B. Engineering College, Karur, Tamil Nadu, India

Pratyush Kumar Das Centre for Biotechnology, Siksha 'O' Anusandhan (Deemed to be University), Bhubaneswar, Odisha, India

Bidyut Prava Das Department of Botany, Sailabala Women's Autonomous College, Cuttack, Odisha, India

Sovik Das Department of Civil Engineering, Indian Institute of Technology Kharagpur, Kharagpur, West Bengal, India

Patitapaban Dash Centre for Biotechnology, Siksha 'O' Anusandhan (Deemed to be University), Bhubaneswar, Odisha, India

Thiruselvi Devaraj Department of Applied Science and Technology, Alagappa College of Technology, Anna University, Chennai, Tamil Nadu, India

Fifi Khoirul Fitriyah Institute of Research and Community Service (LPPM), Universitas Nahdlatul Ulama Surabaya, Surabaya, East Java, Indonesia

M.M. Ghangrekar Department of Civil Engineering, Indian Institute of Technology Kharagpur, Kharagpur, West Bengal, India; PK Sinha Centre for Bioenergy and Renewables, Indian Institute of Technology Kharagpur, Kharagpur, West Bengal, India

Rajat Giri Department of Biosciences, Manipal University Jaipur, Jaipur, Rajasthan, India

Marttin Paulraj Gundupalli Department of Civil Engineering, Indian Institute of Technology Hyderabad, Kandi, Telangana, India

Sathish Paulraj Gundupalli Department of Design and Automation, Vellore Institute of Technology, Vellore, Tamil Nadu, India

Goutam Kishore Gupta Department of Chemical Engineering and Technology, Indian Institute of Technology, Banaras Hindu University, Varanasi, Uttar Pradesh, India

Baskar Gurunathan Department of Biotechnology, St. Joseph's College of Engineering, Chennai, Tamil Nadu, India

Mani Jayakumar Department of Chemical Engineering, Haramaya Institute of Technology, Haramaya University, Dire Dawa, East Hararghe, Ethiopia

T. Jayasree Joshi Department of Food Technology, Saintgits College of Engineering, Kottayam, Kerala, India

Nelson Libardi Junior Federal University of Santa Catarina – UFSC, Florianópolis, Brazil

Luiz Alberto Junior Letti Federal University of Paraná, Centro Politécnico, Curitiba, Paraná, Brazil

G. Kalavathy Department of Biotechnology, St. Joseph's College of Engineering, Chennai, Tamil Nadu, India

Ramani Kandasamy Biomolecules and Biocatalysis Laboratory, Department of Biotechnology, School of Bioengineering, SRM Institute of Science and Technology, Kattankulathur, Kanchipuram, Tamil Nadu, India

Susan Grace Karp Federal University of Paraná, Centro Politécnico, Curitiba, Paraná, Brazil

G. Karthigadevi Sri Venkateswara College of Engineering, Chennai, Tamil Nadu, India

Rupam Kataki Department of Energy, Tezpur University, Tezpur, Assam, India

Manika D. Kataki Department of Biotechnology, Darrang College, Tezpur, Assam, India

Santosh Kumar PK Sinha Centre for Bioenergy and Renewables, Indian Institute of Technology Kharagpur, Kharagpur, West Bengal, India

Huei Yeong Lim Biomass Processing Lab, National Higher Institution, Centre for Biofuel and Biochemical Research, Institute for Self-sustainable Building, Chemical Engineering Department, Universiti Teknologi PETRONAS, Bandar Seri Iskandar, Perak, Malaysia

M. Grace Lydia Phoebe Department of Biotechnology, St. Joseph's College of Engineering, Chennai, Tamil Nadu, India

Hilman Ibnu Mahdi PT Dover Chemical Plant B, Formaldehyde Production Department, Cilegon City, Banten, Indonesia

Sachin A. Mandavgane Visvesvaraya National Institute of Technology, Nagpur, Maharashtra, India

Yuvarani Mani Department of Applied Science and Technology, Alagappa College of Technology, Anna University, Chennai, Tamil Nadu, India

Mpumelelo Thomas Matsena Water Utilization and Environmental Engineering Division, Department of Chemical Engineering, University of Pretoria, Pretoria, Gauteng, South Africa

Rahulkumar Maurya School of Energy Science and Engineering, Indian Institute of Technology Guwahati, Guwahati, India

Kaustubha Mohanty School of Energy Science and Engineering, Indian Institute of Technology Guwahati, Guwahati, India

Monoj Kumar Mondal Department of Chemical Engineering and Technology, Indian Institute of Technology, Banaras Hindu University, Varanasi, Uttar Pradesh, India

Ariane Fátima Murawski de Mello Federal University of Paraná, Centro Politécnico, Curitiba, Paraná, Brazil

Sanjana Narayanan Biomolecules and Biocatalysis Laboratory, Department of Biotechnology, School of Bioengineering, SRM Institute of Science and Technology, Kattankulathur, Kanchipuram, Tamil Nadu, India

Nurfadhila Nasya binti Ramlee School of Chemical and Energy Engineering, Faculty of Engineering, Universiti Teknologi Malaysia (UTM) Skudai, Johor, Malaysia

Evans Martin Nkhalambayausi Chirwa Water Utilization and Environmental Engineering Division, Department of Chemical Engineering, University of Pretoria, Pretoria, Gauteng, South Africa

Mamta Pal Department of Biosciences, Manipal University Jaipur, Jaipur, Rajasthan, India

Shanmugam Palaniyandi Environmental Technology Division, CSIR-Central Leather Research Institute, Chennai, Tamil Nadu, India

Radha Panjanathan Biomolecules and Biocatalysis Laboratory, Department of Biotechnology, School of Bioengineering, SRM Institute of Science and Technology, Kattankulathur, Kanchipuram, Tamil Nadu, India

R. Praveenkumar Department of Biotechnology, Arunai Engineering College, Tiruvannamalai, Tamil Nadu, India

Mahendra Ram Indian Institute of Technology (Banaras Hindu University), Varanasi, Uttar Pradesh, India; BIT Sindri, Dhanbad, Jharkhand, India

E. Raja Sathendra Department of Biotechnology, Arunai Engineering College, Tiruvannamalai, Tamil Nadu, India

R. Sathya School of Life Sciences, B.S. Abdur Rahman Crescent Institute of Science and Technology, Chennai, Tamil Nadu, India

P. Senthil Kumar Department of Chemical Engineering, Sri Sivasubramaniya Nadar College of Engineering, Chennai, Tamil Nadu, India; Centre of Excellence in Water Research (CEWAR), Sri Sivasubramaniya Nadar College of Engineering, Chennai, Tamil Nadu, India

B. Senthil Rathi Department of Chemical Engineering, St. Joseph's College of Engineering, Chennai, Tamil Nadu, India

Rakesh Kumar Sharma Department of Biosciences, Manipal University Jaipur, Jaipur, Rajasthan, India

Carlos Ricardo Soccol Federal University of Paraná, Centro Politécnico, Curitiba, Paraná, Brazil

Anne Sahithi Somavarapu Thomas Department of Biomedical Sciences, Vellore Institute of Technology, Vellore, Tamil Nadu, India

Sivanesan Subramanian Department of Applied Science and Technology, Alagappa College of Technology, Anna University, Chennai, Tamil Nadu, India

Achmad Syafiuddin Department of Public Health, Faculty of Health, Universitas Nahdlatul Ulama Surabaya, Surabaya, East Java, Indonesia; Institute of Research and Community Service (LPPM), Universitas Nahdlatul Ulama Surabaya, Surabaya, East Java, Indonesia

N. Thajuddin National Repository for Microalgae and Cyanobacteria - Freshwater (DBT Sponsored), Bharathidasan University, Tiruchirappalli, Tamil Nadu, India

Kim Kley Valladares-Diestra Federal University of Paraná, Centro Politécnico, Curitiba, Paraná, Brazil

Luciana Porto de Souza Vandenberghe Federal University of Paraná, Centro Politécnico, Curitiba, Paraná, Brazil

C.V. Vidhya Department of Microbiology, Ethiraj College for Women, Chennai, Tamil Nadu, India

V. Vinoth Kumar Department of Biotechnology, SRM Institute of Science and Technology, Chennai, Tamil Nadu, India

Nur Izyan Wan Azelee School of Chemical and Energy Engineering, Faculty of Engineering, Universiti Teknologi Malaysia (UTM) Skudai, Johor, Malaysia; Institute of Bioproduct Development (IBD), Universiti Teknologi Malaysia, UTM Skudai, Johor, Malaysia

Anita R. Warrier Nanophotonics Research Laboratory, Department of Physics, Academy of Maritime Education and Training, Chennai, Tamil Nadu, India

Suzana Yusup Biomass Processing Lab, National Higher Institution, Centre for Biofuel and Biochemical Research, Institute for Self-sustainable Building, Chemical Engineering Department, Universiti Teknologi PETRONAS, Bandar Seri Iskandar, Perak, Malaysia

Zainul Akmar Zakaria School of Chemical and Energy Engineering, Faculty of Engineering, Universiti Teknologi Malaysia, UTM Johor Bahru, Johor, Malaysia

Abdul Hakim Zakkiy Fasya Institute of Research and Community Service (LPPM), Universitas Nahdlatul Ulama Surabaya, Surabaya, East Java, Indonesia

Preface

Biofuels and bioenergy become inevitable alternative fuels to meet the ever-escalating global growing fuel demand and to increase environmental sustainability. This book *Biofuels and Bioenergy—Opportunities and Challenges* focuses on opportunities and challenges associated in commercial production of biofuels and bioenergy such as bioethanol, biodiesel, biogas, and biohydrogen by utilizing various sources ranging from microbial biomass, weeds, and waste and achieved through biochemical and thermochemical processes such as fermentation, transesterification, anaerobic digestion, pyrolysis, gasification, microbial electrochemical reaction, and wastewater treatment. Direct and indirect impacts of biofuel on the environment and country's economy of the overall process also influence the commercialization of biofuels. Technoeconomic analysis and life cycle assessment are powerful tools for examining environmental impacts and economic viability to support the commercialization of biofuels and bioenergy. This book comprises 24 chapters distributed in eight sections as detailed in the following:

Section I briefly discusses the opportunities and challenges in the production of biofuels and bioenergy. **Section I** highlights the biomass conversion technologies available, notably conversion of biomass to biooil, nonwaste technology in the bioethanol and biodiesel industries as well as hydrothermal liquefaction of lignocellulosic biomass to produce biooil and other byproducts. **Section III** comprises five different chapters that focus on the refining of lignocellulose of second-generation waste for bioethanol production, various methods to produce biodiesel and types of catalysts involved, biodiesel production using enzymatic catalyst, the importance of nanocatalyst and its role in biofuel production and potentials, and challenges on biodiesel production from algae. **Section IV** is mainly about the production of biogas that highlights among others sustainable approaches for the production of biohydrogen from corn wastes, critical review on biogas production potential in India, future outlook on biogas and its potential application in a fuel cell, anaerobic digestion, and feasibility of using the gasification approach. Further thermal treatment approaches were also covered in **Section V** where pyrolysis is discussed as an alternative approach for utilization of biomass into bioenergy generation. **Section VI** revolves on the latest trends in bioenergy production using microbial fuel cell technology and bioenergy generation from waste and wastewater treatment processes. **Section VII** focuses on the production of biofuels and bioenergy using the microalgae route. Critical parameters affecting large-scale production of microalgal biomass in open raceway pond, the use of microalgae to produce various biofuels and bioenergy in wastewater treatment, single-cell oil synthesis by de novo and ex novo pathways in oleaginous microorganisms, production of biofuel from genetically modified microalgal biomass, and weeds as a renewable bioresource for bioconversion to biofuels and biomaterials are covered.

Section VIII features coverage on the state-of-the-art of the third-generation biofuel production through broader understanding on technoeconomic analysis and life cycle assessment.

This book will be a useful reference material for students, scientists, researchers, working professionals, and general public who are interested in the current progress of research, development, and commercialization works on biofuels and bioenergy. This book is also envisaged to also directly benefit teaching and learning activities in institution of higher learning notably that involves topics such as biofuels, renewable energy, and bioenergy (both at undergraduate and postgraduate levels) for various academic programs including, but not limited to, industrial biotechnology, chemical engineering, energy engineering, and environmental engineering.

Baskar Gurunathan
Renganathan Sahadevan
Zainul Akmar Zakaria

SECTION I

Biofuels and bioenergy

CHAPTER

1

Opportunities and challenges in industrial production of biofuels

Renu Bala, Monoj Kumar Mondal

Department of Chemical Engineering and Technology, Indian Institute of Technology, Banaras Hindu University, Varanasi, Uttar Pradesh, India

1.1 Introduction

Biofuels are the solid, liquid, and gaseous fuel produced by renewable resources through the biochemical (fermentation, anaerobic digestion) or thermochemical process (pyrolysis, gasification, transesterification). Biofuels are gaining scientific attention due to several advantages associated with them over fossil fuels. The excessive consumption of fossil fuel directs to resource depletion, energy crisis, and a hike in the price of energy sources. Also, their combustion produces greenhouse gas (GHG) emissions that result in acid rain, global warming, and climate change (Sharma et al., 2020). Biofuels are a renewable, economical, and eco-friendly alternative that can be reserved, and energy conservation techniques can be developed to fulfill the broad range of energy requirements (Sharma et al., 2020). They can be generated by a vast variety of biomass such as lignocellulosic residue, algae, seaweeds, noncrop plants, municipal solid waste (MSW), industrial waste, and so on. Biofuel also facilitates waste minimization besides energy production. The production and combustion of ethanol and biodiesel reduce GHG emissions by 12% and 40%, respectively as compared with fossil fuels. Bioenergy production also contributes to improving the economic status in rural and developing areas by providing job opportunities.

The use of biomass for energy production (cooking, heating, and lighting) in the form of firewood has been in existence since human development. Biomass such as wood, straw, cattle dung, and so on has been used after drying and burning as an energy source for cooking purposes. The seasoned air-dried firewood has an energy content of 15 MJ/kg that is one-third to half of the fossil fuel (Guo et al., 2015). The other processed forms of biomass used for energy are wood chips and pellets. Wood fuels contain lower energy content and higher combustion emissions as compared with fossil fuels (Guo et al., 2015). Additionally, the heat of the wood fire cannot melt many metals due to the low temperature of 850°C. Therefore, the transformation of wood into charcoal by the pyrolysis process was developed. The primary use of biofuels includes the transportation sector.

https://doi.org/10.1016/B978-0-323-85269-2.00003-4

Bioethanol, biodiesel, and biogas are the three main biofuels used in the automotive industries. Bioethanol is a substitute for gasoline and can be used to fuel petrol engines. Bioethanol is produced from the fermentation process. Fermentation was first discovered by a French chemist Louis Pasteur in 1857. The first internal combustion engine was designed by a German engineer Nicolaus Otto in 1876 (Guo et al., 2015). Henry Ford designed model T to run on ethanol in 1908. In 1979, the first bioethanol pilot plant was constructed at the South Dakota University. The commercial bioethanol production worldwide was 4 billion gallons in 1990. It increased to 29.03 billion gallons in 2019. The main contributors were the United States (54%), Brazil (30%), European Union (5%), China (3%), and India (2%), as represented in Table 1.1 (RFA, 2019).

Biodiesel is called the monoalkyl esters of fatty acids derived from vegetable oil through the transesterification process. It is used as an alternative to petrol diesel. A German engineer Rudolph Diesel had invented the compression-ignited diesel engine in 1893 (Guo et al., 2015). The first successful engine was exhibited in Paris in the year 1900, based on 100% peanut oil. The major biodiesel producing countries in 2019 were Indonesia (7.9 billion liters), the United States (6.5 billion liters), Brazil (5.9 billion liters), and Germany (3.8 billion liters).

TABLE 1.1 Global bioethanol production in 2019 (RFA, 2019).

Country	Production (million gallons)	% of world production
United States	15,776	54
Brazil	8570	30
European Union	1440	5
China	900	3
India	530	2
Canada	500	2
Thailand	420	1
Argentina	290	1
Rest of the world	600	2
Total	29,030	

Biogas based on sewage sludge was firstly used in 1890 in New Zealand to fuel street lamps (Bond and Templeton, 2011). The world attained more focus on the development of alternative energy fuels during raised oil prices in the 1970s, which increased the spread of biogas technology. The National Project on Biogas Development (NPBD; 1981–82) and National Biogas and Manure Management Program (NBMMP) (2006) have promoted biogas in India (Mittal et al., 2018). NPBD scheme came up with 28.63-lakh family-type and 2674 community-type biogas plants in 1998–99, and 47.5-lakh biogas plants were established under the NBMMP scheme up to March 2014.

The contribution of biomass energy to the total energy consumption was 13% in 2017 (Fig. 1.1). The global biofuel production was increased by 21.9% in 2019 as compared with 2015 (Statista, 2020a). The major biofuel producing countries in 2019 were the United States, Brazil, Indonesia, Germany, France, and China with the production of 1557.1, 992.2, 275.5, 143.4, 113, and 111.3 PJ, respectively (Statista, 2020b). This chapter describes the opportunities and challenges associated with biofuel production at the industrial level.

1.2 Utilization of biomass

1.2.1 Biomass harvesting, storage, and transportation

Biomass cultivation requires proper water availability, land availability, weed control, proper light, and temperature condition. Harvesting is divided into four parts, including direct chip, entire rod harvesting, harvesting in bales, and billets (Balaman, 2019). Some points

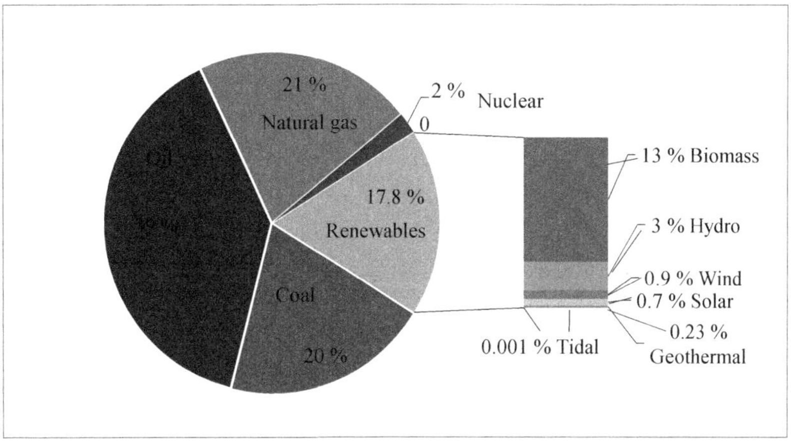

FIGURE 1.1 Contribution of energy resources in energy consumption in 2017 (Global bioenergy statistics, 2019).

being considered during harvesting are (1) the use of the optimal machine that can match spot and operation, (2) low-force harvesting techniques to prevent the soil from compaction and rutting, and (3) incorporation of agricultural processes such as trimming along with harvesting to prevent soil compaction and postharvesting regeneration (Gracia et al., 2014). After harvesting, biomass drying is needed because high moisture can lead to microbial degradation and biomass loss. The moisture content should be less than 20% for longer storage in bale form (Balaman, 2019). Covering the biomass with plastic sheets is also necessary during drying to prevent degradation and losses.

Woody biomass in the form of harvesting chips needs drying facilities, whereas the rods and billets dry naturally (Balaman, 2019). However, drying using facilities is a costly operation because excess heat is required. The agricultural residues along with forest residue are converted into chips, bales, and bundles for easy transportation (Balaman, 2019). The studies regarding biomass to biofuel supply chain include biomass growing, harvesting, preprocessing, storage, and transportation to bioenergy plant associated with 35%–65% of the total biofuel production cost (Albashabsheh and Stamm, 2021). It should not be exceeding 25% for economically feasible biofuel production (Albashabsheh and Stamm, 2021). Biomass densification such as in the form of bundles reduces the cost of the biomass supply chain. Preprocessing of MSW is required to produce physicochemical consistent feed material that can be handled, transported, and stored efficiently. It includes sorting, grinding, blending with other chemicals, drying, baling, and palletization of MSW (Balaman, 2019). Preprocessing methods widely used for the conversion of sewage sludge are thickening, dewatering, and composting (Balaman, 2019). The sewage sludge storing facilities include lagoons, drying beds, and separate tanks.

Microalgae are cultivated in an open system (open ponds, scrubbers, raceway ponds, and tanks) and closed systems such as photobioreactor (PBR) (Yin et al., 2020). Cultivation of algae in a closed system is costly as compared with an open system. The drawbacks associated with the open system are poor light utilization and land requirement. PBR offers better growth conditions of temperature, pH, O_2, and CO_2 (Yin et al., 2020). The different culturing mode of the batch, semicontinuous, and continuous is used for microalgae cultivation. The microalgae growth and high yield are achieved in semicontinuous and continuous cultivation (Yin et al., 2020). The different approaches of microalgae

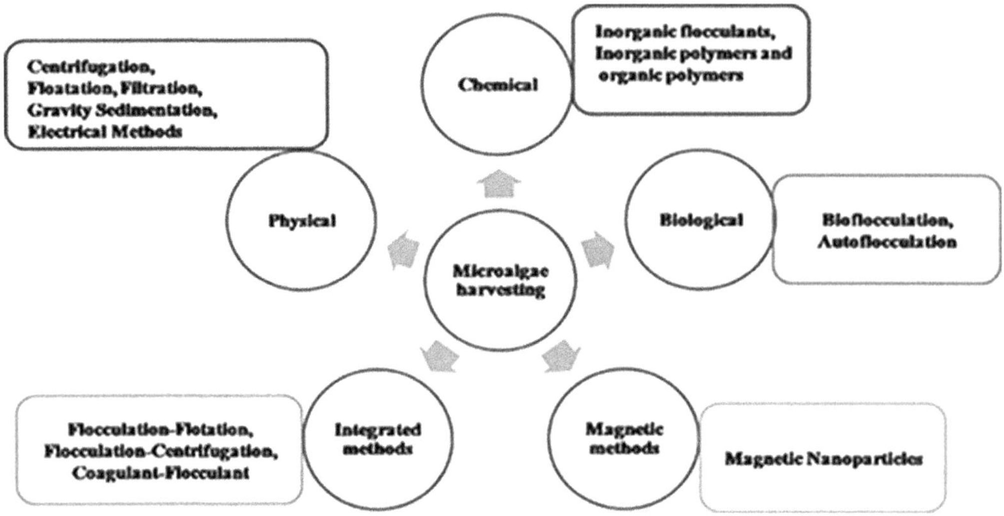

FIGURE 1.2 Different harvesting methods for microalgae.

harvesting are divided into physical, chemical, biological, and magnetic methods as represented in Fig. 1.2 (Yin et al., 2020). Microalgae harvesting is costly due to its small size of 2–20 μm. Flocculation is the economically feasible, efficient, and widely used method for microalgae harvesting (Yin et al., 2020). The major challenges linked with microalgae cultivation and harvesting are as follows: (1) contamination of microalgae species with others, (2) design of a low-budget photobioreactor with efficient yields is difficult, (3) the growth condition of microalgae should be optimized for better lipid accumulation, and (4) a large-scale harvesting method still under investigation considering the product specificity, low cost, separation of additives, and high microalgae recovery (Yin et al., 2020). The biofuels are divided into four generations according to the substrate used.

1.2.2 Generations of biofuels

1.2.2.1 The first generation of biofuel

The first-generation biofuels are produced from food crops (Fig. 1.3). Bioethanol and

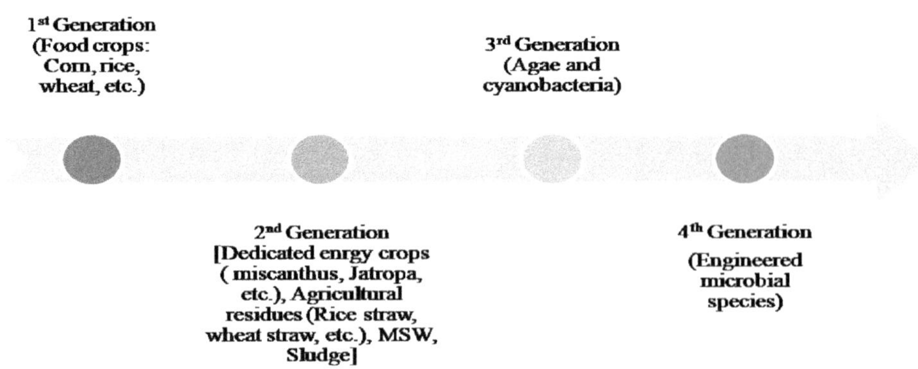

FIGURE 1.3 The generations of biofuels.

biodiesel are the two main types of biofuels used in the transportation sector. Bioethanol is an alternative to gasoline produced by the fermentation process. The substrate for bioethanol production is classified as sucrose-containing biomass (sugarbeet, sweet sorghum, sugarcane, etc.) and starchy feedstocks such as wheat, corn, barley, rice, and so on (Hirani et al., 2018). Biodiesel is the substitute for diesel, generated by the transesterification process using edible vegetable oil crops such as soybean, coconut, palm, peanut, rapeseed, and so on. First-generation biofuels raise the food versus fuel conflict. The number of crops used to produce 1 TJ of biodiesel and bioethanol can feed 110 and 90 people, respectively (Rulli et al., 2016). Moreover, an increased in deforestation rate, large CO_2 emissions, loss of habitat, and threat to biodiversity were noticed in Malaysia and Indonesia due to the replacement of old forests with oil palm plantations (Rulli et al., 2016).

1.2.2.2 *The second generation of biofuel*

The second-generation biofuels are produced either by the nonedible crop (miscanthus, switchgrass, bamboo, sweet sorghum, wheatgrass, etc.), specially grown on the land that is not suitable for growing food crops or by the nonedible part of ordinary crops (rice straw, wheat straw, etc.). It also includes other substrates such as forest residue (lignocellulosic biomass), MSW, industrial effluent, and food waste. The chemical properties of the cell wall present in lignocellulosic biomass create hindrance in the biofuel production process. Therefore, biomass complexity is reduced by altering the cell wall composition through different pretreatment techniques to augment their consumption for biofuel production. The problems associated with this generation of biofuel are sophisticated downstream processes, land and water use competition, and high production cost (Leong et al., 2018).

1.2.2.3 *The third generation of biofuel*

The third-generation biofuels are produced by using algae as a substrate. Algae are characterized as macroalgae and microalgae according to their size and morphology. Macroalgae are also called seaweeds. These are multicellular algae found in marine habitats. The family of macroalgae includes red, green, and brown algae. The blue-green algae (cyanobacteria) are also considered macroalgae (Suganya et al., 2016). Microalgae are the unicellular algae found in fresh or marine habitats. Algae are a promising feedstock as they are devoid of lignin and enriched with polysaccharides. Microalgae can produce special chemicals and nutritional products under the photosynthesis process. It has the ability of oxygenic photosynthesis and hydrogen production. The growth of microalgae can occur in both arable and nonarable land without the need for pesticides or herbicides (Rashid et al., 2013). Microalgae need less space to grow, and also they can grow in a natural and artificial environment in an eco-friendly manner (Chen and Wu, 2011). The main limitations associated with this type of biofuel are insufficient biomass production for commercialization and high production cost and stabilization up to a large scale.

1.2.2.4 *The fourth generation of biofuel*

The fourth-generation biofuels are produced by genetically modified biomass to achieve a high production yield. The substrates used are engineered microalgae, macroalgae, and cyanobacteria. Several improvement mechanisms are used to increase biofuel production such as lipid and carbohydrate mechanism, improved efficiency of nutrient use, improved photosynthesis efficiency, increased stress tolerance, improved cell disintegration, and bioflocculation (Abdullah et al., 2019; Das et al., 2020; Khan and Fu, 2020).

1.2.3 Biomass structural components

Biofuel production yield is dependent on biomass composition. Most of the biomass is composed of cellulose, hemicellulose, and lignin. Cellulose is formed by D-glucopyranose units linked with β-1,4-glycosidic bonds in the form of microfibrils, which in turn entwined into fibrils in such a way that cellulose forms crystalline properties (Chen, 2014). It presents in two forms including amorphous and crystalline. Besides cellulose, lignin is the other main component of the cell wall. It is a complex, irregular, and amorphous structure of phenylpropane by the irregular pairing of C—O and C—H bonds (Chen, 2014). The basic structural monomers of lignin are p-phenyl monomer (H type) obtained from coumaryl alcohol, guaiacyl monomer (G type) obtained from coniferyl alcohol, and syringyl monomer (S type) obtained from sinapyl alcohol (Chen, 2014; Novaes et al., 2010). Hemicellulose is the amorphous component of the cell wall. Xylan, xyloglucan, glucomannan, galactomannan, callose, and so on are present as hemicelluloses in cell matrix (Yin and Fan, 1999). Cellulose, lignin, and hemicelluloses are connected by hydrogen bonds. Other chemical bonds are also present as side chains in between hemicelluloses and lignin such as galactose residues and arabinose residues (Yang, 2008). The presence of cross-links among lignin—cellulose—hemicellulose, the degree of polymerization, and the protecting nature of lignin are responsible for the recalcitrance of biomass. The proportion of cell wall components significantly differ among different feedstocks making the conversion process challenging. Lignin is considered as a hindrance in the biochemical conversion process (cellulosic ethanol production, biohydrogen production, anaerobic digestion, etc.) as compared with cellulose and hemicellulose. However, woody biomass with high lignin content is suitable for thermochemical conversion such as pyrolysis and gasification (Abdullah et al., 2017). The lignin reduction by using different pretreatment methods is the primary step when dealing with lignocellulosic waste for biofuel production.

1.3 Pretreatment techniques

Pretreatment methods are broadly divided into physical, chemical, biological, and combinational approaches. These are generally used due to several advantages such as the increased surface area of the substrate, decreased complexity, increased delignification, and increased porosity that leads to better substrate utilization and better productivity (Abraham et al., 2020). An ideal pretreatment reduces the lignin and conserves the cellulose (Bala and Mondal, 2018). The development of a cost-effective and time-saving treatment strategy is an important challenge in biofuel production. Table 1.2 shows the advantages and disadvantages of different pretreatment methods.

1.3.1 Influence of pretreatment on biomass structure

The cellulose, hemicellulose, and lignin contents are measured before and after the pretreatment to determine its effect on the substrate. The accuracy of measuring methods is a challenging task in biofuel production. Scanning electron microscopy (SEM) is used to study the surface morphology of the substrate. The sample preparation by drying in the oven can lead to deformation and collapsing of the surfaces. The appropriate method to avoid this problem is freeze-drying. Transmission electron microscopy (TEM) and atomic force microscopy (AFM) analysis are the other two ways to detect morphological changes after pretreatment. SEM and TEM are used to capture the two-dimensional images, whereas three-dimensional images could be captured by using AFM analysis. Measurement of biomass crystallinity by X-ray diffraction is also carried out to determine the extent of pretreatment. The crystallinity index (CI) is the ratio calculated by using crystalline and amorphous regions of the XRD curve. The following equation is used to calculate CI (Segal et al., 1959):

$$\mathrm{CI}(\%) = \left(\frac{I_{002} - I_{am}}{I_{002}}\right) \times 100 \tag{1.1}$$

where I_{002} and I_{am} indicate the intensity of the highest peak of the crystalline region, i.e.,

TABLE 1.2 Advantages and disadvantages of common pretreatment methods (Jönsson and Martín, 2016; Haldar and Purkait, 2021).

Pretreatment		Advantage	Disadvantage
Physical	Steam explosion	Hemicellulose conversion to sugars, lignin transformation	Degradation of reducing sugars
	Liquid hot water	Improve cellulose degradability, lignin removal, less inhibitors than acid-based treatments	Furfural and HMF formation
	Microwave	Energy saving, reduce the treatment time, can handle large volume	Phenolic compound and furfural formation at high temperature
	Ultrasound	Disrupt cell wall structure, improve surface area, reduce the degree of polymerization	Expensive, cannot handle large volume
Chemical	Acid	Hemicellulose hydrolysis, increase cellulose susceptibility to microbe and enzymatic hydrolysis	Inhibitor (furfural, HMF, phenolic compound) formation, expensive
	Alkali	Decrease degree of polymerization, cause swelling of biomass, disruption of lignin structure	Neutralization step required, chemical cost, cause water pollution, and soil salinization
	Ionic liquids	Dissolve cellulose, low toxicity, decreased crystallinity, increased porosity	High cost of ionic liquid, regeneration requirement, lack of knowledge about inhibitors, action mode on hemicellulose/lignin
	Hydrogen peroxide	Break lignin and hemicellulose	Losses of cellulose, costly, inhibitor generation
Biological	Fungal	Low energy input, no chemicals, degrade lignin and hemicellulose	Long treatment time, need of microbial growth continuously, large space requirement
	Microbial consortium	Degrade cellulose and hemicellulose	

$2\theta \approx 22$ degrees and lowest intensity of amorphous region $2\theta \approx 18$ degrees, respectively.

In some studies, the CI was found to be decreased after pretreatment (Zhang et al., 2014; Ostovareh et al., 2015), and substrate was considered as more degradable. While, in other cases more degradability was associated with increasing CI after pretreatment (Kristiania et al., 2015; Barman et al., 2014). Hence, measuring crystallinity is somewhat confusing to evaluate the pretreatment effect. Other methods to calculate crystallinity are Fourier transform infrared spectroscopy, infrared spectroscopy, nuclear magnetic imaging, Raman spectroscopy, and so on.

1.3.2 Impact of pretreatment on liquid hydrolysate

Hydrolysate characteristics such as reducing sugar concentration, volatile fatty acid (VFA), soluble chemical oxygen demand (sCOD), and phenolic content are used to determine the influence of pretreatment. The accuracy of the estimating method is required. Reducing sugar concentration is often measured when the substrate is directed to liquid biofuel production such as bioethanol, biobutanol, and so on. Dinitrosalicylic (DNS) method and high-pressure liquid chromatography (HPLC) are commonly used methods reported in the literature for reducing sugar measurement. Hydrolysate color due to the presence of pretreatment chemicals or presence of hydroxymethyl furfural/furfural due to acid treatment or thermal hydrolysis at elevated temperature can interfere with the DNS assay that can lead to false quantification of glucose. HPLC method is a high precision method for sugar estimation but requires a costly refractive index detector, meticulous sample preparation, and skilled handling. VFA and sCOD estimation are considered for anaerobic digestion. Methanogens utilize VFAs and convert them into methane and CO_2 in the methanogenesis phase of anaerobic digestion. sCOD indirectly measures the formation of simpler compounds such as sugars, amino acids, and fatty acids from the breakdown of carbohydrate, protein, and lipid, respectively, due to pretreatment. sCOD enhancement was associated with better pretreatment effectiveness in the literature (Wang et al., 2015). Phenolic content estimation is an indirect method to detect lignin reduction in biomass. Lignin degradation results in the formation of phenolic by-products in the hydrolysate. These are used to measure the efficiency of those pretreatments that are directed to lignin reduction in biomass.

1.4 Enzymatic hydrolysis

Enzymatic hydrolysis is the necessary step after pretreatment when second-generation biomass is subjected to bioethanol production. Pretreatment alters the cell structure and chemical properties of biomass, and then the biomass is treated with enzymes to hydrolyze available cellulose and hemicellulose for fermentable sugar production. These sugars are converted to bioethanol/biobutanol by the fermentation process. Enzymes for cellulose hydrolysis are a group of cellulases that includes endoglucanases, cellobiohydrolases, exoglucohydrolases, and β-glucosidases (Maitan-Alfenas et al., 2014). The most appropriate temperature and pH of cellulose-degrading enzymes reported in the literature are 40–50°C and 4–5, respectively (Maitan-Alfenas et al., 2014). Hemicellulase group of the enzyme for xylan degradation includes endo-β-1,4-xylanase, β-xylosidase, acetylxylan esterase, α-glucuronidase, ferulic acid esterase,α-L-arabinofuranosidase, and α-1 galactosidase (Maitan-Alfenas et al., 2014). Other enzymes derived from white rot fungi such as laccase and peroxidases have the potential to degrade lignin. The hydrolysis time depends upon the substrate concentration, the lignin content of biomass, pretreatment efficiency, cellulose crystallinity, and enzyme activity. Lignin acts as a structural barrier and unproductively binds to the enzyme that reduces the enzyme activity and affects the hydrolysis yield. Enzymatic hydrolysis is also affected by end product inhibition by cellobiose and glucose (Volynets et al., 2017). Enzyme denaturation can be caused due to elevated temperature and shear forces of mechanical mixing. Enzyme improvement can be done by the selection of novel enzymes from alternative microorganisms and using enzyme engineering. The engineering techniques such as site-directed mutagenesis, alteration of the linker region of bimodular cellulases, and domain shuffling are used for enzyme modification (Østby et al., 2020). The high cost of the enzyme is also a critical factor in the commercialization of bioethanol production from lignocelluloses. Recycling of enzymes can reduce its cost. Recycling of adsorbed exo- and endoglucanases from solid substrate could reduce the 30% enzyme dose (Volynets et al., 2017).

1.5 Formation of inhibitory compounds

The inhibitory compound formation depends on the pretreatment method and condition of execution. The acid pretreatment of biomass mainly dissolves cellulose and hemicelluloses, whereas this partially dissolves lignin. The pentose and uronic acid produced from hemicellulose hydrolysis are dehydrated into 2-furaldehyde (furfural). Moreover, the hexoses are dehydrated into 5-hydroxymethyl-2-furaldehyde (HMF) (Jönsson and Martín, 2016). HMF can further be converted into formid and levulinic acid at high temperature and long pretreatment time. Phenolic derivatives could also be present in hydrolysate due to degradation of lignin component. Acid can be corrosive to pretreatment apparatus and cause release of metal ions such as copper, iron, and so on that are inhibitory to microorganisms.Alkaline pretreatments are generally directed to dissolve lignin. Peeling reaction in such type of pretreatment could be responsible for the production of saccharinic acid, formic acid, lactic acid, and dicarboxylic acid (Jönsson and Martín, 2016). Saponification of acetyl groups results in the formation of acetic acid. Phenolic compounds and carboxylic acids could also present in hydrolysate of alkaline pretreatments. Hydrothermal pretreatment could also result in products similar to acidic pretreatment when the condition proceeds from neutral to acidic side (formation of uronic and acetic acid). Detoxification of hydrolysate can be done to mitigate the inhibition problems by using chemical additives (alkali, polymers, and reducing agents), vaporization, enzymatic treatment, liquid–liquid extraction, and liquid–solid extraction (Jönsson and Martín, 2016).

1.6 Conversion of biomass to biofuel

1.6.1 Different routes and associated problems

1.6.1.1 Biochemical route

Microbial conversion of substrate to bioethanol, biobutanol, biohydrogen, and biogas through fermentation and anaerobic digestion is known as the biochemical route of biofuel production (Fig. 1.4). The first-generation crops such as corn, sugarcane, and so on are the well-established substrate for bioethanol production. Most of the bioethanol in the United States and

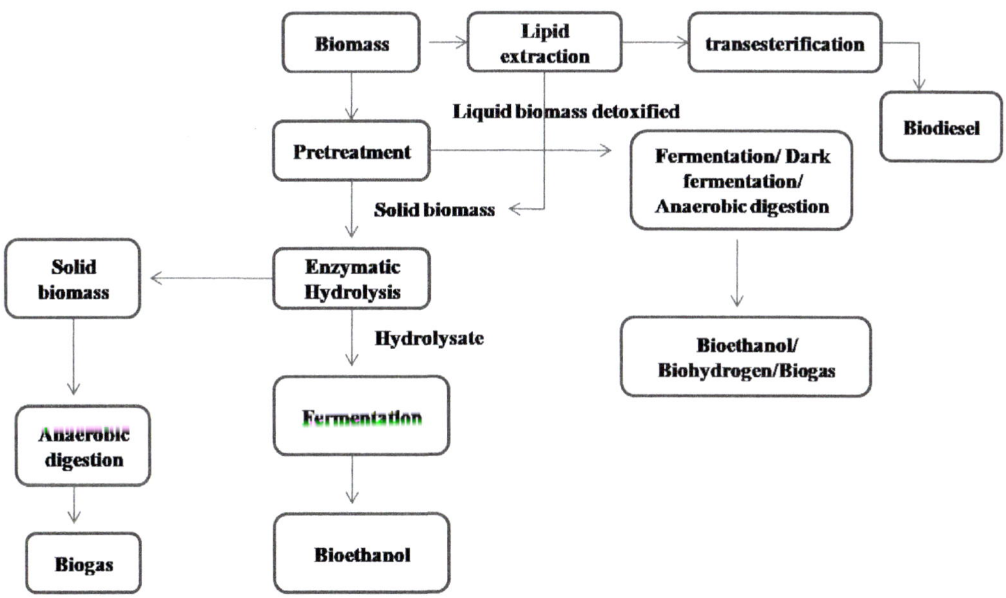

FIGURE 1.4 Biochemical route of biofuel production.

Brazil is produced from corn and sugarcane, respectively. The second- and third-generation bioethanol is still under development. The robust nature of lignocellulosic biomass results in poor yields. Pretreatment techniques and enzymatic hydrolysis are obligatory steps before proceeding to bioethanol production. Furthermore, hemicellulose generally degrades into xylose that cannot be utilized by ethanol-producing yeast such as *Saccharomyces cerevisiae, Zymomonas mobilis*, and so on. Some yeast such as *Candida shehatae* and *Pichia stiplis* can potentially utilize pentose sugars such as xylose but has a low tolerance to ethanol (Baeyens et al., 2015). Bioethanol is produced in three modes including separate hydrolysis and fermentation (SHF), simultaneous saccharification and fermentation (SSF), and consolidated bioprocessing (CBP), as represented in Fig. 1.5. The hydrolysis temperature is 50°C, and the fermentation temperature varies in the range of 28–37°C (Dey et al., 2020). SHF is advantageous as the process parameters can be optimized for both of the processes, due to the occurrence of process in different vessels. However, the process is limited by the loss of enzyme activity due to high sugar yield, and process is costly due to use of different vessels (Dey et al., 2020). SSF is a more promising method of ethanol production than SHF due to less cost, low requirement of enzymes, lower time duration, high ethanol yield and minimized enzymatic inhibition due to product removal, and lower sterile conditions required (Dey et al., 2020). The process has been further improved by the development of short presaccharification (PSSF), semisimultaneous saccharification (SSSF), and semisimultaneous saccharification and cofermentation (SSCF) to deal with the temperature difference of hydrolysis and ethanol production. SSCF is the best method among the existing variants of SSF, as hexose and pentose sugars can be successively

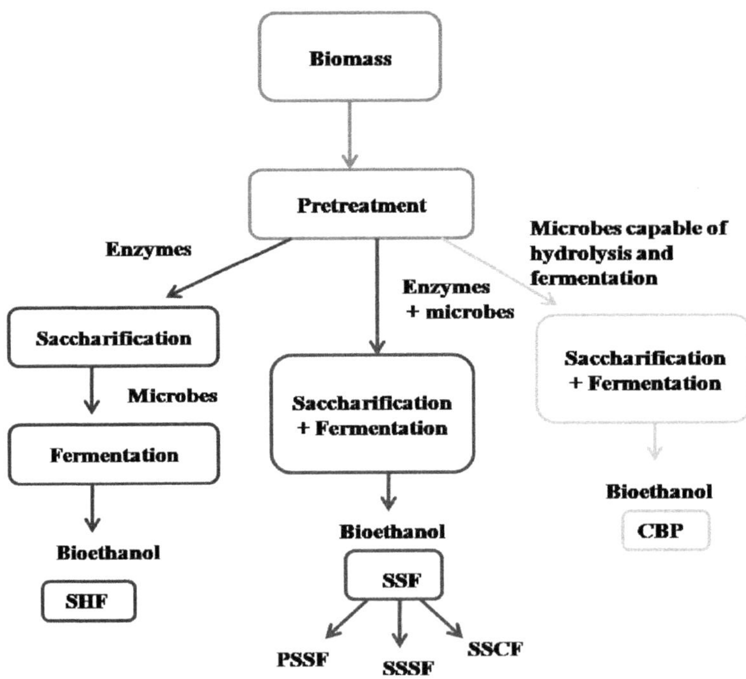

FIGURE 1.5 Different biochemical routes of bioethanol production.

utilized in this process (Dey et al., 2020). CBP is an attractive ethanol production process due to several associated benefits such as hexose and pentose fermentation, low cost, and lower risk of contamination (Singh et al., 2017). The process is carried out by microbial species that has both hydrolysis and fermentation capability. The major challenge associated with the process is the requirement of microbial species that can tolerate elevated temperature, as both the saccharification and ethanol production take place at high temperature (Dey et al., 2020). *Clostridium thermocellum* can be used excellently in CBP ethanol production (Singh et al., 2017). In general, the development of genetically modified species that are capable of producing a large volume of cellulase, hemicellulases, and xylanases enzymes and tolerant to inhibitors is needed to achieve a high amount of ethanol yield.

Biohydrogen could be produced by photolysis, photofermentation, dark fermentation, and microbial fuel cell. Challenges associated with light associated biohydrogen production include low yield. The limitations of photo biohydrogen production are as follows: (1) the liberation of oxygen molecule takes place in photosynthesis process during production by microalgae that inhibits the hydrogenases, (2) different intensity light-absorbing capacity of phototrophs, and (3) the rate of CO_2 absorption to build biomass and then its further translation to biohydrogen is low (Rodionova et al., 2016). Biohydrogen is also produced by the dark fermentation process using anaerobic microorganisms such as *Clostridium acetobutylicum*. Dark fermentation consumes less energy than photofermentation. The environmental parameters such as pH of the medium and the availability of metal cofactors are essential in dark fermentation.

Biogas, a mixture of CH_4 (50%–70%), CO_2 (30%–50%), and traces, is the other gaseous fuel produced from anaerobic digestion (Srivastava et al., 2014). The process includes hydrolysis, acidogenesis, acetogenesis, and methanogenesis steps. A wide variety of feedstocks can be utilized for biogas production. A diverse group of microorganisms participates in each step. Hydrolysis and methanogenesis are the rate-determining step of anaerobic digestion. The main challenge includes the development of a pretreatment technique to alter the complexity of biomass, enhancement of methane concentration in biogas, and process parameters optimization such as temperature, pH, the substrate-to-inoculum ratio, and substrate loading rate.

1.6.1.2 Transesterification and integrated biorefinery

Biodiesel is produced from the transesterification of vegetable oils and animal fats. In this process, the first step is to extract oil and fat from biomass, and the second step is to convert fats into biodiesel in the presence of a catalyst and alcohol. The substrates used for biodiesel production are rapeseed, soybean oil, *Jatropha curcas*, microalgae, and so on. Separation of biomass from fermentation broth after fat accumulation is a bottleneck during biodiesel production. Microalgae have potential in energy industries as it has 50% lipid that can be used in biodiesel formation. The remaining part can be used for bioethanol production. The main challenge of biodiesel production is the availability of substrate throughout the year because of biomass competition for different purposes. The concept of integrated biorefinery consists of the processing of biomass in bio-based products (food/feed) and bioenergy products (biofuels). By using this approach, the substrate availability concern can be conquered as not only crop but also other parts of biomass such

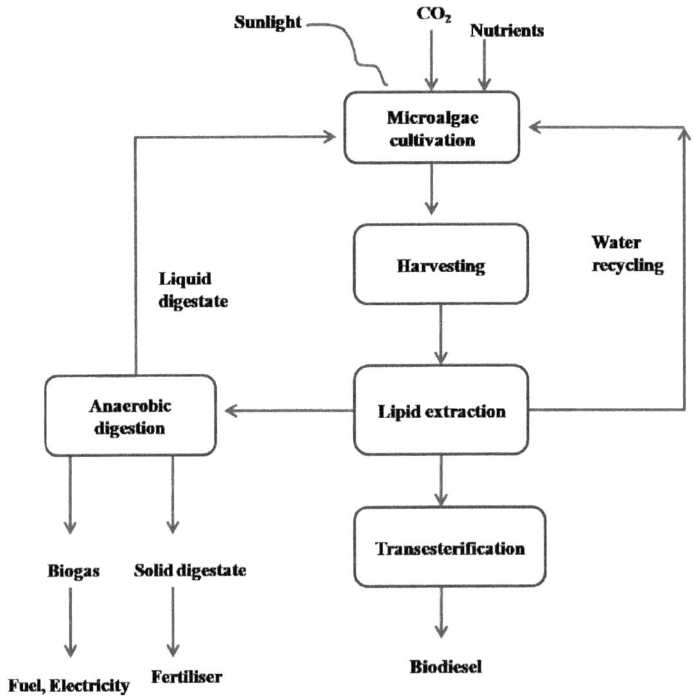

FIGURE 1.6 Integrated approach of biodiesel and biogas production.

as stem, leaves, defruited bunches, roots, and so on can be used for other purposes. Moreover, the wastewater generated can be utilized in microalgal cultivation. Therefore, an integrated biorefinery is a sustainable approach. The integrated production of biodiesel and biogas is represented in Fig. 1.6.

1.6.1.3 Thermochemical route

Biomass is converted into biochar, biooil, flue gas, and syngas via thermochemical route. The conversion processes under this category are pyrolysis, gasification, combustion, liquefaction, carbonization, and torrefaction (Fig. 1.7). Pyrolysis is the formation of biochar, biooil, and flue gas when biomass is heated in the temperature range of 400–700°C in absence of oxygen, whereas gasification is occurred in the presence of a gasifying medium to produce syngas (a mixture of H_2, CH_4, N_2, CO, and CO_2). The degradation ranges for hemicellulose, cellulose, and lignin are 200–380°C, 320–400°C, and >400°C, respectively (Zhang and Zhang, 2019). Temperature, heating rate, and residence time are the main process parameters for all thermochemical processes (Zhang and Zhang, 2019). Feedstock properties such as ash content, heating value, moisture content, and high nitrogen content affect the thermochemical biofuel production. Low moisture substrates are directed to thermochemical pathways for biofuel generation, unlike biochemical processes where high moisture feedstock requires. High lignin is beneficial to the energy balance of thermochemical pathways because it has a high heating value (HHV) than cellulose and starch, contrasting to the biochemical pathway where lignin is undesirable. However, the high lignin content in pyrolysis results in less

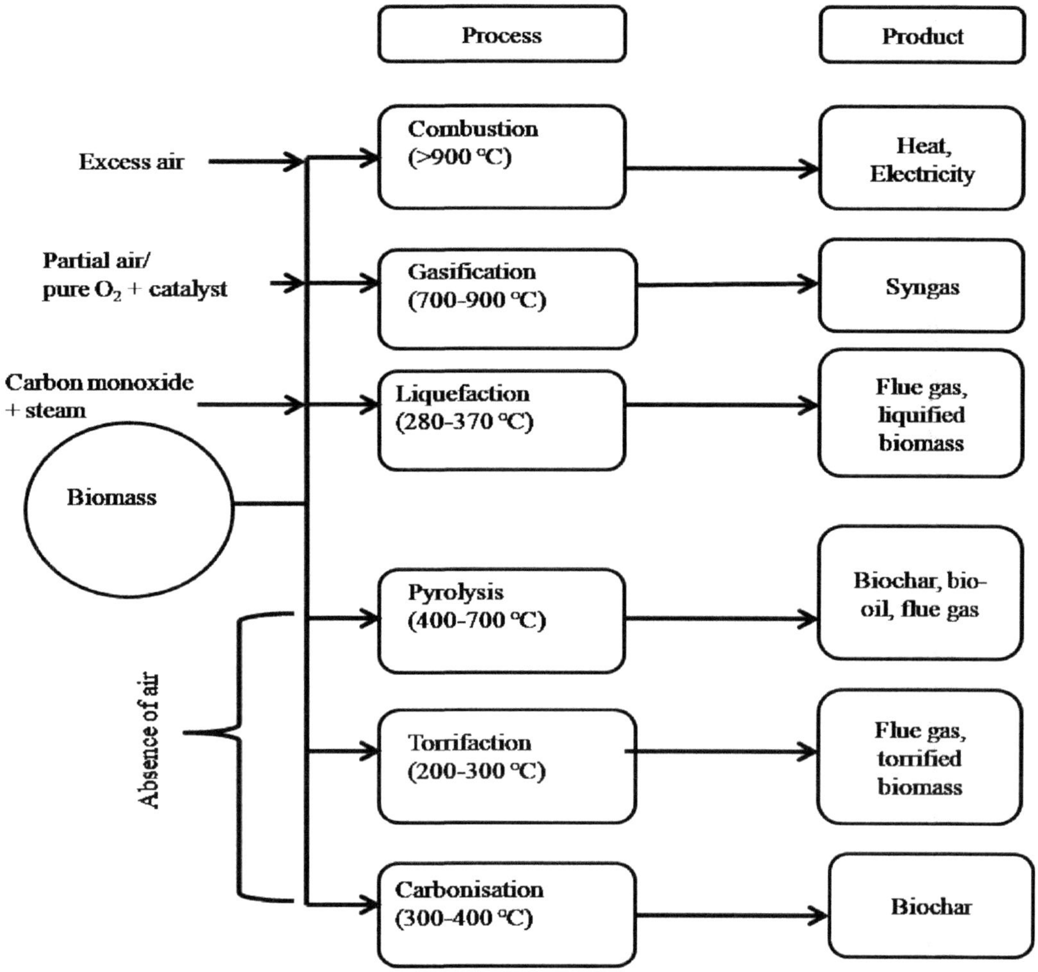

FIGURE 1.7 Thermochemical route of biofuel production.

gas and more char yield. Tar formed during gasification and pyrolysis cause fouling to turbines/engines and environmental pollution. Low product yield and tar formation is a challenge in these processes.

1.6.2 Separation and downstream processing

Downstream processing is defined as the steps involved in the final purification/extraction of biofuel from fermentation broth. The separation techniques account for the high capital cost in biofuel technology. Bioethanol is separated using the distillation process that requires intensive energy input. Ethanol forms an azeotrope with water during the distillation process that is the main limitation of the process. Several process developments such as feed splitting, heat integrated, membrane technology, ohmic-assisted hydrodistillation, and pinch analysis have been done to alleviate this problem by utilizing equipment configuration and physicochemical properties of ethanol mixture and water (Sharma et al., 2020). Glycerol and other impurities such as metal ions, acid, water, and soap have to be separated from biodiesel for better fuel efficiency. Salt-assisted gravitation settling or centrifugation has been reported to separate glycerol from biodiesel. The downstream purification techniques for biodiesel are wet washing, dry washing, and membrane separation technology (Yellapu et al., 2018). Water pollution and large energy input during centrifugation are the main limitations of wet washing separation. Furthermore, high water and acid content in biodiesel decrease its fuel value (Yellapu et al., 2018). Methane is the main component of biogas that determines its fuel value. As biogas contains several impurities such as water vapor, H_2S, CO_2, N_2, and so on, it requires purification before its utilization in combustion engines. The main upgradation techniques used are water scrubbing, membrane separation, pressure swing adsorption, and chemical absorption. In a life cycle assessment and life cycle costing study, membrane separation was found to be advantageous over other techniques (Ardolino et al., 2020). However, water scrubbing is the simplest and most applied method as compared with others. The requirement of a large amount of water and approximately 5% methane loss are drawbacks of this process (Nguyen et al., 2020). Pyrolysis product biooil is also needed to be purified before its use as fuel in transport, boilers, and engines. The downstream techniques for biooil upgradation are solvent addition, emulsification, filtration, electrochemical hydrogenation, and hydrotreatment (Kumar and Streziov, 2021). However, there are some challenges in the purification of biooil. The solvent addition adds water to biooil and decreases pH. Also there is a lack of research regarding the reaction of the solvent to biooil. Emulsification is advantageous but -requires other additional steps of ultrasonication and stirring. Filtration is used to separate the char from biooil, but the membranes used in filtration are highly expensive and need regular solvent washing that further increases the cost of the process (Kumar and Strezov, 2021). Electrocatalytic hydrogenation requires a high-priced membrane, hydrogen, and electricity that make the process costly. The major challenges associated with hydrotreatment are the requirement of hydrogen, its storage, and transportation (Kumar and Strezov, 2021). Syngas is also upgraded by removing impurities such as tar, sulfur-containing compounds (H_2S, COS), nitrogenous compounds (HNC, NH_3), hydrogen halides, and trace metals. The purification techniques are divided into cold gas cleanup that includes wet scrubbing, chemical absorption, physical absorption, and hot gas cleanup that consists of thermal cracking, dry reforming, and catalytic cleanup (Abdoulmoumine et al., 2015). The cold gas clean-up is costly due to its complexity and effluent treatment, whereas catalyst-based hot cleanup is challenging due to the rapid deactivation of catalysts.

1.6.3 Coproduct formation

Coproduct formation depends upon the type of feedstock used in biofuel production. The bioethanol industries produce bagasse and vinasse in sugarcane bioethanol production, dried distillers' grain with soluble (DDGS) from corn/wheat ethanol production, and lignin in lignocellulosic ethanol production. Sugarcane bagasse is the fibrous material remaining after extraction of juice. It is generally used in sugar mills to accomplish its needs of heat and energy. It consists of 48% fiber, 50% water, and 2% soluble solids such as sugars. It has lignin, hemicellulose, and cellulose content of 20%, 28%, and 42%, respectively. It can also be used in the production of single-cell protein, second-generation ethanol, biogas and other fuels, enzymes, pulp and paper, boards, furfural, and animal feeds (Martinez-Hernandez et al., 2017). Vinasse is a liquid by-product of biorefinery generated during distillation and can be used as fertilizer, livestock feed, biogas production, and yeast production (Christofoletti et al., 2013). Distillers' grain is highly nutritious and used as animal feed. Lignin is processed to form various chemical compounds such as phenols, benzene, xylene, toluene, and so on. It is also explored as fertilizer, surfactant, wood adhesives, synthesis of carbon material, and composites in thermoplastics. Protein meal and glycerine is the coproduct of biodiesel industries. Rapeseed and soybean seed produce a protein-rich meal that can be used as animal feed. Glycerine is used in personal care, food, pharmaceutical, and other industrial use. It is also used as a feed ingredient for livestock rations (Popp et al., 2016). The nutrient-rich digestate from anaerobic digestion can be used as manure in agricultural purposes. Coproducts from different biofuel producing process are summarized in Table 1.3.

Coproducts generated from microalgae are divided into three parts as high value, medium value, and low value. High-value products consist of neutraceuticals (β-carotene, polyunsaturated omega-3 fatty acids, carotenoids, lutein, etc.), pharmaceuticals (antimicrobial, antiviral, antifungal, nutraprotective products, etc.), and cosmetics (alguronic acid, anticellulite, etc.). Medium-value products are agar, alginate, paints, dye, colorants, and so on. Low- to medium-grade products are animal feed, bioplastics, and glycerol (Kroumov et al., 2017).

1.7 Water requirements

Water is needed in almost all steps of bioenergy production through the biochemical route. Water is required to grow feedstock crops, cultivation of microalgae, growth media formulation for microbial species, fermentation media, primary washing of biomass, and washing after pretreatment. The shifting of water to biofuel production usage can alter the availability of potable water by putting extra pressure on local water resources (Joshia et al., 2017). Therefore, setting of biorefinery should be done carefully considering

TABLE 1.3 Coproducts of different biofuel production process (Martinez-Hernandez et al., 2017; Christofoletti et al., 2013; Popp et al., 2016).

Process	Coproduct	Uses
Fermentation	Distillers' grain	Animal feed
	Bagasse	Biofuel, single cell protein, furfural production, paper and pulp, animal feed
	Vinasse	Fertiliser, livestock's feed, biogas production
	Lignin	Chemical production (phenol, benzene, xylene, and toluene), synthesis of carbon material, wood adhesive, and fertilizer
Transesterification	Seed meal	Animal feed
	Glycerine	Pharmaceutical, food, and personal care
Anaerobic digestion	Digestate/manure	Fertilizer
Pyrolysis	Biooil	Heating, upgraded biooil use to operate boilers and turbine, chemical synthesis (fertilisers, adhesives, etc.)

the proper water supply. Recycling water can minimize the load of local water sources, but it needs to clarify before use. Algal cultivation and anaerobic digestion is an excellent integrated approach to treat wastewater generated in biorefineries.

1.8 Environmental impact and biofuel economy

Biofuel is considered environmentally friendly due to its low emissions and carbon-neutral nature as compared with fossil fuels. The emitted CO_2 during biofuel combustion goes back to the atmosphere and is captured by feedstock crops for the photosynthesis process. However, there are some negative concerns related to biofuels. Biofuel production technologies use chemicals in substrate pretreatment that can cause water pollution. Climate change, metal deposition, fossil depletion, and freshwater eutrophication were observed due to chemical use in various steps of pretreatment, hydrolysis, and fermentation in bioethanol production (Demichelis et al., 2020). The bioethanol can produce toxic compounds such as formaldehyde and acetaldehyde due to their highly oxidizing nature. Biodiesel can also generate aldehydes and aromatic compounds during their combustion. Moreover, the land required to grow oil-producing crops is a serious issue in biofuel technology. It can lead to deforestation, soil erosion, loss of habitats, and loss of biodiversity. Additionally, it can also cause a drop in food availability and security while increasing the food prices because of the shift of farming trend from food-producing crops to biofuel producing crops to make more income by farmers. The main cost associated with biofuel production is on storage and transportation of biomass, pretreatment techniques, enzyme usage in saccharification, development of equipment/reactors, and downstream processing. The downstream processing of biodiesel contains 60% of the production cost. The severe demand for water, land, crops, energy, and so on is responsible for the poor economics of biofuel technology and limits the sustainable development of biofuel (Joshia et al., 2017).

1.9 Future prospects

The utilization of food crops and arable land for biofuel production is constantly a controversy. The development of efficient alternative feedstocks that do not compete with agriculture is required. It took the world's focus toward algae-derived fuels. However, the third-generation biofuel produced from algae is highly expensive as the manufacturing processes are not cost-effective on an industrial scale (Lin and Lu, 2021). Microalgae are considered as promising biofuel production feedstock in the future due to the high biomass yield, high-fat content, carbon fixation, short growth period, and eco-friendly nature (Lin and Lu, 2021).

The bioethanol production in the United States was 59.5 billion L in 2019. It has been dropped (12%) to 52 billion L in 2020 due to the COVID-19 pandemic (IEA report, 2020). It is anticipated to stabilize at 55 billion L in 2023–25. In Brazil, sugarcane-based ethanol has been dropped in 2020 due to maximization of sugar production, Hence, corn-based ethanol production is expected to increase 2.5 billion L by 2021(IEA report, 2020). An average of 37 billion L ethanol production is expected by the year 2023–25. The RenovaBio policy improves the biofuel economy by incorporating the low emission goals (such as CO_2) in the country. It possibly stimulates the production of biodiesel from low carbon waste oils and animal fats and also increases the biofuel blending. Biodiesel and biomethane (in natural gas) are to be blended by 15% and 5%, respectively, by 2022 (Barr et al., 2021). Biodiesel production can rise to 7 billion L by 2025 in Brazil. The global output for biodiesel production is forecast to 63 billion L in the year 2023–25(IEA report, 2020). Biofuels are struggling despite having eco-friendly nature. The production can be increased by establishing advanced production technologies, developing efficient biomass feedstocks and revising the energy policies (Barr et al., 2021).

1.10 Conclusions

Biofuels are sustainable energy resources. First-generation crops are successively used in bioethanol and biodiesel production. However, they end up with the debate of the necessity of food over fuel. The biomass-to-bioenergy supply chain should be more improved in terms of cost-friendly harvesting and preprocessing methods and the generation of highly specific biomass without harming the soil quality for regeneration. Second- and third-generation biofuels still needed improvements and further investigation for process developments. The several challenges can be concluded as developments of cost-effective pretreatment methods, high temperature withstanding enzymes, genetically engineered microbial species that can utilize the substrate in a better manner, economically feasible reactors to carry out the production, and budget-friendly, highly efficient downstream processing. The competition of land, water, and energy supply also acts as a barrier in the commercialization of biofuels. Bioproducts generated during the biofuel production process are widely advantageous in different sectors of animal feed, biofuel production, and chemical generation, further adding value to the biofuel technology.

References

Abdoulmoumine, N., Adhikari, S., Kulkarni, A., Chattanathan, S., 2015. A review on biomass gasification syngas cleanup. Appl. Energ. 155, 294–307.

Abdullah, B., Muhammad, S.A.F.A.S., Mahmood, N.A.N., 2017. Production of biofuel via hydrogenation of lignin from biomass. In: New Advances in Hydrogenation Processes- Fundamentals and Applications. IntechOpen. https://doi.org/10.5772/66108.

Abdullah, B., Muhammad, S.A.F.S., Shokravic, Z., Ismail, S., Kassime, K.A., Mahmood, A.N., Aziz, M.M.A., 2019. Fourth generation biofuel: a review on risks and mitigation strategies. Renew. Sustian. Energ. Rev. 107, 37–50.

Abraham, A., Mathew, A.K., Parka, H., Choia, O., Sindhu, R., Parameswaran, B., Pandey, A., Park, J.H., Sang, B.I., 2020. Pretreatment strategies for enhanced biogas production from lignocellulosic biomass. Bioresour. Technol. 301, 122725.

Albashabsheh, N.T., Stamm, J.L.H., 2021. Optimization of lignocellulosic biomass-to-biofuel supply chains with densification: literature review. Biomass Bioenergy 144, 105888.

Ardolino, F., Cardamone, G.F., Parrillo, F., Arena, U., 2020. Biogas-to-biomethane upgrading: a comparative review and assessment in a life cycle perspective. Renew. Sustain. Energ. Rev. https://doi.org/10.1016/j.rser.2020.110588.

Baeyens, J., Kang, Q., Appels, L., Dewil, R., Lv, Y., Tan, T., 2015. Challenges and opportunities in improving the production of bio-ethanol. Prog. Energ. Combust. Sci. 47, 60–88.

Bala, R., Mondal, M.K., 2018. Exhaustive characterization on chemical and thermal treatment of sawdust for improved biogas production. Biomass Convers. Bioref. 8, 991–1003.

Balaman, S.Y., 2019. Chapter 1- introduction to biomass- resources, production, harvesting, collection, and storage. In: Decision-Making for Biomass-Based Production Chains, the Basic Concepts and Methodologies. Elsevier, pp. 1–23.

Barman, D.N., Haque, M.A., Kang, T.H., Kim, G.H., Kim, T.H., Kim, M.K., Yun, H.D., 2014. Effect of mild alkali pretreatment on structural changes of reed (Phragmites communis trinius) straw. Environ. Technol. 35, 232–241.

Barr, M.R., Volpe, R., Kandiyoti, R., 2021. Liquid biofuels from food crops in transportation-a balance sheet of outcomes. Chemi. Eng. Sci.: X, 100090.

Bond, T., Templeton, M.R., 2011. History and future of domestic biogas plants in the developing world. Energ. Sustain. Dev. 15, 347–354.

Chen, H., 2011. Biotechnology of Lignocelluloses. Theory and Practice. Chemical Industry Press Beijing, Springer.

Chen, Y., Wu, Q., 2011. Production of biodiesel from algal biomass: current perspectives and future. In: Biofuels: Alternative Feedstocks and Conversion Processes. Elsevier, pp. 399–413.

Christofoletti, C.A., Escher, J.P., Correia, J.E., Marinho, J.F.U., Fontanetti, C.S., 2013. Sugarcane vinasse: environmental implications of its use. Waste Manage 33, 2752–2761.

Das, M., Patra, P., Ghosh, A., 2020. Metabolic engineering for enhancing microbial biosynthesis of advanced biofuels. Renew. Sustain. Energ. Rev. 119, 109562.

Demichelis, F., Laghezza, M., Chiappero, M., Fiore, S., 2020. Technical, economic and environmental assessement of bioethanol biorefinery from waste biomass. J. Clean. Prod. 277, 124111.

Dey, P., Pal, P., Kevin, J.D., Das, D.B., 2020. Lignocellulosic bioethanol production: prospects of emerging membrane technologies to improve the process - a critical review. Rev. Chem. Eng. 36 (3), 333–367.

Global bioenergy statistics, 2019. https://worldbioenergy.org/uploads/191129%20WBA%20GBS%202019_HQ.pdf.

Gracia, C., Velazquez-Martı, B., Estornell, J., 2014. An application of the vehicle routing problem to biomass transportation. Biosyst. Eng. 124, 40–52.

Guo, M., Song, W., Buhain, J., 2015. Bioenergy and biofuels: history, status, and perspective. Renew. Sustain. Energ. Rev. 42, 712–725.

Haldar, D., Purkait, M.K., 2021. A review on the environment-friendly emerging techniques for pretreatment of lignocellulosic biomass: mechanistic insight and Advancements. Chemosphere 264, 128523.

Hirani, A.H., Javed, N., Asif, M., Basu, S.K., Kumar, A., 2018. A review on first- and second generation biofuel productions. In: Biofuels: Greenhouse Gas Mitigation and Global Warming. Springer, pp. 141–154.

IEA, Renewables, 2020. Analysis and Forcast to 2025, Fuel Report- November 2020. https://www.iea.org/reports/renewables-2020/transport-biofuels.

Jönsson, L.J., Martín, C., 2016. Pretreatment of lignocellulose: formation of inhibitory by-products and strategies for minimizing their effects. Bioresour. Technol. 199 (2016), 103–112.

Joshia, G., Pandey, J.K., Rana, S., Rawat, D.S., 2017. Challenges and opportunities for the application of biofuel. Renew. Sustian. Energ. Rev. 79, 850–866.

Khan, S., Fu, P., 2020. Biotechnological perspectives on algae: a viable option for next generation biofuels. Curr. Opin. Biotechnol. 62, 146–152.

Kristiania, A., Effendib, N., Aristiawana, Y., Auliaa, F., Sudiyania, Y., 2015. Effect of combining chemical and irradiation pretreatment process to characteristic of oil palm's empty fruit bunches as raw material for second generation bioethanol. Energy Procedia 68, 195–204.

Kroumov, A.D., Scheufele, F.B., Trigueros, D.E.G., Modenes, A.N., Zaharieva, M., Najdenski, H., 2017. Chapter 11- modeling and technoeconomic analysis of algae for bioenergy and coproducts. In: Algal Green Chemistry: Recent Progress in Biotechnology. Elsevier. https://doi.org/10.1016/B978-0-444-63784-0.00011-4.

Kumar, R., Strezov, V., 2021. Thermochemical production of bio-oil: a review of downstream processing technologies for bio-oil upgrading, production of hydrogen and high value-added products. Renew. Sustain. Energ. Rev. 135, 110152.

Leong, W.H., Lima, J.W., Lam, M.K., Uemura, Y., Ho, Y.C., 2018. Third generation biofuels: a nutritional perspective in enhancing microbial lipid production. Renew. Sustain. Energy Rev. 91, 950–961.

Lin, C.Y., Lu, C., 2021. Development perspectives of promising lignocellulose feedstocks for production of advanced generation biofuels: a review. Renew. Sustain. Energ. Rev. 136, 110445.

Maitan-Alfenas, G.P., Visser, E.M., Guimaraes, V.M., 2014. Enzymatic hydrolysis of lignocellulosic biomass: converting food waste in valuable products. Curr. Opin. Food Sci. https://doi.org/10.1016/j.cofs.2014.10.001.

Martinez-Hernandez, E., Amezcua-Allieri, M.A., Sadhukhan, J., Anell, J.A., 2017. Sugarcane Bagasse Valorization Strategies for Bioethanol and Energy Production. Sugarcane - Technology and Research. https://doi.org/10.5772/intechopen.72237.

Mittal, S., Ahlgen, E., Shukla, P., 2018. Barriers to biogas dissemination in India: a review. Energy Policy 122, 361–370.

Nguyen, L.N., Kumar, J., Vu, M.T., Mohammed, J.A.H., Pathak, N., Commault, A.S., Sutherland, D., Zdarta, J., Tyagi, V.K., Nghiem, L.D., 2020. Biomethane production from anaerobic co-digestion at wastewater treatment plants: a critical review on development and innovations in biogas upgrading techniques. Sci. Total Environ. https://doi.org/10.1016/j.scitotenv.2020.142753.

Novaes, E., Kirst, M., Chiang, V., Winter-Sederoff, H., Sederoff, R., 2010. Lignin and biomass: a negative correlation for wood formation and lignin content in trees. Plant Physiol 154, 555–561.

Østby, H., Hansen, L.D., Horn, S.J., Eijsink, V.G.H., Várnai, A., 2020. Enzymatic processing of lignocellulosic biomass: principles, recent advances and perspectives. J. Ind. Microbiol. Biotechnol. 47, 623–657.

Ostovareh, S., Karimi, K., Zamani, A., 2015. Efficient conversion of sweet sorghum stalks to biogas and ethanol using organosolv pretreatment. Ind. Crops Prod. 66, 170–177.

Popp, J., Harangi-Rákos, M., Gabnai, Z., Balogh, P., Antal, G., Bai, A., 2016. Biofuels and their co-products as livestock feed: global economic and environmental implications. Molecules 21, 285.

Rashid, N., Rehman, M.S., Memon, S., Rahman, Z., Lee, K., Han, J., 2013. Current status, barriers and developments in biohydrogen production by microalgae. Renew. Sustain. Energ. Rev. 22, 571–579.

RFA, 2019. Ethanol Industry Statistics. Renewable Fuel Association, Washington, DC. http://www.ethanolrfa.org/pages/statistics.

Rodionova, M.V., Poudyal, R.S., Tiwari, I., Voloshin, R.A., Zharmukhamedov, S.K., Nam, H.G., Zayadan, B.K., Bruce, B.D., Hou, H.J.M., Allakhverdiev, S.I., 2016. Biofuel production: challenges and opportunities. Int. J. Hydrog. Energ. https://doi.org/10.1016/j.ijhydene.2016.11.125.

Rulli, M.C., Bellomi, D., Cazzoli, A., Carolis, G.D., D'Odorico, P., 2016. The water-land-food nexus of first generation biofuels. Sci. Rep. 6, 22521.

Segal, L., Creely, J.J., Martin, A.E., Conrad, C.M., 1959. An empirical method for estimating the degree of crystallinity of native cellulose using the X-ray diffractometer. Text Res. J. 29, 786–794.

Sharma, S., Kundu, A., Basu, S., Shetti, N.P., Aminabhavi, T.M., 2020. Sustainable environmental management and related biofuel technologies. J. Environ. Manage. 273, 111096.

Singh, N., Mathur, A.S., Tuli, D.K., Gupta, R.P., Barrow, C.J., Puri, M., 2017. Cellulosic ethanol production via consolidated bioprocessing by a novel thermophilic anaerobic bacterium isolated from a Himalayan hot spring. Biotechnol Biofuels 10 (73), 1–18.

Srivastava, R., Krishna, V., Sonkar, I., 2014. Characterization and management of municipal solid waste: a case study of Varanasi city, India. Int. J. Curr. Res. Acad. Rev. 2 (8), 10–16.

Statista, 2020a. Global Biofuel Production from 2000 to 2019. https://www.statista.com/statistics/274163/global-biofuel-production-in-oil-equivalent/.

Statista, 2020b. https://www.statista.com/statistics/274168/biofuel-production-in-leading-countries-in-oil-equivalent/.

Suganya, T., Varman, M., Masjukia, H.H., Renganathan, S., 2016. Macroalgae and microalgae as a potential source for commercial applications along with biofuels production: a biorefinery approach. Renew. Sustain. Energ. Rev. 55, 909–941.

Volynets, B., Ein-Mozaffari, F., Dahman, Y., 2017. Biomass processing into ethanol: pretreatment, enzymatic hydrolysis, fermentation, rheology, and mixing. Green Process Synth 6, 1–22.

Wang, D., Ai, P., Yu, L., Tan, Z., Zhang, Y., 2015. Comparing the hydrolysis and biogas production performance of alkali and acid pretreatments of rice straw using two stage anaerobic fermentation. Biosyst. Eng. 132, 47–55.

Yang, S., 2008. Plant Fiber Chemistry. China Light Industry Press, Beijing.

Yellapu, S.K., Bharti, Kaur, R., Kumar, L.R., Tiwari, B., Zhang, X., Tyagi, R.D., 2018. Recent developments of downstream processing for microbial lipids and conversion to biodiesel. Bioresour. Technol 256, 515–528.

Yin, Z., Fan, R., 1999. The research progress of plant cell wall. Bull Bot. Res. 19 (4), 407–414.

Yin, Z., Zhu, L., Li, S., Hu, T., Chu, R., Mo, F., Hu, D., Liu, C., Li, B., 2020. A comprehensive review on cultivation and harvesting of microalgae for biodiesel production: environmental pollution control and future directions. Bioresour. Technol. 301, 122804.

Zhang, J., Wang, Y., Zhang, L., Zhang, R., Liu, G., Cheng, G., 2014. Understanding changes in cellulose crystalline structure of lignocellulosic biomass during ionic liquid pretreatment by XRD. Bioresour. Technol. 151, 402–405.

Zhang, J., Zhang, X., 2019. Chapter 15: The thermochemical conversion of biomass into biofuels. In: Biomass, Biopolymer-Based Materials, and Bioenergy. Elsevier, pp. 327–328.

SECTION II

Biomass into biofuels and bioenergy

CHAPTER

2

Technology to convert biomass to biooil: challenges and opportunity

Huei Yeong Lim, Suzana Yusup

Biomass Processing Lab, National Higher Institution, Centre for Biofuel and Biochemical Research, Institute for Self-sustainable Building, Chemical Engineering Department, Universiti Teknologi PETRONAS, Bandar Seri Iskandar, Perak, Malaysia

2.1 Introduction

Biooil is a dark brown liquid, generally high in water content, and containing hundreds of organic compounds included acids, phenols, and oxygenates (Hu and Gholizadeh, 2019). At room temperature, biooil is also unstable due to the existence of some reactive molecules and oligomers with extremely large molecular weight. Aging of biooil leads to more water formed, phase separation, and increasing viscosity (Hu and Gholizadeh, 2019). In addition, the high water content and acidity also prohibits biooil from being applied into the existing petroleum infrastructure (Wang et al., 2017). Biooil has two layers: organic oil phase and aqueous phase (Chang, 2018). The organic oil phases possess darker color and higher energy density than the aqueous phase. Thus, liquid–liquid separation of the organic oil phase out from the biooil is needed prior to characterization. However, characterization of biooil with the mixed two phases has been conducted (Chang, 2018). Through ultimate analysis, the elemental composition of biooil is determined, which gives the carbon, nitrogen, hydrogen, sulfur, and oxygen contents. Other physiochemical properties evaluated for the biooil included water content, solid content, ash content, density, kinematic viscosity, pH, and total acid number (TAN) (Chang, 2018). These properties are able to show the heat of combustion, fouling and slagging tendencies, ease of delivery, and corrosiveness of the biooil, which are important in designing the suitable combustion system for it.

2.1.1 Economic of biooil

Economic potential is one of the important parameters for biooil to be commercialized. Consumers are typically cost driven, and therefore, if the selling price of biofuels manufactured from biooil is much higher compared with the conventional fuels, it will be difficult to penetrate the market and lead to lack in its demand. In the effort to reduce the production

Biofuels and Bioenergy
https://doi.org/10.1016/B978-0-323-85269-2.00025-3

costs of biomass biooil plant, one can tackle two factors: energy consumption of the plant and by-products sales (Patel et al., 2016a). The major by-product of pyrolysis plant is the char residue, namely biochar, from the process. Some applications of biochar included absorbent (Carrier et al., 2012; Liu et al., 2012), activated carbon (Choi et al., 2015), and catalyst (Li et al., 2020; Qiu et al., 2020; Yang et al., 2020). It is estimated that if biochar is being considered as a by-product and sold, it can help to reduce the total operational cost of the plant for up to 18% (Patel et al., 2016a).

One useful tool to evaluate the economic potential is the technoeconomic analysis (TEA). Various studies had been made to evaluate the economic potential of biooil produced from several types of feedstocks using TEA. Minimum fuel selling price (MFSP) is a good indicator of how cheap the process can be to produce biooil or biofuels. The MFSP can also be used to compare with other similar processes, as well as with conventional fuels (e.g., fossil fuels). It should be noted that MFSP is the minimum price the product should be sold to reach zero net present value for the plant; therefore the actual selling price of the product will be higher than MFSP depending on the profit margins desired by the manufacturers (Nguyen and Clausen, 2019). The unit is in USD per gasoline gallon equivalent ($/gge), where the prices are adjusted to match with the energy content per gallon of gasoline, giving more comparable values across different fuels. Table 2.1 summarizes TEA studies on pyrolysis and biooil from literatures. It can be observed that the MFSP of fuels produced from biooil differs largely with the process routes the authors had chosen. Comparing with the market price of gasoline, for example, according to Energy Information Administration (EIA), the average gasoline price in the United State for the year 2019 is 2.50 $/gallon (U.S. Energy Information Administration, 2020). Therefore, the MFSP shown in Table 2.1 has the potential to compete with the conventional gasoline in the market economically.

2.1.2 Feedstocks available for biooil production

There are multiple choices of feedstocks that had been investigated to produce biooil. Among them, lignocellulosic biomass attracted a lot of attention due to its properties: abundancy and carbon neutrality. Global lignocellulosic biomass production from various sources (e.g., forestry and timber industry, agricultural activities) is estimated to be around 146 billion tons annually (Pang, 2019). The large number of feedstocks available and the ease to obtain them had provided a large potential for the sustainability for the operation of the industry and process plants. On the other hand, the carbon neutrality of lignocellulosic biomass releases the same amount of CO_2 when combusted, with the absorbed amount during photosynthesis, therefore resulting in carbon neutrality throughout the life cycle (Ibarra-Gonzalez and Rong, 2019; Pang, 2019). In addition, unlike other renewable energy sources, versatility of biomass is another advantage in the energy sector. Biomass can be processed into various products in solid (e.g., biochar), liquid (e.g., biogasoline, biodiesel), and gaseous (e.g., biohydrogen) forms for different applications (Alauddin et al., 2010). Thus, all these advantages made biomass attractive to researchers for the development of renewable and sustainable energy source.

Lignocellulosic biomass constitutes majorly three items: cellulose, hemicellulose, and lignin, where they can make up for about 40%–60%, 15%–30%, and 10%–25%, respectively (Wang et al., 2017). Other than that, there will also be a slight number of extractives and ashes in the lignocellulosic biomass, which in often are not part of the plant cells. Individually, all the components

TABLE 2.1 Technoeconomic analysis studies of biooil production and processing plants, and their minimum fuel selling price.

Feedstocks	Products	Technology/Process	Capacity (MT/day)	Location	Base year	MFSP ($/gge°)	Reference
Switchgrass	Gasoline and diesel	Fast pyrolysis + hydrotreating	2000.00	USA	2014	5.10	Meyer et al. (2020)
Corn stover						4.20	
Hybrid poplar						4.00	
Tulip poplar						4.20	
Clean pine						3.80	
Maple wood	Biooil	Catalytic fast hydropyrolysis + ex situ hydrodeoxygenation	2000.00	Denmark	2017	2.60	Nguyen and Clausen (2019)
Corn stover	Clean naphtha and diesel range products	Fast pyrolysis + hydroprocessing with hydrogen self-generation	2000.00	Not specified	Not specified	3.09	Wright et al. (2010)
		Fast pyrolysis + hydroprocessing with hydrogen purchased				2.11	
Corn stover	Gasoline and diesel	Fast pyrolysis + hydroprocessing	2000.00	Not specified	Not specified	2.57	Brown et al. (2013)
Woody biomass	Gasoline and diesel	Catalytic fast pyrolysis (in situ) + hydrotreating	2000.00	Texas, USA	2007–2012	1.11	Li et al. (2015)
		Catalytic fast pyrolysis (ex situ) + hydrotreating				1.13	
Pulpwood	Gasoline and diesel	Catalytic fast pyrolysis + FCC vacuum gas oil coprocessing	1038.96	Not specified	Not specified	2.60	Wu et al. (2019)
		Fast pyrolysis + hydrotreating + FCC vacuum gas oil coprocessing	989.52			2.63	
Corncob	Bio-oil	Fast pyrolysis + hydrotreating	2323.44	USA	2017	1.47	Brigagão et al. (2019)

have their own characteristics and degradation properties under high temperature condition. Cellulose generally degrades between the range of 315 and 400°C with varying activation energy at 166–250 kJ/mol (Chen et al., 2019b). After that, hemicellulose is easier to decompose than cellulose. The typical degradation temperature ranged at 220–315°C with almost half of cellulose's activation energy at 95.39 kJ/mol (Chen et al., 2019b). Lastly, lignin degradation happens through a large range of temperature, from 44 to 824°C in multiple stages (weak functional group bond breaking, major β-O-4 linkage breakage, and other bond cleavages), with 174.4 kJ/mol activation energy (Yeo et al., 2019).

Lignocellulosic biomass comes from various sources, mainly from the forestry and timber industry, and agricultural-related industry such as plantations and processing mills. In this regard, database such as Food and Agriculture Organization of the United Nation (FAOSTAT) that extensively collects the production data of various crops across the world would be a great tool (Food and Agriculture Organization of the United Nation, 2020). For example, Malaysia is a main oil palm producer in the world since 1960s at about 32% of the total market (Iskandar et al., 2018). From FAOSTAT, the annual production of fresh oil palm fruit in Malaysia for year 2018 is at around 98.5 million tons. The residues generated from palm oil mills are 13.5% mesocarp fibers, 5.5% palm kernel shells, and 22.0% empty fruit bunches from the total mass of fresh oil palm fruit processes (Loh, 2017). Therefore, one can estimate that from the oil palm industry in Malaysia alone can produce up to 40.4 million tons of biomass residues. This example showed the availability of biomass from one agricultural sector in one nation, and if to imagine the entire agriculture and forestry activities globally, the potential to power the world with biomass could be a reality. Due to the carbon neutral nature of biomass, the carbon emission could be greatly reduced and provides a solution to the issues of global warming and climate change.

Other than lignocellulosic biomass, other wastes are also being studied as a feedstock to produce biooil. Plastic, for example, is a type of major wastes disposed in the modern daily life and possesses disposal issues due to its low biodegradability. In year 2018, the total plastic produced around the globe reached 359 million tons, with half of it being manufactured in Asia (Europe and EPRO, 2019). Among various types of plastic, polypropylene (PP), high-density polyethylene (HDPE), and low-density polyethylene (LDPE) constituted almost half of the European demand for plastic resin, mainly from the packaging industry (Europe and EPRO, 2019). The rate of recycling of plastic wastes is also low at only about 10%, and the typical processes of plastic waste recycle also featured with high cost stages such as sorting, cleaning, grinding, and extrusion (Hahladakis and Iacovidou, 2019). Therefore, it would be attractive if the plastic wastes can be recycled or manufactured into value-added products via a simpler and cost-effective method, such as pyrolysis, to recover the energy content.

Transportation sector is one of the largest sectors in the world and, from the automobiles being driven everyday, produces a lot of tire wastes. It is approximated that about 1 billion tires were discarded annually around the globe, and only at about 10% of them were being recycled as the tires were specifically designed to be tough and robust, creating challenges in recycling processes (Goldstein Market Intelligence, 2020). Due to its resistances, waste tires commonly require at least 80 years to degrade; therefore, landfilling is also an undesired disposal method for tires (Chao et al., 2020). Incineration is also not an interesting solution for waste tires disposal, as multiple hazardous matters will be released to the atmosphere, such as dioxins, volatile organic compounds (VOCs), and polycyclic aromatic hydrocarbons (Danon et al., 2015). Thus, thermochemical method might be a suitable disposal solution for the waste tires to produce fuels or other value-added products from the waste tires, while also limiting the harmful emissions.

2.2 Thermochemical biomass conversion method for biooil

Biomass can be exploited into energy in various methods. Direct combustion is the easiest way to obtain heat from woods, which had been existed since ancient time. However, this method is inefficient and produces various pollutants such as dusts and acid rain gases, making it no difference in the effort of reducing the environmental pollution (Alauddin et al., 2010). Hence, conversion technologies for biomass into biofuels are much needed for the efficient and environmentally friendly of extracting energy from biomass. Due to different biomass sources and the desired biofuels, the conversion technologies and pathways could be different with each other (Ibarra-Gonzalez and Rong, 2019). Generally, biomass conversion technologies can be categorized into three main pathways: physical, thermochemical, and biochemical. Among them, pyrolysis and hydrothermal liquefaction (HTL) are the common thermochemical pathway to produce biooil from biomass feedstock, where the general reaction pathway of both processes can be seen in Fig. 2.1.

2.2.1 Pyrolysis

Pyrolysis is categorized into thermochemical pathway of biomass conversion, together with liquefaction, gasification, etc., which uses heat to produce biofuels. By definition, pyrolysis is a process of thermally converting a biomass feedstocks into char, biooil, and noncondensable gases under oxygen-free environment at elevated temperature (Ibarra-Gonzalez and Rong, 2019; Ong et al., 2019; Patel et al., 2016b). Among various types of pyrolysis, the most promising technology is fast pyrolysis, which provides high yields of biooil within couple of seconds (Chen et al., 2019a,b). Fast pyrolysis distinguished itself as having high heating rate, short reaction time, moderate reaction temperature, and a last quenching for vapor condensation (Chen et al., 2019a,b; Ibarra-Gonzalez and Rong, 2019). In this method, biomass would decompose into two products: solid biochar and volatile matters, carried away by a flow of inert carrier gas. Then, separation of biochar would be performed, typically by a gas–solid cyclone. The hot gas was quenched to condensate the volatile compounds,

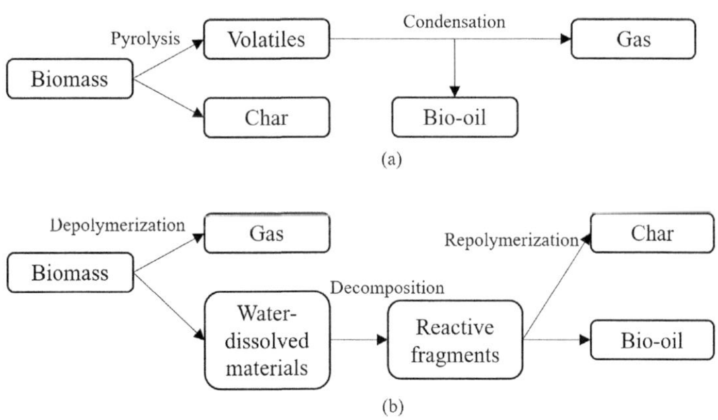

FIGURE 2.1 General reaction pathway of (A) pyrolysis and (B) hydrothermal liquefaction.

producing biooil, leaving noncondensable gases flowing out and collected. Noncondensable gases normally included H_2, CH_4, CO, and CO_2 (Ong et al., 2019). Among the three forms of products of pyrolysis, biooil had particularly received large interest and had been extensively studied for its wide range of applications via upgrading into liquid biofuels or biochemicals (Patel et al., 2016).

The yields and qualities of each product (e.g., biochar, biooil, and gas) are affected by the parameters of pyrolysis (Ibarra-Gonzalez and Rong, 2019). Generally, pyrolysis is affected by reaction temperature, particle size, and the presence of catalyst. Reaction temperature is important in pyrolysis, as it influences mainly the yields of different outputs. Typically, low temperature (e.g., below 450°C) produces more solid biochar, moderate temperature (e.g., 450–800°C) yields more biooil, and temperature beyond that leads to more gas formed (Ong et al., 2019). At lower temperature, the biomass is unable to be completely decomposed and released as volatiles, therefore leaving larger portion of biomass as solid residue, increasing the biochar yields. Due to this, the biochar produced at lower temperature also contains lower carbon contents (Sharma et al., 2015). Furthermore, at the higher temperature, the condensable volatile matters that constituted the biooil would start to have secondary decomposition, turning into noncondensable gases (Sharma et al., 2015). Therefore, pyrolysis is temperature sensitive and optimal at moderate temperature for maximizing biooil production.

Particle size influences as well the pyrolysis product yields. Generally, a bigger particle size increases the solid biochar and gas yields until a certain critical limit where the changes in particle sizes become insignificant toward affecting the yields (Hu and Gholizadeh, 2019). This is due to with bigger particle size, the mass and energy transfer resistance is bigger within the particles, creating a thermal gradient to protect the inner part from getting thermally decomposed, while the decomposed volatile matters might be trapped sufficiently long to trigger secondary decomposition reactions into gases (Sharma et al., 2015). Other than yields, the particle size also influences the compositions of the biooil. One interesting finding is that the biooil water content increases with larger particle size (Hu and Gholizadeh, 2019; Sharma et al., 2015). Hence, particle size of the biomass fed should be sufficiently low (>1 mm) for the effects to be significant, and the particle size should be minimal to avoid the high water content, which reduces the quality of biooil as biofuels.

Catalyst addition into the pyrolysis system to enhance the biooil properties had received considerable interests among research community. This is mainly because of inferior properties of biooil produced from noncatalytic pyrolysis such as low calorific value, highly viscous, highly acidic, and instability, hindering biooil from being used as biofuels (Rahman et al., 2018). Therefore, a wide range of catalysts (e.g., acidic catalysts, basic catalyst, inorganic mineral) had been studied with the aims of maximizing the biooil yields and qualities. For example, abundant of oxygenates decrease the heating value of the biooil, where it can be reduced via deoxygenation by using zeolite-based catalysts (Sharma et al., 2015). Deoxygenation reactions included decarboxylation, decarbonylation, and dehydration. Other than that, acidic catalyst is also capable for cracking, such as C–C bonds cleavages and aromatic side chain scission. Catalyst loading is also another crucial factor to consider, as it affects the catalytic reaction rates in the system. Generally, higher catalyst to biomass ratio promotes the formation of tar as a result of secondary reactions, which reduces the yields of biooil (Rahman et al., 2018). Thus, selection of a suitable catalyst according to the biomass feedstock and the loading feed is vital for producing biooil of high quality.

2.2.2 Hydrothermal liquefaction

HTL is another common thermochemical biomass conversion technology to turn biomass feedstock into liquid product, namely crude oil or biooil. HTL is basically the decomposition of biomass thermally in water medium, in which some described the process as "pyrolysis in water" (Ibarra-Gonzalez and Rong, 2019; Ong et al., 2019). Unlike pyrolysis, in HTL the biomass feedstock is not necessarily needed to be completely dried; therefore, it is suitable for biomass with high moisture content such as sewage sludges and microalgae, where the drying step is energy intensive and costly. Commonly, HTL operates at a lower temperature as pyrolysis within the range of 270–370°C but at elevated pressure of 4–20 MPa (Yang et al., 2019). Therefore, as compared with pyrolysis, HTL provides the advantages of reduced energy required for biomass feedstock drying and operational temperature. To date, the exact mechanism of HTL to convert biomass into biooil is unclear; however, commonly accepted reaction pathway consists of (1) defragmentation via hydrolysis and depolymerization, (2) chemically decomposition, and (3) repolymerization of reactive fragments to form biooil and char (Gollakota et al., 2018). Process parameters that influence the product yields include reaction temperature and residence time.

As one of the thermochemical methods, reaction temperature in the process is one of the important parameters in HTL. In subcritical and supercritical water condition, the effect of rising temperature differs. In subcritical water medium, increasing reaction temperature tends to promote biooil yield due to the increasing activities of biomass repolymerization and decomposition; however, once the medium went into supercritical water region, the yield of char and gas increases with increasing temperature due to higher rate of repolymerization and breakdown of intermediate products (Xue et al., 2016). For example, barley straw HTL at temperature range from 280 to 400°C was performed, which included subcritical and supercritical condition (Zhu et al., 2015). From the study, maximum biooil yield was measured with 34.9% at 300°C, and the yield continued to drop with increasing temperature until reaching 19.9% at 400°C (Zhu et al., 2015). Similarly, Yuan et al. (2007) also performed the HTL of rice straw using alcohol–water mixture under both subcritical and supercritical condition. The study found that the biooil yields were obtained at maximum amount between the temperature 300–310°C across different alcohol:water ratio. Therefore, the relationship between the reaction temperature and the biooil yield showed a negative parabolic, or n-shaped, curve.

Residence time of HTL generally referred as the reaction time of the process at the maximum or desired temperature, excluding the time for heating and cooling (Xue et al., 2016). Similar with reaction temperature, HTL seemed to have an optimum range of residence time for maximum yield for biooil. Eboibi et al. (2014) studied the microalga *Tetraselmis* sp. in HTL with varying reaction temperatures and residence times. In the study, it is found that the optimum residence time for maximum biooil yield is different at different reaction temperatures. For instance, at reaction temperature of 330°C, the optimum residence time is 45 min, obtaining a yield of 62%, while at 350°C, the optimum time shifted greatly to only 5 min with 65% yield. The results have shown that the effect of residence time is highly associated to the reaction temperature, and the great difference of optimum residence time between 330 and 350°C is yet able to be explained clearly due to the highly complex chemistry behind the liquefaction process (Eboibi et al., 2014).

Amount of water being used in HTL is another key parameter in affecting the process output. The water medium can act as solvent and hydrogen donor to enhance hydrolysis for the large carbohydrates in the feedstock, promoting the repolymerization process (Anastasakis and Ross, 2011; Xue et al., 2016). The ratio of biomass-to-water is given in the unit of g/mL. From the study by Anastasakis and Ross (2011), it is observed that the biomass-to-water ratio is increased from 2:10 to 5:30. Yield of biooil increases steadily until reaching a maximum of 17.5% at 3:30, and the yield becomes stagnant afterward. However, in another study of HTL of cornstalk, Liu et al. (2013) observed a different trend. From the highest ratio of 1:6 to 1:10 and 1:14, the biooil yields obtained are 19.6%, 33.9%, and 31.2%, respectively. The decrease in biooil yield between biomass-to-water ratio of 1:10 and 1:14 is related with the increasing volatile organic yields, in which it is speculated that the excess water promotes the evolution of biooil to volatile organics (Liu et al., 2013). Furthermore, excess water in the system might also cause the reaction of hydrolysis and repolymerization to compete with each other, which causes lower biooil yields (Xue et al., 2016).

2.2.3 Comparisons between pyrolysis and hydrothermal liquefaction

Tables 2.2 and 2.3 list out some recent studies on pyrolysis and HTL on various biomass sources focusing on the production of biooil and its properties. By comparing both of the technologies, one may find that the biooil yields typically fall around 30–60 wt%. Therefore, it is not possible to give a simple conclusion that one of the technologies is always better than the other and that they are suitable for different scenarios and cases. For instance, microalgae biomass, which contains high water content, is normally better suited with HTL for conversion into value-added products, as this will save the energy and cost to dry the microalgae for pyrolysis otherwise. In terms of the process parameters, one can see that pyrolysis typically has higher reaction temperature than HTL; however, HTL would require an elevated pressure to operate. This indicates that both of the technologies have their own pros and cons in cost and safety, as higher temperature or pressure means that more energy and higher safety risks are required to operate. On the other hand, fast pyrolysis is definitely on the lead in time saving, as other processes including slow pyrolysis and HTL take quite some time up to 2 h for the process. In a nutshell, extensive analysis should be conducted to make informed decision between the different technologies for biooil production depending on the scenarios one has encountered.

2.3 Biooil quality and testing standard

Fast pyrolysis biooil (FPBO) had entered the stage of production at commercial scale with the commission of production plants by Fortum and Empyro, located in Finland and the Netherlands, respectively, as well as Brazil (Ensyn), which is at the designing stage (Oasmaa et al., 2015). To assist the introduction of biooil into the market, establishment of standards for the quality of biooil is an important factor. For this, ASTM International (formerly American Society for Testing and Materials) had set up a standard for FPBO to be used as burner fuel, namely ASTM D7544. Table 2.4 showed the parameters and their limits for the two grades of FPBO, as well as the corresponding testing standards for the parameters.

TABLE 2.2 Recent studies on pyrolysis and hydrothermal liquefaction of biomass for biooil production.

Feedstocks	Process	Main parameters	Catalyst	Biooil yield (wt%)	Reference
Oil palm empty fruit bunches	Fast pyrolysis	Reaction temperature: 550 °C Particle size: <2000 μm Equipment: Fixed bed reactor	None	43.49	Thangalazhy-Gopakumar et al. (2018)
			CaO Loading: 10 wt%	58.49	
Rice husk	Fast pyrolysis	Reaction temperature: 550 °C Particle size: Not specified Equipment: Two stage fixed bed reactor	None	38.00	Zhang et al. (2018)
			Fe-modified ZSM-5 Loading: 83 wt%	28.00−32.00	
Corn stover	Fast pyrolysis	Reaction temperature: 500 °C Particle size: 300−750 μm Equipment: Free fall pyrolyzer	None	45.00−55.00	Shah et al. (2012)
Sugarcane straw	Fast pyrolysis	Reaction temperature: 450 °C Particle size: 500 μm Equipment: Continuous tube reactor	Dolomite Loading: 10 wt%	36.15	Charusiri and Vitidsant (2017)
Rice husk	Slow pyrolysis	Reaction temperature: 650 °C Particle size: 250−400 μm Equipment: Fixed bed quartz reactor Heating rate: 100 °C/min	None	15.80−17.10	Lazzari et al. (2019)
Sugarcane bagasse	Slow pyrolysis	Reaction temperature: 600 °C Particle size: 2000 μm Equipment: Fixed bed reactor Heating rate: 50 °C/min	None	46.78	Rabiu et al. (2018)
			HZSM-5 Loading: 5 wt%	40.83	
Rice straw + LDPE	Microwave copyrolysis	Reaction temperature: 450 °C Particle size: Not specified Equipment: Microwave fixed bed reactor LDPE loading: 30 wt%	None	22.75	Bu et al. (2018)
			ZSM-5 Loading: 11.04 wt%	28.40	
Food waste	Hydrothermal liquefaction	Reaction temperature: 320 °C Reaction pressure: 150 bar Residence time: 30 min Particle size: 300−750 μm Water added: None Equipment: Stainless steel batch reactor	None	32.90	Chen et al. (2019a)

Continued

TABLE 2.2 Recent studies on pyrolysis and hydrothermal liquefaction of biomass for biooil production.—cont'd

Feedstocks	Process	Main parameters	Catalyst	Biooil yield (wt%)	Reference
Microalgae (*Scenedesmus* sp.)	Hydrothermal liquefaction	Reaction temperature: 250–350°C Reaction pressure: 50–150 bar Residence time: 7–30 min Particle size: Not specified Water used: 17.7 mL/g feedstock Equipment: Stainless steel tubular reactor	None	20.00–36.00	Wądrzyk et al. (2018)
Microalgae (*Coelastrum* sp.)	Single step hydrothermal liquefaction	Reaction temperature: 320 °C Reaction pressure: 200 bar Residence time: 120 min Particle size: Not specified Water used: 6 mL/g feedstock Equipment: Stainless steel tubular reactor	None	29.50	Prapaiwatcharapan et al. (2015)
	Two step hydrothermal liquefaction	Residence time: 120 min Particle size: Not specified Water used: 6 mL/g feedstock Equipment: Stainless steel tubular reactor First step Reaction temperature: 200 °C Reaction pressure: 70 bar Second step Reaction temperature: 320 °C Reaction pressure: 200 bar	None	32.00	
Rice straw	Hydrothermal liquefaction	Reaction temperature: 200–300°C Reaction pressure: Not specified Residence time: 120 min Particle size: 200–1000 μm Water used: 15 mL/g feedstock Equipment: Stainless steel autoclave	None NiO nanoplates Loading: 9 wt%	37.60–62.10 36.50–61.20	Younas et al. (2017)

TABLE 2.3 Physiochemical properties of biooil produced from fast pyrolysis and hydrothermal liquefaction from recent literatures.

Feedstock	Elemental composition (wt%) C	H	N	S	O[a]	Heat of combustion (MJ/kg)	Water content (wt%)	Solids content (wt%)	Ash content (wt%)	Viscosity @40 °C (mm^2/s)	Density @20 °C (kg/dm^3)	pH	TAN (mg KOH/g)	Reference
Fast pyrolysis														
Corn stover	78.00	9.02	1.86	0.38	10.60	33.80	12.80	–	0.18	–	–	–	–	Capunitan and Capareda (2012)
Oil palm empty fruit bunches	69.35	9.61	0.74	–	20.02	36.06	7.90	–	–	–	–	–	–	Abdullah et al. (2010)
Oil palm empty fruit bunches	25.33	9.32	0.32	0.45	64.59	10.43	50.00	–	–	45.47	1.07	2.69	–	Chan et al. (2014)
Oil palm empty fruit bunches	44.87	5.89	0.92	–	48.32	16.26	8.62	–	8.24	–	–	–	–	Park et al. (2019)
Oil palm empty fruit bunches	58.65	7.02	2.74	<0.10	30.14	24.90	–	–	–	–	1.18	–	110.00	Kim et al. (2013)
Palm kernel shell	71.5	10.3	2.82	0.12	15.9	30.90	–	–	–	–	1.23	–	113.00	
Jatropha seed shell cake	65.8	8.92	5.62	0.19	18.80	30.20	–	–	–	–	1.13	–	153.70	
Rice husk	31.63	8.73	0.53	0.34	57.27	17.20	32.95	0.08	0.19	–	1.17	3.22		Cai et al. (2019)
Rice husk	23.38	10.39	0.51	<0.10	65.63	28.71	52.60			1.68	1.07	3.00	55.54	Abu Bakar and Titiloye (2013)
Sugarcane bagasse	45.24	7.47	0.36	–	46.93	18.48	40.21	–	2.73	24.54	1.27	2.40	–	Treedet and Suntivarakorn (2018)

Continued

TABLE 2.3 Physiochemical properties of biooil produced from fast pyrolysis and hydrothermal liquefaction from recent literatures.—cont'd

Feedstock	Elemental composition (wt%) C	H	N	S	O[a]	Heat of combustion (MJ/kg)	Water content (wt%)	Solids content (wt%)	Ash content (wt%)	Viscosity @40 °C (mm^2/s)	Density @20 °C (kg/dm^3)	pH	TAN (mg KOH/g)	Reference
Hydrothermal liquefaction														
Food waste	72.20	9.69	4.28	0.95	12.88	34.79	0.30	-	1.00	–	–	–	–	Chen et al. (2019a)
Litsea cubeba seed	76.20	11.90	1.60	–	10.40	40.80	8.00	–	0.10	–	–	–	100	Wang et al. (2013)
Microalgae *Nannochlorosis* sp.	74.00	10.20	5.40	–	9.50	31.50	–	–	–	–	–	–	–	Li et al. (2014)
Microalgae *Chlorella* sp.	75.60	12.00	0.30	–	11.50	34.20	–	–	–	–	–	–	–	
Microalgae *Dunaliella Tertiolecta*	63.55	7.66	3.71	–	25.08	30.74	8.73		0.20–0.50		1.18	3.5–4.2		Shuping et al. (2010)
Sewage sludge	71.90	6.50	3.30	3.50	4.80	33.00	0.60	–	–	–	–	–	26.10	Mujahid et al. (2020)
Corn straw	72.43	7.81	0.99	0.28	13.29	33.24	–	–	–	–	–	–	–	Tian et al. (2020)
Soybean straw	73.16	7.38	2.19	0.22	12.68	32.98	–	–	–	–	–	–	–	
Peanut straw	72.28	8.09	2.07	0.24	12.26	33.78	–	–	–	–	–	–	–	
Rice straw	72.68	7.96	1.69	0.33	12.40	33.70	–	–	–	–	–	–	–	

[a] *Calculated by difference.*

TABLE 2.4 ASTM D7544 burner fuel standard for FPBO and the suitable testing standards.

Parameter	FPBO grade		Testing standard
	Grade G	Grade D	
Min heat of combustion (MJ/kg)	15	15	ASTM D240
Max water content (wt%)	30	30	ASTM E203
Max solids content (wt%)	2.5	0.25	ASTM D7579
Max kinetic viscosity @ 40°C (mm^2/s)	125	125	ASTM D445
Density @ 20°C (kg/dm^3)	1.1–1.3	1.1–1.3	ASTM D4052
Max sulfur content (wt%)	0.05	0.05	ASTM D5453
Max ash content (wt%)	0.25	0.15	ASTM D482
pH	Report	Report	ASTM E70
Min. flash point (°C)	45	45	ASTM D93B
Max. pour point (°C)	−9	−9	ASTM D97

2.4 Conclusion

Biooil can be treated as a substitute to the current crude petroleum oil, where it can be refined into various valuable products. In this current situation, biooil has a large potential to replace the crude oil while still maintaining similar process chains as petroleum industry, in which biooil refining might be able to also reuse the same facilities with minimum retrofit. The TEA showed that the prices of biooil and its value-added products are getting more competitive with conventional fuels such as gasoline and diesel. In addition, the biomass feedstocks are also abundant and readily available for the process plants to be sustainable, while also providing environmentally friendly disposal methods with certain economical returns. Pyrolysis and hydrothermal liquefaction are two common technologies to convert biomass into biooil, in which both are thermochemical methods. Various factors, for example, the reaction temperature, affect the quality and quantity of the biooil produced from the process. Researchers around the world had studied the biooil production with many different conditions and feedstocks, and the technologies have progress into commercialization stage. To market the biooil, standards have also been established to provide the guideline and reference for the manufacturers and the buyers.

Acknowledgment

The authors of this chapter would like to acknowledge the support given by the Ministry of Higher Education, Malaysia to the HICoE-CBBR, and Universiti Teknologi PETRONAS that enable the research to be conducted.

References

Abdullah, N., Gerhauser, H., Sulaiman, F., 2010. Fast pyrolysis of empty fruit bunches. Fuel 89, 2166–2169.

Abu Bakar, M.S., Titiloye, J.O., 2013. Catalytic pyrolysis of rice husk for bio-oil production. In: Journal of Analytical and Applied Pyrolysis.

Alauddin, Z.A.B.Z., Lahijani, P., Mohammadi, M., Mohamed, A.R., 2010. Gasification of lignocellulosic biomass in fluidized beds for renewable energy development: a review. Renew. Sustain. Energy Rev. 14, 2852–2862.

Anastasakis, K., Ross, A.B., 2011. Hydrothermal liquefaction of the brown macro-alga *Laminaria Saccharina*: effect of reaction conditions on product distribution and composition. Bioresour. Technol. 102, 4876–4883.

Brigagão, G.V., de Queiroz Fernandes Araújo, O., de Medeiros, J.L., Mikulcic, H., Duic, N., 2019. A techno-economic analysis of thermochemical pathways for corncob-to-energy: fast pyrolysis to bio-oil, gasification to methanol and combustion to electricity. Fuel Process. Technol. 193, 102–113.

Brown, T.R., Thilakaratne, R., Brown, R.C., Hu, G., 2013. Techno-economic analysis of biomass to transportation fuels and electricity via fast pyrolysis and hydroprocessing. Fuel 106, 463–469.

Bu, Q., Liu, Y., Liang, J., Morgan, H.M., Yan, L., Xu, F., Mao, H., 2018. Microwave-assisted co-pyrolysis of microwave torrefied biomass with waste plastics using ZSM-5 as a catalyst for high quality bio-oil. J. Anal. Appl. Pyrolysis. 134, 536–543.

Cai, W., Kang, N., Jang, M.K., Sun, C., Liu, R., Luo, Z., 2019. Long term storage stability of bio-oil from rice husk fast pyrolysis. Energy.

Capunitan, J.A., Capareda, S.C., 2012. Assessing the potential for biofuel production of corn stover pyrolysis using a pressurized batch reactor. Fuel 95, 563–572.

Carrier, M., Hardie, A.G., Uras, Ü., Görgens, J., Knoetze, J., 2012. Production of char from vacuum pyrolysis of South-African sugar cane bagasse and its characterization as activated carbon and biochar. J. Anal. Appl. Pyrolysis 96, 24–32.

Chan, Y.H., Dang, K.V., Yusup, S., Lim, M.T., Zain, A.M., Uemura, Y., 2014. Studies on catalytic pyrolysis of empty fruit bunch (EFB) using Taguchi's L9 Orthogonal Array. J. Energy Inst. 87, 227–234.

Chang, S.H., 2018. Bio-oil derived from palm empty fruit bunches: fast pyrolysis, liquefaction and future prospects. Biomass and Bioenergy 119, 263–276.

Chao, L., Zhang, C., Zhang, L., Gholizadeh, M., Hu, X., 2020. Catalytic pyrolysis of tire waste: impacts of biochar catalyst on product evolution. Waste Manag 116, 9–21.

Charusiri, W., Vitidsant, T., 2017. Upgrading bio-oil produced from the catalytic pyrolysis of sugarcane (*Saccharum officinarum* L) straw using calcined dolomite. Sustain. Chem. Pharm 6, 114–123.

Chen, W.H., Lin, Y.Y., Liu, H.C., Chen, T.C., Hung, H.C., Chen, C.H., 2019a. Analysis of physicochemical properties of liquefaction bio-oil from food waste. In: Energy Procedia, 158, pp. 61–66.

Chen, X., Che, Q., Li, S., Liu, Z., Yang, H., Chen, Y., Wang, X., Shao, J., Chen, H., 2019b. Recent developments in lignocellulosic biomass catalytic fast pyrolysis: strategies for the optimization of bio-oil quality and yield. Fuel Process. Technol. 196, 106180.

Choi, G.G., Oh, S.J., Lee, S.J., Kim, J.S., 2015. Production of bio-based phenolic resin and activated carbon from bio-oil and biochar derived from fast pyrolysis of palm kernel shells. Bioresour. Technol. 178, 99–107.

Danon, B., van der Gryp, P., Schwarz, C.E., Görgens, J.F., 2015. A review of dipentene (DL-limonene) production from waste tire pyrolysis. J. Anal. Appl. Pyrolysis 112, 1–13.

Eboibi, B.E., Lewis, D.M., Ashman, P.J., Chinnasamy, S., 2014. Effect of operating conditions on yield and quality of biocrude during hydrothermal liquefaction of halophytic microalga *Tetraselmis* sp. Bioresour. Technol. 170, 20–29.

Europe Plastics, EPRO, 2019. Plastics - the Facts 2019.

Food and Agriculture Organization of the United Nation, 2020. FAOSTAT [WWW Document]. URL. http://www.fao.org/faostat/en/#data/QC. (Accessed 26 September 2020).

Goldstein Market Intelligence, 2020. Global Tire Recycling Industry Analysis by Rubber Type, by Product Type, by End User and by Geography & COVID-19 Impact with Market Outlook 2017–2030.

Gollakota, A.R.K., Kishore, N., Gu, S., 2018. A review on hydrothermal liquefaction of biomass. Renew. Sustain. Energy Rev. 81, 1378–1392.

Hahladakis, J.N., Iacovidou, E., 2019. An overview of the challenges and trade-offs in closing the loop of post-consumer plastic waste (PCPW): focus on recycling. J. Hazard. Mater. 380, 120887.

Hu, X., Gholizadeh, M., 2019. Biomass pyrolysis: a review of the process development and challenges from initial researches up to the commercialisation stage. J. Energy Chem. 39, 109–143.

Ibarra-Gonzalez, P., Rong, B.-G., 2019. A review of the current state of biofuels production from lignocellulosic biomass using thermochemical conversion routes. Chinese J. Chem. Eng. 27, 1523–1535.

Iskandar, M.J., Baharum, A., Anuar, F.H., Othaman, R., 2018. Palm oil industry in South East Asia and the effluent treatment technology—a review. Environ. Technol. Innov. 9, 169–185.

Kim, S.W., Koo, B.S., Ryu, J.W., Lee, J.S., Kim, C.J., Lee, D.H., Kim, G.R., Choi, S., 2013. Bio-oil from the pyrolysis of palm and Jatropha wastes in a fluidized bed. Fuel Process. Technol. 108, 118–124.

Lazzari, E., Polidoro, A., dos, S., Onorevoli, B., Schena, T., Silva, A.N., Scapin, E., Jacques, R.A., Caramão, E.B., 2019. Production of rice husk bio-oil and comprehensive characterization (qualitative and quantitative) by HPLC/PDA and GC × GC/qMS. Renew. Energy.

Li, B., Ou, L., Dang, Q., Meyer, P., Jones, S., Brown, R., Wright, M., 2015. Techno-economic and uncertainty analysis of in situ and ex situ fast pyrolysis for biofuel production. Bioresour. Technol. 196, 49–56.

Li, C., Zhang, C., Gholizadeh, M., Hu, X., 2020. Different reaction behaviours of light or heavy density polyethylene during the pyrolysis with biochar as the catalyst. J. Hazard. Mater. 399, 123075.

Li, H., Liu, Z., Zhang, Y., Li, B., Lu, H., Duan, N., Liu, M., Zhu, Z., Si, B., 2014. Conversion efficiency and oil quality of low-lipid high-protein and high-lipid low-protein microalgae via hydrothermal liquefaction. Bioresour. Technol.

Liu, H., Li, M., Sun, R., 2013. Hydrothermal liquefaction of cornstalk : 7-Lump distribution and characterization of products. Bioresour. Technol. 128, 58–64.

Liu, P., Liu, W.J., Jiang, H., Chen, J.J., Li, W.W., Yu, H.Q., 2012. Modification of bio-char derived from fast pyrolysis of biomass and its application in removal of tetracycline from aqueous solution. Bioresour. Technol. 121, 235–240.

Loh, S.K., 2017. The potential of the Malaysian oil palm biomass as a renewable energy source. Energy Convers. Manag. 141, 285–298.

Meyer, P.A., Snowden-Swan, L.J., Jones, S.B., Rappé, K.G., Hartley, D.S., 2020. The effect of feedstock composition on fast pyrolysis and upgrading to transportation fuels: techno-economic analysis and greenhouse gas life cycle analysis. Fuel 259, 116218.

Mujahid, R., Riaz, A., Insyani, R., Kim, J., 2020. A centrifugation-first approach for recovering high-yield bio-oil with high calorific values in biomass liquefaction: a case study of sewage sludge. Fuel.

Nguyen, T. Van, Clausen, L.R., 2019. Techno-economic analysis of polygeneration systems based on catalytic hydropyrolysis for the production of bio-oil and fuels. Energy Convers. Manag. 184, 539–558.

Oasmaa, A., Beld, B. Van De, Saari, P., Elliott, D.C., Solantausta, Y., 2015. Norms, Standards, and Legislation for Fast Pyrolysis Bio-Oils from Lignocellulosic Biomass. Energy Fuels 29 (4), 2471–2484.

Ong, H.C., Chen, W.-H., Farooq, A., Gan, Y.Y., Lee, K.T., Ashokkumar, V., 2019. Catalytic thermochemical conversion of biomass for biofuel production: a comprehensive review. Renew. Sustain. Energy Rev. 113, 109266.

Pang, S., 2019. Advances in thermochemical conversion of woody biomass to energy, fuels and chemicals. Biotechnol. Adv. 37, 589–597.

Park, J.-W., Heo, J., Ly, H.V., Kim, J., Lim, H., Kim, S.-S., 2019. Fast pyrolysis of acid-washed oil palm empty fruit bunch for bio-oil production in a bubbling fluidized-bed reactor. Energy 179, 517–527.

Patel, M., Zhang, X., Kumar, A., 2016. Techno-economic and life cycle assessment on lignocellulosic biomass thermochemical conversion technologies: a review. Renew. Sustain. Energy Rev. 53, 1486–1499.

Prapaiwatcharapan, K., Sunphorka, S., Kuchonthara, P., Kangvansaichol, K., Hinchiranan, N., 2015. Single- and two-step hydrothermal liquefaction of microalgae in a semi-continuous reactor: effect of the operating parameters. Bioresour. Technol.

Qiu, Z., Zhai, Y., Li, S., Liu, X., Liu, X., Wang, B., Liu, Y., Li, C., Hu, Y., 2020. Catalytic co-pyrolysis of sewage sludge and rice husk over biochar catalyst: bio-oil upgrading and catalytic mechanism. Waste Manag 114, 225–233.

Rabiu, S.D., Auta, M., Kovo, A.S., 2018. An upgraded bio-oil produced from sugarcane bagasse via the use of HZSM-5 zeolite catalyst. Egypt. J. Pet. 27 (4), 589–594.

Rahman, M.M., Liu, R., Cai, J., 2018. Catalytic fast pyrolysis of biomass over zeolites for high quality bio-oil – a review. Fuel Process. Technol. 180, 32–46.

Shah, A., Darr, M.J., Dalluge, D., Medic, D., Webster, K., Brown, R.C., 2012. Physicochemical properties of bio-oil and biochar produced by fast pyrolysis of stored single-pass corn stover and cobs. Bioresour. Technol.

Sharma, A., Pareek, V., Zhang, D., 2015. Biomass pyrolysis—a review of modelling, process parameters and catalytic studies. Renew. Sustain. Energy Rev. 50, 1081–1096.

Shuping, Z., Yulong, W., Mingde, Y., Kaleem, I., Chun, L., Tong, J., 2010. Production and characterization of bio-oil from hydrothermal liquefaction of microalgae *Dunaliella tertiolecta* cake. Energy.

Thangalazhy-Gopakumar, S., Wei Lee, C., Gan, S., Kiat Ng, H., Yee Lee, L., 2018. Comparison of bio-oil properties from non-catalytic and in-situ catalytic fast pyrolysis of palm empty fruit bunch. Mater. Today Proc. 5, 23456–23465.

Tian, Y., Wang, F., Djandja, J.O., Zhang, S.L., Xu, Y.P., Duan, P.G., 2020. Hydrothermal liquefaction of crop straws: effect of feedstock composition. Fuel.

Treedet, W., Suntivarakorn, R., 2018. Design and operation of a low cost bio-oil fast pyrolysis from sugarcane bagasse on circulating fluidized bed reactor in a pilot plant. Fuel Process. Technol. 179, 17–31.

U.S. Energy Information Administration, 2020. Gasoline and Diesel Fuel Update [WWW Document]. URL. https://www.eia.gov/petroleum/gasdiesel/. (Accessed 26 September 2020).

Wądrzyk, M., Janus, R., Vos, M.P., Brilman, D.W.F., 2018. Effect of process conditions on bio-oil obtained through continuous hydrothermal liquefaction of *Scenedesmus* sp. microalgae. J. Anal. Appl. Pyrolysis. 134, 415–426.

Wang, F., Chang, Z., Duan, P., Yan, W., Xu, Y., Zhang, L., Miao, J., Fan, Y., 2013. Bioresource technology hydrothermal liquefaction of *Litsea cubeba* seed to produce bio-oils. Bioresour. Technol. 149, 509–515.

Wang, S., Dai, G., Yang, H., Luo, Z., 2017. Lignocellulosic biomass pyrolysis mechanism: a state-of-the-art review. Prog. Energy Combust. Sci. 62, 33–86.

Wright, M.M., Daugaard, D.E., Satrio, J.A., Brown, R.C., 2010. Techno-economic analysis of biomass fast pyrolysis to transportation fuels. Fuel 89, S2–S10.

Wu, L., Wang, Y., Zheng, L., Wang, P., Han, X., 2019. Techno-economic analysis of bio-oil co-processing with vacuum gas oil to transportation fuels in an existing fluid catalytic cracker. Energy Convers. Manag. 197, 111901.

Xue, Y., Chen, H., Zhao, W., Yang, C., Ma, P., Han, S., 2016. A review on the operating conditions of producing bio-oil from hydrothermal liquefaction of biomass. Int. J. Energy Res. 40, 865–877.

Yang, H., Chen, Z., Chen, W., Chen, Y., Wang, X., Chen, H., 2020. Role of porous structure and active O-containing groups of activated biochar catalyst during biomass catalytic pyrolysis. Energy 210, 118646.

Yang, J., Sophia He, Q., Yang, L., 2019. A review on hydrothermal co-liquefaction of biomass. Appl. Energy 250, 926–945.

Yeo, J.Y., Chin, B.L.F., Tan, J.K., Loh, Y.S., 2019. Comparative studies on the pyrolysis of cellulose, hemicellulose, and lignin based on combined kinetics. J. Energy Inst. 92, 27–37.

Younas, R., Hao, S., Zhang, L., Zhang, S., 2017. Hydrothermal liquefaction of rice straw with NiO nanocatalyst for bio-oil production. Renew. Energy. 113, 532–545.

Yuan, X.Z., Li, H., Zeng, G.M., Tong, J.Y., Xie, W., 2007. Sub- and supercritical liquefaction of rice straw in the presence of ethanol–water and 2-propanol–water mixture. Energy 32, 2081–2088. https://doi.org/10.1016/j.energy.2007.04.011. In this issue.

Zhang, S., Zhang, H., Liu, X., Zhu, S., Hu, L., Zhang, Q., 2018. Upgrading of bio-oil from catalytic pyrolysis of pretreated rice husk over Fe-modified ZSM-5 zeolite catalyst. Fuel Process. Technol.

Zhu, Z., Rosendahl, L., Sohail, S., Yu, D., Chen, G., 2015. Hydrothermal liquefaction of barley straw to bio-crude oil : effects of reaction temperature and aqueous phase recirculation. Appl. Energy 137, 183–192.

CHAPTER

3

Nonwaste technology in the bioethanol and biodiesel industries

Luciana Porto de Souza Vandenberghe[1], *Nelson Libardi Junior*[2], *Kim Kley Valladares-Diestra*[1], *Gustavo Amaro Bittencourt*[1], *Ariane Fátima Murawski de Mello*[1], *Susan Grace Karp*[1], *Luiz Alberto Junior Letti*[1], *Carlos Ricardo Soccol*[1]

[1]Federal University of Paraná, Centro Politécnico, Curitiba, Paraná, Brazil; [2]Federal University of Santa Catarina – UFSC, Florianópolis, Brazil

3.1 Introduction

The well-established advantages of first-generation biofuel industry are constantly competing with the production of traditional fossil fuels with lower costs. However, the planet no more can absorb the hazardous impacts of their processing. So, biofuels and bioenergy appear as an environmentally friendly solution. In addition, different countries have been working, with the support of governmental programs, to establish new goals to reduce the emission of greenhouse gases (GHG). The biorefinery concept provides a wide perspective to reach these goals, with concomitant use of biomass conversion to new value-added products (Santos et al., 2016). Sugarcane, corn, maize, cotton, soy, wheat, and rice dominate culture crops and are the main produce crops in the world, and the processing generates large amounts of wastes/by-products. Beyond the traditional use for electricity and thermal generation, there is the possibility of a more economic and sustainable exploitation of these by-products to produce different products such as biofuels, organic acids, biopolymers, enzymes, biofertilizers, and others.

This chapter brings a new perspective of no-waste technology development under the biorefinery concept with the focus on bioethanol and biodiesel production using some examples of sugarcane, corn, soybean, maize, oil palm, and cassava processing chains. Information about the world biorefineries, their localization, and type of processed crop will be described. An interesting analysis of the new developed technologies and the main innovation developers and patent depositors will also be presented.

Biofuels and Bioenergy
https://doi.org/10.1016/B978-0-323-85269-2.00019-8

3.2 The biorefinery concept

The biorefinery is an efficient platform where there are different processes, which employ biomass and its wastes, to produce different bioproducts such as bioenergy, biofuels, biopolymers, biofertilizers, animal feeds, and other biomolecules (Kumar et al., 2020). These integrated processes involve sustainable approaches with the application of circular bioeconomy and environmentally friendly solutions with less carbon emissions (Ubando et al, 2020, 2021).

The actual situation of biofuels shows that there is still a strong competition with the conventional fossil fuels (Chowdhury et al., 2018). In fact, the high operating costs, poor valorization of biomass, difficulties with feedstock supply chain, scale-up problems, and low technology maturity (Chandel et al., 2018) limit the implantation of biorefineries (Rosales-Calderon and Arantes, 2019; Gonzalez-Contreras et al., 2021). To surpass these limitations, the implementation of simultaneous production of biofuels, heat, food, feed, and other bioproducts, in a single unity, could be achieved. So, the biorefinery concept has quickly gained acceptance with technological and regional variants depending on the applied feedstock (Chowdhury et al., 2018).

Biorefineries can be classified as first-generation (1G), second-generation (2G), and third-generation (3G) biorefinery. 1G biorefineries employ sugar, cereal starch, vegetable oil, or animal fat, and 2G biorefineries use lignocellulosic feedstock, including agroindustrial wastes, woody forestry residues, and municipal solid wastes (MSWs), respectively, as material for biofuels and other biomolecules production (Demirbas, 2010). The 3G biorefinery uses algae (Chowdhury et al., 2018) (Fig. 3.1). In the past

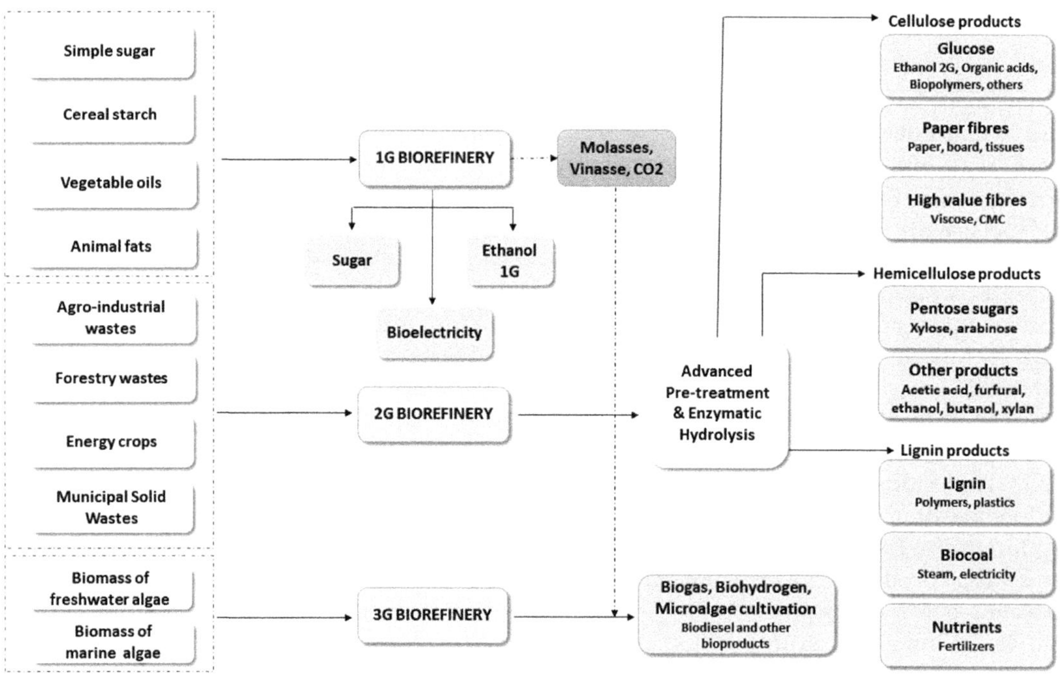

FIGURE 3.1 Classification of biorefineries based on feedstock.

years, 2G biorefineries have received attention in different countries, where numerous large-scale plants have been inaugurated, because they use agroindustrial, forest, or MSWs that are low cost and available.

The 1G biorefinery employs a single feedstock, which limits the portfolio of products that are generated in these unities (Karp and Shield, 2008; Liu et al., 2021) with a high consumption of processed biomass. This fact may cause the increase of food prices and lower incentives for economic development. On the other hand, the 2G biorefineries use different lignocellulosic material as feedstocks (Miao et al., 2004), which amplifies the diversity of products with less dependence of food crops (Liu et al., 2021; Valentine et al., 2012; Thompson et al., 2012). This is a great advantage of 2G biorefineries over the 1G unities, which contributes for 1G technology unsustainability (Valentine et al., 2012; Thompson et al., 2012; Chowdhury et al., 2018).

The next generation are the 3G biorefineries that will employ marine algal biomass and freshwaters and as feedstocks for the production of biofuels and bioproducts. So, the 3G biorefineries represent a significant improvement compared with the 1G and 2G biorefineries. This technology makes use of different fractions of biomass, improving the efficiency and economic value, due to the diversification of produced bioproducts in the operational unities with lower environmental impacts (Liu et al., 2021).

The prevalence and type of biorefineries differ considerably between countries and are clearly dependent on the locally available biomass (Bio-based News, 2017). The classification of biomass is based on the origin or composition, including agroindustrial wastes (derived from plants, animals, or homes), or aquatic biomass (algae and microalgae) (de Jong and Jungmeier, 2015). The production of biomass waste, which are derived from plants, in the world is about 200 × 109 tons, where almost 90% has lignocellulosic composition (Saini et al., 2015; Poveda-Giraldo et al., 2021).

Crops processing and harvesting generate agricultural residues such as bagasse, corncobs, straw, husk, leaves, pulp, and others that can be used for biofuels and chemicals production worldwide (Hattori and Morita, 2010; Kumar and Verma, 2021). The forest residues are divided into primary (short twigs, leaves, and tops) and secondary forestry (barks, sawdust, scrap woods, mill wastes) residues (Balaman, 2019; Woo et al., 2019). New potential sources of biomass are algae, microalgae, and Cyanobacteria due to their fast growth rate (Srivastava, 2019). They are considered the major actors of 3G biofuels and bioproducts generation such as biofuels, bioactive compounds, and biochemicals (Kumar and Verma, 2021).

The developing countries, including Brazil, Russia, India, China, and South Africa, with growing industrial activities and economy, have expressive requirements of energy to sustain this rapid development (Sasana and Ghozali, 2017). That is why governmental programs and incentives for biofuels and bioenergy production have recently boosted the installation of new biorefinery unities (Chowdhury et al., 2018). With this purpose, in Brazil, the National Biofuel Policy (RenovaBio) established mechanisms for decarbonization credits, which is linked to the carbon intensity of biofuels. With these incentives and benefits, new possibilities and prospects for the study and implementation of integrated biorefineries will come to light (Klein et al., 2019).

In 2015, the production of 2G biofuels started at full commercial scale when 67 facilities operated around the world (Nguyen et al., 2017). Actually, in the European Union, 803 biorefineries were identified, which produce liquid biofuels, bio-based chemicals (solvents, polymers, paints, and others), and composites and fibers (wood—plastic and natural fibers composites and others) (Parisi, 2018). The most part of them are in Belgium and the Netherlands and employ agricultural wastes as feedstock.

Bioethanol and biodiesel are largely demanded as biofuels worldwide. That is why the biorefinery solution for these processes is applied with great environment and economic advantages (Longati et al., 2020). Bioethanol is currently the world's most abundant biofuel, which is mainly produced by the Unites States and Brazil, from corn and sugarcane, respectively, representing almost 84% of global ethanol production (110.6 billion liters) (Statista, 2020). Biodiesel is produced from vegetable oils, animal fats, or vegetable oils. The main feedstock for biodiesel production is soybean, representing 67% of the world production. Bioethanol and biodiesel biorefineries will be described in the next sections.

3.3 Biodiesel biorefinery

Biodiesel is a monoalkyl ester that is produced from fatty acids. This biofuel is presented as an eco-friendly alternative with great potential in the energy sector (Singh et al., 2020). The main advantages of biodiesel over fossil diesel are that it has high flash point and biodegradability, it reduces GHG emissions, and it is free of chemicals such as polycyclic aromatic hydrocarbons, sulfur, and most metals. In addition, it can reduce around 85% of the emissions of carcinogenic compounds that are generated by its combustion (Almeida et al., 2012; Valladares-Diestra et al., 2020).

Biodiesel production worldwide is mainly distributed between the United States, the European Union, and Brazil. It is estimated that by 2024 the production of biodiesel will reach 39 million m^3 (Mardhiah et al., 2017; Valladares-Diestra et al., 2020).The renewable and eco-friendly nature of biodiesel are the main characteristics that drive the growth of its production worldwide. So, different governments promote biodiesel production in their energy sectors applying different public policies and giving economic incentives that improve the competitiveness of biodiesel production with a high socioeconomic impact. The use and production of biodiesel contribute to the energy independence of the countries, avoiding the import of crude oil and generating greater development of their productive sectors (Balat and Balat, 2009).

Biodiesel can be obtained through transesterification, pyrolysis, and supercritical fluid methods. Transesterification is the most widely used process at industrial scale, producing glycerol as one of its main by-products (Singh et al., 2020). Depending on the region/country, the main sources that are employed for biodiesel production are soy, palm, rapeseed, and sunflower oils (Almeida et al., 2012). The use of this type of source generates a conflict of "food against fuel," which causes the increase of raw materials' costs. These production costs can be reduced with the use of alternative materials such as industrial wastes with the application of biorefinery concept, aiming at the revaluation of by-products of biodiesel process chain to obtain value-added biomolecules, concomitantly.

3.3.1 Sources for biodiesel production

Lipids are the main component of the oils that serve as raw material in biodiesel production. According to the origin of the raw material, the biodiesel can be classified as 1G, 2G, or 3G.

3.3.1.1 *First-generation biodiesel*

The 1G biodiesel is highly produced worldwide from vegetable oils as main feedstock. Countries such as the United States and Brazil base their biodiesel production on soybean oil. The European Union for its part uses rapeseed and sunflower oils, while regions of the Southeast Asia base their biodiesel production on palm oil, which presents high oil production yields per hectare (Valladares-Diestra et al., 2020). However, the use of feedstocks generates a conflict between the energy industry and the food industry, which compete for the same sources, generating a reduction in the amount of edible food, causing global food prices increase (Mofijur et al., 2020). Because of this, improvements in crop productivity with the introduction of new varieties are commonly studied. In recent years, oilseed crops that produce inedible oils (jatropha, karanja, and polanga) have been evaluated for exclusive energy production, thus avoiding the competition with the food industry (Sharma and Singh, 2010; Suresh et al., 2018).

3.3.1.2 *Second-generation biodiesel*

2G biodiesel emerges as an alternative to avoid the dependence of the food crops in the biofuels production, and it is obtained from inedible raw materials (Mofijur et al., 2020; Rahman et al., 2017). By-products and/or lignocellulosic residues from the agroindustry are commonly employed due to their rich composition in polysaccharides, which are used as carbon sources for oleaginous microorganisms' cultivation. Yeast and bacteria are known to accumulate large amounts of lipids. So, intracellular oils are extracted and then pass through transesterification process for 2G biodiesel production. However, the employed feedstocks need previous pretreatments to facilitate the release of sugars due to the recalcitrant nature of the lignocellulosic materials (Rezania et al., 2020).

Some oleaginous yeasts can accumulate large amounts of lipids (20%–80% of dry weight), mainly C16 and C18 fatty acids, which are used in biodiesel production (Cho and Park, 2018; Severo et al., 2019). However, the lipid accumulation mechanism in these microorganisms has not yet elucidated. Oil yeasts are used due to their rapid growth, high efficiency in the accumulation of lipids, and their ability to consume different organic substrates (Cho and Park, 2018). Some examples are *Yarrowia* sp., *Cryptococcus* sp., *Lipomyces* sp., *Rhodotorula* sp., and *Rhodosporidium* sp. There are also oleaginous bacteria such as *Rhodococcus* sp., *Bacillus* sp., *Gordonia* sp., *Arthrobacter* sp., and *Acinetobacter* sp. with high oil accumulation capacity that can be used in biodiesel production. However, the oleaginous yeast is the most studied compared with bacteria (Cho and Park, 2018; Wei et al., 2015).

Consequently, the use of waste with a high content of organic material as a recyclable source in the production of lipids by oleaginous microorganisms generates great interest due to its high efficiency. Additionally, residential and commercial oil wastes, as well as animal-based oils or baits, are also considered as source for 2G biodiesel. Therefore, the production of 2G biodiesel is a more efficient, profitable, and more environmentally friendly process compared with the 1G biodiesel production (Pinzi et al., 2009; Severo et al., 2019).

3.3.1.3 *Third-generation biodiesel*

One source of biodiesel production is the photosynthetic conversion of solar energy into chemical energy that is performed by plants and algae (Singh et al., 2020). This process allows both plants and algae to fix CO_2, thus accumulating sugars and lipids as energy reserves. Recently, microalgae have shown their high potential as feedstock for the production of 3G biodiesel, reducing food problems and the use of arable land (Chia et al., 2018). Due to their high photosynthesis efficiency and accelerated growth rate, microalgae cultivation can produce 15 to 300 times yield compared with traditional oilseed crops in relation to cultivated area (Mofijur et al., 2020). In addition, microalgae can grow in different and extreme environmental conditions such as sewage, saline water, and brackish water, decreasing the consumption of freshwater. However, the most outstanding characteristic is its robustness in capturing GHGs, especially CO_2, managing to fix it and biotransform it into bioproducts of industrial interest (Deprá et al., 2018). In this way, biodiesel produced from microalgae emerges as a potentially viable alternative compared with 1G and 2G production (Hossain et al., 2020; Mofijur et al., 2020).

Although the main disadvantage in the microalgae cultivation is its relatively new technology, for which a greater economical investment is necessary, it needs a greater amount of light and presents difficulties at large-scale production, and the oil extraction process is complex (Singh et al., 2020). For these reasons, recent research focuses on the definition of ideal culture media composition and selection of high lipid-producing strains. Several species of algae such as *Botryococcus braunii, Chaetoceros calcitrans, Chlorella, and others* have been studied with optimal projections for their use as possible sources of biofuel production (Dutta et al., 2014).

3.3.1.4 *Fourth-generation biodiesel*

This category encompasses the production of biodiesel by means of genetically modified microorganisms. Yeasts, bacteria, and microalgae are designed or modified by molecular and synthetic biology tools for the construction of small biofactories within the same cells (Lü et al., 2011; Singh et al., 2020). The objective of this technology is to reach greater efficiency in the production of both liquid and gaseous biofuels, improving substrate consumption and providing adaptation skills to microorganisms so as tolerance to extreme environments (Abdullah et al., 2019). Modifications within the microbial genome can range from the insertion or deletion of a single gene to the modification, insertion, or inactivation of entire metabolic pathways with the use of metabolomic engineering.

The most outstanding advantages of this technology, which are used in microalgae, are the minimization of photosynthesis process steps for a better efficiency of CO_2 and energy capture in the fungible production of biofuels (Lü et al., 2011). On the other hand, synthesis's systems are also coupled for the coproduction of high added value biomolecules with the design of small biofactories with the capacity of extracellular secretion of products, avoiding high costs of recovery and separation processes.

Although this technology is emerging as a tangible option in the near future, the current investments are still very high. So, different scientists continue to develop their research to improve production and minimize costs. Other points to take into account are the possible environmental and health risks that the use of transgenic microorganisms can cause. Waste disposal of processes involving genetically modified microorganisms may contain chromosomal remains, plasmids, or genes that can cause horizontal gene transfer in contact with other organisms. Therefore, appropriate measures must be taken to mitigate possible risks with the use of this type of technology for sustainable production and development (Abdullah et al., 2019).

3.3.2 Biodiesel biorefinery approach

The search of cost reduction in the production of biodiesel and the growing environmental problems, which are caused by fossil fuels, leads to biorefinery approaches. A biodiesel biorefinery can be defined as a set of integrated technological processes that allow the efficient use of by-products for the production of biodiesel and other bioproducts. These strategies reintegrate the wastes of the process chain into the production cycle in a no-waste technology process unity, thus reducing the environmental impacts compared with conventional production technologies. Biodiesel biorefinery approach provides the advantages in the diversification of production, generating multiple coverage of markets, reducing costs within the production process, energy self-sufficiency, and reducing the generation of GHGs (Boldrin et al., 2013; Parsons et al., 2020; Valladares-Diestra et al., 2020). With the application of biorefinery concept, the production of biodiesel could respond to current energy demand, allowing the sustainable production of various value-added materials, biofuels, biochemicals, and marketable energies (Cherubini, 2010; Hassan et al., 2018).

As it was observed previously, the production of biodiesel can be classified according to the employed source and production methods. However, a biodiesel biorefinery synergistically generates higher production efficiencies, producing large volumes of biofuel with lower costs and with efficient use of raw materials. Although the implementation of this type of technology is in an emerging stage, further studies are still necessary for adequate justification of the implementation of a large-scale plant. In addition, the price of oil and the government policies are determinant for the feasibility of new projects. The combined work of innovation and development is necessary to achieve the implementation of a biodiesel biorefinery (Severo et al., 2019).

Fig. 3.2 shows the ideal biodiesel production process under the concept of biorefinery approach. As it can be seen, the first part is the cultivation of oleaginous plants to obtain oil and the production of 1G biodiesel. Within the different stages of this process, solid, liquid, and gaseous by-products are generated. Each by-product can be used or generates new products: Oil meal derived from seeds can be used in animal feed (Oláh et al., 2017); liquid effluent from the oil refining can serve as an organic substrate in the production of biogas (Choong et al., 2018); and lignocellulosic biomass can generate energy through its combustion (Osman et al., 2017) or be physiochemically pretreated to release fermentable sugars and, thus, obtain 2G bioethanol (Rezania et al., 2020), biomolecules

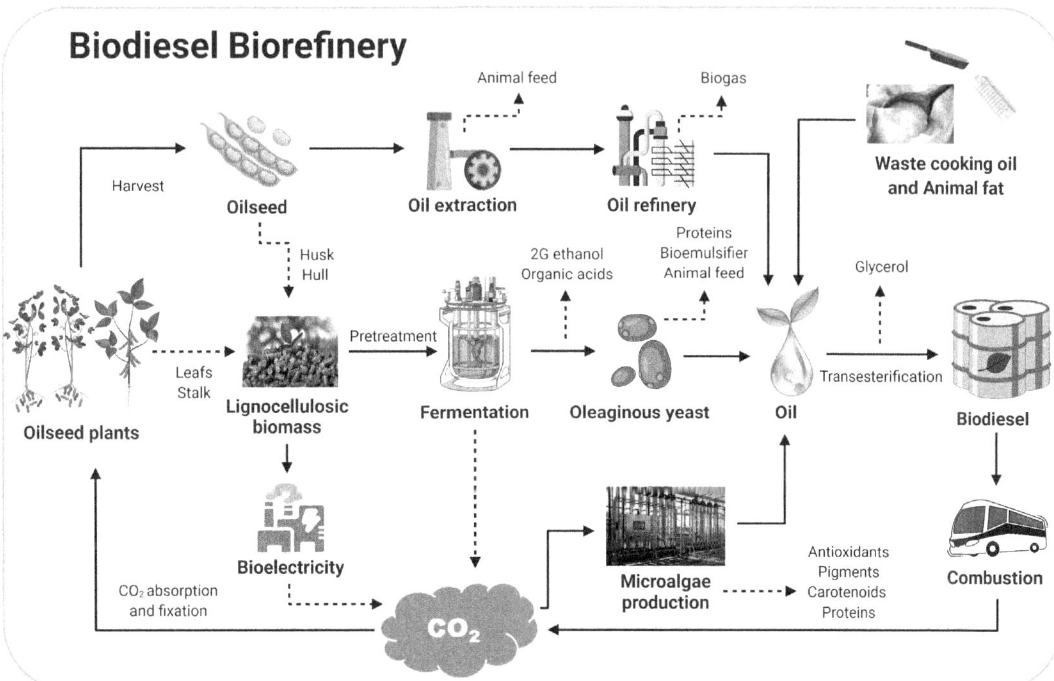

FIGURE 3.2 Biodiesel biorefinery approach. *Created with Biorender.com.*

such as organic acids (Woiciechowski et al., 2020) or cultivate oleaginous yeasts to obtain 2G biodiesel (Patel et al., 2017). In most of these stages, CO_2 is generated as a gaseous by-product, which can be absorbed and fixed by plants or used in the cultivation of microalgae for the generation of pigments, antioxidants, or the production of 3G biodiesel (Bauer et al., 2017). Genetically modified microorganisms, or designed by synthetic biology, can also be applied within the process for 4G biodiesel production and other biofuels (Das et al., 2020). Domestic and/or industrial wastes, such as used oils, as well as fats from animals, can be integrated into a biodiesel biorefinery, considerably increasing the production volumes (Habib et al., 2020; Ibrahim et al., 2020). Finally, the increase in the production of biodiesel generates a great increase in the production of its main by-product, glycerol. This by-product is obtained after transesterification, and its use has been widely studied in the production of biochemicals, biopolymers, and biofuels (Monteiro et al., 2018).

Due to the complexity of biorefineries, efforts are currently being made to improve the feasibility and efficiency of production. Different methods and new technologies are applied with this objective. In addition, new strategies are being sought to make the best use of the biodiesel production cycle.

3.4 Bioethanol biorefineries

Bioethanol biorefineries take hand of physical, chemical, and biological processes to fractionate the biomass feedstock into ethanol, heat, electricity, and other bio-based building blocks (Chandel et al., 2019).

The bioethanol biomass can be classified in four categories: sugar (sugarcane, beetroot, molasse); starch (corn, wheat, cassava); lignocellulosic (agricultural waste, organic fraction of MSW); and algae biomass (Demichelis et al., 2020). In a review of wastes used in biorefineries, Liu et al. (2021) found that agricultural crops industrial residues are the most widely used biorefinery feedstocks, accounting for 32.6% and 29.3%, followed by MSW, woody residues, aquatic organisms, herbs, and manures.

The mains steps involved in 2G biorefineries involve the (1) conditioning of lignocellulosic biomass; (2) biomass pretreatment; (3) saccharification; (4) fermentation, and (5) ethanol recovery (Gonzalez-Contreras et al., 2021). The bioethanol biorefinery concept encompasses the production and recovery of low-added-value products, i.e., ethanol, energy carriers, soil improvers, and fertilizers. However, advanced bioethanol biorefineries can produce high complexity bioproducts, with more specific technical standards (Alibardi et al., 2020). Even if chemicals and polymers continue to be produced from petroleum, hundreds of chemicals can be obtained from cellulosic biomass as a side process/product from the bioethanol biorefinery. To better organize bio-based products, they can be categorized according to the biorefinery platform that includes the key intermediates of raw feedstock processing and bioethanol production. Bio-based chemicals, which are obtained from bioethanol production using sugar feedstock, starch feedstock, and lignocellulosic feedstock, are presented in Fig. 3.3.

3.4.1 Commercial bioethanol biorefineries

Brazil (26.8 billion of L/year) and the United States (55.6 billion of L/year) are the world leaders of 1G large-scale ethanol producers, using sugar- and starch-containing as feedstocks. Also, their high cost was comparable with gasoline prices in January 2020 (gasoline 9.8 USD/L; ethanol 8.6 USD/L) (Demichelis et al., 2020). Bioethanol world production was 94.6 million liters in 2015 and increased to around 109 million liters

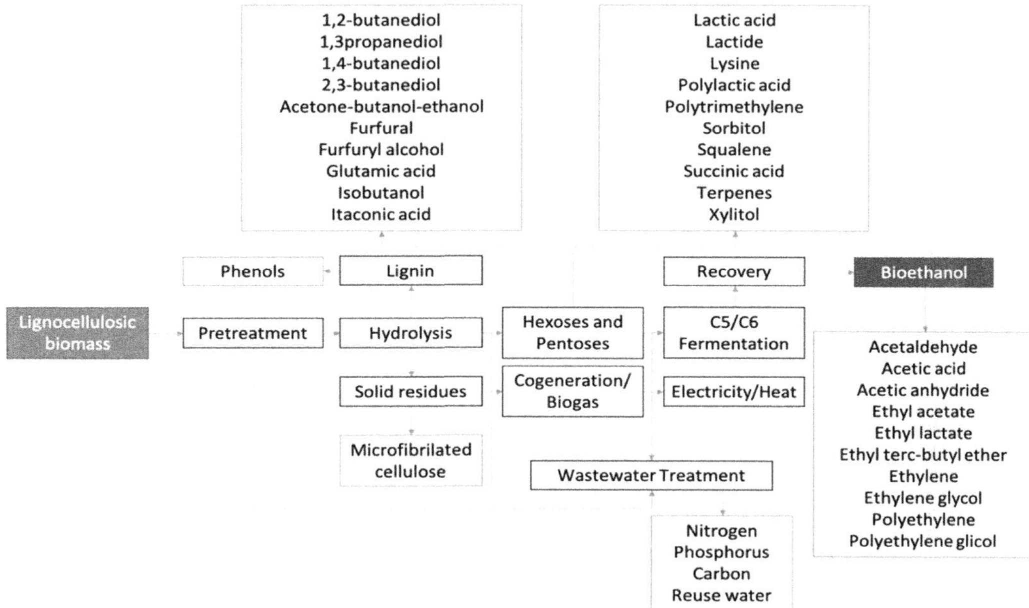

FIGURE 3.3 Biorefinery diagram of the production of bioethanol, biochemicals, and biomaterials. Orange rows and boxes for recovered bio-based chemicals. *Adapted from Chandel, A.K., Albarellu, J.Q., Santos, D.T., Chundawat, S.P., Puri, M. Meireles, M.A.A., 2019. Comparative analysis of key technologies for cellulosic ethanol bagasse at a commercial scale. Biofuel Bioprod. Biorefin. 13 (4), 994–1014. https://doi.org/10.1002/bbb.1990; Rosales-Calderon, O., Arantes, V., 2019. A review on commercial-scale high-value products that can be produced alongside cellulosic ethanol. Biotechnol. Biofuels 12, 1–58. https://doi.org/10.1186/s13068-019-1529–1.*

in 2020 (Rosales-Calderon and Arantes, 2019). The major 2G cellulosic ethanol production (Poet-DSM, DuPont, Abengoa, BetaRenewables, Raizen, Granbio, Praj, and others) have different business models based on the feedstock, processing technology, and bio-based products. The commercial 2G bioethanol biorefinery plants according to the company country are summarized in Table 3.1.

3.4.2 Involved lignocellulosic biomasses

Lignocellulosic biomass is widely available, has low costs, and does not compete with 1G biomass feedstocks such as sugar and starch (Chandel et al., 2019; Rosales-Calderon and Arantes, 2019). Lignocellulosic biomass is a potential alternative since biofuel production from food and energy crops represent potential food insecurity. The worldwide lignocellulosic biomass productivity accounts for 1.5×10^{10} tons/year. Considering the cellulosic composition (30%–60% cellulose, 20%–40% hemicellulose, and 15%–25% lignin) and their structural monomers (glucan, galactan, mannan, arabinan, and xylan), the theoretical lignocellulosic ethanol production is estimated in 442 billion L/year (Rosales-Calderon and Arantes, 2019).

Among lignocellulosic biomass, sugarcane bagasse is one of the most promising biomass, and the main sugarcane producers include India, China, South Africa, and Brazil (Chandel et al., 2019). Around 13.5% of Brazil's renewable energy matrix comes from sugarcane. Brazil processed in 2018 around 700 million tons of sugarcane that generated 98 million tons of bagasse (Chandel et al., 2019). In general, sugarcane bagasse contains 53.5% of carbohydrates

TABLE 3.1 The main installed bioethanol biorefineries.

Location	Company	Feedstock	Pretreatment	Fermentation	Production capacity (million L per year)
Brazil	GranBio and Beta Renewables	Sugarcane bagasse and straw	Proesa steam explosion	Celere-2L recombinant yeast	81.8
Brazil	Raizen Project (Iogen Corp.)	Sugarcane bagasse and straw	Proesa steam explosion	Recombinant yeast	40
Brazil	Centro de Tecnologia Canavieira (CTC)	Sugarcane bagasse	Steam explosion, hydrolysis	Bacterial fermentation	3
Canada	Enerkem and Greenfield	Municipal solid waste	–	Catalytic transformation	38
Canada	Enerkem	Municipal solid waste	Fermentation	Catalytic transformation	38
China	New Tianlong Industry and Dupont	–	–	Fermentation	–
China	M&G Chemicals (Gruppo Mossi Ghisolfi) and Anhui Guozhen	–	–	Fermentation	100
China	COFCO Zhaodong	Dedicated energy crops	–	Fermentation	45
China	Henan Tianguan	Agricultural waste (corn and wheat stalks)	Fermentation	–	38
China	Shandong Longlive	Agricultural waste (corn and wheat stalks)	Fermentation	–	63
Denmark	Maabjerg Energy Center (MEC)	Agricultural waste (straw)	–	–	73
Finland	St1, SOK (NEB)	Woody biomass	Hydrolysis	–	10
India	Hindustan Petroleum Corporation Limited and Institute of Chemical Technology	Agricultural waste	–	Fermentation	32
India	Paraj Industries	Mixed	Hydrolysis	Bacterial fermentation	1
India	Numaligarh refinery Limited and Chempolis	Dedicated energy crops (bamboo)	–	Fermentation	63
India	Indian Oil Corporation Limited and Praj	Agricultural waste	–	Fermentation	32
India	Indian Oil Corporation Limited and Praj	Agricultural waste	–	Fermentation	32

India	Bharat Petroleum Corporation Limited and Praj	Agricultural waste	–	Fermentation	32
India	Bharat Petroleum Corporation Limited and Institute of Chemical Technology	Agricultural waste	–	Fermentation	32
Italy	Beta Renewables	*Arundo donax* (giant cane)	Proesa ssteam explosion	Recombinant yeast	50.7
Macedonia	DuPont and Ethanol Europe Renewables	–	–	Fermentation	100
Norway	St1, Vikeng skog SA	Woody biomass	–	–	50
Norway	Borregaard	Wood	–	–	20
Romania	Clariant	Wheat straw and other cereals	–	–	63
Slovakia	Enviral	Wheat straw	Hydrolysis	–	63
Slovakia	Energochemica, Beta Renewables	Wheat straw, rapeseed straw, corn stover	Hydrolysis	–	70
Sweden	Domsjö Fabriker AB	Wood	–	-	22
USA	Poet DSM	Corn stover	Diluted acid	Recombinant *saccharomyces*	75.7
USA	Aemetis	Agricultural waste (orchard waste and nutshells).	–	Fermentation	38
USA	Abengoa	Corn stover	Diluted acid	Recombinant *saccharomyces*	94.6
USA	New Energy Investors	Agricultural waste (corn stover and wheat straw).	–	Fermentation	49
USA	DowDuPont	Corn stover	Ammonia and steam	Recombinant *saccharomyces*	113.6
USA	Beta Renewables	Dedicated energy crops.	–	–	75
USA	Fiberight	Municipal solid waste	–	–	–
USA	INEOS Bio	Agricultural waste	Fermentation	–	30

References: Morán, D., 2017. BioRefineries.blog, 2015. https://biorrefineria.blogspot.com/2015/08/cellulosic-ethanol-biorefineries.html (Accessed 30 January 2021.); Chandel, A. K., Albarellu, J.Q., Santos, D.T., Chundawat, S.P., Puri, M. Meireles, M.A.A., 2019. Comparative analysis of key technologies for cellulosic ethanol bagasse at a commercial scale. Biofuel Bioprod. Biorefin. 13 (4), 994–1014. https://doi.org/10.1002/bbb.1990.

(cellulose and hemicellulose) and 27.6% of lignin (Chandel et al., 2019). The cost of sugarcane bagasse and corn stover was estimated in $36.36/ton and 58.50/ton, respectively. On the other hand, sugarcane and corn prices were estimated in $60.0/ton and 185.9/ton, respectively. Even so, the complex nature of lignocellulosic biomass requires expensive processes to convert cellulose to fermentable sugars (Rosales-Calderon and Arantes, 2019).

3.4.3 Environmental assessment of bioethanol biorefineries

The main biorefineries' pollutant streams are wastewater, solid wastes, and gaseous emissions (Gonzalez-Contreras et al., 2021). The sustainability of bioethanol biorefineries starts with the feedstock selection, process integration, manufacture, and end-of-life criteria (Katakojwala and Mohan, 2021).

Biofuels represent an alternative to fossil fuels, resulting in less GHG emissions when compared with petroleum-based fuels. The ethanol itself is already considered a closed CO_2 cycle due to its recycling from the atmosphere through the photosynthesis, producing new vegetable material for ethanol production. In addition, the substitution of 1G ethanol to 2G ethanol (bioethanol) in the biorefinery process concept is an advance regarding sustainability (Chandel et al., 2007).

Although the sustainability is in the core concept of biorefineries, they have their own environmental challenges. Some methodologies may be applied to evaluate the environmental impacts. One of them is the life cycle assessment (LCA) (Katakojwala and Mohan, 2021), which is mainly focused on GHG emissions and energy, as a conceptual framework and methodology for assessment of environmental impacts of products and systems on a cradle-to-crave basis (Chandel et al., 2007). LCA approach is a comprehensive tool based on the International Organization for Standardization (ISO 14044: 2006), which considers the criteria such as allocation, functional unit, cutoff criteria, and system boundaries (Katakojwala and Mohan, 2021).

There are some potential environmental impact categories that are considered in biorefinery LCA such as global warming impact (GWI), ozone depletion potential (ODP), acidification or acid rain potential (AP), photochemical oxidation or smog formation (PCOP), human toxicity potential by ingestion (HTPI) and inhalation or dermal exposure (HTPE), aquatic toxicity potential (ATP) terrestrial toxicity potential (TTP), and eutrophication potential (EP). The most explored indicators of biorefinery processes are the EP, HTPI, and AP (Liu et al., 2021). Biorefineries LCA studies use the definition of boundaries, the cradle-to-gate system that includes the well-to-wheel, gate-to-gate, and cradle-to-wheel subtypes. Cradle-to-gate system boundary considers the feedstock cultivation until the factory gate. Wheel considers the product use by combustion vehicles, and grave is linked to the final use of the biorefinery outputs. According to the system boundary of choice, advantages and disadvantages can emerge, and different results can be achieved (Liu et al., 2021).

LCA studies use ton of feedstocks or bioethanol as functional unit (FU). Depending on the chosen FU, bioethanol production promotes different scenarios. Bioethanol usually performs better about GHG emissions, while the upstream processes are controversial (Demichelis et al., 2020).

Each process of a lignocellulosic bioethanol biorefinery has a number of environmental impacts. The use of chemicals for pretreatments, hydrolysis, and fermentation has significant influence on climate change and may generate toxicity, fossil depletion, metal deposition, and freshwater eutrophication. Uncontrolled disposal of agricultural wastes, industrial wastes, or MSWs, such as incineration or landfill destination, consumes a large number of environmental resources, so biorefineries account for a general positive environmental impact. There are also other factors that represent environmental impacts, which are related to the land use changes, fertilizers consumption, equipment, manpower utilization, and others (Katakojwala and Mohan, 2021).

The use of industrial wastes such as sugarcane bagasse in a biorefinery contributes to positive economic and environmental benefits. The use of grass and straw has positive ecological impacts and no competition with the food sector. General lignocellulosic biomass does not compete with feeding sources (Liu et al., 2021). Zhao et al. (2018) applied the LCA methodology in different bioethanol-producing technologies using corn stover, and they concluded that a net saving of the GWI ranging from 900 to 1200 kg CO_2/ton fry corns stover would be achieved. LCA studies reveal that each feedstock accounts in a specific way to the environmental impacts. Demichelis et al. (2020) performed an LCA for bioethanol-producing processes using sugarcane, potato, rice straw, cattle manure, and MSW. Climate change was the most important indicator, especially for rice straw as substrate. Papadaskalopoulou et al. (2019) compared the LCA of an ethanol biorefinery that uses MSW with conventional waste management methods such as landfilling, composting, anaerobic digestion, and incineration. A negative GWI was estimated in −15 kg CO_2 eq/ton of MSW, when compared with 238 kg CO_2 eq/ton when landfilling was analyzed.

Physical, chemical, and biological pretreatment processes have different impacts on the sustainability of ethanol biorefineries that should be evaluated for each case. In general, chemical processes result in high yield but have negative impacts. Diluted acid in comparison with enzymatic hydrolysis has been proposed as a better pretreatment in terms of GHG potential, natural resource depletion, acidification, and eutrophication potential. Diluted acid pretreatment yields more biomass emissions reduction. Contrarily, concentrated acid results in higher GHG emissions (Chandel et al., 2007). Biological processes have been claimed as the ideal pretreatment, since enzymes have biological origin, are biodegradable, and can hydrolyze under mild temperatures. However, the LCA approach reveals other constraints for the use of enzymes as the ideal biocatalyst. An LCA approach can consider the energy and chemicals consumption for enzymes production process, which can cause considerable environmental impact. The slow rate of enzymatic hydrolysis that depends, per example, on accessible cellulose crystallinity, lignin content, enzymes' activity, synergy, and effectiveness is still bottleneck for its use as a pretreatment step (Liu et al., 2021). According to Papadaskalopoulou et al. (2019), the GWI resulting from the use of enzymes has a wide range (11%−62%) and is highly dependent on several factors such as enzyme loading, ethanol yield, and the GWI resulting from the enzyme production process.

3.4.3 Waste treatment—integration to the ethanol production chain

In the biorefinery context, waste streams are integrated to ethanol production chain, reducing the global environmental impact. The generation of wastes is greatly influenced by the technologies adopted in each step of 2G bioethanol production.

The wastewater treatment should guarantee the pollutants removal with a minimum standard efficiency in accordance with the environmental legislation for treated wastewater disposal or reuse, according to the country and process requirements. The carbon concentration is measured by parameters such as total organic carbon (TOC), chemical oxygen demand (COD), and biochemical oxygen demand (BOD). Also, phosphorus and nitrogen are usually used for wastewater analysis. A wastewater treatment plant, which consists of several physical–chemical and biological processes, may release streams that can be considered in a sustainability analysis (Gonzalez-Contreras et al., 2021). The wastewater treatment processes may release GHG from biological reactors such as CO_2, CH_4, and N_2O (Kampschreur et al., 2009). The biogas produced in anaerobic treatment processes, including the CH_4, can be sent to cogeneration processes for heat and energy recovery. In addition, the solid residue from aerobic or anaerobic treatment processes (sludge) also can be sent to cogeneration processes (Gonzalez-Contreras et al., 2021).

3.5 Advancements and innovation

As explored throughout the chapter, biofuels production and biorefineries systems are surrounded by innovation: New technologies are being constantly developed to reduce production costs and make biofuels economically competitive. To add information on biorefineries and reveal the state of art in this subject, a patent search and analysis was conducted. Therefore, the words "biofuel," "bioethanol," "biodiesel," and "biorefinery" were combined and searched in Lens.org and Latipat databases. After manual cleaning of the results, 1890 documents were obtained and analyzed with data science tools, using the programming language Python3 and Microsoft excel software. Energy supply from selected countries was also analyzed. The data were obtained from IEA Energy (total energy supply by source data set from China, Brazil, and the United States).

Throughout the years, the number of filed patents in this subject has grown, especially between the years of 2005 and 2015 (Fig. 3.4). The development of biofuels worldwide has been guided by two main factors: reducing the dependence of fossil fuels and mitigation of global warming effects. Several countries established blend mandates of biofuels in transportation

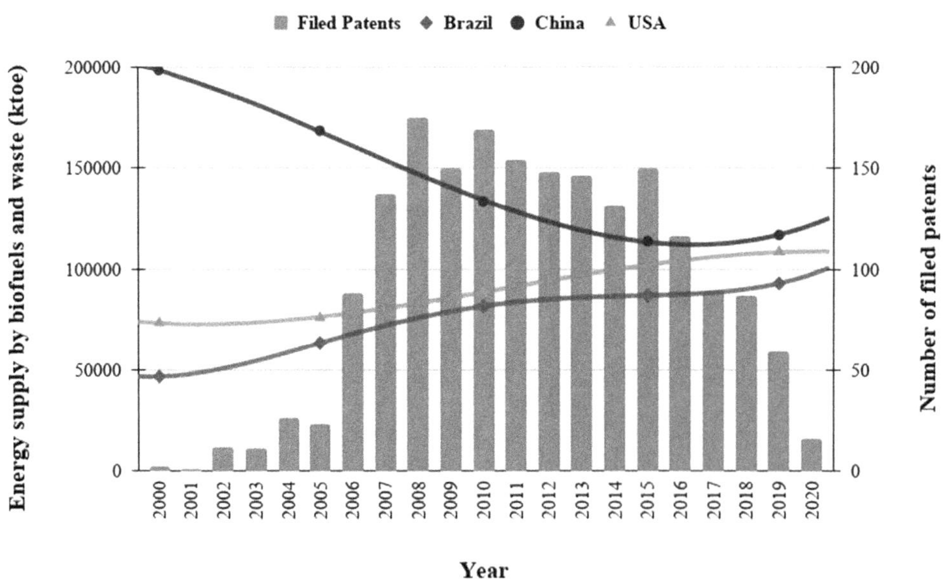

FIGURE 3.4 Evolution of filled patents by China, the United States, and Brazil and energy supply by biofuels and waste from 2000 to 2020.

fuels and promoted the development of vehicles that are moved with alternative energy sources. China, the United States, and Brazil are the countries that have more filed patents (with 30.22%, 20.05% and 14.44% of the filed documents, respectively) and are also among the greatest biofuel producers worldwide (Statista. 2020).

China is one of the largest economies in the world nowadays, and its GDP is growing every year, generating a higher energy demand in the country, which is mainly supplied by coal. However, China has been acting for the development and implementation of biofuels. In 2003, the first ethanol production plant became functional, and in 2004, a pilot project of E10 blend was implemented in the country. China also stands out for its 5-year plans, in which biofuels became an important guideline since the beginning of the 2000s. In 2010, Chinese government also took action favoring biodiesel production, such as a B5 blend national project and the elimination of consumption taxes of 5% (Dyk et al., 2016). On the other hand, the United States seeks biofuel development for energetic independence. The main actions of this country focus on the bioethanol from corn. In 2005, the United States launched an Energy Act that set a bioethanol minimum consumption rate and, in 2007, launched the Energy Independence and Security Act, which determined standards for renewable fuels (Hochman et al., 2017).

Brazil is a country that has been developing biofuels since the past century, starting in 1975 with the foundation of Próalcool, a national ethanol implementation program that had measures such as fixation of ethanol and gasoline blends and lower prices of the biofuel for both the producer and consumer. In 2003, flex passenger cars arrived in Brazil, which instigated ethanol consumption—and therefore, generation (Nogueira and Capaz, 2013)—and in 2008, Petrobras Biocombustíveis was founded with the focus of promoting sugarcane bioethanol production, signing important partnerships such as the one with the Danish company Novozymes in 2010 (Exame, 2010). Petrobras is a great state-owned company that produces a variety of fuels. However, the company suffered from the consequences of the economic and political crisis that have taken place in Brazil, which led to the sale of Petrobrás Biocombustíveis in 2020. Brazil also took actions throughout the years for enhancing biodiesel production in the country. Among them are the creation of the Research and Development Network for biodiesel production in 2002 and the national program of production and use of biodiesel in 2005 (Nogueira and Capaz, 2013). In recent years, the Brazilian government launched RenovaBio (end of 2016), a new program for expansion of biofuel production in the country, and Petrobras announced a new renewable diesel, which will be one of the new focuses of the company.

These countries have a diverse energetic matrix, and biofuels play a critical role in energy supply (Fig. 3.4). Brazil and the United States have fossil oil as their main energy source. However, biofuels and waste have been gaining space, and their energy supply has been growing throughout the years—they are the second and fifth main energy sources in Brazil and the United States, respectively. This can be related to the growing numbers of filed patents. When compared with other countries, China stands out for having a great energy supply from biofuels and waste since the Chinese energy demand is higher due to the substantial industrial activity in the country. Though, when compared with other energy sources, the use of biofuels and waste in China still needs to be encouraged through assertive governmental actions.

The nine most cited patent documents are briefly reviewed as follows. The document US 2008/0057555 (206 citations) describes an integrated process to fractionate various types of lignocellulosic biomass, i.e., hardwoods, softwoods, agroindustrial residues, grasses, recycled papers, and pulp and paper mill residues and convert the fermentable sugar fractions to ethanol and other bioproducts, with an estimated yield of 100 ethanol gallons per ton of dry wood. The stages include size reduction, prehydrolysis (soaking of wood chips in acid solution) and application of heat and pressure to separate hemicellulose in liquid phase, addition of lime to neutralize and detoxify the hemicellulosic prehydrolysate, and delignification of the remaining solid fraction by chemical treatment (sodium chlorite/acetic acid and chlorine/chlorine dioxide systems) under heat and pressure. These last steps generate two other process streams: a liquid lignin-rich fraction and a solid cellulose-rich fraction. Then, the cellulose-containing fraction is subjected to washing, saccharification, and fermentation to produce ethanol. The hemicellulosic prehydrolysate is also fermented, either separately or together with the cellulosic hydrolysate. The lignin fraction can be converted to hydrocarbon fuel, surfactant, drilling aid, or, alternatively, it can be burned as a fuel.

The document US 2010/0041119 (69 citations) reports a method and equipment for the conversion of cellulosic material (straw, corn stover, household waste) to ethanol without the use of chemical agents. The solid material is soaked in liquid before entering a pressurized reactor, at temperatures between 170 and 230°C. Continuous hydrothermal pretreatment is employed and fractionates the material into a liquid stream and a solid fraction (still containing 80% of the lignin). After, the solid fibrous fraction is subjected to enzymatic saccharification and fermentation. The liquid stream is used to produce enzymes and C5 fermenting microorganisms. Another method based on physical pretreatment processes was proposed in the patent document WO 2009/018469 (30 citations). The cellulosic material is explosively dried and pulverized, and then electrically degraded to disrupt lignocellulosic bonds.

The document US 2010/0297749 (114 citations) proposed an integrated biorefinery system to produce biofuel, especially from oleaginous photosynthetic algae cultivated in open ponds. The oil extracted from the algae can be subjected to cracking, transesterification, hydroprocessing, and isomerization to produce jet fuel, diesel fuel, or gasoline. The document US 7135308 (84 citations) describes a process for ethanol production from starch-accumulating algae. Ethanol is produced by yeast fermentation of decaying algal biomass, and the carbon dioxide produced during alcoholic fermentation is fed to the algal aqueous culture as a carbon source.

A different pathway to produce biofuels and chemicals from lignocellulosic biomass was proposed in the patent document US 2012/0283493 (47 citations), the condensation of levulinates derived from biomass containing C6 sugars. The solid material is thermocatalytically reacted in the presence of acid in water and/or alcohol yielding a solution containing levulinic acid and/or levulinic acid ester, which can be extracted using a cyclic ether. Then, the levulinate molecules are condensed with C4–C11 aldehydes, ketones, esters, or ketoacids and hydrotreated, depending on the targeted fuel/chemical hydrotreated compound.

Finally, three patent documents describe the production of biodiesel from lignocellulosic biomass and wastes. The document WO 2010/042842 (47 citations) reports a method based on the simultaneous saccharification and fermentation of lignocellulosic biomass using microorganisms to synthesize lignocellulolytic enzymes and algae to synthesize fatty acids, which can be extracted and transesterified to biodiesel. The document WO 2008/105518 (38 citations) proposes a method based on the use of rare earth elements and Group VIII noble metals, in a low-pressure hydrogen atmosphere, as hydrogenation catalysts for the transesterification of waste cooking oil or oils from other sources. And the document US 2005/0006290 (29 citations) describes a mobile biodiesel biorefinery to process waste cooking oil. The computer-controlled, automated system is built on a semitruck trailer and contains a collection tank, a filter to separate contaminants, a reaction chamber, a glycerin separator, a wash chamber, a pH adjustment system, and a delivery tank.

3.6 Conclusions and perspectives

The world is continuously searching to minimize the fossil fuels dependence. Biofuels appear as a powerful alternative in the energy matrix panorama of crop-producing countries. However, 1G production of biofuels is not economically and environmentally sustainable, as it competes directly with fossil fuels. However, the exploitation of different feedstocks for biofuels production, including bioethanol and biodiesel, generates different by-products that can be reused as feedstock to produce 2G, 3G, and 4G biofuels and other bioproducts. Independently of the type of processing chain, research on economic and sustainable solutions has been stimulated by different governments worldwide. In this way, the biorefinery strategy represents an efficient and viable way to optimize the competitiveness of produced biofuels, bioenergy, and other bioproducts with fossil fuels. The future of bioethanol and biodiesel biorefineries holds numerous questions, and the answers are complex and depend on several factors. Even so, the production of traditional products through mature technologies remains with significant expansion potential.

References

Abdullah, B., Syed Muhammad, S.A.F., Shokravi, Z., Ismail, S., Kassim, K.A., Mahmood, A.N., Aziz, M.M.A., 2019. Fourth generation biofuel: a review on risks and mitigation strategies. Renew. Sustain. Energ. Rev. 107, 37–50. https://doi.org/10.1016/j.rser.2019.02.018.

Alibardi, L., Astrup, T.F., Asunis, F., Clarke, W.P., De Gioannis, G., Dessì, P., Lens, P.N.L., Lavagnolo, M.C., Lombardi, L., Muntoni, A., Pivato, A., Polettini, A., Pomi, R., Rossi, A., Spagni, A., Spiga, D., 2020. Organic waste biorefineries: looking towards implementation. Waste Manag 114, 274–286. https://doi.org/10.1016/j.wasman.2020.07.010.

Almeida, J.R.M., Fávaro, L.C.L., Quirino, B.F., 2012. Biodiesel biorefinery : opportunities and challenges for microbial production of fuels and chemicals from glycerol waste. Biotechnol. Biofuels 5, 1–16. https://doi.org/10.1186/1754-6834-5-48.

Balaman, S.Y., 2019. Introduction to biomass—resources, production, harvesting, collection, and storage. In: Balaman, S.Y. (Ed.), Decision-Making for Biomass-Based Production Chains - Basic Concepts and Methodologies. Academic Press, pp. 1–23. https://doi.org/10.1016/B978-0-12-814278-3.00001-7.

Balat, M., Balat, M., 2009. Political, economic and environmental impacts of biomass-based hydrogen. Int. J. Hyd. Energ. 34, 3589–3603. https://doi.org/10.1016/j.ijhydene.2009.02.067.

Bauer, L.M., Costa, J.A.V., da Rosa, A.P.C., Santos, L.O., 2017. Growth stimulation and synthesis of lipids, pigments and antioxidants with magnetic fields in *Chlorella kessleri* cultivations. Bioresour. Technol. 244, 1425–1432. https://doi.org/10.1016/j.biortech.2017.06.036.

Bio-based News, 2017. Map of 224 European Biorefineries Published by BIC and Nova-Institute. https://news.bio-based.eu/map-of-224-european-biorefineries-published-by-bic-and-nova-institute/. (Accessed 26 January 2021).

Boldrin, A., Balzan, A., Astrup, T., 2013. Energy and environmental analysis of a rapeseed biorefinery conversion process. Biomass Convers. Biorefin. 3, 127–141. https://doi.org/10.1007/s13399-013-0071-9.

Chandel, A.K., Garlapati, V.K., Singh, A.K., Fernandes Antunes, F.A., Silvério da Silva, S., 2018. The path forward for lignocellulose biorefineries: bottlenecks, solutions, and perspective on commercialization. Bioresour. Technol. 264, 370–381. https://doi.org/10.1016/j.biortech.2018.06.004.

Chandel, A.K., Albarellu, J.Q., Santos, D.T., Chundawat, S.P., Puri, M., Meireles, M.A.A., 2019. Comparative analysis of key technologies for cellulosic ethanol bagasse at a commercial scale. Biofuel Bioprod. Biorefin. 13 (4), 994–1014. https://doi.org/10.1002/bbb.1990.

Chandel, A.K., Ravinder, S.E.E., Rudravaram, R., Narasu, M., Venkateswar, R., Ravindra, P., 2007. Economics and environmental impact of bioethanol production technologies: an appraisal. Biotechnol. Mol. Biol. Rev. 2 (1), 14–32. http://www.academicjournals.org/bmbr/PDF/pdf2007/Feb/Chandel et al.pdf. (Accessed 26 January 2021).

Cherubini, F., 2010. The biorefinery concept: using biomass instead of oil for producing energy and chemicals. Energ. Convers. Manag. 51, 1412–1421. https://doi.org/10.1016/j.enconman.2010.01.015.

Chia, S.R., Ong, H.C., Chew, K.W., Show, P.L., Phang, S.-M., Ling, T.C., Nagarajan, D., Lee, D.-J., Chang, J.-S., 2018. Sustainable approaches for algae utilisation in bioenergy production. Renew. Energ. 129, 838–852. https://doi.org/10.1016/j.renene.2017.04.001.

Cho, H.U., Park, J.M., 2018. Biodiesel production by various oleaginous microorganisms from organic wastes. Bioresour. Technol. 256, 502–508. https://doi.org/10.1016/j.biortech.2018.02.010.

Choong, Y.Y., Chou, K.W., Norli, I., 2018. Strategies for improving biogas production of palm oil mill effluent (POME) anaerobic digestion: a critical review. Renew. Sustain. Energy Rev. 82, 2993–3006. https://doi.org/10.1016/j.rser.2017.10.036.

Chowdhury, R., Ghosh, S., Debnath, B., Manna, D., 2018. Indian agro-wastes for 2G biorefineries: strategic decision on conversion processes. In: De, S., Bandyopadhyay, S., Assadi, M., Mukherjee, D. (Eds.), Suistainable Energy Technology and Policies. Green Energy and Technology. Springer, Singapore, pp. 353–373. https://doi.org/10.1007/978-981-10-7188-1_16.

Das, M., Patra, P., Ghosh, A., 2020. Metabolic engineering for enhancing microbial biosynthesis of advanced biofuels. Renew. Sustain. Energ. Rev. 119, 109562. https://doi.org/10.1016/j.rser.2019.109562.

de Jong, E., Jungmeier, G., 2015. Biorefinery concepts in comparison to petrochemical refineries. Industrial Biorefineries & White Biotechnology, pp. 3–33. https://doi.org/10.1016/B978-0-444-63453-5.00001-X.

Demichelis, F., Laghezza, M., Chiappero, M., Fiore, S., 2020. Technical, economic and environmental assessement of bioethanol biorefinery from waste biomass. J. Clean. Prod. 277, 124111. https://doi.org/10.1016/j.jclepro.2020.124111.

Demirbas, A. (Ed.), 2010. Biorefineries for Biomass Upgrading Facilities. Springer-Verlag, London. https://doi.org/10.1007/978-1-84882-721-9.

Deprá, M.C., dos Santos, A.M., Severo, I.A., Santos, A.B., Zepka, L.Q., Jacob-Lopes, E., 2018. Microalgal biorefineries for bioenergy production: can we move from concept to industrial reality? BioEnergy Res 11, 727–747. https://doi.org/10.1007/s12155-018-9934-z.

Dutta, K., Daverey, A., Lin, J.-G., 2014. Evolution retrospective for alternative fuels: first to fourth generation. Renew. Energ. 69, 114–122. https://doi.org/10.1016/j.renene.2014.02.044.

Dyk, S.V., Li, L., Leal, D.B., Hu, J., Zhang, X., Tan, T., Saddler, J., 2016. The Potential of Biofuels in China, IEA Bioenergy: Task 39. http://task39.sites.olt.ubc.ca/files/2020/02/IEA-Bioenergy-Task-39-Implementation-Agendas-Final-Draft-Feb-4-2020.pdf. (Accessed 30 January 2021).

Exame, 2010. Novozymes Faz Parceria Com Petrobras Em Biocombustíveis. https://exame.com/negocios/novozymes-faz-parceria-petrobras-biocombustiveis-604244/. (Accessed 27 January 2021).

Gonzalez-Contreras, M., Lugo-Mendez, H., Sales-Cruz, M., Lopez-Arenas, T., 2021. Synthesis, design and evaluation of intensified lignocellulosic biorefineries - case study: ethanol production. Chem. Eng. Process. – Process Intensification 159, 108220. https://doi.org/10.1016/j.cep.2020.108220.

Habib, M.S., Tayyab, M., Zahoor, S., Sarkar, B., 2020. Management of animal fat-based biodiesel supply chain under the paradigm of sustainability. Energ. Convers. Manag. 225, 113345. https://doi.org/10.1016/j.enconman.2020.113345.

Hassan, S.S., Williams, G.A., Jaiswal, A.K., 2018. Lignocellulosic biorefineries in Europe: current state and prospects. Trends Biotechnol 37 (3), 231–234. https://doi.org/10.1016/j.tibtech.2018.07.002.

Hattori, T., Morita, S., 2010. Energy crops for sustainable bioethanol production; which, where and how? Plant Prod. Sci. 13, 221–234. https://doi.org/10.1626/pps.13.221.

Hochman, G., Traux, M., Zilberman, D., 2017. US biofuel policies and markets. In: Khanna, M., Zilberman, D. (Eds.), Handbook of Bioenergy Economics and Policy: Volume II. Springer New York, New York, NY, pp. 15–38.

Hossain, N., Hasan, M.H., Mahlia, T.M.I., Shamsuddin, A.H., Silitonga, A.S., 2020. Feasibility of microalgae as feedstock for alternative fuel in Malaysia: a review. Energ. Strateg. Rev. 32, 100536. https://doi.org/10.1016/j.esr.2020.100536.

Ibrahim, M.L., Nik Abdul Khalil, N.N.A., Islam, A., Rashid, U., Ibrahim, S.F., Sinar Mashuri, S.I., Taufiq-Yap, Y.H., 2020. Preparation of Na_2O supported CNTs nanocatalyst for efficient biodiesel production from waste-oil. Energ. Convers. Manag. 205, 112445. https://doi.org/10.1016/j.enconman.2019.112445.

Kampschreur, M.J., Temmink, H., Kleerebezem, R., Jetten, M.S.M., Loosdrecht, M.C.M., 2009. Nitrous oxide emission during wastewater treatment. Water Res 43 (17), 4093–4103. https://doi.org/10.1016/j.watres.2009.03.001.

Karp, A., Shield, I., 2008. Bioenergy from plants and the sustainable yield challenge. New Phytol 179, 15–32. https://doi.org/10.1111/j.1469-8137.2008.02432.x.

Katakojwala, R., Mohan, S.V., 2021. A critical view on the environmental sustainability of biorefinery systems. Curr. Opin. Green Sustain. Chem. 27, 100392. https://doi.org/10.1016/j.cogsc.2020.100392.

Klein, B.C., Chagas, M.F., Watanabe, M.D.B., Bonomi, A., Maciel Filho, R., 2019. Low carbon biofuels and the New Brazilian National Biofuel Policy (RenovaBio): a case study for sugarcane mills and integrated sugarcane-microalgae biorefineries. Renew. Sustain. Energ. Rev. 115, 109365. https://doi.org/10.1016/j.rser.2019.109365.

Kumar, B., Bhardwaj, N., Agrawal, K., Chaturvedi, V., Verma, P., 2020. Current perspective on pretreatment technologies using lignocellulosic biomass: an emerging biorefinery concept. Fuel Process. Technol. 199, 106244. https://doi.org/10.1016/j.fuproc.2019.106244.

Kumar, B., Verma, P., 2021. Biomass-based biorefineries: an important architype towards a circular economy. Fuel 288, 119622. https://doi.org/10.1016/j.fuel.2020.119622.

Liu, Y., Lyu, Y., Tian, J., Zhao, J., Ye, N., Zhang, Y., Chen, L., 2021. Review of waste biorefinery development towards a circular economy: from the perspective of a life cycle assessment. Renew. Sustain. Energ. Rev. 139, 110716. https://doi.org/10.1016/j.rser.2021.110716.

Longati, A.A., Batistam, G., Cruz, A.J.G., 2020. Brazilian integrated sugarcane-soybean biorefinery: trends and opportunities. Curr. Opin. Green Sustain. Chem. 26, 100400. https://doi.org/10.1016/j.cogsc.2020.100400.

Lü, J., Sheahan, C., Fu, P., 2011. Metabolic engineering of algae for fourth generation biofuels production. Energy Environ. Sci. 4, 2451. https://doi.org/10.1039/c0ee00593b.

Mardhiah, H.H., Ong, H.C., Masjuki, H.H., Lim, S., Lee, H.V., 2017. A review on latest developments and future prospects of heterogeneous catalyst in biodiesel production from non-edible oils. Renew. Sustain. Energ. Rev. 67, 1225–1236. https://doi.org/10.1016/j.rser.2016.09.036.

Miao, X., Wu, Q., Yang, C., 2004. Fast pyrolysis of microalgae to produce renewable fuels. J. Anal. Appl. Pyrolysis 71, 855–863. https://doi.org/10.1016/j.jaap.2003.11.004.

Mofijur, M., Siddiki, S.Y.A., Shuvho, M.B.A., Djavanroodi, F., Fattah, I.M.R., Ong, H.C., Chowdhury, M.A., Mahlia, T.M.I., 2020. Effect of nanocatalysts on the transesterification reaction of first, second and third generation biodiesel sources- A mini-review. Chemosphere 270, 128642. https://doi.org/10.1016/j.chemosphere.2020.128642.

Monteiro, M.R., Kugelmeier, C.L., Pinheiro, R.S., Batalha, M.O., da Silva César, A., 2018. Glycerol from biodiesel production: technological paths for sustainability. Renew. Sustain. Energy Rev. 88, 109–122. https://doi.org/10.1016/j.rser.2018.02.019.

Morán, D., 2017. BioRefineries.blog, 2015. https://biorrefineria.blogspot.com/2015/08/cellulosic-ethanol-biorefineries.html. (Accessed 30 January 2021).

Nogueira, L.A.H., Capaz, R.S., 2013. Biofuels in Brazil: evolution, achievements and perspectives on food security. Glob. Food Sec 2, 117–125. https://doi.org/10.1016/j.gfs.2013.04.001.

Nguyen, Q., Bowyer, J., Howe, J., Bratkovich, S., Groot, H., Pepke, E., Fernholz, K., 2017. Global Production of Second-Generation Biofuels: Trends and Influences. Dovetail Partners Inc. https://www.dovetailinc.org/report_pdfs/2017/dovetailbiofuels0117.pdf. (Accessed 26 January 2021).

Oláh, J., Lengyel, P., Balogh, P., Harangi-Rákos, M., Popp, J., 2017. The role of biofuels in food commodity prices volatility and land use. J. Compet. 9, 81–93. https://doi.org/10.7441/joc.2017.04.06.

Osman, A.I., Abdelkader, A., Johnston, C.R., Morgan, K., Rooney, D.W., 2017. Thermal investigation and kinetic modeling of lignocellulosic biomass combustion for energy production and other applications. Ind. Eng. Chem. Res. 56, 12119–12130. https://doi.org/10.1021/acs.iecr.7b03478.

Papadaskalopoulou, C., Sotiropoulus, A., Novacovic, J., Barabouti, E., Mai, S., Malamis, D., Kekos, D., Loizidou, M., 2019. Comparative life cycle assessment of a waste to ethanol biorefinery system versus conventional waste management methods. Resour. Conserv. Recycl. 149, 130–139. https://doi.org/10.1016/j.resconrec.2019.05.006.

Parisi, C., 2018. Research Brief: Biorefineries Distribution in the EU. European Commission, Join Research Center. https://publications.jrc.ec.europa.eu/repository/bitstream/JRC113216/online_biorefineries_research_brief.pdf. (Accessed 26 January 2021).

Parsons, S., Allen, M.J., Chuck, C.J., 2020. Coproducts of algae and yeast-derived single cell oils: a critical review of their role in improving biorefinery sustainability. Bioresour. Technol. 303, 122862. https://doi.org/10.1016/j.biortech.2020.122862.

Patel, A., Arora, N., Mehtani, J., Pruthi, V., Pruthi, P.A., 2017. Assessment of fuel properties on the basis of fatty acid profiles of oleaginous yeast for potential biodiesel production. Renew. Sustain. Energ. Rev. 77, 604–616. https://doi.org/10.1016/j.rser.2017.04.016.

Pinzi, S., Garcia, I.L., Lopez-Gimenez, F.J., Luque de Castro, M.D., Dorado, G., Dorado, M.P., 2009. The ideal vegetable oil-based biodiesel composition: a review of social, economical and technical implications. Energy Fuels 23, 2325–2341. https://doi.org/10.1021/ef801098a

Poveda-Giraldo, J.A., Solarte-Toro, J.C., Alzate, C.A.C., 2021. The potential use of lignin as a platform product in biorefineries: a review. Renew. Sustain. Energy Rev. 138, 110688. https://doi.org/10.1016/j.rser.2020.110688.

Rahman, M.M., Rasul, M., Hassan, N.M.S., 2017. Study on the tribological characteristics of australian native first generation and second generation biodiesel fuel. Energies 10, 1–16. https://doi.org/10.3390/en10010055.

Rezania, S., Oryani, B., Cho, J., Talaiekhozani, A., Sabbagh, F., Hashemi, B., Rupani, P.F., Mohammadi, A.A., 2020. Different pretreatment technologies of lignocellulosic biomass for bioethanol production: an overview. Energ 199, 117457. https://doi.org/10.1016/j.energy.2020.117457.

Rosales-Calderon, O., Arantes, V., 2019. A review on commercial-scale high-value products that can be produced alongside cellulosic ethanol. Biotechnol. Biofuels 12, 1–58. https://doi.org/10.1186/s13068-019-1529-1.

Saini, J.K., Saini, R., Tewari, L., 2015. Lignocellulosic agriculture wastes as biomass feedstocks for second-generation bioethanol production: concepts and recent developments. 3 Biotech 5, 337–353. https://doi.org/10.1007/s13205-014- 0246-5.

Santos, V.E.N., Ely, R.N., Magrini, S.A.S., 2016. Chemicals, electricity and fuels from biorefineries processing Brazil's sugarcane bagasse: production recipes and minimum selling prices. Renew. Sustain. Energ. Rev. 53, 1443–1458. https://doi.org/10.1016/j.rser.2015.09.069.

Sasana, H., Ghozali, I., 2017. The impact of fossil and renewable energy consumption on the economic growth in Brazil, Russia, India, China and South Africa. Int. J. Energ. Econ. Policy 7 (3), 194–200.

Severo, I.A., Siqueira, S.F., Deprá, M.C., Maroneze, M.M., Zepka, L.Q., Jacob-Lopes, E., 2019. Biodiesel facilities: what can we address to make biorefineries commercially competitive? Renew. Sustain. Energ. Rev. 112, 686–705. https://doi.org/10.1016/j.rser.2019.06.020.

Sharma, Y.C., Singh, B., 2010. A hybrid feedstock for a very efficient preparation of biodiesel. Fuel Process. Technol 91, 1267–1273. https://doi.org/10.1016/j.fuproc.2010.04.008.

Singh, D., Sharma, D., Soni, S.L., Sharma, S., Kumar Sharma, P., Jhalani, A., 2020. A review on feedstocks, production processes, and yield for different generations of biodiesel. Fuel 262, 116553. https://doi.org/10.1016/j.fuel.2019.116553.

Srivastava, R.K., 2019. Bio-Energy production by contribution of effective and suitable microbial system. Material Science Energ. Technol. 2, 308–318. https://doi.org/10.1016/j.mset.2018.12.007.

Statista, 2020. Leading Countries Based on Biofuel Production in 2019 (In Petajoules). https://www.statista.com/statistics/274168/biofuel-production-in-leading-countries-in-oil-equivalent/. (Accessed 29 January 2021).

Suresh, M., Jawahar, C.P., Richard, A., 2018. A review on biodiesel production, combustion, performance, and emission characteristics of non-edible oils in variable compression ratio diesel engine using biodiesel and its blends. Renew. Sustain. Energ. Rev. 92, 38–49. https://doi.org/10.1016/j.rser.2018.04.048.

Thompson, P.B., 2012. The agricultural ethics of biofuels: the food vs. fuel debate. Agriculture 2 (4), 339–358. https://doi.org/10.3390/agriculture2040339.

Ubando, A.T., Felix, C.B., Chen, W.-H., 2020. Biorefineries in circular bioeconomy: a comprehensive review. Bioresour. Technol. 299, 122585. https://doi.org/10.1016/j.biortech.2019.122585.

Ubando, A.T., Del Rosario, A.J.R., Chen, W.-H., Culaba, A.B., 2021. A state-of-the-art review of biowaste biorefinery. Environ. Pollut. 269, 116149. https://doi.org/10.1016/j.envpol.2020.116149.

Valentine, J., Clifton-Brown, J., Hastings, A., Robson, P., Allison, G., Smith, P., 2012. Food vs. fuel: the use of land for lignocellulosic 'next generation' energy crops that minimize competition with primary food production. GCB Bioenergy 4 (1), 1–19.

Valladares-Diestra, K., de Souza Vandenberghe, L.P., Soccol, C.R., 2020. Oilseed enzymatic pretreatment for efficient oil recovery in biodiesel production industry: a review. Bioenergy Res 13, 1016–1030. https://doi.org/10.1007/s12155-020-10132-9.

Valladares-Diestra, K.K., Porto de Souza Vandenberghe, L., Zevallos Torres, L.A., Nishida, V.S., Zandoná Filho, A., Woiciechowski, A.L., Soccol, C.R., 2020. Imidazole green solvent pre-treatment as a strategy for second-generation bioethanol production from sugarcane bagasse. Chem. Eng. J. 420, 127708. https://doi.org/10.1016/j.cej.2020.127708.

Wei, Z., Zeng, G., Kosa, M., Huang, D., Ragauskas, A.J., 2015. Pyrolysis oil-based lipid production as biodiesel feedstock by Rhodococcus opacus. Appl. Biochem. Biotechnol. 175, 1234–1246. https://doi.org/10.1007/s12010-014-1305-4.

Woiciechowski, A.L., Dalmas Neto, C.J., Porto de Souza Vandenberghe, L., de Carvalho Neto, D.P., Novak Sydney, A.C., Letti, L.A.J., Karp, S.G., Zevallos Torres, L.A., Soccol, C.R., 2020. Lignocellulosic biomass: acid and alkaline pretreatments and their effects on biomass recalcitrance – conventional processing and recent advances. Bioresour. Technol. 304, 122848. https://doi.org/10.1016/j.biortech.2020.122848.

Woo, H., Acuna, M., Cho, S., Park, J., 2019. Assessment techniques in forest biomass along the timber supply chain. Forests 10, 1–14. https://doi.org/10.3390/f10111018.

Zhao, Y., Damgaard, A., Christensen, T.H., 2018. Bioethanol from corn stover - a review and technical assessment of alternative biotechnologies. Progress Energy Combust. Sci. 67, 275–291. https://doi.org/10.1016/j.pecs.2018.03.004.

CHAPTER

4

Hydrothermal liquefaction of lignocellulosic biomass for production of biooil and by-products: current state of the art and challenges

Marttin Paulraj Gundupalli[1], *Sathish Paulraj Gundupalli*[2], *Anne Sahithi Somavarapu Thomas*[3], *Mani Jayakumar*[4], *Debraj Bhattacharyya*[1], *Baskar Gurunathan*[5]

[1]Department of Civil Engineering, Indian Institute of Technology Hyderabad, Kandi, Telangana, India; [2]Department of Design and Automation, Vellore Institute of Technology, Vellore, Tamil Nadu, India; [3]Department of Biomedical Sciences, Vellore Institute of Technology, Vellore, Tamil Nadu, India; [4]Department of Chemical Engineering, Haramaya Institute of Technology, Haramaya University, Dire Dawa, East Hararghe, Ethiopia; [5]Department of Biotechnology, St. Joseph's College of Engineering, Chennai, Tamil Nadu, India

4.1 Introduction

Due to rapid population and global economic growth, energy consumption and its demand are constantly increasing. On the other hand, depletion of fossil fuel reservoirs and increased greenhouse gas (GHG) emissions are harmful to the environment. As an alternative to fossil fuels, biomass (nonedible plants and algae) has long been an attractive source of renewable energy sources. Utilization of biomass for energy production reduces GHG emissions and reliance on fossil fuels (Ragauskas, 2006). There are several thermochemical conversion processes such as pyrolysis, combustion, gasification, and hydrothermal liquefaction (HTL). Pyrolysis is a thermochemical decomposition method that can decompose organic matter (biomass) without oxygen to produce products such as bio-oil, syngas, and biochar (Rajendran et al., 2019). Biomass burning refers to burning energy materials in a furnace, boiler, or stove to produce heat and energy. Gasification converts biomass into syngas such as carbon monoxide (CO),

Biofuels and Bioenergy
https://doi.org/10.1016/B978-0-323-85269-2.00001-0

hydrogen (H_2), carbon dioxide (CO_2), and methane (CH_4) (Heidenreich and Foscolo, 2015). HTL is a technique that can generate biocrude oil from biomass with a high moisture content (>50%) (Divyabharathi and Subramanian, 2020). Among these techniques, the HTL technique has proved to be more effective in the processing of wet biomass (Yerrayya et al., 2020). Pyrolysis and HTL processes are considered as two equivalent technologies because they both provide bio-based intermediate products such as biooil and biocrude. The difference between these two processes has been summarized in Table 4.1. HTL can process wet biomass to biooil because the water in the biomass acts as solvent and reactant. The HTL process is carried out at temperature (200–400°C) and pressure (50–250 bar) as displayed in the table.

The pyrolysis process requires raw material to be dried. On the other hand, this process is not required for the HTL process (Tzanetis et al., 2017). This significantly reduces the economic benefits of fuel production due to the wet nature of the raw materials selected. During the HTL process, the biomass gets converted to biochar, biooil, aqueous crude, and gases (Tungal and Shende, 2014). As summarized, the products of the HTL process have higher heating value (HHV), lower oxygen, and moisture content compared with the pyrolysis process. However, since the HTL process is carried out at higher pressure, it increases the production of biooil due to an increase in investment of HTL process units. A rapid increase in the number of research related to the HTL process has been observed in recent years. This chapter provides detailed information related to the HTL process in relation to the study carried out and reported by different researchers all over the world.

TABLE 4.1 Difference between the physical and chemical properties of by-products from pyrolysis and hydrothermal liquefaction biomass conversion.

Sl. No	Properties	Parameters	Pyrolysis	Hydrothermal liquefaction	References
1	Physical properties	Viscosity	Lower	Higher	Haarlemmer et al. (2016)
		Heating value (MJ/kg)	Low (~17)	High (~30)	Babich et al. (2011), Haarlemmer et al. (2016)
		Density	Lower	Higher	Mørup et al. (2012)
		Temperature (°C)	351–526	200–400	Itskos et al. (2016), Zhang and Chen (2018)
		Pressure (bar)	1–5	50–200	Mei Wu et al. (2014)
2	Chemical properties	Water content	Higher	Lower	Zhang et al. (2019)
		Oxygen content	Higher	Lower	Mørup et al. (2012)
		Nitrogen content	Lower	Higher	Mullen and Boateng (2008), Usami et al. (2020)
		Hydrogen content	Lower	Higher	Mørup et al. (2012)
		Carbon content	Lower	Higher	Mørup et al. (2012)
3	Other parameters	Drying	Required	Not required	Mørup et al. (2012)
		Catalyst	Not required	Required occasionally	Dimitriadis and Bezergianni (2017)

4.2 Lignocellulosic biomass

Lignocellulosic biomass is currently considered as an alternative material source for fossil fuels and products worldwide. It is used to produce bioethanol, biological products, electricity, and heat (Gupta and Verma, 2015). Lignocellulosic biomass is carbon-based feedstock, which can be used for the production of bioenergy and biofuel (Gundupalli et al., 2021). Also, it is abundant worldwide and can be used to produce renewable energy. It is mainly composed of 40%–50%, 25%–30%, and 15%–20% of cellulose, hemicellulose, and lignin, respectively. In addition, pectin, inorganic compounds, and nitrogen compounds are also components of lignocellulosic biomass (Mori et al., 2015). The wood materials of lignocellulosic biomass are classified as softwood, hardwood, and grasses (Gundupalli et al., 2020). The plant's components, such as cellulose, hemicellulose, and lignin content, vary for these different forms of wood materials. The content of cellulose, hemicellulose, and lignin for softwood, hardwood, and grasses are summarized in Table 4.2.

Cellulose is a polysaccharide arranged in bundles and in a crystalline structure that allows it to form a stable and robust network (Sankaran et al., 2020) A group of cellulose chains (18 chains), called microfibrils, are connected to the hemicellulose and lignin components of plant cells (Kubicki et al., 2018). The degree of polymerization (DP) of cellulose chains varies between different lignocellulosic biomass and is generally in the range of 1500–5000 (Hallac and Ragauskas, 2011). All the hydrogen atoms of cellulose chains are in the axial position, but all the hydroxyl groups (OH) are equatorial. These equatorial OH groups form hydrogen bonds with the nearest neighbors and adjacent cellulose chains through intrahydrogen and interhydrogen bonds, allowing the cellulose to crystallize (Lindman et al., 2017). The structure of cellulose is usually expressed using the molecular formula $(C_6H_{10}O_5)_n$ (Kumar Gupta et al., 2019).

On the other hand, hemicellulose is a heterogeneous polymer composed of C_5 and C_6 sugars such as xylose, arabinose, mannose, galactose, and glucose, respectively (Ahmad and Zakaria, 2019). Xylan is the main component of hemicellulose and contains C_5 and C_6 sugars. It also contains sugar acids such as 4-O-methyl-D-glucuronic acid (Ji et al., 2011). Xylan is composed of xylose monomer units linked via β-(1 → 4) linkages. It is the second most common and abundant polysaccharide after cellulose (Mohanraj, 2019). The structure of C_5 sugar, C_6 sugar, sugar acid, and xylan is shown in Fig. 4.1A–F. The DP of hemicellulose chains consists mostly of 100–200 sugar monomer units (Bajpai, 2018). The usual molecular expression for C_5 and C_6 polysaccharide is $(C_5H_8O_4)_n$ and $(C_6H_{10}O_5)_n$, respectively (Abu Ghalia and Dahman, 2017).

Lignin is a complex primary component of plant biomass that contains many oxygen functional groups such as hydroxyl, carbonyl, and carboxyl groups. The lignin matrix consists of monolignols, which are in the form of hydroxyphenylpropane units. These units include trans-p-coumaryl alcohol (H unit), sinapyl alcohol (S unit), and coniferyl alcohol (G unit). The

TABLE 4.2 Lignocellulosic biomass composition of softwood, hardwood, and grasses (Wang et al., 2013).

Lignocellulosic biomass type	Cellulose (%)	Hemicellulose (%)	Lignin (%)
Softwood	46–50	19–22	21–29
Hardwood	40–46	17–23	18–25
Grasses	28–37	23–29	17–20

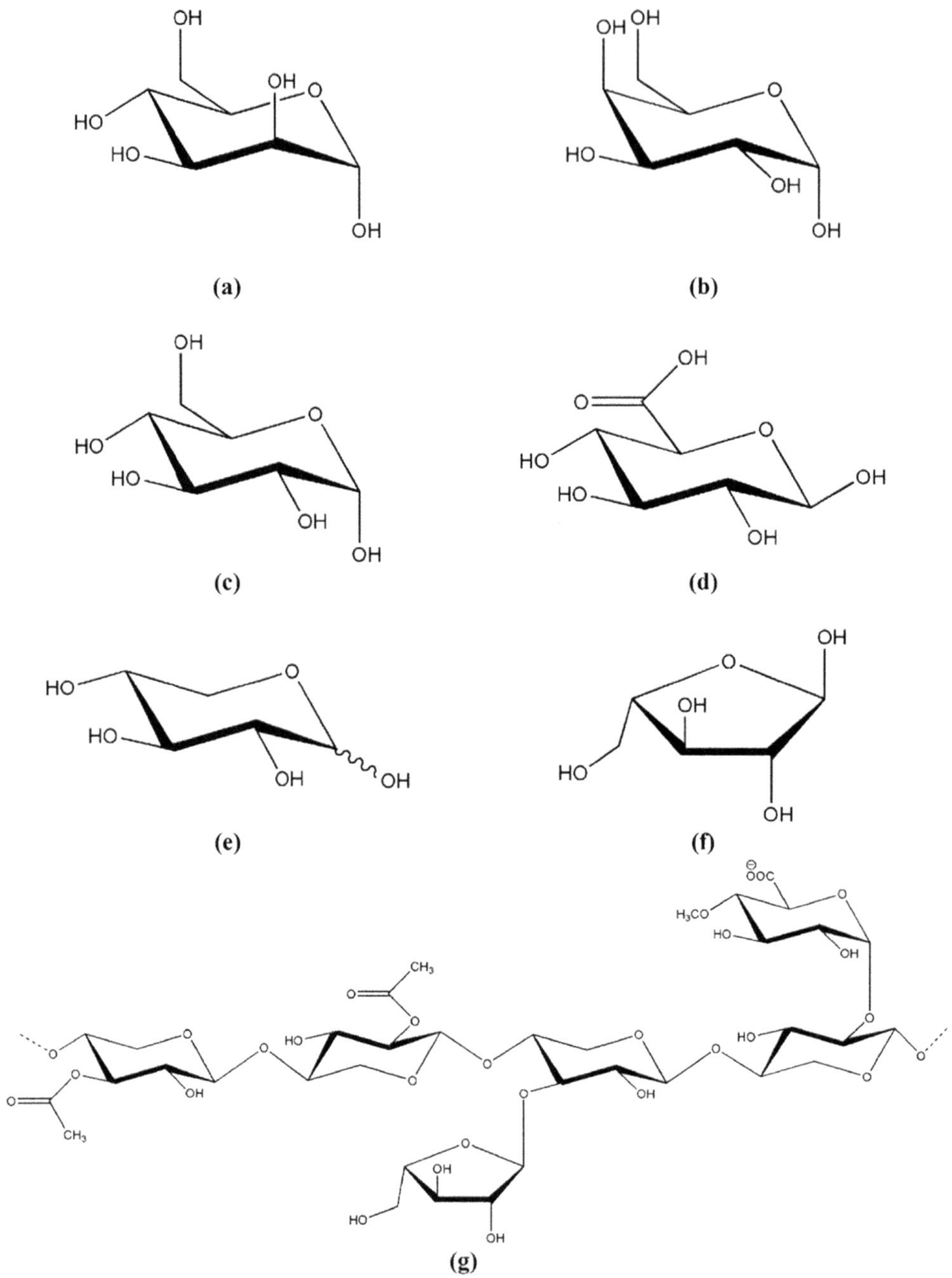

FIGURE 4.1 Hemicellulose monomer units and their structures. (A) mannose; (B) galactose; (C) glucose; (D) 4-O-methyl-D-glucuronic acid; (E) xylose; (F) arabinose; (G) xylan.

nomenclature of the hydroxyphenylpropane unit and the structure of monolignols are shown in Fig. 4.2. The content of lignin in the lignocellulosic biomass ranges from 3% to 30% by weight. They differ based on different types of lignocellulosic biomass, as displayed in Table 4.2.

The monolignols units (G, H, and S) are distributed in range for softwood, hardwood, and grasses, as summarized in Table 4.3. The G unit is higher in softwood (>95%) compared with hardwood and grasses. However, the H unit is higher in grasses (30%–80%) compared with softwood and hardwood. The S unit is usually found in hardwood (45–75) and grasses (20–55), but they are generally not found in softwood. Monolignol units are linked to each other via ether linkages. Ether bonds dominate the lignin matrix (Marchessault, 1994). Some of

TABLE 4.3 Monolignol (G, H, and S units) distribution in lignin for various lignocellulosic biomass materials (Schutyser et al., 2017).

Lignocellulosic biomass	G unit (%)	H unit (%)	S unit (%)
Softwood	>95	<5	0
Hardwood	25–50	0–8	45–75
Grasses	35–80	5–35	20–55

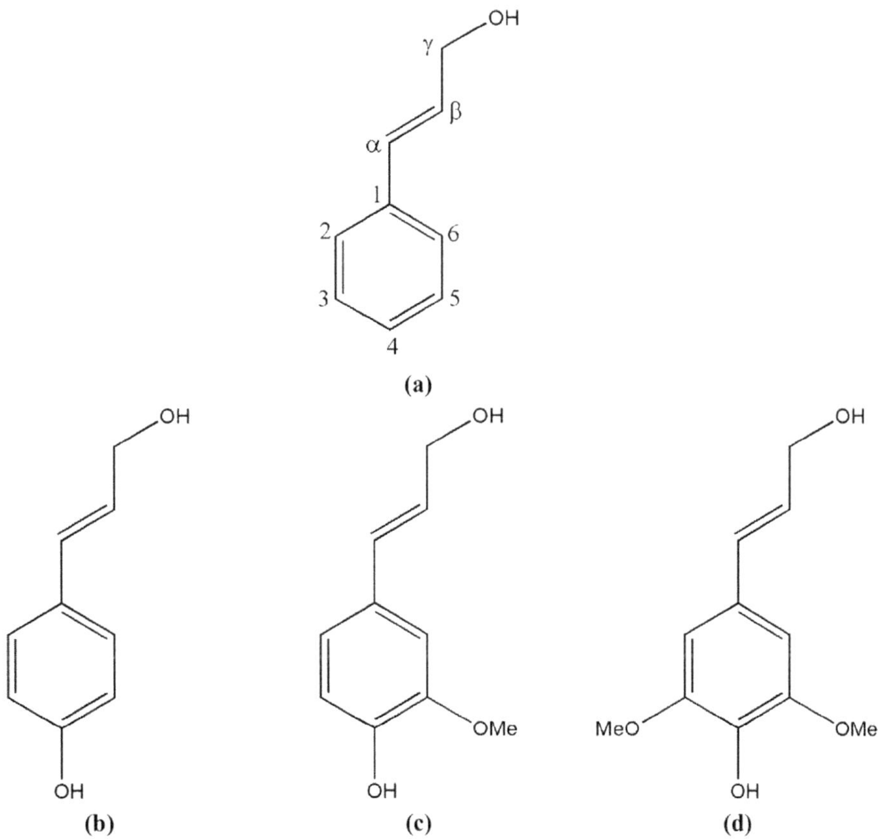

FIGURE 4.2 Lignin matrix with three hydroxyphenylpropane units. (A) Nomenclature; (B) trans-p-coumaryl alcohol; (C) sinapyl alcohol; (D) coniferyl alcohol.

the linkages that are present in the lignin matrix that bind the monolignols units are shown in Fig. 4.3. The most dominant bond found in the lignin matrix that exists between monolignols in softwood and hardwood is the β-O-4 (phenylpropane β-aryl ester) bond of about 45%–50%. The linkages such as β-β bond, the β-5 bond (phenylcumaran), the α-O-4 bond (phenylpropane, α-aryl ether), 5-5 (biphenyl), and β-1 (1,2-diarylpropane) in softwood and hardwood are approximately 3%–5%, 6%–12%, 7%–10%, 5%–11%, and 5%–7%, respectively (Adler, 1977; McDonald and Donaldson, 2001). Due to its complex structure and composition, the decomposition of lignin occurs in a wide range of temperature resulting in different by-products (Brunner, 2014).

4.3 Hydrothermal liquefaction

Biofuels are the main building blocks of bioenergy and can be classified into various forms such as solids, liquids, and gases (Chen et al., 2010). The conversion of lignocellulosic biomass materials into renewable bioenergy can be accomplished through biological and thermochemical processes. Since the cellulose chains, which have a high crystalline structure, are embedded in the lignin-hemicellulose layer through hydrogen bonds, the biochemical process of converting this lignocellulosic biomass into valuable by-products is more complex than the biomass of food crops (Yousuf, 2012). The conversion of biomass to valuable by-products through conventional methods includes (1) pretreatment, (2) enzymatic saccharification, and (3) fermentation. However, these methods require more time, energy, man force, and hardship (Gundupalli and Bhattacharyya, 2019).

Therefore compared with conventional pretreatment, torrefaction, combustion, liquefaction, pyrolysis, and gasification (Chen et al., 2019a) are considered more suitable for the valorization of biomass. These methods are carried out to increase the ease of use of the components in a wide range of applications, especially in the value-added products from lignocellulosic biomass. Thermal and chemical methods change both the microstructure and the macrostructure of the biomass, which leads to a change in the chemical composition of the biomass. As mentioned previously, the HTL technique has proved to be more effective in the processing of biomass with high moisture (Yerrayya et al., 2020). In the HTL process, high moisture content biomass is liquefied in a high-pressure reactor at temperatures of 200–400°C and pressures of 50–250 bar. The reaction is carried out in the presence of water or a mixture of water and a solvent (ethanol, propanol, methanol, phenol) (Dimitriadis and Bezergianni, 2017). The process flow diagram for the HTL process is shown in Fig. 4.4.

In this process, water acts as a reactant, a catalyst, and a solvent at a higher temperature (Medina-Ramos et al., 2014). There are several benefits to using water as a solvent. It naturally contains biomass, is environmentally friendly, and adopts certain properties under HTL conditions. Water near its critical point (374°C and 22.064 MPa) leads to a decrease in dielectric constant and viscosity (Chen et al., 2020). Under these conditions, the water remains liquid due to the increase in pressure during the process. This property of water makes it a polar solvent (weak), leading to increased solubility of organic compounds (hydrophobic) in biomass (Zhu et al., 2015b). In addition, during the HTL process, the enhancement of an iconic product such as hydronium ions (H_3O^+) increases the rate of acid–base reactions (Brunner, 2009). Some advantages of the HTL process include the following: (1) it does not require drying of lignocellulosic biomass, (b) it does not depend on the type of biomass, because conventional methods rely on the type of biomass to increase the yield of by-products, (c) it can also include wet biomass from other processes, and (4) recycle of aqueous products. Therefore, the

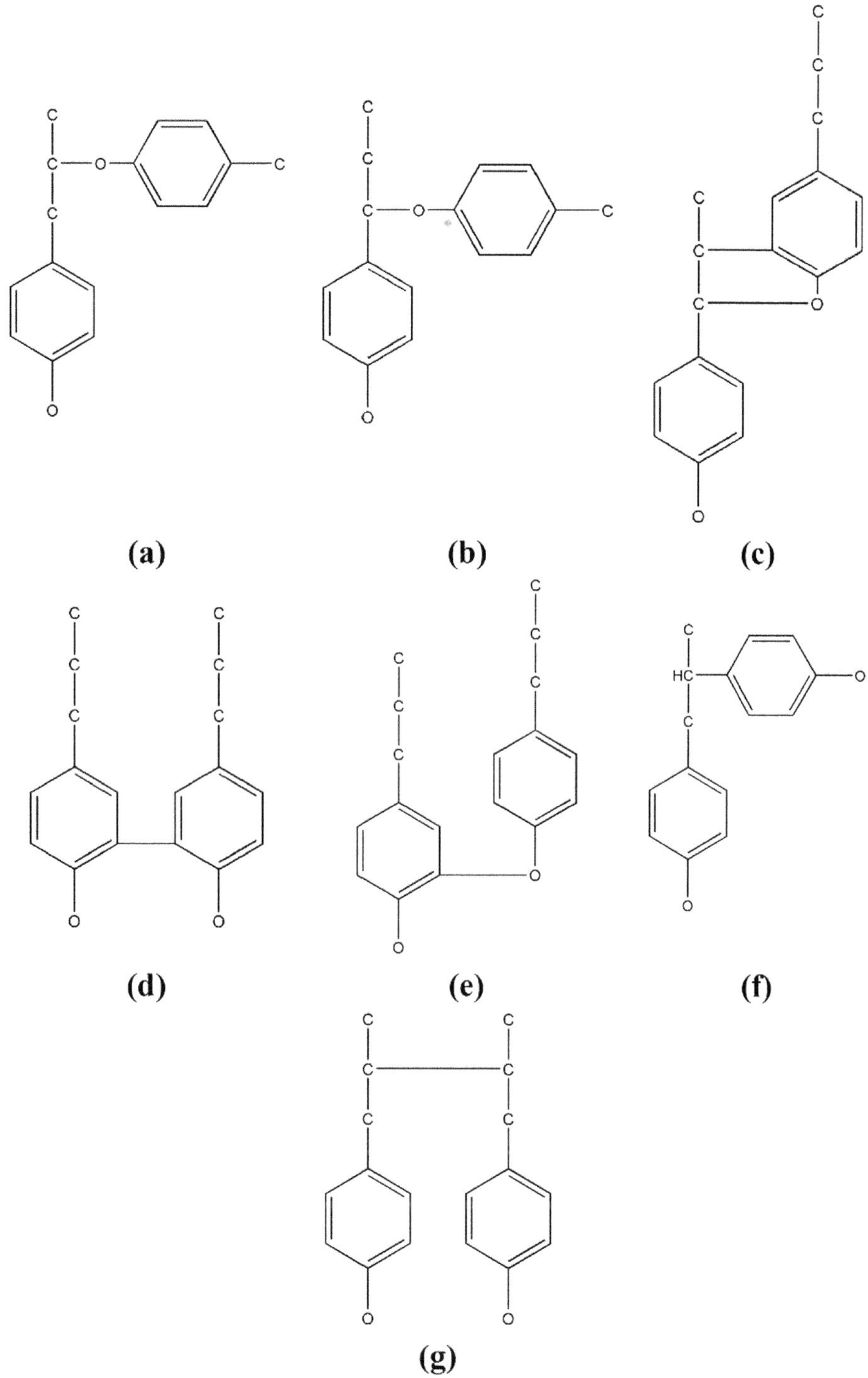

FIGURE 4.3 Different type of linkages connecting the monolignols present in the lignin matrix. (A) β-O-4; (B) α-O-4; (C) β-5; (D) 5-5; (E) 4-O-5; (F) β-1; (G) β-β

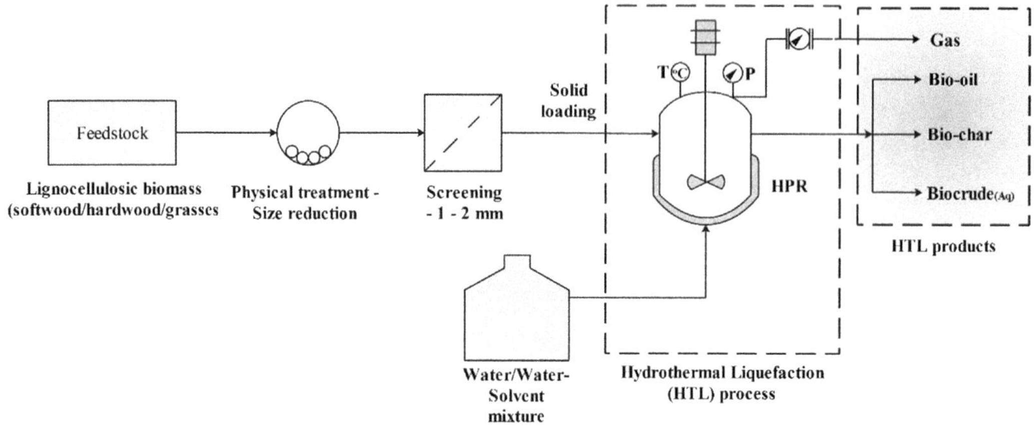

FIGURE 4.4 Schematic outline of the hydrothermal liquefaction (HTL) process. *HPR*, high-pressure reactor; *P*, pressure; *T*, temperature.

HTL process is a robust technology that is gaining more and more attention from researchers and is the focus of global biomass energy research (Xue et al., 2016). During the HTL process, as the temperature rises slowly toward the subcritical points, cellulose gets hydrolyzed to glucose or oligomers. Similarly, hemicellulose gets hydrolyzed to monomers or oligomers. However, as the temperature exceeds 160°C, glucose and oligomers further degrade to different by-products. Near the critical point, the rate of hydrolysis and decomposition increases. Thus, products with a high yield are obtained (Sasaki et al., 1998). On the other hand, lignin is a complex structure that is bonded to cellulose and hemicellulose. Under subcritical conditions, lignin gets degraded to different phenolic compounds that are the main constituents in biooil (Pourali et al., 2010). In the HTL process, cellulose and hemicellulose are hydrolyzed due to the increased formation of OH^- and H^+ ions, exhibiting weak basicity and weak acidity, respectively. Therefore, these properties attribute to the high reactivity that occurs during the HTL process (Möller et al., 2013). As the HTL process temperature rises nearer to 230°C, the cellulose and hemicellulose monomers and oligomers further degrade to formic acid, levulinic acid, furan compounds, and acetic acid. These acids decrease the pH of the mixture to 3 due to an increase in hydronium ions (H_3O^+). Therefore, the by-products soluble in the water further degrade, dehydrate, and condense to form organic compounds that form biochar (Qi et al., 2016).

Biochar is a product formed by the repolymerization of unsaturated compounds due to lack of hydrogen atoms. This product is formed as the reaction time and temperature during the HTL process (Dimitriadis and Bezergianni, 2017). Biochar is considered to be an undesirable product of the HTL process because its increased formation will reduce the production of biooil. Biocrude (Aq) is a water-soluble by-product formed in the dehydration step of the HTL process (Akhtar and Amin, 2011). The gaseous product is a mixture of carbon dioxide (CO_2), carbon monoxide (CO), methane, alkanes, alkenes, etc. Similarly, the gas production rate also increases with the reaction time and temperature of the HTL process. The biooil produced in the HTL process is rich in carbon and hydrogen content. In addition, biooil has low oxygen and moisture content. The oil produced has a high viscosity and is unstable and corrosive in nature. The quantity of HTL products formed depends on the type of lignocellulosic biomass, HTL process operating parameters, lignin content, and catalyst type (Mazaheri et al., 2010). However, to increase the biooil yield, it is necessary to optimize the operational parameters such as temperature and reaction time to lower the production of biochar.

The path and mechanism of the HTL process have not been explained and clearly demonstrated due to the complexity that exists in the process. However, in the existing literature, the proposed mechanism includes three major steps: (1) depolymerization, (2) decomposition, and (3) recombination (Nanda et al., 2016). The plant components made of three major polymers cellulose, hemicellulose, and lignin are rich in carbon and hydrogen. Although some mechanisms may provide insights into possible reaction pathways for these plant components, the exact chemical pathway of HTL is not yet known (Blaschek et al., 2010). At high temperatures, polymer chains depolymerize, which is the process of breaking down macromolecular compounds into monomers. Then, due to the dehydration, the monomer decomposes since the oxygen atoms react with the hydrogen atoms to form water. The main chemical reaction that occurs during the HTL process is aldol condensation, ketonation, secarboxylation, and esterification. As the most abundant component of woody biomass, cellulose is widely used in hydrolysis research. During cellulose hydrolysis, the cleavage of the β-1,4-glycosidic bonds of polysaccharides opens up the possibility of converting cellulose into value-added products (Huang and Fu, 2013). In the HTL process, cellulose is depolymerized to form glucose monomer and soluble oligosaccharides. Since cellulose is a natural polymer with a strong crystalline structure, it has high chemical stability properties. In this regard, cellulose depolymerization requires more energy. Methods are known for depolymerizing the cellulose chain naturally using cellulase catalytic enzymes. However, due to low efficiency, long residence time, and high cost of enzymes, this method is not feasible (Lehmann et al., 2012). In general, biooil and other byproducts are determined using Eqs. (4.1)–(4.4) (Gundupalli and Bhattacharyya, 2021):

$$\text{Bio}-\text{oil yield},\% = \frac{\text{Bio}-\text{oil weight, g}}{\text{Biomass weight, g}} \times 100 \tag{4.1}$$

$$\text{Biochar},\% = \frac{\text{Weight of solid residue, g}}{\text{Weight of biomass, g}} \times 100 \tag{4.2}$$

$$\text{Biocrude}_{\text{aqueous}},\% = \frac{\text{Weight of solids obtained after eaporation, g}}{\text{Weight of biomass, g}} \times 100 \tag{4.3}$$

$$\text{Gas},\% = 100 - \left(\text{Bio}-\text{oil yield},\% + \text{Biochar},\% + \text{Biocrude}_{\text{aqueous}},\%\right) \tag{4.4}$$

4.4 Biooil production from lignocellulosic biomass—state of the art and current challenge

4.4.1 Biomass composition

One of the most promising biomass-reforming methods for processing biofuels is the HTL process. This method has been under development for several years, but due to its difficulties with process parameters, it has always been difficult to generalize data about optimum conditions (Arturi et al., 2016). One of the most important process parameters is the choice of biomass and its composition. To obtain a higher yield of by-products, the selection of lignocellulosic biomass is important for the HTL process. The selection of biomass depends on many factors, such as availability and simplicity during transport. However, before processing of the biomass, it is necessary to determine the composition of the biomass (cellulose, hemicellulose, and lignin). In addition, it is also necessary to determine the elemental composition of biomass such as carbon (C), hydrogen (H), nitrogen (N), oxygen (O), and sulfur (S). Lignocellulosic biomass, which includes wood, forestry residues, and agricultural waste, contains varying amounts of lignin, hemicellulose, and cellulose (Peterson et al., 2008). Under high temperature and pressure, these complex polymers decompose into chemical mixtures, mainly composed of C, H, and O. The composition of different lignocellulosic biomass with elemental composition is summarized in Table 4.4. Tian (2020) studied the effect of feedstock composition on biooil yield. The study was carried out using corn straw (CS), peanut straw (PS), soybean straw (SS), and rice straw (RS). HTL process was carried out under 320°C for 60 min. Biooil yields of 15.8%, 15.1%, 14.6%, and 7.9% were reported for SS, RS, PS, and CS, respectively. The study was concluded with a biooil yield higher for SS. These results were attributed to the higher content of cellulose and ratio of cellulose and hemicellulose. These parameters contributed to increasing the biooil yield (Tian et al., 2020). In another study, the effect of different lignocellulosic biomass on biooil yield was studied. The liquefaction process was carried out at 300°C, and 15 min under an initial pressure of 2.0 MPa using N_2 gas. Feng (2014) highlighted the effect of different biomass such as white spruce bark, white birch bark, and white pine bark on biooil yield. The biooil yields for white spruce bark, white birch bark, and white pine

TABLE 4.4 Proximate and ultimate analysis of different lignocellulosic biomass feedstock.

Lignocellulosic biomass	Cellulose (%)	Hemicellulose (%)	Lignin (%)	C (%)	H (%)	N (%)	O (%)	S (%)	References
Spruce wood	39.95	35.3	23.01	47.79	5.84	0.09	46.28	–	Alper et al. (2019)
Empty fruit bunch	26.6	26.9	18.6	43.62	4.03	1.96	50.22	0.17	Chan et al. (2014)
Palm mesocarp fiber	23.1	22.2	30.6	46.29	4.67	1.42	47.37	0.24	
Palm kernel shell	24.5	22.9	33.5	47.77	4.06	0.46	47.55	0.16	
Barley straw	46	23	15	44.66	6.34	0.46	47.97	0.57	Zhu et al. (2015b)
Water hyacinth	18	48	3.5	34.98	6.46	0.76	56.4	1.4	Nigam (2002), Singh et al. (2015a)
Birch sawdust	45.3	24.2	22.9	48.5	6.3	0.3	45.2	–	Malins (2017)
Pine wood	41	19	31	50.5	6.7	0.03	42.8	–	Hardi et al. (2017)
Poplar	51.91	18.8	25.5	47.04	5.6	0.91	43.20	–	Wu et al. (2018)
Sorghum bagasse	37.68	17.02	13.84	43.2	5.8	0.7	41.4	0.8	Bi et al. (2017)
Sugarcane bagasse	39	24.9	23.1	44.7	5.3	0.4	47.6	–	Yan et al. (2017)
Rice husk	31.3	24.3	14.3	40.20	4.68	1.2	31.77	0.14	Hu et al. (2018)
Aspen wood	47.14	19.64	22.11	50.39	6.19	0.19	43.23	–	Pedersen et al. (2016)
Corn straw	30.81	25.52	16.76	44.57	5.53	0.93	33.7	0.1	Tian et al. (2020)
Peanut straw	36.56	20.27	18.36	41.42	5.51	1.27	35.21	0.15	
Rice straw	46.33	31.09	10.17	41.34	5.33	1.12	34.29	0.14	
Soybean straw	42.39	22.05	18.93	45.99	6.07	1.38	39	0.11	

bark were 67%, 58%, and 36%, respectively. The HHV of the biooil derived from different barks varied from 25 to 39 MJ/kg. The molecular weight of the aromatic/phenolic compounds determined was lower ($M_w = 800-1700$), thus making the biooil suitable for biofuel applications or as a substitute material (Feng et al., 2014).

On the other hand, forest biomass (deodar, pinewood) and agricultural biomass (sugarcane bagasse, wheat straw) were studied for biooil production. The HTL process was carried out under 280°C for 15 min. In this study, two types of biooil yield, such as ether soluble biooil and acetone soluble biooil, were studied and reported. It was reported that ether soluble biooil was light oil, and acetone soluble oil was heavy oil. Light biooil yields of 4%, 6%, 12%, and 8% were reported for deodar, pinewood, wheat straw, and sugarcane bagasse, respectively. On the other hand, the heavy biooil yields of 18% and 17% were observed for forest biomass and agricultural biomass, respectively. It can be observed that the biooil yield for different biomass varies on the basis of light biooil fraction compared with heavy bioyield. This attributes to the different composition of hemicellulose, cellulose, and lignin composition. Furthermore, this study concluded that the highest biomass conversion of 95% was reported for sugarcane bagasse (Singh et al., 2015b). Although there have been several studies of the HTL process using thermal energy, a study highlighted the importance of the HTL process using microwave irradiation for the yield of biooil from a mixture of biomass (pine and spruce). It was reported that biomass conversion, gas yield, and biooil yield from the biomass mixture varied as 13%–77%, 7%–67%, and 1%–29%, respectively. The HHV value of biooil was 20 MJ/kg when the temperature of the HTL process was set at 250°C (Remón et al., 2019). Similarly, different studies have reported biooil production from different lignocellulosic biomass such as barley straw (Zhu et al., 2018), rice straw (Cao et al., 2016), corn straw (Chen et al., 2019c) spent coffee grounds (Yang et al., 2016), birch sawdust (Malins, 2017), waste furniture sawdust (Kumar and Jha, 2016), duckweed (Yan et al., 2016), eucalyptus (Wu et al., 2019), blackcurrant pomace (Anouti et al., 2016), cornstalk (Zhu et al., 2017), mulberry bark (Chen et al., 2018), sorghum bagasse (Bi et al., 2017), coocnut (*Cocos nucifera*) residues such as coir and pith (Gundupalli and Bhattacharyya, 2021) and many more.

4.4.2 Cosolvent mixture

In HTL processes, water is considered an important reactant that also acts as a solvent. Under subcritical conditions, the properties of water change, which leads to homogeneous, fast, and efficient reactions (Zhang and Chen, 2018). Compared with other fractionation methods that use chemicals, the HTL process is done using only water. Water is abundant in nature and is very cheap compared with other solvents. Water at a higher temperature near to subcritical point has a more adverse effect on lignin, hemicellulose, and cellulose. The thermal and physical properties of water change with increase in temperature. This leads to depolymerization and gasification of the cellulose, hemicellulose, and lignin present in the lignocellulosic biomass. However, in recent years, attempts have been made to increase the biooil yield from biomass by introducing different solvents (cosolvents). These solvents act as capping agents enhancing the depolymerization process and reducing the rate of repolymerization and production of biochar. In recent years, alcohols have been utilized as cosolvents/capping agents for enhancing the biooil from lignocellulosic biomass.

On the otherh hand, a novel study was carried out to evaluate the effect of methanol as solvent for HTL of rice straw. In this study, reaction time, temperature, biomass-to-solvent ratio, alkali catalysts, and solvent composition were varied and studied to increase the biooil yield. The temperature, reaction time, and fed ratio were varied as 300–350°C, 30–60 min, and 1:1–1:15

w:w, respectively. Further different alkali catalysts such as KOH, NaOH, and K_2CO_3 were studied, and the performance of each was evaluated. The methanol and water were mixed in different ratios such as 0:100, 25:75, 50:50, and 75:25 v/v. A maximum biooil of 12.3% was obtained when the temperature, time, and solid loading were maintained at 300°C, 60%, and 10%, respectively. However, it was reported that a ratio of water and methanol (50:50) increased the biooil from 12.3 to 36.8 wt%. In addition, using the KOH catalyst, the biooil increased from 36.8 to 40 wt%. Furthermore, through a detailed study of the biooil composition, it was further reported that water increased the phenol and its derivative compounds in biooil. However, addition of methanol increased the ester in the biooil due to an increase in esterification of the phenol compounds in the biooil. Therefore, it was concluded that, change in solvents can influence the biooil composition (Yerrayya et al., 2020).

Though several studies on HTL have been carried out using water as reactant and solvent, in recent years, water and cosolvents (alcohol, glycerol, etc.) have been applied to the HTL process to increase the yield of biooil. In addition to increase the biomass conversion and HHV of the biooil produced from lignocellulosic biomass, Cao et al. (2016) examined the effect of glycerol on the HTL of rice straw for biooil production (Cao et al., 2016). It was reported that the biooil yield for rice straw liquefied using glycerol:water (1:1 ratio) was twice compared with the previously reported study by Minowa (1998) (Minowa et al., 1998). A biooil yield of 50.31% was obtained from rice straw under 260°C and 1 h. Furthermore, it was noted that the process temperature reduced when cosolvent mixtures were utilized for the HTL process. While the reaction temperature was lower, the biooil yield and biomass conversion were higher for rice straw.

In another study, water, black liquor, water/ethanol, and water/black liquor were utilized as the solvent for the HTL of bagasse. The HTL process was carried out under a reaction temperature of 300°C and reaction time of 30 min. A maximum biooil yield of 61% for bagasse liquefied using pure black liquor was reported. However, the highest HHV value for biooil was obtained for bagasse liquefied using ethanol/black liquor solvent (50:50). Therefore, it is inferred from the results that alcohol and black liquor mixture increase the quality of biooil compared with other studied solvents and mixtures (Kosinkova et al., 2015). Similarly, the effect of ethanol effect on Mulberry bark HTL process reported biooil yield of 30.32%. The study was concluded with the highest biooil yield obtained for water and ethanol in 1:1 ratio, the temperature of 300°C and time of 1 h (Chen et al., 2018). Some of the other studies related to water and cosolvents are summarized in Table 4.5.

4.4.3 Operating conditions

Increase in biooil yield through the maximum conversion of biomass is obtained through determining the optimal conditions for temperature, reaction time, and pressure. It has been reported in several works of literature that temperature and reaction tie vary for different biomass due to its different composition. Therefore, it is necessary to determine the optimum conditions of time and temperature to obtain maximum biooil yield with lower biochar production.

4.4.2.1 Role of temperature and reaction time

De Caprariis (2017) studied the influence of temperature and time on biooil yield from three different biomasses. The temperature was varied from 240 to 340°C for the HTL of natural hay, oak wood, and walnut shell. The study was concluding with temperature considered as an important factor to maximize biooil yield. The maximum biooil yield was observed between 260 and 320°C, and the HHV varied between 27 and 31 MJ/kg. The study also highlighted that the increase in HTL temperature attributed

TABLE 4.5 A literature study on the effect of different solvents on biooil yield from different lignocellulosic biomass.

S. No	Lignocellulosic biomass	Type of solvent	Solvent ratio	Biooil yield (%)	HHV, MJ/kg	References
1	Rice straw	Water and ethanol	1:1	36.8	35.3	Yerrayya et al. (2020)
2	Aspen wood	Water and glycerol	1:1	>30	34.3	Pedersen et al. (2016)
3	Cornstalk	Water and ethanol	1:1	~50	30.86	Zhao et al. (2021)
4	Spent coffee grounds	Water and ethanol	1:1	47.3	31	Yang et al. (2016)
5	Rice straw	Water and isopropyl alcohol	5:5	39.7	35.86	Yuan et al. (2007)
6	White pine sawdust	Water and ethanol	1:1	65	–	Cheng et al. (2010)
7	Wheat straw	Water and ethanol	1:1	30	28	Patil et al. (2014)

to the presence of the higher lignin content in biomass (de Caprariis et al., 2017). It can be understood from the study that, since lignin is highly complicated matrix, higher lignin content can increase the structure complexity. Therefore, higher energy is required to break the bonds that are present between the monolignols (G, H, and S unit). In another study, along with temperature, other factors such as pressure, retention time, feed ratio, feed particle size, and so on were investigated for the yield of biooil and other by-products from corn stover. In this study, temperature and reaction time varied from 250 to 375°C and 0–60 min, respectively. The highest yield of biooil was reported at 300°C at a pressure of 2200 psi and a reaction time of 0 min. It was observed that corn stover depolymerization occurred when the temperature reached to the set point, thus increasing the yield of biooil. However, as the reaction time increased from 0 to 60 min, the biochar content increased. Therefore, it is understood that the reaction occurs when the temperature rises to the set value. These factors need to be considered in the HTL process. The best biooil of 29.25 wt% was reported with a carbon conversion rate of 86.62%. In addition, HHV values of 35.13 MJ/kg was reported for biooils obtained from corn stover liquefied at 300°C, 2200 psi, and 0 min (Mathanker et al., 2020).

In addition to the factors, a study was carried to understand the role of temperature and solvent on biooil production from *Azolla filiculoides*. In this study, the temperature and time of the HTL process were varied from 260 to 300°C and 15–60 min. Furthermore, the HTL process was carried out using different solvents such as ethanol, methanol, and water. The study reported a maximum biooil yield of 28.8 wt% for *A. filiculoides* liquefied with ethanol at a temperature and reaction time of 280°C and 60 min, respectively. On the other hand, maximum biooil yield of 28.7 and 21.5 wt% was reported for methanol and water under 260°C, 15 min and 280°C, 15 min, respectively. It can be observed that the temperature, time, and solvent have greater influence on biooil yield from *A. filiculoides* (Biswas et al., 2021). Zhu et al. (2015a) studied that effect of temperature on HTL of barley straw to increase the biooil yield. In this study, the temperature was varied from 280 to 400°C. It is observed that the biooil yield was maximum when the temperature was maintained at 280°C. In addition, the solid residue (biochar) and others (organic compounds in aqueous crude and gas) were lower. It was reported that as the temperature increased to 400°C, the biooil decreased, and other by-products increased. The maximum biooil yield of 34.9% was obtained at 300°C. Therefore, the study concluded that as the temperature increased, the biooil composed of lower-molecular-weight compounds. This study clearly showed that temperature had greater influence on quantity and quality of biooil produced from barley straw (Zhu et al., 2015a).

In another study, Wang (2014) investigated the effect of temperature on HTL of various feedstocks (lignite, wheat straw, and plastic waste) mixed in different ratios. The temperature was varied from 260 to 320°C. The study showed that maximum biooil yield was achieved for feedstock ratio of 5:4:1. A similar pattern on biooil yield was observed compared with previous literature. It was reported that the conversion of feedstock biomass mixture to by-products increased, as the temperature increased (260–300°C). However, the biooil yield was maximum only when the temperature was varied between 280 and 320°C. A lower and higher temperature than 280 and 320°C resulted in lower yield (Wang et al., 2014). The lower yield of biooil during lower temperature can be due to the reduced biomass conversion to necessary by-products. On the other hand, biooil yield at a higher temperature, as discussed previously, is due to the increase in the gas and biochar content. Therefore, it is necessary to optimize the reaction temperature of HTL. A similar influence of temperature on HTL process was carried out and reported for different biomass such as *Pteris vittata* L (Chen, 2018), palm kernel shell (Chan et al., 2018), corn straw (Chen et al., 2019c), wheat stalk (Chen et al., 2019b), corn stalk (Liu et al., 2013; Zhu et al., 2017), cornelian cherry stones (Akalin et al., 2012), poplar (Wu et al., 2018), rice straw (Cao et al., 2016), and more.

Though temperature plays an essential role, reaction time for the HTL process is also considered as the most important factor to convert the biomass to biooil. Lower reaction time and higher temperature can cause cleavage of cellulose, hemicellulose, and lignin to monomers and oligomers. On the other hand, longer reaction time for lower and higher temperature can increase the biochar content in the system reducing the biooil content. Henceforth, it is essential that temperature and reaction time are optimized to maximize the biooil yield. In a case study, Singh (2015a) studied the effect of reaction time on the HTL process of water hyacinth. The reaction time and reaction temperature were varied from 15 to 60 min and from 250 to 300°C, respectively. The study concluded that maximum biooil yield of 23 wt% from water hyacinth was obtained when reaction time and the temperature were maintained at 15 min and 280°C (Singh et al., 2015a). Similarly, another study reported a maximum biooil yield of 38.72% when the HTL conditions were maintained at 304.8°C and 15.5 min (Zhu et al., 2018). In another study, reaction time, temperature, and feedstock concentration were varied, and the biooil was studied. It was reported that maximum biooil yield of 18% from lemon peels was obtained when reaction temperature, time, and feedstock concentration were maintained at 336°C, 50 min, and 9.6% w/w, respectively (Zhang et al., 2019). Though the yield was lower, it is understood that the biooil is influenced by temperature, time, and biomass concentration or loading. Some of the studies carried out for maximizing the biooil yield by varying and optimizing the HTL conditions were displayed in Table 4.6.

4.4.2.2 Role of catalyst

In previous years, HTL of lignocellulosic biomass was carried out considering only operational parameters such as temperature and time. However, in recent years, the thermochemical process of converting biomass to different valuable by-product under high temperature is carried out in the presence of a catalyst (Gollakota et al., 2018). Catalyst enhances the reaction that occurs during aldol condensation, ketonation,

TABLE 4.6 Influence of HTL conditions on biooil yield from different lignocellulosic biomass.

S. No	Biomass type/solvent	Biomass loading (% w/v)	Time, min	Temperature (°C)	Catalyst loading	Biooil (%)	References
1	Barley straw/water	10	15.5	304.8	11.7% w/w K_2CO_3	38.72	Zhu et al. (2018)
2	Corn straw/NaOH	10	60	190	2% w/v CuO	38.2	Chen et al. (2019c)
3	Pinewood sawdust/water	10	30	300	5% Na_2CO_3	48	Hu et al. (2020)
4	Wheat straw/water	20	15	350	K_2CO_3	32.34	Seehar et al. (2020)
5	Soybean/water	30	60	320	–	15.8	Tian et al. (2020)
6	Rice straw/water	30	60	320	–	15.17	
7	Corn straw/water	30	60	320	–	7.9	
8	Birch sawdust/water	25	5	300	5% w/w NaOH	54.1	Malins (2017)
9	Oak wood/water	20	15	320	15% w/v Fe_3O_4	37	de Caprariis et al. (2019)
10	Waste bamboo chopsticks/water	10	30	300	20% w/v K_2CO_3	33.1	Chang et al. (2016)
11	Oak wood/water	20	10	330	Nano-spiked Ni	32.88	de Caprariis et al. (2020)

decarboxylation, and esterification processes. Beckman and Elliott (1985) carried out the study to understand the effect of different operational parameters on biooil yield. The study concluded with the importance of optimizing temperature, reaction time, pressure, and catalyst loading to increase the biooil yield (Beckman and Elliott, 1985). Therefore, in recent years, several studies have been carried out, and attempts have been made by researchers to utilize a different type of alkali and other catalysts to enhance the quantity and quality of biooil yield. Different types and their effect on biooil yield from lignocellulosic biomass can be seen in Table 4.6. The most common alkali catalyst that is utilized to enhance the biooil yield and biomass conversion in the HTL process includes K_2CO_3, Na_2CO_3, KOH, and NaOH. The catalytic activity depends on the type of biomass and its composition, temperature, time, and pressure.

For instance, Jindal and Jha (2016) investigated the effect of different alkali catalysts on the biooil yield from furniture sawdust. The study reported a maximum biooil yield of 34.9% when the temperature and time were maintained at 280°C and 15 min. Furthermore, the study also reported that the reactivity order for alkali catalyst to achieve maximum conversion of biomass to biooil was $K_2CO_3 > KOH > Na_2CO_3 > NaOH$ (Jindal and Jha, 2016). Similar results were published by Zhu (2015b) reporting the effectiveness of achieving maximum biooil yield from barley straw when using K_2CO_3 as a catalyst. A biooil yield of 34.85% was obtained when barley straw was liquefied under 300°C. Furthermore, the study also concluded with achieving the twice the quantity of biooil compared with study carried out without the addition of catalyst (Zhu et al., 2015b). Likewise, in another study, the effectiveness of the alkali catalyst (K_2CO_3) was studied by varying the catalyst loading from 0, 0.1, 0.5, and 1.0% w/v. Two different biomasses such as mongolian oak and larch were liquefied under HTL conditions (300°C and 30 min). The study reported a biooil yield of 27.7% and 25.6% from Mongolian oak and larch, respectively. The catalyst loading of 0.5% w/v enhanced the yield and HHV of biooil. Addition of catalyst improved the HHV value of biooil to 31.9 M/kg (Hwang et al., 2019). This clearly explains the effect of a catalyst on the quality of biooil. An increased value of HHV infers to the presence of higher-molecular-weight organic compounds in the biooil mixture.

On the other hand, a study was carried out for liquefaction of cumin seed cake with and without catalyst. The effect and significance of catalyst in a liquefaction process was investigated. In this study, black cumin seed cake was liquefied under different temperatures such as 300, 325, and 350°C. However, the HTL process was carried out under these conditions with and without catalyst. The catalysts utilized in this study were Al_2O_3, $Ca(OH)_2$, and $SnCl_4 \cdot 5H_2O$. In this study, the biooil was categorized as light biooil (LBO) and heavy biooil (HBO). These oils were separated on the basis of different extraction methods. It was reported that a maximum LBO of 3.18 wt% and HBO of 37.37 wt% were obtained for reaction carried out at 300°C with catalyst $SnCl_4 \cdot 5H_2O$ and Al_2O_3, respectively. The study concluded with the significance of synthesizing lower and higher molecular compounds in the biooil for different catalyst using gas chromatography—mass spectroscopy (GC-MS) profile. Furthermore, an HHV of 21.87 and 31.78 MJ/kg as reported for LBO and HBO produced from liquefaction of cumin seed cake, respectively (Durak and Genel, 2020a). A study was carried out by Bi (2017) to understand the importance and the influence of alkali catalyst on biooil yield and its quality. In this study, the sorghum bagasse was liquefied under 300°C and studied for three different catalysts (K_2CO_3, KOH, and Ni/Si-Al). The study was concluded with maximum biooil yield obtained for sorghum bagasse liquefied in the presence of K_2CO_3 catalyst. However, a detailed study on the composition of biooil obtained from sorghum bagasse revealed that K_2CO_3 enhanced the quality of the biooil compared with another catalyst. The effect of a different catalyst on biooil composition can be observed clearly and understood. This study was concluded with K_2CO_3 being a suitable catalyst to enhance the yield with lower production of biochar (Bi et al., 2017).

As discussed related to the reactivity order for different catalyst, a study was carried out to understand the influence of K_2CO_3 and KOH on bioyield from wheat straw, sugarcane bagasse, pinewood, and deodar. Singh (2015b) studied the effect of K_2CO_3 and KOH on four different biomasses. The study reported that in all four cases, K_2CO_3 enhanced the biooil yield compared with KOH (Singh et al., 2015b). Therefore, it is clearly understood that among the alkali catalysts, K_2CO_3 has more effect during the HTL of lignocellulosic biomass. Though several studies have reported K_2CO_3 to be effective, a different approach to catalytic activity was studied. Effect of individual and mixture catalyst on biooil yield was studied and reported (Chen et al., 2019b). Since NaOH is cost-effective compared with K_2CO_3, an attempt was made to utilize NaOH with other catalysts to enhance its catalytic activity. Cupric oxide (CuO) (heterogeneous catalyst), NaOH (homogeneous catalyst), and its mixture were utilized as a catalyst to enhance the biooil yield during the HTL process from corn straw. The study reported that CuO + NaOH had more effect on increasing the quality and quantity of biooil from corn straw compared with other catalysts. A maximum biooil yield of 38.2% was obtained from corn straw liquefied at 190°C and 30 min. This study showed that the combination of heterogeneous and homogeneous catalyst could enhance the catalytic activity (Chen et al., 2019c). Similar attempts were made by Wu (2019) to utilize homogeneous catalyst (NaOH and KOH) and heterogeneous catalyst (Pd/C). The study reported that the utilization of NaOH resulted in maximum biooil yield (61.2%). In this study, the biooil was recovered as HBO and LBO through different extraction procedures. Higher content of phenol compounds was observed in LBO compared with HBO (Wu et al., 2019).

Though several studies were focused on catalysts such as K_2CO_3 and NaOH, Chen (2019a) investigated the catalytic effect of Na_2CO_3, Iron (Fe), and its mixture. The wheat stalk was liquefied under different HTL conditions by varying the type of catalyst and its concentration. The study reported that the combination of Na_2CO_3 and Fe could assist in increasing the biomass conversion compared with individual catalyst. Furthermore, a maximum biooil yield of 24.25% was achieved for catalyst mixture compared with the individual catalyst (Chen et al., 2019a). Therefore, it is understood that types of catalyst have different effects on biooil yield and its composition. In another study, Nazari (2015) screened and studied the different catalytic activity of KOH, K_2CO_3, $FeSO_4 \cdot 7H_2O$, synthetic hydrotalcite (HT), MgO, and ground colemanite on biooil yield from birch sawdust. It was reported that the alkali catalyst (KOH, K_2CO_3, and colemanite) increased the biooil yield compared with other catalysts. Furthermore, the study showed that the catalyst with lower catalytic activity was $FeSO_4$ and MgO (Nazari et al., 2015). Some of the other catalysts that have been utilized for enhancing the depolymerization and increasing the biooil from lignocellulosic biomass are summarized in Table 4.7. Therefore, it is necessary that a suitable catalyst is selected for carrying out the HTL process for different lignocellulosic biomass. In general, the alkali catalyst has been proved to be more effective in increasing and decreasing the biooil yield and biochar, respectively.

4.4 Future prospects

An increase in population can cause an increase in demand for food and fuel in the coming years. Fuels are produced from fossil crude oils through a chemical process. On the other hand, fuels are also produced via a biochemical process from different lignocellulosic biomass. However, currently, substrates such as sugarcane and corn are utilized for the production of fuel and other biochemicals. However, an increase in world population increases the demand for food and land. Also, fossil fuel in coming years will be exhausted completely until an alternative source is not determined. Therefore, in recent years, alternative lignocellulosic biomass has been utilized and studied for the production of fuels. In this context, lignocellulosic biomass is considered redundant and renewable since it is considered an abundant carbon source on earth. As discussed, there are several methods (pyrolysis, combustion, gasification, and HTL) to convert lignocellulosic biomass to valuable byproducts. In this chapter, detailed aspects of HTL were discussed. HTL process offers an advantage of utilizing biomass with high moisture content compared with other methods. Therefore, the HTL process eliminates the need to dry biomass before proceeding with the thermochemical process. Though several studies have been carried out to date, there are still many challenges that are yet to be addressed in future HTL studies, which are as follows.

Literature reported related to the HTL process was carried out in lab scale. Therefore, it is necessary to scale up the process to pilot scale and industrial scale. Through scale-up, the effect of the HTL process on biomass conversion and biooil yield can be studied in higher quantity. Cost–benefit analysis of the HTL process is yet to be addressed. Since in several works of literature, the biooil yield has been discussed in detail, the production cost of biooil from different biomass has not been discussed. Few kinds of literature

TABLE 4.7 Biooil yield from different lignocellulosic biomass with different catalyst.

Lignocellulosic biomass	Solvent	Temperature (°C)	Time (min)	Catalyst	Bio-oil (%)	HHV (MJ/kg)	References
Lactuca scariola	Water	300	15	Fe	24.25	30.66	Durak and Genel (2020b)
Korean native kenaf	Water	350	30	NaOH	23.74	37.06	Kim and Um (2020)
Construction wood waste	Water	400	15	K_2CO_3	36.35	31.64	Seehar et al. (2021)
Glycyrrhiza glabra L.	Water	350	–	H_3BO_3	35.73	31.78	Durak (2020)
Sweet sorghum bagasse	Water	320	30	NaOH	53.2	–	Sehume et al. (2020)
Ricinus communis residue	Water	300	60	K_2CO_3	21.20	33.61	Kaur et al. (2020)
Pteris vittata L	Water	275	30	Na_2CO_3/ HCOOH/ CH_3COOH	30.32	28.51	Jiang et al. (2020)

have taken the initiation to address and compare the fuel properties of crude obtained from fossil fuels. It is necessary that these properties are addressed for biooil obtained from different biomass. Through this, suitable biomass and HTL process can be commercialized for the production of fuel on a larger scale. Recycle and reuse—few literatures have addressed the possibility of recycling the solvents used in the extraction and recovery of biooil after HTL process. It is necessary to address this in the studies to reduce the production cost of biooil from lignocellulosic biomass. Few known facts in mechanism and pathways related to the conversion of polysaccharides and lignin to biooil, gas, biochar, and aqueous crude are known. However, a detailed study on the pathway in understanding the HTL process and the system is still a challenge for many researchers. A detailed study on the interaction effect of hemicellulose, cellulose, and lignin on increasing the biooil from lignocellulosic biomass during the HTL process is yet a challenge to understand. Several studies concluded with biooil yield and its HHV. However, it is necessary that a detailed quantitative analysis of the biooil and aqueous crude is addressed using GC-MS and LC-MS (liquid chromatography–mass spectroscopy) profile, respectively. Furthermore, the same is related to the HTL operational parameter conditions.

Few studies have reported the importance and effect of cosolvents during HTL process. It has been concluded in fewer studies that combination of water with organic solvents can lower the reaction time and reaction temperature. Therefore, this contributes to reducing the production cost of biooil from lignocellulosic biomass.

4.5 Conclusion

In this chapter, a detailed discussion on the HTL process, type of lignocellulosic biomass, composition, cosolvents, operational conditions, and type of catalyst were discussed. The pieces of literature enlighten us with the effect of different parameters on the quantity and quality of biooil from lignocellulosic biomass. Furthermore, the advantages and disadvantages of HTL process were also discussed briefly.

References

Abu Ghalia, M., Dahman, Y., 2017. Synthesis and utilization of natural fiber-reinforced poly (lactic acid) bionanocomposites. Lignocellul. Fibre Biomass-Based Compos. Mater. 313–345. https://doi.org/10.1016/B978-0-08-100959-8.00015-9. Elsevier.

Adler, E., 1977. Lignin chemistry—past, present and future. Wood Sci. Technol. 11, 169–218. https://doi.org/10.1007/BF00365615.

Ahmad, N., Zakaria, M.R., 2019. Oligosaccharide from Hemicellulose. Lignocellul. Futur. Bioecon. 135–152. https://doi.org/10.1016/B978-0-12-816354-2.00008-6. Elsevier.

Akalin, M.K., Tekin, K., Karagöz, S., 2012. Hydrothermal liquefaction of cornelian cherry stones for bio-oil production. Bioresour. Technol. 110, 682–687. https://doi.org/10.1016/j.biortech.2012.01.136.

Akhtar, J., Amin, N.A.S., 2011. A review on process conditions for optimum bio-oil yield in hydrothermal liquefaction of biomass. Renew. Sustain. Energy Rev. 15, 1615–1624. https://doi.org/10.1016/j.rser.2010.11.054.

Alper, K., Tekin, K., Karagöz, S., 2019. Hydrothermal liquefaction of lignocellulosic biomass using potassium fluoride-doped alumina. Energy Fuels 33, 3248–3256. https://doi.org/10.1021/acs.energyfuels.8b04381.

Anouti, S., Haarlemmer, G., Déniel, M., Roubaud, A., 2016. Analysis of physicochemical properties of bio-oil from hydrothermal liquefaction of blackcurrant pomace. Energy Fuel 30, 398–406. https://doi.org/10.1021/acs.energyfuels.5b02264.

Arturi, K.R., Kucheryavskiy, S., Søgaard, E.G., 2016. Performance of hydrothermal liquefaction (HTL) of biomass by multivariate data analysis. Fuel Process. Technol. 150, 94–103. https://doi.org/10.1016/j.fuproc.2016.05.007.

Babich, I.V., van der Hulst, M., Lefferts, L., Moulijn, J.A., O'Connor, P., Seshan, K., 2011. Catalytic pyrolysis of microalgae to high-quality liquid bio-fuels. Biomass Bioenergy 35, 3199–3207. https://doi.org/10.1016/j.biombioe.2011.04.043.

Bajpai, P., 2018. Wood and fiber fundamentals. In: Biermann's Handbook of Pulp and Paper. Elsevier, pp. 19–74. https://doi.org/10.1016/B978-0-12-814240-0.00002-1.

Beckman, D., Elliott, D.C., 1985. Comparisons of the yields and properties of the oil products from direct thermochemical biomass liquefaction processes. Can. J. Chem. Eng. 63, 99–104. https://doi.org/10.1002/cjce.5450630116.

Bi, Z., Zhang, J., Peterson, E., Zhu, Z., Xia, C., Liang, Y., Wiltowski, T., 2017. Biocrude from pretreated sorghum bagasse through catalytic hydrothermal liquefaction. Fuel 188, 112–120. https://doi.org/10.1016/j.fuel.2016.10.039.

Biswas, B., Arun Kumar, A., Bish, Y., Krishna, B.B., Kumar, J., Bhaskar, T., 2021. Role of temperatures and solvents on hydrothermal liquefaction of *Azolla filiculoides*. Energy 217, 119330. https://doi.org/10.1016/j.energy.2020.119330.

Blaschek, H.P., Ezeji, T.C., Scheffran, J., 2010. Biofuels from Agricultural Wastes and Byproducts. Wiley-Blackwell, Oxford, UK. https://doi.org/10.1002/9780813822716.

Brunner, G., 2009. Near critical and supercritical water. Part I. Hydrolytic and hydrothermal processes. J. Supercrit. Fluids 47, 373–381. https://doi.org/10.1016/j.supflu.2008.09.002.

Brunner, G., 2014. Hydrothermal and supercritical water processes. Supercrit. Fluid Sci. Technol. 511–523. https://doi.org/10.1016/B978-0-444-59413-6.00009-1.

Cao, L., Zhang, C., Hao, S., Luo, G., Zhang, S., Chen, J., 2016. Effect of glycerol as co-solvent on yields of bio-oil from rice straw through hydrothermal liquefaction. Bioresour. Technol. 220, 471–478. https://doi.org/10.1016/j.biortech.2016.08.110.

Chan, Y.H., Yusup, S., Quitain, A.T., Uemura, Y., Sasaki, M., 2014. Bio-oil production from oil palm biomass via subcritical and supercritical hydrothermal liquefaction. J. Supercrit. Fluids 95, 407–412. https://doi.org/10.1016/j.supflu.2014.10.014.

Chan, Y.H., Quitain, A.T., Yusup, S., Uemura, Y., Sasaki, M., Kida, T., 2018. Optimization of hydrothermal liquefaction of palm kernel shell and consideration of supercritical carbon dioxide mediation effect. J Supercrit Fluids 133, 640–646. https://doi.org/10.1016/j.supflu.2017.06.007.

Chang, C.C., Chen, C.P., Yang, C.S., Hau, C.Y., Huang, M., Chang, C.Y., et al., 2016. Conversion of waste bamboo chopsticks to bio-oil via catalytic hydrothermal liquefaction using K_2CO_3. Sustain. Environ. Res. 26, 262–267. https://doi.org/10.1016/j.serj.2016.08.002.

Chen, J., 2018. Bio-oil production from hydrothermal liquefaction of *Pteris vittata* L.: effects of operating temperatures and energy recovery. Bioresour. Technol. 265, 320–327. https://doi.org/10.1016/j.biortech.2018.06.019.

Chen, W.-H., Tu, Y.-J., Sheen, H.-K., 2010. Impact of dilute acid pretreatment on the structure of bagasse for bioethanol production. Int. J. Energy Res. 34, 265–274. https://doi.org/10.1002/er.1566.

Chen, C., Zhu, J., Jia, S., Mi, S., Tong, Z., Li, Z., et al., 2018. Effect of ethanol on Mulberry bark hydrothermal liquefaction and bio-oil chemical compositions. Energy 162, 460–475. https://doi.org/10.1016/j.energy.2018.08.026.

Chen, W.-H., Lin, Y.-Y., Liu, H.-C., Chen, T.-C., Hung, C.-H., Chen, C.-H., et al., 2019a. A comprehensive analysis of food waste derived liquefaction bio-oil properties for industrial application. Appl. Energy 237, 283–291. https://doi.org/10.1016/j.apenergy.2018.12.084.

Chen, Y., Cao, X., Zhu, S., Tian, F., Xu, Y., Zhu, C., et al., 2019b. Synergistic hydrothermal liquefaction of wheat stalk with homogeneous and heterogeneous catalyst at low temperature. Bioresour. Technol. 278, 92–98. https://doi.org/10.1016/j.biortech.2019.01.076.

Chen, Y., Dong, L., Miao, J., Wang, J., Zhu, C., Xu, Y., et al., 2019c. Hydrothermal liquefaction of corn straw with mixed catalysts for the production of bio-oil and aromatic compounds. Bioresour. Technol. 294, 122148. https://doi.org/10.1016/j.biortech.2019.122148.

Chen, W.-H., Lin, Y.-Y., Liu, H.-C., Baroutian, S., 2020. Optimization of food waste hydrothermal liquefaction by a two-step process in association with a double analysis. Energy 199, 117438. https://doi.org/10.1016/j.energy.2020.117438.

Cheng, S., D'cruz, I., Wang, M., Leitch, M., Xu, C., 2010. Highly efficient liquefaction of woody biomass in hot-compressed alcohol–water co-solvents. Energy Fuels 24, 4659–4667. https://doi.org/10.1021/ef901218w.

de Caprariis, B., De Filippi, s P., Petrullo, A., Scarsella, M., 2017. Hydrothermal liquefaction of biomass: influence of temperature and biomass composition on the bio-oil production. Fuel 208, 618–625. https://doi.org/10.1016/j.fuel.2017.07.054.

de Caprariis, B., Bavasso, I., Bracciale, M.P., Damizia, M., De Filippis, P., Scarsella, M., 2019. Enhanced bio-crude yield and quality by reductive hydrothermal liquefaction of oak wood biomass: effect of iron addition. J. Anal. Appl. Pyrolysis 139, 123–130. https://doi.org/10.1016/j.jaap.2019.01.017.

de Caprariis, B., Bracciale, M.P., Bavasso, I., Chen, G., Damizia, M., Genova, V., Marra, F., Paglia, L., Pulci, G., Scarsella, M., Tai, L., De Filippis, P., 2020. Unsupported Ni metal catalyst in hydrothermal liquefaction of oak wood: effect of catalyst surface modification. Sci. Total Environ. 709, 136215. https://doi.org/10.1016/j.scitotenv.2019.136215.

Dimitriadis, A., Bezergianni, S., 2017. Hydrothermal liquefaction of various biomass and waste feedstocks for biocrude production: a state of the art review. Renew. Sustain. Energy Rev. 68, 113–125. https://doi.org/10.1016/j.rser.2016.09.120.

Divyabharathi, R., Subramanian, P., 2020. Hydrothermal liquefaction of paddy straw for biocrude production. Mater. Today Proc. 45, 603–606. https://doi.org/10.1016/j.matpr.2020.02.390.

Durak, H., 2020. Hydrothermal liquefaction of *Glycyrrhiza glabra* L. (Liquorice): effects of catalyst on variety compounds and chromatographic characterization. Energy Sour. A Recover. Util. Environ. Eff. 42, 2471–2484. https://doi.org/10.1080/15567036.2019.1607947.

Durak, H., Genel, S., 2020a. Characterization of bio-oil and bio-char obtained from black cumin seed by hydrothermal liquefaction: investigation of potential as an energy source. Energy Sour. A Recover. Util. Environ. Eff. 1–11. https://doi.org/10.1080/15567036.2020.1714821.

Durak, H., Genel, S., 2020b. Catalytic hydrothermal liquefaction of lactuca scariola with a heterogeneous catalyst: the investigation of temperature, reaction time and synergistic effect of catalysts. Bioresour. Technol. 309, 123375. https://doi.org/10.1016/j.biortech.2020.123375.

Feng, S., Yuan, Z., Leitch, M., Xu, C.C., 2014. Hydrothermal liquefaction of barks into bio-crude – effects of species and ash content/composition. Fuel 116, 214–220. https://doi.org/10.1016/j.fuel.2013.07.096.

Gollakota, A.R.K., Kishore, N., Gu, S., 2018. A review on hydrothermal liquefaction of biomass. Renew. Sustain. Energy Rev. 81, 1378–1392. https://doi.org/10.1016/j.rser.2017.05.178.

Gundupalli, M.P., Bhattacharyya, D., 2019. Sequential acid hydrolysis and enzymatic saccharification of coconut coir for recovering reducing sugar: Process evaluation and optimization. Bioresour. Technol. Rep 6, 70–80. https://doi.org/10.1016/j.biteb.2019.01.015.

Gundupalli, M.P., Bhattacharyya, D., 2021. Hydrothermal liquefaction of residues of *Cocos nucifera* (coir and pith) using subcritical water: Process optimization and product characterization. Energy 121466. https://doi.org/10.1016/j.energy.2021.121466.

Gundupalli, M.P., Cheng, Y.-S., Chuetor, S., Bhattacharyya, D., Sriariyanun, M., et al., 2021. Effect of dewaxing on saccharification and ethanol production from different lignocellulosic biomass. Bioresour. Technol. 339, 125596. https://doi.org/10.1016/j.biortech.2021.125596.

Gundupalli, M.P., Kajiura, H., Ishimizu, T., Bhattacharyya, D., et al., 2020. Alkaline hydrolysis of coconut pith: process optimization, enzymatic saccharification, and nitrobenzene oxidation of Kraft lignin. Biomass Convers. Biorefin. 1–19. https://doi.org/10.1007/s13399-020-00890-z.

Gupta, A., Verma, J.P., 2015. Sustainable bio-ethanol production from agro-residues: a review. Renew. Sustain. Energy Rev. 41, 550–567. https://doi.org/10.1016/j.rser.2014.08.032.

Haarlemmer, G., Guizani, C., Anouti, S., Déniel, M., Roubaud, A., Valin, S., 2016. Analysis and comparison of bio-oils obtained by hydrothermal liquefaction and fast pyrolysis of beech wood. Fuel 174, 180–188. https://doi.org/10.1016/j.fuel.2016.01.082.

Hallac, B.B., Ragauskas, A.J., 2011. Analyzing cellulose degree of polymerization and its relevancy to cellulosic ethanol. Biofuels Bioprod. Biorefining 5, 215–225. https://doi.org/10.1002/bbb.269.

Hardi, F., Mäkelä, M., Yoshikawa, K., 2017. Non-catalytic hydrothermal liquefaction of pine sawdust using experimental design: material balances and products analysis. Appl. Energy 204, 1026–1034. https://doi.org/10.1016/j.apenergy.2017.04.033.

Heidenreich, S., Foscolo, P.U., 2015. New concepts in biomass gasification. Prog. Energy Combust. Sci. 46, 72–95. https://doi.org/10.1016/j.pecs.2014.06.002.

Hu, Y., Wang, S., Li, J., Wang, Q., He, Z., Feng, Y., et al., 2018. Co-pyrolysis and co-hydrothermal liquefaction of seaweeds and rice husk: comparative study towards enhanced biofuel production. J. Anal. Appl. Pyrolysis 129, 162–170. https://doi.org/10.1016/j.jaap.2017.11.016.

Hu, Y., Gu, Z., Li, W., Xu, C., 2020. Alkali-catalyzed liquefaction of pinewood sawdust in ethanol/water co-solvents. Biomass Bioenergy 134, 105485. https://doi.org/10.1016/j.biombioe.2020.105485.

Huang, Y.-B., Fu, Y., 2013. Hydrolysis of cellulose to glucose by solid acid catalysts. Green Chem. 15, 1095–1111. https://doi.org/10.1039/c3gc40136g.

Hwang, H., Lee, J.H., Choi, I.-G., Choi, J.W., 2019. Comprehensive characterization of hydrothermal liquefaction products obtained from woody biomass under various alkali catalyst concentrations. Environ. Technol. 40, 1657–1667. https://doi.org/10.1080/09593330.2018.1427799.

Itskos, G., Nikolopoulos, N., Kourkoumpas, D.-S., Koutsianos, A., Violidakis, I., Drosatos, P., et al., 2016. Energy and the environment. Environ. Dev. 363–452. https://doi.org/10.1016/B978-0-444-62733-9.00006-X. Elsevier.

Ji, X.-J., Huang, H., Nie, Z.-K., Qu, L., Xu, Q., Tsao, G.T., 2011. Fuels and chemicals from hemicellulose sugars. Adv. Biochem. Eng. Biotechnol. 199–224. https://doi.org/10.1007/10_2011_124. Springer.

Jiang, H., Fan, L., Cai, C., Hu, Y., Zhao, F., Ruan, R., et al., 2020. Study on the bio-oil characterization and heavy metals distribution during the aqueous phase recycling in the hydrothermal liquefaction of As-enriched *Pteris vittata* L. Bioresour. Technol. 317, 124031. https://doi.org/10.1016/j.biortech.2020.124031.

Jindal, M.K., Jha, M.K., 2016. Catalytic hydrothermal liquefaction of waste furniture sawdust to bio-oil. Indian Chem. Eng. 58, 157–171. https://doi.org/10.1080/00194506.2015.1006145.

Kaur, R., Biswas, B., Kumar, J., Jha, M.K., Bhaskar, T., 2020. Catalytic hydrothermal liquefaction of castor residue to bio-oil: effect of alkali catalysts and optimization study. Ind. Crops Prod. 149, 112359. https://doi.org/10.1016/j.indcrop.2020.112359.

Kim, S.J., Um, B.H., 2020. Biocrude production from Korean native kenaf through subcritical hydrothermal liquefaction under mild alkaline catalytic conditions. Ind. Crops Prod. 145, 112001. https://doi.org/10.1016/j.indcrop.2019.112001.

Kosinkova, J., Ramirez, J.A., Nguyen, J., Ristovski, Z., Brown, R., Lin, C.S.K., et al., 2015. Hydrothermal liquefaction of bagasse using ethanol and black liquor as solvents. Biofuels Bioprod. Biorefining 9, 630–638. https://doi.org/10.1002/bbb.1578.

Kumar, M.J., Jha, M.K., 2016. Effect of process parameters on hydrothermal liquefaction of waste furniture sawdust for bio-oil production. RSC Adv. 6, 41772–41780. https://doi.org/10.1039/c6ra02868c.

Kubicki, J.D., Yang, H., Sawada, D., O'Neill, H., Oehme, D., Cosgrove, D., et al., 2018. The Shape of Native Plant Cellulose Microfibrils. Sci. Rep. 8 (1), 13983. https://doi.org/10.1038/s41598-018-32211-w.

Kumar Gupta, P., Sai Raghunath, S., Venkatesh Prasanna, D., Venkat, P., Shree, V., Chithananthan, C., et al., 2019. An Update on Overview of Cellulose, its Structure and Applications. Cellulose, IntechOpen. https://doi.org/10.5772/intechopen.84727.

Lehmann, C., Sibilla, F., Maugeri, Z., Streit, W.R., Domínguez de María, P., Martinez, R., et al., 2012. Reengineering CelA2 cellulase for hydrolysis in aqueous solutions of deep eutectic solvents and concentrated seawater. Green Chem. 14, 2719. https://doi.org/10.1039/c2gc35790a.

Lindman, B., Medronho, B., Alves, L., Costa, C., Edlund, H., Norgren, M., 2017. The relevance of structural features of cellulose and its interactions to dissolution, regeneration, gelation and plasticization phenomena. Phys. Chem. Chem. Phys. 19, 23704–23718. https://doi.org/10.1039/C7CP02409F.

Liu, H.-M., Li, M.-F., Sun, R.-C., 2013. Hydrothermal liquefaction of cornstalk: 7-Lump distribution and characterization of products. Bioresour. Technol. 128, 58–64. https://doi.org/10.1016/j.biortech.2012.09.125.

Malins, K., 2017. Production of bio-oil via hydrothermal liquefaction of birch sawdust. Energy Convers. Manag. 144, 243–251. https://doi.org/10.1016/j.enconman.2017.04.053.

Marchessault, R.H., 1994. Wood chemistry, fundamentals and applications. Carbohydr. Res. 252, C1. https://doi.org/10.1016/0008-6215(94)90030-2.

Mathanker, A., Pudasainee, D., Kumar, A., Gupta, R., 2020. Hydrothermal liquefaction of lignocellulosic biomass feedstock to produce biofuels: parametric study and products characterization. Fuel 271, 117534. https://doi.org/10.1016/j.fuel.2020.117534.

Mazaheri, H., Lee, K.T., Bhatia, S., Mohamed, A.R., 2010. Subcritical water liquefaction of oil palm fruit press fiber in the presence of sodium hydroxide: an optimisation study using response surface methodology. Bioresour. Technol. 101, 9335–9341. https://doi.org/10.1016/j.biortech.2010.07.004.

McDonald, A.G., Donaldson, L.A., 2001. Wood, constituents of. Encycl. Mater. Sci. Technol. 9612–9615. https://doi.org/10.1016/B0-08-043152-6/01739-3. Elsevier.

Medina-Ramos, W., Mojica, M.A., Cope, E.D., Hart, R.J., Pollet, P., Eckert, C.A., et al., 2014. Water at elevated temperatures (WET): reactant, catalyst, and solvent in the selective hydrolysis of protecting groups. Green Chem. 16, 2147–2155. https://doi.org/10.1039/C3GC42569J.

Mei Wu, L., Hui Zhou, C., Shen Tong, D., Hua Yu, W., 2014. Catalytic thermochemical processes for biomass conversion to biofuels and chemicals. Bioenergy Res. Adv. Appl. 243–254. https://doi.org/10.1016/B978-0-444-59561-4.00015-2. Elsevier.

Minowa, T., Kondo, T., Sudirjo, S.T., et al., 1998. Thermochemical liquefaction of indonesian biomass residues. Biomass Bioenergy 14 (5–6), 517–524. https://doi.org/10.1016/S0961-9534(98)00006-3.

Mohanraj, R., 2019. Plant-derived resorbable polymers in tissue engineering. Mater. Biomed. Eng. 19–40. https://doi.org/10.1016/B978-0-12-818415-8.00002-4. Elsevier.

Möller, M., Harnisch, F., Schröder, U., 2013. Hydrothermal liquefaction of cellulose in subcritical water—the role of crystallinity on the cellulose reactivity. RSC Adv. 3, 11035. https://doi.org/10.1039/c3ra41582a.

Mori, T., Tsuboi, Y., Ishida, N., Nishikubo, N., Demura, T., Kikuchi, J., 2015. Multidimensional high-resolution magic angle spinning and solution-state NMR characterization of 13C-labeled plant metabolites and lignocellulose. Sci. Rep. 5, 11848. https://doi.org/10.1038/srep11848.

Mørup, A.J., Christensen, P.R., Aarup, D.F., Dithmer, L., Mamakhel, A., Glasius, M., et al., 2012. Hydrothermal liquefaction of dried distillers grains with solubles: a reaction temperature study. Energy Fuel 26, 5944–5953. https://doi.org/10.1021/ef3008163.

Mullen, C.A., Boateng, A.A., 2008. Chemical composition of bio-oils produced by fast pyrolysis of two energy crops. Energy Fuel 22, 2104–2109. https://doi.org/10.1021/ef700776w.

Nanda, S., Dalai, A.K., Gökalp, I., Kozinski, J.A., 2016. Valorization of horse manure through catalytic supercritical water gasification. Waste Manag. 52, 147–158. https://doi.org/10.1016/j.wasman.2016.03.049.

Nazari, L., Yuan, Z., Souzanchi, S., Ray, M.B., Xu, C., 2015. Hydrothermal liquefaction of woody biomass in hot-compressed water: catalyst screening and comprehensive characterization of bio-crude oils. Fuel 162, 74–83. https://doi.org/10.1016/j.fuel.2015.08.055.

Nigam, J., 2002. Bioconversion of water-hyacinth (*Eichhornia crassipes*) hemicellulose acid hydrolysate to motor fuel ethanol by xylose–fermenting yeast. J. Biotechnol. 97, 107–116. https://doi.org/10.1016/S0168-1656(02)00013-5.

Patil, P.T., Armbruster, U., Martin, A., 2014. Hydrothermal liquefaction of wheat straw in hot compressed water and subcritical water–alcohol mixtures. J. Supercrit. Fluids 93, 121–129. https://doi.org/10.1016/j.supflu.2014.01.006.

Pedersen, T.H., Grigoras, I.F., Hoffmann, J., Toor, S.S., Daraban, I.M., Jensen, C.U., et al., 2016. Continuous hydrothermal co-liquefaction of aspen wood and glycerol with water phase recirculation. Appl. Energy 162, 1034–1041. https://doi.org/10.1016/j.apenergy.2015.10.165.

Peterson, A.A., Vogel, F., Lachance, R.P., Fröling, M., Antal Jr., M.J., Tester, J.W., 2008. Thermochemical biofuel production in hydrothermal media: a review of sub- and supercritical water technologies. Energy Environ. Sci. 1, 32. https://doi.org/10.1039/b810100k.

Pourali, O., Asghari, F.S., Yoshida, H., 2010. Production of phenolic compounds from rice bran biomass under subcritical water conditions. Chem. Eng. J. 160, 259–266. https://doi.org/10.1016/j.cej.2010.02.057.

Qi, Y., Zhang, M., Qi, L., Qi, Y., 2016. Mechanism for the formation and growth of carbonaceous spheres from sucrose by hydrothermal carbonization. RSC Adv. 6, 20814–20823. https://doi.org/10.1039/C5RA26725K.

Ragauskas, A.J., 2006. The path forward for biofuels and biomaterials. Science 311, 484–489. https://doi.org/10.1126/science.1114736.

Rajendran, K., Lin, R., Wall, D.M., Murphy, J.D., 2019. Influential aspects in waste management practices. Sustain. Resour. Recover. Zero Waste Approaches 65–78. https://doi.org/10.1016/B978-0-444-64200-4.00005-0. Elsevier.

Remón, J., Randall, J., Budarin, V.L., Clark, J.H., 2019. Production of bio-fuels and chemicals by microwave-assisted, catalytic, hydrothermal liquefaction (MAC-HTL) of a mixture of pine and spruce biomass. Green Chem. 21, 284–299. https://doi.org/10.1039/C8GC03244K.

Sankaran, R., Parra Cruz, R.A., Pakalapati, H., Show, P.L., Ling, T.C., Chen, W.H., Tao, Y., et al., 2020. Recent advances in the pretreatment of microalgal and lignocellulosic biomass: a comprehensive review. Bioresour. Technol. 298, 122476. https://doi.org/10.1016/j.biortech.2019.122476.

Sasaki, M., Kabyemela, B., Malaluan, R., Hirose, S., Takeda, N., Adschiri, T., et al., 1998. Cellulose hydrolysis in subcritical and supercritical water. J. Supercrit. Fluids 13, 262–268. https://doi.org/10.1016/S0896-8446(98)00060-6.

Schutyser, W., Renders, T., Van den Bossche, G., Van den Bosch, S., Koelewijn, S.-F., Ennaert, T., et al., 2017. Catalysis in lignocellulosic biorefineries: the case of lignin conversion. Nanotechnol. Catal. https://doi.org/10.1002/9783527699827.ch23.

Seehar, T.H., Toor, S.S., Shah, A.A., Pedersen, T.H., Rosendahl, L.A., 2020. Biocrude production from wheat straw at sub and supercritical hydrothermal liquefaction. Energies 13, 3114. https://doi.org/10.3390/en13123114.

Seehar, T.H., Toor, S.S., Shah, A.A., Nielsen, A.H., Pedersen, T.H., Rosendahl, L.A., 2021. Catalytic hydrothermal liquefaction of contaminated construction wood waste for biocrude production and investigation of fate of heavy metals. Fuel Process. Technol. 212, 106621. https://doi.org/10.1016/j.fuproc.2020.106621.

Sehume, T.Z., Strydom, C.A., Bunt, J.R., Schobert, H.H., 2020. Bio-oil production from sweet sorghum bagasse via liquefaction using alkaline solutions and identification of phenolic products. Waste Biomass Valoriz. 11, 3593–3607. https://doi.org/10.1007/s12649-019-00893-6.

Singh, R., Balagurumurthy, B., Prakash, A., Bhaskar, T., 2015a. Catalytic hydrothermal liquefaction of water hyacinth. Bioresour. Technol. 178, 157–165. https://doi.org/10.1016/j.biortech.2014.08.119.

Singh, R., Prakash, A., Balagurumurthy, B., Singh, R., Saran, S., Bhaskar, T., 2015b. Hydrothermal liquefaction of agricultural and forest biomass residue: comparative study. J. Mater. Cycles Waste Manag. 17, 442–452. https://doi.org/10.1007/s10163-014-0277-3.

Tian, Y., Wang, F., Djandja, J.O., Zhang, S.-L., Xu, Y.-P., Duan, P.-G., 2020. Hydrothermal liquefaction of crop straws: effect of feedstock composition. Fuel 265, 116946. https://doi.org/10.1016/j.fuel.2019.116946.

Tungal, R., Shende, R.V., 2014. Hydrothermal liquefaction of pinewood (*Pinus ponderosa*) for H2, biocrude and bio-oil generation. Appl. Energy 134, 401–412. https://doi.org/10.1016/j.apenergy.2014.07.060.

Tzanetis, K.F., Posada, J.A., Ramirez, A., 2017. Analysis of biomass hydrothermal liquefaction and biocrude-oil upgrading for renewable jet fuel production: the impact of reaction conditions on production costs and GHG emissions performance. Renew. Energy 113, 1388–1398. https://doi.org/10.1016/j.renene.2017.06.104.

Usami, R., Fujii, K., Fushimi, C., 2020. Improvement of bio-oil and nitrogen recovery from microalgae using two-stage hydrothermal liquefaction with solid carbon and HCl acid catalysis. ACS Omega 5, 6684–6696. https://doi.org/10.1021/acsomega.9b04468.

Wang, H., Male, J., Wang, Y., 2013. Recent advances in hydrotreating of pyrolysis bio-oil and its oxygen-containing model compounds. ACS Catal. 3, 1047–1070. https://doi.org/10.1021/cs400069z.

Wang, B., Huang, Y., Zhang, J., 2014. Hydrothermal liquefaction of lignite, wheat straw and plastic waste in sub-critical water for oil: product distribution. J. Anal. Appl. Pyrolysis 110, 382–389. https://doi.org/10.1016/j.jaap.2014.10.004.

Wu, X.-F., Zhou, Q., Li, M.-F., Li, S.-X., Bian, J., Peng, F., 2018. Conversion of poplar into bio-oil via subcritical hydrothermal liquefaction: structure and antioxidant capacity. Bioresour. Technol. 270, 216–222. https://doi.org/10.1016/j.biortech.2018.09.032.

Wu, X.-F., Zhang, J.-J., Li, M.-F., Bian, J., Peng, F., 2019. Catalytic hydrothermal liquefaction of eucalyptus to prepare bio-oils and product properties. Energy Convers. Manag. 2019 (199), 111955. https://doi.org/10.1016/j.enconman.2019.111955.

Xue, Y., Chen, H., Zhao, W., Yang, C., Ma, P., Han, S., 2016. A review on the operating conditions of producing bio-oil from hydrothermal liquefaction of biomass. Int. J. Energy Res. 40, 865–877. https://doi.org/10.1002/er.3473.

Yan, W.-H., Duan, P.-G., Wang, F., Xu, Y.-P., 2016. Composition of the bio-oil from the hydrothermal liquefaction of duckweed and the influence of the extraction solvents. Fuel 185, 229–235. https://doi.org/10.1016/j.fuel.2016.07.117.

Yan, X., Ma, J., Wang, W., Zhao, Y., Zhou, J., 2017. The effect of different catalysts and process parameters on the chemical content of bio-oils from hydrothermal liquefaction of sugarcane bagasse. BioResources 13, 997–1018. https://doi.org/10.15376/biores.13.1.997-1018.

Yang, L., Nazari, L., Yuan, Z., Corscadden, K., Xu, C.C., He, Q.S., 2016. Hydrothermal liquefaction of spent coffee grounds in water medium for bio-oil production. Biomass Bioenergy 86, 191–198. https://doi.org/10.1016/j.biombioe.2016.02.005.

Yerrayya, A., Shree Vishnu, A.K., Shreyas, S., Chakravarthy, S.R., Vinu, R., 2020. Hydrothermal liquefaction of rice straw using methanol as co-solvent. Energies 13, 2618. https://doi.org/10.3390/en13102618.

Yousuf, A., 2012. Biodiesel from lignocellulosic biomass - prospects and challenges. Waste Manag. 32, 2061–2067. https://doi.org/10.1016/j.wasman.2012.03.008.

Yuan, X., Li, H., Zeng, G., Tong, J., Xie, W., 2007. Sub- and supercritical liquefaction of rice straw in the presence of ethanol–water and 2-propanol–water mixture. Energy 32, 2081–2088. https://doi.org/10.1016/j.energy.2007.04.011.

Zhang, Y., Chen, W.-T., 2018. Hydrothermal liquefaction of protein-containing feedstocks. Direct Thermochem. Liq. Energy Appl. 127–168. https://doi.org/10.1016/B978-0-08-101029-7.00004-7. Elsevier.

Zhang, B., Chen, J., He, Z., Chen, H., Kandasamy, S., 2019. Hydrothermal liquefaction of fresh lemon-peel: parameter optimisation and product chemistry. Renew. Energy 143, 512–519. https://doi.org/10.1016/j.renene.2019.05.003.

Zhang, S., Yang, X., Zhang, H., Chu, C., Zheng, K., Ju, M., et al., 2019. Liquefaction of biomass and upgrading of bio-oil: a review. Molecules 24, 2250. https://doi.org/10.3390/molecules24122250.

Zhao, B., Hu, Y., Qi, L., Gao, J., Zhao, G., Ray, M.B., et al., 2021. Promotion effects of metallic iron on hydrothermal liquefaction of cornstalk in ethanol-water mixed solvents for the production of biocrude oil. Fuel 285, 119150. https://doi.org/10.1016/j.fuel.2020.119150.

Zhu, Z., Rosendahl, L., Toor, S.S., Yu, D., Chen, G., 2015a. Hydrothermal liquefaction of barley straw to bio-crude oil: effects of reaction temperature and aqueous phase recirculation. Appl. Energy 137, 183–192. https://doi.org/10.1016/j.apenergy.2014.10.005.

Zhu, Z., Toor, S.S., Rosendahl, L., Yu, D., Chen, G., 2015b. Influence of alkali catalyst on product yield and properties via hydrothermal liquefaction of barley straw. Energy 80, 284–292. https://doi.org/10.1016/j.energy.2014.11.071.

Zhu, Z., Si, B., Lu, J., Watson, J., Zhang, Y., Liu, Z., 2017. Elemental migration and characterization of products during hydrothermal liquefaction of cornstalk. Bioresour. Technol. 243, 9–16. https://doi.org/10.1016/j.biortech.2017.06.085.

Zhu, Z., Rosendahl, L., Toor, S.S., Chen, G., 2018. Optimizing the conditions for hydrothermal liquefaction of barley straw for bio-crude oil production using response surface methodology. Sci Total Environ. 630, 560–569. https://doi.org/10.1016/j.scitotenv.2018.02.194.

SECTION III

Bioethanol and biodiesel production

CHAPTER

5

Refining lignocellulose of second-generation biomass waste for bioethanol production

E. Raja Sathendra[1], R. Praveenkumar[1], Baskar Gurunathan[2], S. Chozhavendhan[3], Mani Jayakumar[4]

[1]Department of Biotechnology, Arunai Engineering College, Tiruvannamalai, Tamil Nadu, India; [2]Department of Biotechnology, St. Joseph's College of Engineering, Chennai, Tamil Nadu, India; [3]Department of Biotechnology, V.S.B. Engineering College, Karur, Tamil Nadu, India; [4]Department of Chemical Engineering, Haramaya Institute of Technology, Haramaya University, Dire Dawa, East Hararghe, Ethiopia

5.1 Introduction

Plant biomass is the most consistent carbon-based feedstock on earth (Parisutham et al., 2014). The plant biomass of lignocellulosic material procured from second-generation energy crops such as agricultural residues and wood is the most abundant global source of waste renewable biomass (Lin and Tanaka, 2006). Lignocellulose contains polysaccharides of hemicellulose, cellulose, phenol, lignin, and soluble polar compounds and nonpolar materials. Pretreatment of lignocellulose changes the natural structure into a simple soluble monomer that is accessible to enzymes (Kurien et al., 2010). The skill in conversion is to release cellulose and hemicelluloses from hard lignin, a material that hinders the processing of biomass and makes pretreatment expensive. The morphologic structures of hemicellulose and cellulose are associated with lignin material, which make it hard for polysaccharide to use enzymes. Physical pretreatment decreases the plant biomass particle size, increasing the surface area. Chemical pretreatment overcomes the hard nature of lignin present in lignocellulose. Major types of agrowaste that are used as feedstock for bioethanol production are rice straw, wheat straw, corn straw, and sugarcane bagasse, owing to their availability throughout the year in terms of quantity (Kim and Dale, 2004). Possible lignocellulosic plant biomass sources and chemical constituents (percent dry weight) are, respectively: **agricultural residues** (hemicellulose, cellulose, lignin, and other ash content: 25–50, 37–50, 5–15, and 12–10);

Biofuels and Bioenergy
https://doi.org/10.1016/B978-0-323-85269-2.00016-2

hardwood (25–40, 45–47, 20–25, and 0.80); **softwood** (25–40, 40–45, 30–60, and 0.50); **grasses** (35–50, 25–40, no traces of lignin content and others, and 2–5); **waste paper from chemical pulp** (12–40, 50–70, 6–10, and 5); **newspaper** (25–40, 40–55, 18–30, and 5–8); and **switchgrass** (30–35, 40–45, 12, and other ash content 4–5, respectively) (Michalska et al., 2012; Palmqvist and Hahn-Hägerdal, 2000).

Pretreatment converts the chemical composition of lignocellulose through its microstructure and macrostructure (An et al., 2015). During the degradation of lignin, hemicellulose covering cellulose is degraded and hydrolyzed, and the cellulose structure becomes improved for release and availability. Thus, such a process can promote and improve the substrate and enzyme interaction and stimulate the hydrolysis of monomer (Chandra et al., 2015).

5.1.1 Lignocellulose

The lignocellulosic component mainly consists of three polymers, cellulose (40%–50%), hemicellulose (25%–30%), and lignin (15%–20%), and a trace amount of pectin, nitrogen, and inorganic ingredients (Mori et al., 2015; Knauf, 2004). These polymers are arranged so that each forms a hetero-matrix to unusual degrees with varying relative compositions depending on the source of biomass (Carere et al., 2008; Chandra et al., 2007).

The main component of the plant cell wall is cellulose. It confers structural rigidity and exists as an unbranched homopolymer. It is a β-D-glucopyranose polymer moiety linked via β-(1,4)glycosidic bonds. The cellulose chain consists of a repeating structure of disaccharide cellobiose (Desvaux, 2005). These cellulose structures are grouped into microfibrils, each about 3–6 nm in diameter. They contain 36 glucan long chains with thousands of glucose molecules, and they form cellulose fiber through hydrogen bonding and van der Waals force. Hemicellulose is the second most abundant polymer (20%–50% of lignocellulose biomass). It differs from cellulose, it is not chemically homogeneous. The ultra complex nature of lignocellulose consist of number of components which makes it woody and hard nature shown in Fig. 5.1.

Hemicelluloses are branched, heterogeneous polymers of pentoses (xylose and arabinose), hexoses (mannose, glucose, and galactose), and acetylated sugars (Saha, 2003).They have a low molecular weight with respect to cellulose. Polymerization of this polysaccharide is low without crystalline regions, so it can be quickly degraded into monomers. Lignin is a highly complex structure with a hydrophobic cross-linked aromatic compound. It is the third most abundant polymer in nature and confers impermeability, rigidity, and microbial resistance to invasion and stress by oxidation. Lignin is an amorphous heteropolymer network of phenyl propane units (coniferyl, sinapyl alcohol, and p-coumaryl) held together by different linkages (Hendricks and Zeeman, 2009). Cellulose, hemicelluloses, and lignin in

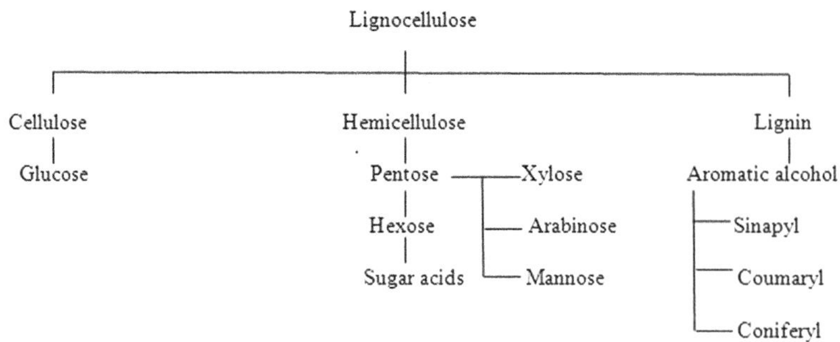

FIGURE 5.1 Schematic representation of lignocellulosic material.

the cell wall of plant have a highly ordered crystalline structure that constrains degradation. Thus, pretreatment of the lignocellulose network structure of lignin improves the use of cellulose efficiently.

5.1.2 Pretreatment strategy

The pretreatment strategy has an important part in lignocellulose processing for bioethanol production. Extensive research was conducted to improve the breakdown of lignocellulose material. Pretreatments change the structural and physicochemical properties of the original lignocellulose by opening the biomass recalcitrance structure, and then render the direct enzymatic hydrolysis of cellulose to produce monomers (Mosier et al., 2005). In the native form of C5 and C6 sugar-based hemicelluloses, the major part of the component is lignin in an intertwined and complicated form (Wang et al., 2015). The pretreatment strategies include physical, physicochemical (Sun and Cheng, 2002), chemical (Galbe and Zacchi, 2007), and biological methods (Itoh et al., 2003). A combination pretreatment method is also used, with more than one or two techniques from the same or different strategy.

5.1.2.1 Physical pretreatment

Physical methods of pretreatment require relatively powerful high energy to process lignocellulosic material, and thus increase the production cost.

5.1.2.1.1 Mechanical pretreatment

Mechanical methods of pretreatment primary break the cellulose morphology (crystalline form), increase the surface area, and reduce the polymerization of cellulosic biomass into small particles and then render the substrate to become more amenable for hydrolysis through enzymes. The material is reduced from 10–30 mm (chipping) to 0.2–2 mm (grinding or milling) The energy requirement to break the agricultural material using a mechanical method depends on the ultimate size of the particle (Sun and Cheng, 2002). For hardwood, the final particle size would be 1.60 mm; energy consumption using a knife mill is 130 kWh/ton, and using a hammer mill is also 130 kWh/ton. For rice straw, the final particle size would be 1.60 mm; energy consumption using a knife mill is 7.5 kWh/ton, and using a hammer mill is 42 kWh/ton. For corn stover, the final particle size would be 1.60 mm; energy consumption using hammer mill is 14 kWh/ton (Cadoche and Lopez, 1989). It is energy-intensive, expensive, and time-consuming. The mechanical method is much less efficient than pretreatment using chemicals because it does not remove lignin effectively.

Studies on mechanical milling effectiveness and metal salt pretreatment of lignocellulose plant material also researched the results of metal salt with various corrosion numbers as well as the milling method of pretreatment on crystalline biomass and polymerization. The effect of $Al(NO_3)_3$ and milling is promising and desirable compared with different pretreatment methods (Zhang et al., 2015).

5.1.2.1.2 Uncatalyzed steam explosion

Uncatalyzed steam explosion is also called autohydrolysis. It this method, only steam water is used. Biomass particles are rapidly heated using high-pressure saturated steam for a period to promote the hydrolysis of hemicellulose. By quickly releasing pressure, the process is terminated and then causes the lignocellulosic biomass to undergo explosive decomposition, which causes the hydrolysis of hemicellulose by organic acids released from biomass by acetyl or other functional groups. Thus, the degradation of sugar is due to acidic conditions in uncatalyzed steam explosion (Cantarella et al., 2004; Garcia et al., 2006). Rapid heating of biomass by steam causes less moisture to exists inside the reactor, resulting a high sugar content. Important factors for uncatalyzed steam explosion treatment are the temperature, time, moisture content, and particle size (Ballesteros et al., 2008; Negro et al., 2003). In uncatalyzed steam explosion, hemicellulose is removed and lignin is altered, enhancing the hydrolysis of enzyme on biomass. Uncatalyzed steam explosion is performed at 160 to 270°C for a few minutes before pretreated materials are taken into a cooling vessel. Low heating and an extended residence time are more encouraging than high heating and a shorter time (Wright, 1988). Speedy thermal expansion induces the plant morphology to decrease the size of particles and enhance the pore size volume (Michalowicz et al., 1991).

5.1.2.1.3 Pyrolysis

Pyrolysis is thermochemical process for changing biomass into chemical products and energy. Pyrolysis maximizes the biomass yield with respect to the residence time and heating rate. Normally, pyrolysis involves dehydration, primary, and secondary (dehydrogenation, depolymerization and fragmentation) are considered to be the most important reactions in the primary decomposition of biomass. Thermal decomposition for lignocellulosic biomass proceeds at a low rate of heating with adequate time for repolymerization to maximize the solid yield (Demirbas and Arin, 2002). The decomposition of cellulosic material takes place at a temperature greater than 300°C, rapidly producing residue. With the mild acidic condition (1N H_2SO_4, 97°C, 2.5 h) of hydrolysis, this yields 80%–85% of cellulose to be converted to reducing sugar with a maximum glucose concentration of more than 50% (Fan et al., 1987). Thermal pretreatment causes an decrease in the essential moisture content and oxygen in the lignocellulosic biomass and uses the excess heat in pyrolysis to improve the quality of the product and produce energy efficiently (Tao et al., 2016).

5.1.2.1.4 Pretreatment with high-energy radiation

Radiation with high energy is carried through the irradiation of rays to lignocellulosic material. High-energy radiation improves the degradation of cellulosic biomass using gamma rays, which break at the β-1–4 glycosidic bond, giving a large surface area and decreasing crystalline cellulose. This method is expensive on a large scale and creates safety concerns for other methods including microwave heating, ultraviolet rays, ultrasound, pulsed electrical field, and electronic beams.

High-energy radiation alters biomass, including crystalline cellulose , increasing the surface area and degree of polymerization, degrading hemicellulose and the fractional consumption of lignin content, and loosening the cellulose structure by improving degradation and the conversion rates of substrate. It lowers pollution and is eco-friendly. The high intensity of radiation leads to damage to the structure of cellulose. High-energy radiation reaction is energy-intensive, slow processing and highly expensive and they are intense substrate-specific (Chang et al., 1981). In studies on ultraviolet pulsed laser beams and pretreatment of bamboo using electron beams, samples underwent enzymatic hydrolysis for 3 h at 37°C. Result showed that ultraviolet pulsed laser beams increased enzymatic hydrolysis speed by 38% and pretreatment with electron beams enhance monomer (sugar) yield by 40% (Ota et al., 2010). Gamma rays combined with mechanical milling for wheat straw pretreatment obtained 10.24% sugar yield (Yang et al., 2008).

5.1.2.1.5 Pretreatment with ultrasonication

Ultrasonic pretreatment opens cellulose crystalline regions, degrading lignin molecules. Importantly, it improve the ease of access and cellulose reactivity. Its effects on cellulose microstructure are small. Ultrasonication of biomass can effectively improve cellulose saccharification by increasing the surface area ratio (Yachmenev et al., 2009). Studies on different lignocellulosic materials and their relative saccharification rate before and after ultrasonic pretreatment were conducted and resulted in small changes to surface conformation (Zhang et al., 2008). Ultrasound assisted with alkaline pretreatment degraded the hydrogen bond between lignocellulosic molecules and increased crystalline cellulose to render enzymatic saccharification more effectively. The impact of mechanical milling produces cavitation bubbles, permitting enzyme hydrolysis on the substrate surface. The introduction of ultrasound can significantly improve high enzymatic hydrolysis by enhancing enzyme transport toward the surface of substrate through the effect of cavitation.

5.1.2.1.6 Pretreatment using microwave radiation

Microwave pretreatment is an easy method. The efficiency of energy can produce a high temperature in a small time, less than that for nuclear thermal power, but the investment cost in equipment is high. However, acid and alkali pretreatment of lignocellulosic material requires a relatively high energy as well high-pressure and high-temperature conditions. Enzymatic activity increases in lignocellulosic material pretreated after microwave treatment, enhancing the specificity of the enzyme and substrate. Microwave pretreatment on bagasse material with glycerol was carried out at >160°C. A considerable change was identified in the physicochemical characteristics of the material and a large volume of sugar was released in the hydrolysis of the enzyme. Studies on the pretreatment of bagasse in a short time with microwaves along with acid, alkali, and enzymatic saccharification on pretreated substrate enhanced the removal of lignin and improved the fermentable sugar yield. Lignin removal was significantly affected in the microwave-alkali pretreatment process (Xue et al., 2012). Switchgrass material was drenched at different proportions in a solution of alkali (0.05%–0.3% w/v). Microwave irradiation of this substrate produced a total sugar content of 99% with a 70% increase in sugar (Hu and Wen, 2008).

5.1.2.2 *Physicochemical pretreatment*

Physicochemical pretreatment affects both the physical and chemical properties of the biomass. Various physicochemcial pretreatments and thier effects on physical and chemical properties of the biomass are described in the following sections.

5.1.2.2.1 Catalyzed steam explosion or autohydrolysis

Catalyzed steam explosion is similar in action to uncatalyzed steam explosion, but some chemicals are added (gases and liquids). The lignocellulosic biomass is soaked in chemicals such as SO_2, CO_2, oxalic acid, and H_2SO_4 before steam explosion can effectively improve enzymatic digestion, leading to complete hemicellulose removal. It is hydrothermal with extreme steam temperature pretreatment for several seconds. Removal of steam and liquid from the reactor cools rapidly owing to pressure (Chen and Liu, 2015). Catalyzed steam explosion can enhance the complete removal of hemicellulose, rendering increased enzymatic hydrolysis on biomass with no or less generation of toxic inhibitory compounds compared with uncatalyzed steam explosion. The physically preconditioned biomass is treated with saturated steam at a high pressure, at 160 to 240°C and from 0.7 to 4.8 MPa. Pressure inside the reactor is retained for a few minutes to promote hydrolytic activity on hemicellulose and release it. Both mechanical force (steam heat and the size of the particle) and chemical effects (residence time and humidity) cause the acetyl hydrolysis group in the hemicellulose (Melin et al., 2015). Catalyzed steam explosion can be an effectively pretreatment for agricultural materials and hardwood. It require less energy input for processing. It has a low environmental impact and results in the complete recovery of monomer.

5.1.2.2.2 Liquid hot water

Liquid hot water (LHW) is the same as steam explosion; however, water in a liquid form at an extreme temperature is used rather than steam. The terms "hydrothermolysis" and "solvolysis" are used for LHW. LHW causes the hydrolysis of hemicellulose, lignin removal, and enhanced accessibility of biomass cellulose while preventing the development of inhibitors at an extreme temperature (Yang and Wyman, 2004). Studies on LHW pretreatment reported the potential enhancement of digestibility of cellulose, extraction of sugar, and recovery of pentose with the advantage of generating no or less inhibitor during the fermentation of sugar. Enriched sugar hydrolysate is used for ethanol production. Eighty percent removal of hemicellulose was achieved and enhanced corn fiber substrate on enzymatic hydrolysis. LHW pretreatment enhanced the splitting and decomposition of uronic acid substitution and O-acetyl on hemicellulose, which then produced other organic acids and acetic acid. These acidic compounds enhance raw substrate hydrolysis efficiently. In LHW, a low temperature (180–190°C) for corn stover and low dry matter (1%–8%) leads to better production of cellulose oligosaccharides. In LHW, the biomass and hot water are exposed for 15 min at 200 to 230°C. The extreme hot liquid breaks the linkage between hemiacetal and release acid upon biomass hydrolysis, which enhances the cleavage of the ether bond on biomass (Mosier et al., 2005). Generally, 40%–60% of materials can be dissolved with a reaction time of 2–15 min, and 4%–22% of cellulose and 35%–60% of lignin and hemicellulose were removed using a relatively low solid material (<20%), achieving a high use of energy and less yield (Laser et al., 2002).

5.1.2.2.3 Carbon dioxide explosion

Carbon dioxide pretreatment of lignocellulosic material generates toxins during degradation. It is not flammable, recovery is easy during extraction, it uses a low temperature, and large amount of solid material can be formed. CO_2 explosion uses supercritical carbon dioxide (SC–CO_2) at high pressure (1000–4000 psi) at around 200°C. It is maintained for several minutes to improve the degradation of biomass (Kim and Hong, 2001). It is the same as CO_2 explosion steam and ammonia explosion, forming carbonic acid and enhancing the degradation rate. Studies on the pretreatment on alfalfa fiber (4 kg CO_2/kg) and pressure (5.62 MPa) produced 75% of sugar released for 24 h in the hydrolysis of enzyme. Ammonia and steam explosion are relatively low, but fast compared to enzymatic hydrolysis without preconditioning. In this method, CO_2 released at a high pressure rapidly disrupts the hemicellulosic and cellulosic structure and enhances the surface area of substrate, making it more accessible to enzymatic hydrolysis. This supercritical fluid can degrade lignin effectively. Studies on ammonia, steam, and CO_2 explosion pretreatment on sugarcane bagasse, recycled paper mix, and repulping waste of recycled paper compared CO_2 explosion, which was more cost-effective than pretreatment with ammonia and had a negative impact on the inhibitory substituent produced in the steam processing explosion (Zheng et al., 1998).

5.1.2.2.4 Ammonia fiber/freeze explosion

Lignocellulosic pretreatment with ammonia as a catalyst is an alkaline method that is an effective alternative to increase the degradation of agricultural crop residues. Ammonia can be easily recycled through evaporation. The ammonia residue provides a vital nitrogen source for biological processes (Mosier, 2005). Pretreatment of lignocellulosic biomass with ammonia removes the lignin content (ammonolysis), and hemicellulose can be partially solubilized (high temperature) while preserving the high cellulose solution (Kim and Lee, 2005). Glucan is retained at about 100%, and the treated residue of lignocellulosic biomass has a high amount of cellulose oligosaccharides. Pretreatment of ammonia causes the cleavage of lignin complex linkages between hemicellulose side chains and lignin phenol, which allows the partial removal of hemicellulose and lignin. The influence of cellulose crystallization depends on the duration of treatment and temperature. Treatment of anhydrous liquid ammonia on cellulose results in crystallinity and different degrees of polymerization (Mittal et al., 2011).

5.1.2.2.5 SO_2 explosion

SO_2 explosion is a type of explosion using the CO_2 method with the addition of acid, catalyzing the hemicellulose to solubilize it. It requires an optimal low temperature for pretreatment and enhances the partial degradation of cellulose hydrolysis (Brownell and Saddler, 1986). SO_2 catalyzed steam explosion is considered an effective approach for the pretreatment of soft wood substrates (Tengborg et al., 1998). Studies on the processing of corn straw using explosion SO_2 at 190°C with the addition of 3% SO_2 for 5 min and then cellulase and enzyme hemicellulase supplement pretreatment on corn straw during degradation. The sugar yield increased to near to the theoretical conversion efficiency. Hydrolysis with enzyme and xylose generated 70%–74% of yield. Drawbacks of this method include high equipment costs and the use of acid for large quantities of degraded compound.

5.1.2.2.6 Pulsed electric energy

Pulsed electric energy avoids undesirable changes in biological material that could occur in treatment including chemical, thermal, and enzymatic (Donsi et al., 2010). It is consumes energy by securing unit operations and uses renewable plant resources. Difficulties in processing lignocellulosic material are poor digestibility and the ineffectiveness of the enzyme for quick degradation; in addition, it alters into monomer of fermentable sugars. Lignocellulosic materials are heterogeneous in the natural form and require an incorporated approach to pretreatment for various types of biomass. Pulsed electric energy uses high-frequency (1.2-MHz) cold plasma discharge to alter and improve structural and chemical changes of the lignocellulosic material (Ioanid et al., 2013). It is clean and avoids contamination. Studies on pulsed electric energy at a field strength of 1–10 kV/cm were useful for pretreating switchgrass and wood chips (Kumar et al., 2011); they measured the result of pulsed electric energy on diffusion into plant tissue. The use of dye neutral red was calculated. The substrate is resistant to changes in structure at a low field strength. Conversely, the useful intake of a dye for pulsed electric energy field pretreatment at 10 kV/cm was predicted, reflecting enhanced plant tissue porosity. Pulsed electric energy field can improve the hydrolysis of cellulose to fermentable sugar.

5.1.2.2.7 Electrical catalysis

This electrical catalytic method attracts attention owing to its cleanliness, lack of secondary pollution, and advantage of oxidation. Studies on the hydrolysis of cellulose processing used titanium-based SnO_2 and CeO_2 as an electrode substance and observed the hydrolysis of cellulose with the properties on oxidation through electric catalysis. Research also enhanced lignocellulosic digestibility by applying an acid two-step approach method such as lignocellulose hydrolysis with concentrated acid and then following electrocatalysis with diluted acid.

5.1.2.3 *Chemical pretreatment*

Chemical pretreatment is the widely considered method. It was originally and primarily developed exclusively for the paper industry to remove lignin from lignocellulosic material to create high-quality paper products. However, the chemical method is a promising alternative. Combined pretreatment with sonication resulted in enhanced reactivity (Masanao et al., 2003).

5.1.2.3.1 Acids

Acid concentration is most important parameter affecting the effective degradation of lignocellulosic materials. It can be used at a low concentration with extreme temperatures or a high concentration with a lower temperature. Dilute acid treatment is the most widely used technique because it can liberate sugar efficiently from part of hemicellulose and produce a solid material with rich cellulose, which is subject to hydrolysis using enzymes. However, pretreatment with concentrated acid is also striking by affecting in breaking of carbon bond, hydrogen bond and ester bond which provides the highly ordered complex nature of lignocellulose into solubilizing sugar from cellulose and hemicellulose fraction and eliminating subsequent enzymatic hydrolysis step. Acid acts as a catalyst by striking intramolecular and intermolecular bonds on cellulose, lignin, and hemicellulose and then hydrolyzing carbohydrate.

5.1.2.3.1.1 Dilute acid Dilute acid pretreatment on biomass is carried out with a low concentration acid (<5% w/v) at 120 to 210°C at 10 atm, to yield a low glucose content from cellulose (Ghatak, 2011). A dilute acid solution cleaves the hemicellulose fraction in the biomass; then, it is easily hydrolyzed using enzyme owing to the side chain, linkage type, and short length. In dilute acid pretreatment, nearly 80%–90% of hemicellulose is obtained (Yang and Wyman, 2008). Pretreatment of lignocellulosic biomass on dilute acid with enhanced cellulose hydrolysis and the rate of reaction were developed compared with traditional dilute acid based on the hydrolysis of cellulose (Esteghalian et al., 1997). Dilute acid pretreatment can selectively enhance the removal of hemicellulose in the preliminary step during the fractionation of lignocellulosic biomass overall processing and subsequent remove lignin by alkaline pretreatment, rendering the solid substrate for pure cellulose. High temperature in dilute acid pretreatment favorably improves cellulose hydrolysis compared with a moderate temperature on saccharification, which results in a low yield owing to the decomposition of sugar (McMillan et al., 1994). Before treatment on dilute H_2SO_4, the structure of biomass shows a rupture owing to the removal of hemicellulose content. A change in biomass color to dark brown is obtained as a result of lignin residue enhanced in the solid. Lignin content is separated on subsequent treatment with alkaline solution in the second stage, rendering clear and separate cellulose fiber. Primarily two different types of dilute acid pretreatment exist that is from high temperature (>160°C) for continuous flow processing with to low solid loading of (5–10% wt of substrate/wt of reaction mixture) shown in Table 5.1.

5.1.2.3.1.2 Concentrated acid Concentrated acid is used to pretreat lignocellulosic biomass (approximately 30% w/v) at optimum atmospheric pressure and temperature (<100°C) (Jung and Kim, 2015). The high acid

TABLE 5.1 Dilute alkaline pretreatment processing conditions of rice straw, cotton linter, and wheat straw and monomer yields.

Pretreatment method	Lignocellulosic materials used	Pretreatment conditions	Output obtained	References
Alkaline pretreatment	Wheat straw and rice straw	5 g wheat straw and rice straw with sodium hydroxide solution concentration (1%–5%) sodium chlorite (1%–5%) and H_2O_2 (1%–5%), deionized using water, dried with hot air oven for 24 h at 70°C.	Effect of alkali (NaOH), chlorite (Na-chlorite), and oxidative (H_2O_2) treatment on lignocellulosic substrates (rice straw and wheat straw). Maximum results for lignin loss and holocellulose gain were obtained with 5% NaOH, 5% with Na-chlorite, and 2% with H_2O_2.	Singh et al. (2014)
Alkaline pretreatment	Cotton linter	Cotton linter, soaked, sodium hydroxide solution (0%, 2%, 4%, 6%, 8%, 10%, 12%, 14%, 16%, 18% and 20%) at 0, 23, and 100°C. The ratio of liquid/dry solid was 20 mL/g.	Degradation output was determined at 12% NaOH and 0°C, in which 99.1% of cotton was digested to glucose within 4 days.	Azamet al. (2009)

concentration breaks the lignocellulosic structure into both cellulose and hemicellulose and releases a high sugar yield from the hydrolysate. Common concentrated acid solutions are nitric, trifluoroacetic, sulfuric, hydrochloric, and phosphoric acid to pretreat the lignocellulosic biomass of sulfuric acid for all types of lignocellulosic materials. Concentrated acid can remove hemicellulose in combination with cellulose to yield sugar. Otherwise, it is applied before dilute acid cellulose hydrolysis. Concentrated sulfuric acid pretreatment on lignocellulosic biomass (1) initiates solubilization, which causes the lignocellulosic structure to decrystallize; and (2) hydrolyzation, which hydrolyzes the fragment of hemicellulose and cellulose to monosaccharide (Liu et al., 2012). The recovery of sugar efficiency in this pretreatment method includes operating variables such as the temperature, solid loading, acid concentration, and reaction time.

Main advantaged of concentrated acid pretreatment are that the hydrolysis of cellulose is effective and avoids the further use of enzyme for hydrolysis, because acid itself can hydrolyze the cellulose and hemicellulose fraction to fermentable monomer. The result is an elevated sugar yield and low pressure and temperature, which can affect the economics of the process. Disadvantage are that it is corrosive, highly toxic, and hazardous, the construction of reaction involves expensive resistant material, high energy is required for acid recycling and recovery, the acid should be recovered or a large volume of neutralizing substance will be necessary to regulate the pH of the treated substance early to the fermentation mixer, and degradation of sugar causes the formation of inhibitors such as hydroxyl methyl furfural and furfural. This affects the growth of the microorganism in fermentation, requiring detoxification. Figs. 5.2 and 5.3 show the lignocellulosic plant biomass before and after pretreatment.

5.1.2.3.2 Combination of acid with other methods

Combining acid with other pretreatment methods obtain efficiency and economy with lower toxic compound formation, a shorter reaction time, and high sugar recovery. Studies reported combining acid pretreatment with others, including microwave heating, extrusion, steam explosion, and ultrasound assisted pretreatment. A study of treatment on a pine sawdust twin screw extruder reactor used concentrated acid pretreatment on this lignocellulosic biomass. The process conditions included concentrated sulfuric acid (70%), a sawdust–acid ratio of 1:0.8, screw speed of 110 rpm, temperature of 60°C, and head pressure of 780 psi. Under this optimum condition, cellulose (44.4%) and hemicellulose (73.8%) were converted to sugar and oligosaccharide (Miller and Hester, 2007). Pretreatment of rice straw on diluted sulfuric acid using a twin screw extruder reactor with

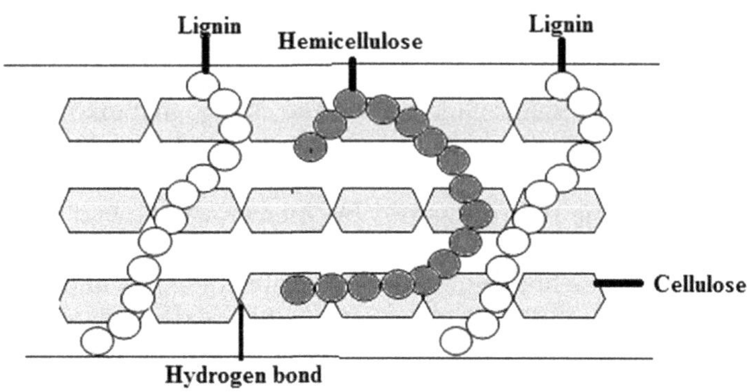

FIGURE 5.2 Ultra and complex nature of lignocellulose before pretreatment.

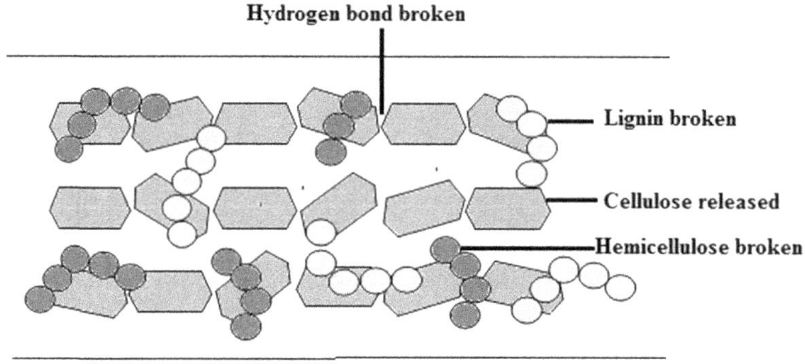

FIGURE 5.3 Release of cellulose molecule after pretreatment showing broken hydrogen bond, lignin, and hemicellulose.

various screw speeds (30–150 rpm), acid concentrations (1%–3%), and temperatures (80 to 160°C) under this process condition resulted in the highest exchange of xylan to monomer xylose (83.7%) yield at an acid concentration of 3%, 40 rpm, and 120°C (Chen et al., 2011).

5.1.2.3.3 Alkaline

Pretreatment using alkaline is a chief strategy based on several studies. Alkaline pretreatment can take place at ambient temperature for a few days. Various bases employed in lignocellulosic biomass pretreatment include calcium hydroxide (lime), sodium hydroxide, ammonium hydroxide, aqueous ammonia, and potassium hydroxide to improve the efficient removal of lignin. Other types of alkaline include barium, a lithium reactant with respect to the biomass category. Among them, sodium hydroxide is the most well-studied one after lime because it is inexpensive and easily recoverable. Alkaline pretreatment is generally a delignification method. It affects the solubility of hemicellulose significantly. The subsequent delignification of feedstock is increased by oxygen or surplus air supply (Hu et al., 2008). Alkaline pretreatment disrupt the lignocellulosic biomass structure by saponifying ester bonds of cross-linking intermolecular (lignin and other hemicellulose) xylan hemicellulose. Added substances remove uronic acid substitution and acetyl groups on hemicellulose that hinder or reduce cellulose and hemicellulose accessibility for enzymatic hydrolysis (Chang and Holtzapple, 2000).

5.1.2.3.3.1 Sodium hydroxide Sodium hydroxide in dilute alkaline pretreatment is useful for hydrolyzing low lignin in straw at 10%–18% (Bjerre et al., 1996). The hydrolysis of sodium hydroxide pretreated hardwood enhanced from 14–55% with decrease in lignin content between 24–55% to 20% and subsequently no effect on pretreating with dilute sodium hydroxide to softwood with lignin more than 26% (Millet et al., 1976). Studies on pretreatment used alkali such as sodium, potassium, and calcium hydroxide to pretreat rice straw at 25°C for 24 h. Sodium hydroxide of (6% loading g/g dry rice straw) enhances 85% increase of sugar yield on enzyme hydrolysis (Yi et al., 2009). NaOH dilute pretreatment enhanced efficiency for straw but was less for hardwood and very little for softwood.

The change that takes place in the structure of cellulose by NaOH during treatment causes an alteration in crystalline distribution, crystalline by mercerization, and amorphous regions as well as partial depolymerization. The effect of pretreatment on hot and cold NaOH on the crystalline index depends on the substrate residue (Nelson and O'Connor, 1964). For pretreatment involving Na^+ cations in the sodium hydroxide method, cellulose swelling is studied in the form of $(C_6H_{10}O_5)_6(NaOH)$ $(H_2O)_3$. Na^+ cations are small ions able to penetrate the smallest pore present in cellulose, so they can transit across cellulose (Deshpande et al., 2008). Higher crystalline cellulose is achievable only upon complete decrystallization.

5.1.2.3.3.2 Ammonia Alkaline ammonia pretreatment on different lignocellulosic biomass has been extensively studied and is a promising alternative (Yang et al., 2003). Pretreatment using ammonia greatly enhances the digestibility of low-grade forage and crops residue in animal feed (Selim et al., 2004). Studies on various publications describe ammonia pretreatment using lignocellulosic biomass for subsequent biogas generation and biofuel production (Taherzadeh and Karimi, 2008). Pretreatment on ammonia enhances lignin removal and partial hemicellulose solubilization although high cellulose content is maintained (Kim and Lee, 2005). Ammonia can cleave lignin by ammonolysis while solubilizing hemicellulose at a high temperature (Gupta and Lee, 2009). Using ammonia, almost all glucan is retained at about 100% and a high amount of short-length (oligosaccharide) cellulose is present in the treated solid residue. Pretreatment using ammonia breaks the carbohydrate-lignin complex linkage between the hemicellulose side chain and lignin phenol, which allows lignin removal and partial hemicellulose. Studies also show that the influence on cellulose crystallization depends on the duration of treatment and temperature. Samples of cellulose with various degrees of polymerization and crystallinity indexes were subject to anhydrous liquid ammonia treatment (Mittal et al., 2011). An important aspect of pretreatment using ammonia is the economic and high quality of lignin content productivity as a secondary product (Bouxin et al., 2014).

5.1.2.3.3.3 Alkaline peroxides Peracetic acid and hydrogen peroxide are commonly used as oxidizing agents (Xu and Tschirner, 2012). Hydrogen peroxide pretreatment is becoming popular. Hydrogen peroxide requires the pH to be in an alkaline state to produce oxidizing radicals for lignin degradation. Thus, at an alkaline pH, powerful radicals of hydroxyl are formed that attack the lignin content. Simultaneously, hemicellulose is dissolved through the action of hydrogen peroxide. Important parameters that increase efficiency are the concentration of hydrogen peroxide, the temperature, the pH of the reaction mixture, and the duration; these should be optimized for various types of lignocellulosic biomass (Michalska and Ledakowicz, 2014). The advantage of using alkaline hydrogen peroxide (AHP) is that the final product is water and oxygen, so no residual hydrogen peroxide is expected in biomass degrade samples. Oxidation on biomass with hydrogen peroxide degrades and removes lignin and uronic groups in hemicellulose through solvation and saponification on biomass during alkaline peroxide hydrolysis. Optimum processing conditions are a concentration of hydrogen peroxide of 5%, temperature of 25°C, and duration of 24 h (Gould, 1984). Studies on AHP pretreatment on cashew apple bagasse pretreatment altered the surface structure of lignocellulosic biomass, leading to efficient delignification of biomass with operating conditions of 35°C, 11.5 pH, and a concentration of hydrogen peroxide equal to 5% v/v (Costa et al., 2013).

5.1.2.3.3.4 Other alkaline types Pretreatment of lignocellulosic material removes the acetyl group from hemicellulose using $Ca(OH)_2$ treatment and enhances the hydrolysis and reduction of steric hindrance on enzymes (Reilly et al., 2015). Studies of lignocellulosic biomass on $Ca(OH)_2$ enhance the crystalline index through amorphous removal (Alvira et al., 2010). High methane production from wheat straw can be achieved upon handling with $Ca(OH)_2$ (7.4% w/w) at 35°C and a duration of 2 h. Pretreatment using $Ca(OH)_2$ is inexpensive and less safe with sodium hydroxide or potassium hydroxide, but it is easily recovered by reaction with CO_2 from hydrolysate (Mosier et al., 2005). Alkali primary removes lignin from the lignocellulosic material by solubilization. KOH, $Ca(OH)_2$, NaOH, and NH_4OH are good alkali solutions for lignocellulosic biomass pretreatment (Mirmohamadsadeghi et al., 2016).

5.1.2.3.3.5 Alkali with oxidizing agents Adding oxidizing agents with alkali for alkaline pretreatment such as H_2O_2 enhances lignin content removal. In papermaking, H_2O_2 is chiefly used for oxidizing cellulose bleaching. Combination alkaline peroxide has an excellent effect on lignin removal with reduced energy under mild pretreatment conditions. The method involves the formation of hydrogen peroxide anions in alkali media. This acts as an energetic chemical in hydrogen peroxide through bleaching. Hydrogen peroxide–derived radicals (superoxide anion radicals and hydroxyl radicals) from the oxidative action contribute to lignin content depolymerization by breaking side chains on lignin and fragment the macrostructure of lignin into a low–molecular weight compound, leading to carbonyl group formation, cleavage of some interbonds, and dissolution of hemicelluloses and lignin (Umezawa et al., 2000). For the alkaline peroxide method, the pH is most important for efficient peroxide application (Maziero et al., 2012). The oxidizing agents attack the aliphatic region of macromolecules (Sun et al., 2000). For pretreatment of lignocelluloses using the AHP method, such as for corncob, processing conditions were 55°C and low alkaline (2% w/w H_2O_2) and pH (11.5) to separate three components efficiently. The ratios of hemicellulose and lignin removal after 6 h of treatment were 38.7% and 75.4%, respectively, and cellulose residue improved to 81.3%. Enzymatic hydrolysis on AHP residue after 24 h cellulose conversion was 80%. For subsequent desalination and precipitation using ethanol, 89% was recovered hemicelluloses as a white solid and 77.5% of lignin was a brown solid (Su et al., 2015).

5.1.2.3.3.6 Combination of alkaline with ultrasound, solar radiation, and microwave Studies reported that a small amount of lignin was released when lignocellulose softwood was treated with a dilute NaOH aqueous solution. Microwave treatment of an alkali medium enhanced the property of the component to be separated. There were also studies on the effect of heating using microwaves on lignin separation. The powdered wood sample was soaked in a sodium hydroxide medium heated using microwaves at different intervals (Xue et al., 2012). Lignin extraction with microwave-dependent heating had higher molecular weight and polydispersity compared with that without microwave heating. Thus, the high molecular weight of lignin was released under microwave irradiation. The pretreated component obtained using ultrasound and alkali dependent on sonication pretreatment can enhance the release of monomer molecules from biomass in the hydrolysis of enzymes and increase ethanol produced in fermentation (Tang and Sivakumar, 2015). In ultrasound-assisted alkali pretreatment, the lignin content increased in a short time using sonication. Ultrasound and sonication-assisted alkaline (NaOH concentrations of 1% and 3%) pretreatment increased sugar to 5.78, 9.11, and 14.84 g/L, respectively. The efficiency of this hydrolysate was examined for ethanol production; yields reached 0.38 g/g of glucose after 72 h.

Radiant light such as solar energy and heat energy from the sun can be harnessed using devices such as radiation collectors, solar concentrators, and wavelength shifting radiation convertors. Solar energy can be used to degrade the liquor content in paper industries photocatalytically (Kansal et al., 2008). Subsequently, solar energy using can be used effectively for lignocellulose preprocessing.

5.1.2.3.3.7 Alkali with dilute acid combination Studies on dilute acid and alkali pretreatment in techniques using ammonia and sulfuric acid for rice straw hydrolysis to improve xylose and glucose extraction obtained results through the sequential extraction of hemicellulose and lignin. In the first step, hemicellulose was extracted along with the lignin content, yielding a lignin–xylose blend as a black liquor. This resulted in a complete loss of monomer for fermentation, because xylose in the black liquor strongly inhibited microbes in xylose fermentation. To overcome this, sequential dilute acid and alkaline pretreatment maintained high xylose recovery upon enzymatic digestibility from pretreated residue. Various combinations of alkali acid catalyze successfully hydrolyzed various lignocellulose biomasses, including corn and rice straw (Kim and Han, 2012). Dilute sulfuric acid pretreatment of rice hull and then sodium hydroxide led to delignification and the removal of lignin. The cellulose extract was enhanced to 68.5% –70.5%. The combination pretreatment resulted in a ninefold increased enhancement in the hydrolysis of enzyme (Martin et al., 2007).

5.1.2.3.3.8 Alkali with steam explosion Lignocellulosic biomass is pretreated using a complex method combining steam explosion and alkaline pretreatment, causing hemicelluloses to be extracted. This exposes the cellulose surface. Alkaline delignification increases the accessibility of enzyme to cellulose. Explosion using steam can enhance cellulose by 90% in the pretreated hydrolysate. The quantity of reducing sugar achieved was 460 mg/g for early dry hydrolysate using bamboo after treatment with steam explosion and 10% for sodium hydroxide.

With combined alkali sodium hydroxide and steam explosion preprocessing, the cellulose content in the wheat straw was enhanced to 67.2%, the monomer quantity in hydrolysate was 110.9 g/L, the total quantity of ethanol produced in the solid-state fermentation of wheat straw was 51.5 g/L, and a total yield of 81.1% was achieved (Chen et al., 2008a). Studies on the pretreatment of sugarcane bagasse using steam explosion and then alkaline pretreatment lignin removal resulted in 91%; 72% hemicellulose was removed by combining pretreatment with steam explosion along with alkali and alkaline/ethanol. Recovery resulted in 28.9%–60.1% of lignin from biomass (Sun et al., 2014).

5.1.2.4 Metal salt

Metal salt can be used to hydrolyze lignocellulose material to generate monomer (Liu et al., 2009). Metal salt pretreatment is preceded by a low pH and high catalytic reaction system compared with inorganic acids. Metal salts were segregated based on the periodic table, such as alkaline earth metals (Ca and Mg), transition metals (Ca, Fe, Cu, Mn, Co, and Fe), alkali metals (Li, Na, and K), and (A1) group III A metals. They can be seen in the form of phosphate nitrate and chloride (Kang et al., 2013). Metal salt pretreatment enhances hemicellulose hydrolysis and enzymatic saccharification and results in a higher reaction rate and higher recyclability solution. It is less corrosive, with moderate sugar degradation, which subsequently leads to a low amount of toxic compound generated (Monavari et al., 2011). Metal salt upon lignocellulose biomass pretreatment is based on the Lewis acid characteristic of being able to attract pairs of electrons.

5.1.2.5 Organosolv

The organosolv method of pretreatment includes organic solvents such as methanol, phenol, ethanol, acetone, and glycol by adding a smaller amount of minerals acids such as HCL or H_2SO_4 to enhance porous pretreatment (Hergert, 1998). The processing temperature is 100–250°C. This solubilizes lignin in certain organic solvents. The recovered lignin can be used for various purposes. The organosolv lignin content has a molecular weight of 700–1500 g/mol, transition temperature of 70–170°C, a high amount of phenol hydroxyl groups, low polydispersity, and a very low degree of changes in chemical structure (Lora and Glasser, 2002). Organosolv is done at a high temperature between 100 and 250°C with a solvent of methanol and ethanol at a low boiling point, glycerol tetrahydrofurfuryl alcohol, ethylene glycol at a high boiling point, and other organic compounds such as ketones, dimethylsulfoxide, organic acid, phenol, and ether (Thring et al., 1990).

5.1.2.6 Biological pretreatment

Biological pretreatment is achieved using enzymes and microorganism under ambient temperature and pressure. Biological processes require low chemical and energy inputs with negligible or reduced generation of inhibiting agents. The rate of reaction in the biological method is low, at 10–14 days. It requires optimum growth conditions with a large space to carry out biological pretreatment. Because of this, biological pretreatment has technical challenges with less commercial attraction.

5.1.2.6.1 Microorganism

Biological pretreatment uses microorganisms that secrete ligninolytic enzymes to fractionate the lignin content in biomass and thus disrupt the native structure and subsequently hydrolyze holocellulose using enzymes to liberate monomers. Microorganisms that degrade lignocellulosic biomass are either fungi or bacteria. The former one has a major role in degrading wood. The most common and effective one for lignin and hemicellulosic degrading microorganisms is brown and white rot fungi (Chen et al., 2010).

White rot fungi, mostly of basidiomycetes are used for lignocelluloses material pretreatment. They can degrade both cellulose and lignin. Brown rot fungi mainly degrade the cellulose content in biomass (Fan et al., 1987). Studies on pretreating wheat straw used white rot fungi; nearly 35% of wheat straw was converted into monomers by *Pleurotus ostreatus* in 5 weeks (Ander and Eriksson, 1977). Some soil bacteria that grow faster than fungi have the ability to degrade lignin content into water-soluble compounds for potential enzyme production. Many bacteria has proved from the gene of *Streptomyces* have capability to hydrolyze lignin content in the lignocellulose of plant biomass.

5.1.2.6.1.1 Fungi Fungal pretreatment of biomass degrades lignin content by converting it into water and carbon dioxide. White rot fungi degrades all components in the lignocellulose biomass and that derived from wood rot basidiomycetes (Martınez et al., 2005). Fungi produce different patterns of rot with selective delignification and simultaneous rotting. The former degrades the lignin content selectively over cellulose, whereas the latter degrades holocellulose while degrading lignin (Otjen and Blanchette, 1986). White rot fungi is mostly seen in hardwood rather than softwood. It selectively degrades S units rather than G units. Studies have been done on several fungi to pretreat lignocellulosic plant biomass (Hatakka, 1983).

5.1.2.6.1.2 Bacteria The most widely characterized bacteria for lignin depolymerization is the α-proteobacteria *Sphingobium* Sp. SYK-6. Other bacteria such as *Rhodococcus jostii* [RHA1] and *Pseudomonas putida* have been found to hydrolyze lignin content in lignocellulose. The ligninolytic system for depolymerization is through laccases or some other nonperoxidase (Ahmad et al., 2010). Studies on the pretreatment of paddy straw using *Streptomyces griseorubens* for 10 days produced an enzymatic hydrolysis of 97.8% (Saritha et al., 2013). *Trichoderma reesei* and *Cellulomonas uda* were used to pretreat sugarcane bagasse (Singh et al., 2007).

5.1.2.6.2 Enzyme

The extraction of enzymes can be directly added to the substrate for lignin depolymerization. Hydrolysis using enzymes is achieved at the optimum temperature of 45–50°C and pH of 4.8 (Duff and Murray, 1996), which produces no corrosion; the product is the monomers. This method is not promising for microorganisms because the ligninolytic is a complex enzymatic system that requires a synergic action of various enzymes and mediators to degrading the lignin effectively. The extract of crude enzyme from *Coriolus versicolor* and *Phanerochaete chrysosporium* enhanced the yield of sugar upon hydrolysis on corn stover by 50.2% with subsequent partial lignin removal (Wang et al., 2013). The cellulase enzyme contains three groups, each of which performs a specific action on lignocelluloses biomass: (1) endoglucanase, which attaches cellulose fiber in the region of low crystallization and treats the free end chain; (2) exoglucanase, which degrades further by removing the free chain end from the cellobiose unit; and (3) β−-glucosidase, which hydrolyzes cellobiose to glucose (Coughlan and Ljungdahl, 1988). These are apart from several mediator enzymatic systems that can attack and cleave hemicelluloses such as acetyl esterase, glucomannanases, glucuronidase, and galactomannanaase (Duff and Murray, 1996). Cellulose hydrolysis using enzyme is degraded to produce glucose monomers by the action of enzyme cellulase, which can be used to produce ethanol using microorganisms.

Biological pretreatment was conducted with microorganism that can secrete ligninolytic enzyme or enzyme to degrade and remove lignin content and subsequent destruction in morphologic structures, which allows enzymatic hydrolysis on substrate to liberate more fermentable sugar for biorefinery processes (Chen et al., 2010). Biological pretreatment on lignocellulosic biomass has great potential. There are hurdles to overcome for a more feasible process on a large scale, such as the time of pretreatment. Types of biological pretreatment on rice straw, wheat straw, sugarcane bagasse, and corn straw and their respective output are:

Pretreatment method: biological pretreatment (*T. reesei.*), **Lignocellulosic materials used:** rice straw and wheat straw. **Pretreatment conditions:** straw saccharification; time (7 days); substrate (5%); microorganism concentration (5 mL); temperature (30°C); rate (200 rpm); pH (5.5); and G Tween 80 (1% v/v). **Output obtained:** microbial saccharification (using fungi) of delignified substrate was studied. The optimum sugar released (175.27 ± 3.34 mg/g and 98.68 ± 2.43 mg/g with rice straw and wheat straw, respectively) was obtained on the seventh day of saccharification (Srivastava et al., 2014).

Pretreatment method: balling milling (4 h) after enzymatic hydrolysis. **Lignocellulosic materials used:** sugarcane bagasse. **Pretreatment conditions:** enzymatic (acremonium cellulase at 5 filter paper units/g substrate of cellulase and 20 U/g substrate of xylanase from Optimash BG at 45°C, pH 5.0, for 72 h. **Output obtained:** 89.2 ± 0.7% (glucose) and 77.2 ± 0.9% (xylose) (Buaban et al., 2010).

Pretreatment method: alkaline pretreated with enzymatic hydrolysis. **Lignocellulosic materials used:** corn straw. **Pretreatment conditions:** 2% NaOH, 80°C, 1 h after enzymatic hydrolysis by cellulase of *T. reesei* ZU-02 and cellobiose of *Aspergillus niger* ZU-07. **Output obtained:** xylose 23.6 g/L, glucose56.7 g/L, and arabinose 5.7 g/L (Chen et al., 2008b).

5.1.3 Fermentation strategy for bioethanol production

The most vital step in fermentation is to finding the optimum working condition to achieve the maximum yield from the given carbon source of substrate (Baeyens, 2015). In fermentation, the hydrolysate of hemicellulose has hexose as well as pentose, in which xylose may be the chief monomer in the hydrolysate in hemicellulose of hardwood. After saccharification, the broth is neutralized and centrifuged after filtration to remove the cell mass and the hydrolysate is concentrated then inoculated with microorganisms of actively grown cells after a new fermentation medium is added. It is sterilized then incubated and shaken at different process optimization conditions of the substrate to achieve the maximum yield, such as pH 4.5, 5.0, or 5.5, a temperature of 35, 40, or 45°C, an agitation rate of 150, 160, or 170 rpm, a substrate concentration of 4%, 6%, or 8% (v/v), and an inoculum size of 3%, 4%, or 5% (v/v) for 96 h according to one factor design using batch fermentation. The sample of 5 mL is taken to study the cell mass, substrate, and ethanol concentration in the fermentation broth. The bioethanol produced is determined using potassium dichromate for every interval of 6 h (William, 1950). The various factors that influence bioethanol production are temperature, pH, inoculum size, substrate concentration, agitation rate, and fermentation time (Zabed et al., 2014). At the end of fermentation, the medium is filtrated to remove the suspended solids and distillation is performed on the round-bottom flask at 80°C (Prasad et al., 2009). A simplified representation of glucose degrading to the ethanol pathway by anaerobic bacteria is shown in Fig. 5.4.

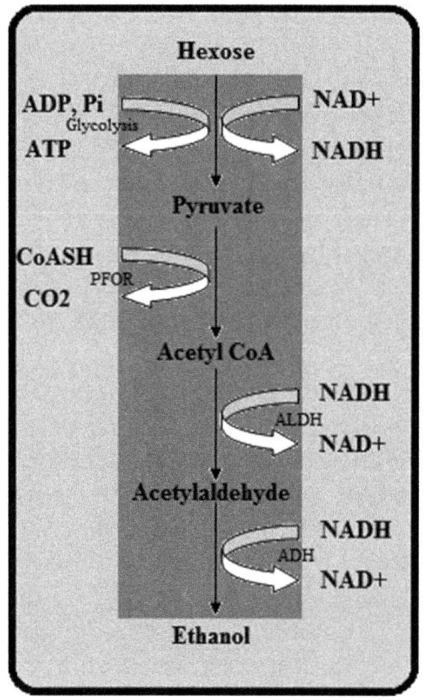

FIGURE 5.4 Simplified representation of glucose degradation to end product pathway by anaerobic bacteria. *ADH*, alcohol dehydrogenase; *ALDH*, acetaldehyde dehydrogenase; *PFOR*, pyruvate ferredoxin oxidoreductase (Scully and Orlygsson, 2014).

For an ideal microorganism for ethanol production,

1. During fermentation, rapid growth rate is achieved.
2. The maximum ethanol production is produced with optimal growth of the microorganism during fermentation conditions.
3. The organism should withstand high tolerance to ethanol concentration produced in fermentation.
4. The microorganism should have the ability to ferment the substrate on the available substrate.
5. The microorganism should have the ability to tolerate inhibitors produced during hydrolysis.
6. The organism should produce a maximum ethanol yield with a high fermentation rate.
7. The microorganism should have the ability to tolerate any extreme chemical condition during fermentation (e.g., low pH).

Different microorganisms for ethanol fermentation with various lignocellulosic substrates are listed in Table 5.2.

5.1.4 Purification of bioethanol

The most common method of distillation is the standard industrial purification practice for bioethanol. It is performed with a variation in the volatility of mixers present in the crude component. The method is by heating the mixer; thus, lesser boiling point constituents

TABLE 5.2 Bioethanol production from different microorganisms and substrates under optimized fermentation conditions.

Microbial species	Substrate	Fermentation condition (°C)	Yield (g/L)	References
Saccharomyces cerevisiae D5A	Switchgrass	37	21.9	Faga et al. (2010)
Kluyveromyces marxianus IMB3	Kanlow switchgrass	45	22.5	Pessani et al. (2011)
S. cerevisiae TJ 14	161 g/L paper sludge material	42	40	Prasetyo et al. (2011)
S. cerevisiae JZ1C	Jerusalem artichoke	40	65.2	Hu et al. (2012)
K. marxianus OT-1	Jerusalem artichoke	40	73.6	Hu et al. (2012)
Pichia kudriavzevii HOP-1	Rice straw	45	24.25	Oberoi et al. (2012)
Blastobotrys adeninivorans RCKP 2012	Sugarcane bagasse	50	14.05	Antil et al. (2015)
K. marxianus TISTR 5925	Palm sap	40	45.4	Murataet al. (2015)
K. marxianus DBKKU-Y102	Jerusalem artichoke	40	97.46	Charoensopharat et al. (2015)
S. cerevisiae ZM1-5	Sugarcane bagasse	40	18.79	Huang et al. (2015)
K. marxianus MTCC 1389	Banana peduncle	40	21.89	Raja Sathendra et al. (2019)
K. marxianus MTCC 1389	Palm wood	40	23	Raja Sathendra et al. (2018)
Crude cellulase enzyme	Sugarcane leaves	30	18	Dharmendira Kumar et al. (2014)
Zymomonas mobilis	*Prosopis juliflora*	35	16.40	Sivarathnakumar et al. (2017)
Immobilization of cellulase onto MnO_2 nanoparticles	Agricultural waste	–	21.96	Dharmendirakumar et al. (2015)
K. marxianus	*Moringa oleifera*	42	20.91	Sivarathnakumar et al. (2014)

are determined in the volatile phase. and by condensing this vapor content with cold water, more concentrated and less volatile compounds are obtained in the liquid phase. It is the most well-organized bioseparation method. A fermentation medium was distilled and 40 mL of distilled ethanol was collected to analyze the concentrated ethanol. The concentration of ethanol was calibrated using a standard dichromate test by spectrophotometric method (Caputi et al., 1968).

5.1.5 Techno-economic strategy

The cost of ethanol production can be reduced by lessening investments in either the untreated substrate or the cellulase enzyme. It was predicted that the use of genetically engineered substrate with high carbohydrate content together combined with the improvement in conversion technology could reduce the cost of ethanol by $0.11 per liter over the next 10 years (Wooley et al., 1999).

The conversion of xylose/pentose to ethanol can enhance the production of ethanol efficiently and reduce the value of designing the species metabolic pathway using genetically modified species (Wood et al., 1997). The reconstructed operon encoding xylose assimilation and pentose phosphate pathway enzymes were transformed into the bacterium *Zymomonas mobilis* for the effective fermentation of xylose to produce ethanol (Dien et al., 2000). The recombinant DNA strain of *Escherichia coli* with the genes from *Z. mobilis* to adapt the pyruvate pathway into ethanol was published by Dipardo (2000).

The hurdle faced during the production of bioethanol was the constant action of yeasts modified genetically in the stream processing operations on a large saleable scale (Hendriks and Zeeman, 2009), enhanced with an efficient pretreatment method for plant lignocellulosic biomass, and incorporated with substituent into a production of ethanol systems with financial feasibility. The economic assessment of physical and chemical pretreatment processes is:

Physical pretreatment method

Mechanical: monomer yield: very low; waste: low; outlay: low. **Explosion on steam**: monomer yield: high; wastes: low; outlay: high. **Ammonia fiber explosion**: monomer yield: moderate; wastes: very low; outlay: high. **Carbonic acid:** monomer yield: very high; waste: very low; outlay: low.

Chemical pretreatment method

Acid dilute: monomer yield: very high; waste: high; outlay: moderate; **Acid concentrated**: monomer yield: very high; waste: high; outlay: high; **Alkaline**: monomer yield: very high; waste: high; outlay: low. **Wet oxidation**: monomer yield: high; waste: low; outlay: low. **Organosolv**: monomer yield: very high; waste: low: outlay: very high (Gurgel et al., 2014; Behera et al., 2014; Zeng et al., 2014; Phan et al., 2014; Prado et al., 2014; Singh et al., 2014, 2015; Biswas et al., 2014; Sharma et al., 2014).

5.1.6 Sustainable improvement of lignocellulosic biomass in the environment

Lignocellulosic biomass feedstocks gained worldwide attention as a potential renewable energy source. The integration of a combined conversion method and process is determined to be a sustainable approach. The lignocellulosic biomass feedstock for biorefinery is an useful technique for wide-ranging use for lignocellulosic components. This will have a major part in the energy industry and chemical industry in future developments. Owing to the current system of feedstock logistics, biomass volarization has many drawbacks, which makes biorefinery operations unsuccessful throughout the world. To overcome this, lignocellulosic feedstock is being pursued, from nonfood crops to agroforest residue and waste. The lignocellulosic biofuel of agricultural crops is interesting worldwide for numerous approaches such as energy security, climate change mitigation, and development in the rural economy. Lignocellulosic plant biomasses in the form of waste crops are geographically rich globally and may sustain the transportation of liquid fuel. The second generation of lignocellulosic plant biomass containing cellulose-released monomer molecules is hydrolyzed using microorganisms or enzymes despite the cost. Cellulose-containing agricultural by-products such as sawdust, bamboo, and straw could be developed using this technology (Demirbas et al., 2009).

5.1.7 Conclusion

The use of lignocellulosic substrate as a renewable source for carbon is successful and depends on the economic perspective. The various ranges of pretreatment options with chemical, physical, physicochemical, and biological processes were analyzed and developed for lignocellulosic biomass pretreatment. The effective use of lignocellulosic plant biomass for biorefinery through the fractionation of its constituents, such as cellulose, hemicellulose, and lignin, was achieved using efficient pretreatment techniques. Alkali pretreatment can solubilize hemicellulose and lignin in a process operating under ambient conditions, which gives a greater advantage for lignocellulose to be fractionated compared with others. Lignocellulosic pretreatment for biorefinery is cost-effective. The different methods of biomass pretreatment listed for biorefinery development for various products can be generated from the lignocellulose of plant biomass. Pretreatment of lignocellulosic biomass and the breakdown of three-dimensional morphologic structures modify the degree of polymerization and enhance the surface area and pore size, promoting fermentable sugar for maximum ethanol production.

References

Ahmad, M., Taylor, C.R., Pink, D., Burton, K., Eastwood, D., Bending, G.D., 2010. Development of novel assays for lignin degradation: comparative analysis of bacterial and fungal lignin degraders. Mol. Biosyst. R. Soc. Chem. 6 (5), 815–821.

Alvira, P., Tomás-Pejó, E., Ballesteros, M., Negro, M., 2010. Pretreatment technologies for an efficient bioethanol production process based on enzymatic hydrolysis: a review. Bioresour. Technol. 101, 4851–4861.

An, Y.-X., Zong, M.-H., Wu, H., Li, N., 2015. Pretreatment of lignocellulosic biomass with renewable cholinium ionic liquids: biomass fractionation, enzymatic digestion and ionic liquid reuse. Bioresour. Technol. 192, 165–171.

Ander, P., Eriksson, K.-E., 1977. Selective degradation of wood components by white-rot fungi. Physiol. Plant. 41, 239–248.

Antil, P., Gupta, R., Kuhad, R., 2015. Simultaneous saccharification and fermentation of pretreated sugarcane bagasse to ethanol using a new thermotolerant yeast. Ann. Microbiol. 65, 423–429.

Azam, J., Mohammad, J., Taherzadeh, 2009. Ethanol production from cotton-based waste textiles. Bioresour. Technol. 100, 1007–1010.

Baeyens, J., Kang, Q., Appels, L., Dewil, R., Lv, Y., Tan, T., 2015. Challenges and opportunities in improving the production of bio-ethanol. Prog. Energy Combust. Sci. 4, 60–88.

Ballesteros, I., Ballesteros, M., Manzanares, P., 2008. Dilute sulfuric acid pretreatment of cardoon for ethanol production. Biochem. Eng. J. 42, 84–91.

Behera, S., Arora, R., Nandhagopal, N., Kumar, S., 2014. Importance of chemical pretreatment for bioconversion of lignocellulosic biomass. Renew. Sustain. Energy Rev. 36, 91–106.

Biswas, R., Uellendahl, H., Ahring, B.K., 2014. Wet explosion pretreatment of sugarcane bagasse for enhanced enzymatic hydrolysis. Biomass Bioenergy 61, 104–113.

Bjerre, A.B., Olesen, A.B., Fernqvist, T., 1996. Pretreatment of wheat straw using combined wet oxidation and alkaline hydrolysis resulting in convertible cellulose and hemicellulose. Biotechnol. Bioeng. 49, 568–577.

Bouxin, F.P., Jackson, S.D., Jarvis, M.C., 2014. Isolation of high quality lignin as a by-product from ammonia percolation pretreatment of poplar wood. Bioresour. Technol. 162, 236–242.

Brownell, H., Yu, E., Saddler, J., 1986. Steam-explosion pretreatment of wood: effect of chip size, acid, moisture content and pressure drop. Biotechnol. Bioeng. 28, 792–801.

Buaban, B., Inoue, H., Yano, S., Tanapongpipat, S., Ruanglek, V., Champreda, V., 2010. Bioethanol production from ball milled bagasse using an on-site produced fungal enzyme cocktail and xylose-fermenting *Pichia stipitis*. J. Biosci. Bioeng. 110 (1), 18–25.

Cadoche, L., Lopez, G.D., 1989. Assessment of size reduction as a preliminary step in the production of ethanol from lignocellulosic wastes. Biol. Wastes. 30, 153–157.

Cantarella, M., Cantarella, L., Gallifuoco, A., 2004. Effect of inhibitors released during steam-explosion treatment of poplar wood on subsequent enzymatic hydrolysis and SSF. Biotechnol. Prog. 20, 200–206.

Caputi Jr., A., Ueda, M., Brown, T., 1968. Spectrophotometric determination of ethanol in wine. Am. J. Enol. Vitic. 19, 160–165.

Carere, C.R., Sparling, R., Cicek, N., Levin, D.B., 2008. Third generation biofuels via direct cellulose fermentation. Int. J. Mol. Sci. 9, 1342–1360.

Chandra, R.P., Bura, R., Mabee, W.E., Berlin, A., Pan, X., Saddler, J.N., 2007. Substrate pretreatment: the key to effective enzymatic hydrolysis of lignocellulosics. Adv. Biochem. Eng. Biotechnol. 108, 67–93.

Chandra, R.P., Arantes, V., Saddler, J., 2015. Steam pretreatment of agricultural residues facilitates hemicellulose recovery while enhancing enzyme accessibility to cellulose. Bioresour. Technol. 185, 302–307.

Chang, V., Holtzapple, M., 2000. Fundamental factors affecting biomass enzymatic reactivity. Appl. Biochem. Biotechnol. 84 (86), 5–37.

Chang, M.M., Chou, T.Y.C., Tsao, G.T., 1981. Structure, pretreatment and hydrolysis of cellulose. Adv. Biochem. Eng. 20, 15–42.

Charoensopharat, K., Thanonkeo, P., Thanonkeo, S., Yamada, M., 2015. Ethanol production from Jerusalem artichoke tubers at high temperature by newly isolated thermotolerant inulin-utilizing yeast kluyveromyces marxianus using consolidated bioprocessing. Antonie Van Leeuwenhoek 108, 173–190.

Chen, H.Z., Liu, Z.H., 2015. Steam explosion and its combinatorial pretreatment refining technology of plant biomass to bio-based products. Biotechnol. J. 10, 866–885.

Chen, H., Han, Y., Xu, J., 2008a. Simultaneous saccharification and fermentation of steam exploded wheat straw pretreated with alkaline peroxide. Process Biochem. 43, 1462.

Chen, M., Zhao, J., Xia, L., 2008b. Enzymatic hydrolysis of maize straw polysaccharides for the production of reducing sugars. Carbohydr. Polym. 71, 411–415.

Chen, S., Zhang, X., Singh, D., Yu, H., Yang, X., 2010. Biological pretreatment of lignocellulosics: potential, progress and challenges. Biofuels 1 (1), 177–199. Taylor & Francis.

Chen, W.-H., Xu, Y.-Y., Hwang, W.-S., Wang, J.-B., 2011. Pretreatment of rice straw using an extrusion/extraction process at bench-scale for producing cellulosic ethanol. Bioresour. Technol. 102, 10451–10458.

Cherian, E., Dharmendirakumar, M., Baskar, G., 2015. Immobilization of cellulase onto MnO_2 nanoparticles for bioethanol production by enhanced hydrolysis of agricultural waste. Chin. J. Catal. 36 (2015), 1223–1229.

Coughlan, M.P., Ljungdahl, L.G., 1988. Comparative biochemistry of fungal and bacterial cellulolytic enzyme system. In: Aubert, J.-P., Beguin, P., Millet, J. (Eds.), Biochemistry and Genetics of Cellulose Degradation, pp. 11–30.

da Costa Correia, J.A., Marques Jr., J.E., Gonçalves, L.R.B., Rocha, M.V.P., 2013. Alkaline hydrogen peroxide pretreatment of cashew apple bagasse for ethanol production: study of parameters. Bioresour. Technol. 139, 249–256.

Demirbas, A., Arin, G., 2002. An overview of biomass pyrolysis. Energy Source 24, 471–482.

Demirbas, M.F., Balat, M., Balat, H., 2009. Potential contribution of biomass to the sustainable energy development. Energy Convers. Manag. 50 (7), 1746–1760.

Deshpande, M., Scheicher, R.H., Ahuja, R., Pandey, R., 2008. Binding strength of sodium ions in cellulose for different water contents. J. Phys. Chem. B. 112, 8985.

Desvaux, M., 2005. Clostridium cellulolyticum: model organism of mesophilic cellulolytic clostridia. FEMS Microbiol. Rev. 29, 741–764.

Dien, B.S., Nochols, N.N., O'Bryan, P.J., Bothast, R.J., 2000. Development of new ethanologenic *Escherichia coli* strains for fermentation of lignocellulosic biomass. Appl. Biochem. Biotechnol. 84 (86), 181–196.

Dipardo, J., 2000. Outlook for biomass ethanol production. Energy Inf. Adm. Available from: http://www.eia.doe.gov/oiaf/analysispaper/biomass.html.

Donsi, F., Ferrari, G., Pataro, G., 2010. Applications of pulsed electric field treatments for the enhancement of mass transfer from vegetable tissue. Food Eng. Rev. 2, 109–130.

Duff, S.J.B., Murray, W.D., 1996. Bioconversion of forest products industry waste cellulosics to fuel ethanol: a review. Bioresour. Technol. 55, 1–33.

Elsa Cherian, M., Kumar, D., Baskar, G., 2014. Cellulosic bioethanol production by sequential fermentation using agricultural waste. Chemtech 6 (14), 5653–5660.

Esteghalian, A., Hashimoto, A.G., Fenske, J.J., Penner, M.H., 1997. Modelling and optimization of dilute-sulfuric-acid pretreatment of corn stove, poplar and switch grass. Bioresour. Technol. 59, 129–136.

Faga, B.A., Wilkins, M.R., Banat, I.M., 2010. Ethanol production through simultaneous saccharification and fermentation of switchgrass using *Saccharomyces cerevisiae* D5A and thermotolerant *Kluyveromyces marxianus* IMB strains. Bioresour. Technol. 101, 2273–2279.

Fan, L.T., Gharpuray, M.M., Lee, Y.-H., 1987. In: Cellulose Hydrolysis Biotechnology Monographs, vol. 57. Springer, Berlin.

Galbe, M., Zacchi, G., 2007. Pretreatment of lignocellulosic materials for efficient bioethanol production. Biofuels 108, 41–65.

Garcia-Aparicio, M.P., Ballesteros, I., Gonzalez, A., 2006. Effect of inhibitors released during steam-explosion pretreatment of barley straw on enzymatic hydrolysis. Appl. Biochem. Biotechnol. 129, 278–288.

Ghatak, H.M., 2011. Biorefineries from the perspective of sustainability: feedstocks, products, and processes. Renew. Sustain. Energy Rev. 15, 4042–4052.

Gould, J.M., 1984. Alkaline peroxide delignification of agricultural residues to enhance enzymatic saccharification. Biotechnol. Bioeng. 26, 46–52.

Gupta, R., Lee, Y., 2009. Pretreatment of hybrid poplar by aqueous ammonia. Biotechnol. Prog. 25 (2), 357–364.

Gurgel, L.V.A., Pimenta, M.T.B., Curvelo, A.A.S., 2014. Enhancing liquid hot water (LHW) pretreatment of sugarcane bagasse by high pressure carbon dioxide (HP-CO_2). Ind. Crops Prod. 57, 141–149.

Hatakka, A.I., 1983. Pretreatment of wheat straw by white-rot fungi for enzymic saccharification of cellulose. Eur. J. Appl. Microbiol. Biotechnol. 18 (6), 350–357.

Hendricks, A.T., Zeeman, G., 2009. Pretreatments to enhance the digestibility of lignocellulosic biomass. Bioresour. Technol. 100, 10–18.

Hendriks, A.T.W.M., Zeeman, G., 2009. Pretreatments to enhance the digestibility of lignocellulosic biomass. Bioresour. Technol. 100 (1), 10–18.

Hergert, H.L., 1998. Developments in organosolv pulping-an overview. In: Akhtar, M., Young, R.A. (Eds.), Environmentally Friendly Technologies for the Pulp and Paper Industry. Wiley, New York, pp. 5–68.

Hu, Z., Wen, Z., 2008. Enhancing enzymatic digestibility of switchgrass by microwave-assisted alkali pretreatment. Biochem. Eng. J. 38, 369–378.

Hu, G., Heitmann, J.A., Rojas, O.J., 2008. Feedstock pretreatment strategies for producing ethanol from wood, bark, and forest residues. BioResources 3, 270.

Hu, N., Yuan, B., Sun, J., Wang, S.A., Li, F.L., 2012. Thermotolerant Kluyveromyces marxianus and *Saccharomyces cerevisiae* strains representing potentials for bioethanol production from Jerusalem artichoke by consolidated bioprocessing. Appl. Microbiol. Biotechnol. 95, 1359–1368.

Huang, Y., Qin, X., Luo, X.M., Nong, Q., Yang, Q., Zhang, Z., 2015. Efficient enzymatic hydrolysis and simultaneous saccharification and fermentation of sugarcane bagasse pulp for ethanol production by cellulase from *Penicillium oxalicum* EU2106 and thermotolerant *Saccharomyces cerevisiae* ZM1-5. Biomass Bioenergy 77, 53–63.

Imai, M., ikari, K., Suzuki, I., 2003. High Performance Hydrolysis of Cellulose Using Mixed Cellulase Species and Ultra-sonication Pretreatment.

Ioanid, E.G., Rusu, D., Ursescu, M., Totolin, M., Vlad, A.M., 2013. HF plasma discharge for surface modification of cellulose-based materials. IEEE Trans. Plasma Sci. 41, 3201–3206.

Itoh, H., Wada, M., Honda, Y., Kuwahara, M., 2003. Bioorganosolve pretreatments for simultaneous saccharification and fermentation of beech wood by ethanolysis and white-rot fungi. J Biotechnol 103, 273–280.

Jung, Y.H., Kim, K.H., 2015. Pretreatment of Biomass. Processes and Technologies. Elsevier, pp. 27–50.

kan, T., Vladimir, Strozov, T., Evans, J., 2016. Lignocellulosic biomass pyrolysis: a review of product properties and effects of pyrolysis parameter. RSER 57, 1126–1140.

Kang, K.E., Park, D.-H., Jeon, G.-T., 2013. Effects of inorganic salts on pretreatment of Miscanthus straw. Bioresour. Technol. 132, 160–165.

Kansal, S., Singh, M., Sud, D., 2008. Effluent quality at kraft/soda agro-based paper mills and its treatment using a heterogeneous photocatalytic system. Desalination 228, 183.

Kim, S., Dale, B.E., 2004. Global potential bioethanol production from wasted crops and crop residues. Biomass Bioenergy 26, 361–375.

Kim, I., Han, J.-I., 2012. Optimization of alkaline pretreatment conditions for enhancing glucose yield of rice straw by response surface methodology. Biomass Bioenergy 46, 210.

Kim, H.K., Hong, J., 2001. Supercritical CO_2 pretreatment of lignocellulose enhances enzymatic cellulose hydrolysis. Bioresour. Technol. 77, 139–144.

Kim, T.H., Lee, Y.Y., 2005. Pretreatment of corn stover by soaking in aqueous ammonia. Appl. Biochem. Biotechnol. 124 (1–3), 1119–1132.

Knauf, M., Moniruzzaman, M., 2004. Lignocellulosic biomass processing: a perspective. Int. J. 106, 147–150.

Kumar, P., Barrett, D.M., Delwiche, M.J., Stroeve, P., 2011. Pulsed electric field pretreatment of switchgrass and wood chip species for biofuel production. Ind. Eng. Chem. Res. 50, 10996–11001.

Kurien, N., Sukumaran, R.K., Pandey, A., 2010. Bioethanol production from rice straw: an overview. Bioresour. Technol. 101, 4767–4774.

Laser, M., Schulman, D., Allen, S.G., Lichwa, J., Antal, M.J., Lynd, L.R., 2002. A comparison of liquid hot water and steam pretreatments of sugar cane bagasse for bioconversion to ethanol. Bioresour. Technol. 81, 33–44.

Lin, Y., Tanaka, S., 2006. Ethanol fermentation from biomass resources: current state and prospects. Appl. Microbiol. Biotechnol. 69, 627–642.

Liu, L., Sun, J., Cai, C., Wang, S., Pei, H., Zhang, J., 2009. Corn stover pretreatment by inorganic salts and its effects on hemicellulose and cellulose degradation. Bioresour. Technol. 100, 5865–5871.

Liu, Z.-S., Wu, X.-L., Kida, K., Tang, Y.-Q., 2012. Corn stover saccharification with concentrated sulfuric acid: effects of saccharification conditions on sugar recovery and by-product generation. Bioresour. Technol. 119, 224–233.

Lora, J.H., Glasser, W.G., 2002. Recent industrial applications of lignin: a sustainable alternative to nonrenewable materials. J. Polym. Environ. 10, 39–48.

Martin, C., De Moraes Rocha, G.J., Perez, M., Lopez, Y., Hernandez, E., Plasencia, Y., 2007. Acid prehydrolysis, alkaline delignification and enzymatic hydrolysis of rice hulls. Cellulose Chem. Technol. 41, 129.

Martınez, A.T., Speranza, M., Ruiz-Duenas, F.J., Ferreira, P., Camarero, S., Guillen, F., 2005. Biodegradation of lignocellulosics: microbial, chemical, and enzymatic aspects of the fungal attack of lignin. Int. Microbiol. 8 (3), 195–204.

Maziero, P., Neto, M.D.O., Machado, D., Batista, T., Cavalheiro, C.C.S., Neumann, M.G., 2012. Structural features of lignin obtained at different alkaline oxidation conditions from sugarcane bagasse. Ind. Crops Prod. 35, 61.

Pretreatment of lignocellulosic biomass. In: McMillan, J.D., Himmel, M.E., Baker, J.O., Overend, R.P. (Eds.), 1994. Enzymatic Conversion of Biomass for Fuels Production. American Chemical Society, Washington, DC, pp. 292–324.

Melin, K., Kohl, T., Koskinen, J., Hurme, M., 2015. Performance of biofuel processes utilising separate lignin and carbohydrate processing. Bioresour. Technol. 92, 397–409.

Michalowicz, G., Toussaint, B., Vignon, M.R., 1991. Ultrastructural-changing in poplar cell wall during steam explosion treatment. Holzforschung 45, 175–179.

Michalska, K., Ledakowicz, S., 2014. Alkaline hydrogen peroxide pretreatment of energy crops for biogas production. Chem. Pap. 68 (7), 913–922.

Michalska, K., Miazek, K., Krzystek, L., Ledakowicz, S., 2012. Influence of pretreatment with Fenton's reagent on biogas production and methane yield from lignocellulosic biomass. Bioresour. Technol. 119, 72–78.

Miller, S., Hester, R., 2007. Concentrated acid conversion of pine softwood to sugars. Part I: use of a twin-screw reactor for hydrolysis pretreatment. Chem. Eng. Commun. 194, 85–102.

Millet, M.A., Baker, A.J., Scatter, L.D., 1976. Physical and chemical pretreatment for enhancing cellulose saccharification. Biotech. Bioeng. Symp. 6, 125–153.

Mirmohamadsadeghi, S., Chen, Z., Wan, C., 2016. Reducing biomass recalcitrance via mild sodium carbonate pretreatment. Bioresour. Technol. 209, 386–390.

Mittal, A., Katahira, R., Himmel, M.E., Johnson, D.K., 2011. Effects of alkaline or liquid-ammonia treatment on crystalline cellulose: changes in crystalline structure and effects on enzymatic digestibility. Biotechnol. Biofuels 4 (1), 41.

Monavari, S., Galbe, M., Zacchi, G., 2011. The influence of ferrous sulfate utilization on the sugar yields from dilute-acid pretreatment of softwood for bioethanol production. Bioresour. Technol. 102, 1103–1108.

Mori, T., Tsuboi, Y., Ishida, N., Nishikubo, N., Demura, T., Kikuchi, J., 2015. Multidimensional high-resolution magic angle spinning and solution-state NMR characterization of 13C-labeled plant metabolites and lignocellulose. Sci. Rep. 5, 1–12.

Mosier, N., 2005. Features of promising technologies for pretreatment of lignocellulosic biomass. Bioresour. Technol. 96 (6), 673–686.

Mosier, N.S., Wyman, C.E., Dale, B.E., Elander, R., Lee, Y.Y., Holtzapple, M.T., Ladisch, M., 2005. Features of promising technologies for pretreatment of lignocellulosic biomass. Bioresour. Technol. 96, 673–686.

Murata, Y., Danjarean, H., Fujimoto, K., Kosugi, A., Arai, T., Ibrahim, W.A., 2015. Ethanol fermentation by the thermotolerant yeast, Kluyveromyces marxianus TISTR5925, of extracted sap from old oil palm trunk. AIMS Energy 3, 201–213.

Negro, M.J., Manzanares, P., Oliva, J.M., 2003. Changes in various physical/chemical parameters of *Pinus pinaster* wood after steam explosion pretreatment. Biomass Bioenergy 25, 301–308.

Nelson, M.L., O'Connor, R.T., 1964. Relation of certain infrared bands to cellulose crystallinity and crystal latticed type. Part I. Spectra of lattice types I, II, III and of amorphous cellulose. J. Appl. Polym. Sci. 8, 1311.

Oberoi, H.S., Babbar, N., Sandhu, S.K., Dhaliwal, S.S., Kaur, U., Chadha, B., 2012. Ethanol production from alkali-treated rice straw via simultaneous saccharification and fermentation using newly isolated thermotolerant *Pichia kudriavzevii* HOP-1. J. Ind. Microbiol. Biotechnol. 39, 557–566.

Ota, M., Suzuki, H., Iguchi, H., Fugita, K., Yanagisawa, Y., Suzuki, T., et al., 2010. Promotion of enzymatic saccharification of lignocellulose by irradiation with pulsed UV laser beams or low-energy electron beams. In: Pacifichem 2010, International Chemical Congress of Pacific Basin Societies. Pacifichem, Honolulu, HI, United States, pp. 15–20.

Otjen, L., Blanchette, R.A., 1986. A discussion of microstructural changes in wood during decomposition bywhite rot basidiomycetes. Cancer J. Bot. 64 (5), 905–911. NRC Research Press, Ottawa, Canada.

Palmqvist, E., Hahn-Hägerdal, B., 2000. Fermentation of lignocellulosic hydrolysates. II: inhibitorsand mechanisms of inhibition. Bioresour. Technol. 74, 25–33.

Parisutham, V., Kim, T.H., Lee, S.K., 2014. Feasibilities of consolidated bioprocessing microbes:from pretreatment to biofuel production. Bioresour. Technol. 161, 431–440.

Pessani, N.,K., Atiyeh, H.K., Wilkins, M.,R., Bellmer, D.D., Bana, t I.M., 2011. Simultaneous saccharification and fermentation of Kanlow switchgrass by thermotolerant Kluyveromyces marxianus IMB3: the effect of enzyme loading, temperature and higher solid loadings. Bioresour. Technol. 102, 10618–10624.

Phan, D.T., Tan, C.S., 2014. Innovative pretreatment of sugarcane bagasse using supercritical CO_2 followed by alkaline hydrogen peroxide. Bioresour. Technol. 167, 192–197.

Prado, J.M., Follegatti-Romero, L.A., Forster-Carneiro, T., Rostagno, M.A., Maugeri Filho, F., Meireles, M.A.A., 2014. Hydrolysis of sugarcane bagasse in subcritical water. J. Supercrit. Fluids 86, 15–22.

Prasad, M.P., Rekha, S., Tamilarasan, M., Subha, K.S., 2009. Production of bioethanol using various agricultural raw materials by two step enzymatic process. J. Adv. Biotechnol. 9, 41–43.

Prasetyo, J., Naruse, K., Kato, T., Boonchird, C., Harashima, S., Park, E.Y., 2011. Bioconversion of paper sludge to biofuel by simultaneous saccharification and fermentation using a cellulase of paper sludge origin and thermotolerant *Saccharomyces cerevisiae* TJ14. Biotechnol. Biofuels 4, 1–13.

Raja Sathendra, E., Baskar, G., Praveenkumar, R., 2018. Optimization of bioethanol production from palm wood using *Trichoderma reesei* and *Kluveromyces marxianus*. Bioresour. Technol. 271, 345–352.

Raja Sathendra, E., Baskar, G., Praveenkumar, R., 2019. Production of bioethanol from lignocellulosic banana peduncle waste using *Kluveromyces marxianus*. J. Environ. Biol. 40, 769–774.

Reilly, M., Dinsdale, R., Guwy, A., 2015. Enhanced biomethane potential from wheat straw by low temperature alkaline calcium hydroxide pre-treatment. Bioresour. Technol. 189, 258–265.

Saha, B.C., 2003. Hemicellulose bioconversion. J. Ind. Microbiol. Biotechnol. 30, 279–291.

Saritha, M., Arora, A., Singh, S., Nain, L., 2013. Streptomyces griseorubens mediated delignification of paddy straw for improved enzymatic saccharification yields. Bioresour. Technol. 135, 12–17.

Scully, S., Orlygsson, J., 2014. Recent advances in second generation ethanol production by thermophilic bacteria. Energies 8 (1), 1–30.

Selim, A.J., Pan, T., Takano, T., Suzuki, S., Koike, Y., Kobayashi, Tanaka, K., 2004. Effect of ammonia treatment on physical strength of rice straw, distribution of straw particles and particle-associated bacteria in sheep rumen. Anim. Feed Sci. Technol. 115 (1–2), 117–128.

Sharma, S., Kumar, R., Gaur, R., 2014. Pilot scale study on steam explosion and mass balance for higher sugar recovery from rice straw. Bioresour. Technol. 175, 350–357.

Singh, P., Suman, A., Tiwari, P., Arya, N., Gaur, A., Shrivastava, A.K., 2007. Biological pretreatment of sugarcane trash for its conversion to fermentable sugars. World J. Microbiol. Biotechnol. 24 (5), 667–673.

Singh, R., Shukla, A., Tiwari, S., Srivastava, M., 2014. A review on delignification of lignocellulosic biomass for enhancement of ethanol production potential. Renew. Sustain. Energy Rev. 32, 713–728.

Singh, J., Suhag, M., Dhaka, A., 2015. Augmented digestion of lignocellulose by steam explosion, acid and alkaline pretreatment methods: a review. Carbohydr. Polym. 117, 624–631.

Sivarathnakumar, S., Baskar, G., Praveenkumar, R., Bharathiraja, B., 2017. Simultaneous saccharification and fermentation of woody stem *Prosopis juliflora* by *Zymomonas mobilis* for the production of cellulosic ethanol. IJMPT 55 (1/2/3), 238–253.

Sivarathnakumar, S., Bharathiraja, B., Jayamuthunagai, J., Praveen Kumar, R., Baskar, G., et al., 2014. Simultaneous saccharification and fermentation of bioethanol from softwood moringa oleifera using thermo-tolerant yeast kluyveromyces marxianus MTCC 1388. Int. J. Chemtech Res. 6 (12), 5064–5070.

Srivastava, A.K., Agrawal, P., Abdul Rahiman, 2014. Pretreatment and production of bioethanol from different Lignocellulosic biomass. Int. J. Adv. Res. 2 (4), 888–896.

Su, Y., Du, R., Guo, H., Cao, M., Wu, Q., Su, R., 2015. Fractional pretreatment of lignocellulose by alkaline hydrogen peroxide: characterization of its major components. Food Bioprod. Process. 94, 322.

Sun, Y., Cheng, J., 2002. Hydrolysis of lignocellulosic materials for ethanol production: a review. Bioresour. Technol. 83, 1–11.

Sun, R.C., Tomkinson, J., Ma, P.L., Liang, S.F., 2000. Comparative study of hemicelluloses from rice straw by alkali and hydrogen peroxide treatments. Carbohydr. Polym. 42, 111.

Sun, S.-L., Wen, J.-L., Ma, M.-G., Sun, R.-C., 2014. Enhanced enzymatic digestibility of bamboo by a combined system of multiple steam explosion and alkaline treatments. Appl. Energy 136, 519.

Taherzadeh, M.J., Karimi, K., 2008. Pretreatment of lignocellulosic wastes to improve ethanol and biogas production: a review. IJMS 9 (9), 1621–1651.

Tang, S.Y., Sivakumar, M., 2015. Ultrasound as a green processing technology for pretreatment and conversion of biomass into biofuels. In: Production of Biofuels and Chemicals with Ultrasound. Springer, p. 189.

Tengborg, C., Stenberg, K., Galbe, M., Zacchi, G., Larsson, S., Palmqvist, E., Hahn-Hägerdal, B., 1998. Comparison of SO_2 and H_2SO_4 impregnation of softwood prior to steam pretreatment on ethanol production. Biotechnol. Fuels Chem. 70 (72), 3–15. Springer.

Thring, R.W., Chornet, E., Overend, R., 1990. Recovery of a solvolytic lignin: effects of spent liquor/acid volume ratio, acid concentrated and temperature. Biomass 23, 289–305.

Umezawa, K., Kumabe, T., Shirane, R., Yoshimoto, T., Sun, R., Tomkinson, J., 2000. Characterization of lignins from wheat straw by alkaline peroxide treatment. Polym. Degrad. Stab. 67, 101.

Wang, F.Q., Xie, H., Chen, W., Wang, E.T., Du, F.G., Song, A.D., 2013. Biological pretreatment of corn stover with ligninolytic enzyme for high efficient enzymatic hydrolysis. Bioresour. Technol. 144, 572–578. Elsevier Ltd.

Wang, G., Zhang, S., Xu, W., Qi, W., Yan, Y., Xu, Q., 2015. Efficient saccharification by pretreatment of bagasse pith with ionic liquid and acid solutions simultaneously. Energy Convers. Manag. 89, 120–126.

William, M.B., Reese, D., 1950. Colorimetric determination of ethyl alcohol. Anal 22, 1556–1561.

Wood, B.E., Beall, D.S., Ingram, L.O., 1997. Production of recombinant bacterial endoglucanase as a co-product with ethanol during fermentation using derivatives of *Escherichia coli* KO11. Biotechnol. Bioeng. 55 (3), 547–555.

Wooley, R., Ruth, M., Glassner, D., Sheehan, J., 1999. Process design and costing of bioethanol technology: a tool for determining the status and direction of research and development. Biotechnol. Prog. 15, 794–803.

Wright, J.D., 1988. Ethanol from biomass by enzymatic hydrolysis. Chem. Eng. Prog. 84, 62–74.

Xu, L., Tschirner, U., 2012. Peracetic acid pretreatment of alfalfa stem and aspen biomass. BioResources 7 (1), 203–216.

Xue, B.L., Li, M.F., Xu, F., Sun, R.C., Jones, G., 2012. Microwave-enhanced alkali treatment of *Pinus yunnanensis*, Bagasse: physiochemical characterization of the dissolved lignins. Ind. Crops Prod. 36, 209.

Yachmenev, V., Condon, B., Klasson, A., 2009. Lambert. Acceleration of the enzymatic hydrolysis of corn stover and sugar cane bagasse celluloses by low intensity uniform ultrasound. J. Biobaased Mater. Bioenergy 3, 25–31.

Yang, B., Wyman, C.E., 2004. Effect of xylan and lignin removal by batch and flow through pretreatment on enzymatic digestibility of corn stover cellulose. Biotechnol. Bioeng. 86, 88–95.

Yang, B., Wyman, C.E., 2008. Pretreatment: the key to unlocking low-cost cellulosic ethanol. Biofuels Bioprod. Biorefin. 2, 26–40.

Yang, D., Li, X., Gao, Z., Wang, Y., 2003. Improving biogas production of corn stalk through chemical and biological pretreatment: a preliminary comparison study. Trans. Chin. Soc. Agric. Eng. 19 (5), 209.

Yang, C., Shen, Z., Yu, G., Wang, J., 2008. Effect and aftereffect of γ radiation pretreatment on enzymatic hydrolysis of wheat straw. Bioresour. Technol. 99, 6240–6245.

Yi, Z., Zhongli, P., Ruihong, Z., 2009. Overview of biomass pretreatment for cellulosic ethanol production. Int. J. Agric. Biol. Eng. 2 (3), 51–68.

Zabed, H., Faruq, G., Sahu, J.N., Azirun, M.S., Hashim, R., Boyce, A.M., 2014. Bioethanol production from fermentable sugar juice. Sci. World J. Article ID 957102.

Zeng, J.J., Tong, Z.H., Wang, L.T., Zhu, J.Y., Ingram, L., 2014. Isolation and structural characterization of sugarcane bagasse lignin after dilute phosphoric acid plus steam explosion pretreatment and its effect on cellulose hydrolysis. Bioresour. Technol. 154, 274–281.

Zhang, Y., Fu, E., Liang, J., 2008. Effect of ultrasonic waves on the saccharification processes of lignocellulose. Chem. Eng. Technol. 31, 1510–1515.

Zhang, Y., Li, J., Su, Q., Lin, Y., Huang, Z., Lu, Y., Sun, G., Yang, M., Huang, A., Hu, H., 2015. A green and efficient technology for the degradation of cellulosic materials: structure changes and enhanced enzymatic hydrolysis of natural cellulose pretreated by synergistic interaction of mechanical activation and metal salt. Bioresour. Technol. 177, 176–181.

Zheng, Y.Z., Lin, H.M., Tsao, G.T., 1998. Pretreatment for cellulose hydrolysis by carbon dioxide explosion. Biotechnol. Prog. 14, 890–896.

CHAPTER

6

Various methods of biodiesel production and types of catalysts

V. Amirthavalli[1], Anita R. Warrier[2], Baskar Gurunathan[3]

[1]Department of Petroleum Engineering, Academy of Maritime Education and Training, Chennai, Tamil Nadu, India; [2]Nanophotonics Research Laboratory, Department of Physics, Academy of Maritime Education and Training, Chennai, Tamil Nadu, India; [3]Department of Biotechnology, St. Joseph's College of Engineering, Chennai, Tamil Nadu, India

6.1 Introduction

Depletable fossil fuel resources meet 81.1% of world's energy requirement. Hydro power, nuclear energy, biofuel, and other renewable energy sources account for only 18.9%. Energy procured from depletable sources has limited reserve of around 50 years only. Urbanization and industrialization have led to increased usage of petroleum-derived fuels. The main problems associated with petroleum-derived fuels are global warming, ozone depletion, and increased emission of carbon monoxide. The energy demand and associated climatic changes have fueled the necessity for finding an alternate energy resource. Biodiesel is a renewable source of energy, best suited for direct usage in engines. It outperforms the ordinary diesel due to its reportedly high efficiency and low emissivity of unburned hydrocarbons (68%), particulates (40%), sulfur oxide (100%), carbon monoxide (44%), and polycyclic aromatic hydrocarbons (80%–90%), and also it can be easily produced and stored safely and provides better engine lubrication (Leduc et al., 2009).

Biodiesel has high flash point and high cetane number. Apart from that, it is nontoxic and biodegradable. Biodiesel contains 11% of oxygen in its chemical structure, which permits the complete combustion of fuel and lowers the emission of oxides of sulfur, carbon, and other hydrocarbons and hence protects the earth from global warming. Biodiesel can be used either by blending it with diesel or in pure form (commercially available B70 is 70% biodiesel and 30% diesel). Biodiesel blends are referred as B_{yy}, where yy refers to the amount of biodiesel in the blend. The lubricating properties of the blended fuel are enhanced due to formation of methyl esters.

Biodiesel is a fatty acid alkyl ester obtained from vegetable oils or animal fats. Vegetable oils consist of triglycerides consisting of glycerol and esters of fatty acids with different number of bonds and carbon chain. More than 300 types of animal or plant lipids serve as feedstock for

Biofuels and Bioenergy
https://doi.org/10.1016/B978-0-323-85269-2.00020-4

biodiesel generation, as they are a source of good fatty acid (Qiu et al., 2011; Wen et al., 2010; Lin et al., 2009; Pleanjai et al., 2009; Yee et al., 2009). They react with alcohol (with/without catalyst) to generate fatty acid methyl ester. The main drawback in the usage of animal fats or plant lipids is that the food chain gets disturbed. Raw materials such as feedstock and alcohol are responsible for 70%–80% cost of biodiesel (Zhang et al., 2003). Biodiesel production is hindered when the market price of edible vegetable oil starts to increase, and the profitability of the process is affected. Hence, nonedible sources such as algae, microalgae, jatropha seeds, and grease or used oil are widely used as the raw material. Usage of fuels obtained from any of these feedstocks is highly promoted due to the factors such as reduction in crude oil reserves and fluctuations in crude oil prices due to the geopolitical tension in Middle East. Oil crops, which give high oil yield, are mostly preferred in the biodiesel production since they reduce the production cost. However, large capital investment is required for mass production of feedstock, harvesting, recovering the oil, and converting it into biodiesel. Crude oil and natural gas production costs are comparatively less than that of alternative renewable fuels, which causes stagnation in the production of biofuels. Nonedible sources act as an alternative to animal fat and vegetable oil. Among the nonedible sources, jatropha, *Paullinia pinnata*, and castor oil give high yield. The yield obtained is affected by the plantation type and oil extraction techniques. Biodiesel production cost reduces by 60%–90% when waste edible oils are used as the feedstock. Researchers are trying to produce high-quality biodiesel at low cost employing different techniques to reduce the volume of alcohol, quantity of catalyst, reaction time, and temperature.

Many researchers have reported that waste cooking oil will be an ideal and eco-friendly feedstock (Mazubert et al., 2014), as it is nonedible, and when ingested by the human, it will create health problems such as diarrhea, growth retardation, obesity, increased weight of the liver and kidneys, and other biochemical reactions. When this oil is discarded, it clogs the pipelines. When waste cooking oil is released into the water bodies, it affects the aquatic and marine organisms. Therefore, instead of draining into sewage and water bodies, we can recycle them into biodiesel, which in turn reduces the water treatment cost.

6.2 Methods of biodiesel production

Biodiesel is produced using any of the commonly used methods such as follows: Table 6.1 shows the various methods of biodiesal generation at different operating conditions along with their conversion yield.

1. Direct use and mixing
2. Pyrolysis
3. Microemulsion
4. Transesterification

Using vegetable oils directly is not feasible, as their viscosity and free fatty acid content is very high. Further, the issues such as (1) gum formation due to oxidation and polymerization, (2) carbon deposition, and (3) thickening of the lubricants are major constraints. But when these vegetable oils are blended along with diesel, viscosity is reduced and can be used as fuel. By reducing the viscosity, issues such as injector choking and carbon deposits are minimized. Vegetable oil can be treated at high temperature by a process called pyrolysis (thermal cracking) and converted with the help of a catalyst in the absence of air or oxygen. Pyrolysis involves the breakage of chemical bonds to give hydrocarbon molecules such as alkanes, alkenes, alkadienes, carboxylic acids, aromatics, and small amounts of gaseous products. The resulting product is in liquid state and has chemical composition matching with that of conventional diesel (Amit, 2012).

Microemulsion technique is one of the ways in which the viscosity of the feedstock can be

TABLE 6.1 Various methods for biodiesel generation along with the conversion yield and operating conditions.

S. No.	Name of the method	Source of feed stock	Operating conditions	Biodiesel yield (%)	References
1	Microwave-assisted transesterification	Corncobs	Catalyst: oil mass is 20%, alcohol:oil is 6:1, duration: 20 min, and MW power of 0–600 W	88.7	Rocha et al. (2019)
2	Microwave-assisted transesterification	Papaya oil	Amount of catalyst: 0.95 wt%, methanol:oil is 9.5:1, reaction temperature is 62.33°C, duration: 3.30 min	99.3	Nayak et al. (2019)
3	Ultrasonic-assisted oil extraction followed by transesterification	Spent coffee grounds	Amount of catalyst is 4 wt% the of KOH, methanol:oil is 30:1, 30% ultrasonic amplitude (500 W), and duration: 3 h	97.11	Goh et al. (2020)
4	Ultrasonic-assisted esterification process	Rubber seed	Amount of catalyst: 7.50 wt% sulfuric acid, reaction temperature is 50°C, methanol:oil is 23:1, and duration: 30 min	98	Trinh et al. (2018)
5	Ultrasonic transesterification in situ transesterification	Almonds of *Syagrus cearensis*	Amount of catalyst: 5 wt% KOH catalyst, methanol:oil (molar ratio): 1:6, and duration: 30 min	99.99	Pascoal et al. (2019)
6	In situ transesterification	Industrial spent coffee grounds	Reaction temperature: 50°C, duration: 3 h	77	Tuntiwiwattanapun et al. (2017)
7	Transesterification	Sardine fish	Amount of catalyst is 1.25 wt% KOH, methanol:oil is 20:1, reaction temperature: 150°C, duration: 25 min	96.57	Anand Kumar et al. (2019)
8	Transesterification	Waste edible oil	Catalyst concentration of 4.571 wt%, methanol:oil is 16.7:1, reaction temperature: 69.37°C, duration: 7.08 h	98.37	Foroutan et al. (2020)
9	Noncatalytic supercritical methanol transesterification	Tobacco (*Nicotiana tabacum*) seed	Methanol:oil is 43:1, reaction temperature: 300°C, and duration: 90 min	92.8	Garcia et al. (2017)
10	Noncatalytic supercritical methanol transesterification	*Schizochytrium limacinum* microalgae	Reaction temperature: 543K, reaction pressure: 20 MPa, methylating agent:algae is 10:1, and duration: 40 min	>90	Rathnam et al. (2019)
11	Catalytic supercritical methanol transesterification	Jatropha seeds and Krating oil	Reaction temperature: 320°C, methanol:oil ratio is 40:1, reaction pressure:15 MPa, and duration: 5 min Reaction temperature: 260°C, 16 MPa, and 10 min, molar ratio of methanol to oil 40:1.	84.6% for jatropha oil 90.4% for Krating oil	Samniang et al. (2014)
12	Catalytic supercritical methanol transesterification	Palm fatty acid distillate	Methanol:palm fatty acid distillate is 6:1, amount of catalyst is 2 wt% of sulfated calcined angel wing shell (CAWS), reaction temperature: 290°C, duration: 15 min	98	Nur Syazwani et al. (2017)

lowered considerably. Feedstocks in combination with ester and dispersant (cosolvent) or with alcohol and surfactant with can be used to produce microemulsion. It is a transparent, isotropic, colloidal dispersion of oil, water, surfactant, and amphiphilic molecule. The size of the dispersed particles is approximately 100 nm. Due to presence of alcohol, they have low calorific value, combustion chamber remains cooler, and further nozzle choking is reduced, due to their high latent heat of vaporization (Yusuf et al., 2011).

Transesterification is one of the simplest processes that help to decrease the viscosity of the oil by transforming it to biodiesel. The rate of transesterification can be increased by using ultrasonic sound waves or high-frequency microwave irradiation or by performing both extraction and transesterification simultaneously (in situ). Considering the advantages of this technique, in this chapter, we have elaborated on the various transesterification mechanisms that are being adopted for improving the biodiesel production.

6.3 Transesterification

Transesterification is also called as alcoholysis, as it involves chemical reaction of the feedstock with alcohol and a catalyst to release glycerol and esters. Biodiesel is produced more commonly by transesterification process, which takes place in three consecutive steps. In transesterification, ester is obtained in every step, by transformation of triglycerides to diglycerides, followed by monoglycerides and eventually glycerol. The transesterification reaction can be carried out with feedstocks such as fresh or used vegetable oil, animal fat such as beef tallow, fish oil, poultry waste, and methanol. Used cooking oil is more preferred as the feedstock due to its low cost and abundant availability from various food processing units, hotels, restaurants, and street sellers.

Various types of alcohols can be used for transesterification process; however, due to the low cost of methanol; it is normally preferred in the transesterification reaction. Being an equilibrium reaction, excess amount of alcohol is required to displace and complete the reaction to yield fatty acid methyl esters. The reaction rate and yield of biodiesel can be enhanced by addition of catalyst. The quantity of the catalyst required for the reaction depends on the pH of the feedstock.

The entire transesterification process can be specified in three consecutive and reversible steps given in Eqs. (6.1–6.3).

$$\text{Triglyceride} + \text{ROH} \overset{\text{Catalyst}}{\rightleftharpoons} \text{Diglyceride} + \text{R}^{\text{I}}\text{COOR} \tag{6.1}$$

$$\text{Diglyceride} + \text{ROH} \overset{\text{Catalyst}}{\rightleftharpoons} \text{Monoglyceride} + \text{R}^{\text{II}}\text{COOR} \tag{6.2}$$

$$\text{Monoglyceride} + \text{ROH} \overset{\text{Catalyst}}{\rightleftharpoons} \text{Glycerol} + \text{R}^{\text{III}}\text{COOR} \tag{6.3}$$

In Eqs. (6.1–6.3), R^{I}, R^{II}, and R^{III} are long-chain hydrocarbons, normally fatty acids, and R is an alkyl group. Transesterification reaction can be used to produce biodiesel in few liters (laboratory scale) to few million liters.

There are several transesterification processes such as noncatalytic supercritical methanol transesterification, catalytic supercritical methanol transesterification, ultrasound- and radio frequency–assisted transesterification, microwave irradiation–assisted transesterification, and in situ transesterification techniques.

6.3.1 Noncatalytic supercritical alcohol transesterification

Supercritical methanol is considered to be an ecofriendly process for transesterification even without catalyst. Supercritical fluid is defined as a fluid, which exists beyond the critical temperature and pressure, where distinct liquid phase and gas phase do not exist. Methanol has critical temperature and pressure of 512.2K and 8.1 MPa; in this state, transesterification occurs very quickly. Kusdiana and Saka have reported that 95% conversion can be achieved for rapeseed oil at a temperature of 350°C, 30 MPa, at methanol:oil as 30:1 in a duration of only 4 min (Kusdiana et al., 2001). In the supercritical state, dielectric constant of methanol reduces, which gives hydrophobic nature of the system. At the supercritical methanol condition, the conversion efficiency remains unaffected by the presence of water. It has been found that enhanced yield is obtained at temperature lower than 300°C and pressure greater than 8.09 MPa.

The reaction of feedstock with supercritical alcohol, which occurs at high reaction temperature of 222–400°C and pressure of 80–650 bar in reaction time of 222 s to 6 min, results in a yield of 95%–100%. The main advantages of this process, due to the absence of catalyst, are as follows: washing and neutralization are not needed, and further products and by-products can be separated easily after the transesterification process. The main disadvantages of this process are that costlier apparatus is required to maintain high reaction temperature and pressure. Several researchers have worked on the production of biodiesel using supercritical alcohols (Santana et al., 2012; De Boer et al., 2011).

Biodiesel can be produced from castor oil without the catalyst using supercritical methanol. Yield of 96.5% is obtained at reaction temperature of 300°C, 21 MPa, and duration of 90 min (Roman-Figueroa et al., 2016). Under these conditions, the yield of intermediate is very low, and fatty acid methyl ester does not undergo thermal decomposition. However, for temperature of >350°C and reaction time of ~90 min, rapid decomposition has been reported. Methyl ricinoleate fatty acid, which is mostly available in castor oil, will be unstable at 300°C and 60 min, leading to decreased yield.

Biodiesel can be produced from tobacco seeds (*Nicotiana tabacum*), with maximum yield (92.8%) at 300°C, with methanol:oil at 43:1 in 90 min (Garcia-Martinez et al., 2017). Thermal decomposition of biodiesel is seen above reaction temperature of 325°C and duration of 60 min. Transesterification of jatropha oil using supercritical methanol without a catalyst results in 100% yield at 593 K, 8.4 MPa, 43:1 (methanol: oil) in 4 min (Hawash et al., 2009). Palm olein oil on converting into biodiesel shows a yield of 94.96% at a temperature of 350°C, 35 MPa, 40:1 (methanol: oil), and 20 min (Choi et al., 2011). Oil extracted from microalgae under noncatalytic supercritical methanol-assisted transesterification (at 240–300 °C, in 15–45 min and 15: 1 to 45:1 (methanol: oil)) has 99.16% conversion efficiency (Srivastava et al., 2018). Biodiesel obtained from microalgae oil is unstable at temperature more than 300°C and duration time of 60 min.

6.3.2 Catalytic supercritical methanol transesterification

In this technique, feedstock transforms into biodiesel by the addition of catalyst and supercritical methanol. CaO has high activity in sub- and supercritical conditions, whereas it has weaker catalytic activity at ambient temperature. On increasing the amount of CaO from 1% to 3%, there was an increase in transesterification rate, and upon increasing the reaction temperature to 525K, reaction ended within 6 min (Demirbas et al., 2007). *Jatropha curcas* seed on reaction with supercritical methanol and calcium lanthanum mixed oxide ($CaO-La_2O_3$) (at 240°C, 8.2 MPa, methanol:oil of 21:1, 1 wt% of catalyst in

60 min) gives an efficiency of 93% (Teo et al., 2015). Calcium lanthanum mixed oxide catalyst reduces the reaction temperature, pressure, and duration of supercritical conditions. Decreased concentration of calcium lanthanum mixed oxide is sufficient to give high yield in a shorter duration with the help of supercritical conditions. Biodiesel yield of more than 97% is obtained at a reaction temperature of 250°C, 23:1 (methanol: oil), 1.0 wt% CH_3ONa, and 20 min from soybean oil (Zeng et al., 2017). $Cs_{2.5}PW_{12}O_{40}$ (Cs-doped heteropolyacid) during transesterification shows enhanced stability in supercritical methanol. FAME yield of around 92% was achieved at 260°C, 20 MPa, 40:1(methanol: oil), 3% of catalyst, and duration of 40 min. Supercritical methanol makes up for the low reaction rate of solid acid catalytic transesterification, whereas Cs-doped HPA catalyst reduces the severe operation conditions of the supercritical methanol process.

Yoo et al. have widely studied the supercritical methanol-assisted transesterification using various metal oxide catalysts. They have reported that SrO and CaO get converted into strontium methoxide and calcium methoxide during the reaction, whereas ZnO shows improved activity in a duration of 10 min (Yoo et al., 2010). A yield of 95% was obtained from palm oil when the catalyst used in this technique was supported by solid base alumina (CaO/KI/-alumina) (Asri et al., 2013). Transesterification of *Madhuca indica* oil at 523K, 10 MPa, 40:1 (methanol:oil), and 1 wt% of ZnO catalyst shows a conversion of more than 90% in 50 min, as activation energies for ZnO catalyst-assisted transesterification are lower when compared with that of a noncatalytic supercritical methanol-assisted transesterification (Lamba et al., 2019). This technique has faster reaction rate and requires comparatively low reaction temperature and less amount of catalyst.

6.3.3 Ultrasonic wave–assisted transesterification

Biodiesel production cost is much higher in conventional transesterification methods, as they require mechanical stirring to be carried out at high temperature (70–200°C) and pressure (6–10 atm) for duration of 70 h to achieve a conversion of around 90%–95%. By using nonconventional approaches such as ultrasonication, due to their energy efficiency, cost can be reduced. Ultrasonic wave irradiation promotes three types of reactions such as (1) acoustic mixing, (2) rapid movement of fluids, and (3) formation of cavitational bubbles. These ultrasonic waves generate large heat (>5000K) and pressure (>1000 atm) energy that influences the rate of reaction, mass transfer, and active catalytic surface area (Yu et al., 2010). Comparatively higher yield is obtained by ultrasonication due to complete miscibility of oil and alcohol, formation of smaller droplets, increase in contact area, and formation of new active sites in the catalyst.

The highlighted advantages of this process are reduced reaction time, alcohol-to-oil ratio, energy requirement, amount of catalyst required, and increased reaction rate and conversion efficiency. The simple and cost-effective setup along with ease of separation and purification process has made this technique highly suitable for large-scale production. The main drawbacks of this process are high heat generation when compared with conventional process and maintenance of the ultrasonic power and frequency, as it may lead to formation of soap. Hence, yield will be affected for frequencies greater than 40 kHz.

Using this technique, biodiesel (yield of 95%) is obtained from palm oil by reducing the amount of catalyst and duration to 2.8% wt and 50 min (Salmatinia et al., 2010). By assisting the ultrasonic transesterification with vibration, the reaction rate and enzyme activity can be enhanced, as found in the case of biodiesel generated from soybean oil using an immobilized form of Novozym 435 as catalyst (Yu et al., 2010). By optimizing the parameters such as water content, solvent-to-oil ratio, power, and temperature can enhance the reaction efficiency of 96% (Stavarache et al., 2007; Singh et al., 2006; Hanh et al., 2008; Kelkar et al., 2008; Thanh et al., 2010).

Figs. 6.1 and 6.2 depict pictorial representation of the experimental set up and flowchart of the ultrasonic-assisted transesterification process, respectively.

6.3.4 Microwave-assisted transesterification

Microwave irradiation is considered to be a standard method, which is used to accelerate the chemical reaction as it delivers the energy directly to the reactant. Heat transfer is more powerful than conventional heating, and the transesterification process can be completed in much shorter reaction time. This technique does not require any intermediate heating phase, as the temperature of the polar solvent increases rapidly, and this method is more preferred when compared with other warm up methods. Yield of biodiesel achieved using microwave irradiation is higher when compared with other methods since the reactants are heated using thermal energy in the form of conduction, convection, and radiation from the reactor surface. Polar/nonpolar molecules on absorbing the microwave energy accelerate the chemical reaction rate, and microwave energy is considered to be nonionizing radiation. Molecular movements are affected due to the microwaves, but it cannot alter the molecular structure. Boiling point will be easily attained when low-molecular-weight and high-dielectric-constant polar solvents are used.

Microwaves, during their interaction with molecules, rise the temperature without

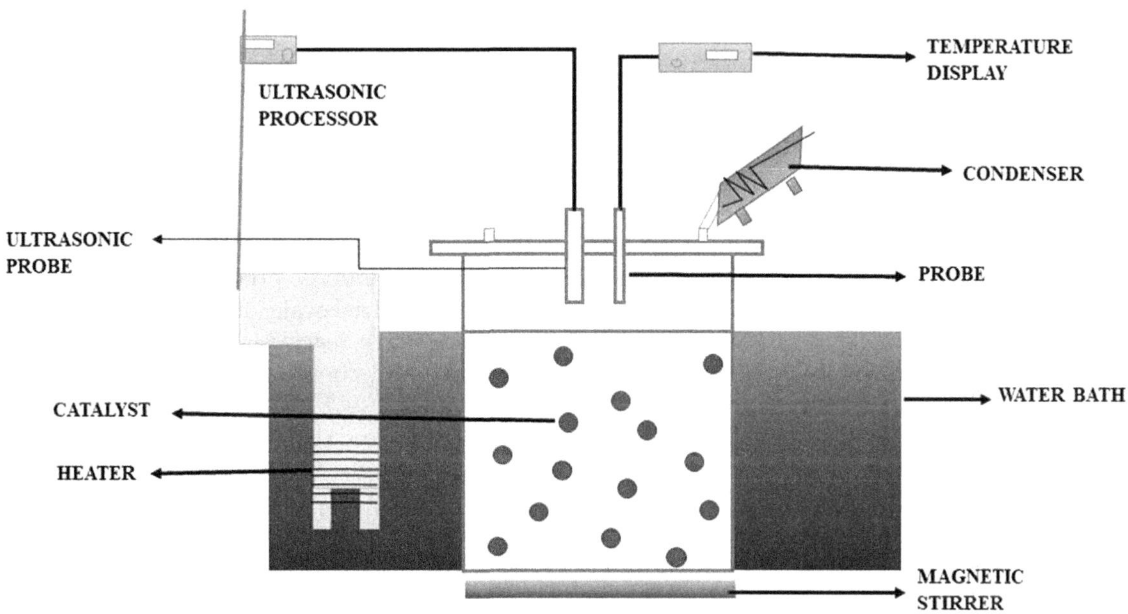

FIGURE 6.1 Pictorial representation of the experimental setup used for ultrasonic-assisted transesterification.

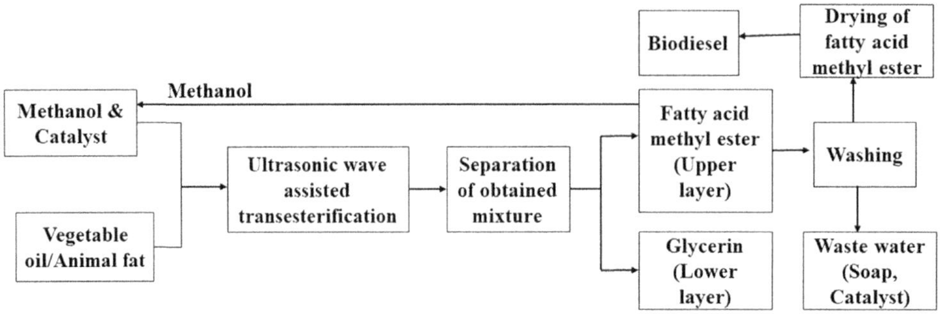

FIGURE 6.2 Flowchart of the ultrasonic-assisted transesterification process.

breaking the bonds. The main advantages of this microwave-assisted transesterification process are that it is ecofriendly and requires only minimum energy, low reaction time, less amount of solvent, decreased by-products formation, and more effective heat transfer when compared with conventional modes such as conduction, convection, and radiation. Preheating of the reactants is not needed since microwave directly supplies the energy. Microwave-assisted transesterification is faster and gives enhanced yield and reaction time, which is 10 times shorter when compared with other conventional methods. Biodiesel is produced from waste palm oil using sodium hydroxide as catalyst with a domestic microwave (800 W) in a continuous transesterification reaction unit and has a conversion efficiency of ~97% while using 12:1 methanol-to-oil ratio. The reaction attains complete conversion of 100% in a duration of only 10 s when it is carried out with a 9:1 methanol–to-oil ratio.

Microwave heating is a simple and fast method for the production of biodiesel from various renewable feedstocks in batch process (Koberg et al., 2011; Tangy et al., 2016). Microwave heating enhances the rate of reactions and improves the selectivity. The time involved for the conversion is decreased due to the enhanced interaction between the molecules in the reaction medium due to electromagnetic field. The efficiency of microwave process enhanced due to high polarity and high dielectric loss of methanol in transesterification reaction. The microwave irradiation improved the rate of reaction by thermal effect and methanol evaporation. The strong microwave interaction with triglycerides and methanol results in decreased activation energy due to the improved dipolar polarization. Transesterification reaction using a homogenous catalyst under microwave irradiation in a continuous reactor conditions gives high yield or conversion in reduced time. For a continuous-flow process, microwave irradiation serves as the possible method to accelerate the transesterification reaction. This technique gives large quantities of biodiesel in a reduced reaction time. Implementation of this technique in large scale is difficult due to the fact that uniform heating at industrial scale is not possible without internal mixing. Disadvantages of microwave processes can be decreased by using continuous process approach as a possible industrial-scale strategy. The continuous microwave reactors are a combination of microwave and continuous flow enabled technique and serve as a new platform for industrial study. These reactors operate by passing the reactants through a flow cell present in the microwave cavity. The drawbacks of these continuous flow reactors are difficult to process solids, highly viscous liquids, or heterogenous reaction mixtures. Bowman et al. have reported that it will be time-consuming to adapt to the conditions from small scale to industrial scale (Bowman et al., 2008). A report compares the production of biodiesel under the conventional heating and microwave irradiation using two different catalysts NaOH (homogenous) and Ba $(OH)_2$ H_2O (heterogenous) (Mazzocchia et al., 2004). Both the catalysts gave similar results on the usage of microwave irradiation. Yield and activity were higher when compared with conventional heating methods.

Biodiesel is produced from waste palm oil using microwave with a power of 800 W using sodium hydroxide as catalyst; 97% conversion efficiency is obtained with ethanol-to-oil ratio of 12:1 for a reaction time of 30 s; and 100% conversion is obtained with ethanol-to-oil ratio of 9:1 and 10 s of reaction time (Lertsathapornsuk et al., 2008). Microwave irradiation needs reaction time, which is 10 times shorter than other conventional methods. Yield of ~88% is obtained for both jatropha oil and waste cooking oil while using microwave irradiation, with methanol-to-oil ratio of 12:1, 1% of NaOH, 7 min of reaction time, and temperature ~65°C (Yaakob et al., 2009). Reports on production of biodiesel from waste cooking oil using NaOH and CH_3ONa using microwave irradiation suggest that fatty acid methyl ester yield is higher for CH_3ONa when compared with NaOH when same weight percentage of catalyst is used for the reaction. It has also been confirmed that increasing the microwave irradiation power to 750 W and reaction time from 1 to 3 min enhances the yield of biodiesel with 6:1 methanol-to-oil ratio and 0.75 wt% of CH_3ONa catalyst amount (Chen et al., 2012). However, reports on the production of ethyl ester using 0.5% of NaOH, 100% excess ethanol, and 750 W microwave irradiation suggest that 4 min is sufficient for the completion of the reaction when compared with 75 min for conventional method (Dias et al., 2008). In this technique, quantity of by-products and separation time is reduced. It enables a comfortable, safe, and clean way of working with chemical reactions. Due to this property, it is being used in many areas such as assisting transesterification of vegetable oils in the production of biodiesel.

6.3.4 In situ transesterification

Biodiesel produced from in situ transesterification is quite costlier than biodiesel produced by conventional transesterification, since large quantity of methanol is required for the process. The yield of ester production from in situ transesterification method using sunflower seeds and excess methanol in the presence of sulfuric acid as catalyst is 20% greater when compared with transesterification using preextracted oil. Biodiesel is produced by in situ transesterification of cottonseed oil using alkaline catalyst. The effect of moisture content and catalyst concentration involved in the conversion of triacylglycerols into biodiesel in a nontoxic cottonseed meal is studied. The conversion of oil to methyl esters shows a steep increase from 80% to 98%, as the moisture content is decreased from 8.7% to 1.9%. Similarly, drastic increase in conversion of methyl esters is seen from 33% to 98% when the alkaline catalyst concentration is increased from 0.05 to 0.1 mol/L. In in situ alkaline-catalyzed transesterification when carried out with sunflower seeds and NaOH as catalyst, a yield of 96% methyl esters when methanol is used and 90% of yield of ethyl esters when ethanol is used. The mixture is refluxed and stirred using either mechanical stirrer or ultrasonicator with low frequency.

Yield is obtained from in situ transesterification of sunflower seeds using acid catalyst and methanol and, the obtained yield is compared with that of conventional transesterification. Crude methyl esters yield obtained from in situ process are 38.9% when compared with the yield of 33.2% from conventional treatment methods. In situ transesterification is carried out using soybean oil, methanol, ethanol, n-propanol, and n-butanol, and sulfuric acid as the catalyst for the reaction. Ethanol and higher alcohols give improved yield of alkyl esters for soybean oil when compared with methanol. 80%–85% of conversion is obtained at reaction temperature of 78°C and reaction time of 3 h.

Biodiesel can be produced from rape seed by alkaline transesterification. Koutsouk et al. have studied about in situ and conventional transesterification under sonication without using external heat and the results were compared with mechanical stirring. Improved conversion is obtained with in situ transesterification due to ultrasonication and mechanical stirring in a duration of 120 min. Biodiesel yield of around 37% is obtained when only mechanical stirring is carried out, whereas 80.6% yield is obtained from ultrasonication at 7.5% NaOH w/w of oil, methanol-to-oil molar ratio of 400:1 (Koutsouki et al., 2016).

Biodiesel is obtained from almonds of *Syagrus cearensis* using ultrasonic-assisted in situ transesterification. Conversion efficiency of around 99.99% is obtained when ultrasonication is combined along with in situ transesterification using 5 wt% catalyst (KOH), 1:6 methanol:oil (molar ratio), and duration time of 30 min (Pascoal et al., 2019). Biodiesel produced by in situ transesterification of cotton seed oil shows that a conversion of 98% at 40°C, 13:1 methanol-to-oil ratio, 0.1 mol/L of NaOH in a duration of 3 h, and moisture content in the cottonseed flours should be less than 2% (Qian et al., 2008). Biodiesel yield of around 95% is obtained by conventional alkaline transesterification along with ultrasonication and methanol in a duration of 20 min, which is similar to the conventional alkaline transesterification with mechanical stirring produced from sunflower seed oil. In situ transesterification along with methanol gives same yield as that of conventional alkaline transesterification. When ethanol is used instead of methanol in in situ alkaline transesterification along with ultrasonication, it gives yield of 98% in 40 min of duration time, whereas conventional alkaline transesterification with mechanical stirring gives a yield of 88% in 4 h. Methanol gives improved yield in a short duration of time when compared with ethanol (Georgogianni et al., 2008).

In situ transesterification and conventional alkaline transesterification of *Cynara cardunculus* L. seed oil are carried out with methanol employing ultrasonication and mechanical stirring. Increased fatty acid content of 96% and 93% is seen for in situ transesterification with ultrasonication and mechanical stirring with 550:1 methanol-to-oil ratio, 9.5% NaOH, and duration of 20 min, whereas conventional transesterification with ultrasonication and mechanical stirring shows fatty acid content of 97% in 20 min and 95.8% in 1 h with 7:1 methanol-to-oil ratio and 1% (w/w) oil NaOH. Biodiesel yield obtained from in situ–assisted transesterification along with mechanical stirring (50.4%) is less when compared with that of ultrasonication (85.1%) (Koutsouki et al., 2015).

6.4 Significance of catalyst

Catalyst accelerates the rate of the reaction without being consumed in the transesterification reaction. Catalyst is generally classified as acid, base, bio-, and nanocatalyst, which accelerates the conversion of feedstock into biodiesel using methanol. Acid and base catalyst are further divided into homogenous and heterogenous.

6.4.1 Acid catalyst

Sulfuric acid, hydrochloric acid, phosphoric acid, and sulfonated organic acids are the most prevalent homogenous acid catalysts. For the nonedible oil and used cooking oil, acid catalyst is preferred since their acid value is considered to be high. The drawbacks of acid catalyst are that it requires more volume of alcohol, high reaction temperature and pressure, low reaction rate, corroding and environmental problems, and the overall cost of the process increases due to the need for reactors to be made of special materials.

Research carried out in this field from feedstocks with high acid value uses homogenous acid catalyst by two-step esterification (Cao et al., 2008). Esterification is carried out to reduce acid value using liquid acid (catalyst) and methanol, and the feedstock with reduced acid value further undergoes transesterification with methanol in the presence of a base catalyst. The main advantages of acid catalyst over base catalyst are that the cost of acid catalyst is lesser.

Heterogeneous acid catalyst does not disintegrate in the solution containing feedstock and methanol; hence, it is easier to separate the solution by filtration and can be reused. These catalysts are more suitable for both esterification and transesterification reactions. Heterogeneous acid catalyst is further classified into liquid acid catalyst and solid acid catalyst such as zirconium oxide (ZrO_2), sulfated oxides, and cation-exchange resins, which have the ability to replace liquid acid catalyst. Being nonreactive to free fatty acid content, the solid acid catalyst can accelerate the reaction; hence the washing step of the product is excluded. Due to the easy separation, reduced corrosion problems even in the presence of acid species, regeneration, and reusage of solid acid catalyst are easier in contrast to liquid acid catalyst (Dalai et al., 2006).

6.4.2 Base catalyst

Base catalyst possesses the following advantages over acid catalyst: ability to complete the transesterification at higher speed, decreased reaction temperature and pressure with less volume of methanol, and high conversion efficiency. Base catalyst has few side reactions, and it does not undergo oxidation. They are readily available at low cost. The rate of transesterification reaction is very much fast when base catalyst is used. Base catalyst gives good result when the acid value (free fatty acid content) of the source is less than two. The disadvantages of base catalyst are that it does not give good results when water is present in the source and acid value of the source is high. Base catalyst is highly susceptible to water content, and it causes formation of soap, which in turn makes the separation difficult.

Homogenous base catalysts such as potassium hydroxide, sodium hydroxide, sodium methoxide, and potassium methoxide serve as effective catalyst, but removal of these catalysts from methyl esters requires more washing. Hence, lot of water, energy, and time are used, and it cannot be recycled. Moreover, obtained biodiesel is contaminated with metal ions (Alaei et al., 2020). Most widely used heterogeneous base catalysts are alkaline earth metal oxides such as MgO, BaO, SrO, CaO, zeolites, and KNO_3 loaded on Al_2O_3, KNO_3/Al_2O_3, BaO, SrO, CaO, and MgO (Baskar and Aiswarya, 2016). Heterogenous catalyst can carry out both esterification and transesterification simultaneously. Since heterogeneous base catalysts are insoluble in the solution, i.e., they do not mix with methanol or ethanol, it is easier to separate and reuse in the transesterification reaction many numbers of times, and obtained biodiesel quality is high. Heterogeneous base catalysts are environmentally friendly and find usage in continuous and noncontinuous processes, and it shows less sensitivity to FFA, as the surface microcrystal structure is stabilized (Yan et al., 2010; Bournay et al., 2005).

MgO and CaO are widely used catalysts in the biodiesel production due to low cost and require mild reaction temperature (Zhang et al., 2003; Issariyakul et al., 2007; Mahdavi et al., 2014). MgO is considered to be the first heterogeneous base catalyst, which is used in the biodiesel production in the pilot scale (Umdu et al., 2009). Various researchers reported that MgO shows high potency in noncontinuous reactions. Similarly, CaO is more commonly used basic heterogenic catalyst because of its improved basicity and low solubility in methanol. CaO can be synthesized using lime stone and calcium hydroxide, as they are inexpensive and contribute less to the environmental pollution. MgO and CaO catalysts are loaded on Al_2O_3 (Wu et al., 2013) and ZSM-5 zeolites as supports, and these supports also take part in the reaction. The surface area of these supports helps in the uniform distribution of the active phase. These ceramic supports are preferred for the catalysis of biodiesel due to their large surface area, high mechanical strength, and improved thermal resistance (Keane et al., 2003). Various researchers have started to focus biodiesel production using spinels such as $CoFe_2O_4$ (Zhang et al., 2014), $SrAl_2O_4$ (Mierczynski et al., 2015), $ZnAl_2O_4$ (Liu et al., 2013), $ZnFe_2O_4$ (Sankaranarayanan et al., 2013), $Zn_xCu_{(1-x)}Al_2O_4$(Hashemzehi et al., 2016), and $MgAl_2O_4$.

CaO, when used as a catalyst in the production of biodiesel, gives FAME yield of 98% in the first cycle of reaction (Demirbas et al., 2005). Since reusability is difficult, it has been modified into organometallic in nature, e.g., $Ca(OCH_3)$ or $Ca(C_3H_7O_3)_2$, which shows high reusability. Waste materials such as eggshell, mollusc shell, bones, and so on can be used as feedstock for the synthesis of CaO, which in turn is used in the production of biodiesel. The usage of these waste materials will reduce the wastage, and it is a very cost-effective way to produce the catalyst CaO. SrO is also used as a heterogenous catalyst for transesterification reaction (Yoo et al., 2010) and can be reused 10 times.

6.4.3 Biocatalyst or enzyme catalyst

Biocatalyst or enzyme catalyst can be used for feedstocks with high free fatty acid content to undergo transesterification. 90% of oil can be converted into biodiesel. Problems involved during purification, washing, and neutralization are eliminated since soap is not formed during their usage. The main disadvantages of this catalyst are that it is expensive and it requires longer reaction time (Leung et al., 2010). Various catalysts such as lipozyme, novozym 435, lipozyme TLIM, and lipozyme RMIM show excellent results in biodiesel production. Lipase from *Pseudomonas fluorescens* has been reported to give an enhanced yield of more than 90% biodiesel.

6.4.4 Nanocatalyst

Nanomaterials are either natural or human-made with a size of 100 nm or less and find applications in fields of medical devices, optoelectronics, drug delivery, Q3 electronics, chemical catalysts, and cosmetics. Recently, nanocatalysts are gaining importance due to their high catalytic activity, easy to separate after the reaction, and high surface-to-volume ratio. It has good reusability and simple operational procedure, and it contaminates the environment at a minimal level.

Among various heterogeneous catalysts, nanocatalysts are used recently in the production of biodiesel. Various researches are being conducted in the field of biodiesel using nanocatalyst (Amirthavalli and Warrier, 2019). Nanocatalysts possess increased surface area and porosity, which are important properties of the catalyst to get attached with the underlayer, thereby improving the effectiveness of transesterification reaction. Nanoparticles/nanocomposites as catalyst in the transesterification reaction are preferred due to their nanodimension and possess large specific area, and also they show enhanced selective catalytic activity. Biodiesel is produced from soybean oil as the feedstock using various alkali metal oxides such as MgO/SiO_2, CaO/SiO_2, and BaO/SiO_2. Biodiesel is produced using CaO/Au nanoparticles as a heterogenous catalyst (Bet-Moushoul et al., 2016).

6.4.4.1 MgO nanocatalyst

MgO has been used by many researchers in the biodiesel production due to its ease of synthesis, ready availability, cost-effectiveness, and high yield. MgO possesses high basicity and mild solubility in methanol and is eco-friendly; due to these properties, it is used in transesterification process. MgO is considered to be a wide bandgap semiconductor and finds applications in various fields such as catalysis. Biodiesel production cost decreases in a discontinuous reactor using nano-MgO. MgO shows an enhanced yield in critical conditions with increased proportion of methanol-to-oil ratio. Nano-MgO shows enhanced yield for soy oil transesterification with supercritical methanol. Wang et al. have reported biodiesel generation from soybean oil in the presence of nano-MgO (Wang et al., 2007). MgO nanocatalyst does not catalyze the transesterification reaction alone due to its weak basic affinity, whereas when it is used in combination with nano-CaO, there is an enhancement in the basic properties and can be used as a suitable base for the catalyst so that there is an increase in CaO contact surface and hence increase in transesterification yield. A yield of 98.95% of biodiesel is achievable in a duration of 6 h, with 1:7 methanol-to-oil ratio and amount of nano-CaO and nano-MgO being 0.7 and 0.5 gm, respectively. CaO–MgO nanocatalyst shows improved reusability than pure nano-CaO (Mahdavi et al., 2014). When the transesterification reaction is carried out using cotton seed oil with 14.4 wt%, CaO–MgO catalyst loaded on Al_2O_3 at 95.63°C reaction temperature, with 12.24 ethanol-to-oil ratio, is able to obtain a yield of ~97.62% (Tahvildari et al., 2015). The improved yield is obtained from transesterification reaction when nano-MgO is used along with nano-CaO.

6.4.4.2 CaO nanocatalyst

Waste shells of egg, oyster, mud crab, mussel, and shrimp when used in the biodiesel production give good yield only when catalyst is calcinated prior to usage in transesterification process. Calcite exists in nature in the form of CaO and consists of calcium carbonate and traces of alumina, silica, and ferrite. It finds applications in reforming and gasification since it is inexpensive and ecofriendly and has high basicity (Suppes et al., 2001; Yoosuk et al., 2010). Dolomite is made up of calcium carbonate and magnesium carbonate when it is decomposed in two steps at temperature between 350–550°C and 825°C to obtain CaO and MgO (Ngamcharussrivichai et al., 2007; Maes et al., 1997).

The transesterification process of canola oil to produce biodiesel is studied using two types of calcium oxide nanopowders, i.e., nanopowders that exhibit high surface area (HAS nano-CaO) and moderate surface area (nano-CaO) (Tahvildari et al., 2015). HAS nano-CaO shows enhanced activity than moderate surface area (nano-CaO) due to large surface area. 99.85% yield of biodiesel is obtained when 9:1 methanol-to-oil molar ratio and 3 wt% of the moderate surface area (nano-CaO) are used in a duration of 2 h. When HSA nano-CaO catalyst is used in the transesterification, the amount of catalyst/oil weight ratio needed to attain the same yield under the same set of conditions is 10 times less than moderate surface area (nano-CaO). Biodiesel yield of 96% is obtained from waste cooking oil using CaO nanocatalyst at optimized conditions of 8:1 methanol-to-oil molar ratio, 1 wt% of CaO nanocatalyst, 50°C reaction temperature, and 90 min of reaction time (Degfie et al., 2019). Even though CaO is reported to be effective in carrying out transesterification process to produce biodiesel, it has the main drawbacks such as absorption of water and carbon dioxide that leads to the difficulty in separation due to enhanced viscosity, which retards the kinetics of the reaction (Bet-Moushoul et al., 2016).

6.4.4.3 ZnO nanocatalyst

ZnO nanocatalyst shows improved catalytic activity for improved biodiesel production (Varghese et al., 2017). The main drawback is that the separation of ZnO catalyst is tedious and this becomes a difficult task in the biodiesel industry. Copper-doped zinc oxide (CZO) heterogenous nanocatalysts have been used in the production of biodiesel from neem oil, which is considered to be a nonedible source of feedstock (Gurunathan and Ravi, 2015). Biodiesel yield of 97.18% is obtained by transesterification of neem oil at a reaction temperature of 55°C using 10:1 methanol-to-oil molar ratio and 10% (w/w) CZO nanocatalyst in a reaction time of 60 min. Biodiesel is produced from the cheapest feedstock like waste cooking oil using copper doped zinc oxide nanocomposite as a heterogeneous catalyst which gives a yield of 97.71% (w/w) for the reaction temperature of 55°C, 1:8 oil to methanol molar ratio, 12% (w/w) amount of nanocatalyst in a duration of 50 min. Zn-doped CaO nanocatalyst is used for transesterification of waste cooking oil conversion efficiency of 96.74% using 20:1 methanol-to-oil molar ratio, 5 wt% of catalyst loading, 65°C reaction temperature, and reaction time of 4 h (Borah et al., 2019).

Recently, many researchers have started to focus on bifunctional heterogenous catalysts that accelerate esterification of FFA and transesterification of triglycerides into biodiesel. The drawback of the other catalysts can be overcome by using heterogenous catalyst with both acidic and basic sites. The main important feature of bifunctional catalyst is that it can act as acid and base simultaneously. These catalysts can be modified to bring out the needed physicochemical properties so that FFA or presence of water does not influence the reaction steps during the transesterification process. The performance of the heterogenous catalyst can be increased by doping and nanocomposites and also shows improved catalytic activity and reusability in transesterification reaction. The main problem in the implementation of production in the industrial scale is the high production cost; almost 70%–95% of the production cost is contributed by the raw material cost (Zhang et al., 2003). The production cost can be decreased by the usage of inexpensive materials as the source and catalyst (Azocar et al., 2010). Table 6.2 shows various nanocatalyst used for production of biodiesel from different feed stocks and their operating conditions.

6.4.5 Nanocomposites

It has been reported by various researchers that composite metal oxides have high efficiency and reusability in comparison with bulk metal oxides. CaO/MgO composite is a nanocomposite used in biodiesel generation. The diatoms skeleton offers large surface area with high porosity. Biodiesel yield of around 96.47% is obtained using 6 wt% of catalyst, 1:15 (oil: methanol) at 90°C (Shaban et al., 2019). This nanocomposite acts as an efficient heterogenous catalyst, which gives high output in a cost-effective way. Metal oxides in the chemical structure of the nanocomposite impact the alkalinity. Calcium and magnesium improve the negative charges on the oxygen atom, which enhances the catalytic activity and stability.

Mg/Al/Zn-based hydrotalcite/SBA-15 has enhanced catalytic efficiency and stability due to addition of hydrotalcite. Nanocomposite shows a constant catalyst performance when used for more than 200 h without any deactivation of the catalyst. Nanocomposite catalytic efficiency is connected with the basic site density, which can be inferred from TPD graph. Prabu et al. (2019) have reported a yield of 90% when MAZ-1/SBA-15 and 66% when MAZ-1.5/SBA-15 used. Increasing the content of Mg or decreasing the content of Zn shows a reduction in the catalytic activity due to formation of AlO structure on the surface of Mg–O pairs, which inhibits the concentration of surface O^{2-}, Lewis basic sites. SBA-15 stands for Santa Barbara Amorphous-15 which is considered to be a highly stable mesoporous silica sieve structure. The highest yield shown by MAZ-1/SBA-15 nanocomposite gives surface basicity, which is due to more MgO getting dissolved in the surface

TABLE 6.2 The operating conditions and catalyst used in the production of biodiesel.

S. No.	Nanocatalyst	Source	Temperature (°C)	Alcohol-to-oil ratio	Reaction time (min)	Catalyst amount wt%	Biodiesel yield (%)	References
1	Cs_2O loaded onto a nanomagnetic core	Tannery waste	65	21:1	300	7	97.1	Booramurthy et al. (2020)
2	Cu-impregnated TiO_2	Palm oil	45	20:1	45	2	90.93	De et al. (2020)
3	Zinc-doped calcium oxide	*Calophyllum inophyllum* oil	55	9:1	80	6	89	Naveenkumar et al. (2019)
4	Calcium oxide (CaO) nanocatalyst	Used cooking oil	50	8:1	90	1	96	Degfie et al. (2019)
5	Sodium oxide impregnated on carbon nanotubes	Used cooking oil	65	20:1	180	3	97	Ibrahim et al. (2020)
6	$TiO_2/PrSO_3H$ solid acid nanocatalyst	Used cooking oil	60	15:1	540	4.5	98.3	Gardy et al. (2017)
7	Boehmite nanoparticles-chlorocholine hydroxide (BNPs-CCH)	Soybean oil	60	11:1	264	4.13	95.2	Hosseini et al. (2019)
8	CaO nanocatalyst	Soybean oil	60	11:1	120	3.675	97.61	Bharti et al. (2019)
9	$MgO/MgAl_2O_4$ nanocatalyst	Sunflower oil	110	12:1	180	3	95	Rahmani et al. (2017)
10	Magnesium oxide nanocatalyst	Used cooking oil	65	24:1	60	2	93.3	Ashok et al. (2018)
11	Nanomagnetic bifunctional catalyst (RHC/K_2O-20%/Ni-5%)	Rice husk	65	12:1	120	4	98.2	Hazmi et al. (2021)
12	Magnesium oxide nanocatalyst	Goat fat	70	12:1	180	1	93.12	Rasouli et al. (2019)
13	CaO nanocatalyst	Sugar beet agroindustrial residue	75	4.5:1	60	1	93	Abdelhady et al. (2020)

bringing about the basicity. Zn^{2+} ions in the Mg^{2+} lattice result in network of acidic and basic centers, which leads to increase in the amount of Lewis basicity. The increase of Zn or Mg content reduces the yield due to the formation of more surface acidic and basic sites. Here, SBA-15 alters the selectivity of fatty acid methyl ester and shows good dispersion of catalytically active centers and mass transfer. Comparatively enhanced conversion, selectivity, and stability are observed in MAZ-1/SBA-15. SBA-15 provides a way for SBO components to move to the active sites, which are available at the surface of MAZ-1, hence improving the selectivity.

Zinc-doped calcium oxide shows enhanced catalytic activity and stability due to the improved surface area, which assists in the conversion of feedstock into biodiesel. $CaO-Al_2O_3$-mixed oxides (calcium aluminate) exhibit improved pore size, basicity, and activity in the production of biodiesel. ZnO considered to be a heterogenous catalyst is readily available, nontoxic and plentiful in nature. $ZnO-SiO_2$ nanocomposite shows improved specific surface area, catalytic activity, and selectivity. Nanocomposite increases the product quality in the transesterification reaction. SrO/MgO nanocatalyst has weak or moderate $Mg^{2+}-O^{2-}$, strong $Sr^{2+}-O^{2-}$, and few new basic sites on the surface, which is connected with isolated O^{2-} sites. The catalyst synthesized by coprecipitation has higher basicity than that synthesized by impregnation method. A yield of 87.49% is obtained when 0.1 g of SrO/MgO nanocatalyst is used at 50.16°C and 7.77% (methanol:oil) for a duration of 1.37 h (Shahbazi et al., 2019).

Two important metal oxides (Zn and Cu give Zn–CuO catalyst) in combination result in synergy of their physical, chemical, and textural properties (Kuniyil et al., 2021). Carbon in their different forms is used to enhance the surface morphology, catalytic efficiency, and shelf life of the catalyst. Graphene-based material gives additional benefit of high surface area, improved porosity, electrical conductivity, and surface chemistry. Nitrogen (N)-doped graphene exhibits enhanced catalytic activity by building N bondage, which gives impurity sites by altering conjugation and aromaticity of graphene.

N-doped graphene mixed with Zn and Cu shows enhanced catalytic activity under optimized conditions when compared with the mixed metal oxides without NDG at the same conditions. Nanocomposite shows an enhanced activity due to the splitting of N-doped graphene and further improves the mixed metal oxide properties. Nanocomposite can be easily separated after the reaction and reusable, which makes it possible to be used in industries.

6.4.6 Ionic liquids

Ionic liquid (IL) is a saltlike material, which is made of cations, anions, and short-lived ion pairs, and exists in liquid phase at room temperature, as their melting point is less than 100°C (Mohammad Fauzi and Amin, 2012). ILs such as Brønsted acidic ILs (BAILs) help in the conversion of free fatty acids into biodiesel without pretreatment of feedstock. Physicochemical properties get affected by the cations and anions, which are present in ILs, and also it imparts important characteristic for addition of functional groups for catalysis/reactivity, maintaining solubility, hydrophobicity, and hydrophilicity. ILs can be classified as acidic, basic, or task specific and have functional group attached by covalent bond to the cation or anion or both. ILs, which are used for the production of biodiesel, are sulfonic acid–functionalized ILs (SFILs) (Han et al., 2016; Qiu et al., 2016; Feng et al., 2017), choline-based ILs (Fan et al., 2013; Reddy et al., 2014), hydroxide-based ILs (Khiratkar et al., 2018; Fang et al., 2013; Liang et al., 2010), polyionic-liquids (PILs) (Zhang et al., 2018; Jiang et al., 2017; Boakye et al., 2016), and hydrophobic ILs enzyme (Fan et al., 2018; Su et al., 2016; Cheng et al., 2014). ILs improve the molecular interaction between cation and anion and combines with enzymatic catalyst, mixed acid catalyst (BLAILs), or combinations of them. Acidic ILs are more commonly used than basic ILs. Brønsted acidic ILs have multiacid sites, and BLAILs show improved catalytic efficiency when compared with Brønsted acidic liquids with sole acid site. In acidic and basic ILs, the improved acidity/basicity of the ILs increases the catalytic activity in both esterification and transesterification reactions. When ILs are used as a catalyst in microwave and ultrasound process, they improve the mixing of reactants and reduce the duration of transesterification when compared with other conventional methods (Yang et al., 2016). The choice of ILs is considered to be efficient since it surpasses the hurdles in using homogenous catalyst and heterogenous catalyst (Tiong et al., 2018).

ILs help in the forward shift of the transesterification. ILs are being used in supercritical conditions for biodiesel generation. The main aim of using this in supercritical conditions is to overcome the problems, which are faced in conventional transesterification reaction. Caldas et al. (2016) obtained 97.6% of conversion efficiency with supercritical ethanol using BAILs. Ortiz-Martínez et al. (2016) obtained 61.6% and 95% of conversion efficiency with supercritical methanol using two different acidic anion-based BAILs, i.e., 1-butyl-3-methylimidazolium hydrogen sulfate ([BMIM] [HSO_4]) and choline dihydrogen phosphate ([Chol][H_2PO_4]). Thus, we can infer that the catalytic efficiency is influenced by the acidity order of the acidic anion (Tadesse and Luque, 2011).

6.5 Future prospects

Nanocatalysts in composite and hybrid forms in combination with various metal oxides, polymers, and metal nanoparticles have to be widely researched to bring out the synergetic properties of the different materials. Exploring the incorporation of nanocatalyst in solid matrix is important that requires immediate attention to enhance the reusability of the nanocatalyst. While there are several methods and feedstocks from which one can produce biodiesel, the most efficient way would be the way in which the cost and time required for processing the reaction is considerably reduced. An integrated approach of using a suitable nanocatalyst embedded in a solid matrix with microwave/ultrasound-assisted transesterification may be followed to enhance the conversion yield and produce high-quality biodiesel suitable for commercialization.

6.6 Conclusion

Biodiesel serves as the best source of energy for reducing carbon footprint and reducing global warming. This chapter elaborates on various methods by which we can produce biodiesel with focus on transesterification process, since it is a simple and cost-effective technique, which is very efficient while using cheap feedstocks such as nonedible seed oils, used cooking oils, and methanol. 100% conversion efficiency has been achieved by using microwave- and ultrasound-assisted transesterification. Further choosing the catalyst has a significant role in improving the efficiency of the transesterification process. Cost-effective production of biodiesel has been achieved by using nanocatalysts such as MgO, CaO, ZnO, and nanocomposites such as ZnO$-$$SiO_2$, CaO$-$$Al_2O_3$ mixed oxides, and SrO/MgO are widely used for obtaining high yield.

References

Abdelhady, H.H., Elazab, H.A., Ewais, E.M., Saber, M., El-Deab, M.S., 2020. Efficient catalytic production of biodiesel using nano-sized sugar beet agro-industrial waste. Fuel 261, 116481.

Alaei, S., Haghighi, M., Vahid, B.R., Shokrani, R., Naghavi, H., 2020. Conventional vs. hybrid methods for dispersion of MgO over magnetic Mg$-$Fe mixed oxides nanocatalyst in biofuel production from vegetable oil. Renew. Energy 154, 1188$-$1203.

Amirthavalli, V., Warrier, A.R., 2019. Production of biodiesel from waste cooking oil using MgO nanocatalyst. AIP Conf. Proc. 2115, 030609-1$-$030609-4.

Amit, S., 2012. Biodiesel: Production and Properties. Royal Society of Chemistry, UK.

Anand Kumar, S.A., Sakthinathan, G., Vignesh, R., Rajesh Banu, J., Al-Muhtaseb, A.H., 2019. Optimized transesterification reaction for efficient biodiesel production using Indian oil sardine fish as feedstock. Fuel 253, 921–929.

Ashok, A., Kennedy, L.J., Vijaya, J.J., Aruldoss, U., 2018. Optimization of biodiesel production from waste cooking oil by magnesium oxide nanocatalyst synthesized using coprecipitation method. Clean Technol. Environ. Policy 20 (6), 1219–1231.

Asri, N.P., Machmudah, S., Wahyudiono, S., Budikarjono, K., Roesyadi, A., Goto, M., 2013. Palm oil transesterification in sub- and supercritical methanol with heterogeneous base catalyst. Chem. Eng. Process. Process Intensif. 72, 63–67.

Azocar, L., Ciudad, G., Heipieper, H., Navia, R., 2010. Biotechnological processes for biodiesel production using alternative oils. Appl. Microbiol. Biotechnol. 88, 621–636.

Baskar, G., Aiswarya, R., 2016. Trends in catalytic production of biodiesel from various feedstocks. Renew. Sustain. Energy Rev. 57, 496–504.

Bet-Moushoul, E., Farhadi, K., Mansourpanah, Y., Nikbakht, A.M., Molaei, R., Forough, M., 2016. Application of CaO-based/Au nanoparticles as heterogeneous nanocatalysts in biodiesel production. Fuel 164, 119–127.

Bharti, P., Singh, B., Dey, R., 2019. Process optimization of biodiesel production catalyzed by CaO nanocatalyst using response surface methodology. J. Nanostruct. Chem. 9, 269–280.

Boakye-Boaten, N.A., Xiu, S., Shahbazi, A., Wang, L., Li, R., Schimmel, K., 2016. Uses of miscanthus press juice within a green biorefinery platform. Bioresour. Technol. 207, 285–292.

Booramurthy, V.K., Kasimani, R., Subramanian, D., Pandian, S., 2020. Production of biodiesel from tannery waste using a stable and recyclable nano-catalyst: an optimization and kinetic study. Fuel 260, 116373.

Borah, M.J., Das, A., Das, V., Bhuyan, N., Deka, D., 2019. Transesterification of waste cooking oil for biodiesel production catalyzed by Zn substituted waste egg shell derived CaO nanocatalyst. Fuel 242, 345–354.

Bournay, L., Casanave, D., Delfort, B., Hillion, G., Chodorge, J.A., 2005. New heterogeneous process for biodiesel production: a way to improve the quality and the value of the crude glycerin produced by biodiesel plants. Catal. Today 106, 190–192.

Bowman, M.D., Holcomb, J.L., Kormos, C.M., Leadbeater, N.E., Williams, V.A., 2008. Approaches for scale-up of microwave-promoted reactions. Org. Process. Res. Dev. 12, 41–57.

Caldas, B.S., Nunes, C.S., Souza, P.R., Rosa, F.A., Visentainer, J.V., Júnior, O.O.S., Muniz, E.C., 2016. Supercritical ethanolysis for biodiesel production from edible oil waste using ionic liquid [HMim][HSO_4] as catalyst. Appl. Catal. B. 181, 289–297.

Cao, F., Chen, Y., Zhai, F., Li, J., Wang, J., Wang, X., Zhu, W., 2008. Biodiesel production from high acid value waste frying oil catalyzed by superacid heteropolyacid. Biotechnol. Bioeng. 101 (1), 93–100.

Chen, K.S., Lin, Y.C., Hsu, K.H., Wang, H.K., 2012. Improving biodiesel yield from waste cooking oil by using sodium methoxide and microwave heating system. Energy 38, 151–156.

Cheng, J., Li, T., Huang, R., Zhou, J., Cen, K., 2014. Optimizing catalysis conditions to decrease aromatic hydrocarbons and increase alkanes for improving jet biofuel quality. Bioresour. Technol. 158, 378–382.

Choi, C.-S., Kim, J.-W., Jeong, C.-J., Kim, H., Yoo, K.-P., 2011. Transesterification kinetics of palm olein oil using supercritical methanol. J. Supercrit. Fluids 58 (3), 365–370.

Dalai, K.M.G., Meher, L.C., 2006. Presented at the IEEE EIC Climate Change Technology Conference, Article No. 4057358, Ottawa.

De Boer, K., Bahri, P.A., 2011. Supercritical methanol for fatty acid methyl ester production: a review. Biomass Bioenergy 35 (3), 983–991.

De, A., Boxi, S.S., 2020. Application of Cu impregnated TiO_2 as a heterogeneous nanocatalyst for the production of biodiesel from palm oil. Fuel 265, 117019.

Degfie, T.A., Mamo, T.T., Mekonnen, Y.S., 2019. Optimized biodiesel production from waste cooking oil (WCO) using calcium oxide (CaO) nanocatalyst. Sci. Rep. 9 (18982), 1–8.

Demirbas, A., 2005. Biodiesel production from vegetable oils via catalytic and non-catalytic supercritical methanol transesterification methods. Prog. Energy Combust. Sci. 31, 466–487.

Demirbas, A., 2007. Biodiesel from sunflower oil in supercritical methanol with calcium oxide. Energy Convers. Manag. 48 (3), 937–941.

Dias, J.M., Alvim-Ferraz, M.C.M., Almeida, M.F., 2008. Comparison of different homogeneous alkali catalysts during transesterification of waste and virgin oils and evaluation of biodiesel quality. Fuel 87, 3572–3578.

Fan, M., Huang, J., Yang, J., Zhang, P., 2013. Biodiesel production by transesterification catalyzed by an efficient choline ionic liquid catalyst. Appl. Energy 108, 333–339.

Fan, M., Liu, H., Zhang, P., 2018. Ionic liquid on the acidic organic-inorganic hybrid mesoporous material with good acid-water resistance for biodiesel production. Fuel 215, 541–550.

Fang, D., Jiang, C., Yang, J., 2013. Preparation of biodiesel from castor oil catalyzed by novel basic ionic liquid. Energy Technol. 1, 135–138.

Feng, Y., Qiu, T., Yang, J., Li, L., Wang, X., Wang, H., 2017. Transesterification of palm oil to biodiesel using Brønsted acidic ionic liquid as high-efficient and eco-friendly catalyst. Chin. J. Chem. Eng. 25, 1222–1229.

Foroutan, R., Mohammadi, R., Esmaeili, H., Mirzaee Bektashi, F., Tamjidi, S., 2020. Transesterification of waste edible oils to biodiesel using calcium oxide@magnesium oxide nanocatalyst. Waste Manag. 105, 373–383.

García-Martínez, N., Andreo-Martínez, P., Quesada-Medina, J., de los Ríos, A.P., Chica, A., Beneito-Ruiz, R., Carratalá-Abril, J., 2017. Optimization of non-catalytic transesterification of tobacco (*Nicotiana tabacum*) seed oil using supercritical methanol to biodiesel production. Energy Convers. Manag. 131, 99–108.

Gardy, J., Hassanpour, A., Lai, X., Ahmed, M.H., Rehan, M., 2017. Biodiesel production from used cooking oil using a novel surface functionalised TiO_2 nano-catalyst. Appl. Catal. B: Environ. 207, 297–310.

Georgogianni, K.G., Kontominas, M.G., Pomonis, P.J., Avlonitis, D., Gergis, V., 2008. Conventional and in situ transesterification of sunflower seed oil for the production of biodiesel. Fuel Process. Technol. 89 (5), 503–509.

Goh, B.H.H., Ong, H.C., Chong, C.T., Chen, W.-H., Leong, K.Y., Tan, S.X., Lee, X.J., 2020. Ultrasonic assisted oil extraction and biodiesel synthesis of spent coffee ground. Fuel 261, 116121.

Gurunathan, B., Ravi, A., 2015. Process optimization and kinetics of biodiesel production from neem oil using copper doped zinc oxide heterogeneous nanocatalyst. Bioresour. Technol. 190, 424–428.

Han, X., Yan, W., Hung, C.-T., He, Y., Wu, P.-H., Liu, L.-L., 2016. Transesterification of soybean oil to biodiesel by tin-based Brønsted-Lewis acidic ionic liquid catalysts. Korean J. Chem. Eng. 33, 2063–2072.

Hanh, H.D., Dong, N.T., Starvarache, C., Okitsu, K., Maeda, Y., Nishimura, R., 2008. Methanolysis of triolein by low frequency ultrasonic irradiation. Energy Convers. Manag. 49 (2), 276–280.

Hashemzehi, M., Saghatoleslami, N., Nayebzadeh, H., 2016. A study on the structure and catalytic performance of $Zn_xCu_{1-x}Al_2O_4$ catalysts synthesized by the solution combustion method for the esterification reaction. C. R. Chim. 19, 955–962.

Hawash, S., Kamal, N., Zaher, F., Kenawi, O., Diwani, G.E., 2009. Biodiesel fuel from Jatropha oil via non-catalytic supercritical methanol transesterification. Fuel 88 (3), 579–582.

Hazmi, B., Rashid, U., Ibrahim, M.L., Nehdi, I.A., Azam, M., Al-Resayes, S.I., 2021. Synthesis and characterization of bifunctional magnetic nano-catalyst from rice husk for production of biodiesel. Environ. Technol. Innov. 21, 101296.

Hosseini, S., Moradi, G.R., Bahrami, K., 2019. Synthesis of a novel stabilized basic ionic liquid through immobilization on boehmite nanoparticles: a robust nanocatalyst for biodiesel production from soybean oil. Renew. Energy 138, 70–78.

Ibrahim, M.L., Nik Abdul Khalil, N.N.A., Islam, A., Rashid, U., Ibrahim, S.F., Sinar Mashuri, S.I., Taufiq-Yap, Y.H., 2020. Preparation of Na_2O supported CNTs nanocatalyst for efficient biodiesel production from waste-oil. Energy Convers. Manag. 205, 112445.

Issariyakul, T., Kulkarne, M.G., Meher, L.C., Dalai, A.K., Bakhshi, N.N., 2007. Biodiesel production from mixtures of canola oil and used cooking oil. Chem. Eng. J. 140, 77–85.

Jiang, B., Wang, Y., Zhang, L., Sun, Y., Yang, H., Wang, B., 2017. Biodiesel production via transesterification of soybean oil catalyzed by superhydrophobic porous poly (ionic liquid) solid base. Energy Fuel 31, 5203–5214.

Keane, M.A., 2003. Ceramics for catalysis. J. Mater. Sci. 38, 4661–4475.

Kelkar, M.A., Gogate, P.R., Pandit, A.B., 2008. Intensification of esterification of acids for synthesis of biodiesel using acoustic and hydrodynamic cavitation. Ultrason. Sonochem. 15 (3), 188–194.

Khiratkar, A.G., Balinge, K.R., Patle, D.S., Krishnamurthy, M., Cheralathan, K.K., Bhagat, P.R., 2018. Transesterification of castor oil using benzimidazolium based Brønsted acid ionic liquid catalyst. Fuel 231, 458–467.

Koberg, M., Abu-Much, R., Gedanken, A., 2011. Optimization of bio-diesel production from soybean and wastes of cooked oil: combining dielectric microwave irradiation and a SrO catalyst. Bioresour. Technol. 102, 1073–1078.

Koutsouki, A.A., Tegou, E., Badeka, A., Kontakos, S., Pomonis, P.J., Kontominas, M.G., 2016. In situ and conventional transesterification of rapeseeds for biodiesel production: the effect of direct sonication. Ind. Crops Prod. 84, 399–407.

Koutsouki, A.A., Tegou, E., Kontakos, S., Kontominas, M.G., Pomonis, P.J., Manos, G., 2015. In situ transesterification of *Cynara cardunculus* L. seed oil via direct ultrasonication for the production of biodiesel. Fuel Process. Technol. 134, 122–129.

Kuniyil, K., Shanmukha Kumar, J.V., Adil, S.F., Assal, M.E., Shaik, M.R., Al-Warthan, A.R., Siddiqui, M.R.H., 2021. Production of biodiesel from waste cooking oil using ZnCuO/N-doped graphene nanocomposite as an efficient heterogeneous catalyst. Arab. J. Chem. 14 (3), 102982.

Kusdiana, D., Saka, S., 2001. Kinetics of transesterification in rapeseed oil to biodiesel fuel as treated in supercritical methanol. Fuel 80, 693–698.

Lamba, N., Gupta, R., Modak, J.M., Madras, G., 2019. ZnO catalyzed transesterification of *Madhuca indica* oil in supercritical methanol. Fuel 242, 323–333.

Leduc, S., Natarajan, K., Dotzauer, E., McCallum, I., Obersteiner, M., 2009. Optimizing biodiesel production in India. Appl. Energy 86, S125–S131.

Lertsathapornsuk, V., Pairintra, R., Aryusuk, K., Krisnangkura, K., 2008. Microwave assisted in continuous biodiesel production from waste frying palm oil and its performance in a 100-kW diesel generator. Fuel Process. Technol. 89, 1330–1336.

Leung, D.Y.C., Wu, X., Leung, M.K.H., 2010. A review on biodiesel production using catalyzed transesterification. Appl. Energy 87, 1083.

Liang, J.-H., Ren, X.-Q., Wang, J.-T., Jinag, M., Li, Z.-J., 2010. Preparation of biodiesel by transesterification from cottonseed oil using the basic dication ionic liquids as catalysts. J. Fuel Chem. Technol. 38, 275–280.

Lin, L., Ying, D., Chaitep, S., Vittayapadung, S., 2009. Biodiesel production from crude rice bran oil and properties as fuel. Appl. Energy 86 (5), 681–688.

Liu, Q., Wang, L., Wang, C., Qu, W., Tian, Z., Ma, H., Wang, D., Wang, B., Xu, Z., 2013. The effect of lanthanum doping on activity of Zn-Al spinel for transesterification. Appl. Catal. B: Environ. 136–137, 210–217.

Maes, I.I., Gryglewicz, G., Yperman, J., Franco, D.V., Mullens, J., Van Poucke, L.C., 1997. Effect of calcium and calcium minerals in coal on its thermal analysis. Fuel 76 (2), 143–147.

Mahdavi, V., Monajemi, A., 2014. Optimization of operational conditions for biodiesel production from cottonseed oil on CaO-MgO/Al_2O_3 solid base catalysts. J. Taiwan Instit. Chem. Eng. 45, 2286–2292.

Mazubert, A., Taylor, C., Aubin, J., Poux, M., 2014. Key role of temperature monitoring in interpretation of microwave effect on transesterification and esterification reactions for biodiesel production. Bioresour. Technol. 161, 270–279.

Mazzocchia, C., Modica, G., Kaddauri, A., Nannicini, R., 2004. Fatty acid methyl ester synthesis from triglycerides over heterogeneous catalyst in the presence of microwave. C. R. Chim. 7, 601–605.

Mierczynski, P., Chalupka, K.A., Maniukiewicz, W., Kubicki, J., Szynkowska, M.I., Maniecki, T.P., 2015. $SrAl_2O_4$ spinel phase as active phase of transesterification of rapeseed oil. Appl. Catal. B Environ. 164, 176–183.

Mohammad Fauzi, A.H., Amin, N.A.S., 2012. An overview of ionic liquids as solvents in biodiesel synthesis. Renew. Sustain. Energy Rev. 16, 5770–5786.

Naveenkumar, R., Baskar, G., 2019. Biodiesel production from *Calophyllum inophyllum* oil using Zinc doped Calcium oxide (Plaster of Paris) nanocatalyst. Bioresour. Technol. 280, 493–496.

Nayak, M.G., Vyas, A.P., 2019. Optimization of microwave-assisted biodiesel production from Papaya oil using response surface methodology. Renew. Energy 138, 18–28.

Ngamcharussrivichai, C., Wiwatnimit, W., Wangnoi, S., 2007. Modified dolomites as catalysts for palm kernel oil transesterification. J. Mol. Catal. A 276, 24–33.

Nur Syazwani, O., Lokman Ibrahim, M., Wahyudiono, Kanda, H., Goto, M., Taufiq-Yap, Y.H., 2017. Esterification of high free fatty acids in supercritical methanol using sulfated angel wing shells as catalyst. J. Supercrit. Fluids 124, 1–9.

Ortiz-Martínez, V.M., Salar-García, M.J., Hernandez-Fernandez, F.J., Olivares Carrillo, P., de los Ríos, A.P., Quesada-Medina, J., 2016. Ionic liquids in supercritical methanol greatly enhance transesterification reaction for high-yield biodiesel production. AIChE J. 62, 3842–3846.

Pascoal, C.V.P., Oliveira, A.L.L., Figueiredo, D.D., Assunção, J.C.C., 2019. Optimization and kinetic study of ultrasonic-mediated in situ transesterification for biodiesel production from the almonds of *Syagrus cearensis*. Renew. Energy 147, 1815–1824.

Pleanjai, S., Gheewala, S.H., 2009. Full chain energy analysis of biodiesel production from palm oil in Thailand. Appl. Energy 86 (S1), S209–S214.

Prabu, M., Manikandan, M., Kandasamy, P., Kalaivani, P.R., Rajendiran, N., Raja, T., 2019. Synthesis of biodiesel using the Mg/Al/Zn hydrotalcite/SBA-15 nanocomposite catalyst. ACS Omega 4 (2), 3500–3507.

Qian, J., Wang, F., Liu, S., Yun, Z., 2008. In situ alkaline transesterification of cottonseed oil for production of biodiesel and nontoxic cottonseed meal. Bioresour. Technol. 99 (18), 9009–9012.

Qiu, F., Li, Y., Yang, D., Li, X., Sun, P., 2011. Biodiesel production from mixed soybean oil and rapeseed oil. Appl. Energy 88 (6), 2050–2055.

Qiu, T., Guo, X., Yang, J., Zhou, L., Li, L., Wang, H., 2016. The synthesis of biodiesel from coconut oil using novel Brønsted acidic ionic liquid as green catalyst. Chem. Eng. J. 296, 71–78.

Rahmani Vahid, B., Haghighi, M., 2017. Biodiesel production from sunflower oil over MgO/$MgAl_2O_4$ nanocatalyst: effect of fuel type on catalyst nanostructure and performance. Energy Convers. Manag. 134, 290–300.

Rasouli, H., Esmaeili, H., 2019. Characterization of MgO nanocatalyst to produce biodiesel from goat fat using transesterification process. 3 Biotech 9, 429.

Rathnam, V.M., Madras, G., 2019. Conversion of *Shizochitrium limacinum* microalgae to biodiesel by non-catalytic transesterification using various supercritical fluids. Bioresour. Technol. 288, 121538.

Reddy, E.R., Sharma, M., Chaudhary, J.P., Bosamiya, H., Meena, R., 2014. One-pot synthesis of biodiesel from high fatty acid *Jatropha curcas* oil using bio-based basic ionic liquid as a catalyst. Curr. Sci. 106, 1394–1400.

Rocha, P.D., Oliveira, L.S., Franca, A.S., 2019. Sulfonated activated carbon from corn cobs as heterogeneous catalysts for biodiesel production using microwave-assisted transesterification. Renew. Energy 143, 1710–1716.

Román-Figueroa, C., Olivares-Carrillo, P., Paneque, M., Palacios-Nereo, F.J., Quesada-Medina, J., 2016. High-yield production of biodiesel by non-catalytic supercritical methanol transesterification of crude castor oil (*Ricinus communis*). Energy 107, 165–171.

Salamatinia, B., Mootabadi, H., Bhatia, S., Abdullah, A.Z., 2010. Optimization of ultrasonic-assisted heterogeneous biodiesel production from palm oil: a response surface methodology approach. Fuel Process. Technol. 91, 441–448.

Samniang, A., Tipachan, C., Kajorncheappun-ngam, S., 2014. Comparison of biodiesel production from crude Jatropha oil and Krating oil by supercritical methanol transesterification. Renew. Energy 68, 351–355.

Sankaranarayanan, T.M., Shanthi, R.V., Thirunavukkarasu, K., Pandurangan, A., Sivasanker, S., 2013. Catalytic properties of spinel-type mixed oxides in transesterification of vegetable oils. J. Mol. Catal. A: Chem. 379, 234–242.

Santana, A., Maçaira, J., Larrayoz, M.A., 2012. Continuous production of biodiesel using supercritical fluids: a comparative study between methanol and ethanol. Fuel Process. Technol. 102, 110–115.

Shaban, M., Abukhadra, M.R., Hosny, R., Rabie, A.M., Ahmed, S.A., Negm, N.A., 2019. Diatomite supported by CaO/MgO nanocomposite as heterogeneous catalyst for biodiesel production from waste cooking oil. J. Mol. Liq. 279, 224–231.

Shahbazi, F., Mahdavi, V., Zolgharnein, J., 2019. Preparation and characterization of SrO/MgO nanocomposite as a novel and efficient base catalyst for biodiesel production from waste cooking oil: a statistical approach for optimization. J. Iran Chem. Soc. 17, 333–349.

Singh, A.K., Fernando, S.D., 2006. Catalyzed Fasttransesterification of Soybean Oil Using Ultrasonication. American Society of Agricultural Engineers, ASAE Annual Meeting, Portland, Oregon, USA.

Srivastava, G., Paul, A.K., Goud, V.V., 2018. Optimization of non-catalytic transesterification of microalgae oil to biodiesel under supercritical methanol condition. Energy Convers. Manag. 156, 269–278.

Stavarache, C., Vinatoru, M., Maeda, Y., 2007. Aspects of ultrasonically assisted transesterification of various vegetable oils with methanol. Ultrason. Sonochem. 14, 380–386.

Su, F., Peng, C., Li, G.-L., Xu, L., Yan, Y.-J., 2016. Biodiesel production from woody oil catalyzed by *Candida rugosa* lipase in ionic liquid. Renew. Energy 90, 329–335.

Suppes, G.J., Bockwinkel, K., Lucas, S., Botts, J.B., Mason, M.H., Heppert, J.A., 2001. Calcium carbonate catalyzed alcoholysis of fats and oils. J. Am. Oil Chem. Soc. 78, 139–145.

Tadesse, H., Luque, R., 2011. Advances on biomass pretreatment using ionic liquids: an overview. Energy Environ. Sci. 4, 3913–3929.

Tahvildari, K., Anaraki, Y.N., Fazaeli, R., Mirpanji, S., Delrish, E., 2015. The study of CaO and MgO heterogenic nano-catalyst coupling on transesterification reaction efficacy in the production of biodiesel from recycled cooking oil. J. Environ. Health Sci. Eng. 13, 73.

Tangy, A., Pulidindi, I.N., Gedanken, A., 2016. SiO_2 beads decorated with SrO nanoparticles for biodiesel production from waste cooking oil using microwave irradiation. Energy Fuel 30, 3151–3160.

Teo, S.H., Goto, M., Taufiq-Yap, Y.H., 2015. Biodiesel production from *Jatropha curcas* L. oil with Ca and La mixed oxide catalyst in near supercritical methanol conditions. J. Supercrit. Fluids 104, 243–250.

Thanh, L.T., Kenji, O., Yasuhiro, S., Norimichi, T., Yasuaki, M., Hiroshi, B., 2010. Ultrasonicassisted production fuel from vegetable oils in a small-scale circulation process. Bioresour. Technol. 101, 639–645.

Tiong, Y.W., Yap, C.L., Gan, S., Ya, P.W.S.P., 2018. Conversion of biomass and its derivatives to levulinic acid and levulinate esters via ionic liquids. Ind. Eng. Chem. Res. 57, 4749–4766.

Trinh, H., Yusup, S., Uemura, Y., 2018. Optimization and kinetic study of ultrasonic assisted esterification process from rubber seed oil. Bioresour. Technol. 247, 51–57.

Tuntiwiwattanapun, N., Monono, E., Wiesenborn, D., Tongcumpou, C., 2017. In-situ transesterification process for biodiesel production using spent coffee grounds from the instant coffee industry. Ind. Crops Prod. 102, 23–31.

Umdu, E.S., Tuncer, M., Seker, E., 2009. Transesterification of nannochloropsis oculata microalga's lipid to biodiesel on Al_2O_3 supported CaO and MgO catalysts. Bioresour. Technol. 100, 2828–2231.

Varghese, R., Henry, J.P., Irudayaraj, J., 2017. Ultrasonication-assisted transesterification for biodiesel production by using heterogeneous ZnO nanocatalyst. Environ. Prog. Sustain. Energy 37 (3), 1176–1182.

Wang, L., Yang, J., 2007. Transesterification of soybean oil with nano-MgO or not in supercritical and subcritical methanol. Fuel 86 (3), 328–333.

Wen, Z., Yu, X., Tu, S.T., Yan, J., Dahlquist, E., 2010. Synthesis of biodiesel from vegetable oil with methanol catalyzed by Li-doped magnesium oxide catalysts. Appl. Energy 87 (3), 743–748.

Wu, H., Zhang, J., Wei, Q., Zheng, J., Zhang, J., 2013. Transesterification of soybean oil to biodiesel using zeolite supported CaO as strong base catalysts. Fuel Process. Technol. 109, 13–18.

Yaakob, Z., Ong, B.H., Kumar, M.N.S., Kamarudin, S.K., 2009. Microwave-assisted transesterification of Jatropha and waste frying palm oil. Int. J. Sustain. Energy 28, 195–201.

Yan, S., DiMaggiom, C., 2010. Advancements in heterogeneous catalsysis for biodiesel synthesis. Top. Catal. 53, 721–736.

Yang, W., Wang, H., Zhang, M., Zhu, J., Zhou, J., Wu, S., 2016. Fuel properties and combustion kinetics of hydrochar prepared by hydrothermal carbonization of bamboo. Bioresour. Technol. 205, 199–204.

Yee, K.F., Tan, K.T., Abdullah, A.Z., Lee, K.T., 2009. Life cycle assessment of palm biodiesel: revealing facts and benefits for sustainability. Appl. Energy 86 (S1), 189–196.

Yoo, S.J., Lee, H., Veriansyah, B., Kim, J., Kim, J.-D., Lee, Y.-W., 2010. Synthesis of biodiesel from rapeseed oil using supercritical methanol with metal oxide catalysts. Bioresour. Technol. 101 (22), 8686–8689.

Yoosuk, B., Udomsap, P., Puttasawat, B., Krasae, P., 2010. Modification of calcite by hydration–dehydration method for heterogeneous biodiesel production process: the effects of water on properties and activity. Chem. Eng. J. 162, 135–141.

Yu, D., Tian, L., Wu, H., Wang, S., Wang, Y., Na, D., 2010. Ultrasonic irradiation with vibration for biodiesel production from soybean oil by Novozym 435. Process. Biochem. 45, 519–525.

Yusuf, N.N.A.N., Kamarudin, S.K., Yaakub, Z., 2011. Overview on the current trends in biodiesel production. Energy Convers. Manag. 52 (7), 2741–2751.

Zeng, D., Yang, L., Fang, T., 2017. Process optimization, kinetic and thermodynamic studies on biodiesel production by supercritical methanol transesterification with CH_3ONa catalyst. Fuel 203, 739–748.

Zhang, H., Li, H., Pan, H., Wang, A., Souzanchi, S., Xu, C., 2018. Magnetically recyclable acidic polymeric ionic liquids decorated with hydrophobic regulators as highly efficient and stable catalysts for biodiesel production. Appl. Energy 223, 416–429.

Zhang, P., Han, Q., Fan, M., Jiang, P., 2014. Magnetic solid base catalyst $CaO/CoFe_2O_4$ for biodiesel production: influence of basicity and wettability of the catalyst in catalytic performance. Appl. Surf. Sci. 317, 1125–1130.

Zhang, Y., Dubé, M.A., McLean, D.D., Kates, M., 2003. Biodiesel production from waste cooking oil: 2. Economic assessment and sensitivity analysis. Bioresour. Technol. 90, 229–240.

CHAPTER

7

Biodiesel production using enzymatic catalyst

Nurfadhila Nasya binti Ramlee[1], *Hilman Ibnu Mahdi*[2], *Nur Izyan Wan Azelee*[1,3]

[1]School of Chemical and Energy Engineering, Faculty of Engineering, Universiti Teknologi Malaysia (UTM) Skudai, Johor, Malaysia; [2]PT Dover Chemical Plant B, Formaldehyde Production Department, Cilegon City, Banten, Indonesia; [3]Institute of Bioproduct Development (IBD), Universiti Teknologi Malaysia, UTM Skudai, Johor, Malaysia

7.1 Introduction

Lipases (triacylglycerol acylhydrolases, EC 3.1.1.3) are enzymes that catalyze the hydrolysis of triglycerides (TAGs), via diacylglycerols (DAGs) and monoacylglycerols (MAGs), to produce glycerol and free fatty acids (FFAs) (Almyasheva et al., 2018). Lipases belong to the group of esterases that catalyze the hydrolytic cleavage (Iboyo et al., 2017). The versatility of lipases as biocatalysts makes them one of the most useful enzymes in the industry thanks to the wide range of substrate specificity and high stability in nonaqueous media. In addition, lipases also catalyze the synthesis of esters through esterification, transesterification, interesterification, thioesterification, alcoholysis, acidolysis, and aminolysis (López-Fernández et al., 2020; Zeng et al., 2018). Microbial lipases production has received a great deal of attention because of their industrial applications for biodiesel production (David et al., 2016), valorization of waste (Louhasakul et al., 2020), emulsification (Furini et al., 2018), lyophilization of milk fat (Fraga et al., 2018), and ester synthesis.

Biodiesel is a renewable and biodegradable energy that can be produced either by esterification of FFA or transesterification of TAG to FAME. Conventional methods of biodiesel production include catalyzing via transesterification using alkali catalysts. The advantages of the alkali-catalyzed process are a high conversion rate in a short reaction time, however, this process consumes high energy, generates wastewater, and complicates the separation of glycerol and the catalyst. Alkali catalyst is unsuitable for oil with high FFA content because the reaction favors the saponification reaction. Thus, reducing the FFA content via a deacidification or esterification reaction is prominent during the pretreatment step. After completion of the esterification reaction using acid catalysts, a

Biofuels and Bioenergy
https://doi.org/10.1016/B978-0-323-85269-2.00017-4

neutralization process must be carried out, which generates a large quantity of wastewater (Gupta and Rathod, 2020).

The enzyme-catalyzed reaction has received a great deal of attention as this process is simple, energy efficient, and nontoxic due to the biodegradability properties of the enzymes. Various types of lipases have been studied for biodiesel production including intracellular, extracellular, and membrane-bound lipases, however free lipases are expensive and cannot be recycle, thereby limiting large-scale biodiesel production. Immobilized lipase could be used to overcome the free enzyme limitations, and therefore, future studies should develop this process to facilitate biodiesel production. This chapter describes the sources and properties of lipase, the enzymatic reaction of biodiesel production, and factors that have influenced lipase-catalyzed biodiesel production. At the end of the chapter, a summary of the challenges and opportunities for lipase in biodiesel production is given.

7.2 Sources and properties of lipases for biodiesel production

Lipases can be found in animals, plants, and microorganisms. Lipase from animals usually is obtained from the pancreas, and on the surface of mucous cells of the gastric mucosa (Pahoja and Sethar, 2002). Due to easier cultivation of microorganisms, most lipases are commercially produced from bacteria, fungi, and yeasts. Microbial lipases have been widely exploited for various applications owing to their robustness, ease of production on a large scale, and the low cost of the final product. The cost of enzyme production depends on downstream processes and postharvesting processes, which affect the total enzyme yield at the end of the process. Several methods to improve enzyme yield include by choosing the most suitable fermentation technique and optimizing the culture conditions of the microorganisms.

The selection of these groups for biodiesel production is dependent on the production systems, which can be either a homogeneous or heterogeneous system, and the types of substrates used. Lipases are characterized by their regioselectivity and enantioselectivity to synthesize the desired product. *Aspergillus niger* exhibits 1,3-regiospecificity toward a triglyceride for transesterification to fatty acid methyl ester (Almyasheva et al., 2018). Meanwhile *Aspergillus oryzae* is more suitable for synthesis of 1,3-dioleoyl-2-palmitoylglycerol via the acidolysis route of palm stearin and oleic acid because the enzymes are characterized by sn-1,3-regiospecificity. Enzymes are classified into three major groups: intracellular, extracellular, and membrane-bound. The sources of lipases and their applications are tabulated in Table 7.1.

7.2.1 Intracellular lipases

Intracellular enzymes, also known as endoenzymes, are defined as enzymes that perform their functions within the cell that produces them. Microorganisms such as *Rhizopus oryzae*, *Aspergillus niger*, and *Yarrowia lipolytica* are reported to be the best at producing intracellular lipase. Optimization of medium growth is important to improve biomass yield and activity of lipase. The selection of the best carbon and nitrogen sources included yeast extract, soybean meal, and sunflower oil resulting in 3.2- and 2.4-fold increased intracellular enzyme activity and cell mass concentration, respectively, compared to the initial medium (Almyasheva et al., 2018). David et al. (2016) developed a novel technology consisting of magnetically supported bioreactors to generate biomass and intracellular lipase from *Yarrowia lipolytica* using glycerol as a carbon source. The study integrated a biorefinery concept by reusing the glycerol formed from biodiesel production as a fermentation

TABLE 7.1 The applications of enzymes from various sources.

Groups	Microorganisms	Systems	Process	References
Intracellular	*Rhizopus oryzae*	Whole cells	Transesterification	Navarro Lopez et al. (2016)
	Aspergillus niger	Immobilization	Methanolysis	Almyasheva et al. (2018)
	Yarrowia lipolytica	Free enzymes	Biodiesel production	David et al. (2016)
	Brevibacterium sp. and *Pseudomonas* sp.	Whole cells	Biodiesel production	Haq et al. (2020)
	Sphingomonas sp.	Free enzymes	Aminolysis	Zeng et al. (2018)
Extracellular	*Acinetobacter baylyi*	Free enzymes	Emulsification	Furini et al. (2018)
	Yarrowia lipolytica	Free enzymes	Valorization of waste	Louhasakul et al. (2020)
	Aspergillus niger	Free enzymes	–	Jia et al. (2015)
	Yarrowia lipolytica	Immobilized enzymes	Synthesis of L-ascorbyl palmitate	Ping et al. (2018)
	Fusarium solani	Free enzymes	Esterification and transesterification	Geoffry and Achur (2017)
	Serratia marcescens	Free enzymes	Detergent formulation and biodiesel production	Garcia-Silvera et al. (2018)
	Enterobacter cloacae	Free enzymes	Waste management	Iboyo et al. (2017)
	Candida rugosa	Immobilized enzymes	Biodiesel production	Iuliano et al. (2020)
Membrane bound	*Yarrowia lipolytica*	Free enzymes	Valorization of waste	Louhasakul et al. (2020)
	Yarrowia lipolytica	Immobilized enzymes	Lipolyzed milk fat	Fraga et al. (2018)

substrate for lipase production. Then, lipase was applied as a biocatalyst for biodiesel production while the remaining biomass can be formulated into feed production.

Purification of lipase free from contaminants is an expensive and tedious technique. An alternative to substitute for the use of purified lipase is using whole-cell biocatalysis. This approach involves the employment of enzymes to generate directly species as biocatalysts in the reaction. In this case, intracellular enzymes expressed on the cell wall or on the membrane are used rather than extracellular enzymes. Almyasheva et al. (2018) conducted an optimization study of *Aspergillus niger* immobilized on polyurethane foam for the production of intracellular lipase. The maximum lipase activity of 1.12 U/mg was observed when the medium was supplemented with sunflower oil and yeast extract. The application of genetic engineering toward increasing lipase production has been proven to increase the catalytic activity of lipase. Zeng and co-workers observed that the expression of intracellular lipase from *Sphingomonas* sp. into *Escherichia coli* as a whole-cell biocatalyst shows a remarkable increase in catalytic activity during amide synthesis (Zeng et al., 2018).

7.2.2 Extracellular lipases

In the cell, extracellular enzymes are produced, then packaged and secreted from the cell. Extracellular enzymes, also called exoenzymes, catalyze their reactions outside the cell. The production of bacterial extracellular lipases is influenced by nutritional and physiological factors such as pH, carbon source, and temperature (Furini et al., 2018). Thus, the optimization of these factors is crucial for high production of lipase. Jia and colleagues found that the source of nitrogen, source of carbon, and lipid were the most influential factors in the production of lipase. The optimized conditions of 0.5 g/L corn starch, 35.4 g/L soybean meal, and 10.9 g/L soybean oil resulted in the highest lipase activity of 2171 U/mL, which was 16.4% higher than for the initial medium used (Jia et al., 2015). Researchers have investigated various methods to increase the production of extracellular lipase. Biosurfactant enhanced the performance of the yeast by modifying hydrophobic substrates and cell membrane permeability, leading to an increase in substrate entry and consequently enhanced lipase production.

Instead of producing lipase using a single bacterial culture, the study of mixed bacterial culture can help increase lipase production. Willsey and Wargo (2015) cultivated mixed-culture bacteria isolated from drinking water to study the interaction of the bacterial community on lipase production. This study showed that each bacteria species produces varying enzyme activities but the absence of one bacteria species from the mixed culture showed no significant difference in lipase activity for short-chain lipases (ρ-nitrophenylbutyrate) and long-chain lipases (ρ-nitrophenylpalmitate). Meanwhile, the absence of *Chryseobacterium gleum* shows an increment of phosphatidylcholine-specific phospholipases C (ρ-nitrophenylphosphorylcholine). The regulatory processes, such as regulation of feedback and mechanisms of chemical signaling, are likely to integrate together to guide the production of hydrolytic enzymes within microbial communities (Willsey and Wargo, 2015). Extracellular lipase production also can be improved significantly by mutating wild-type bacteria either using chemical or physical mutagenesis. The mutant strain of *Rhizopus oligosporus* was developed by treating the culture with ethyl methanesulfonate and methyl methanesulfonate as mutagenesis chemicals, showing an increment in the maximum lipase activity up to 124 U/g (Waseem et al., 2018). Physical mutagenesis methods, such as low-energy ion beam implantation, were applied to *Yarrowia lipolytica* and gave outstanding results by increasing the lipase activity by up to 5.5-fold compared to the wild type (Ping et al., 2018).

7.2.3 Membrane-bound lipases

Membrane-bound lipases also play an important role in catalyzing chemical reactions. Membrane-bound enzymes are referred to as enzymes that are located in the membranes of cells. Membrane-bound enzymes such as *Yarrowiya lipolytica* are naturally immobilized on cell debris, showing high catalytic activity to produced lipolyzed milk fat. While showing a high catalytic activity of 181 U/g, the membrane-bound lipase exhibited good operational stability by retaining 70% of the lipolytic activity after the second cycle (Fraga et al., 2018). Immobilization of lipases may improve the catalytic activity and the operational stability to synthesize products in a repeated cycle. Another way to enhance the catalytic activity of membrane-bound lipase is through the addition of surfactant. The addition of biosurfactant shows an increment in lipase activity of *Yarrowiya lipolytica* up to 555 U/L, which is higher than the culture without the addition of surfactant with 361 U/L of lipase activity (Louhasakul et al., 2020).

7.3 Enzymatic reaction for biodiesel production

Homogeneous catalysts such as KOH, NaOH, H_2SO_4, K_3PO_4, CH_3ONa, CaO, and CH_3OK have been widely utilized in biodiesel production under esterification and transesterification reactions as reported widely in the literature (Sakai et al., 2009; Atadashi et al., 2013; Singh et al., 2019; Agarwal et al., 2012; Guan et al., 2009), whereupon the addition of sulfuric acid and methanol in the reactions is, at 70°C and 4 bar, conducted to mitigate free fatty acid (FFA) content (Knothe, 2010; Knothe and Razon, 2017; Konwar et al., 2014). Despite being applied in many processes for years, the use of the homogeneous catalysts requires high energy consumption in purification and separation processes, removal of catalyst in the last process, and nonrecyclability of the catalysts (Long et al., 2014; Konwar et al., 2014). Also, it can cause serious corrosion in the reactor and form soap when using edible and nonedible oils with high water and FFA contents and thus lower biodiesel yield (Mansir et al., 2017; Farooq et al., 2013). These drawbacks have encouraged many researchers to seek other potential catalysts such as heterogeneous acid and base catalysts and enzymes in biodiesel production, as observed in recent reports (Sani et al., 2014; Mansir et al., 2017; Borges and Díaz, 2012; Akhabue et al., 2020; Karmakar et al., 2020).

The heterogeneous solid catalysts contain alkaline earth atoms such as Ca, Na, Ba, Mg, and K which can increase the active catalyst sites, thus improving the catalyst activity (Mansir et al., 2017). Furthermore, some of the heterogeneous catalysts have highly active sites and reusability during several cycles with less deactivation (Teo et al., 2015; Galadima and Muraza, 2014; Borges and Díaz, 2012). However, their ability and stability are also greatly decreased when using oils with high FFA and water contents (Atadashi et al., 2012; Atapour et al., 2014; Encinar et al., 2005; Nayak et al., 2016; Vieitez et al., 2014) causing the formation of more soap. Accordingly, oils with low FFA (<1%) and water (<0.3%) contents must be properly employed to lower the soap formation (Encinar et al., 2005; Amini et al., 2017). On the other hand, the process using heterogeneous catalysts is carried out at high temperature, longer reaction time, and elevated methanol:oil molar ratio, leading to corrosion in the main reactor (Romdhane et al., 2013; Lotero et al., 2005, 2006). It is noteworthy to say that either homogeneous or heterogeneous solid catalysts are not suitable for application in biodiesel production using oils possessing high FFA and water contents.

Therefore, enzymes are prospective compatible biocatalysts for biodiesel production using oils with high FFA and water contents, especially immobilized enzymes (Razack and Duraiarasan, 2016; Bateni et al., 2014; Poppe et al., 2018; Ching-Velasquez et al., 2020; Andrade et al., 2019). Guo et al. (2020) employed *Candida antarctica* lipase A (CALA) enzyme in biodiesel production from waste palm frying oil with high FFA and water contents and claimed that the enzyme emerged as a cleaner and cheaper biocatalyst. In addition, the immobilized enzymes have perfect water-solubility (Jegannathan et al., 2008; Lima et al., 2015), thus reducing the mass transfer limitedness and resolving the problem of by-product adsorption (Andrade et al., 2016, 2019; Pollardo et al., 2018). The utilization of enzymes as the biocatalyst in biodiesel production under the esterification and transesterification reactions is outlined in Table 7.2.

Apart from troubleshooting for the high FFA and water contents and the adsorption of by-product, biodiesel production using enzymes as biocatalysts can be also performed at low temperature (Rial et al., 2020), has no environmental emissions (Tan et al., 2010; Li et al., 2011; Atabani et al., 2012; Verma et al., 2013), and gives high production efficiency (Andrade et al., 2016). For example, Taher et al. (2020) used

TABLE 7.2 Biodiesel production using various optimized enzymatic catalysts.

Pathways	Various oils	Operating conditions				Y (%)	Type of biocatalysts	Novelty	References
		Temp (°C)	RPM	Time (h)	Enzyme content				
Esterification	Waste frying acid oil (WFAO)	30–50	100–200	4–8	2%	75–78.33	Gelatin-alginate lipase (GAL) and pectin-alginate lipase (PAL)	The biocatalysts could be reused over seven cycles and became a promising catalyst to replace the solid catalysts due to biodiesel properties similar to commercial diesel fuel	Muanruksa et al. (2020)
Esterification	Sludge palm oil	40	200	4	2%	91.30	*Rhizopus oryzae* lipase immobilized with alginate-polyvinyl alcohol (PVA)	The low-budget production was conducted using low-price oil and the biocatalyst was reused over 15 batches with less deactivation	Muanruksa and Kaewkannetra (2020)
Esterification	Soybean oil	50	200	$^1/_2$	10 g	98	Lipozyme *thermomyces lanuginosus* (TL 100L) lipase synthesized over resin of CLEA-TLL NKA	The biocatalyst activity revealed much better performance than the commercial enzyme lipozyme TLIM activity and then had good stability with 20 consecutive batches	Zheng et al. (2020)
Esterification	Residual chicken oil (RCO)	30	Na	3	15%	92.4	*Candida antarctica* lipase A and B (CALA and CALB) and *Rhizomucor miehei* (RML)	The biodiesel production was optimized over the Taguchi method at the best operating conditions	Rocha et al. (2020)

Esterification	Soybean oil	40	200	NA	15%	94.1	*Pseudomonas fluorescens* lipase (PFL) immobilized with SDB	The high-grade quality of biodiesel was produced from waste cooking oil with a good yield	Costa et al. (2020)
Esterification	Palm fatty acid distillate (PFAD)	45	Na	72	8.7 U/g	59	Lipozyme RM IM lipase impregnated by dry fermented solid (DFS)	The DFS-enhanced biocatalyst reusability and solid-state fermentation (SSF) played a significant influence toward the enzyme yield	Collaço et al. (2020)
Esterification	Soybean oil industrial waste	37	200	12	3 wt.%	96	NS - 40,116 (Novozymes, Denmark)	The utilization of industrial oil waste supplied many benefits not only for reducing production cost but also for decreasing environmental pollution	Mibielli et al. (2020)
Esterification	Palm oil	40	250	>2	5%	Na	*Candida antarctica* lipase B (CALA B)	The two-step reaction of transes-terification process followed by the esterification process was viable technology at large-scale production	Hansen et al. (2020)

Continued

TABLE 7.2 Biodiesel production using various optimized enzymatic catalysts.—cont'd

Pathways	Various oils	Operating conditions				Y (%)	Type of biocatalysts	Novelty	References
		Temp (°C)	RPM	Time (h)	Enzyme content				
Esterification	Hydrolyzed microbial oil (HMO)	45	Na	NA	4%	85	Lipomod™ 34 MDP lipase	The oils like *Rhodosporidium toruloides* NRRL Y-27012 and *R. kratochvilovae* Y-43 were produced over the fermentation method from cane sugar or sugarcane molasses	Boviatsi et al. (2020)
Esterification	Soybean oil	35	300	8	1 and 1.45 wt.%	85.08	*Thermomyces lanuginosus* lipase	Feeding higher oil: methanol ratio and lower enzyme content played a significant role in increasing the biodiesel yield	Wancura et al. (2019b)
Esterification	Soybean oil	35	500	5	70 wt.%	97.1	*Thermomyces lanuginosus* lipase	The two-step reaction was a low-cost potential technology and water content was applied in optimization of productivity	Wancura et al. (2019a)
Esterification	Vegetable oils	50–100	200–1000	1–4	1–10 wt.%	95.08–98.6	Various catalysts	The review declared that biodiesel produced over the two-step reaction could be optimized over various methods and the esterification step reaction could be significantly reduced to less [illegible]	Thoai et al. (2019)

Esterification	Soybean oil	35	300	24	1.45 wt.%	96.9	Callera^{TM} Trans L lipase	The biodiesel yield was optimized using the central composite statistical design (CCSD) method	Wancura et al. (2018b)
Esterification	Soybean oil	35	300	8	1.45 wt.%	84.6	*Thermomyces lanuginosus* lipase	The highest yield was achieved when using 100% of fats with soluble lipase	Wancura et al. (2018a)
Esterification	Cinnamic acid	45	120	12	30 mg	90.4	Porcine pancreatic lipase (PPL)	The lipase immobilized with PW/MIL-100(Fe) showed better stability than soluble enzymes with stability of 80% after eight cycles	Nobakht et al. (2018)
Esterification	Waste cooking oil (WCO)	30	300	12	3 g	91.8	*Thermomyces lanuginosus lipase* and *Candida antarctica* lipase B (CALB)	The enzymes were reused over five cycles with less deactivation and produced high yield	Vescovi et al. (2016)
Esterification	*Nannochloropsis gaditana* microalga	25	200	4	0.1 g	92.6	*Candida antarctica* lipase A (CALA)	Less methanol and energy consumptions were accomplished and the purity of biodiesel was 83 wt.%	López et al. (2015)

Continued

TABLE 7.2 Biodiesel production using various optimized enzymatic catalysts.—cont'd

Pathways	Various oils	Operating conditions				Y (%)	Type of biocatalysts	Novelty	References
		Temp (°C)	RPM	Time (h)	Enzyme content				
Transesterification	Fish waste	35	216	24	10 wt.%	75.3	Immobilized *Thermomyces lanuginosus* lipase (TLL)	Increasing biocatalyst content played a crucial role toward higher biodiesel yield. However, higher reaction temperatures and agitation speeds affected lower yield	Ching-Velasquez et al. (2020)
Transesterification	Fungal oil	50	200	7	2%	83.6	*Aspergillus niger*	The immobilized lipase was reused within five runs with a stable yield of 70%	El-Ghonemy et al. (2020)
Transesterification	Waste frying palm oil (WFPO)	30	400	22	5.5 wt.%	94.6	Free *Candida antarctica* lipase A (CALA)	The cleaner and more cost-efficient biodiesel production was carried out using waste oil and inexpensive biocatalyst	Guo et al. (2020)
Transesterification	Waste cooking oil (WCO)	Na	Na	4	1.5%	88	Lipase enzyme	The biodiesel blended with diesel fuel at compositions of 20%:80% and 40%:60% yielded lower gas emissions (i.e., CO, HC, and NOx) and could be applied in the diesel engine very well	Jayaraman et al. (2020)

Transesterification	*Serratia* sp. ISTD04 lipids	30	200	3	30%	97.41	*Pseudomonas* sp. ISTPL3 lipase	The biocatalysts were strongly influenced over different supports (i.e., activated biochar, pure biochar, and glass-ceramic) and the activated biochar-immobilized biocatalyst was recycled over 10 runs	Kumar and Thakur (2020)
Transesterification	Sunflower oil	40	200	25	5 wt.%	~73	*Eversa transform* (ET) and *Thermomyces lanuginosus* lipase (TLL)	Either pure ET or TLL immobilized over octyl agarose beads indicated the same remarkable activity and stability and also different pH gave an important effect toward stability of the biocatalysts	Martínez-Sanchez et al. (2020)
Transesterification	Waste cooking oil (WCO)	40	NA	48	15.2 wt.%	93.1	*Rhizomucor miehei* lipase (RML) and *Thermomyces lanuginosus* lipase (TLL)	The TLL resulted in higher biodiesel yield (93.1%) than the RML (57.5%) and these biocatalysts were recycled over five runs	Ashjari et al. (2020)

Continued

TABLE 7.2 Biodiesel production using various optimized enzymatic catalysts.—cont'd

Pathways	Various oils	Operating conditions				Y (%)	Type of biocatalysts	Novelty	References
		Temp (°C)	RPM	Time (h)	Enzyme content				
Transesterification	*Chlorella vulgaris* microalgae	45	NA	24	8.6 mg	~78	*Rhizopus oryzae* lipase (ROL) synthesized over Fe_3O_4 superpara magnetic nanoparticles (MNPs)	Biocatalyst capability and stability were improved by adding covalent bonding of 3-amino propyl triethylene silane-glutaraldehyde (MNP-AP-GA), and thus could be reused over five cycles	Nematian et al. (2020)
Transesterification	Used cooking oil	45	200	28	30 kUnit	96.5	Nile tilapia viscera (*Oreochromis niloticus*) lipase	By-products of oils and catalyst were utilized to produce biodiesel possessing physic and chemical properties similar to commercial diesel fuel in EN 14214 and ASTM D6751	Patchimpet et al. (2020)
Transesterification	Crude plant oils	30–40	35	72	0.2 g	98.8	A recombinant *Fusarium heterosporum* lipase	A harmless biodiesel solvent significantly enhanced the enzyme performance and yield produced and thus could be reused for nine runs	Quayson et al. (2020b)

Transesterification	Sunflower oil	30	1350	4	0.33 mg	>99	*Pseudomonas fluorescens* (L-AK) immobilized with carbon nanotubes	An amino-doped carbon nanotubes and GDE were applied in the reaction to escalate the enzymatic activity and stability with a reusability of 20 cycles	Bartha-Vári et al. (2020)
Transesterification	Yellow horn seed oil	35–60	NA	Up to 2	1%–6%	92.3	*Candida antarctica* lipase B (CALB) immobilized with polyporous magnetic cellulose beads (PMCBs)	PMCBs were a good carrier for immobilizing the enzyme to jack up the enzymatic activity and stability with a high reusability of five runs	Zhang et al. (2020)
Transesterification	Triolein, sunflower oil, and castor oils	25–100	NA	12	5 wt.%	99	Novozym 435 enzyme immobilized with biocompatible and hydrophilic ionic liquid (choline; H_2PO_4	The ionic liquid played a crucial role in increasing the enzyme activity and stability and the ionic liquid-immobilized enzyme was reused over five cycles	Zhu et al. (2020)
Transesterification	Phoenix tree seed oil	30	300	6.98	1, 8, and 15 wt.%	98.8	Liquid lipozyme *Thermomyces laguginosus* lipase TL100L	The lipozyme TL100L was the best enzyme among the other free liquid lipases and the performance was optimized over RSM	Sun and Li (2020)

Continued

TABLE 7.2 Biodiesel production using various optimized enzymatic catalysts.—cont'd

Pathways	Various oils	Operating conditions				Y (%)	Type of biocatalysts	Novelty	References
		Temp (°C)	RPM	Time (h)	Enzyme content				
Transesterification	Palm oil	37	250	24	25 mg	95	*Aspergillus oryzae* ST11 lipase covered as magnetic cross-linked enzyme aggregates (mCLEAs)	The mCLEAs showed excellent performance and were stably recycled over five runs with a conversion rate of more than 60%	Paitaid and Aran (2020)
Transesterification	Wet spent coffee grounds (SCGs)	60.2	6000	0.5	14.46 mL/g	97.18	1,8-Diazabicyclo [5.4.0] undec-7-ene (DBU)	The DBU biocatalyst showed perfect activity, was recycled within 10 batches, and was easily separated from the products in the last process	Nguyen et al. (2020)
Transesterification	Castor oil	50	120	24	10 wt.%	78	*Pseudomonas cepacia* lipase immobilized on a hybrid BSM	The biocatalyst resulted in high biodiesel conversion and was reused over 12 cycles	Gupta and Rathod (2020)
Transesterification	Soybean oil	37	80	48	400 mg	~79	*Pseudomonas fluorescens* lipase immobilized on sodium-modified SBA-15	Various vegetable oils produced different biodiesel yields based on different FFA contents	Ferrero et al. (2020)

Transesterification	Jatropha oil	37	80	48	400 mg	~ 85	*Pseudomonas fluorescens* lipase immobilized on sodium-modified SBA-15	Various vegetable oils produced different biodiesel yields based on different FFA contents	Ferrero et al. (2020)
Transesterification	Frying waste oil	37	80	48	400 mg	~ 94	*Pseudomonas fluorescens* lipase immobilized on sodium-modified SBA-15	Various vegetable oils produced different biodiesel yields based on different FFA contents	Ferrero et al. (2020)
Transesterification	Sunflower oil	37	80	48	400 mg	~ 96	*Pseudomonas fluorescens* lipase immobilized on sodium-modified SBA-15	Various vegetable oils produced different biodiesel yields based on different FFA contents	Ferrero et al. (2020)
Transesterification	Castor oil	35	750	8	5 wt.%	>90	Liquid enzyme *Eversa transform* lipase	Water content positively contributed to lipase stability and performance due to enlarged interfacial area between the aqueous and lipase active surface sites	Andrade et al. (2020)

Continued

TABLE 7.2 Biodiesel production using various optimized enzymatic catalysts.—cont'd

Pathways	Various oils	Operating conditions				Y (%)	Type of biocatalysts	Novelty	References
		Temp (°C)	RPM	Time (h)	Enzyme content				
Transesterification	Oleaginous microalgae	50	300	24	15 mg	97.2	*Candida antarctica* lipase B (CALB) and *Pseudomonas cepacia* (PC) lipase	The CALB and PC lipase were combined to enhance the enzyme ability, thus resulting in higher biodiesel yield than the yield obtained from the single enzymes	Sánchez-Bayo et al. (2019)
Transesterification	*Nannochloropsis* strains (i.e., *N. oculata*)	80	5000	1	0%–20%	92.59	*Thermomyces lanuginosus* (TL) lipase	The biodiesel yield was affected by different reactants and the used oil was a clean and inexpensive feedstock	He et al. (2020)
Transesterification	Waste cooking oil (WCO)	45	250	24	1%	96.5	*Candida rugosa* and *Rhizomucor miehei* lipases	The used lipases were reused over six cycles and produced biodiesel yield of 96.5%	Binhayeeding et al. (2020)
Transesterification	Oleaginous fungal lipid	45	200	4	300 mg	84.8	Lipase was immobilized over nanosized magnetic	The immobilized enzyme lipase was first optimized over magnetic nanoparticles and reused over five cycles	Jambulingam et al. (2019)

Transesterification	Soybean oil	40	200	4–12	4–14 wt.%	95.6	*Rhizomucor miehei* lipase (RML) immobilized with RML-ZIF-8	A synergistic effect between RML and ZIF-8 was a crucial factor in increasing enzymatic activity and stability with 10 batches	Adnan et al. (2018)
Transesterification	Soybean oil	45	200	24	0.2–0.5 g	88	*Aspergillus niger* lipase (ANL) immobilized with UiO866 coming from polydimeth ylsiloxane 12 (ANL-UiO-66-PDMS)	The ANL-UiO-66-PDMS could increase the reaction rate and produced high biodiesel yield and reusability of 10 cycles with less deactivation	Hu et al. (2018)

BSM, biosupport material; *CLEA*, cross-linked enzyme aggregates; *NKA*, on macroporous resin; *NN*, not mentioned; *RML-ZIF-8*, X-shaped zeolitic imidazolate frameworks; *RSM*, response surface methodology; *SDB*, styrene-divinylbenzene; *Y*, Yield.

freeze-dried *Nannochloropsis gaditana* enzyme via the supercritical CO_2 method and compressed natural gas (CNG) method to produce 1000 ton/year biodiesel for 16 years. The authors declared that the biodiesel process using enzymes was more profitable and resulted in fewer gas emissions than those using CNG (Taher et al., 2020). Due to the predominance of enzymatic processes for biodiesel production, the enzymes are widely applied in the process under the esterification reaction of FFAs and transesterification reaction of triglycerides with novel developments by 2020 (Arif et al., 2020; Awais et al., 2020; Foteinis et al., 2020; Shrivastava et al., 2020; Simsek, 2020; Rial et al., 2020).

Biodiesel production is mostly enforced over a single-step reaction, which is transesterification using oils with low FFA content (Pisarello and Querini, 2013; Loong and Idris, 2017), however a two-step reaction is also necessary when using vegetable oils with a high FFA content (Ashraful et al., 2014; Dubey et al., 2015; Kumar et al., 2014). The two-step reaction is the esterification reaction in the first step reaction (Cai et al., 2015; Suppalakpanya et al., 2010; Nakpong and Wootthikanokkhan, 2010; Jaliliannosrati et al., 2013; Wang et al., 2010; Patil et al., 2010) followed by the transesterification reaction in the second step (Rizwanul Fattah et al., 2020), where the esterification reaction in the first step is aimed at reducing the FFA content by less than 2 wt.% (Canakci and Van Gerpen, 2001) so that the lower formation of soap in the second process of the step—transesterification—is reached (Thoai et al., 2019; Math and Irfan, 2007). The lower amount of soap formed during the process facilitates the separation process between biodiesel as the main product and glycerol as the by-product in the final process and requires a small amount of water in the washing column (Thoai et al., 2017). Meanwhile, the second step reaction has been developed to decrease the biodiesel production budget from different reactants (Thoai et al., 2017, 2019).

Yahya et al. (2020) performed an esterification–transesterification reaction for biodiesel production from WCO using Fe-montmorillonite K10 catalyst in a batch reactor with a speed of 300 rpm within 6 h and the biodiesel yield obtained was 95.26%. Thoai et al. (2019) studied a two-step reaction in the production of biodiesel and announced that the two-step process was the most effective, efficient, and promising technology applied in various vegetable oils to produce biodiesel. This had already been proven by other authors (Cai et al., 2015; Canakci and Van Gerpen; 1999; Jaliliannosrati et al., 2013; Mendow et al., 2012).

Cai et al. (2015) esterified the FFA content of waste cooking oil (WCO) into acylglycerols with conversion of 99.6% reached at a catalyst content of 0.5 wt.% within 4 h; further, the acylglycerols was transesterified into biodiesel (FAME content) with a yield of 93.1% and FAME content of 98.6%. The authors claimed that the two-step process was a low-budget technology which is suitable to apply in biodiesel production using high FFA content oils. Meanwhile, Jaliliannosrati et al. (2013) successfully reduced the FFA content from 14% to less than 1% using a microwave irradiation heating method in the first step reaction (the esterification process) and implemented the response surface methodology (RSM) method to optimize the resulting biodiesel yield to 90.01% with a FAME content of 97.29% in the second step (the transesterification process).

Due to the extreme reduction rate of the FFA content of oils, the microwave irradiation heating (MIH) method has also been employed by other researchers. Suppalakpanya et al. (2010) utilized the MIH method for conversion of crude palm oil using an ethanol:FFA molar ratio of 1:24 and catalyst content of 4 wt.% in the first step process. During the esterification process, the FFA content was dramatically decreased from 7.5 wt.% to less than 2 wt.%, which was reached at a microwave power of 70 W that could be next transformed into biodiesel in the second step. This experiment resulted in a yield of 80 wt.% with a FAME content of 97.4 wt.%. Suppalakpanya et al. (2011) optimized the reduction of FFA content of crude palm oil from 7.5 wt.% to 1.4 wt.% reached at a microwave power of 78 W, oil:ethanol molar ratio of 1:6, and 1.25 wt.% of catalyst content within 1.5 h.

Although, the microwave irradiation heating (MIH) method could significantly decrease the reduction of FFA content in oils, this method needs a long reaction time. Alternatively, the ultrasonic irradiation method is a compatible and potential technology to address this problem (Worapun et al., 2010; Deng et al., 2010). Worapun et al. (2010) investigated the ultrasonic irradiation method in the esterification process to degrade FFA content only within 20 min (Worapun et al., 2010). Moreover, Deng et al. (2010) needed just 30 min to obtain a biodiesel yield of 96.4% with a stable and clean reaction in the transesterification process (Deng et al., 2010). In addition, Malani et al. (2019) employed the ultrasonic irradiation method for biodiesel production using immobilized *Thermomyces lanuginosus* lipase with an oil:alcohol ratio of 1: 7.64 and enzyme content of 3.55 wt.% by adding water. The experiment increased the biodiesel yield from 90% to 94% and the enzyme was reused over six runs.

Biodiesel production can be not only produced using esterification–transesterification reactions, but can be done using esterification–esterification reactions (Thoai et al., 2019) and transesterification–transesterification reactions (Thoai et al., 2017; Mendow et al., 2011; Okitsu et al., 2010; Yahya et al., 2020; Cai et al., 2015). In The authors highlight biodiesel production conducted over the esterification reaction and the transesterification reaction using enzymatic catalysts that have been commonly investigated in the recent literature.

7.3.1 Esterification reaction using enzymes

The hydroesterification process (combination of hydrolysis and esterification) of high FFA containing oils and fats with alcohols (mostly methanol and ethanol) has been conducted using various enzymes. (Mibielli et al., 2020; Jayaraman et al., 2020; Malani et al., 2019; Muanruksa and Kaewkannetra, 2020; Muanruksa et al., 2020). The biodiesel production carried out using the esterification process has many advantages, including: (1) average reaction temperatures; (2) no need for vegetable oil treatment; (3) toleration of high FFA-containing oils; and (4) use of low-grade oils (Zheng et al., 2020; Brusamarelo et al., 2010). Furthermore, the utilization of enzymes as biocatalysts is widely implemented by dint of removing saponification reactions (no forming soap), producing high-quality by-product (glycerol), and enabling low-cost production (Vescovi et al., 2016). Hence, the implementation of enzymes in the esterification process supplies great benefits from an economic point of view and productivity efficiency. The flowchart of the hydroesterification process to produce biodiesel using enzymes as the biocatalysts is portrayed in Fig. 7.1.

The hydroesterification process consists of a two-step reaction in which the first step results in light products containing a FAME content, and the light phase is transferred into the second step reaction by adding fresh methanol, enzymes, and water to pressurize the reaction equilibrium into the product phases. The vegetable oils are first heated up to the required temperatures to facilitate the reaction using water-diluted methanol and an enzyme catalyst in the reactor (Wancura et al., 2019a). The reactor most often used is the batch reactor performed in the range of 30–80°C, speed agitation of 200–1000 rpm, and enzyme content of over 2 wt.%, whereupon the addition of water plays a significant role in the enzymatic performance

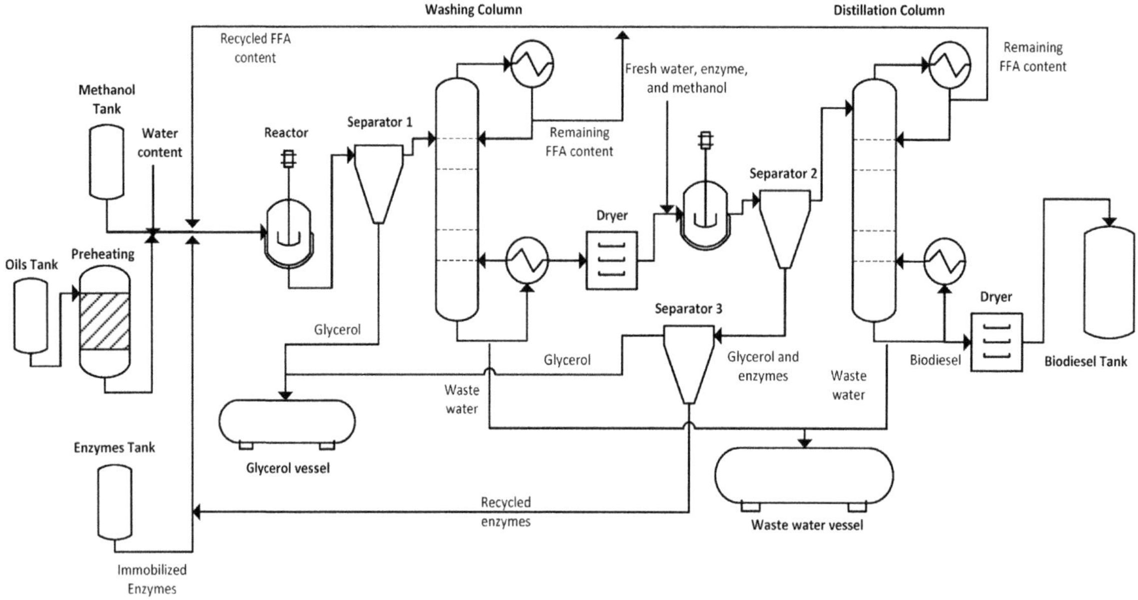

FIGURE 7.1 Hydroesterification reaction for biodiesel production using enzymes. *Adapted from Wancura, J. H., Rosset, D. V., Mazutti, M. A., Ugalde, G. A., De Oliveira, J. V., Tres, M. V., Jahn, S. L., 2019a. Improving the soluble lipase–catalyzed biodiesel production through a two-step hydroesterification reaction system. App. Microbiol. Biotechnol. 103, 7805–7817 and Hansen, R., Agerbaek, M., Nielsen, P., Rancke-Madsen, A. & Woodley, J. 2020. Esterification using a liquid lipase to remove residual free fatty acids in biodiesel. Process Biochem. 97, 213–221.*

and stability (Bartha-Vári et al., 2020; Santos et al., 2020; Valladares-Diestra et al., 2020; Aguieiras et al., 2020; Collaço et al., 2020; Binhayeeding et al., 2020). The separator used in the first and second step reactions is used to separate the main product containing biodiesel and by-products consisting of glycerol. The glycerol can be then utilized to produce other intermediate products (Samoilov et al., 2020; Brock et al., 2020; Czapka et al., 2020; Sun et al., 2020; Xu et al., 2021), and the biodiesel is used as a substitute for petroleum/diesel fuel in the transportation sector (Chandran, 2020; Abomohra et al., 2020; de Jesus et al., 2020; Hoang et al., 2020). Meanwhile, the reflux condenser installed in the top distillation column is used to remove the lost methanol, and the remaining FFA content in the last process can be recycled into the process system for reuse and the enzymes can be recycled in the reaction. The wastewater produced is stored in a wastewater tank.

The addition of different amounts of methanol and enzymes strongly affects the biodiesel yield as observed by Wancura et al. (2019b), who employed oil:methanol ratios of 1:4 and 1:4.5 and *Thermomyces lanuginosus* lipase contents of 1 wt.% and 1.45 wt.%. The maximum yield was 85.08%, achieved with an oil:methanol molar ratio of 1:4.5 and lipase content of 1 wt.% with the methanol and enzyme flow rate of 3 g/h (Wancura et al., 2019b). The use of an oil:methanol molar ratio of 1:4.5 was also investigated by Wancura et al. (2018b) using 1.45 wt.% of Callera Trans L lipase at 35°C and resulting in a biodiesel yield of 96.9% within 24 h. Meanwhile, in their previous study, Wancura et al. (2018a) stated that the maximum biodiesel yield was attained when using 100% fats with a soluble lipase after reaction time of

8 h. The productivity was also optimized with a water content with the highest yield of 97.1%, as illustrated by Wancura et al. (2019a).

By 2020, novel developments had been undertaken to discover the best method for the optimization of biodiesel yield, reusability of enzymes, productivity efficiency, lower environmental pollution, and reduced production cost for biodiesel production carried out using an esterification process. For instance, Mibielli et al. (2020) utilized soybean oil obtained from industrial waste and NS-40116 enzyme provided by Novozymes (Denmark), where the reaction was undertaken at 37°C, 200 rpm, and with an enzyme content of 3 wt.% within 12 h. The highest yield obtained was 96%. The utilization of industrial waste oils created a large profit not only by reducing production costs but also by decreasing environmental pollution (Mibielli et al., 2020).

Hansen et al. (2020) harnessed a recyclable and cost-efficient enzyme, namely *Candida antarctica* lipase B (CALB), in the reaction carried out at 40°C, 250 rpm and an enzyme content of 5% within more than 2 h as well as palm oil possessing an FFA content of 3 wt.%. The two-step transesterification process was followed by an esterification process and was a viable technology for large-scale production (Hansen et al., 2020). In contrast, Zheng et al. (2020) focused on the improvement of reusability, stability, and worthiness of the used enzymes by applying an immobilization method of cross-linked enzyme aggregates (CLEA) on *Thermomyces lanuginosus* lipase (TLL) to produce biodiesel from soybean oil. The CLEA immobilization method resulted in a 98% biodiesel yield and exhibited a good stability of 90% over 20 cycles. The CLEA-TLL performance was much better than the commercial TLL activity.

Optimization of enzymes was also developed by Rocha et al. (Rocha et al., 2020) who combined *Candida antarctica* lipase A and B (CALA and CALB) and *Rhizomucor miehei* lipase (RML) in conversion of the FFA content in residual chicken oil (RCO) at 30°C, oil:methanol ratio of 1:5%, and 15% enzyme content for 3 h. The use of 67% CALB and 33% RML generated a conversion rate of 89.95%, while the Taguchi method contributed to enhancing the biodiesel yield to 92.4%. Muanruksa et al. (2020) affirmed that the lower production cost could be accomplished when using low-price nonedible oils and recycled enzymes. In this study, the esterification reaction was executed using sludge palm oil (SPO) and *Rhizopus oryzae* lipase immobilized with alginate-polyvinyl alcohol (PVA) beads at 30–50°C, oil:methanol ratios of 1:1–1:5, and speed of 100–300 rpm within 4 h. The optimum yield was 91.30%, reached at 40°C, oil: methanol ratio of 1:3, and speed of 200 rpm. The superior conversion of low price oil (SPO) and the reuse of the lipase (15 batch runs) enabled the production cost to be lowered and thus it could be applied in wide-scale industry.

Generally, the immobilization method consists of four types, which are entrapment, physical adsorption, covalent binding, and cross-linking of the enzymes (Muanruksa et al., 2020; Pilipenko et al., 2019; Nobakht et al., 2018; Takei et al., 2020; İnal et al., 2017). However, the entrapment method is known as the best immobilization method due to its low-cost, simple process, low operating conditions, good recovery, simple diffusion, and stable enzymes (Fernandez-Arrojo et al., 2013; Zhao et al., 2015; Kahveci and Xu, 2011). For this reason, lipases immobilized with the entrapment method were also observed by Muanruksa et al. (2020) using hydrogels such as gelatin alginate (GA) and pectin alginate (PA) as a carrier for the lipase synthesis (i.e., GAL and PAL). GAL and PAL were novel biocatalysts, with waste frying acid oil (WFAO) as the reactant. These biocatalysts gave a biodiesel yield in the range of 75%–78.33% and were reused over seven cycles. Meanwhile, Rehman et al. (2019) harnessed pectin as the carrier to increase the enzyme stability, and the pectin combined with proteins and lipids was the highest potential biocatalyst.

In addition to the entrapment method, a covalent method of enzyme immobilization was, in 2020, exploited by Nezhad and Aghaei using tosylated cloisite (TCL) as the carrier on *Candida rugosa* lipase to produced biodiesel from waste frying oil (WFO). The highest yield was 97.1%, reached at 50°C, oil:methanol ratio of 1:15, and water content of 10 wt.% for 8 h. The lipase was recycled over 10 batch runs (Nezhad and Aghaei, 2020). In contrast, Pinheiro et al. (2019) used chitosan-divinyl sulfone (CHDVS) as the carrier in covalent immobilization of *Candida antarctica* lipase B (CALB), where the CHDVS-CALB showed extraordinary performance and stability. In addition, Ashjari et al. (2020) employed a covalent method for the immobilization of *Rhizomucor miehei* lipase (RML) and *Thermomyces lanuginosa* lipase (TLL) with an immobilization yield of 81%–100%. The authors declared that the lipases immobilized under the covalent method revealed extraordinary activity and stability, resulting in a biodiesel yield of 93.1% and reusability for five cycles (Ashjari et al., 2020).

Lower methanol and energy consumptions were accomplished and the purity of the biodiesel was 83 wt.% as reported by López et al. (2015) using *Candida antarctica* lipase A (CALA) to produce biodiesel using the esterification process. The highest biodiesel yield was 92.6%, reached at 25°C, 200 rpm, oil:methanol ratio of 1:1.5, and 4 g of FFA content during 4 h. Meanwhile, *Candida antarctica* lipase B (CALB) was employed by Vescovi et al. (2016) for the esterification process at 30°C, oil:water ratio of 1:4, oil: methanol ratio of 1:7, and 10 wt.% enzyme content. The highest biodiesel yield was 91.8 wt.% and the conversion rate was 90% with five cycles of the enzyme. Furthermore, Nobakht et al. (2018) used porcine pancreatic lipase (PPL) immobilized with iron-carboxylate [MIL-100(Fe)] and Keggin phosphotungstic acid $H_3PW_{12}O_{40}$ (PW), resulting in a yield of 90.4% over five cycles and the PW-MIL-100(Fe)/PPL biocatalyst maintained stability over eight runs with a constant yield of 80%.

The application of low-grade oils having high FFA content can be completely addressed using the esterification process. Also, it requires low energy consumption due to the average operating conditions and no need enzyme treatments (less chemical and water consumption). Thus, it can answer the problems raised by the use of poor-quality edible and nonedible oils with high FFA contents without producing soap and causing the correction of the reactor. Less soap forming in the reaction requires a small amount of water washing, and so water consumption is reduced. All of these solutions help in creating low-cost production and a cleaner environmental impact.

7.3.2 Enzyme-catalyzed transesterification reaction

The transesterification reaction consists of two types, which are catalytic and noncatalytic (supercritical conditions) reactions, where the catalytic reaction is performed using various catalysts (Danner et al., 2016; MacLeod et al., 2008; Ong et al., 2019; Cao et al., 2009; Sotoft et al., 2010), particularly enzymes as biocatalysts (Ngoie et al., 2020; Rafiei et al., 2018; Santos et al., 2020). The enzymatic reaction is operated at low operating parameters and is more appropriate for the reaction using low-grade oils (Marchetti et al., 2007; Al-Zuhair, 2007). Lipases also require less energy and have more simple product recovery (Balat and Balat, 2010; Fjerbaek et al., 2009). There are two enzyme types, namely liquid enzymes and immobilized enzymes. The liquid enzymes are much cheaper than the immobilized enzymes (Andrade et al., 2020) and the addition of water is required to activate the liquid enzymes and to prevent the deactivation of the lipases from methanol exposure (Poppe et al., 2015; Firdaus et al., 2016). However, too much water can lead to mitigation of the biodiesel yield and so the addition of water must be properly controlled.

To overcome the problems of enzyme ability and stability, various enzyme immobilization methods are often implemented in the enzyme treatments (Marchetti et al., 2007; Al-Zuhair, 2007; Demirbas et al., 2009; Balat and Balat, 2010; Fjerbaek et al., 2009) in order to increase the enzyme performance thus resulting in high yield and reusability of more than two cycles with less deactivation, as illustrated by other studies (Rafiei et al., 2018; Lee et al., 2006; Xie and Wang, 2012; Moreira et al., 2007). Also, the immobilized enzymes have a more remarkable performance than the liquid enzymes. Different immobilization methods applied in enzyme synthesis to produce biodiesel and the general immobilization methods used include physical adsorption (Yagiz et al., 2007; Shah and Gupta, 2007; Katiyar and Ali, 2012; Hartmann and Jung, 2010), entrapment (Noureddini et al., 2005; Kuan et al., 2013; Hsu et al., 2001; Kennedy et al., 1990), covalent bonding (Mendes et al., 2011; Rodrigues et al., 2016; Yücel, 2011; Yücel et al., 2011), and cross-linking (Shah et al., 2006; Abdulla and Ravindra, 2013; Dizge and Keskinler, 2008; Kumari et al., 2007; Han and Kim, 2011). Moreover, the advantages and disadvantages of nanoparticle-immobilized biocatalysts have been discussed by Zhong et al. (2020).

Also, excess methanol in the reaction has a negative impact in deactivation of the enzymes. Therefore, to minimize the deactivation of enzymes caused by excess methanol, the application of organic solvents plays a significant role in enhancing the methanol solubility and reducing the methanol content in the reactor (Andrade et al., 2020). In addition, glycerol negatively contributes to lipase ability and stability due to the formation of a hydrophilic circumference. Hence, glycerol removal is necessary (Guldhe et al., 2015; Lotti et al., 2015). Notwithstanding, the transesterification reaction is widely applied in biodiesel production, and some of drawbacks such as extensive reaction time, slow rate of reaction, lost activity after operation over several days, and high production cost are problems that apply when used at an industry scale (Fjerbaek et al., 2009; Zhang et al., 2003). Moreover, the high cost of the enzymes is also becomes one of the obstacles to use of the reaction, leading to increased biodiesel production cost caused by the high biocatalyst price. The reusability of the enzymes and maintaining the ability and stability of enzymes during batch runs are the best solutions to the high production expense (Nielsen et al., 2008; Nordblad et al., 2014; Amoah et al., 2016).

For these main reasons, improvements employed in the transesterification reaction have been continuously sought (Abrahamsson et al., 2015; Chen et al., 2009; De Diego et al., 2011; Guan et al., 2010; Guldhe et al., 2015; Halim et al., 2011; Hossain et al., 2020; Kawakami et al., 2011; Kumari et al., 2007; Shah et al., 2006; Lai et al., 2012; Lee et al., 2013; Li et al., 2007, 2009). For instance, El-Ghonemy et al. (2020) modified *Aspergillus niger* lipase immobilized by agarose (2%) with the highest immobilization yield of 81.7%. The improved lipase resulted in a FAME yield of 83.6% and it was recycled over five batch runs while maintaining the biodiesel yield of 70%. The flowchart process for biodiesel production conducted under the transesterification reaction using enzymes and lipases is illustrated in Fig. 7.2.

There are five main process units, namely the pretreatment unit, the main transesterification reaction unit, the recovered methanol process unit, the separation process unit between the main product (biodiesel) and by-product (glycerol), and the purified biodiesel process unit (Andrade et al., 2020). In the pretreatment unit, oils are preheated up to the required temperature to facilitate the reaction and the used enzymes are immobilized with the carriers to activate the active surface sites. Then, oils, methanol, the immobilized enzymes, and water are combined in the process system, where the

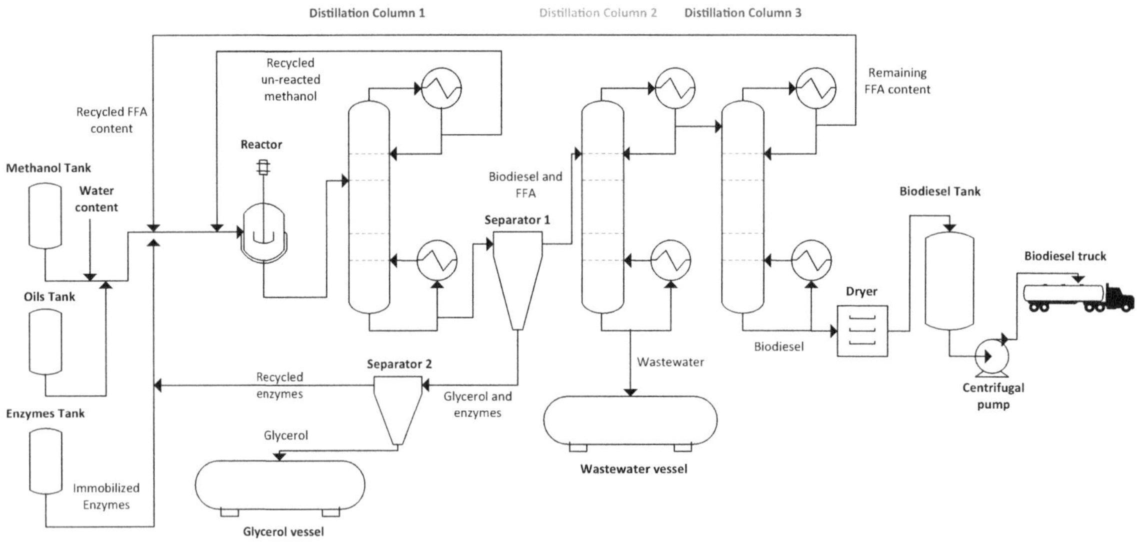

FIGURE 7.2 Transesterification reaction to produce biodiesel using enzymes. *Adapted from Andrade, T. A., Errico, M., Christensen, K. V., 2020. Implementation of biodiesel production process using enzyme-catalyzed routes. Process. Systems Eng. Biofuels Dev. 191–220 and Hansen, R., Agerbaek, M., Nielsen, P., Rancke-Madsen, A., Woodley, J., 2020. Esterification using a liquid lipase to remove residual free fatty acids in biodiesel. Process Biochem. 97, 213–221.*

addition of water content is around 5 wt.% with enzyme contents of 0.3–10 wt.%. The oil:methanol molar ratio used is in the range of 1:3–1:9 (Andrade et al., 2016, 2017c). To concatenate the different streams to create perfectly mixed solutions, a mixer static pipe is sometimes employed. Furthermore, the perfectly mixed solution is reacted in a semibatch reactor at 30–50°C, agitation speed of 200–1000 rpm and within about 2 h the immobilized enzymes and lipases are harnessed as the biocatalysts (Modi et al., 2007; Kumari et al., 2007; Ranganathan et al., 2008; Kumar et al., 2014; Karmee et al., 2015; Nayak et al., 2016; Collaço et al., 2020). The resulting products are biodiesel (FAME), glycerol (by-product), unreacted methanol, unreacted glycerides (including diglycerides and monoglycerides), water, and the remaining enzymes (Andrade et al., 2017d).

The resulting products are then sent into the first distillation column at 50 kPa to recover the nonreacted methanol that will be recycled back in the reaction. Therefore, a heat exchanger is installed in the top distillation column for methanol recovery. Meanwhile, the heat exchanger installed in the bottom distillation column keeps the temperatures under 150°C to prevent thermal decomposition of glycerol (Zhang et al., 2003). From the first distillation column, the separation process is conducted in two different separators.

Separator 1 is implemented to separate between the organic phases (biodiesel, the unreacted glycerides, the unreacted methanol, and water content) and the side-products (glycerol and the remaining enzymes). Hereinafter, the remaining enzymes and glycerol are, based on density, separated in separator 2, whereupon the enzymes are recycled into the process system and the produced glycerol is stored in a glycerol tank (Andrade et al., 2017b). The purity of glycerol should be over 92% (Andrade et al., 2020), which is why the water content is added to (Andrade et al., 2019). The high glycerol purity can be utilized to produce other intermediate chemical products to increase the factory sale value with a cleaner industry owing to the use by the waste industry into more worthy chemical products such as solketal, glycerol carbonate, glycerol monostearate (GMS), lecithin, etc. (Okoye et al., 2020; Chotchuang et al., 2020; Samoilov et al., 2020; Brock et al., 2020; Kaur et al., 2020).

Meanwhile, the products (biodiesel, unreacted glycerides, unreacted methanol, and water) produced from the top separator 1 are transferred into the second distillation column at 2.3 kPa to separate those products. The bottom distillation column is used to minimize the unreacted glycerides and to produce the waste water that is next saved in the wastewater vessel where the bottom distillation column temperature is significantly reduced by a cooler to less than 350°C to prevent the decomposition of the unreacted oils (Andrade et al., 2017a). Meanwhile, the top distillation column produces biodiesel and the remaining FFAs, thus the temperature of the top distillation column must be carefully controlled to not more than 250°C to avoid decomposition of the resulting biodiesel (Andrade et al., 2019). The temperature can be restrained using a heat exchanger installed in the top distillation column (Andrade et al., 2020).

The biodiesel and remaining FFAs are then separated in the third distillation column at 3.5 kPa, where the remaining FFAs obtained in the top column are recycled into the process system. Meanwhile, the biodiesel produced in the bottom column is dried by a dryer to obtain the desired biodiesel purity and the bottom temperature is reduced by a heat exchanger to less than 250°C to prevent disintegration of biodiesel and FFAs (Andrade et al., 2020). The biodiesel and FFA purities achieved are 99.97% and 98.2%, respectively, and thus they can be applied in the transportation sectors to substitute diesel fuel (Andrade et al., 2020). The whole process is conducted at the atmospheric pressures beside the reactor and heat exchangers so that the pressure drop during the process is ignored (Andrade et al., 2017c, 2017d, 2019). Meanwhile, the high sulfide content in the biodiesel can be removed by high vacuum distillation to obtain 84.1% biodiesel yield with a sulfur content of less than 10 ppm (Xie et al., 2019). The high sulfur content in the biodiesel has an adverse effect on the combustion of the diesel engine, resulting in the release of toxic compounds such as SOx and H_2SO_4 to the environment (Xie et al., 2019).

To optimize the productivity and to minimize the environmental effects, researchers have undertaken several modifications and accomplishments for the transesterification process such as various immobilization methods for the enzymes, as reviewed by Zhong et al. (2020), whereupon improvements in the productivity are commonly focused on the enzyme synthesis (Gupta and Rathod, 2020; Santaraite et al., 2020; Rial et al., 2020; Zhang et al., 2020). The effect of different immobilization carriers was studied by Kumar and Thakur (2020) using *Pseudomonas* sp. ISTPL3 lipase immobilized with activated biochar supported by calcite (L1), with biochar (L2) and glass-ceramic (L3). The highest biodiesel yield was 97.41%, achieved by L1, participated by L2 (94.91%) and L3 (91.50%). Moreover, the immobilized lipases generated higher FAME yield than NaOH catalyst and could be reused over 5 and 10 cycles with less deactivation (Kumar and Thakur, 2020). As can be seen, the immobilization of enzymes plays a critical position in enzyme strength and the formation of products (Andrade et al., 2019; Zhang et al., 2012).

In 2020, Bartha-Vári and co-workers improved the enzyme activity and stability by immobilizing *Pseudomonas fluorescens* lipase (L-AK) using the covalent method and by using carbon nanotubes for biodiesel production in the transesterification process. The immobilized lipase generated a conversion rate of over 99%, maintained stability of over 99%, and was recycled after 20 runs (Bartha-Vári et al., 2020). Ching-Velasquez et al. (2020) used a response surface methodology (RSM) method to optimize the octadecyl metacrylate beads-immobilized

Thermomyces lanuginosus lipase activity, resulting in the highest biodiesel yield of 75.3% reached after a 1-day reaction (Ching-Velasquez et al., 2020).

Optimization using the RSM method was also performed by Guo et al. (2020) employing frying palm oil (FPO) as the reactant and free *Candida antarctica* lipase A (CALA) as the biocatalyst. The biocatalyst was used in the esterification process of FFA content followed by the transesterification process of FPO at 30°C, water content of 16.6 wt.%, enzyme content of 5.5 wt.%, and oil:methanol molar ratio of 1:7 within 22 h. The highest FAME yield was 94.6%. The cheap FPO and biocatalyst supported by a green method positively contributed to the cost-effective production and minimal environmental effects (Guo et al., 2020). The green biocatalyst and low price oil were also implemented by He et al. (2020) exploiting *Thermomyces lanuginosus* lipase (TLL) as the biocatalyst and *N. oculata* cells as the cheap reactant. The highest biodiesel yield was 90.24% (He et al., 2020). On the other hand, the immobilized TLL activity was much better than the immobilized Eversa transform lipase performance, as observed by Martínez-Sanchez et al. (2020).

Generally, biodiesel production has been commercially conducted using the transesterification process by reacting edible and nonedible oils with alcohols (mostly methanol and ethanol) and producing biodiesel and glycerol. In contrast, the resulting glycerol can be utilized to produce other worthy intermediate chemical products. The transesterification process consists of five main process units which are the pretreatment process unit, the transesterification reaction process unit, the methanol recovery process unit, the separation process unit, and finally the purification of biodiesel process unit. All process units are carried out at atmospheric pressure, whereas the high vacuum distillation unit is conducted to reduce the sulfur content. The capability and stability of enzymes and lipases can be resolved using the immobilization method with the carriers. Thus, the immobilized enzymes and lipases could be reused in many cycles and result in a high biodiesel yield as well as contributing to a greener environment and lowering production costs.

7.4 The application of immobilized lipases for biodiesel production

The lipase catalyst reaction is very convenient to obtain highly purified product due to mild operating conditions, hence minimizing energy consumption and reducing wastewater generation. Thus, it can reduce the whole operational cost. The application of free lipase as a homogeneous catalyst for biodiesel production is restricted to a limited extent due to low reusability and recovery issue that leads to a significant increment in biodiesel production cost (Dhawane et al., 2018). Therefore, researchers are looking for a solution to address this issue by immobilizing enzymes on an inert support. The advantages of enzyme immobilization are increased catalyst stability and reusability, and reduced downstream processing costs as it removes the catalyst separation step. Despite these advantages, immobilized enzymes may be less active than free enzymes due to conformational changes at the active sites of the enzymes, and the immobilization procedure can be costly and time-consuming (Hommes et al., 2019).

There are various conventional immobilization techniques that have been reported in recent decades for biodiesel production, such as adsorption, covalent bonding, cross-linking, encapsulation, and entrapment. Colin-Luna et al. (2018) conducted a study on biodiesel production using WCO by immobilizing the PPL on a zirconium-pillared clay (Zr-PILC) using adsorption and cross-linking. The results show that the adsorption method yielded the highest methyl esters with 1000 ppm in contrast to cross-linking with 500 ppm (Colín-Luna et al., 2018). The results indicate that the cross-linker such as glutaraldehyde may become toxic to enzymes, thus reducing its activity. The immobilization of *Burkholderia cepacia* lipase and *Candida antarctica* lipase via encapsulation in a silica aerogel was investigated in the synthesis of biodiesel from sunflower seed oil with methyl acetate. In such aerogels, similar protein specific activity was achieved with *Burkholderia cepacia* lipase and *Candida antarctica* lipase, with ≈0.39 and ≈0.30 U/mg protein, respectively. The study suggested that the aerogel encapsulation technique helps in maintaining enzymes in a dispersion state with similar properties to the dispersion prevailing in an aqueous solution, even when applied in organic hydrophobic media (Orçaire et al., 2006).

Advancements in technology have led to the development of modern immobilization techniques such as cross-linked enzyme aggregates (CLEAs) that have been employed and eliminated the use of support during enzyme immobilization. The mixture of lipases was immobilized using CLEAs that could be reused for up to five cycles in the ethanolysis of soybean oil resulting in more than 80% biodiesel production (Ramos et al., 2019). The improvement of CLEA recovery and its recyclability can be achieved by introducing magnetite nanoparticles and it can functionalized with 3-aminopropyl triethoxysilane (APTES) by a silanization process in order to acquire magnetic CLEAs. The experiment managed to obtain satisfactory results for biodiesel production at 71%, whereas free lipase produced 42% biodiesel from WCO after 36 h (Badoei-dalfard et al., 2019). The fabrication of a new and advanced immobilization method received a great deal of attention in order to reduce the economic value by using enzymes as biocatalyst. Hopefully, the emerging new technologies, especially in the immobilization area, will pave the way for the production of high-quality and high-yield biodiesel.

7.5 Challenges and opportunities of lipase for biodiesel production

Dealing with a biological catalyst such as lipase promises good results for future generations. They offer unique benefits over conventional chemical processes with regard to sustainability and efficiency. These will gradually reduce the world's dependence on fossil fuels and simultaneously reduce harmful emissions that cause environmental pollution. Animal fats, vegetable oils, and waste cooking oil are seen some of the alternatives that can be employed as raw materials to produce biodiesel. However, in-depth research needs to be carried out to achieve the best conditions with the best catalysts for high biodiesel yield. Researchers worldwide have carried out and there are still on-going experiments with the aim of achieving the highest biodiesel yield with the most economical value and sustainable process. Several challenges faced and future opportunities when undergoing enzymatic biodiesel production are discussed in the next section.

7.5.1 Challenges

Among the main challenges when dealing with free lipase is the high cost in maintaining the overall process and the final production. The challenge is greater when biodiesel is produced on a larger scale. As is generally known, biological catalysts are very expensive and have a slower reaction rate compared with chemical catalysts. However, due to their environmental-friendly properties and excellent performance, researchers are putting great interest and investment into searching for the best conditions and parameters for high biodiesel yield. Immobilization of lipase on several inert supports is being explored widely with the aim of achieving high recyclability of the lipase for multiple reactions. Additionally, lipase is prone to be deactivated by the excess solvent (methanol/ethanol) available in the reaction. In order to make the biodiesel e suitable for current diesel engines, several parameters and properties of the biodiesel need to be aligned with the regulatory standards.

7.5.2 Opportunities

The use of lipase on low-priced feedstock such as waste cooking oil can reduce the production cost and thus improve the environment, whereby, instead of disposing of the waste, it can be reused to produce valuable biodiesel. The lipase-catalyzed reaction for biodiesel production is highly suitable for low-grade oil containing high impurities such as high FFA and high moisture contents compared with a chemical catalyst (Wancura et al., 2020). This is because of saponification when the chemical catalyst is used with oil that has high FFA and moisture contents. The occurrence of saponification will reduce the biodiesel yield and make the separation process more complicated. The ability of lipase to process a high amount of residual raw materials may compensate for the high cost of the biological catalyst. An interesting fact when dealing with lipase is the possibility of the enzymatic reaction being combined with alkaline transesterification, where this enables high-acidity feedstocks to be processed for biodiesel production (Nordblad et al., 2016). A more sustainable process can be achieved when immobilized lipase is used. Moreover, a multienzymatic system could enable a great conversion efficiency, which allows for more effective bioconversion. Coimmobilization of the enzyme with a chemical catalyst is also seen as a possible route to enhance the overall catalytic performance. Overall, the lipase-catalyzed reaction is a more environmental-friendly process compared to the current technique applied industrially. Employing lipase in the biodiesel production process will emphasize the importance of using biocatalyst in addressing resource efficiency, low toxicity, safe and milder reaction, and energy-efficiency. With a deeper investigation into the factors influencing enzymatic biodiesel production, enzymes are projected to have a significant contribution in improving overall food processing sustainability.

7.6 Conclusion

Lipase is known as a versatile enzyme due to its ability to catalyze various types of chemical reactions using a wide array of substrates. The integration of fermentation technology and genetic engineering help in increasing the efficiency of lipase production for various applications ranging from the food, pharmaceutical, and biofuel sectors. The application of enzymes as a catalyst has paved the way toward the development of sustainable and environmental-friendly industrial processes by generating less waste and by-products, while consuming less energy compared to chemical catalysts. The lipase-catalyzed esterification or transesterification reaction provides tremendous opportunities in enhancing the efficiency of biodiesel production using cleaner and greener processes. In this context, the advancements in enzyme technology, such as enzyme immobilization, are great achievements in increasing the stability, activity, and robustness of the enzymes to meet the industrial objectives.

Acknowledgments

This work was supported by the Ministry of Higher Education Malaysia under the Fundamental Research Grant Scheme (FRGS/1/2018/STG01/UTM/01/4) and Universiti Teknologi Malaysia (UTM).

References

Abdulla, R., Ravindra, P., 2013. Immobilized *Burkholderia cepacia* lipase for biodiesel production from crude *Jatropha curcas* L. oil. Biomass Bioenergy 56, 8–13.

Abomohra, A.E.-F., Elsayed, M., Esakkimuthu, S., El-Sheekh, M., Hanelt, D., 2020. Potential of fat, oil and grease (FOG) for biodiesel production: a critical review on the recent progress and future perspectives. Prog. Energy Combust. Sci. 81, 100868.

Abrahamsson, J., Andreasson, E., Hansson, N., Sandström, D., Wennberg, E., Maréchal, M., Martinelli, A., 2015. A Raman spectroscopic approach to investigate the production of biodiesel from soybean oil using 1-alkyl-3-methylimidazolium ionic liquids with intermediate chain length. App. Energy 154, 763–770.

Adnan, M., Li, K., Xu, L., Yan, Y., 2018. X-shaped ZIF-8 for immobilization rhizomucor miehei lipase via encapsulation and its application toward biodiesel production. Catalysts 8, 96.

Agarwal, M., Chauhan, G., Chaurasia, S., Singh, K., 2012. Study of catalytic behavior of KOH as homogeneous and heterogeneous catalyst for biodiesel production. J. Taiwan Inst. Chem. Eng. 43, 89–94.

Aguieiras, É.C., Cavalcanti, E.D., Da Silva, P.R., Soares, V.F., Fernandez-Lafuente, R., Assunção, C.L.B., Da Silva, J.A.C., Freire, D.M., 2020. Enzymatic synthesis of neopentyl glycol-bases biolubricants using biodiesel from soybean and castor bean as raw materials. Renew. Energy 148, 689–696.

Akhabue, C.E., Osa-Benedict, E.O., Oyedoh, E.A., Otoikhian, S.K., 2020. Development of a bio-based bifunctional catalyst for simultaneous esterification and transesterification of neem seed oil: modeling and optimization studies. Renew. Energy 152, 724–735.

Almyasheva, N.R., Shuktueva, M.I., Petrova, D.A., Kopitsyn, D.S., Kotelev, M.S., Vinokurov, V.A., Novikov, A.A., 2018. Biodiesel fuel production by Aspergillus Niger whole-cell biocatalyst in optimized medium. Mycoscience 59, 147–152.

Al-Zuhair, S., 2007. Production of biodiesel: possibilities and challenges. Biofuels Bioprod. Biorefin. 1, 57–66.

Amini, Z., Ilham, Z., Ong, H.C., Mazaheri, H., Chen, W.-H., 2017. State of the art and prospective of lipase-catalyzed transesterification reaction for biodiesel production. Energy Convers. Manag. 141, 339–353.

Amoah, J., Ho, S.-H., Hama, S., Yoshida, A., Nakanishi, A., Hasunuma, T., Ogino, C., Kondo, A., 2016. Lipase cocktail for efficient conversion of oils containing phospholipids to biodiesel. Bioresour. Technol. 211, 224–230.

Andrade, M.F., Parussulo, A.L., Netto, C.G., Andrade, L.H., Toma, H.E., 2016. Lipase immobilized on polydopamine-coated magnetite nanoparticles for biodiesel production from soybean oil. Biofuel Res. J. 3, 403–409.

Andrade, T., Errico, M., Christensen, K., 2017a. Castor oil transesterification catalysed by liquid enzymes: feasibility of reuse under various reaction conditions. Chem. Eng. Trans. 57, 913–918.

Andrade, T.A., Errico, M., Christensen, K.V., 2017b. Evaluation of reaction mechanisms and kinetic parameters for the transesterification of castor oil by liquid enzymes. Ind. Eng. Chem. Res. 56, 9478–9488.

Andrade, T.A., Errico, M., Christensen, K.V., 2017c. Influence of the reaction conditions on the enzyme catalyzed transesterification of castor oil: a possible step in biodiesel production. Bioresour. Technol. 243, 366–374.

Andrade, T.A., Errico, M., Christensen, K.V., 2017d. Transesterification of castor oil catalyzed by liquid enzymes: optimization of reaction conditions. Comput. Aided Chem. Eng. 40, 2863–2868.

Andrade, T.A., Martin, M., Errico, M., Christensen, K.V., 2019. Biodiesel production catalyzed by liquid and immobilized enzymes: optimization and economic analysis. Chem. Eng. Res. Des. 141, 1–14.

Andrade, T.A., Errico, M., Christensen, K.V., 2020. Implementation of biodiesel production process using enzyme-catalyzed routes. Process. Systems Eng. Biofuels Dev. 191–220.

Arif, M., Bai, Y., Usman, M., Jalalah, M., Harraz, F.A., Al-Assiri, M., Li, X., Salama, E.-S., Zhang, C., 2020. Highest accumulated microalgal lipids (polar and non-polar) for biodiesel production with advanced wastewater treatment: role of lipidomics. Bioresour. Technol. 298, 122299.

Ashjari, M., Garmroodi, M., Asl, F.A., Emampour, M., Yousefi, M., Lish, M.P., Habibi, Z., Mohammadi, M., 2020. Application of multi-component reaction for covalent immobilization of two lipases on aldehyde-functionalized magnetic nanoparticles; production of biodiesel from waste cooking oil. Process Biochem. 90, 156–167.

Ashraful, A.M., Masjuki, H.H., Kalam, M.A., Fattah, I.R., Imtenan, S., Shahir, S., Mobarak, H., 2014. Production and comparison of fuel properties, engine performance, and emission characteristics of biodiesel from various non-edible vegetable oils: a review. Energy Convers. Manag. 80, 202–228.

Atabani, A.E., Silitonga, A.S., Badruddin, I.A., Mahlia, T., Masjuki, H., Mekhilef, S., 2012. A comprehensive review on biodiesel as an alternative energy resource and its characteristics. Renew. Sustain. Energy Rev. 16, 2070–2093.

Atadashi, I., Aroua, M.K., Aziz, A.A., Sulaiman, N., 2012. The effects of water on biodiesel production and refining technologies: a review. Renew. Sustain. Energy Rev. 16, 3456–3470.

Atadashi, I., Aroua, M., Aziz, A.A., Sulaiman, N., 2013. The effects of catalysts in biodiesel production: a review. Ind. Eng. Chem. Res. 19, 14–26.

Atapour, M., Kariminia, H.-R., Moslehabadi, P.M., 2014. Optimization of biodiesel production by alkali-catalyzed transesterification of used frying oil. Process Safety Environ. Protect. 92, 179–185.

Awais, M., Musmar, S.E.A., Kabir, F., Batool, I., Rasheed, M.A., Jamil, F., Khan, S.U., Tlili, I., 2020. Biodiesel production from *Melia azedarach* and *Ricinus communis* oil by transesterification process. Catalysts 10, 427.

Badoei-Dalfard, A., Malekabadi, S., Karami, Z., Sargazi, G., 2019. Magnetic cross-linked enzyme aggregates of Km12 lipase: a stable nanobiocatalyst for biodiesel synthesis from waste cooking oil. Renew. Energy 141, 874–882.

Balat, M., Balat, H., 2010. Progress in biodiesel processing. App. Energy 87, 1815–1835.

Bartha-Vári, J.-H., Moisă, M.E., Bencze, L.C., Irimie, F.-D., Paizs, C., Tosa, M.I., 2020. Efficient biodiesel production catalyzed by nanobioconjugate of lipase from *Pseudomonas fluorescens*. Molecules 25, 651.

Bateni, H., Karimi, K., Zamani, A., Benakashani, F., 2014. Castor plant for biodiesel, biogas, and ethanol production with a biorefinery processing perspective. App. Energy 136, 14–22.

Binhayeeding, N., Klomklao, S., Prasertsan, P., Sangkharak, K., 2020. Improvement of biodiesel production using waste cooking oil and applying single and mixed immobilised lipases on polyhydroxyalkanoate. Renew. Energy 162, 1819–1827.

Borges, M.E., Díaz, L., 2012. Recent developments on heterogeneous catalysts for biodiesel production by oil esterification and transesterification reactions: a review. Renew. Sustain. Energy Rev. 16, 2839–2849.

Boviatsi, E., Papadaki, A., Efthymiou, M.N., Nychas, G.J.E., Papanikolaou, S., Da Silva, J.A., Freire, D.M., Koutinas, A., 2020. Valorisation of sugarcane molasses for the production of microbial lipids via fermentation of two *Rhodosporidium* strains for enzymatic synthesis of polyol esters. J. Chem. Technol. Biotechnol. 95, 402–407.

Brock, D., Koder, A., Rabl, H.-P., Touraud, D., Kunz, W., 2020. Optimising the biodiesel production process: implementation of glycerol derivatives into biofuel formulations and their potential to form hydrofuels. Fuel 264, 116695.

Brusamarelo, C.Z., Rosset, E., De Cesaro, A., Treichel, H., De Oliveira, D., Mazutti, M.A., Di Luccio, M., Oliveira, J.V., 2010. Kinetics of lipase-catalyzed synthesis of soybean fatty acid ethyl esters in pressurized propane. J. Biotechnol. 147, 108–115.

Cai, Z.-Z., Wang, Y., Teng, Y.-L., Chong, K.-M., Wang, J.-W., Zhang, J.-W., Yang, D.-P., 2015. A two-step biodiesel production process from waste cooking oil via recycling crude glycerol esterification catalyzed by alkali catalyst. Fuel Process. Technol. 137, 186–193.

Canakci, M., Van Gerpen, J., 1999. Biodiesel production via acid catalysis. Transac. ASAE 42, 1203.

Canakci, M., Van Gerpen, J., 2001. Biodiesel production from oils and fats with high free fatty acids. Transac. ASAE 44, 1429.

Cao, P., Tremblay, A.Y., Dubé, M.A., 2009. Kinetics of canola oil transesterification in a membrane reactor. Ind. Eng. Chem. Res. 48, 2533–2541.

Chandran, D., 2020. Compatibility of diesel engine materials with biodiesel fuel. Renew. Energy 147, 89–99.

Chen, Y., Xiao, B., Chang, J., Fu, Y., Lv, P., Wang, X., 2009. Synthesis of biodiesel from waste cooking oil using immobilized lipase in fixed bed reactor. Energy Convers. Manag. 50, 668–673.

Ching-Velasquez, J., Fernández-Lafuente, R., Rodrigues, R.C., Plata, V., Rosales-Quintero, A., Torrestiana-Sánchez, B., Tacias-Pascacio, V.G., 2020. Production and characterization of biodiesel from oil of fish waste by enzymatic catalysis. Renew. Energy 153, 1346–1354.

Chotchuang, A., Kunsuk, P., Phanpitakkul, A., Chanklang, S., Chareonpanich, M., Seubsai, A., 2020. Production of glycerol carbonate from glycerol over modified sodium-aluminate-doped calcium oxide catalysts. Catalysis Today. https://doi.org/10.1016/j.cattod.2020.06.007 (in Press).

Colín-Luna, J.A., Zamora-Rodea, E.G., González-Brambila, M.M., Barrera-Calva, E., Rosas-Cedillo, R., Medina-Mendoza, A.K., García-Martínez, J.C., 2018. Biodiesel production using immobilized lipase supported on a zirconium-pillared clay. Effect of the immobilization method. Int. J. Chem. React. 16.

Collaço, A.C.A., Aguieiras, E.C., Santos, J.G., De Oliveira, R.A., De Castro, R.D.P.V., Freire, D.M., 2020. Experimental study and preliminary economic evaluation of enzymatic biodiesel production by an integrated process using co-products from palm (*Elaeais guineensis Jaquim*) industry. Ind. Crops. Prod. 157, 112904.

Costa, M.J., Silva, M.R., Ferreira, E.E., Carvalho, A.K.F., Basso, R.C., Pereira, E.B., De Castro, H.F., Mendes, A.A., Hirata, D.B., 2020. Enzymatic biodiesel production by hydroesterification using waste cooking oil as feedstock. Chem. Eng. Process. 108131.

Czapka, A., König, S., Pergola, C., Grune, C., Vougogiannopoulou, K., Skaltsounis, A.-L., Fischer, D., Werz, O., 2020. The indirubin derivative 6-bromoindirubin-3′-glycerol-oxime ether (6BIGOE) potently modulates inflammatory cytokine and prostaglandin release from human monocytes through GSK-3 interference. Biochem. Pharmacol. 180, 114170.

Danner, T., Brown, P.N., Mauss, M., 2016. Method of Alcoholisis of Fatty Acids and Fatty Acid Glycerides. Google Patents.

David, G.F., Perez, V.H., Justo, O.R., Cubides, D.C., Cardona, C.A., Hristov, J., 2016. Glycerol bioconversion in unconventional magnetically assisted bioreactor seeking whole cell biocatalyst (intracellular lipase) production. Chem. Eng. Res. Des. 111, 243–252.

De Diego, T., Manjón, A., Lozano, P., Iborra, J.L., 2011. A recyclable enzymatic biodiesel production process in ionic liquids. Bioresour. Technol. 102, 6336–6339.

De Jesus, S.S., Ferreira, G.F., Moreira, L.S., Maciel Filho, R., 2020. Biodiesel production from microalgae by direct transesterification using green solvents. Renew. Energy 160, 1283–1294.

Demirbas, M.F., Balat, M., Balat, H., 2009. Potential contribution of biomass to the sustainable energy development. Energy Convers. Manag. 50, 1746–1760.

Deng, X., Fang, Z., Liu, Y.-H., 2010. Ultrasonic transesterification of *Jatropha curcas* L. oil to biodiesel by a two-step process. Energy Convers. Manag. 51, 2802–2807.

Dhawane, S.H., Kumar, T., Halder, G., 2018. Process optimisation and parametric effects on synthesis of lipase immobilised carbonaceous catalyst for conversion of rubber seed oil to biodiesel. Energy Convers. Manag. 176, 55–68.

Dizge, N., Keskinler, B., 2008. Enzymatic production of biodiesel from canola oil using immobilized lipase. Biomass Bioenergy 32, 1274–1278.

Dubey, S.M., Gole, V.L., Gogate, P.R., 2015. Cavitation assisted synthesis of fatty acid methyl esters from sustainable feedstock in presence of heterogeneous catalyst using two step process. Ultrason. Sonochem. 23, 165–173.

El-Ghonemy, D.H., Ali, T.H., Hassanein, N.M., Abdellah, E.M., Fadel, M., Awad, G.E., Abdou, D.A., 2020. Thermo-alkali-stable lipase from a novel *Aspergillus niger*: statistical optimization, enzyme purification, immobilization and its application in biodiesel production. Prep. Biochem. Biotech. 1–16.

Encinar, J.M., Gonzalez, J.F., Rodríguez-Reinares, A., 2005. Biodiesel from used frying oil. Variables affecting the yields and characteristics of the biodiesel. Ind. Eng. Chem. Res. 44, 5491–5499.

Farooq, M., Ramli, A., Subbarao, D., 2013. Biodiesel production from waste cooking oil using bifunctional heterogeneous solid catalysts. J. Clean. Prod. 59, 131–140.

Fernandez-Arrojo, L., Rodriguez-Colinas, B., Gutierrez-Alonso, P., Fernandez-Lobato, M., Alcalde, M., Ballesteros, A.O., Plou, F.J., 2013. Dried alginate-entrapped enzymes (DALGEEs) and their application to the production of fructooligosaccharides. Process Biochem. 48, 677–682.

Ferrero, G.O., Faba, E.M.S., Rickert, A.A., Eimer, G.A., 2020. Alternatives to rethink tomorrow: biodiesel production from residual and non-edible oils using biocatalyst technology. Renew. Energy 150, 128–135.

Firdaus, M.Y., Guo, Z., Fedosov, S.N., 2016. Development of kinetic model for biodiesel production using liquid lipase as a biocatalyst, esterification step. Biochem. Eng. J. 105, 52–61.

Fjerbaek, L., Christensen, K.V., Norddahl, B., 2009. A review of the current state of biodiesel production using enzymatic transesterification. Biotechnol. Bioeng. 102, 1298–1315.

Foteinis, S., Chatzisymeon, E., Litinas, A., Tsoutsos, T., 2020. Used-cooking-oil biodiesel: life cycle assessment and comparison with first-and third-generation biofuel. Renew. Energy 153, 588–600.

Fraga, J.L., Penha, A.C.B., Da, S.P.A., Silva, K.A., Akil, E., Torres, A.G., Amaral, P.F.F., 2018. Use of Yarrowia lipolytica lipase immobilized in cell debris for the production of lipolyzed milk fat (LMF). Int. J. Mol. Sci. 19.

Furini, G., Berger, J.S., Campos, J.a.M., Sand, S.T.V., Germani, J.C., 2018. Production of lipolytic enzymes by bacteria isolated from biological effluent treatment systems. An. Acad. Bras. Cienc. 90, 2955–2965.

Galadima, A., Muraza, O., 2014. Biodiesel production from algae by using heterogeneous catalysts: a critical review. Energy 78, 72–83.

Garcia-Silvera, E.E., Martinez-Morales, F., Bertrand, B., Morales-Guzman, D., Rosas-Galvan, N.S., Leon-Rodriguez, R., Trejo-Hernandez, M.R., 2018. Production and application of a thermostable lipase from *Serratia marcescens* in detergent formulation and biodiesel production. Biotechnol Appl. Biochem. 65, 156–172.

Geoffry, K., Achur, R.N., 2017. A novel halophilic extracellular lipase with both hydrolytic and synthetic activities. Biocatal. Agricult. Biotechnol. 12, 125–130.

Guan, G., Kusakabe, K., Yamasaki, S., 2009. Tri-potassium phosphate as a solid catalyst for biodiesel production from waste cooking oil. Fuel Process. Technol. 90, 520–524.

Guan, F., Peng, P., Wang, G., Yin, T., Peng, Q., Huang, J., Guan, G., Li, Y., 2010. Combination of two lipases more efficiently catalyzes methanolysis of soybean oil for biodiesel production in aqueous medium. Process Biochem. 45, 1677–1682.

Guldhe, A., Singh, B., Rawat, I., Permaul, K., Bux, F., 2015. Biocatalytic conversion of lipids from microalgae Scenedesmus obliquus to biodiesel using Pseudomonas fluorescens lipase. Fuel 147, 117–124.

Guo, J., Sun, S., Liu, J., 2020. Conversion of waste frying palm oil into biodiesel using free lipase A from Candida antarctica as a novel catalyst. Fuel 267, 117323.

Gupta, A.R., Rathod, V.K., 2020. Biodiesel synthesis from palm fatty acid distillate using enzyme immobilized on magnetic nanoparticles. SN App. Sci. 2.

Halim, R., Gladman, B., Danquah, M.K., Webley, P.A., 2011. Oil extraction from microalgae for biodiesel production. Bioresour. Technol. 102, 178–185.

Han, J.-Y., Kim, H.-K., 2011. Transesterification using the cross-linked enzyme aggregate of Photobacterium lipolyticum lipase M37. J. Microbiol. Biotechnol. 21, 1159–1165.

Hansen, R., Agerbaek, M., Nielsen, P., Rancke-Madsen, A., Woodley, J., 2020. Esterification using a liquid lipase to remove residual free fatty acids in biodiesel. Process Biochem. 97, 213–221.

Haq, A., Adeel, S., Khan, A., Rana, Q.U.A., Khan, M.a.N., Rafiq, M., Ishfaq, M., Khan, S., Shah, A.A., Hasan, F., Ahmed, S., Badshah, M., 2020. Screening of lipase-producing bacteria and optimization of lipase-mediated biodiesel production from *Jatropha curcas* seed oil using whole cell approach. BioEnergy Res. 13, 1280–1296.

Hartmann, M., Jung, D., 2010. Biocatalysis with enzymes immobilized on mesoporous hosts: the status quo and future trends. J. Mat. Chem. 20, 844–857.

He, Y., Zhang, B., Guo, S., Guo, Z., Chen, B., Wang, M., 2020. Sustainable biodiesel production from the green microalgae Nannochloropsis: novel integrated processes from cultivation to enzyme-assisted extraction and ethanolysis of lipids. Energy Convers. Manag. 209, 112618.

Hoang, A.T., Tabatabaei, M., Aghbashlo, M., Carlucci, A.P., Ölçer, A.I., Le, A.T., Ghassemi, A., 2020. Rice bran oil-based biodiesel as a promising renewable fuel alternative to petrodiesel: a review. Renew. Sustain. Energy Rev. 135, 110204.

Hommes, A., De Wit, T., Euverink, G.J.W., Yue, J., 2019. Enzymatic biodiesel synthesis by the biphasic esterification of oleic acid and 1-butanol in microreactors. Ind. Eng. Chem. Res. 58, 15432–15444.

Hossain, S.Z., Razzak, S.A., Al Shater, A.F., Moniruzzaman, M., Hossain, M.M., 2020. Recent advances in enzymatic conversion of microalgal lipids into biodiesel. Energy Fuels 34 (6), 6735–6750.

Hsu, A.-F., Jones, K., Marmer, W.N., Foglia, T.A., 2001. Production of alkyl esters from tallow and grease using lipase immobilized in a phyllosilicate sol-gel. J. Am. Oil Chem. Soc. 78, 585–588.

Hu, Y., Dai, L., Liu, D., Du, W., 2018. Rationally designing hydrophobic UiO-66 support for the enhanced enzymatic performance of immobilized lipase. Green Chem. 20, 4500–4506.

Iboyo, A., David Asitok, A., George Ekpenyong, M., Peter Antai, S., 2017. Selection of *Enterobacter cloacae* Strain POPE6 for fermentative production of extracellular lipase on palm kernel oil processing effluent. Int. J. Sci. 3, 1–17.

Iuliano, M., Sarno, M., De Pasquale, S., Ponticorvo, E., 2020. Candida rugosa lipase for the biodiesel production from renewable sources. Renew. Energy 162, 124–133.

İnal, M., Işıklan, N., Yiğitoğlu, M., 2017. Preparation and characterization of pH-sensitive alginate-g-poly (N-vinyl-2-pyrrolidone)/gelatin blend beads. Ind. Eng. Chem. Res. 52, 128–137.

Jaliliannosrati, H., Amin, N.a.S., Talebian-Kiakalaieh, A., Noshadi, I., 2013. Microwave assisted biodiesel production from *Jatropha curcas* L. seed by two-step in situ process: optimization using response surface methodology. Bioresour. Technol. 136, 565–573.

Jambulingam, R., Shalma, M., Shankar, V., 2019. Biodiesel production using lipase immobilised functionalized magnetic nanocatalyst from oleaginous fungal lipid. J. Clean. Prod. 215, 245–258.

Jayaraman, J., Alagu, K., Appavu, P., Joy, N., Jayaram, P., Mariadoss, A., 2020. Enzymatic production of biodiesel using lipase catalyst and testing of an unmodified compression ignition engine using its blends with diesel. Renew. Energy 145, 399–407.

Jegannathan, K.R., Abang, S., Poncelet, D., Chan, E.S., Ravindra, P., 2008. Production of biodiesel using immobilized lipase—a critical review. Crit. Rev. Biotech. 28, 253–264.

Jia, J., Yang, X., Wu, Z., Zhang, Q., Lin, Z., Guo, H., Lin, C.S., Wang, J., Wang, Y., 2015. Optimization of fermentation medium for extracellular lipase production from Aspergillus Niger using response surface methodology. Biomed. Res. Int. 2015, 497462.

Kahveci, D., Xu, X., 2011. Enhancement of activity and selectivity of *Candida rugosa* lipase and *Candida antarctica* lipase A by bioimprinting and/or immobilization for application in the selective ethanolysis of fish oil. Biotechnol. Lett. 33, 2065–2071.

Karmakar, B., Samanta, S., Halder, G., 2020. Delonix regia heterogeneous catalyzed two-step biodiesel production from *Pongamia pinnata* oil using methanol and 2-propanol. J. Clean. Prod. 255, 120313.

Karmee, S.K., Linardi, D., Lee, J., Lin, C.S.K., 2015. Conversion of lipid from food waste to biodiesel. Waste Manag. 41, 169–173.

Katiyar, M., Ali, A., 2012. Immobilization of *Candida rugosa* lipase on MCM-41 for the transesterification of cotton seed oil. J. Oleo Sci. 61, 469–475.

Kaur, J., Gera, P., Jha, M., Sarma, A.K., 2020. A study on conversion of glycerol into solketal using rice husk-derived catalyst. In: Advances in Energy Research, Vol. 2. Springer.

Kawakami, K., Oda, Y., Takahashi, R., 2011. Application of a *Burkholderia cepacialipase*-immobilized silica monolith to batch and continuous biodiesel production with a stoichiometric mixture of methanol and crude Jatropha oil. Biotechnol. Biofuels 4, 42.

Kennedy, J., Melo, E.M., Jumel, K., 1990. Immobilized enzymes and cells. Chem. Eng. Prog. 86, 81–89.

Knothe, G., 2010. Biodiesel and renewable diesel: a comparison. Prog. Energy Combust. Sci. 36, 364–373.

Knothe, G., Razon, L.F., 2017. Biodiesel fuels. Prog. Energy Combust. Sci. 58, 36–59.

Konwar, L.J., Boro, J., Deka, D., 2014. Review on latest developments in biodiesel production using carbon-based catalysts. Renew. Sustain. Energy Rev. 29, 546–564.

Kuan, I.-C., Lee, C.-C., Tsai, B.-H., Lee, S.-L., Lee, W.-T., Yu, C.-Y., 2013. Optimizing the production of biodiesel using lipase entrapped in biomimetic silica. Energies 6, 2052–2064.

Kumar, V., Thakur, I.S., 2020. Biodiesel production from transesterification of *Serratia* sp. ISTD04 lipids using immobilised lipase on biocomposite materials of biomineralized products of carbon dioxide sequestrating bacterium. Bioresour. Technol. 123193.

Kumar, V., Muthuraj, M., Palabhanvi, B., Ghoshal, A.K., Das, D., 2014. Evaluation and optimization of two stage sequential in situ transesterification process for fatty acid methyl ester quantification from microalgae. Renew. Energy 68, 560–569.

Kumari, V., Shah, S., Gupta, M.N., 2007. Preparation of biodiesel by lipase-catalyzed transesterification of high free fatty acid containing oil from *Madhuca indica*. Energy Fuels 21, 368–372.

Lai, J.-Q., Hu, Z.-L., Wang, P.-W., Yang, Z., 2012. Enzymatic production of microalgal biodiesel in ionic liquid [BMIm] [PF6]. Fuel 95, 329–333.

Lee, D.H., Kim, J.M., Shin, H.Y., Kang, S.W., Kim, S.W., 2006. Biodiesel production using a mixture of immobilized *Rhizopus oryzae* and *Candida rugosa* lipases. Biotechnol. Bioprocess Eng. 11, 522–525.

Lee, O.K., Kim, Y.H., Na, J.-G., Oh, Y.-K., Lee, E.Y., 2013. Highly efficient extraction and lipase-catalyzed transesterification of triglycerides from Chlorella sp. KR-1 for production of biodiesel. Bioresour. Technol. 147, 240–245.

Li, X., Xu, H., Wu, Q., 2007. Large-scale biodiesel production from microalga *Chlorella protothecoides* through heterotrophic cultivation in bioreactors. Biotechnol. Bioeng. 98, 764–771.

Li, N.-W., Zong, M.-H., Wu, H., 2009. Highly efficient transformation of waste oil to biodiesel by immobilized lipase from *Penicillium expansum*. Process Biochem. 44, 685–688.

Li, S.-F., Fan, Y.-H., Hu, R.-F., Wu, W.-T., 2011. *Pseudomonas cepacia* lipase immobilized onto the electrospun PAN nanofibrous membranes for biodiesel production from soybean oil. J. Mol. Catal. B 72, 40–45.

Lima, L.N., Oliveira, G.C., Rojas, M.J., Castro, H.F., Da Rós, P.C., Mendes, A.A., Giordano, R.L., Tardioli, P.W., 2015. Immobilization of *Pseudomonas fluorescens* lipase on hydrophobic supports and application in biodiesel synthesis by transesterification of vegetable oils in solvent-free systems. J. Ind. Microbiol. 42, 523–535.

Long, Y.-D., Fang, Z., Su, T.-C., Yang, Q., 2014. Co-production of biodiesel and hydrogen from rapeseed and Jatropha oils with sodium silicate and Ni catalysts. App. Energy 113, 1819–1825.

Loong, T.C., Idris, A., 2017. One step transesterification of biodiesel production using simultaneous cooling and microwave heating. J. Clean. Prod. 146, 57–62.

López, B.C., Cerdán, L.E., Medina, A.R., López, E.N., Valverde, L.M., Peña, E.H., Moreno, P.A.G., Grima, E.M., 2015. Production of biodiesel from vegetable oil and microalgae by fatty acid extraction and enzymatic esterification. J. Biosci. Bioeng. 119, 706–711.

López-Fernández, J., Benaiges, M.D., Valero, F., 2020. Rhizopus oryzae lipase, a promising industrial enzyme: biochemical characteristics, production and biocatalytic applications. Catalysts 10.

Lotero, E., Liu, Y., Lopez, D.E., Suwannakarn, K., Bruce, D.A., Goodwin, J.G., 2005. Synthesis of biodiesel via acid catalysis. Ind. Eng. Chem. Res. 44, 5353–5363.

Lotero, E., Goodwin Jr., J.G., Bruce, D.A., Suwannakarn, K., Liu, Y., Lopez, D.E., 2006. The catalysis of biodiesel synthesis. Catalysis 19, 41–83.

Lotti, M., Pleiss, J., Valero, F., Ferrer, P., 2015. Effects of methanol on lipases: molecular, kinetic and process issues in the production of biodiesel. Biotechnol. J. 10, 22–30.

Louhasakul, Y., Cheirsilp, B., Intasit, R., Maneerat, S., Saimmai, A., 2020. Enhanced valorization of industrial wastes for biodiesel feedstocks and biocatalyst by lipolytic oleaginous yeast and biosurfactant-producing bacteria. Int. Biodeterior. Biodegrad. 148, 104911.

Macleod, C.S., Harvey, A.P., Lee, A.F., Wilson, K., 2008. Evaluation of the activity and stability of alkali-doped metal oxide catalysts for application to an intensified method of biodiesel production. Chem. Eng. J. 135, 63–70.

Malani, R.S., Umriwad, S.B., Kumar, K., Goyal, A., Moholkar, V.S., 2019. Ultrasound–assisted enzymatic biodiesel production using blended feedstock of non–edible oils: kinetic analysis. Energy Convers. Manag. 188, 142–150.

Mansir, N., Taufiq-Yap, Y.H., Rashid, U., Lokman, I.M., 2017. Investigation of heterogeneous solid acid catalyst performance on low grade feedstocks for biodiesel production: a review. Energy Convers. Manag. 141, 171–182.

Marchetti, J., Miguel, V., Errazu, A., 2007. Possible methods for biodiesel production. Renew. Sustain. Energy Rev. 11, 1300–1311.

Martínez-Sanchez, J.A., Arana-Peña, S., Carballares, D., Yates, M., Otero, C., Fernandez-Lafuente, R., 2020. Immobilized biocatalysts of Eversa® Transform 2.0 and lipase from Thermomyces lanuginosus: comparison of some properties and performance in biodiesel production. Catalysts 10, 738.

Math, M., Irfan, G., 2007. Optimization of restaurant waste oil methyl ester yield. J. Sci. Ind. Res. 66 (9), 772–776.

Mendes, A.A., Giordano, R.C., De Lc Giordano, R., De Castro, H.F., 2011. Immobilization and stabilization of microbial lipases by multipoint covalent attachment on aldehyde-resin affinity: application of the biocatalysts in biodiesel synthesis. J. Mol. Catal. B 68, 109–115.

Mendow, G., Veizaga, N., Sánchez, B., Querini, C., 2011. Biodiesel production by two-stage transesterification with ethanol. Bioresour. Technol. 102, 10407–10413.

Mendow, G., Veizaga, N., Sánchez, B., Querini, C., 2012. Biodiesel production by two-stage transesterification with ethanol by washing with neutral water and water saturated with carbon dioxide. Bioresour. Technol. 118, 598–602.

Mibielli, G.M., Fagundes, A.P., Bohn, L.R., Cavali, M., Bueno, A., Bender, J.P., Oliveira, J.V., 2020. Enzymatic production of methyl esters from low-cost feedstocks. Biocat. Agr. Biotechnol. 101558.

Modi, M.K., Reddy, J., Rao, B., Prasad, R., 2007. Lipase-mediated conversion of vegetable oils into biodiesel using ethyl acetate as acyl acceptor. Bioresour. Technol. 98, 1260–1264.

Moreira, A.B., Perez, V.H., Zanin, G.M., De Castro, H.F., 2007. Biodiesel synthesis by enzymatic transesterification of palm oil with ethanol using lipases from several sources immobilized on silica–PVA composite. Energy Fuels 21, 3689–3694.

Muanruksa, P., Kaewkannetra, P., 2020. Combination of fatty acids extraction and enzymatic esterification for biodiesel production using sludge palm oil as a low-cost substrate. Renew. Energy 146, 901–906.

Muanruksa, P., Dujjanutat, P., Kaewkannetra, P., 2020. Entrapping immobilisation of lipase on biocomposite hydrogels toward for biodiesel production from waste. Fry. Acid Oil. Catalysts 10, 834.

Nakpong, P., Wootthikanokkhan, S., 2010. High free fatty acid coconut oil as a potential feedstock for biodiesel production in Thailand. Renew. Energy 35, 1682–1687.

Navarro Lopez, E., Robles Medina, A., Gonzalez Moreno, P.A., Esteban Cerdan, L., Martin Valverde, L., Molina Grima, E., 2016. Biodiesel production from *Nannochloropsis gaditana* lipids through transesterification catalyzed by *Rhizopus oryzae* lipase. Bioresour. Technol. 203, 236–244.

Nayak, P.K., Dash, U., Rayaguru, K., Krishnan, K.R., 2016. Physio-chemical changes during repeated frying of cooked oil: a review. J. Food Biochem. 40, 371–390.

Nematian, T., Salehi, Z., Shakeri, A., 2020. Conversion of bio-oil extracted from *Chlorella vulgaris* micro algae to biodiesel via modified superparamagnetic nano-biocatalyst. Renew. Energy 146, 1796–1804.

Nezhad, M.K., Aghaei, H., 2020. Tosylated cloisite as a new heterofunctional carrier for covalent immobilization of lipase and its utilization for production of biodiesel from waste frying oil. Renew. Energy 164, 876–888.

Ngoie, W.I., Oyekola, O.O., Ikhu-Omoregbe, D., Welz, P.J., 2020. Valorisation of edible oil wastewater sludge: bioethanol and biodiesel production. Waste Biomass Valori. 11, 2431–2440.

Nguyen, H.C., Nguyen, M.L., Wang, F.-M., Juan, H.-Y., Su, C.-H., 2020. Biodiesel production by direct transesterification of wet spent coffee grounds using switchable solvent as a catalyst and solvent. Bioresour. Technol. 296, 122334.

Nielsen, P.M., Brask, J., Fjerbaek, L., 2008. Enzymatic biodiesel production: technical and economical considerations. Eur. J. Lipid Sci. Technol. 110, 692–700.

Nobakht, N., Faramarzi, M.A., Shafiee, A., Khoobi, M., Rafiee, E., 2018. Polyoxometalate-metal organic framework-lipase: an efficient green catalyst for synthesis of benzyl cinnamate by enzymatic esterification of cinnamic acid. Int. J. Biol. Macromol. 113, 8–19.

Nordblad, M., Silva, V.T., Nielsen, P.M., Woodley, J.M., 2014. Identification of critical parameters in liquid enzyme-catalyzed biodiesel production. Biotechnol. Bioeng. 111, 2446–2453.

Nordblad, M., Pedersen, A.K., Rancke-Madsen, A., Woodley, J.M., 2016. Enzymatic pretreatment of low-grade oils for biodiesel production. Biotechnol. Bioeng. 113, 754–760.

Noureddini, H., Gao, X., Philkana, R., 2005. Immobilized *Pseudomonas cepacia* lipase for biodiesel fuel production from soybean oil. Bioresour. Technol. 96, 769–777.

Okitsu, K., Sadanaga, Y., Takenaka, N., Maeda, Y., Bandow, H., 2010. A two-step continuous ultrasound assisted production of biodiesel fuel from waste cooking oils: a practical and economical approach to produce high quality biodiesel fuel. Bioresour. Technol. 101, 5394–5401.

Okoye, P.U., Wang, S., Khanday, W.A., Li, S., Tang, T., Zhang, L., 2020. Box-Behnken optimization of glycerol transesterification reaction to glycerol carbonate over calcined oil palm fuel ash derived catalyst. Renew. Energy 146, 2676–2687.

Ong, H.C., Milano, J., Silitonga, A.S., Hassan, M.H., Wang, C.-T., Mahlia, T.M.I., Siswantoro, J., Kusumo, F., Sutrisno, J., 2019. Biodiesel production from *Calophyllum inophyllum-Ceiba pentandra* oil mixture: optimization and characterization. J. Clean. Prod. 219, 183–198.

Orçaire, O., Buisson, P., Pierre, A.C., 2006. Application of silica aerogel encapsulated lipases in the synthesis of biodiesel by transesterification reactions. J. Mol. Cat. B 42, 106–113.

Pahoja, V.M., Sethar, M.A., 2002. A review of enzymatic properties of lipase in plants, animals and microorganisms. Pak. J. App. Sci. 2, 474–484.

Paitaid, P., Aran, H., 2020. Magnetic Cross-Linked Enzyme Aggregates of Aspergillus oryzae ST11 Lipase using polyacrylonitrile coated magnetic nanoparticles for biodiesel production. App. Biochem. Biotechnol. 190, 1319–1332.

Patchimpet, J., Simpson, B.K., Sangkharak, K., Klomklao, S., 2020. Optimization of process variables for the production of biodiesel by transesterification of used cooking oil using lipase from Nile tilapia viscera. Renew. Energy 153, 861–869.

Patil, P., Deng, S., Rhodes, J.I., Lammers, P.J., 2010. Conversion of waste cooking oil to biodiesel using ferric sulfate and supercritical methanol processes. Fuel 89, 360–364.

Pilipenko, N., Gonçalves, O.H., Bona, E., Fernandes, I.P., Pinto, J.A., Sorita, G.D., Leimann, F.V., Barreiro, M.F., 2019. Tailoring swelling of alginate-gelatin hydrogel microspheres by crosslinking with calcium chloride combined with transglutaminase. Carbohydr. Polym. 223, 115035.

Ping, L., Yuan, X., Zhang, M., Chai, Y., Shan, S., 2018. Improvement of extracellular lipase production by a newly isolated Yarrowia lipolytica mutant and its application in the biosynthesis of L-ascorbyl palmitate. Int. J. Biol. Macromol. 106, 302–311.

Pinheiro, B.B., Rios, N.S., Aguado, E.R., Fernandez-Lafuente, R., Freire, T.M., Fechine, P.B., Dos Santos, J.C., Gonçalves, L.R., 2019. Chitosan activated with divinyl sulfone: a new heterofunctional support for enzyme immobilization. Application in the immobilization of lipase B from Candida antarctica. Int. J. Biol. Macromol. 130, 798–809.

Pisarello, M.L., Querini, C.A., 2013. Catalyst consumption during one and two steps transesterification of crude soybean oils. Chem. Eng. J. 234, 276–283.

Pollardo, A.A., Lee, H.-S., Lee, D., Kim, S., Kim, J., 2018. Solvent effect on the enzymatic production of biodiesel from waste animal fat. J. Clean. Prod. 185, 382–388.

Poppe, J.K., Fernandez-Lafuente, R., Rodrigues, R.C., Ayub, M.a.Z., 2015. Enzymatic reactors for biodiesel synthesis: present status and future prospects. Biotechnol. Adv. 33, 511–525.

Poppe, J.K., Matte, C.R., Fernandez-Lafuente, R., Rodrigues, R.C., Ayub, M.a.Z., 2018. Transesterification of waste frying oil and soybean oil by combi-lipases under ultrasound-assisted reactions. App. Biochem. Biotechnol. 186, 576–589.

Quayson, E., Amoah, J., Rachmadona, N., Hama, S., Yoshida, A., Kondo, A., Ogino, C., 2020b. Biodiesel-mediated biodiesel production: a recombinant Fusarium heterosporum lipase-catalyzed transesterification of crude plant oils. Fuel Process. Technol. 199, 106278.

Rafiei, S., Tangestaninejad, S., Horcajada, P., Moghadam, M., Mirkhani, V., Mohammadpoor-Baltork, I., Kardanpour, R., Zadehahmadi, F., 2018. Efficient biodiesel production using a lipase@ ZIF-67 nanobioreactor. Chem. Eng. J. 334, 1233–1241.

Ramos, M.D., Miranda, L.P., Fernandez-Lafuente, R., Kopp, W., Tardioli, P.W., 2019. Improving the yields and reaction rate in the ethanolysis of soybean oil by using mixtures of lipase CLEAs. Molecules 24.

Ranganathan, S.V., Narasimhan, S.L., Muthukumar, K., 2008. An overview of enzymatic production of biodiesel. Bioresour. Technol. 99, 3975–3981.

Razack, S.A., Duraiarasan, S., 2016. Response surface methodology assisted biodiesel production from waste cooking oil using encapsulated mixed enzyme. Waste Manag. 47, 98–104.

Rehman, A., Ahmad, T., Aadil, R.M., Spotti, M.J., Bakry, A.M., Khan, I.M., Zhao, L., Riaz, T., Tong, Q., 2019. Pectin polymers as wall materials for the nano-encapsulation of bioactive compounds. Trends Food Sci. Technol. 90, 35–46.

Rial, R.C., De Freitas, O.N., Nazário, C.E.D., Viana, L.H., 2020. Biodiesel from soybean oil using Porcine pancreas lipase immobilized on a new support: p-nitrobenzyl cellulose xanthate. Renew. Energy 149, 970–979.

Rizwanul Fattah, I., Ong, H., Mahlia, T., Mofijur, M., Silitonga, A., Rahman, S., Ahmad, A., 2020. State of the art of catalysts for biodiesel production. Front. Energy Res. 8, 101.

Rocha, T.G., Pedro, H.D.L., De Souza, M.C., Monteiro, R.R., Dos Santos, J.C., 2020. Lipase cocktail for optimized biodiesel production of free fatty acids from residual chicken. Oil. Catal. Lett. 1–12.

Rodrigues, J., Canet, A., Rivera, I., Osório, N., Sandoval, G., Valero, F., Ferreira-Dias, S., 2016. Biodiesel production from crude Jatropha oil catalyzed by non-commercial immobilized heterologous *Rhizopus oryzae* and Carica papaya lipases. Bioresour. Technol. 213, 88–95.

Romdhane, I.B.-B., Romdhane, Z.B., Bouzid, M., Gargouri, A., Belghith, H., 2013. Application of a chitosan-immobilized *Talaromyces thermophilus* lipase to a batch biodiesel production from waste frying oils. App. Biochem. Biotechnol. 171, 1986–2002.

Sakai, T., Kawashima, A., Koshikawa, T., 2009. Economic assessment of batch biodiesel production processes using homogeneous and heterogeneous alkali catalysts. Bioresour. Technol. 100, 3268–3276.

Samoilov, V., Borisov, R., Stolonogova, T., Zarezin, D., Maximov, A., Bermeshev, M., Chernysheva, E., Kapustin, V., 2020. Glycerol to renewable fuel oxygenates. Part II: gasoline-blending characteristics of glycerol and glycol derivatives with C3-C4 alkyl (idene) substituents. Fuel 280, 118585.

Sánchez-Bayo, A., Morales, V., Rodríguez, R., Vicente, G., Bautista, L.F., 2019. Biodiesel production (FAEEs) by heterogeneous combi-lipase biocatalysts using wet extracted lipids from microalgae. Catalysts 9, 296.

Sani, Y.M., Daud, W.M.a.W., Aziz, A.A., 2014. Activity of solid acid catalysts for biodiesel production: a critical review. App. Catal. A. Gen. 470, 140–161.

Santaraite, M., Sendzikiene, E., Makareviciene, V., Kazancev, K., 2020. Biodiesel production by lipase-catalyzed in situ transesterification of rapeseed oil containing a high free fatty acid content with ethanol in diesel fuel media. Energies 13, 2588.

Santos, S., Puna, J., Gomes, J., 2020. A review on bio-based catalysts (immobilized enzymes) used for biodiesel production. Energies 13, 3013.

Shah, S., Gupta, M.N., 2007. Lipase catalyzed preparation of biodiesel from Jatropha oil in a solvent free system. Process Biochem. 42, 409–414.

Shah, S., Sharma, A., Gupta, M.N., 2006. Preparation of cross-linked enzyme aggregates by using bovine serum albumin as a proteic feeder. Analytical Biochem. 351, 207–213.

Shrivastava, P., Verma, T.N., Samuel, O.D., Pugazhendhi, A., 2020. An experimental investigation on engine characteristics, cost and energy analysis of CI engine fuelled with Roselle, Karanja biodiesel and its blends. Fuel 275, 117891.

Simsek, S., 2020. Effects of biodiesel obtained from Canola, sefflower oils and waste oils on the engine performance and exhaust emissions. Fuel 265, 117026.

Singh, D., Sharma, D., Soni, S., Sharma, S., Kumari, D., 2019. Chemical compositions, properties, and standards for different generation biodiesels: a review. Fuel 253, 60–71.

Sotoft, L.F., Rong, B.-G., Christensen, K.V., Norddahl, B., 2010. Process simulation and economical evaluation of enzymatic biodiesel production plant. Bioresour. Technol. 101, 5266–5274.

Sun, S., Li, K., 2020. Biodiesel production from phoenix tree seed oil catalyzed by liquid lipozyme TL100L. Renew. Energy 151, 152–160.

Sun, T., Zhang, H., Dong, Z., Liu, Z., Zheng, M., 2020. Ultrasonic-promoted enzymatic preparation, identification and multi-active studies of nature-identical phenolic acid glycerol derivatives. RSC Adv. 10, 11139–11147.

Suppalakpanya, K., Ratanawilai, S., Tongurai, C., 2010. Production of ethyl ester from crude palm oil by two-step reaction with a microwave system. Fuel 89, 2140–2144.

Taher, H., Giwa, A., Abusabiekeh, H., Al-Zuhair, S., 2020. Biodiesel production from Nannochloropsis gaditana using supercritical CO_2 for lipid extraction and immobilized lipase transesterification: economic and environmental impact assessments. Fuel Process. Technol. 198, 106249.

Takei, T., Yoshihara, R., Danjo, S., Fukuhara, Y., Evans, C., Tomimatsu, R., Ohzuno, Y., Yoshida, M., 2020. Hydrophobically-modified gelatin hydrogel as a carrier for charged hydrophilic drugs and hydrophobic drugs. Int. J. Biol. Macromol. 149, 140–147.

Tan, T., Lu, J., Nie, K., Deng, L., Wang, F., 2010. Biodiesel production with immobilized lipase: a review. Biotechnol. Adv. 28, 628–634.

Teo, S.H., Goto, M., Taufiq-Yap, Y.H., 2015. Biodiesel production from Jatropha curcas L. oil with Ca and La mixed oxide catalyst in near supercritical methanol conditions. J. Supercrit. Fluids 104, 243–250.

Thoai, D.N., Tongurai, C., Prasertsit, K., Kumar, A., 2017. A novel two-step transesterification process catalyzed by homogeneous base catalyst in the first step and heterogeneous acid catalyst in the second step. Fuel Process. Technol. 168, 97–104.

Thoai, D.N., Tongurai, C., Prasertsit, K., Kumar, A., 2019. Review on biodiesel production by two-step catalytic conversion. Biocataly. Agricult. Biotechnol. 18, 101023.

Valladares-Diestra, K., De Souza Vandenberghe, L.P., Soccol, C.R., 2020. Oilseed enzymatic pretreatment for efficient oil recovery in biodiesel production industry: a review. BioEnergy Res. 13, 1016–1030.

Verma, M.L., Barrow, C.J., Puri, M., 2013. Nanobiotechnology as a novel paradigm for enzyme immobilisation and stabilisation with potential applications in biodiesel production. App. Microbiol. Biotechnol. 97, 23–39.

Vescovi, V., Rojas, M.J., Baraldo, A., Botta, D.C., Santana, F.a.M., Costa, J.P., Machado, M.S., Honda, V.K., Giordano, R.D.L.C., Tardioli, P.W., 2016. Lipase-catalyzed production of biodiesel by hydrolysis of waste cooking oil followed by esterification of free fatty acids. J. Am. Oil Chem. Soc. 93, 1615–1624.

Vieitez, I., Callejas, N., Irigaray, B., Pinchak, Y., Merlinski, N., Jachmanián, I., Grompone, M.A., 2014. Acid value, polar compounds and polymers as determinants of the efficient conversion of waste frying oils to biodiesel. J. Am. Oil Chem. Soc. 91, 655–664.

Wancura, J.H., Rosset, D.V., Brondani, M., Mazutti, M.A., Oliveira, J.V., Tres, M.V., Jahn, S.L., 2018a. Soluble lipase-catalyzed synthesis of methyl esters using a blend of edible and nonedible raw materials. Bioprocess Biosyst. Eng. 41, 1185–1193.

Wancura, J.H., Rosset, D.V., Tres, M.V., Oliveira, J.V., Mazutti, M.A., Jahn, S.L., 2018b. Production of biodiesel catalyzed by lipase from Thermomyces lanuginosus in its soluble form. Can. J. Chem. Eng. 96, 2361–2368.

Wancura, J.H., Rosset, D.V., Mazutti, M.A., Ugalde, G.A., De Oliveira, J.V., Tres, M.V., Jahn, S.L., 2019a. Improving the soluble lipase–catalyzed biodiesel production through a two-step hydroesterification reaction system. App. Microbiol. Biotechnol. 103, 7805–7817.

Wancura, J.H., Rosset, D.V., Ugalde, G.A., Oliveira, J.V., Mazutti, M.A., Tres, M.V., Jahn, S.L., 2019b. Feeding strategies of methanol and lipase on eversa® transform-mediated hydroesterification for FAME production. Can. J. Chem. Eng. 97, 1332–1339.

Wancura, J.a.H.C., Tres, M.V., Jahn, S.E.L., Oliveira, J.E.V.D., 2020. Lipases in liquid formulation for biodiesel production: current status and challenges. Biotechnol. Appl. Biochem. 67, 648–667.

Wang, Y., Nie, J., Zhao, M., Ma, S., Kuang, L., Han, X., Tang, S., 2010. Production of biodiesel from waste cooking oil via a two-step catalyzed process and molecular distillation. Energy Fuels 24, 2104–2108.

Waseem, A., Ali, S., Khalid, S.W., 2018. Enhanced production of an extracellular lipase by EMS and MMS-induced mutant strain of *Rhizopus oligosporus* EM-7 using almond meal as a basal substrate. Pak. J. Zool. 50.

Willsey, G.G., Wargo, M.J., 2015. Extracellular lipase and protease production from a model drinking water bacterial community is functionally robust to absence of individual members. PLoS One 10, e0143617.

Worapun, I., Pianthong, K., Thaiyasuit, P., 2010. Synthesis of biodiesel by two-step transesterification from crude *Jatropha curcus* L. oil using ultrasonic irradiation assisted. Eng. App. Sci. Res. 37, 169–179.

Xie, W., Wang, J., 2012. Immobilized lipase on magnetic chitosan microspheres for transesterification of soybean oil. Biomass Bioenergy 36, 373–380.

Xie, Q., Cai, L., Xia, F., Liang, X., Wu, Z., Liu, Y., Li, X., Lu, M., Nie, Y., Ji, J., 2019. High vacuum distillation for low-sulfur biodiesel production: from laboratory to large scale. J. Cleaner Prod. 223, 379–385.

Xu, Z., Pan, C., Li, X., Hao, N., Zhang, T., Gaffrey, M.J., Pu, Y., Cort, J.R., Ragauskas, A.J., Qian, W.-J., 2021. Enhancement of polyhydroxyalkanoate production by Co-feeding lignin derivatives with glycerol in *Pseudomonas putida* KT2440. Biotechnol. Biofuels 14 (1), 11. https://doi.org/10.1186/s13068-020-01861-2.

Yagiz, F., Kazan, D., Akin, A.N., 2007. Biodiesel production from waste oils by using lipase immobilized on hydrotalcite and zeolites. Chem. Eng. J. 134, 262–267.

Yahya, S., Wahab, S.K.M., Harun, F.W., 2020. Optimization of biodiesel production from waste cooking oil using Fe-Montmorillonite K10 by response surface methodology. Renew. Energy 157, 164–172.

Yücel, Y., 2011. Biodiesel production from pomace oil by using lipase immobilized onto olive pomace. Bioresour. Technol. 102, 3977–3980.

Yücel, Y., Demir, C., Dizge, N., Keskinler, B., 2011. Lipase immobilization and production of fatty acid methyl esters from canola oil using immobilized lipase. Biomass Bioenergy 35, 1496–1501.

Zeng, S., Liu, J., Anankanbil, S., Chen, M., Guo, Z., Adams, J.P., Snajdrova, R., Li, Z., 2018. Amide synthesis via aminolysis of ester or acid with an intracellular lipase. ACS Catalysis 8, 8856–8865.

Zhang, Y., Dube, M., Mclean, D., Kates, M., 2003. Biodiesel production from waste cooking oil: 1. Process design and technological assessment. Bioresour. Technol. 89, 1–16.

Zhang, B., Weng, Y., Xu, H., Mao, Z., 2012. Enzyme immobilization for biodiesel production. App. Microbiol. Biotechnol. 93, 61–70.

Zhang, H., Liu, T., Zhu, Y., Hong, L., Li, T., Wang, X., Fu, Y., 2020. Lipases immobilized on the modified polyporous magnetic cellulose support as an efficient and recyclable catalyst for biodiesel production from Yellow horn seed oil. Renew. Energy 145, 1246–1254.

Zhao, X., Qi, F., Yuan, C., Du, W., Liu, D., 2015. Lipase-catalyzed process for biodiesel production: enzyme immobilization, process simulation and optimization. Renew. Sustain. Energy Rev. 44, 182–197.

Zheng, J., Wei, W., Wang, S., Li, X., Zhang, Y., Wang, Z., 2020. Immobilization of Lipozyme TL 100L for methyl esterification of soybean oil deodorizer distillate. 3 Biotech 10, 1–10.

Zhong, L., Feng, Y., Wang, G., Wang, Z., Bilal, M., Lv, H., Jia, S., Cui, J., 2020. Production and use of immobilized lipases in/on nanomaterials: a review from the waste to biodiesel production. Int. J. Biol. Macromol. 152, 207–222.

Zhu, A., Feng, W., Li, Z., Cheng, S., Chen, Q., Fan, D., Guo, Y., Li, L., Wang, J., 2020. Cleaner enzymatic production of biodiesel with easy separation procedures triggered by a biocompatible hydrophilic ionic liquid. Green Chem. 22, 1944–1951.

CHAPTER

8

Importance of nanocatalyst and its role in biofuel production

Elsa Cherian[1], G. Kalavathy[2], T. Jayasree Joshi[1], M. Grace Lydia Phoebe[2], Baskar Gurunathan[2]

[1]Department of Food Technology, Saintgits College of Engineering, Kottayam, Kerala, India; [2]Department of Biotechnology, St. Joseph's College of Engineering, Chennai, Tamil Nadu, India

8.1 Introduction to nanocatalysis

Nanotechnology and nanoscience are an important area of research that is growing at a very faster rate. Nanotechnology is the field of research in which nanoscale (1–100 μm) molecules and atoms are used. Reducing the size can alter the structure, so that many functional properties may be activated (Mahmoud et al., 2021). These nanostructured materials impart a wide range of properties with innovative applications in many fields including health care, food processing and packaging sector, construction, defense, cosmetics, and energy manufacturing (Xu et al., 2019). Through nanotechnology, molecular properties like strength, durability, conduction, and reactivity may be modified, which can add many benefits to the area of research. It also helps in maintaining the environmental balance through bioremediation, and supports the filtration of water and desalination. There are many biosensors working on this principle. Most currently available products rely on some aspects of nanotechnology. Therefore, in near future, nanotechnology will become an irreplaceable part of research and development.

Catalysis is a process of enhancing the rate of a chemical reaction by the addition of catalyst. This is achieved by lowering the activation of energy. The reaction can be homogeneous or heterogeneous. A catalytic reaction where all the reactants are in the same phase is homogeneous catalysis. When the reactants are in different phases, it is called a heterogeneous catalysis reaction. When enzymes are used as a catalyst, the reaction also can be called a biocatalytic reaction or biocatalysis. In most cases, biocatalysis is considered to be an intermediate between homogeneous and heterogeneous reactions.

Nanocatalysis is considered to be an emerging scientific field as it opens up a new way to meet the problems of energy, environmental stability,

Biofuels and Bioenergy
https://doi.org/10.1016/B978-0-323-85269-2.00022-8

and sustainability. These are the problems faced currently around the world which can be resolved through nanocatalysis. Nanocatalysis can be homogeneous or heterogeneous.

In homogeneous catalysis, mostly nanoparticles in transition metal form which are available in colloidal form are utilized as catalysts. Here the nanoparticles are finely disseminated in an organic or aqueous solution, or a solvent mixture to prevent agglomeration. The use of stabilizers plays a major role in the prevention of agglomeration. Also, stabilizers are selected in such a way that they will enhance the activity of the catalyst. The main advantage of this type of catalyst is the availability of a huge number of atoms on the surface.

Some of the benefits of homogeneous catalysts include manageable operating conditions, increased catalytic activity, availability, etc. Along with the advantages, one of the main disadvantages in the usage of homogeneous catalyst is the difficulty in separation (Kim et al., 2008). The advantages of heterogeneous catalysts include easier separation and improved reusability (Nayebzadeh et al., 2018), along with the difficulty in leaching and long reaction time (Wang et al., 2019). One of the important facets of the nanocatalyst that helps in this activity is the nature of the catalyst. Other than the nature of the catalyst, choosing a suitable synthesis method has an important effect on the physicochemical properties and performance (Todorović et al., 2019; Mierczynski et al., 2015).

Out of the different synthesis methods for the homogeneous transition metal complexes, the chemical reduction method, is most commonly used. Some of the methods for the synthesis include thermal, sonochemical, or photochemical, reduction of the precursor transition metal salts, electrochemical reduction of transition metal precursor salts, ligand reduction, and displacement from organometallic precursors and metal vapor synthesis. Heterogeneous nanocatalysts can be synthesized by the adsorption of nanoparticles onto the support. $CuO/ZnO/Al_2O_3$ mixed oxide catalysts were produced by hydrothermal, gel oxalate precipitation, and nonthermal glow discharge plasma synthesis methods (Seyed and Mohammad, 2020). Novel nanoparticles and technologies for the synthesis of these are being discovered currently.

8.2 Role of nanoparticles as catalyst

Nanoparticles are described as nanoclusters, nanorods, nanofibers, and nanowires. Materials in the nano range can quicken any process by creating reaction sites for the substrates in solid, liquid, or gaseous phase. The effects of nanoparticles make a difference in moving forward in the general adequacy of the method by increasing bioethanol productivity (Deepika et al., 2019). Nanoparticles enhance the production of bioethanol through the following steps.

- Pretreatment: The raw biomass is pretreated with metal nanoparticles which are intensively productive by piercing the cell wall of the biomass to release carbohydrates, which can be used for biofuel production.
- Enzymatic hydrolysis: The addition of a smaller quantity of nanoparticles when added to the media can increase the activity of the produced enzyme by 40%. In close proximity to metal nanoparticles, enzymes are thermally stable for up to a few hours. Metal nanoparticles are used as cofactors to increase the stability of enzymes and the enzyme immobilization of a backing material which then enhances the activity of enzymes also.
- Fermentation: Magnetic nanoparticles play a major role in enhancement of the reaction rate and efficiency of entrapping yeast cells.

The major variables that have to be considered while producing the final products are allowing active factors affecting the rate of the reaction, such as size of the particle, external shape, surface area, and the sort of biomass used to be at the optimum level (Santos et al., 2016).

8.3 Effect of nanocatalysts in the biofuel industry

8.3.1 Bioethanol production

With expanded accessibility and reduced usage as fuel, ethanol is a potential base molecule for the manufacture of an assortment of necessary chemicals. There have been many recent advances in the conversion of ethanol utilizing a catalytic method into an extensive range of chemicals, fuels, etc. A variety of lignocellulosic biomasses are considered to be the most significant and most apt feedstocks for ventures related to the production of a diverse range of high-value products which include bioethanol. However, the perplexing structure of these biomasses supports their pretreatment to discharge the carbohydrates and other polymers in them.

Pretreatment of biomass has been found as one of the most vital and first step in bioethanol synthesis. Acids such as butyl carboxylic acid (Fe_3O_4–MNPs–Si–BCOOH) and alkyl sulfonic acid (Fe_3O_4–MNPs–Si–AS) MNPs were examined for their efficiency in the pretreatment. In addition, recovery and subsequent utilization of catalyst, and the feasibility of nanocatalysts to pretreat different lignocellulosic biomass was successfully completed by the conjoint effect of magnetic nanocatalysts. Nanocatalysts used in the pretreatment can be easily recovered and utilized in numerous cycles which then reduces the cost associated with the process. Two distinctive nanocatalysts formed through the fictionalization of MNPs were used for numerous cycles of pretreatments and were found to be the foremost part of this approach (Avinash et al., 2020).

The nanoparticles influence the change in enzymatic activity and mass transfer rate during the biochemical conversion of feedstock. Numerous metal oxide catalysts such as iron, cobalt, and manganese oxides act in an effective way to catalyze various energy production processes. In syngas fermentation, a higher bioethanol production (166.1%) was observed using silica nanoparticles and the only disadvantage that was observed was poor reusability of catalyst (Kim et al., 2014). A study was conducted with six different nanoparticles which are unique in type, being alumina, silica, carbon nanotubes, iron, and its oxides. Among these, silica was found to have a significant effect in improving the bioethanol conversion, on further modification of the silica structure by adding functional groups (methyl and isopropyl). Another study was conducted using magnetic nanoparticles that are functionalized using methyl group which were found to increase the bioethanol production, with an almost 213.5% increase being observed (Kim and Lee, 2016).

Another method of improving the activity of catalyst is by modifying its surface, thereby increasing the number of active sites and thus increasing the interactions between molecules and nanoparticles (Zhao et al., 2000). A number of modifications in nanoparticles were carried out to enhance their activity through various methods such as coprecipitation, hydrothermal activations, functionalization, microwave irradiation, and impregnation method (Deepika et al., 2019). A composite of SiO and SrO nanopowder was developed to enhance the activity of catalyst in the production of biofuels such as bioethanol and biodiesel (Naor et al., 2017). Modified nanoparticles were found to have increased the activity and improved the yield in many studies (Ivanova et al., 2011). Carbon nanotubes (CNTs) are not regular nanoparticles, where catalysts can be immobilized and they provide a high specific surface area for the reaction and also have high tensile strength and stability at higher temperatures (Chaturvedi et al., 2012). In one study, rhodium nanoparticles were loaded on CNTs which were used in a reactor to improve ethanol production (Pan et al., 2007).

The most important step in the production of bioethanol from lignocellulosic biomass is the saccharification reaction followed by fermentation. The first step is usually the hydrolytic reaction for the release of glucose. This instantaneous reaction has to be carried out in optimized conditions. Current research activities support the use of nanosize biocatalysts. This will support the heat and mass transfer rate and enzymatic activities along with having other benefits (Elsa et al., 2015).

8.3.2 Biodiesel production

Nanocatalysts are known to have exceptional properties, such as selective reactivity, recyclability, exact rate of reaction, recoverability, and high activation energy. Nanocatalysts with a high catalytic activity and high specific surface area may address the foremost common issues of heterogeneous catalysts, such as time consumption, mass transfer resistance, inefficiency, and fast deactivation. Biodiesel is considered to be one of the favorable renewable fuels and is found to be sparkling, aromatic free sulfur, and nontoxic when equated with petroleum based diesel. Wide varieties of nanocatalysts affect the production of biodiesel in different ways (Baskar and Aiswarya, 2016).

The production of biodiesel can be carried out using any fatty sources such as algal, vegetable oil, or animal fat in a blend with alcohol in the presence of a catalyst. Calcium oxide nanocatalysts showed enhanced reaction efficacy of soybean oil to biodiesel under reasonable reaction circumstances. They have been commonly utilized within the synthesis processes for their low cost, high catalytic activity, nontoxicity, and accessibility. Their catalytic nature and outer surface area of CaO have been progressed to be improved by hydration, dehydration, and calcination processes. It has been stated that nanosized calcium oxide could be utilized either along other complexes using CaO as a carrier or by immobilization of CaO onto a distinctive carrier. The point of interest raised by nanoCaO is its reusable nature and limitations in the leaching process (Priyanka et al., 2019).

Biodiesel production using two microalgae, *Chlorella vulgaris* and *Spirulina plantensis*, utilizing two nanomagnetic base catalysts which are solid in nature, that is, $KOH/CaO{-}Fe_3O_4$ and $KF/KOHFe_3O_4$, was studied for its efficacy. Heterogeneous nanocatalysts with magnetic property are a promising choice for better recyclability. The magnetic property helps to ease the separation of nanocatalyst and reduces the catalyst loss. The use of magnetic nanocatalysts resulted in higher process effectiveness, thereby saving raw materials and accelerating the reaction. Efficient utilization of microbial fuel cells modified with nanoparticles for the generation of bioelectricity rate reduces the processing time. As magnetic nanocatalysts are heterogeneous and can be reused for many cycles, they are more affordable (Tahereh et al., 2020).

The process of extracting the lipid utilizing autoflocculated wet microalgae biomass with diverse pretreatment strategies was efficiently carried out. To reduce the antagonistic impact of utilizing an acid–base catalyst within the transesterification process, heterogeneous catalysis was illustrated utilizing Fe_2O_3 nanoparticles. The Fe_2O_3 magnetic nanocatalyst was synthesized using a greener process with algae extract. Furthermore, the effect of Fe_2O_3 magnetic nanocatalyst in the generation of biodiesel on the environment was measured. This Fe_2O_3 nanocatalyst was synthesized using extract of *Hibiscus rosasinensis*. The transesterification parameters such as reaction time, catalyst load, and reaction temperature were optimized for maximum biodiesel yield. Integration of a biorefinery approach for the production of bioethanol using residual algal biomass made biodiesel production more economical. This process implies the utilization of Fe_2O_3 nanocatalyst for the production of biodiesel using lipids from microalgae that had less natural results compared to conventional biodiesel production methods (Srijoni et al., 2019).

Unwanted biomass from agricultural areas can be utilized to create biofuels using nanocatalysts that improve the gasification process. The lignocellulosic part of this biomass is a potential alternative feedstock for economic generation of bioethanol. Nanomaterials can be used as a catalyst to improve the energy effectiveness and product quality. Biodiesel was produced from mixed biomass of weeds such as *Carthamous oxyacantha*, *Asphodelus tenuifolius*, and *Chenopodium* using nanocatalysts. Nickel and cobalt were used as nanocatalysts to speed up the reactions for the production of biofuels at lower temperatures, which improved the gasification process efficiency and product quality (Safdar et al., 2020). More findings on biodiesel production using nanocatalysts are listed in Table 8.1.

8.3.3 Biobutanol production

Biofuels are considered as a green and sustainable alternative to petroleum-derived fuels. It has been estimated that the global demand for biofuels transportation will increase by up to 51.1% by 2022 (Navigant Research, 2014). Biobutanol has been always of interest in the current era to researchers as a green alternative due to its higher energy level, lower volatility, reduced corrosiveness, and lower moisture sorption properties (Durre, 2007). Foremost companies such as GEVO, Cobalt Technologies, DuPont, Green Biologics, Metabolic Explorer, and BUTALCO have also initiated the production

TABLE 8.1 Usage of nanocatalysts for biodiesel production.

Catalyst	Operating conditions	Key findings	References
Zn-doped CaO catalyst	*Calophyllum inophyllum* oil	Biodiesel conversion of 91.95% was achieved when maintaining the methanol to oil ratio at 9.66:1, concentration of catalyst 5% (w/v), time 81.31 min, and temperature 56.71°C with green chemistry value of 0.873	Naveenkumar and Baskar (2020)
CaO/Au nanocomposites	Soybean oil	Methanol-oil molar ratio at 12: 1 with catalyst content of 1.0 wt.% at 70°C for 3 h	Jiadi et al. (2020)
$SO_4^{2-}/Fe_3O_4-Al_2O_3$ as a magnetic nanocatalyst	Microalgae	Temperature of 120°C, catalyst concentration of 8 wt.%, 9 mL methanol/g microalgae, and 4 h of reaction time	Ebrahim et al. (2020)
Magnetic $ZnO/BiFeO_3$	Canola oil	Molar ratio of methanol/canola oil of 15:1, a reaction temperature of 65°C, and a catalyst amount of 4 wt.%	Zahra et al. (2019)
Zn-doped CaO nanocatalyst	Waste cooking oil	20:1 methanol to oil molar ratio, 5 wt.% catalyst loading, 65°C reaction temperature, and 4 h of reaction time	Manash et al. (2019)
Alumina-doped calcium oxide nanocomposite	Pork fat	Catalyst loading of 8% (w/w), methanol to oil ratio of 9:1, reaction temperature of 65°C, and reaction time of 90 min	Elsa et al. (2019)
Nickel-doped ZnO nanocatalyst	Castor oil	Oil to methanol molar ratio of 1:8, catalyst loading 11% (w/w), reaction temperature of 55°C, for 60 min of reaction time	Baskar et al. (2018)

of butanol (Han et al., 2013). The global trade in butanol is anticipated to reach $247 billion by 2020, owing to its advantages over ethanol and biodiesel (Bharathiraja et al., 2017).

8.3.3.1 *Biosynthesis of biobutanol*

The manufacture of biobutanol is often carried out by acetone-butanol-ethanol (ABE) fermentation (Kumar et al., 2012; Jurgens et al., 2012). A bacterial strain of *Clostridium* is commonly used for this purpose. Sugar beets, glucose, starch-based, cellulosic, and other biomass materials can be used as substrate for this fermentation (Ezeji et al., 2007a,b; Hansen et al., 2005). The selection of microbial strain for butanol synthesis depends on the type of substrates and the concentration, nutrient requirement, tolerance limit, and resistance to other competitive microbes and antibiotics (Bharathiraja et al., 2017). *Clostridium pasteurianum*, *Clostridium acetobutylicum*, *Clostridium* (Elsa et al., 2019) *cadaveris*, *Clostridium beijerinckii*, *Clostridium aurantibutyricum*, *Clostridium saccaroper butylacetonicum*, *Clostridium saccharoacetobutylicum*, *Clostridium sporogenes*, and *Clostridium tetanomorphum* have been found to synthesize butane effectively. The ABE fermentation process is comprised of two phases, namely the solventogenic phase and the acidogenic phase. The raw materials are converted to acetic and butyric acid in the acidogenic phase and as a result the pH of the solution decreases. In the solventogenic phase, the sugars react with the acids and are further transformed into solvents like ethanol, butanol, and acetone, which results in an increase in the pH (Maddox et al., 2000).

8.3.3.2 *Nanocatalysts in biobutanol production*

Researchers have been working on biocatalysts and their modification to aid the transformation of biomass into biofuels (Kim and Lee 2016; Kim et al., 2005). Nanomaterials and nanoparticles have the potential to influence the bioconversion process. The main variables affecting the end product generation and efficient control of the rate of reaction include particle size, nature of nanoparticles, and kind of biomass being used (Chaturvedi et al., 2012). These nanocatalysts differ from conservative catalysts based on several factors such as huge area-to-volume ratio and vigorous surfaces for carrying out chemical reactions (Sirajunnisa and Surendhiran, 2014).

8.3.3.2.1 Metal oxide nanocatalysts

Researchers have studied the potential of metal nanoparticles such as oxides from of metals including iron, silver, gold, cobalt, cerium, ruthenium, palladium, etc. in the production of butanol (Skrotskyi et al., 2019; Quesada et al., 2018; Siqueira et al., 2019). Quesada et al. (2018) studied the potential of ruthenium and palladium nanoparticles supported on magnesium-aluminum mixed oxide for the production of butanol. Ruthenium nanoparticles showed 15 times higher butanol productivity than the parent substance under soft conditions. The behavior of palladium was found to be dependent on the decarbonylation side-product. Palladium nanoparticles showed consistent butanol productivity at low temperatures with a high dehydrogenation rate. Skrotskyi et al. (2019) studied the influence of iron oxide, cerium, silver, gold, and gadolinium nanoparticles on the biosynthesis of butanol. They used the *Clostridium beijerinckii* IMV B-7806 and *C. acetobutylicum* IMV B-780 strains of acetone-butyl bacteria for the biosynthesis process. Silver oxide nanoparticles showed an increase of butanol yield from 9.0 ± 0.6 g/L in a control up to 11.1 ± 1.8 g/L during biosynthesis using *C. acetobutylicum* species. The effect of the nanoparticles on the biosynthesis of butanol by acetone-butyl bacteria is highly strain-specific and dependent on the bacterial growth kinetics and enzyme action (Skrotskyi et al., 2019). The influence of metal oxide nanoparticles in biosynthesis butanol production by different microbial strains is summarized in Table 8.2.

TABLE 8.2 Usage of nanocatalysts for biobutanol production.

Catalyst	Operating conditions	Key findings	References
Mono and bimetallic modified 13× zeolite catalyst	93K, 1 bar, W/F = 5.6 g/hmol (W = catalyst weight and F = flow rate of feedstock)	No n-butanol recovery was showed by monometallic modified zeolites, whereas the bimetallic modified zeolite catalyst showed 40% n-butanol selectivity	Yang and Meng (1993)
Zr-containing zeolites BEA doped with palladium particles	Atmospheric pressure and ethanol feed space velocity of 0.15–6.5 $g_{ethanol}/(g_{cat}\ h)$	The rate of butane formation decreased with increase in the palladium content. The sample with 1.5% palladium showed butanol selectivity of 8.6%	Kots et al. (2019)
ZSM-5 zeolites	35–65 bar	Observed 13% increase in-butanol conversion rate. Alkali-treated HZSM-5 zeolites showed enhanced rate of conversion to hydrocarbons	Souza et al. (2019)
HZSM-5 zeolites	360°C and 300 psig	Catalyst showed selectivity toward the production of ethylene, higher olefins, and aromatics. No trace of butanol was observed	Ramasamy and Wang (2013)
ZSM-5 zeolites exchanged with alkali cations (Li, Na, K, and Cs)	100–400°C; atmospheric pressure	Observed that catalytic activity ranks are in the order: LiZSM-5 > NaZSM-5 > KZSM-5 > CsZSM-5	Jaumin and Su (2003)

Magnesium oxide catalysts showed intrinsic activity toward the production of butanol due to the dominant behavior of basic centers of the catalyst. The performance of magnesium oxide catalyst can be enhanced by the addition of ruthenium or nickel. The addition of the metal particles showed an enhancement in the number and nature of basic sites. Butanol production has also increased fivefold the original volume and the selectivity was doubled at 400°C by the addition of nickel (Cimino et al., 2018).

8.3.3.2.2 Zeolite-based nanocatalysts

Zeolites are microporous crystalline aluminosilicate structures, usually composed of aluminum (Al), silicon (Si), and oxygen. Increasing the Si/Al ratio in the structure increases the strength of the zeolite, and its structure stabilizes at higher temperatures (Corma and Martinez, 1995). Various mechanisms are involved in the synthesis of zeolites in the form of compounds, and nanosized and mesoporous crystals (Cejka and Mintova, 2007 Tosheva and Valtchev, 2005; Tao et al., 2006). Natural zeolites are formed when layers of ash and rocks formed after volcanic eruption crystallize in response to groundwater or alkaline/saline lakes. Examples of natural zeolites are ofrite, chabazite, mordonite (MOR), and foazite (FAU). Synthetic zeolites are produced from clays and industrial residues (Barczyk et al., 2013; Johan et al., 2015). X, A, ZSM-5, Y, and beta are a few of the most potent synthetic zeolites (Yilmaz et al., 2012). Synthetic zeolites are produced under well-regulated conditions. By altering the processing parameters, zeolites with different chemical compositions and structural properties can be obtained. The sol–gel method is one of the most widely used methods for the manufacture of zeolite. The product characteristics depend on the composition, pH, reaction time, operating temperature, and templates used during processing (Rebrov, 2009; Zhang et al., 2019). Some of the major studies carried out in the catalytic conversation of ethanol to butanol using zeolite-based catalyst are shown in Table 8.2. Although many studies has been carried out using zeolite as an adsorbent and catalyst in biofuel production, only a few attempts have been made in the utilization of zeolite nanocatalysts in biobutanol production.

8.3.3.2.3 Zirconia nanocatalysts

Zirconia can be used both as an acid and base heterogeneous catalyst (Cutrufello et al., 2002). Zirconium oxide and its stabilized variants have been used in acid–base heterogeneous catalytic processes (Cutrufello et al., 2002; Jung and Bell, 2002). Yttria-stabilized zirconia is a successful catalyst for conversion of ethanol to n-butanol using the Guerbert coupling reaction (Vlasenko et al., 2019a). The catalyst showed selectivity up to 70% and n-butanol yield up to 16.8% greater than with individual zirconia. Vlasenko et al. (2019a) studied the effect of modification of zirconia with ceria additives on ethanol to butanol conversion. They found a correlation between the rate of butanol formation and the strength (Jaumain and Su, 2003) of base sites on the catalyst surfaces. They found that the addition of oxide of ceria increases the strength of base sites, which resulted in a 13.3% increase in the ethanol conversion rate and a 31.6% increase in n-butanol selectivity.

8.3.3.2.4 Hydroxyapatite catalysts

Hydroxyapatite catalysts [$Ca_{10}(PO_4)_6(OH)_2$] are calcium phosphate-based catalysts that possesses both acidic and basic characteristics. Scalbert et al. (2014) investigated ethanol to butanol conversion using commercial hydroxyapatite catalyst at 350–410°C, and observed butanol selectivity of 50%. Tsuchida et al. (2008) studied the effect of hydroxyapatite catalysts in the conversion of ethanol to butanol. They varied the ratios of calcium and phosphorus content in the catalysts and found that as the Ca/P ratio of HAP increases the density and strength of basic sites on the catalyst surface also increases. The product selectivity also showed a strong correlation with the acid–basic properties of the catalysts. They also obtained 1-butanol selectivity of 29%–70%. Cimino et al. (2018) studied the catalytic conversion of ethanol into butanol at 200–500°C using HAP catalysts. HAP catalysts showed only a slight activity up to 300°C and almost complete conversion of ethanol was observed at the temperature range between 300 and 400°C. They also observed that butanol production increases with temperature and the maximum butanol selectivity of 4.0% was observed at 400°C. Strontium hydroxyapatite has higher selectivity to butanol than calcium hydroxyapatite catalysts (Ogo et al., 2008). The catalytic conversion of ethanol to butanol depends on the ions present in the hydroxyapatite catalyst. Sr–P hydroxyapatite catalyst showed 70% (mol) of 1-butanol selectivity in the catalytic conversion of ethanol (Ogo et al., 2011). Catalysts with higher Sr/P molar ratios were found to have higher densities of strong basic sites and those catalysts exhibited enhanced catalytic activity and 1-butanol selectivity in the conversion of ethanol (Ogo et al., 2012).

8.3.3.2.5 Hydrotalcites

Hydrotalcites are a group of mixed-metal layered hydroxide catalysts used in the ethanol to butanol conversion reaction. Hydrotalcite-derived magnesium-alumina mixed oxide catalyst is one of the most promising catalytic materials for the conversion of ethanol to butanol. Perrone et al. (2018) studied the effect of hydrotalcites modified with copper, palladium, and indium in the conversion of ethanol to butanol. The catalytic reaction was carried out in stainless steel reactors at 403–573K. The modified hydrotalcites with copper, palladium, and indium showed an increase in butanol production and selectivity. Also, the addition of palladium or indium ions increased the number of acid sites. In–CuMgAl catalyst showed a high butanol selectivity of 65% at 443K. Pd–CuMgAl catalyst showed butanol selectivity of 45%. Another research showed that the calcination temperature modifies the chemical properties of the hydrotalcite-derived magnesium-alumina catalyst, which in turn affects the 1-butanol synthesis. As the calcination temperature increases, the butanol selectivity also increases and after reaching a particular point it starts decreasing. At 600°C, calcined hydrotalcite catalyst showed ethanol conversion of 44% and 1-butanol selectivity of 50% (Ramasamy et al., 2015).

8.4 Future applications of nanocatalysts in the biofuel industry

Essential advancements will take place in the area of nanotechnology in the coming years. Nanomaterials have shown enormous projections for the synthesis of biofuels with greener preparation of biomass and increased selection of product and yield in a conservative way. It is clear that nanocatalysts have successfully demonstrated wide applicability in pretreating biomass and its conversion into liquid fuels. Further, the use of nanomaterials for the utilization of diverse intermediates in process and final products remains to be investigated to discover prudent methods of preparation for better control and monitoring of the fermentation process and optimization of economic demand. Even now, a wide variety of related challenges need to be overcome.

Specifically, the following issues merit the consideration of analysts employed in this field:

- Advanced sensors for the discovery of midway metabolites, final products, and commercialization of the synthesized nanoparticles;
- Greener course of nanoparticle amalgamation for utilization of biocatalysts as support so that these can effectively be exploited for fermentation of bioalcohols;
- Production of more flexible nanocatalysts which can be utilized for a variety of steps supporting biomass processing;
- Advancement in the synthesis of more discerning nanocatalysts to change biologically derived sugars into biofuel;
- Understanding the chemistry of nanostructures and the capacity to regulate materials in the nanometer scale will surely guarantee the cost-effective progress of novel and suitable catalysts;
- Commercial application should be investigated for scaled-up reactions.

There is vast inherent potential of this innovation and a more is needed to be done to overcome the obstacles present within the area of biofuel manufacture utilizing nanocatalysts.

8.5 Conclusion

The study and research into bioenergy and biofuel production has a major role to play, because of the drastic fossil fuel depletion and concerns about environmental deterioration. Biofuel can help in the progress toward sustainable development. All types of biofuels and each stage of production have key roles in the successful implementation of biofuel production. Current research is focused on the use of nanocatalysts to improve the production efficiency of biofuels such as bioethanol, biodiesel, and biobutanol. The reusability of the nanocatalyst considerably reduces the economic costs of biofuel production. Thus nanocatalysts for biofuel production have a vital role to play in the current energy crisis.

References

Avinash, P.I., Rafael, R.P., Yasmin, C.S.M., Silvio, S.S., 2020. Acid-functionalized magnetic nano catalysts mediated pretreatment of sugarcane straw: an eco-friendly and cost-effective approach. Cellulose 27, 7069–7078.

Barczyk, K., Mozgawa, W., Krol, M., 2013. Studies of anions sorption on natural zeolites. Spectrochim. Acta Part A Mol. Biomol. Spectrosc. 133, 876–882.

Baskar, G., Aiswarya, R., 2016. Trends in catalytic production of biodiesel from various feedstocks. Renew. Sustain. Energy Rev. 57, 496–504.

Baskar, G., Selvakumari, I.A.E., Aiswarya, R., 2018. Biodiesel production from castor oil using heterogeneous Ni doped ZnO nanocatalyst. Bioresour. Technol. 250, 793–798.

Bharathiraja, B., Jayamuthunagai, J., Sudharsanaa, T., Bharghavi, A., Praveenkumar, R., Chakravarthy, M., Yuvraj, D., 2017. Biobutanol – an impending biofuel for future: a review on upstream and downstream processing techniques. Renew. Sustain. Energy Rev. 68, 788–807.

Cejka, J., Mintova, S., 2007. Perspectives of micro/mesoporous composites in catalysis. Catal. Rev. 49 (4), 457–509.

Chaturvedi, S., Dave, P.N., Shah, N.K., 2012. Application of nano-catalyst in new era. J. Saudi Chem. Soc. 16, 307–325.

Cimino, S., Lisi, L., Romanucci, S., 2018. Catalysts for conversion of ethanol to butanol: Effect of acid-base and redox properties. Catal. Today 304, 58–63.

Corma, A., Martinez, A., 1995. Zeolites and zeotypes as catalysts. Adv. Mater. 7 (2), 137–144.

Cutrufello, M.G., Ferino, I., Monaci, R., Rombi, E., Solinas, V., 2002. Acid-base properties of zirconium, cerium and lanthanum oxides by calorimetric and catalytic investigation. Top. Catal. 19 (3), 225–240.

Deepika, K., Upadhyay, S.N., Mishra, P.K., 2019. Nanotechnology in bioethanol/biobutanol production. Green Nanotechnol. Biofuel Prod. 115–127.

Durre, P., 2007. Biobutanol: an attractive biofuel. Biotechnol. J. Healthc. Nutr. Technol. 2 (12), 1525–1534.

Ebrahim, S., Hamed, N., Naser, S., Sina, K., 2020. Comprehensive assessment of the preparation conditions of a separable magnetic nanocatalyst for biodiesel production from algae. Algal Res. 49, 101949.

Elsa, C., Dharmendirakumar, M., Baskar, G., 2015. Immobilization of cellulase onto MnO_2 nanoparticles for bioethanol production by enhanced hydrolysis of agricultural waste. Chin. J. Catal. 36 (8), 1223–1229.

Elsa, C., Yazhini, D., Merlin, V., Baskar, G., 2019. Production of biodiesel from pork fat using alumina-doped calcium oxide nanocomposite as heterogeneous catalyst. Energy Sour. A Recovery Util. Environ. Eff. https://doi.org/10.1080/15567036.2019.1637971.

Ezeji, T.C., Qureshi, N., Blaschek, H.P., 2007a. Bioproduction of butanol from biomass: from genes to bioreactors. Curr. Opin. Biotechnol. 18 (3), 220–227.

Ezeji, T., Qureshi, N., Blaschek, H.P., 2007b. Production of acetone–butanol–ethanol (ABE) in a continuous flow bioreactor using degermed corn and *Clostridium beijerinckii*. Process Biochem 42 (1), 34–39.

Han, S.H., Cho, D.H., Kim, Y.H., Shin, S.J., 2013. Biobutanol production from 2-year-old willow biomass by acid hydrolysis and acetone-butanol-ethanol fermentation. Energy 61, 1 3–17.

Hansen, A.C., Zhang, Q., Lyne, P.W., 2005. Ethanol–diesel fuel blends—a review. Bioresour. Technol. 96 (3), 277–285.

Ivanova, V., Petrova, P., Hristov, J., 2011. Application in the ethanol fermentation of immobilized yeast cells in matrix of alginate/magnetic nanoparticles, on chitosan-magnetite microparticles and cellulose-coated magnetic nanoparticles. Int. Rev. Chem. Eng. 3 (2), 289–299.

Jaumain, D., Su, B.-L., 2003. Direct catalytic conversion of chloromethane to higher hydrocarbons over a series of ZSM-5 zeolites exchanged with alkali cations. J. Molec. Catal. A: Chem. 197 (1–2), 263–273.

Jiadi, L., Minghuan, L., Shaoyun, C., Bingqing, W., Jin, C., Da-Peng, Y., Shangzhou, Z., Wenxiao, Du, 2020. Conversion of Au(III)-polluted waste eggshell into functional CaO/Au nanocatalyst for biodiesel production. Green Energy Environ. https://doi.org/10.1016/j.gee.2020.07.019. Published online.

Johan, E., Yamada, T., Munthali, M.W., Kabwadza-Corner, P., Aono, H., Matsue, N., 2015. Natural zeolites as potential materials for decontamination of radioactive cesium. Procedia Environ. Sci. 28, 52–56.

Jung, K.T., Bell, A.T., 2002. Effects of catalyst phase structure on the elementary processes involved in the synthesis of dimethyl carbonate from methanol and carbon dioxide over zirconia. Top. Catal. 20 (1–4), 97–105.

Jurgens, G., Survase, S., Berezina, O., Sklavounos, E., Linnekoski, J., Kurkijarvi, A., Vakeva, M., van Heiningen, A., Granstrom, T., 2012. Butanol production from lignocellulosics. Biotechnol. Lett. 34, 1415–1434.

Kim, Y.K., Lee, H., 2016. Use of magnetic nanoparticles to enhance bioethanol production in syngas fermentation. Bioresour. Technol. 204, 139–144.

Kim, B.C., Nair, S., Kim, J., Kwak, J.H., Grate, J.W., Kim, S.H., Gu, M.B., 2005. Preparation of biocatalytic nanofibers with high activity and stability via enzyme aggregate coating on polymer nanofibres. Nanotechnology 16 (7), 82–88.

Kim, M., Salley, S.O., Ng, K.Y.S., 2008. Transesterification of glycerides using a heterogeneous resin catalyst combined with a homogeneous catalyst. Energy Fuels 22, 3594–3599.

Kim, Y.K., Park, S.E., Lee, H., Yun, J.Y., 2014. Enhancement of bioethanol production in syngas fermentation with *Clostridium ljungdahlii* using nanoparticles. Bioresour. Technol. 159, 446–450.

Kots, P.A., Zabilska, A.V., Grigor'ev, Y.V., Ivanova, I.I., 2019. Ethanol to butanol conversion over bifunctional zeotype catalysts containing palladium and zirconium. Petr. Chem. 59 (8), 925–934.

Kumar, M., Goyal, Y., Sarkar, A., Gayen, K., 2012. Comparative economic assessment of ABE fermentation basedon cellulosic and non-cellulosic feedstocks. Appl. Energy 93, 193–204.

Maddox, I., Steiner, E., Hirsch, S., Wessner, S., Gutierrez, N., Gapes, J., Schuster, K., 2000. The cause of acid crash and acidogenic fermentations during the batch acetone-butanol-ethanol (ABE) fermentation process. J. Mol. Microbiol. Biotechnol. 2 (1), 95–100.

Mahmoud, N., Mohaddeseh, S., Siavash, I., Rajender, S.V., 2021. Green-synthesized nanocatalysts and nanomaterials for water treatment: current challenges and future perspectives. J. Hazard. Mater. 401, 123401.

Manash, J.B., Ankur, D., Velentina, D., Nilutpal, B., Dhanapati, D., 2019. Transesterification of waste cooking oil for biodiesel production catalyzed by Zn substituted waste egg shell derived CaO nanocatalyst. Fuel 242, 345–354.

Mierczynski, P., Ciesielski, R., Kedziora, A., Maniukiewicz, W., Shtyka, O., Kubicki, J., Albinska, J., Maniecki, T.,P., 2015. Biodiesel production on MgO, CaO, SrO and BaO oxides supported on (SrO)(Al_2O_3) mixed oxide. Catal. Lett. 145, 1196–1205.

Naor, E.,O., Koberg, M., Gedanken, A., 2017. Nonaqueous synthesis of SrO nanopowder and SrO/SiO_2 composite and their application for biodiesel production via microwave irradiation. Renew. Energy 101, 493–499.

Naveenkumar, R., Baskar, G., 2020. Optimization and techno-economic analysis of biodiesel production from *Calophyllum inophyllum* oil using heterogeneous nanocatalyst. Bioresour. Technol. 315, 123852.

Navigant Research, 2014. https://www.navigantresearch.com/newsroom/biofuels-demand-forroad-transportaion-will-surpass-51-billion-gallons-annually-by-2022.

Nayebzadeh, H., Haghighi, M., Saghatoleslami, N., Tabasizadeh, M., Yousefi, S., 2018. Fabrication of carbonated alumina doped by calcium oxide via microwave combustion method used as nanocatalyst in biodiesel production: influence of carbon source type. Energy Convers. Manag. 171, 566–575.

Ogo, S., Onda, A., Yanagisawa, K., 2008. Proceeding of the 102nd Annual Meeting of the Catalysis Society of Japan, A, p. 311.

Ogo, S., Onda, A., Yanagisawa, K., 2011. Selective synthesis of 1-butanol from ethanol over strontium phosphate hydroxyapatite catalysts. Appl. Catal. A-Gen. 402 (1–2), 188–195.

Ogo, S., Onda, A., Iwasa, Y., Hara, K., Fukuoka, A., Yanagisawa, K., 2012. 1-Butanol synthesis from ethanol over strontium phosphate hydroxyapatite catalysts with various Sr/P ratios. J. Catal. 296, 24–30.

Pan, X., Fan, Z., Chen, W., Ding, Y., Luo, H., Bao, X., 2007. Enhanced ethanol production inside carbon-nanotube reactors containing catalytic particles. Nat. Mater. 6 (7), 507–511.

Perrone, O.M., Lobefaro, F., Aresta, M., Nocito, F., Boscolo, M., Dibenedetto, A., 2018. Butanol synthesis from ethanol over CuMgAl mixed oxides modified with palladium (II) and indium (III). Fuel Process. Technol. 177, 353–357.

Priyanka, B., Bhaskar, S., Dey, R.K., 2019. Process optimization of biodiesel production catalyzed by CaO nanocatalyst using response surface methodology. J. Nanostruct. Chem. 9, 269–280.

Quesada, J., Faba, L., Díaz, E., Ordonez, S., 2018. Enhancement of the 1-butanol productivity in the ethanol condensation catalyzed by noble metal nanoparticles supported on Mg-Al mixed oxide. Appl. Catal. A-Gen. 563, 64–72.

Ramasamy, K.K., Wang, Y., 2013. Catalyst activity comparison of alcohols over zeolites. J. Energy Chem. 22 (1), 65–71.

Ramasamy, K.K., Gray, M., Job, H., Santosa, D., Li, X.S., Devaraj, A., Wang, Y., 2015. Role of calcination temperature on the hydrotalcite derived MgO–Al2O_3 in converting ethanol to butanol. Top. Catal. 59 (1), 46–54.

Rebrov, E.V., 2009. Sol-gel synthesis of zeolite coatings and their application in catalytic microstructured reactors. Catal. Ind. 1, 322–347.

Safdar, A., Obaidullah, S., Shahid, M., Tariq, M., Bashir, A.K., Ijaz, A., 2020. Biofuels production from weed biomass using nanocatalyst technology. Biomass Bioenergy 139, 105595.

Santos, F.C.U., Paim, L.L., Silva, J.L.D., Stradiotto, N.R., 2016. Electrochemical determination of total reducing sugars from bioethanol production using glassy carbon electrode modified with graphene oxide containing copper nanoparticles. Fuel 163, 112–121.

Scalbert, J., Thibault-Starzyk, F., Jacquot, R., Morvan, D., Meunier, F., 2014. Ethanol condensation to butanol at high temperatures over a basic heterogeneous catalyst: how relevant is acetaldehyde self-aldolization? J. Catal. 311, 28–32.

Seyed, B.B., Mohammad, H., 2020. Texture-evolution of copper-based nanocatalyst via hybrid glow discharge plasma-oxalate-precipitation method for hydrogen production from CH_3OH/H_2O mixture. Appl. Catal. A-Gen. 591, 117402.

Siqueira, M.R., Perrone, O.M., Metzker, G., de Oliveira Lisboa, D.C., Thomeo, J.C., Boscolo, M., 2019. Highly selective 1-butanol obtained from ethanol catalyzed by mixed metal oxides: reaction optimization and catalyst structure behavior. Mol. Catal. 476, 110516.

Sirajunnisa, A.R., Surendhiran, D., 2014. Nanosilver fabrication mediated by exopolysaccharides from *Pseudomonas fluorescens* and its biological activities. Am. J. Pharm. Tech. Res. 4, 727–742.

Skrotskyi, S., Voychuk, S., Khomenko, L., Vasyliuk, O., Pidgorskyi, V., 2019. Influence of nanoparticles on the solventogenesis of bacteria *Clostridium beijerinckii* IMV B-7806, *Clostridium acetobutylicum* IMV B-7807. Ukr. Food J. 8 (1), 110–118.

Souza de Carvalho Filho, J.F., Maciel Pereira, M., Gomes Aranda, D.A., Monnerat Araujo Ribeiro de Almeida, J., Falabella Sousa-Aguiar, E., Nothaft Romano, P., 2019. Application of response surface methodology for ethanol conversion into hydrocarbons using ZSM-5 zeolites. Catalysts 9 (7), 617.

Srijoni, B., Swagatika, R., Sanjukta, B., Arnab, A., Debabrata, D., 2019. Fe_2O_3 nanocatalyst aided transesterification for biodiesel production from lipid-intact wet microalgal biomass: a biorefinery approach. Energy Convers. Manag. 195, 844–853.

Tahereh, N., Zeinab, S., Alireza, S., 2020. Conversion of bio-oil extracted from *Chlorella vulgaris* micro algae to biodiesel via modified superparamagnetic nano-biocatalyst. Renew. Energy 146, 1796–1804.

Tao, Y., Kanoh, H., Abrams, L., Kaneko, K., 2006. Mesopore-modified zeolites: preparation, characterization, and applications. Chem. Rev. 106, 896–910.

Todorović, Z.B., Troter, D.Z., Đokić-Stojanović, D.R., Veličković, A.V., Avramović, J.M., Stamenković, O.S., Veselinović, L.M., Veljković, V.B., 2019. Optimization of CaO-catalyzed sunflower oil methanolysis with crude biodiesel as a cosolvent. Fuel 237, 903–910.

Tosheva, L., Valtchev, V.P., 2005. Nanozeolites: synthesis, crystallization mechanism, and applications. Chem. Mater. 17, 2494–2513.

Tsuchida, T., Kubo, J., Yoshioka, T., Sakuma, S., Takeguchi, T., Ueda, W., 2008. Reaction of ethanol over hydroxyapatite affected by Ca/P ratio of catalyst. J. Catal. 259 (2), 183–189.

Vlasenko, N.V., Kyriienko, P.I., Valihura, K.V., Kosmambetova, G.R., Soloviev, S.O., Strizhak, P.E., 2019a. Yttria-stabilized zirconia as a high-performance catalyst for ethanol to n-butanol guerbet coupling. ACS Omega 4 (25), 21469–21476.

Wang, S., Shan, R., Wang, Y., Lu, L., Yuan, H., 2019. Synthesis of calcium materials in biochar matrix as a highly stable catalyst for biodiesel production. Renew. Energy 130, 41–49.

Xu, C., Nasrollahzadeh, M., Sajjadi, M., Maham, M., Luque, R., Puente-Santiago, A.R., 2019. Benign-by-design nature-inspired nanosystems in biofuels production and catalytic applications. Renew. Sustain. Energy Rev. 112, 195–252.

Yang, C., Meng, Z.Y., 1993. Bimolecular condensation of ethanol to 1-butanol catalyzed by alkali cation zeolites. J. Catal. 142 (1), 37–44.

Yilmaz, B., Trukhan, N., Müller, U., 2012. Industrial outlook on zeolites and metal organic frameworks Chinese. J. Catal. 33, 3–10.

Zahra, S., Seyed, A.H., 2019. Study and optimization of conditions of biodiesel production from edible oils using $ZnO/BiFeO_3$ nano magnetic catalyst. Fuel 239, 1204–1212.

Zhang, X., Wang, Y., Wang, X., Chen, H., Li, H., Sun, C., Sun, L., Fan, C., Wang, C., 2019. An efficient route for synthesis of ERI zeolite through conversion of FAU zeolite in the presence of N, N-dimethylpiperidinium hydroxide. Microporous Mesoporous Mater. 279, 407–415.

Zhao, X.S., Lu, G.Q., Hu, X., 2000. Characterization of the structural and surface properties of chemically modified MCM-41 material. Microporous Mesoporous Mater. 41, 37–47.

CHAPTER

9

Potentials and challenges in biodiesel production from algae—technological outlook

S. Chozhavendhan[1], *G. Karthigadevi*[2], *R. Praveenkumar*[3], *M. Aniskumar*[6], *Mani Jayakumar*[4], *Baskar Gurunathan*[5]

[1]Department of Biotechnology, V.S.B. Engineering College, Karur, Tamil Nadu, India; [2]Sri Venkateswara College of Engineering, Chennai, Tamil Nadu, India; [3]Department of Biotechnology, Arunai Engineering College, Tiruvannamalai, Tamil Nadu, India; [4]Department of Chemical Engineering, Haramaya Institute of Technology, Haramaya University, Dire Dawa, East Hararghe, Ethiopia; [5]Department of Biotechnology, St. Joseph's College of Engineering, Chennai, Tamil Nadu, India; [6]V.S.B. Engineering College, Karur, Tamil Nadu, India

9.1 Introduction

The increases in transportation, modernization, and industrialization have augmented destructive environmental pollution and energy demand in consumption (Yin et al., 2020; Addy et al., 2017). The world energy production mainly depends on fossil fuels, which poses multiple threats to society such as fuel demand, climatic change, and biotic health issues (Okoro et al., 2019; Zhu et al., 2019). It is calculated that the CO_2 will reach 500 ppm (parts per million) in 2050 from the current level of 394.5 ppm if the emission is unrestricted (Mathimani and Mallick, 2018). The reduction of carbon dioxide and other pollutants can be accomplished by consuming by using less energy altogether or substituting fossil fuel with renewable fuels. Technological development has paved the way for the transformation of fossil fuels to search for renewable energy sources. The efforts are continuously made to overcome the shortage and reliance on fossil fuels and efficient alternate renewable sources (Mahlia et al., 2020). Among the other alternative energy sources, biofuels are distinguished as primary fuels due to their unique properties. Biofuels are produced from renewable biomass resources such as plants, algae species, and animal wastes (Venturini et al., 2019; Li et al., 2017; Sanchez et al., 2019). The biofuel is also popularly known as either biodiesel or renewable diesel and for its characteristics properties such as biodegradable, sustainable, and clean energy.

Biofuels and Bioenergy
https://doi.org/10.1016/B978-0-323-85269-2.00014-9

In near future, biofuels play a crucial role in maintaining global energy security by targeting large fuel market (Abdullah et al., 2019). The transportation sector occupies third place in energy consumption and greenhouse gas emission after the industry and building sector. Due to an increase in industrialization and transportation sector, utilization of fossil fuel is expected to rise by 60% in 2030. The energy generation without CO_2 from renewable sources has attracted researches to look at work on renewable energy sources. Around 70% of the world's energy productions are accounted for the renewable source for biomass waste as sustainable energy production (Chozhavendhan et al., 2019). Various liquid biofuels can be produced either biologically or thermochemical route. The biodiesel produced from the organic feedstock can be instantly fueled to the vehicles without any modification and reduces the output of pollution and transferable hazards (Bharathiraja et al., 2015; Doshi et al., 2016). Apart from transportation fuel, it can also be used as heating oil, power generation, lubricants, and plasticizer. Crude glycerol obtained as a by-product during biodiesel production has numerous applications. The crude glycerol can be transformed into a high-value product by a biological method (Chozhavendhan et al., 2016).

9.2 The global scenario on biodiesel production

It is firmly believed that the existing fossil fuels will lose with five to six decades if the usage continues as like now. The consumption has been increased by around 1.5% globally due to the intensive development in transportation, industrial activities, and economics. By accounting for 50% of the global biofuel production, the United States, Brazil, and Germany are standing as a leading biofuel producer in the world. The African continent contributes less than 1% of biofuel production. Inevitably, it is also clear that the global production of biofuel gradually increases every year. Biodiesel comprises 82% of the overall biofuel supply and has the potential to expand further based on the political and economic objectives (Mata et al., 2010). The roadmap of the European Union 2050 clearly states that 30% of the energy consumption of the European Union should be from renewable energy in 2050. Similarly, it is also anticipated that the market for algae-derived biodiesel will reach USD around 1200 million by 2024 (Mehta et al., 2018).

9.3 Generation of biodiesel

Biodiesel offers a new opportunity to progress for an extensive standby to fossil fuels and a reduction in greenhouse gas. Biodiesel, a mixture of FAME produced by the transesterification process could satisfy the global demand for energy consumption. The biodiesel is commonly classified into first, second, and third generation based on the origin of feedstock. The quality and efficiency of the biodiesel mainly depend on the properties of feedstock and the conversion process adopted in the production process (Karmakar and Halder, 2019).

9.3.1 First-generation feedstock

The common products of the first generation of biofuels are bioethanol and biodiesel. The first generation of biodiesel is generally produced from the edible feedstock such as grains, sugarcane, oil culture (Bhatia et al., 2017). The usage of edible feedstock for the production of biodiesel by adopting simple technology is pretty common during the initial stage of the biodiesel era. The first generation of biofuel has successfully constituted a firm foundation on biodiesel production by implementing the transesterification process. Vegetable oils, when mixed with the methanol, yield biodiesel and glycerol in the presence of a catalyst (Ullah et al., 2015). The glycerides in oil sources are converted into biodiesel. Similarly, bioethanol is produced by a simple fermentation process of starch, sugar, and other materials by utilizing the microbial sources. The bioethanol produced by this process can be used either in alcoholic beverage industries or as blend in diesel (Velazquez et al., 2018). Continuing the usage of edible food crops as a feedstock led to increasing the cost of food products (Aransiola et al., 2014). High cost, limited area of cultivation, and food versus fuel debut constrained the users to shift to other alternate sources on biodiesel production (Tariq et al., 2012).

9.3.2 Second-generation feedstock

The second generation of biodiesel is produced from nonedible feedstocks such as *Calophyllum inophyllum*, *Jatropha curcas*, *Mahua indica*, karanja, neem, Rubber seed, and so on (Mohamed Shameer et al., 2017). The drawback of first-generation feedstocks attracted the scholars to work on nonedible feedstock, which is cheap, eco-friendly, abundant, and renewable and eradicated the food inequality (Aransiola et al., 2014; Mahdavi et al., 2015). An added advantage of the second generation of biodiesel is that the feedstocks are grown on noncultivable land. The second generation of biofuel can be produced by either thermochemical or biological methods. The thermochemical process can be carried out by either the pyrolysis or gasification process in the presence or absence of oxygen. In the biological route, enzymatic hydrolysis is used. The lower efficiency and lower yield of biodiesel due to the feedstocks led the researcher to emphasize unique feedstock for the effective and economic production of biodiesel.

9.3.3 Third-generation feedstock

Limitation of the first and second generation of biodiesel such as low return on investment, food insecurity, and climate change laid a foundation for the third-generation biodiesel (Verma et al., 2016). Biodiesel delivered from microalgae, waste cooking oil animal fat and fish oil, and so on constitutes the third-generation biodiesel. Algae are the aquatic organism, and its size ranges from micro- to macrolevel. Microalgae can be an impending feedstock for biodiesel generation since it can be produced in all types of water and easily cultivable and fix the atmosphere CO_2 (Saladini et al., 2016). The average lipid content of the majority of algal species is 70%, which can be enhanced when cultivated under optimum condition. The major disadvantage is that it requires sunlight and difficulties in the extraction process (Tariq et al., 2012).

9.4 Algae a potential source of biodiesel

Biodiesel produced from algae seems to be the only biofuel produced from the renewable source without altering the food chain and other crop products. It has numerous advantages when compared with other feedstocks used in biodiesel production. Shorter-generation time, culturing in open or closed vessels, high carbon sequestering capability, high-lipid productivity, and capable of growing in freshwater or brackish water are some of the unique features that made the microalgae as a potential source for biodiesel production. The lipid content in microalgae ranges between 8% and 40%, and the lipid contents are highly influenced by the culture condition (Najafi et al., 2011). The one-cell algal culture could generate a huge amount of lipids and provide more yield per hectare when compared with the sources. Various biofuels such as bioethanol, biodiesel, biohydrogen, and syngas are produced from different species of algae based on the chemical composition and growth habitat of algae (Janaun and Ellis, 2010). The doubling time of microalgae is 24 h, and it can be easily mass-cultivated, and with the aid of genetic engineering tools, the time can be even reduced. The residual biomasses after lipid extraction can be used as a livestock feed. Besides, algae possess great application in pharmaceutical- and cosmetics-based research (Markou and Nerantzis, 2013). The calorific value of algal biodiesel is 41 MJ/kg, whereas calorific value of the biodiesel obtained oilseeds such as canola or soybean is 39.5 MH/kg only.

9.5 Biodiesel production from algae

The biodiesel-produced algae consists of several processes, including selection and cultivation of algae, harvesting and dewatering, pretreatment or lipid extraction process, transesterification process, and product purification process (Gong and You, 2015). The overall processes involved in the biodiesel production from microalgae are shown in Fig. 9.1.

9.5.1 Microalgae selection and cultivation

Microalgae are basic microscopic unicellular autotrophic or heterotrophic photosynthetic organisms that live in saline or freshwater (Singh et al., 2011). These photosynthetic organisms effectively utilize CO_2 as a carbon source, sunlight as an energy source, and water to synthesize proteins, lipids, nucleic acids, and phospholipids (Enamala et al., 2018). Several algal species are distinguished as a potential feedstock for biodiesel production *including Nannochloropsis oculata, Chlorella vulgaris, Botryococcus braunii*, and so on (Mohamadzadeh et al., 2017; Lam et al., 2017). The cultivation method

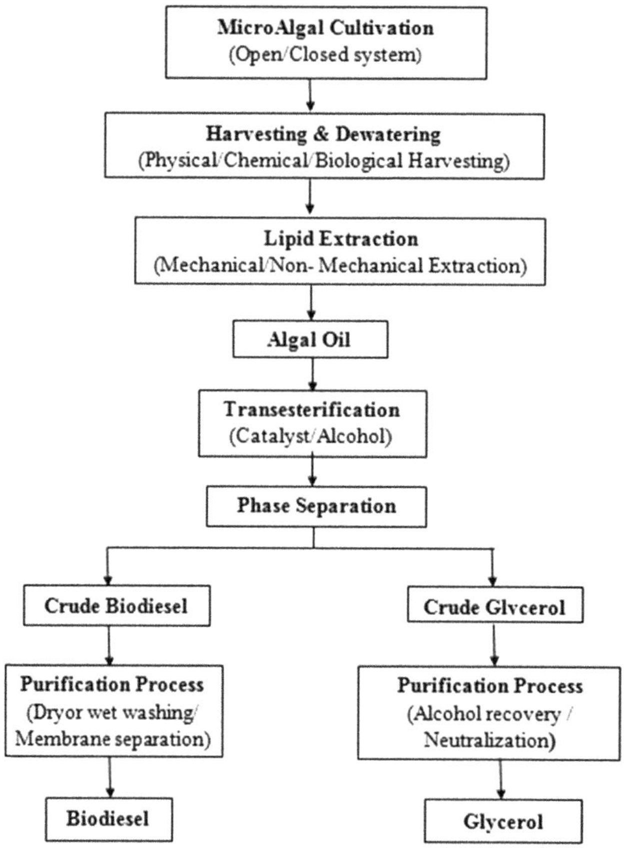

FIGURE 9.1 Overall process involved in biodiesel production from microalgae.

is closely associated with the microbial selection. The algal cultivation can be conducted in either an open pond or a closed pond system. The open pond cultivation is the conventional method for the large-scale cultivation of algae and superior to a closed system. In the open pond system, sunlight and the atmospheric CO_2 are utilized as a carbon source for the production of high algal biomass. The paddle system is used for the circulation and proper mixing of external nutrients with the algal cells. In the open system, there is a chance of contamination by other species of algae and bacteria. The algae can be grown at lower capital cost and operational cost; however, it is challenging to maintain the production stability and the yield of algal biomass.

A closed photobioreactor system is made up of polyethylene bags, plastics and glass in which light can penetrate 7–11 cm from the top and allows the algae to grow. It is designed in such a way to provide liquid circulation and gas exchange process to supply CO_2 to photobioreactor. Though the system is smaller, algae are protected from contamination and can be grown in an arid condition also. Microalgae can be cultivated in wastewater as it contains a load of organic waste along with essential nitrate and phosphate for the algae to produce lipids (Lowrey et al., 2015; Nugroho and Zhu, 2019). During the cultivation process, the operational parameters such as pH, temperature, and gas diffusion are maintained under optimum condition. Algae are easily cultivated anywhere where the light and pH play a crucial role in algal cultivation and lipid conversion. All species of algae contain protein, carbohydrates, and lipids.

Light is a vital factor the microalgal growth and provides the energy source to drive the photosynthesis and to initiate the translation of inorganic matter in the medium (da Fontoura et al., 2017). The microalgae can absorb the wavelength from 400 to 700 nm (Sajjadi et al., 2018). The *B. braunii, Spirogyra, Chlorella, and Picochlorum* microalgae can be cultivated under light and dark conditions for a certain period alternatively (Singh and Singh, 2015). Light distribution and the intensity of every two cells differ due to shading when microalgae are grown in a photobioreactor (Naderi et al., 2017). High acidity or alkalinity affects the biomass growth and lipid accumulation of microalgal metabolism (Qiu et al., 2017). *Chroococcus turgidus, Lyngbya confervoides, Nostoc commune*, and *Chlorella* sp. show the best growth rate of natural pH. Lower pH leads to lower biomass growth. For *Nannochloropsis salina*, pH in the range of 8–9 is considered as suitable for biomass production. *Chlorella* sp. tend to show lipid accumulation in slight alkali condition; however, biomass yield is reduced. Apart from the light intensity, the pH, temperature, culture density, and nutrient composition are the important factors that affect the microalgae biomass and lipid accumulation. High-quality algal biomass is produced under improved cultivation techniques.

9.5.2 Harvesting and dewatering

Harvesting and dewatering after the growth cycle is the second and most important step in algae biodiesel production process (Sharma et al., 2013). This process is a prerequisite for the mass production of biodiesel from the microalgae. The microalgal cell size (2–20 μm) and their colloidal property make the harvesting process more difficult and expensive (Barros et al., 2015; Wan et al., 2015). Harvesting is a solid–liquid separation process and energy-consuming and expensive process, which consumes 20%–30% of total biomass production cost. To reduce the cost of harvesting process typically involves bulk harvesting and dewatering or thickening process (Gao et al., 2014). The bulk harvesting procedure could reduce energy consumption during the dewatering process (Lam and Lee, 2012). In general, the various harvesting procedures are followed to harvest the microalgae such as physical, chemical, and biological methods. Sometimes, the combination of the methods is also employed to increase biomass separation (Mathimani and Mallick, 2018).

9.5.2.1 Physical methods

9.5.2.1.1 Centrifugation

Centrifugation process is a prevalent and the finest selection for harvesting microalgal cells in applying centripetal force. It evades the use of chemical solvents and eradicates the risk of chemical contamination during the separation process (Laamanen et al., 2016). The efficiency of the centrifugation process depends on the concentration of algal biomass and also the types of centrifuge and culture used. Milledge and Heaven (2013) stated that disc stalk centrifuge was found to be efficient in separating the algal cell with a density of 0.02%–0.05% in a medium. The efficiency of algal separation by this process is more than 95%. The major drawbacks such as cell damage may occur because of high gravitational force and shear stress applied. Besides, this process is highly energy expensive and costly, and maintenance is equally difficult (Mathimani and Mallick, 2018).

9.5.2.1.2 Filtration

Filtration is a mechanical or physical process usually employed to separate the solids from liquids by interpolating a medium where liquids alone will pass (Enamala et al., 2018). The culture and algal suspension will allow passing through the filter where the water passed through the filter using the driving force and algal slurry retains. Dewatering is usually achieved by the filtration process. There are several membrane filters available for the filtration process based on the size of the algal cell. Filtration shows 70%–80% efficiency in cell separation, and the major advantage is that it maintains the cell structure. Low-density culture can be easily separated by managing tangential flow to avoid the thick filter cake (Fuad et al., 2018). Microfiltration and ultrafiltration are most commonly employed for the separation process. The prime concerns of the filtration process are the increase in flow resistance due to clogging, high maintenance cost, inefficient cell separation, and extensive washing requirement of screening filters (Bilad et al., 2014).

9.5.2.1.3 Gravity sedimentation

Gravity sedimentation is predominately used to separate the solid–liquid. The algae slurry and clear water are separated under the action of gravity (Nie et al., 2018). The variety of cell provides different setting velocity based on their size and density. The cluster growth and density of microalgae could promote the effective settling velocity (Kumar et al., 2017). The economical gravity sedimentation progression requires less energy. However, Christenson and Sims (2011) concluded the settling rate by this process ranges from 0.1 to 2.6 cm/h. Hence, this method is not found to be reliable on a large scale.

9.5.2.1.4 Floatation

Floatation is an inverted sedimentation process based on the attachment of air or bubbles to the solid particles when they are carried to the liquid surface and more effective and advantageous when compared with the sedimentation process. The air bubble lifting force and self-floating characteristics of microalgae help it to float on the surface of water (Laamanen et al., 2016). Dissolved floatation and dispersed floatation are the most employed types of floatation techniques in the harvesting of microalgae. Both the methods generate an air bubble with a diameter ranging from 10 to 100 μm and 700–1500 μm, respectively. Floatation process depends on the instability of the suspended particles since high air and particle contact can be achieved with lower instability particle. The attachment of air bubbles to algae depends on air and aqueous contact angle between the air bubble and microalgae. It is found to be a promising method of harvesting microalgae and requires less space and a relatively short period. However, the bubble size, surface concentration and pH determine the efficiency of this approach (Barros et al., 2015).

9.5.2.2 Chemical method

9.5.2.2.1 Flocculation

Flocculation is the most conventional and effective method among the aforementioned processes of microalgae harvesting of large-volume of culture. Flocculation is the preparatory step to all other harvesting methods such as filtration, floatation, or gravity sedimentation. Flocculation and coagulation process is often carried out by either chemical or biological flocculants or agents. In all flocculation processes, the coalescence of smaller particles (microalgae) or cells into larger particles by colliding and adhering to each other and allowing for the sedimentation and easy separation. Zhu et al. (2018) reported that the flocculation generally occurs by any one of the mechanisms such as charge neutralism, patching, sweeping, and adsorption bridging. However, the selection of flocculants depends on the microalgal culture and the desired end product. The chemical flocculation can be induced by inorganic flocculants, inorganic polymers, and organic polymers. The major limitation in the flocculation method is that a large concentration of inorganic flocculants is needed to initiate the solid–liquid separation and ultimately results in a large quantity of sludge. Besides, the metal remains in the residue after the lipid extraction when metal salts are used as flocculants.

9.5.2.3 Biological method

9.5.2.3.1 Bioflocculation

The process of bioflocculation is an emerging technique, which is usually carried out by various microorganisms and their secreted biopolymers such as extracellular polymeric substance (EPS) or glutamic acid to harvest the microalgae (Barros et al., 2015; Chen et al., 2018). Bioflocculation can be carried out in any one of the following methods: (1) bacterial flocculation, (2) fungal flocculation, (3) actinomycetes flocculation, and (4) plant-based flocculation. The efficiency of the bioflocculation determines the type of bioflocculant used and the microalgal culture behavior (competency of the microalgae) to attach to them to form floc. Sometimes the bioflocculant organisms are cocultured along with the microalgal culture, which also results in higher lipid production. However, the coculturing of microalgae with other organisms may result in contamination and affect the downstream processing of the biodiesel production process (Vandamme et al., 2013). Bioflocculation process is simple, economic, and energy efficient, and combining with the centrifugation or sedimentation process will enhance the harvesting process.

9.5.2.3.2 Other methods

Apart from the aforementioned methods, magnetic methods and electricity-assisted techniques are employed in the harvesting of microalgae cells. The magnetic method had gained attention due to the simple operation, reusability, and time-saving process. In this process, microalgal cell tags with magnetic particles and are separated effortlessly from the culture by applying an external magnetic field (Markeb et al., 2019). Nowadays, magnetic nanocomposite is also applied for efficient harvesting of microalgae. Electricity employed techniques are found to be more environmental and improved efficiency when compared with flocculation and floatation method. When the culture is exposed to electric fields, microalgae are concentrated in the anode as it is negatively charged. Microalgae are negatively charged (Barros et al., 2015). Each method possesses its own merits and demerits, and the single method cannot effectively harvest the microalgal cells from the culture medium. The integration of the harvesting method could effectively improve the harvesting efficiency and reduce the cost and time under suitable conditions.

9.5.3 Pretreatment or lipid extraction process

The harvesting of algae and lipid extraction is the core process, which can be performed simultaneously. However, to enhance the lipid extraction and reduce the usage of the solvent and reduction of time, many techniques are developed. The ultimate aim of the other techniques will help in cell wall disruption, which allows the solvent to direct accesses of the intracellular lipid and improves the lipid extraction efficiency (Sierra et al., 2017). The extraction of lipids is also carried out by either mechanical or nonmechanical methods as depicted in Fig. 9.2. During extraction, lipids and other carbohydrates from microalgae should be removed effectively for the biodiesel production process (Noraini et al., 2014). The advantages and limitations of various lipids extraction methods are discussed in Table 9.1.

9.5.3.1 *Mechanical methods*

The mechanical method of lipid extraction implies the disintegration of the microalgal cell via shear, wave, or electric forces, before the introduction of solvents. Ultrasonication, microwave, electric pulse mode, and homogenization are mechanical methods assisted for the lipid extraction from microalgae.

9.5.3.1.1 Ultrasonication

Ultrasonic wave energy with the frequencies ranging between 10 kHz and 20 MHz is applied for cell disruption. This technique utilizes the shock wave to propagate fluctuations that induce cavitation and mechanoacoustic effects for the cell wall rupturing (Naveena et al., 2015; Mubarak et al., 2015). The microbubbles and reduced pressure eventually collapse the

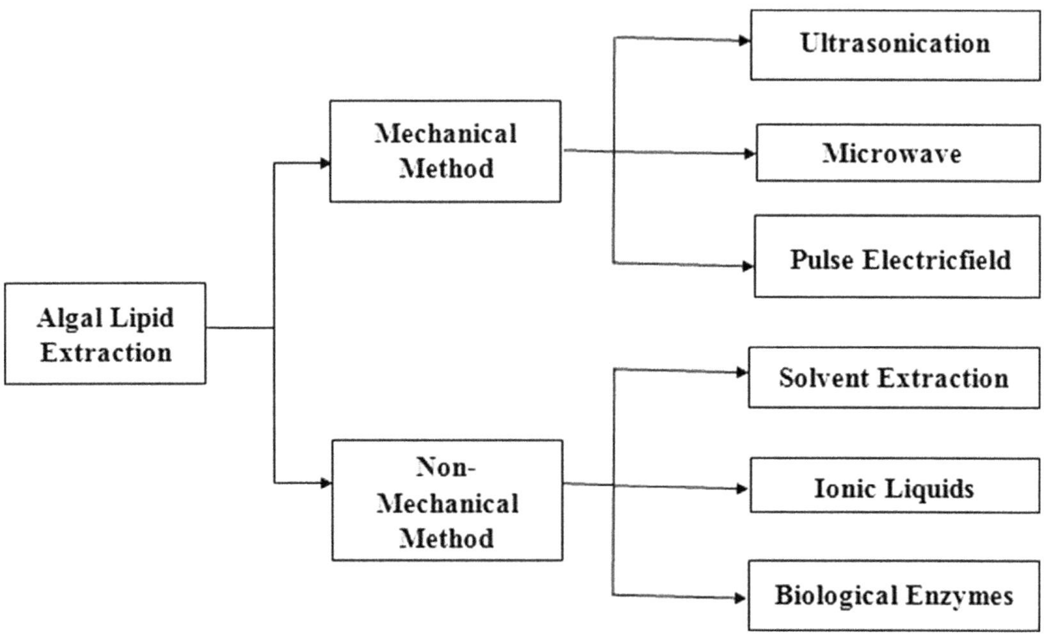

FIGURE 9.2 Mechanical and nonmechanical methods of algal lipid extraction.

TABLE 9.1 Advantages and limitation of various extraction processes.

Methods	Advantages	Limitations
Ultrasonication	• Increases the extraction rate • No solvent required • No effects on fatty acid profile • Eco-extraction process with higher penetration	• High heat generation • Produces free radicals • High power consumption • Scale-up process is difficult
Microwave	• High heat and mass transfer • More economical • Usage of minimal solvent • Improved extraction yield	• Maintenance cost is higher • Additional process is required to remove solid residue • Scale-up process is difficult
Pulsed electric field	• No additional chemicals required • Less energy requirement • Easy scalability • Less time-consuming • Less operating cost	• No Lipids is detected in extracellular medium • Formation of oil content due to the existence of C−C double bonds of unsaturated fatty acids
Solvent Extraction	• Easy processing • Very simple, cheap and reliable • High efficiency	• Few solvents are toxic • Large volume of solvents are required and highly flammable • Time consumption process • Removal of solvent residue after extraction is energy expensive
Acid/alkali	• High yield • Low cost and fast reaction time	• May corrode the reactor vessel • Denaturation of proteins
Ionic liquids	• Stable green solvent • Nonflammable and less hazardous	• Water immiscible ionic liquids have lower extraction • Scale-up process is difficult • Expensive approach • Difficult in recovery and reuse
Enzymatic approach	• Cell disruption with minimal damage • High lipid recovery	• Strongly dependent on pH • Extraction efficiency is also dependent on incubation time • Enzymes are expensive

emitting shock wave, which leads to the opening up of cells by cytoplasmic disruption (Greenly and Tester, 2015). This method reduces extraction time in tenfold along with 50%–500% in enhancement lipid extraction process. The disruption efficiency of the algal cell is highly dependent on the culture's cell size, shape, and age apart from the process parameters such as frequencies and exposure time (Lari et al., 2019). Natarajan et al. (2014) concluded that ultrasonic-assisted methods are highly effective against rigid cell wall algae such as *Chlorella* sp., rather than flexible cell walls species such as *Tetraselmis suecica and Nannochloropsis* sp. The flexible cell wall algae tend to coil up and retain the lipids. However, high-energy consumption and denaturation due to the high temperature generated during the process are the major drawbacks associated with this process (Dong et al., 2016).

9.5.3.1.2 Microwave

The feasibility of microwave extraction of lipids is first developed in the 1980s and found to be more effective when compared with the conventional methods. Microwaves represent electromagnetic waves with the radiation frequencies ranging from 0.3 to 300 GHz and cause cell disruption. In general, for algal cell, disruption takes place at 2450 MHz (D'Hondt et al., 2017). This technology does not require the dewatering of algal biomass, and also it is secure, rapid, and economical in lipid extraction from algae. The rapid oscillating electric field produced by microwave generates heat from inter- and intracellular movements. The generated heats form the water vapor, which ultimately ruptures the cell and opens the cell membrane for the accomplishment of the efficient extraction process. Rokicka et al. (2018) stated that *Botryococcus braunii* yields more lipids from 18% to 38% at 45°C for 15 min under microwave-assisted treatment. The microwave radiation ruptures the microalgal cell wall for easy extraction and conversion technologies to produce the biodiesel at the desired temperature at a particular time. Microwave-assisted techniques offer high disruption efficiency, short time, and even heating, which utilizes a minimal amount of solvent (Dai et al., 2014). It is also reported that in microwave-assisted process, the biodiesel can be recovered from the reaction mixture within 15–20 min, which is much quicker when compared with the conventional heating method.

9.5.3.1.3 Pulsed electric field

Pulsed electric field (PEF) is a promising clean, cheap, and simple extraction process from microalgal cells for biodiesel production. In this process, the cell membranes become permeable, and this process is known as electroporation or electropermeabilization (Silve et al., 2018). Electroporation is used to extract oil from the microalgae. The electroporators for PEF are designed to meet out the specific requirements such as high voltage and short electric pulses for cell disruption in a continuous system. This method of extraction does not produce cell debris and hence makes the downstream processing easier when compared with other methods (Carullo et al., 2018). The PEF can be combined with the solvent for the extraction of lipids from microalgae. On industrial scale, the solvent lose will be less than 1%. Optimization parameter includes the microalgal size, shape, and position in the electric field and the pulse field such as a number of the pulses, voltage intensity, and energy input (Raso et al., 2016).

9.5.3.1.4 Other methods

Sudden depressurization technique is the most common and simple method of cell disruption to get intracellular compounds. Depressurization is initiated, when a diffusible gas (carbon dioxide) is permitted to pass through the cell wall, where the gas expands the cell wall until saturation is accomplished. The pressure gradients rupture high disruption efficiency (Uquiche et al., 2016). The major advantages of using this process are free from toxic solvents and shorten the lipid extraction time. Halim et al. (2011) reported that supercritical carbon dioxide will reduce the lipid extraction time of *Chlorococcum* sp. by 5.6 times when compared with Soxhlet extraction. Similarly, bead mill is one of the simplest and oldest techniques used for the microalgal disruption. The agitation between the beads and algae brings cell disruption (Onumaegbu et al., 2018). The cell wall structure, shape and bead size, shape and agitation speed, and the design play a vital role in the cell. In general, the mechanical method of cell disruption for lipid extraction for biodiesel production is reported as a success. However, most of the methods will be more successful when combined with another nonmechanical extraction process.

9.5.3.2 *Nonmechanical methods*

The nonmechanical method of lipid extraction includes solvent extraction, chemical, and biological method. This method of extraction is scalable.

9.5.3.2.1 Solvent extraction

Lipid extraction proficiency in microalgal is liable for the usage of polar and nonpolar solvents (Abomohra et al., 2016). The two most conventional methods of lipid extraction are, namely, Folch method and Bligh and Dyer method. In both the methods, solvents such as chloroform and methanol are used in the ratio 2:1 and 1:2 v/v, respectively (Guo et al., 2019). The lipid extraction is usually carried out with the solvents such as ethanol, methanol, chloroform, and hexane. The polar solvents coupled with nonpolar solvents ensure the total extraction of all lipids including the free-standing globules. The uses of solvents are based on the class of lipids to be extracted from algae. However, few solvents such as chloroform and hexane cause adverse effects and toxicity to the environment and hence are excluded in large-scale lipid extraction (Mubarak et al., 2015). To accelerate and achieve better lipid extraction from algae in a short time, the solvent extraction process can be combined with heat or pressure. Hence, alternate methods of nonmechanical lipid extractions are developed.

9.5.3.2.2 Acid/alkali methods

Acid/alkali methods of cell disruption of microalgal biomass are usually carried with strong acids and alkalis. Sulfuric acid and sodium hydroxide are the most commonly used chemicals for the extraction of intracellular lipids (Xia et al., 2015). Acid hydrolyzes the cell wall polymer, whereas the alkali saponifies the cell wall lipids, which results in the disintegration of cell structure and cell disruption (Yoo et al., 2015). The chemical method of extraction is preferred for their low energy demand and easy scalability. However, fewer drawbacks such as safe disposal of used chemicals, corrosion of reactor vessels, and denaturation of valuable proteins due to extreme pH (Dixon and Wilken, 2018).

9.5.3.2.3 Ionic liquids

Ionic liquids are the green solvents and are popularly used in the lipid extraction process from microalgae. The ionic liquids contain cations and anions, and typical examples are 1-ethyl-3-methyl imidazolium acetate, 1-ethyl-3-methyl imidazolium diethylphosphate, and 1-ethyl-3-methyl imidazolium tetrafluoroborate. When these ionic liquids are employed in the extraction process, it dissolves microalgal biomass and makes the low-density lipids float in ionic liquids. In general, the ionic liquids are good solubility in an organic and inorganic solvent, thermal stability highly polar, and chemically inert. Olkiewicz et al. (2015) reported that these ionic liquids show a better extraction yield from raw sludge when compared with dried sludge, which could eliminate the cost of the drying process. Similarly, Pan et al. (2016) used ionic liquids in microwave-assisted microalgae lipid extraction process and observed that the increased lipid yields of about 10–15 times.

9.5.3.2.4 Biological methods

Biological methods of lipid extraction primarily involve using enzymatic hydrolysis of the cell wall to alter the cell wall permeability. These enzymes bind to the exact target molecules in the cell wall and hydrolyze the cell wall, which ultimately fallouts in the rupture of the cell wall. This process is superior when compared with other methods and depends on the composition of the microalgal cell wall and its components toward the enzyme catalytic activity (Hom-Diaz et al., 2016). The choice of enzymes such as chitinase, pectinase, and lysozymes varies for each microalgal species. Zabed et al. (2019) said that this is a simple method and does not involve any harmful chemicals and sophisticated equipment. Sierra et al. (2017) reported that a 30% higher lipid yield was achieved when *Chlamydomonas reinhardtii* incubated with autolysin. Many microalgae are reported higher yielding of biodiesel components after enzymatic pretreatment. The efficiency of lipid extraction depends on the temperature, pH biomass lipid content, and choice of enzyme selection. All the extraction processes are chosen based on the microalgal cell wall components and lipid concentration, which would be required for the biodiesel production.

9.5.4 Transesterification process

The extracted lipids from microalgae cannot be used directly for the transportation process as biodiesel. The extracted lipids are converted into fuel by four methods such as pyrolysis, microemulsification, dilution, and transesterification process. The transesterification may be dry esterification process or wet transesterification process. When concerning other processes, biodiesel produced from the transesterification process has the similar properties to diesel fuels and more economical concerning commercial value. The transesterification process is also known as alcoholysis, due to the reaction of triglycerides with alcohol (Mahmudul et al., 2017). In the transesterification process, one part of triglycerides reacts three parts of alcohol to give three parts of ester (fatty acid methyl ester) and glycerol as a by-product (Chozhavendhan et al., 2020a,b). As the process is reversible, the separation process should be fast and uses excess alcohol to maintain a forward reaction. In the transesterification process, ethanol or methanol is used as alcohol, and the catalysts such as homogeneous, heterogeneous, enzyme, and nanocatalyst are most commonly used. The catalyst used, oil to alcohol ratio, reaction temperature, water content, and reaction time determine the quality of biodiesel (Kumar et al., 2014).

9.5.4.1 Catalyst

The catalyst increases the speed of the transesterification process. The common catalysts used in the transesterification process such as homogeneous (Zhang et al., 2015), heterogeneous (Carvalho et al., 2017), enzyme (Huang et al., 2015), and nanocatalyst are commonly used. The homogeneous and heterogeneous catalysts may be in the form of acid or alkali. The choice of a catalyst depends on the fatty acid content of oil (Verma and Sharma, 2016). The detailed study of the catalyst is as follows.

9.5.4.1.1 Homogeneous catalyst

The acid-catalyzed transesterification in the homogeneous reactions uses sulfuric acid and hydrochloric acid and has a high tolerance to free fatty acids (FFAs) and water contents. These acids can simultaneously catalyze the esterification and transesterification processes. The reusability of the catalyst is possible. This type of catalysis process is not generally engaged in the biodiesel production, due to the requirement of the high alcohol molar ratio, slow reaction rate, and high catalyst concentration (Vyas et al., 2010).

In general, sodium hydroxide and potassium hydroxide are used as a base catalyst in the transesterification process. In the most common transesterification process, oil and methanol are stirred in the presence of potassium hydroxide as a catalyst for 3 h at 60°C and left unattended for 18 h for phase separation to occur (Onay et al., 2014). The major advantages are short reaction time, high activity, affordability, availability, and much faster than acid-catalyzed reaction (Avhad and Marchetti, 2015). Saponification due to high fatty acid and the catalyst will be impure after the reaction, and hence, there is no chance of reusability. The byproducts contain soap, which hinders the separation and purification process in biodiesel. The sodium hydroxide catalyzes *Scenedesmus* sp. and yields 55% biodiesel per gram of a lipid in

2 h, whereas the acid catalyzes the same process in 10 h at the same temperature (Alhassan et al., 2015). In general, homogeneous (acid and alkali) catalysts are predominately used in the industrial production process due to the easy conversion at moderate temperature and high yields at a single run.

9.5.4.1.2 Heterogeneous catalyst

A heterogeneous catalyst is developed to overcome the problem that occurs with a homogeneous catalyst such as reusability, recovery of glycerol, and purification of biodiesel (Olkiewicz et al., 2016). The heterogeneous acid catalyst reduces the purification process and produces the purest form of glycerol, whereas the homogeneous catalyst yields around 80%. The heterogeneous catalyst performed by inorganic, sulfonated carbons, and polymeric membranes yields around 99% of methyl ester. The advantages are similar to that of a homogeneous acid catalyst; the major disadvantages are leaching associated with lower activity and more energy consumption (Lee and Wilson, 2015). The heterogeneous base catalysts have chemical steadiness, highest action at encompassing temperature, and by and largely composed metal oxides, clays, and alkaline zeolites. The base catalysts usually have additional active sites when compared with the heterogeneous acid catalyst, and thus, automatically reduce the reaction time (Yang et al., 2018).

9.5.4.1.3 Enzyme catalyst

The usage of biocatalyst in biodiesel production has attracted the researchers, and its performance was equal to the chemical-based catalyst. The aforementioned problems of using chemical catalysts are ruined out by using a biological route (enzyme catalyst) (Gog et al., 2012). The process has several advantages such as high selectivity, lower energy consumption, and low reaction temperature, and it can be immobilized. The enzymes are reused when it is immobilized on the solid supports (Aguieiras et al., 2015). Lipase and Novozymes are found to be effective against various species of microalgae such as *Botryococcus, Chlorella, and Aurantiochytrium* sp. Bacterial and fungal lipase enzymes are generally used in the enzyme-catalyzed reactions. The high-purity enzyme could achieve more than 90% conversion at 30–50°C temperature with the minimum 8 h duration to maximum of 90 h. Apart from the enzyme, alcohol type and reaction temperature have an impact on the biodiesel production and the endurance of the enzyme used. However, on industrial-scale using the enzymes is so expensive and a very slow process.

9.5.4.1.4 Nanocatalyst

The nanoparticles are used as a catalyst and have combined advantages over the homogeneous and heterogeneous catalytic process (Ashu and Anshul, 2017). The biodiesel derived from microalgae using nanocatalytic (nano calcium methoxide) transesterification process yields 99% of FAME at the reaction time of 3 h at 80°C (Siow et al., 2016). Also, few studies revealed that ZnO nanorods show better catalytic performance when compared with conventional ZnO. Many nanocatalysts have developed in the recent and past years and have been showing better efficiency in converting the oil into biodiesel (Baskar and Aiswarya, 2016). At present, nanoporous carbon and other derivatives are developed to establish the adsorbents for the biodiesel separation process. However, the major parameters that affect the nanocatalyst include the size of the catalyst, surface functionality, and surface composition apart from the toxicity toward the human and the environment (Ashu and Anshul, 2017).

9.5.4.2 Alcohol to oil ratio

In general, for the transesterification process, three parts of alcohol react with one part triglycerides to yield three parts of FAME and glycerol. However, excess alcohol is required to drive the reaction in the forward direction and ensure the completion conversion (Hailegiorgis et al., 2015).

The molar ratio of 6:1 is used in a base-catalyzed reaction, and the excess and unused alcohol settles in the glycerol phase during the separation process and makes the glycerol purification difficult. The *C. vulgaris* shows 90% of fatty acid to FAME conversion efficiency in 4 h at the 4:1 of alcohol to ratio. Methanol and ethanol are the most common alcohols used in the transesterification process. The alcohol-to-oil ratio plays a crucial role in determining the purity of biodiesel. The increase in the alcohol-to-oil ratio of more than 6:1 will not increase the biodiesel yield, and on one hand, it complicates the purification process, and on the other hand, it increases the production cost.

9.5.4.3 Water and free fatty acid composition

Water content has a low adverse effect on the biodiesel production process. Kusdiana and Saka (2004) noted the presence of water could aid in the hydrolysis of triglycerides present in the oil and to the formation of FFA. Thus, the formation of FFAs enhances the production of biodiesel (FAME). However, the excess amount of water on biomass makes an energy-intensive drying of biomass. The excess water in FFA on reaction alkali catalysts such as sodium or potassium hydroxides leads to soap formation and thus affects the choice of a catalyst (Suwannakarn et al., 2009). Demirbas (2009) stated that indeed a pint-sized amount of water (0.1%) in the transesterification reaction will decrease the methyl ester translation from vegetable oil. In the conventional catalyst method, the presence of water causes adverse effects on methyl ester formation, but in contrast, the presence of water improves the yield of methyl ester when methanol is replaced by supercritical methanol by modern technologies.

9.5.4.4 Reaction time

The reaction time is the most important factor, which determines the investment cost of any biodiesel production process. The reaction time is determined by the solvent used for the extraction of lipids and the lipids present in the microalgae. Reaction time has also been impacted by the catalyst and strongly influenced by the reaction temperature (Jazzar et al., 2015). The conversion rate is more than 95%; when there is an increase in temperature from 250 to 400°C, the reaction time decreased from 30 to 10 min. The choice of a catalyst and the water content were the critical parameters that affect the reaction time. The reaction time in alkali catalyst is five times faster than the acid catalyst in dry *Scenedesmus* biomass.

9.5.4.5 Temperature

The reaction temperature is one of the vital parameters that affects the yield of biodiesel production. In the conventional method, the reaction temperature should be in the range of 50–60°C or less than the boiling point of alcohol to prevent the evaporation of the alcohol used. An increase in temperature reduces the viscosity of the oil and enhances the interaction between the triglyceride, methanol, and catalyst (Leung et al., 2010). Cao et al. (2013) stated that employing the higher process temperature can break the microalgal cell wall and improves the methanol–lipid interactions. It is observed that biodiesel yield is improved with an increase in temperature from 200 to 400°C in the supercritical process. However, an increase in temperature beyond the optimum temperature reduces the biodiesel production and accelerates the soap formation in triglycerides (Mathiyazhagan and Ganapathi, 2011).

9.5.5 Product purification process

The biodiesel and glycerol mixture is obtained as the result of the transesterification process. The raw biodiesel must be purified to remove the various impurities such as metal ions, water, acid, glycerol, and excess alcohol to have better fuel efficiency and emission characteristics (Shirazi et al., 2013). Glycerol phase is separated from biodiesel by gravitational settling in the pretreatment process. Several downstream processing techniques such as wet washing, dry washing, membrane separation (Saleh et al., 2010), and distillation are employed to maintain the standard quality (Banga et al., 2014).

9.5.5.1 *Wet washing*

Wet washing is the traditional method of removing impurities from biodiesel such as residual impurities such as a catalyst, methanol soap, and glycerol. Mendow et al. (2012) stated that biodiesel purification in wet washing is carried out by two consecutive washing. First washing with 5 wt% of HCl followed by saturated carbon dioxide water, and second washing is with neutral water. The mixture is shaken well and allowed for settling for about 15 min and centrifuged to obtain the final purified biodiesel. Apart from water and acid, other organic solvents are also employed in the biodiesel purification process. The acid washing helps to neutralize the base catalyst used and decompose the soap formed and later washed with water to expel the remaining impurities (Berrios et al., 2011). The major advantages of this method are to remove glycerol, excess alcohol, and soap and could achieve 99% biodiesel purity. The limitation of these methods is the requirement of the centrifugation process and a large amount of water required for multiple washing and supplementary drying steps. This process occupies a large surface area for washing and settling tank and not applicable in continuous operation.

9.5.5.2 *Dry washing*

Dry washing is developed to replace the wet washing process and to make water-free purification process. Dry washing removes the contaminants from crude biodiesel from the adsorption process. This process does not employ the solvent and uses ion exchange adsorbents such as Magnesol, Amberlite, and so on, which are used in industrial plants (Mansir et al., 2018). Many other different types of resins are also available to purify the crude biodiesel. The positively charged resin adsorbs the negative charged water and glycerol (Atadashi, 2015). This process is excellent in removing soap and free glycerol, no wastewater production, less energy-intensive, and no additional requirement of the drying process. Also, the methanol can be reused further in the transesterification process. Few drawbacks associated with these processes are the resins that are positively charged and difficult to predict the composition of resin. The resins are nonreusable and result in frequent replacement of adsorbents. This process does not remove methanol, and hence, additional method should be employed for methanol removal (Veljkovic et al., 2015).

9.5.5.3 *Membrane separation process*

Membrane-based separation is quite familiar in the water purification and protein separation process. At present, these membrane separation technologies are employed in the separation and production of crude biodiesel. Solvent resistance polymeric membranes, microporous ceramic membranes, and nanofiltration membranes were prepared by phase inversion process (Torres et al., 2017). This process has several advantages over the other by high specific mass transfer and minimizes the capital cost. These membranes show resistance in adverse conditions even at pH 12 and temperature up to 60°C. The average pore size of the membrane is 44 μm and with the lower and higher limits of 12 and 400 μm, respectively. However, the soap content in crude biodiesel can foul the membranes, leading to repeated replacements of expensive membranes (Chozhavendhan et al., 2020a,b).

It has to be ensured that the final product should be free from other pollutants and meet out the specification of biodiesel. Various experimental trials are done to increase biodiesel production, economic feasibility, emission reduction, and usage of biodiesel without altering the engine and to develop the easy method for recovery.

9.6 By-product applications

The microalgal feedstock faces numerous challenges such as low productivity, cell harvesting, extraction of lipids, separation of biodiesel, and overall capital cost investment. The microalgae contain many high-value bioproducts such as pigments, vitamins, proteins, and antioxidants that can provide additional income in the short term (Chew et al., 2017). The deoiled algae can be used for their protein content, which can be extracted using ethanol as similar to lipid extraction. Later the solvents are recovered by the evaporation process, and the protein output is used in animal feed. The remaining microalgal biomass can be used in combustion and as fertilizer.

Crude glycerol is the main by-product obtained from biodiesel after the transesterification process of the lipid extracted from algae. For each 40 kg of biodiesel produced, 4 kg of crude glycerol is produced as a by-product with a copious number of impurities scuh as methanol, salt, water, and so on. Generally, glycerol is a versatile, high-value compound and has found its application in food, cosmetics, and pharmaceutical industries. However, the usage of crude glycerol in those industries is restricted due to its low purity. At present, the crude glycerol is generally sold at low cost for animal feed and used as cocombustion fuel in the boiler due to its high energy value (Chozhavendhan et al., 2017). Conversion of crude glycerol to other products through the microbiological route aids the economic development of biodiesel industries and paves way for the biorefinery concept. The crude glycerol is a good carbon and energy source, which is readily utilized by several microorganisms for the production of high platform chemicals such as ethanol, hydrogen, and succinic acid. Monterio et al. (2018) stated that crude glycerol can be used as feedstock for the production of biopolymers and other fuel additive chemicals (Pradima and Kulkarni, 2017). This process of converting waste into wealth leads to zero waste discharge from the biodiesel industries, as microalgae, lipids, protein, pigment, and the by-product crude glycerol are effectively utilized. Thus, the environment is conserved ecofriendly without dumping any waste.

9.7 Conclusions

The first and second generation of biodiesel could not able to meet out the global demand. Biodiesel from microalgae is a potential renewable source and sustainable solution for the energy deficiency around the globe. However, there are a lot of blockades in the commercialization of microalgal biodiesel. Microalgal growth, harvesting, extraction of lipids, and purification process are the key parameters in algal biodiesel production. Many research studies identifying the lipid-rich and fast-growing microalgae for biodiesel production have been completed. The downstream processing contributes to the major portion of the capital investment and the production cost. Technical experience, technological development, and integrated concepts can meet the required amount of biodiesel in a short period time and will promote the economic viability of bio-based industries. In the end, microalgae looked up as a potential source for biodiesel production in near future.

References

Abdullah, B., Muhammad, S.A.F.A.S., Shokravi, Z., Ismail, S., Kassim, K.A., Mahmood, A.N., Aziz, M.M.A., 2019. Fourth generation biofuel: a review on risks and mitigation strategies. Renew. Sust. Energy Rev. 107, 37–50.

Abomohra, A.E.F., Jin, W., El-Sheekh, M., 2016. Enhancement of lipid extraction for improved biodiesel recovery from the biodiesel promising microalga *Scenedesmus obliquus*. Energy Convers. Manag. 108, 23–29.

Addy, M.M., Kabir, F., Zhang, R., Lu, Q., Deng, X., Current, D., Griffith, R., Ma, Y., Zhou, W., Chen, P., 2017. Co-cultivation of microalgae in aquaponic systems. Bioresour. Technol 245, 27–34.

Aguieiras, E.C.G., Cavalcanti-Oliveira, E.D., Freire, D.M.G., 2015. Current status and new developments of biodiesel production using fungal lipases. Fuel 159, 52–67.

Alhassan, F.H., Rashid, U., Taufiq-yap, Y.H., 2015. Synthesis of waste cooking oil-based biodiesel via effectual recyclable bi-functional $Fe_2O_3MnOSO_4^{2-}/ZrO_2$ nanoparticle solid catalyst. Fuel 142, 38–45.

Aransiola, E.F., Ojumu, T.V., Oyekola, O.O., Madzimbamuto, T.F., Ikhu-Omoregbe, D.I.O., 2014. A review of current technology for biodiesel production: state of the art. Biomass Bioenergy 61, 276–297.

Ashu, C., Anshul, S., 2017. Nano-catalysts: a newfangled gem in the catalytic world. Recent Adv. Petrochem. Sci. 3 (5), 555–625.

Atadashi, I.M., 2015. Purification of crude biodiesel using dry washing and membrane technologies. Alexandria Eng. J. 54, 1265–1272.

Avhad, M.R., Marchetti, J.M., 2015. A review on recent advancement in catalytic materials for biodiesel production. Renew. Sust. Energy Rev. 50, 696–718.

Banga, S., Varshney, P., Kumar, N., Pal, M., 2014. Optimization of parameters for purification of *Jatropha curcas* based biodiesel using organic adsorbents. Inter. J. Renew. Energy Res. 4 (3), 598–603.

Barros, A.I., Gonçalves, A.L., Simoes, M., Pires, J.C.M., 2015. Harvesting techniques applied to microalgae: a review. Renew. Sust. Energy Rev. 41, 1489–1500.

Baskar, G., Aiswarya, R., 2016. Trends in catalytic production of biodiesel from various feedstocks. Renew. Sustain. Energy Rev. 57, 496–504.

Berrios, M., Martín, M., Chica, A., Martín, A., 2011. Purification of biodiesel from used cooking oils. Appl. Energy 88 (11), 3625–3631.

Bharathiraja, B., Chakravarthy, M., Ranjith Kumar, R., Yogendran, D., Yuvaraj, D., Jayamuthunagai, J., Praveenkumar, R., 2015. Aquatic biomass (algae) as a future feed stock for biorefineries: a review on cultivation, processing and products. Renew. Sust. Energy Rev. 47, 635–653.

Bhatia, S.K., Kim, S.H., Yoon, J.J., Yang, Y.H., 2017. Current status and strategies for second generation biofuel production using microbial systems. Energy Convers. Manag. 148, 1142–1156.

Bilad, M.R., Discart, V., Vandamme, D., Foubert, I., Muylaert, K., Vankelecom, I.F.J., 2014. Coupled cultivation and pre-harvesting of microalgae in a membrane photobioreactor (MPBR). Bioresour. Techno. 155, 410–417.

Cao, H., Zhang, Z., Wu, X., Miao, X., 2013. Direct biodiesel production from wet microalgae biomass of *Chlorella pyrenoidosa* through in situ transesterification. BioMed Res. Int. 2013, 1–6.

Carullo, D., Abera, B.D., Casazza, A.A., Donsi, F., Perego, P., Ferrari, G., Pataro, G., 2018. Effect of pulsed electric fields and high pressure homogenization on the aqueous extraction of intracellular compounds from the microalgae *Chlorella vulgaris*. Algal Res. 31, 60–69.

Carvalho, A.K.F., da Conceiçao, L.R.V., Silva, J.P.V., Perez, V.H., de Castro, H.F., 2017. Biodiesel production from *Mucor circinelloides* using ethanol and heteropolyacid in one and two-step transesterification. Fuel 202, 503–511.

Chen, J., Leng, L., Ye, C., Lu, Q., Addy, M., Wang, J., Liu, J., Chen, P., Ruan, R., Zhou, W., 2018. A comparative study between fungal pellet and spore-assisted microalgae harvesting methods for algae bioflocculation. Bioresour. Technol. 259, 181–190.

Chew, K.W., Yap, J.Y., Show, P.L., Suan, N.H., Juan, J.C., Ling, T.C., et al., 2017. Microalgae biorefinery: high value products perspectives. Bioresour. Technol. 229, 53–62.

Chozhavendhan, S., Praveenkumar, R., Bharathiraja, B., Jayakumar, M., 2016. Recent progress on transforming crude glycerol into high value chemicals: a critical review. Biofuels 10 (3), 309–314.

Chozhavendhan, S., Praveen Kumar, R., Sivarathnakumar, S., Vinoth Arulraj, J., Elavazhagan, S., Bharathirja, B., Sunita, J.V., 2017. Production of ethanol by *Zymomonas mobilis* using partially purified glycerol. J. Energy Env. Sust. 4, 15–19.

Chozhavendhan, S., Karthiga Devi, G., Subbaiah, R., Vinoth, S., 2019. Production of Dihydroxyacetone by *Gluconobacter oxydans* from crude glycerol. Int. J. Environ. Prot. 39 (2), 110–114.

Chozhavendhan, S., Vijay Pradhap Singh, M., Fransila, B., Praveen Kumar, R., Karthiga Devi, G., 2020a. A review on influencing parameters of biodiesel production and purification processes. Currn. Res. Green Sust. Chem. 2, 1–6.

Chozhavendhan, S., Karthiga Devi, G., Bharathiraja, B., Praveen Kumar, R., Elavazhagan, S., 2020b. Chapter 11—Assessment of crude glycerol utilization for sustainable development of biorefineries. In: Praveen kumar, R. (Ed.), Refining Biomass Residues for Sustainable Energy and Bioproducts. Elsevier, pp. 195–212.

Christenson, L., Sims, R., 2011. Production and harvesting of microalgae for wastewater treatment, biofuels, and bioproducts. Biotechnol. Adv. 29 (6), 686–702.

da Fontoura, J.T., Rolim, G.S., Farenzena, M., Gutterres, M., 2017. Influence of light intensity and tannery wastewater concentration on biomass production and nutrient removal by microalgae *Scenedesmus sp*. Process. Saf. Environ. Prot. 111, 355–362.

Dai, Y.M., Chen, K.T., Chen, C.C., 2014. Study of the microwave lipid extraction from microalgae for biodiesel production. Chem. Eng. J. 250, 267–273.
Demirbas, A., 2009. Progress and recent trends in biodiesel fuels. Energy Convers. Manag. 50, 14–34.
Dixon, C., Wilken, L.R., 2018. Green microalgae biomolecule separations and recovery. Bioresour. Bioprocess. 5 (1), 1–14.
Dong, T., Knoshaug, E.P., Pienkos, P.T., Laurens, L.M.L., 2016. Lipid recovery from wet oleaginous microbial biomass for biofuel production: a critical review. Appl. Energy 177, 879–895.
Doshi, A., Pascoe, S., Coglan, L., Rainey, T.J., 2016. Economic and policy issues in the production of algae-based biofuels: a review. Renew. Sust. Energy Rev. 64, 329–337.
D'Hondt, E., Martin-Juarez, J., Bolado, S., Kasperoviciene, J., Koreiviene, J., Sulcius, S., Elst, K., Bastiaens, L., 2017. 6 - cell disruption technologies. In: Gonzalez-Fernandez, C., Munoz, R. (Eds.), Microalgae-Based Biofuels and Bioproducts. Woodhead Publishing, pp. 133–154.
Enamala, M.K., Enamala, S., Chavali, M., Donepudi, J., Yadavalli, R., Kolapalli, B., et al., 2018. Production of biofuels from microalgae-a review on cultivation, harvesting, lipid extraction, and numerous applications of microalgae. Renew. Sust. Energy Rev. 94, 49–68.
Fuad, N., Omar, R., Kamarudin, S., Harun, R., Idris, A., Wan, W.A.K.H., 2018. Mass harvesting of marine microalgae using different techniques. Food Bioprod. Process. 112, 169–184.
Gao, H., Scherson, Y.D., Wells, G.F., 2014. Towards energy neutral wastewater treatment: methodology and state of the art. Environ. Sci. Process Impacts. 16, 1223–1246.
Gog, A., Roman, M., Tosa, M., Paizs, C., Irimie, F.D., 2012. Biodiesel production using enzymatic transesterification – current state and perspectives. Renew. Energy 39, 10–16.
Gong, J., You, F., 2015. Value-added chemicals from microalgae: Greener, more economical, or both? ACS Sustain. Chem. Eng. 3 (1), 82–96.
Greenly, J.M., Tester, J.W., 2015. Ultrasonic cavitation for disruption of microalgae. Bioresour. Technol. 184, 276–279.
Guo, Y.Y., Sze, Y.L., Pau, L.S., Yang, T., Chung, L.L., Thi, T.C.N., et al., 2019. Recent advances in algae biodiesel production: from upstream cultivation to downstream processing. Bioresou. Technol. Rep. 7, 100227.
Hailegiorgis, S.M., Mahadzir, S., Subbarao, D., 2015. Reaction kinetics study for microwave energy pretreated *Jatropha curcas L* in-situ transesterification. In: Hashim, M.A. (Ed.), ICGSCE 2014. Springer Singapore, pp. 131–138.
Halim, R., Gladman, B., Danquah, M.K., Webley, P.A., 2011. Oil extraction from microalgae for biodiesel production. Bioresour. Technol. 102 (1), 178–185.
Hom-Diaz, A., Passos, F., Ferrer, I., Vicent, T., Blánquez, P., 2016. Enzymatic pretreatment of microalgae using fungal broth from *Trametes versicolor* and commercial laccase for improved biogas production. Algal Res. 19, 184–188.
Huang, J., Xia, J., Jiang, W., Li, Y., Li, J., 2015. Biodiesel production from microalgae oil catalyzed by a recombinant lipase. Bioresour. Technol. 180, 47–53.
Janaun, J., Ellis, N., 2010. Perspectives on biodiesel as a sustainable fuel. Renew. Sust. Energy Rev. 14 (4), 1312–1320.
Jazzar, S., Olivares-Carrillo, P., Perez de los Rios, A., Marzouki, M.N., Acien- Fernandez, F.G., Fernandez-Sevilla, J.M., et al, 2015. Direct supercritical methanolysis of wet and dry unwashed marine microalgae (*Nannochloropsis gaditana*) to biodiesel. Appl. Energy 148, 210–219.
Karmakar, B., Halder, G., 2019. Progress and future of biodiesel synthesis: advancements in oil extraction and conversion technologies. Energy Convers. Manag. 182, 307–339.
Kumar, V., Muthuraj, M., Palabhanvi, B., Ghoshal, A.K., Das, D., 2014. Evaluation and optimization of two stage sequential in situ transesterification process for fatty acid methyl ester quantification from microalgae. Renew. Energy 68, 560–569.
Kumar, V., Nanda, M., Verma, M., 2017. Application of agar liquid-gel transition in cultivation and harvesting of microalgae for biodiesel production. Bioresour. Technol. 243, 163–168.
Kusdiana, D., Saka, S., 2004. Effects of water on biodiesel fuel production by supercritical methanol treatment. Bioresour. Technol. 91, 289–295.
Laamanen, C.A., Ross, G.M., Scott, J.A., 2016. Flotation harvesting of microalgae. Renew. Sust. Energy Rev. 58, 75–86.
Lam, M.K., Lee, K.T., 2012. Microalgae biofuels: a critical review of issues, problems and the way forward. Biotechnol. Adv 30, 673–690.
Lam, M.K., Yusoff, M.I., Uemura, Y., Lim, J.W., Khoo, C.G., Lee, K.T., et al., 2017. Cultivation of *Chlorella vulgaris* using nutrients source from domestic wastewater for biodiesel production: growth condition and kinetic studies. Renew. Energy 103 (Suppl. C), 197–207.
Lari, Z., Ahmadzadeh, H., Hosseini, M., 2019. Chapter 2 – cell wall disruption: a critical upstream process for biofuel production. In: Hosseini, M. (Ed.), Advances in Feedstock Conversion Technologies for Alternative Fuels and Bioproducts. Wood head Publishing, pp. 21–35.
Lee, A.F., Wilson, K., 2015. Recent developments in heterogeneous catalysis for the sustainable production of biodiesel. Catal. Today 242, 3–18.
Leung, D.Y.C., Wu, X., Leung, M.K.H., 2010. A review on biodiesel production using catalyzed transesterification. Appl. Energy 87, 1083–1095.

Li, Y., Zhou, L.W., Wang, R.Z., 2017. Urban biomass and methods of estimating municipal biomass resources. Renew. Sust. Energy Rev. 80, 1017–1030.

Lowrey, J., Brooks, M.S., Mc Ginn, P.J., 2015. Heterotrophic and mixotrophic cultivation of microalgae for biodiesel production in agricultural wastewaters and associated challenges a critical review. J. Appl. Phycol. 27, 1485–1498.

Mahdavi, M., Abedini, E., Darabi, A.H., 2015. Biodiesel synthesis from oleic acid by nanocatalyst (ZrO_2/Al_2O_3) under high voltage conditions. RSC. Adv. 5, 55027–55032.

Mahlia, T.M.I., Syazmi, Z.A.H.S., Mofijur, M., Abas, A.E.P., Bilad, M.R., Ong, H.C., Silitonga, A.S., 2020. Patent landscape review on biodiesel production: technology updates. Renew. Sust. Energy Rev. 118, 1–9.

Mahmudul, H.M., Hagos, F.Y., Mamat, R., Adam, A.A., Ishak, W.F.W., Alenezi, R., 2017. Production, characterization and performance of biodiesel as an alternative fuel in diesel engines – a review. Renew. Sust. Energy Rev. 72, 497–509.

Mansir, N., Teo, S.H., Rashid, U., Saiman, M.I., Tan, Y.P., Alsultan, G.A., et al., 2018. Modified waste egg shell derived bifunctional catalyst for biodiesel production from high FFA waste cooking oil. A review. Renew. Sust. Energy Rev. 82, 3645–3655.

Markeb, A.A., Llimos-Turet, J., Ferrer, I., Blanquez, P., Alonso, A., Sanchez, A., Moral-Vico, J., Font, X., 2019. The use of magnetic iron oxide based nanoparticles to improve microalgae harvesting in real wastewater. Water Res. 159, 490–500.

Markou, G., Nerantzis, E., 2013. Microalgae for high-value compounds and biofuels production: a review with focus on cultivation under stress conditions. Biotechnol. Adv. 31, 1532–1542.

Mata, T.M., Martins, A.A., Caetano, N.S., 2010. Microalgae for biodiesel production and other applications: a review. Renew. Sust. Energy Rev. 14 (1), 217–232.

Mathimani, T., Mallick, N., 2018. A comprehensive review on harvesting of microalgae for biodiesel–key challenges and future directions. Renew. Sust. Energy Rev. 91, 1103–1120.

Mathiyazhagan, M., Ganapathi, A., 2011. Factors affecting biodiesel production. Res. Plant Biol. 1 (2), 1–5.

Mehta, P., Singh, D., Saxena, R., Rani, R., Gupta, R., Puri, S., Mathur, A., 2018. High-value Coproducts from Algae-An Innovational Way to Deal with Advance Algal Industry. Waste to wealth, pp. 343–363.

Mendow, G., Veizaga, N., Sánchez, B., Querini, C., 2012. Biodiesel production by two-stage transesterification with ethanol by washing with neutral water and water saturated with carbon dioxide. Bioresour.Technol. 118, 598–602.

Milledge, J.J., Heaven, S., 2013. A review of the harvesting of microalgae for biofuel production. Rev. Environ. Sci. Bio. 12 (2), 165–178.

Mohamadzadeh, S.H., Karimi-Sabet, J., Ghotbi, C., 2017. Biodiesel production from Spirulina microalgae feedstock using direct transesterification near supercritical methanol condition. Bioresour. Technol. 239 (Suppl. C), 378–386.

Mohamed Shameer, P., Ramesh, K., Sakthivel, R., Purnachandran, R., 2017. Experimental evaluation on oxidation stability of biodiesel/diesel blends with alcohol addition by rancimat instrument and FTIR spectroscopy. J. Mech. Sci. Technol. 31 (1), 455–463.

Monteiro, M.R., Kugelmeier, C.L., Pinheiro, R.S., Batalha, M.O., da Silva César, A., 2018. Glycerol from biodiesel production: technological paths for sustainability. Renew. Sust. Energy Rev. 88, 109–122.

Mubarak, M., Shaija, A., Suchithra, T.V., 2015. A review on the extraction of lipid from microalgae for biodiesel production. Algal Res. 7, 117–123.

Naderi, G., Znad, H., Tade, M.O., 2017. Investigating and modelling of light intensity distribution inside algal photobioreactor. Chem. Eng. Process: Process Intensif. 122, 530–537.

Najafi, G., Ghobadian, B., Yusaf, T.F., 2011. Algae as a sustainable energy source for biofuel production in Iran: a case study. Renew. Sust. Energy Rev. 15, 3870–3876.

Natarajan, R., Ang, W.M.R., Chen, X., Voigtmann, M., Lau, R., 2014. Lipid releasing characteristics of microalgae species through continuous ultrasonication. Bioresour. Technol. 158, 7–11.

Naveena, B., Armshaw, P., Tony Pembroke, J., 2015. Ultrasonic intensification as a tool for enhanced microbial biofuel yields. Biotechnol. Biofuels 8, 140.

Nie, C., Pei, H., Jiang, L., Cheng, J., Han, F., 2018. Growth of large-cell and easily-sedimentation microalgae *Golenkinia* SDEC-16 for biofuel production and campus sewage treatment. Renew. Energy 122, 517–525.

Noraini, M., Ong, H.C., Badrul, M.J., Chong, W., 2014. A review on potential enzymatic reaction for biofuel production from algae. Renew. Sust. Energy Rev. 39, 24–34.

Nugroho, Y.K., Zhu, L., 2019. An integration of algal biofuel production planning, scheduling, and order-based inventory distribution control systems. Biofuel Bioprod. Bior. 13 (4), 920–935.

Okoro, V., Azimov, U., Munoz, J., Hernandez, H.H., Phan, A.N., 2019. Microalgae cultivation and harvesting: growth performance and use of flocculants-A review. Renew. Sust. Energy Rev. 115, 1–16.

Olkiewicz, M., Plechkova, N.V., Fabregat, A., Stüber, F., Fortuny, A., Font, J., et al., 2015. Efficient extraction of lipids from primary sewage sludge using ionic liquids for biodiesel production. Sep. Purif. Technol. 153, 118–125.

Olkiewicz, M., Plechkova, N.V., Earle, M.J., Fabregat, A., Stüber, F., Fortuny, A., et al., 2016. Biodiesel production from sewage sludge lipids catalysed by Bronsted acidic ionic liquids. Appl. Catal. B Environ. 181, 738–746.

Onay, M., Sonmez, C., Oktem, H.A., Yucel, A.M., 2014. Thermo-resistant green microalgae for effective biodiesel production: isolation and characterization of unialgal species from geothermal flora of Central Anatolia. Bioresour. Technol. 169, 62–71.

Onumaegbu, C., Mooney, J., Alaswad, A., Olabi, A.G., 2018. Pre-treatment methods for production of biofuel from microalgae biomass. Renew. Sustain. Energy Rev. 93, 16–26.

Pan, J., Muppaneni, T., Sun, Y., Reddy, H.K., Fu, J., Lu, X., Deng, S., 2016. Microwave-assisted extraction of lipids from microalgae using an ionic liquid solvent [BMIM][HSO_4]. Fuel 178, 49–55.

Pradima, J., Kulkarni, M.R., 2017. Review on enzymatic synthesis of value added products of glycerol, a by-product derived from biodiesel production. Resour. Eff. Technol. 3, 394–405.

Qiu, R., Gao, S., Lopez, P.A., Ogden, K.L., 2017. Effects of pH on cell growth, lipid production and CO_2 addition of microalgae *Chlorella sorokiniana*. Algal Res. 28, 192–199.

Raso, J., Frey, W., Ferrari, G., Pataro, G., Knorr, D., Teissie, J., Miklavcic, D., 2016. Recommendations guidelines on the key information to be reported in studies of application of PEF technology in food and biotechnological processes. Innovative Food Sci. Emerg. Technol. 37, 312–321.

Rokicka, M., Zielinski, M., Debowski, M., Głowacka, K., 2018. Microwave heating impact on the oil yield from *Botryococcus braunii*. Algae Biomass. Proc. 2, 1289.

Sajjadi, B., Chen, W.Y., Raman, A.A.A., Ibrahim, S., 2018. Microalgae lipid and biomass for biofuel production: a comprehensive review on lipid enhancement strategies and their effects on fatty acid composition. Renew. Sust. Energy Rev. 97, 200–232.

Saladini, F., Patrizi, N., Pulselli, F.M., Marchettini, N., Bastianoni, S., 2016. Guidelines for emergy evaluation of first, second and third generation biofuels. Renew. Sust. Energy Rev. 66, 221–227.

Saleh, J., Dub, M.A., Tremblay, A.Y., 2010. Effect of soap, methanol, and water on glycerol particle size in biodiesel purification. Energy Fuels 24, 6179–6186.

Sanchez, J., Curt, M.D., Robert, N., Fernandez, J., 2019. Chapter two – biomass resources. In: Lago, C., Caldes, N., Lechón, Y. (Eds.), The Role of Bioenergy in the Bioeconomy. Academic Press, pp. 25–111.

Sharma, K.K., Garg, S., Li, Y., Malekizadeh, A., Schenk, P.M., 2013. Critical analysis of current microalgae dewatering techniques. Biofuels 4, 397–407.

Shirazi, M.M.A., Kargari, A., Tabatabaei, M., Mostafaeid, B., Akia, M., Barkhi, M., Shirazi, M.J.A., 2013. Acceleration of biodiesel–glycerol decantation through NaCl-assisted gravitational settling: a strategy to economize biodiesel production. Bioresour. Technol. 134, 401–406.

Sierra, L.S., Dixon, C.K., Wilken, L.R., 2017. Enzymatic cell disruption of the microalgae *Chlamydomonas reinhardtii* for lipid and protein extraction. Algal Res. 25, 149–159.

Silve, A., Papachristou, I., Wüstner, R., Strabner, R., Schirmer, M., Leber, K., et al., 2018. Extraction of lipids from wet microalga *Auxenochlorella protothecoides* using pulsed electric field treatment and ethanol-hexane lends. Algal Res. 29, 212–222.

Singh, S.P., Singh, P., 2015. Effect of temperature and light on the growth of algae species: a review. Renew. Sust. Energy Rev. 50, 431–444.

Singh, A., Nigam, P.S., Murphy, J.D., 2011. Renewable fuels from algae: an answer to debatable land based fuels. Bioresour. Technol. 102, 10–16.

Siow, H.T., Aminul, I., Yun, H.T., 2016. Algae derived biodiesel using nanocatalytic transesterification process. Chem. Eng. Res. Design. 111, 362–370.

Suwannakarn, K., Lotero, E., Ngaosuwan, K., Goodwin, J.G., 2009. Simultaneous free fatty acid esterification and triglyceride transesterification using a solid acid Catalyst with in situ removal of water and unreacted methanol. Ind. Chem. Eng. Res. 48, 2810–2818.

Tariq, M., Ali, S., Khalid, N., 2012. Activity of homogeneous and heterogeneous catalysts, spectroscopic and chromatographic characterization of biodiesel: a review. Renew. Sust. Energy Rev. 16, 6303–6316.

Torres, J.J., Toledo Arana, J., Ochoa, N.A., Marchese, J., Pagliero, C., 2017. Biodiesel purification using polymeric nanofiltration composite membranes highly resistant to hard conditions. Chem. Eng. Technol. 41 (2), 253–260.

Ullah, K., Ahmad, M., Kumar, V., Lu, P., Harvey, A., Zafar, M., Sultana, S., 2015. Assessing the potential of algal biomass opportunities for bioenergy industry: a review. Fuel 143, 414–423.

Uquiche, E., Antilaf, I., Millao, S., 2016. Enhancement of pigment extraction from *B. braunii* pretreated using CO_2 rapid depressurization. Braz. J. Microbiol. 47 (2), 497–505.

Vandamme, D., Foubert, I., Muylaert, K., 2013. Flocculation as a low-cost method for harvesting microalgae for bulk biomass production. Trends Biotechnol. 31, 233239.

Velazquez lucio, J., Rodríguez jasso, R.M., Colla, L.M., Sáenz galindo, A., Cervantes, D.E., Aguilar, et al., 2018. Microalgal biomass pretreatment for bioethanol production: a review. Biofuel Res. J. 17, 780–791.

Veljkovic, V.B., BankovicIlic, I.B., Stamenkovic, O.S., 2015. Purification of crude biodiesel obtained by heterogeneously- catalyzed transesterification. Renew. Sust. Energy Rev. 49, 500–516.

Venturini, G., Pizarro-Alonso, A., Munster, M., 2019. How to maximize the value of residual biomass resources: the case of straw in Denmark. Appl. Energy 250, 369–388.

Verma, P., Sharma, M.P., 2016. Review of process parameters for biodiesel production from different feedstocks. Renew. Sust. Energy Rev. 62, 1063–1071.

Verma, P., Sharma, M.P., Dwivedi, G., 2016. Impact of alcohol on biodiesel production and properties. Renew. Sust. Energy Rev. 56, 319–333.

Vyas, A.P., Verma, J.L., Subrahmanyam, N., 2010. A review on FAME production processes. Fuel 89, 1–9.

Wan, C., Alam, M.A., Zhao, X.Q., Zhang, X.Y., Guo, S.L., Ho, S.H., Chang, J.S., Bai, F.W., 2015. Current progress and future prospect of microalgal biomass harvest using various flocculation technologies. Bioresour. Technol. 184, 251–257.

Xia, A., Cheng, J., Song, W., Su, H., Ding, L., Lin, R., et al., 2015. Fermentative hydrogen production using algal biomass as feedstock. Renew. Sust. Energy Rev. 51, 209–230.

Yang, X.X., Wang, Y.T., Yang, Y.T., Feng, E.Z., Luo, J., Zhang, F., et al., 2018. Catalytic transesterification to biodiesel at room temperature over several solid bases. Energy Convers. Manag. 164, 112–121.

Yin, Z., Zhu, L., Li, S., Hu, T., Chu, R., Mo, F., Hu, D., Liu, C., Li, B., 2020. A comprehensive review on cultivation and harvesting of microalgae for biodiesel production: environmental pollution control and future directions. Bioresour. Technol. 301, 1–19.

Yoo, G., Park, M.S., Yang, J.-W., 2015. Chemical pretreatment of algal biomass. In: Pandey, A., Negi, S., Binod, P., Larroche, C. (Eds.), Pretreatment of Biomass. Elsevier, Amsterdam, pp. 227–258 (Chapter 12).

Zabed, H.M., Akter, S., Yun, J., Zhang, G., Awad, F.N., Qi, X., Sahu, J.N., 2019. Recent advances in biological pretreatment of microalgae and lignocellulosic biomass for biofuel production. Renew. Sust. Energy Rev. 105, 105–128.

Zhang, Y., Li, Y., Zhang, X., Tan, T., 2015. Biodiesel production by direct transesterification of microalgal biomass with co-solvent. Bioresour. Technol. 196, 712–715.

Zhu, L., Li, Z., Hiltunen, E., 2018. Microalgae *Chlorella vulgaris* biomass harvesting by natural flocculant: effects on biomass sedimentation, spent medium recycling and lipid extraction. Biotechnol. Biofuels. 11, 183.

Zhu, L., Li, S., Hu, T., Nugroho, Y.K., Yin, Z., Hu, D., Chu, R., Mo, F., Liu, C., Hiltunen, E., 2019. Effects of nitrogen source heterogeneity on nutrient removal and biodiesel production of mono-and mix-cultured microalgae. Energy Convers. Manage. 201, 1–8.

SECTION IV

Biogas production

CHAPTER

10

Sustainable approaches for biohydrogen and biogas production from corn wastes: prospects and challenges

Achmad Syafiuddin[1,2], *Fifi Khoirul Fitriyah*[2], *Difran Nobel Bistara*[2], *Abdul Hakim Zakkiy Fasya*[2], *Zainul Akmar Zakaria*[3]

[1]Department of Public Health, Faculty of Health, Universitas Nahdlatul Ulama Surabaya, Surabaya, East Java, Indonesia; [2]Institute of Research and Community Service (LPPM), Universitas Nahdlatul Ulama Surabaya, Surabaya, East Java, Indonesia; [3]School of Chemical and Energy Engineering, Faculty of Engineering, Universiti Teknologi Malaysia, UTM Johor Bahru, Johor, Malaysia

10.1 Introduction

In the current development, an increase in the population in many countries is projected to escalate the need for energy (Bayomi and Fernandez, 2018; Ganivet, 2020; Sugiawan and Managi, 2019). Globally, the current consumption of energy is approximately 542 quadrillion BTU (QBTU), and the consumption is predicted to rise continuously up to 50% from the current consumption by 2050. Among these, the current energy consumption is dominantly derived from fossil fuels, which are more than 80% of the world's energy demand (Guo et al., 2015). Fossil fuels are categorized as nonrenewable and limited resources. By using the current consumption, the energy supply derived from coal, petroleum, and natural gas will be available only for another 45, 60, and 120 years, respectively (Guo et al., 2015). Without adequate energy production and technology, the energy scarcity in the future seems to be pronounced. Therefore, several alternative energy generations that can be derived from wind, solar, hydro, and biomass are crucial.

Biomasses such as agriculture and forest residues are investigated for bioenergy production (Hosseini Koupaie et al., 2019; Klassen et al., 2020; Pesce et al., 2017; Yan et al., 2016). Agricultural wastes have the potential to be explored for bioenergy production (Almomani, 2020; Mugodo et al., 2017; Prajapati and Singh, 2020; Zhu et al., 2019). In addition, exploration of crop residues also potentially supports the environmental protection. This is because the crop residues left in the field have the potential to produce greenhouse gases or have the potential as air pollutant sources by openly burning in the field (Sukumaran et al., 2010).

Biofuels and Bioenergy
https://doi.org/10.1016/B978-0-323-85269-2.00012-5

Corn wastes have potential for biogas and hydrogen production because they commonly contain cellulose, hemicellulose, and lignin (Bautista et al., 2019; Paul et al., 2019; Rughoonundun et al., 2020; Yang et al., 2020). Cellulose and hemicellulose are considered as substrates that can be easily degraded while lignin is harder to be degraded. For the degradation of lignin, pretreatment procedure is needed such as the solubilization and biodegradation to facilitate biogas energy generation (Fernandes et al., 2009).

Although many pretreatment methods for the improvement of bioenergy production from crop residues are possible, the current trend is focused on how it can be carried out in a greener and cheaper procedure. Physical methods have several drawbacks such as high electricity demand and scalability issues. Chemical methods have adverse effects on humans and environment since some formations of toxic products can be created during or at the end of the process. Hence, biological pretreatment methods such as the use of bacteria, fungi, and enzymes can eliminate toxic chemical use, relatively free from toxic products, and low energy consumption. Therefore, this chapter presents the prospects and challenges of the use biological pretreatment methods to enhance biohydrogen and biogas fabrication from corn wastes. The provided findings and discussions in this chapter are useful in the designing the best performance of biological pretreatment methods for future full-scale bioenergy production and the protection of environment from wastes.

10.2 Effects of pretreatment on chemical composition

The biological pretreatment commonly uses the single or mixed microbes to treat corn wastes for the bioenergy production. It has been well proven that fungi are commonly employed for delignification compared with cellulose and hemicellulose degradation. Several microbes have been explored for the biodegradation of lignocellulosic biomass and found that the use of bacteria is preferable because they perform faster growth rate compared with fungi (Xu et al., 2018). Also, the mechanisms of lignin degradation are quite specific depending on the type of microbes. Several studies have reported that the production of ligninolytic enzyme was higher in fungi compared with bacteria (Abraham et al., 2020). This indicates that more studies are needed for the investigation of bacteria that have better delignification of lignocellulosic biomass.

Biological approach using fungal pretreatment can be effective for the removal of lignin present in corn wastes. For instance, *Ceriporiopsis subvermispora* can be used to treat corn stover for the enhancement of removal of lignin (Huang et al., 2019). Lignin degradation from corn stover by up to 51.99% can be achieved within 15 days using *Ceriporiopsis subvermispora* and can achieve a maximum degradation at 90 days, which was up to 71.54%. Fungus action has been found to have capability to reduce the recalcitrance of the substrate as pretreatment of biomass via oxidative enzymes and lignolytic extracellular systems (Rouches et al., 2016). White-rot fungal pretreatment can also remove the lignin content of corn stover (Saha et al., 2016).

The use of *Phanerochaete chrysosporium* as the pretreatment method for corn stover and corn stover silage affects their chemical contents (Liu et al., 2014). Pretreatment of corn stover silage can perform dry mass loss of 14.2%. As a comparison, increase in dry mass loss for the corn stover (55.3%) after the fungal pretreatment was also ratified. Alternatively, 70% of lignin degradation can also be achieved for corn silage pretreated with *Trametes versicolor* (Tišma et al., 2018). Weight loss and lignin degradation of corn stover pretreated with *Pleurotus eryngii* were 30% and 30%, respectively (Wyman et al., 2018). Another study found that *Phanerochaete chrysosporium* was also capable for the reduction of lignin up to 34.3% from the corn stalk (Zhao et al., 2012).

Beside the use of fungi, bacteria can also be explored for the enhancement of the hydrolysis of corn wastes since they have capability produce various enzymes. For instance, *Bacillus subtilis* as the pretreatment method for corn straw was evaluated by Xu et al. (2018). The degradation of lignin after the pretreatment can achieve 17%–23%; depending on the ratio between biogas liquid and bacteria, solution with the highest lignin degradation was obtained for 0: 1 v/v. Moreover, crystallinity degree analysis also confirmed that the lowest crystallinity index (25.58) was also obtained at the ratio (0:1 v/v), suggesting that more crystalline cellulose was destroyed compared with other ratios. It has been well known that microorganism can be used for decreasing their crystallinity index (Gupta et al., 2016).

The degradation rates of hemicellulose, cellulose, and lignin of corn stover using the mixed microbes (fungi and bacteria) were 44.4%, 34.9%, and 39.2%, respectively (Li et al., 2020). This was also better in terms of these degradation rates compared with the composting pretreatment, which only achieved the degradation rates of hemicellulose, cellulose, and lignin by 19.9%, 14.7% and12.5%, respectively. The mixture of fungi and bacteria can also be successful for the removal of lignin from corn straw by up to 53% within 20 days (Zhong et al., 2011).

10.3 Enhancement of biogas and biohydrogen production

The enhancement of biohydrogen production from corn stalk using fungal pretreatment was proposed using *Trichoderma reesei* Rut C-30 by Cheng and Liu (2012). At the optimum blending ratio (1:5), biohydrogen yield can be improved up to 243% using the fungal pretreatment xcompared with that without pretreatment. The enhancement of hydrogen production can be correlated to the production of cellulase released by *Trichoderma reesei* Rut C-30. Another study proposed *Phanerochaete chrysosporium* as the pretreatment method for corn stover silage to improve the biogas production and methane yield (Liu et al., 2014). The study found that the proposed pretreatment can perform 10.5%–19.7% higher biogas production and 11.7%–21.2% higher methane yield compared with the untreated corn stover silage.

Bacteria can also be alternative in accelerating the hydrolysis of lignocellulosic biomass, thus increasing the bioenergy production. Improvement in methane production from corn straw was also observed when pretreating with *Bacillus subtilis* (Xu et al., 2018). It can be observed from the study that the methane yield can be increased from 0 to 17 days. It was also found that the methane yields for all pretreatment cases were higher compared with the untreated samples (control).

The mixed bacteria for the improvement of biogas and total methane production from corn stalks were proposed by Yuan et al. (2011). It is noted that Yuan et al. (2011) employed 26 different clones assembled into three phyla, which are Clostridiales, Proteobacteria, and Bacteriodetes. The study achieved 68.3% and 87.9% greater the total biogas production and the total methane production, respectively, compared with the untreated samples. In addition, mixed microbes were used with the assumption that the degradation of lignocellulose present in corn stover can be improved facilitated by multiple microorganisms (Li et al., 2020). It was found that the mixed microbes can improve the methane production up to 131.6% compared with the untreated corn stover. This finding was higher compared with that reported by Zhong et al. (2011) who found that the mixed microbes (fungi and bacteria) can only achieve 33.07% higher biogas yield improvement and 75.57% higher methane yield improvement than the untreated samples.

Enzymes such as cellulase (in powder form) can be used for the improvement of biogas production from corn stover (Wang et al., 2018). The study found that up to 36.9% higher biogas production can be obtained compared with untreated corn stover. Moreover, it was also found that increase in enzyme load can improve the biogas production. Alternatively, laccase and mixed enzymes (Mn peroxidase + versatile peroxidase) can also be potential as proposed by Schroyen et al. (2014), who reported 25% and 17% higher methane yield, respectively, than the untreated samples. By employing other enzymes, which are amylase, protease, and cellulase in liquid form, the improvement of methane production from corn straw was tested and evaluated (Wang et al., 2016). The use of amylase and cellulase can improve methane yields by 110.79% and 103.20%, respectively, but the use of protease did not significantly improve the methane production (an increase of only 1.47%) compared with the untreated samples.

Simple biological method using lime pretreatment can also be carried out to improve the biohydrogen production from corn stalk (Cao et al., 2012). The study reported an improvement of the biohydrogen production by 38.1% compared with the untreated sample obtained at the recommended conditions of lime loading 0.10 g/g biomass, ambient temperature, and residence time 96 h. Moreover, it is interesting to study the effect of lime loading on the biohydrogen yield. It is clear from the study that the biohydrogen yield increases with lime loading and achieves the optimum yield at 0.10 g/g raw biomass and starts to decrease but still higher compared with the untreated sample. For a comprehensive overview, different pretreatments by using biological approaches for the enhancement of biogas and biohydrogen production are summarized in Table 10.1.

10.4 Mathematical modeling

Cumulative bioenergy production is commonly modeled using the modified Gompertz equation (He et al., 2019). This is given as Eq. (10.1):

$$P(t) = P_m \times \exp\left\{-\exp\left[\frac{R_m \times e}{P_m} \times (\lambda - t) + 1\right]\right\} \tag{10.1}$$

where $P(t)$ is defined as the methane production, P_m is defined as the potential methane production, R_m is defined as the maximum production rate, and e, λ, and t are the base of the natural logarithms (2.7183), the lag-phase time, and the elapsed time, respectively.

The cumulative bioenergy production can also be simulated using the transfer function (TF) and Cone models, which are given in Eqs. (10.2) and (10.3), respectively, as follows:

$$P(t) = P_m\left(1 - \exp\left(\frac{-R_m(t-\lambda)}{P_m}\right)\right) \tag{10.2}$$

$$P(t) = \frac{P_m}{1 + \left(K_{hyd}t\right)^{-n}} \tag{10.3}$$

where K_{hyd} represents the hydrolysis rate constant (1/day), and n represents the shape factor.

Cumulative methane production from corn straw pretreated with the bacteria, *Bacillus subtilis,* was analyzed using the modified Gompertz equation (Xu et al., 2018). Value of P_m for the untreated corn straw was 225.67 mL CH_4/g_{VS}, which can be enhanced up to 262.84 mL $CH4/g_{VS}$ after the pretreatment depending on the mixture ratio. In addition, value of R_m can be enhanced after the treatment (16.4 mL/(g_{VS}·d)) compared with untreated corn straw (13.6 mL/(g_{VS}·d)).

By using the TF model, accumulated biogas production obtained from corn stover was modeled (Wyman et al., 2018). Three strains of

TABLE 10.1 Different pretreatment methods by using biological approaches for the enhancement of bioenergy production.

Pretreatment method	Microorganism	Biomass	Effect on bioenergy production	References
Fungal pretreatment				
	Trichoderma reesei Rut C-30	CSta	243% higher biohydrogen yield	Cheng and Liu (2012)
	Phanerochaete chrysosporium	CSto and CSSi	10.5%–19.7% higher biogas yield and 11.7%–21.2% higher MY	Liu et al. (2014)
	Trametes versicolor	CSil	Improve the methane generation rate by 0.236 $m^3{}_{CH4}$ $kg_{VS}{}^{-1}$ compared with untreated one by 0.167 $m^3{}_{CH4}$ $kg_{VS}{}^{-1}$	Tišma et al. (2018)
	Pleurotus eryngii	CSto	19% higher biogas production	Wyman et al. (2018)
	Ceriporiopsis subvermispora	CSto	15.2% higher MY	Huang et al. (2019)
Bacterial pretreatment				
	Bacillus subtilis	CStr	17.35% higher MY	Xu et al. (2018)
	Mixed bacteria	CSta	68.3% higher biogas production and 87.9% greater the total methane production	Yuan et al. (2011)
Mixed microbes pretreatment				
	Fungi and bacteria	CStr	33.07% higher biogas yield and 75.57% higher MY	Zhong et al. (2011)
	Fungi and bacteria	CStr	131.6% higher MY	Li et al. (2020)
Enzyme pretreatment				
	Cellulase	CSto	36.9% higher biogas production	Wang et al. (2018)
	Laccase	CSto	25% higher MY	Schroyen et al. (2014)
	Mn peroxidase + versatile peroxidase	CSto	17% higher MY	Schroyen et al. (2014)
	Amylase	CStr	110.79% higher methane production	Wang et al. (2016)
	Protease	CStr	1.47% higher methane production	Wang et al. (2016)
	Cellulase	CStr	103.20% higher methane production	Wang et al. (2016)
Fruit extract pretreatment				
	Lime	CSta	38.1% higher biohydrogen yield	Cao et al. (2012)

CSil, corn silage; *CSSi*, corn stover silage; *CSta*, corn stalk; *CSto*, corn stover; *CStr*, corn straw; *MY*, methane yield.

fungi, which are *Pleurotus eryngii* (*Pe*), *Pleurotus ostreatus* (*Po*), and *Trametes versicolor* (*Tv*), showed different characteristics. In general, the TF model was capable for the modeling cumulative biogas production as indicated by high values of R^2 for all treatments. It is interesting here that the cumulative biogas production was evaluated using different fungi and different corn stover sizes. The study found that the biological treatment time for all cases can improve the P_m compared with the untreated corn stover. At the longest treatment (30 day), the P_m values were higher at the smaller corn stover sizes, exhibiting better biogas production when the smaller sizes were used for the biogas production particularly by using *Pe*, *Po*, and *Tv* as the biological pretreatment (Wyman et al., 2018).

The modified Gompertz model and Cone model were used for the modeling of biogas production from corn stover and found that the modified Gompertz model was better in terms of R^2 values (Wang et al., 2018). The value of P_m of the pretreated corn stover was 262.7 mL, which was higher compared with the untreated corn stover (201.6 mL). Using the Cone model, the K_{hyd} can be used for the estimation of hydrolysis rate, and a higher K_{hyd} value is recommended. The study found that the proposed enzyme pretreatment significantly improved the K_{hyd} value by 42%.

10.5 Advantages and disadvantages

It is also useful to discuss advantages and disadvantages of biological treatment for the biogas and biohydrogen production. Focusing on the disadvantages, it is therefore needed to be tackled before implementing in full scale. One of the main advantages of biological pretreatment is its capability to eliminate the use of hazardous chemicals, which is really hard to handle when chemical methods are used. In addition, it is also noted that the biological method performs mild process conditions compared with chemical process (Xu et al., 2019). The biological pretreatments are considered as low capital cost and have the potential as sustainable approaches because of the elimination of the need for toxic chemical uses (Maurya et al., 2015).

The main disadvantage of biological method is the low rate of hydrolysis (Sun and Cheng, 2002). In addition, this method also needs more longer incubation time for growing the employed microbes (Maurya et al., 2015). In addition, the use of this method has the potential loss of carbohydrates during the pretreatment. This is because employed microbes use nutrients for growing and metabolism processes (Abraham et al., 2020; Liu et al., 2014; Tišma et al., 2018). Although enzymes perform an improvement in biogas production, they still have drawbacks because of their expensive cost (Schroyen et al., 2014). This calls studies to propose a cheaper method for the enhancement of biogas and biohydrogen production such as the assessment of lime pretreatment as proposed by Cao et al. (2012). In the future, the biological pretreatments using other biological substrates rather than microorganisms are recommended.

10.6 Conclusion

This chapter highlights the current state of knowledge concerning biohydrogen and biogas production from corn wastes focusing on the exploration of biological pretreatment method. The use of bacteria, fungi, mixed microbes, enzymes, and lime has been done and reported in literature. The biological pretreatment has been found to be suitable for the biodegradation of lignin present in corn wastes. By using the provided data analysis obtained from literature, the biological method can perform higher biomethane production varying from 1% to 131% and higher biohydrogen production from 38% to 243% compared with the untreated samples. The modified Gompertz, transfer function, and

Cone models are three common models employed for the modeling the accumulative biohydrogen and biogas production, but the Cone model provides the beneficial specific parameter that represents the hydrolysis rate. Moreover, this chapter also provides advantages and disadvantages of the use of biological pretreatment for the improvement of biohydrogen and biogas production from corn wastes that must be considered before implementing in full scale. A study on the evaluation of other biological substrates rather than microorganisms for the improvement of biohydrogen and biogas production needs more research, as only a few reports can be found in literature.

References

Abraham, A., Mathew, A.K., Park, H., Choi, O., Sindhu, R., Parameswaran, B., Pandey, A., Park, J.H., Sang, B.-I., 2020. Pretreatment strategies for enhanced biogas production from lignocellulosic biomass. Bioresour. Technol. 301, 122725.

Almomani, F., 2020. Prediction of biogas production from chemically treated co-digested agricultural waste using artificial neural network. Fuel 280, 118573.

Bautista, K., Unpaprom, Y., Ramaraj, R., 2019. Bioethanol production from corn stalk juice using *Saccharomyces cerevisiae* TISTR 5020. Energy Source A 41, 1615–1621.

Bayomi, N., Fernandez, J.E., 2018. Trends of energy demand in the Middle East: a sectoral level analysis. Int. J. Energy Res. 42, 731–753.

Cao, G.-L., Guo, W.-Q., Wang, A.-J., Zhao, L., Xu, C.-J., Zhao, Q.-l., Ren, N.-Q., 2012. Enhanced cellulosic hydrogen production from lime-treated cornstalk wastes using thermophilic anaerobic microflora. Int. J. Hydrog. Energy. 37, 13161–13166.

Cheng, X.-Y., Liu, C.-Z., 2012. Fungal pretreatment enhances hydrogen production via thermophilic fermentation of cornstalk. Appl. Energy 91, 1–6.

Fernandes, T.V., Klaasse Bos, G.J., Zeeman, G., Sanders, J.P.M., van Lier, J.B., 2009. Effects of thermo-chemical pre-treatment on anaerobic biodegradability and hydrolysis of lignocellulosic biomass. Bioresour. Technol. 100, 2575–2579.

Ganivet, E., 2020. Growth in human population and consumption both need to be addressed to reach an ecologically sustainable future. Environ. Dev. Sustain. 22, 4979–4998.

Guo, M., Song, W., Buhain, J., 2015. Bioenergy and biofuels: history, status, and perspective. Renew. Sustain. Energy Rev. 42, 712–725.

Gupta, V.K., Kubicek, C.P., Berrin, J.-G., Wilson, D.W., Couturier, M., Berlin, A., Filho, E.X.F., Ezeji, T., 2016. Fungal enzymes for bio-products from sustainable and waste biomass. Trends Biochem. Sci. 41, 633–645.

He, C., Zhao, J., Wang, S., Guan, S., Zhang, Z., Zhang, Q., Pan, X., Jiao, Y., 2019. Ammonium bicarbonate pretreatment of corn stalk for improved methane production via anaerobic digestion: kinetic modeling. Bioresour. Technol. 292, 122052.

Hosseini Koupaie, E., Dahadha, S., Bazyar Lakeh, A.A., Azizi, A., Elbeshbishy, E., 2019. Enzymatic pretreatment of lignocellulosic biomass for enhanced biomethane production-A review. J. Environ. Manage. 233, 774–784.

Huang, W., Wachemo, A.C., Yuan, H., Li, X., 2019. Modification of corn stover for improving biodegradability and anaerobic digestion performance by *Ceriporiopsis subvermispora*. Bioresour. Technol. 283, 76–85.

Klassen, V., Blifernez-Klassen, O., Bax, J., Kruse, O., 2020. Wastewater-borne microalga *Chlamydomonas* sp.: a robust chassis for efficient biomass and biomethane production applying low-N cultivation strategy. Bioresour. Technol. 315, 123825.

Li, P., He, C., Li, G., Ding, P., Lan, M., Gao, Z., Jiao, Y., 2020. Biological pretreatment of corn straw for enhancing degradation efficiency and biogas production. Bioengineered 11, 251–260.

Liu, S., Wu, S., Pang, C., Li, W., Dong, R., 2014. Microbial pretreatment of corn stovers by solid-state cultivation of Phanerochaete chrysosporium for biogas production. Appl. Biochem. Biotechnol. 172, 1365–1376.

Maurya, D.P., Singla, A., Negi, S., 2015. An overview of key pretreatment processes for biological conversion of lignocellulosic biomass to bioethanol. 3 Biotech 5, 597–609.

Mugodo, K., Magama, P.P., Dhavu, K., 2017. Biogas production potential from agricultural and agro-processing waste in South Africa. Waste Biomass Valoriz. 8, 2383–2392.

Paul, S., Dutta, A., Thimmanagari, M., Defersha, F., 2019. Techno-economic assessment of corn stover for hybrid bioenergy production: a sustainable approach. Case Stud. Therm. Eng. 13, 100408.

Pesce, G.R., Negri, M., Bacenetti, J., Mauromicale, G., 2017. The biomethane, silage and biomass yield obtainable from three accessions of *Cynara cardunculus*. Ind. Crops Prod. 103, 233–239.

Prajapati, K.B., Singh, R., 2020. Enhancement of biogas production in bio-electrochemical digester from agricultural waste mixed with wastewater. Renew. Energy 146, 460–468.

Rouches, E., Herpoël-Gimbert, I., Steyer, J.P., Carrere, H., 2016. Improvement of anaerobic degradation by white-rot fungi pretreatment of lignocellulosic biomass: a review. Renew. Sustain. Energy Rev. 59, 179–198.

Rughoonundun, H., Fu, Z., Mann, T., Wales, M.E., Holtzapple, M.T., 2020. Oxidative submerged lime pretreatment and high-solids saccharification of corn stover. Biomass Bioenergy 139, 105626.

Saha, B.C., Qureshi, N., Kennedy, G.J., Cotta, M.A., 2016. Biological pretreatment of corn stover with white-rot fungus for improved enzymatic hydrolysis. Int. Biodeterior. Biodegrad. 109, 29–35.

Schroyen, M., Vervaeren, H., Van Hulle, S.W.H., Raes, K., 2014. Impact of enzymatic pretreatment on corn stover degradation and biogas production. Bioresour. Technol. 173, 59–66.

Sugiawan, Y., Managi, S., 2019. New evidence of energy-growth nexus from inclusive wealth. Renew. Sustain. Energy Rev. 103, 40–48.

Sukumaran, R.K., Surender, V.J., Sindhu, R., Binod, P., Janu, K.U., Sajna, K.V., Rajasree, K.P., Pandey, A., 2010. Lignocellulosic ethanol in India: prospects, challenges and feedstock availability. Bioresour. Technol. 101, 4826–4833.

Sun, Y., Cheng, J., 2002. Hydrolysis of lignocellulosic materials for ethanol production: a review. Bioresour. Technol. 83, 1–11.

Tišma, M., Planinić, M., Bucić-Kojić, A., Panjičko, M., Zupančič, G.D., Zelić, B., 2018. Corn silage fungal-based solid-state pretreatment for enhanced biogas production in anaerobic co-digestion with cow manure. Bioresour. Technol. 253, 220–226.

Wang, S., Li, F., Wu, D., Zhang, P., Wang, H., Tao, X., Ye, J., Nabi, M., 2018. Enzyme pretreatment enhancing biogas yield from corn stover: feasibility, optimization, and mechanism analysis. J. Agric. Food Chem. 66, 10026–10032.

Wang, X., Li, Z., Zhou, X., Wang, Q., Wu, Y., Saino, M., Bai, X., 2016. Study on the bio-methane yield and microbial community structure in enzyme enhanced anaerobic co-digestion of cow manure and corn straw. Bioresour. Technol. 219, 150–157.

Wyman, V., Henríquez, J., Palma, C., Carvajal, A., 2018. Lignocellulosic waste valorisation strategy through enzyme and biogas production. Bioresour. Technol. 247, 402–411.

Xu, N., Liu, S., Xin, F., Zhou, J., Jia, H., Xu, J., Jiang, M., Dong, W., 2019. Biomethane production from lignocellulose: biomass recalcitrance and its impacts on anaerobic digestion. Front. Bioeng. Biotech. 7, 1–12.

Xu, W., Fu, S., Yang, Z., Lu, J., Guo, R., 2018. Improved methane production from corn straw by microaerobic pretreatment with a pure bacteria system. Bioresour. Technol. 259, 18–23.

Yan, N., Ren, B., Wu, B., Bao, D., Zhang, X., Wang, J., 2016. Multi-objective optimization of biomass to biomethane system. Green Energy Environ. 1, 156–165.

Yang, Y., Ni, J.-Q., Zhou, S., Xie, G.H., 2020. Comparison of energy performance and environmental impacts of three corn stover-based bioenergy pathways. J. Clean. Prod. 272, 122631.

Yuan, X.F., Li, P., Wang, H., Wang, X., Cheng, X., Cui, Z., 2011. Enhancing the anaerobic digestion of corn stalks using composite microbial pretreatment. J. Microbiol. Biotechnol. 21, 746–752.

Zhao, L., Cao, G.-L., Wang, A.-J., Ren, H.-Y., Dong, D., Liu, Z.-N., Guan, X.-Y., Xu, C.-J., Ren, N.-Q., 2012. Fungal pretreatment of cornstalk with Phanerochaete chrysosporium for enhancing enzymatic saccharification and hydrogen production. Bioresour. Technol. 114, 365–369.

Zhong, W., Zhang, Z., Luo, Y., Sun, S., Qiao, W., Xiao, M., 2011. Effect of biological pretreatments in enhancing corn straw biogas production. Bioresour. Technol. 102, 11177–11182.

Zhu, T., Curtis, J., Clancy, M., 2019. Promoting agricultural biogas and biomethane production: lessons from cross-country studies. Renew. Sustain. Energy Rev. 114, 109332.

CHAPTER

11

Biogas production potential in India, the latest biogas upgradation techniques and future application in a fuel cell

Thiruselvi Devaraj[1], *Yuvarani Mani*[1], *Salma Aathika*[1], *Shanmugam Palaniyandi*[2], *Baskar Gurunathan*[3], *Sivanesan Subramanian*[1]

[1]Department of Applied Science and Technology, Alagappa College of Technology, Anna University, Chennai, Tamil Nadu, India; [2]Environmental Technology Division, CSIR-Central Leather Research Institute, Chennai, Tamil Nadu, India; [3]Department of Biotechnology, St. Joseph's College of Engineering, Chennai, Tamil Nadu, India

11.1 Introduction

All the untreated waste in the water and land is the main cause for all kinds of pollution in environment. Waste to energy technologies may eradicate the problems due to solid and liquid waste (EAI, 2019). Fossil fuel dependency is reduced by using renewable source, as fuel (waste) utilization of waste to energy technique not only reduces the waste but also produces reliable energy, socioeconomic benefits, and monetary benefits (Panwara et al., 2011).

Among the world energy demand, 14% is supplied form the renewable sources such as wind, hydro, solar, biomass, geothermal, and marine power (UNDP, 2000; Dincer, 2001; Bilgen, 2004). Contribution of energy from the renewable sources has expected an increase of 30%–80% by 2100 (Fridleifsson, 2001). Usage of fossil fuels already has negative impact on environment, which leads to energy crisis, and society is focusing toward promotion of clean energy from renewable sources. In India, MNRE (Ministry of New and Renewable Energy) aims in developing legislative reforms to raise to energy share from the renewable sources in the total potential (Debadayita et al., 2014). According to MNRE in 2019, 225 MW of power is generated from the sewage (liquid waste) and 1460 MW of power from municipal solid waste (MSW) (EAI, 2019). India's electricity production targets on less carbon emission policies; clean cooking process has been

Biofuels and Bioenergy
https://doi.org/10.1016/B978-0-323-85269-2.00009-5

adopted with the biogas. National Project on Biogas Development, a manure management initiative, and National Biomass Cook Stove Initiative were the energy service schemes that promoted the efficient and clean cooking process among the rural populations. Hence, India concentrated on the production of biogas from renewable sources for reducing the greenhouse gases (GHGs), global warming, and other pollution.

11.2 Societal benefits

11.2.1 Source for renewable energy

Global energy is supplied from the dead and decayed matters of the plant and animals (fossil sources), which has been under heat and earth pressure for many years. Hence, fossil fuels are nonrenewable and being depleted without recharge for the new ones. For world's better future, among the energy resources such as fossil fuels, nuclear, and renewable sources, renewable energy sources are considered to be one of the main factors (Demirbas, 2000). Sources from which energy can be produced continuously are called renewable resources. It is otherwise called as alternative energy source (wind, solar, biomass, etc.). Eradication of air pollution and GHG is achieved with these kinds of sources. Many critical problems include fuel economy, energy, water, and sustainable lifestyle in the remote zones and can be overwhelmed through renewable resources. Hence, projects with renewable energy will upgrade the rural part of the country by reducing urban migration and creating job opportunities. Biogas from the Anaerobic Digestion (AD) is one of the important renewable energy sources as it requires the wastes (biomass, sludge, industrial waste etc.) as their feedstock. Utilization of biogas will increase the energy balance and also preserves the environment and natural resources.

11.2.2 Proper utilization of waste

During AD process, the possibility of converting the waste into valuable product is achieved. Problems associated with waste from agriculture, industry, and house are the most challenging among all the countries (Rathore and Panwar, 2007). Production of biogas using the organic waste generated and utilizing the digested leftover as an effective fertilizer is the best method throughout the world. Digestion without the presence of air may reduce the sludge volume, time taken for digestion, and also cost for treating waste.

11.2.3 Reduced global warming and greenhouse gas emission

CO_2 is one of the important GHGs; its emission into the atmosphere increases the global warming. Biogas is also composed of carbon dioxide, but it can be taken by the plants through photosynthesis. AD reduces the methane emission into atmosphere, as it can be stored properly and utilized for energy processing. Hence, replacing fossil fuel by biogas reduces the GHG emission, and thereby, positivity toward mitigation of global warming is achieved.

11.2.4 Minimal usage of water

As AD requires very less amount of water for its processing, it is considered to be one of the advantageous methods among other methods of biofuel production. Reduced water usage is the important aspect during shortage of water in future. Hence, AD is more acceptable and effective method worldwide.

11.2.5 Flexibility on the usage of biogas

Gaseous energy carrier is more flexible to utilize in many applications. The best and easiest among the applications of biogas is cooking. But the application of purified biogas extends to power generation, vehicular fuel, and input for fuel cells.

11.2.6 Creating job in rural areas

Manpower requirement is tremendous in the AD process, substrate collection, transport, digester manufacturing, operation, and maintenance of digestion plants. This process is involved with the workers in that area, and hence, rural income and new jobs have been created.

11.3 Power sector position in four regions (IEA bioenergy, 2000)

The current and future forecasting of the renewable power generation in southern, eastern and northeastern, western, and northern regions of India are given in Table 11.1.

11.3.1 Southern Region

The southern part of the country comprises Andhra Pradesh, Karnataka, Tamil Nadu, Kerala, Andaman and Nicobar Islands, Puducherry, and Lakshadweep Islands. The potential of renewable installation will increase by 13% and reach 41% by 2022. Current renewable share will increase by 12% and attains 24% by 2022. These parts of the country focus on the wind and solar energy for which proper scheduling and forecasting on hourly basis may increase the energy demand.

11.3.2 Eastern and northeastern regions

Eastern and northeastern part of the country comprises Bihar, Odisha, West Bengal, Jharkhand,

TABLE 11.1 Forecasting of renewable power generation in region wise (IEA bioenergy, 2000).

Year	Total installed capacity (GW)	Renewable installed capacity (GW)	Total generation (billion kWh)	Renewable generation (billion kWh)
Northern region				
2017	89.5	11.5	290.6	18.0
2022	134.3	45.5	468.1	78.8
Eastern region				
2017	34.4	1.0	185.3	2.5
2022	75.1	13.0	171.2	45.2
Northeeastern region				
2017	3.91	0.28	15.05	0.30
2022	11.0	1.56	23.8	5.42
Western region				
2017	107.0	18.3	422.1	27.6
2022	182	54.3	481.5	96.2
Southern region				
2017	91.8	26.1	241.3	33.1
2022	144.9	59.7	420.7	101.2

Manipur, Assam, Meghalaya, Arunachal Pradesh, Nagaland, and Sikkim. Renewable share of 10% (2017) is expected to increase 31% by 2022. According to Meghalaya State Renewable Energy Department, small hydro power is the widely utilized renewable source for power than the wind and solar energy. Even though there is a greater possibility of utilizing solar source, these regions concentrate only on the hydro power. Hence, proper balancing should be carried out for various renewable sources through forecasting.

11.3.3 Western region

Maharashtra, Goa, Madhya Pradesh, Chhattisgarh, Gujarat, and Diu and Daman are considered to be the western part of the country. Installed potential of renewable energy in these regions is 17%, which may increase to 30% in 2022. Wind and solar energy are the renewable sources in these parts, and hence, proper scheduling must be carried to deal other sources.

11.3.4 Northern region

The northern region states include Haryana, Punjab, Rajasthan, Himachal Pradesh, Delhi, Jammu and Kashmir, Uttarakhand, and Uttar Pradesh (UP). The total geographical area is around 30% and consists of 28% of total population. Current renewable share of 18 billion kilowatt hours (6%) will increase to 78.8 bkWh (17%) by 2022. Coal-based power production is 52.84 GW of which 17.97 GW is from UP and 10.93 GW is from Rajasthan. Least demand of electricity in the region tends to operate thermal power plants with less (58%) plant load factor.

11.4 Current energy production scenario

In India, according to CEA (Central Electricity Authority), renewable energy generation was around 9433.21 MW in January 2019, and it increased by 17.2% during the month of January 2020 as 10,698.91 MW. In comparison with the previous year (2019), wind energy increased by 12.36%, solar energy increased by 18.57%, and in total, renewable energy generation increased by 13.42%.

11.5 Anaerobic digestion

During the anaerobic process, two types of microorganism acid–forming and methane-forming bacteria play the vital role in digestion. Conversion of complex organics into simple compounds is achieved in acid phase and final conversion as acetic acid. The acetic acid is stabilized to form methane using methanogen species. If the mixture of feedstocks is used in the AD, the process is termed as codigestion, which is a successful method and used widely. Biogas and the digested slurry are the two resultant products of AD. Biogas is the mixture of methane and carbon dioxide with little hydrogen sulfide gas. Process of methane production is divided into different stages with the help of several microorganisms such as hydrolysis, acidogenesis, acetogenesis, and methanogenesis. Table 11.2 explains the composition of biogas from various sources.

11.6 Operating parameters during biogas production

To optimize the AD process, parameters such as pH, redox potential, volatile fatty acids (VFAs), hydraulic retention time, dissolved oxygen, sludge retention time, organic loading rate, chemical oxygen demand, alkalinity, and volatile suspended solids are monitored to maintain the system stable and efficient. Characterizations of the biogas are given in Table 11.3. Neutral pH of 6.5–7.5 is active because of the sensitivity of methanogens in acidic media in digester. AD at

TABLE 11.2 Composition of biogas from various sources.

Composition	Biogas from various sources				
	Natural gas	Wastewater	Food waste	Animal waste	Landfill waste
Methane (vol %)	80–100	50–60	50–70	45–60	40–55
Carbon dioxide (vol %)	0–3	30–40	25–45	35–60	35–50
Nitrogen (vol %)	0–3	0–4	0–4	0–4	0–20
Oxygen (vol %)	0–0.2	0–1	0–1	0–1	0–2
H_2S (ppm)	0–0.1	0–400	0–10,000	0–300	0–200
Non-H_2S sulfur (ppm)	0–10	0–1	0–1000	0–30	0–30
Halogens (ppm)	0–0.1	0–0.2	0–0.2	0–0.2	0–100
Moisture (%)	0–0.02	3 ± 0.5	3 ± 0.5	3 ± 0.5	3 ± 0.5

TABLE 11.3 Physical characterization of biogas.

Factors	Values
Biogas composition	50%–75% CH_4, 50%–25% CO_2, traces of other gases
Total energy content	6.0–6.5 kW/m^3
Fuel equivalent	0.6–0.65 L oil/m^3 biogas
Explosive limits	6%–12% biogas in air
Ignition temperature	650–750°C
Critical pressure	75–89 bar
Critical temperature	−82.5°C
Normal density	1.2 kg/m^3
Odor	Bad eggs (due to hydrogen sulfide gas)

temperature of 35–45°C is called mesophilic and 45–60°C is thermophilic. Among these processes, thermophilic is considered to be more efficient. Redox potential indicates the oxidation and reduction process during the digestion process. ORP denotes the oxygen reserve in the AD. VFA, a midway product of the AD process, will be changed as methane at end. Dissolved oxygen is also similar to ORP, which inhibits the oxygen in AD. Alkalinity is the main parameter to keep the process under proper pH. There are two types of retention times: one is sludge retention time (SRT) and second is hydraulic retention time (HRT). The extended retention time will have negative effect on the efficiency of the digester. According to the type of reactor, difference in the retention time may occur. The biological conversion ability of the AD process is termed as organic loading rate (OLR) based on volatile solids (VS) and chemical oxygen demand (COD).

11.7 Biomass resources and catastrophe in India

There are many types of sources available for the biogas production in India, which include agricultural waste, animal waste, industrial waste, sewage sludge, MSW, and so on. For biogas production, wet biomass (moisture content 70%–80%) can be used directly. Biomass from sewage treatment plant (STP), animal waste, and food processing has been utilized in many developed and underdeveloping countries. Currently, many agricultural wastes (maize, rice straw, grains) are utilized for the biogas production. The quality of the feedstock is monitored to use the AD waste as biofertilizer.

11.7.1 Municipal waste

MSW consists of organic and inorganic mixture of waste. The characteristic features of the MSW are shown in Table 11.4.

In big cities with population above 10 million and above 2 million, 58.43% and 49.63% of compostable matters are disposed of, respectively. From Table 11.4, average compostable matter of 52.26% is the efficient composition of MSW for the AD process (NBB, 2017). The carbon–nitrogen ratio ranges between 23.78 and 36.96, which is very vital for methanogenesis (Statista, 2020). Moisture content varies between 39.67% and 58.30%, and calorific values vary between 800 and 1010 kcal/kg.

11.7.2 Agriculture and crop waste

India is an agricultural country, and major people depend on it for their livelihood. Biomass quality and quantity differ from crops to crops significantly. Usually, crop residues are dumped and burned in the cultivated field till next season, and it is utilized as fodder for animals and fire wood. Any biomass can be used as substrate for the biogas production; no specifications are given for the crops and crop residues (Chi et al., 2019; Urja, 2016; Bhuvaneshwari et al., 2019). The estimated biomass amount of 279 Mt/year is available for energy generation, and its biogas production potential is around 46 Mm^3/day.

11.7.3 Waste water

An average of 80% wastewater generated in the urban area is collected for STP. Sewage is treated, and either sludge is utilized as fertilizer or it can be utilized as substrate for biogas production. Total wastewater generated is around 15,390 MLD, and 10,170 MLD is collected for sewage treatment system, which is about 66% (Statista, 2020; Elango et al., 2007; ENVIS, 2020).

11.7.4 Animal manure

In bovine population, India stands as one of the largest with 294 million in the world including cows, buffalos, calves, and bullocks. Total dung production is around 660 metric tons annually from cattle and buffalo (MFAD, 2020; Myint and Nirmalakhandan, 2009). The dung utilizing biogas plants are estimated to be

TABLE 11.4 Characterization of municipal solid waste in 59 Indian cities.

		Percentage composition				
Total population (million)	No. of cities	Moisture Content	Compostable	Recyclable	Carbon nitrogen ratio	Waste quantity (Tons/day)
0.01–0.05	5	39.67	49.16	20.05	23.78	59
0.05–0.1	3	58.30	55.83	22.59	30.17	121
0.1–0.5	11	44.9	51.16	19.60	26.39	1065
0.5–1	16	47.75	54.16	16.24	34.36	5250
1–1.5	11	42.73	47.42	12.29	27.99	5289
2–5	11	49.00	49.63	18.06	34.71	16,005
10 and above	2	51.50	58.43	16.09	36.96	61,242

Central Pollution Control Board (CPCB), 2017. Consolidated Annual Review Report on Implementation of Solid Wastes Management Rules. Accessible at: cpcb.nic.in.

12 million by 2010 and produce a biogas of 3448 Mm^3. Fig. 11.1 illustrates the animal waste generated in India.

11.7.5 Industrial waste

Based on the raw materials and processes in industries, the composition of waste varies from one industry to another. The waste generated from food processing, beverages, and dairy industry can be effectively used as raw material for biogas generation.

11.8 Biogas upgradation techniques

Biogas is a mixture of methane and carbon dioxide, and it contains traces of hydrogen sulfide, nitrogen, and water droplets. Amount of impurities is based on the characteristics of the substrate used for biogas production. Hence, purity of the CH_4 can be increased by many upgradation methods (Aziz et al., 2019; Amir et al., 2019; García-Gutiérrez et al., 2016). Biogas enrichment depends on CO_2 removal and sulfur reduction using microbial cultures. Carbon dioxide and H_2S are removed using NaOH by physical adsorption and chemical absorption. Pressure swing adsorption (PSA) method is used to remove the oxygen and nitrogen from the biogas produced from landfill. Sludge from STP and landfill produces biogas with plenty of siloxanes (silicon and oxygen bond). These siloxanes are separated using adsorption and cryogenic method (Rekia, 2020). Comparative assessments on all the upgradation methods are given in Table 11.5.

11.8.1 Absorption using chemical

Carbon dioxide present in biogas is absorbed using chemical absorbents, which is termed as chemical absorption. Fig. 11.2 illustrates the flow pattern involved in chemical absorption. Amine solutions such as monoethanolamine (MEA), diethanolamine (DEA), monodimethyl ethanolamine (MDEA), and triethanolamine (TEA) are important reagents and can be utilized as absorbents (Rimika et al., 2019; Tippayawong and Thanompongchart, 2010; Kadam and Panwar, 2017). Carbon dioxide in biogas reacts with amine group compounds to produce carbamides and can be decayed by heating (Cebula, 2009; Muñoz et al., 2015). Alkalomines, which include MEA, DEA, and TEA, differs with the ability to absorb the gaseous impurities. MDEA, MEA, and TEA react readily with CO_2, CO, and H_2 present in the biogas, whereas tertiary amines

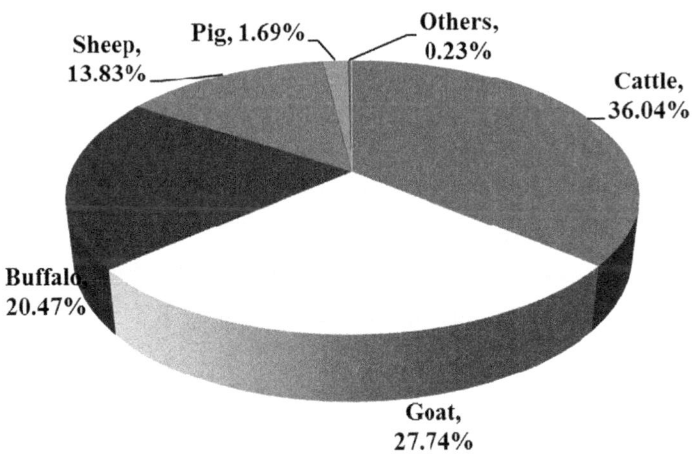

FIGURE 11.1 Animal waste generated in percentage throughout India.

TABLE 11.5 Various biogas upgradation process.

Process	Chemicals/ process procedure	Pretreatment	Temperature (°C)	Percentage of methane content	Percentage of impurities	Percentage of methane loss	Working pressure (bar)	Energy needed in kW h/N m^3 cleaned gas	Operation cost (€/year for 1000m^3)	Benefits	Drawbacks
Absorption process—physical	Polyethylene glycol, water, polluting resisting agents	Required	20.0–40.5	96–99	O_2, N_2 (0.51–1.0), CO_2 (0.51–3.0)	<3	5.0–9.0	0.45–0.91	15,500	Easy operation and economically feasible	Needs large quantity of water, formation of foam, clogging
Scrubbing	Water, solvents	Not required	10.1–20.5	>95–98	CO_2 (2.0–5.0)	<2	4.0–7.0	0.2–0.3 0.49–0.67	39,100	Cost effective, less methane loss, more CO_2 and H_2S removal	Difficult to operate, high operational cost, and external heat for redevelopment
Absorption cchemically	Alkaline and amine solutions	Required	35.1–50.2	>99	CO_2 (0.51–1.0)	0.04–0.1<5 <1	1.0–2.0	<0.151	59,010	Pure methane is obtained, feasible carbon dioxide removal	Very costly, pollution due to chemicals and microbial growth
Pressure swing adsorption	Adsorbing agents, drying solutions, sieves for molecular separation	Required	5.0–30.5	>98	CO_2 (1.0–4.0)	<3.0–4.0	4.0–7.0	0.251	56,010	No microbial growth, water and chemical not needed. Dry process	More operation cost. Hydrogen sulfide is removed by pretreatment

Membrane method	Gas–gas permeability Gas–liquid solubility	Required	25.0–60.0	91–96	CO_2 (2.0–9.0)	<5	20.0–36.0	0.181–0.351	25,000	Less initial investment, maintenance and operation cost. Chemical and heat not needed	High methane loss, fouling and clogging of membrane, costlier due to replacement
Cryogenic separation	Condensation and distillation with compression	Required	−56 to −45	>97–99	CO_2 <3; <1	<1	40	0.21–0.79 0.18–0.251	NA*	High purity of methane with less energy requirement and carbon dioxide is collected separately	Complex, costlier process with pretreatment

Vrbova, V., Ciahotný, K., 2017. Upgrading biogas to biomethane using membrane separation. Energy Fuel. 31, 9393–9401; BIS. Biomethane Standards. IS 16087, 2015. https://archive.org/details/gov.in.is.16087.2013. Accessed March 2020; Central Pollution Control Board (CPCB), 2017. Consolidated Annual Review Report on Implementation of Solid Wastes Management Rules. (Accessible at: cpcb.nic.in); Persson, M., Jönsson, O., Wellinger, A., 2006. Biogas upgrading to vehicle fuel standards and grid injection. In: Braun, R., Holm-Nielsen, J.B., Seadi, T.A., Baxter, D., Jormanainnen, M., Rintala, J., et al., (Eds.), Task 37: Energy from Biogas and Lanfill Gas, Paris, France. Available online: http://www.seai.ie/Renewables/Bioenergy/Biogas_upgrading_to_vehicle_fuel_standards_and_grid_connection_IEA.pdf; Masebinu, S.O., Aboyade, A., Muzenda, E., 2014. Enrichment of biogas for use as vehicular fuel: a review of the upgrading techniques. Int. J. Res. Chem. Metall. Civil Eng. 1, 2349–1442; Niesner, J., Jecha, D., Stehlik, P., 2013. Biogas upgrading techniques: state of art review in European region. Chem. Eng. Trans. 35, 517–522; Sun, Q., Li, H., Yan, J., Liu, L., Yu, Z., Yu, X., 2015. Selection of appropriate biogas upgrading technology-a review of biogas cleaning, upgrading and utilisation. Renew. Sustain. Energy Rev. 51, 521–532.

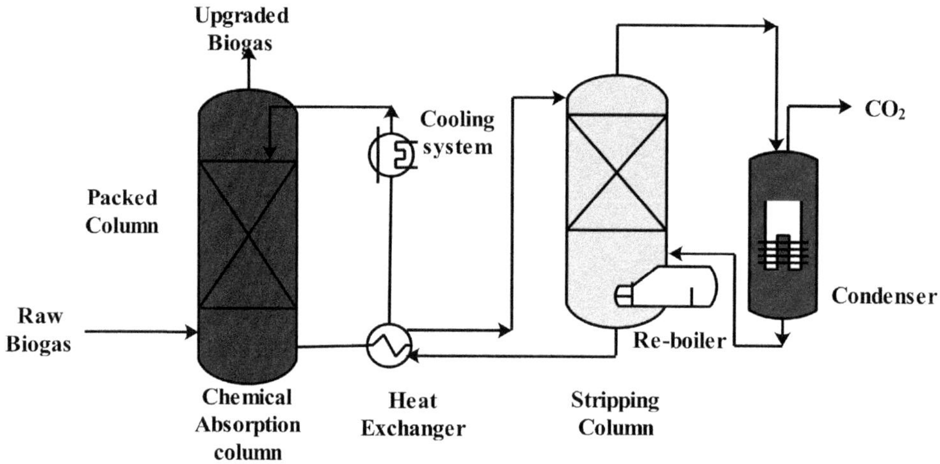

FIGURE 11.2 Flowchart for chemical absorption method.

fit well for energy regeneration due to less heat of reaction (Li et al., 2013; Andriani et al., 2014; Abdeen et al., 2016; Bishnoi and Rochelle, 2000). On comparing with the amine solutions, pure ammonia solution acts as an efficient absorbent for CO removal (Rinprasertmeechaiet al., 2012; Mcleod et al., 2014; Makhloufi et al., 2014). In addition to this, chemical scrubbing, which uses alkaline solutions such as sodium, sodium hydroxide, calcium hydroxide, potassium, and so on, absorbs more carbon monoxide than all other amine solutions (Angelidaki et al., 2018). When comparing physical and chemical processes, chemical scrubbing had shown elevation in CO_2 absorption than physical scrubbing (Nie et al., 2013). Disadvantage of chemical scrubbing is solvent regeneration at high temperature of 160–800°C because the process involves cleavage of chemical bonds between carbon dioxide and absorbent molecules (Maile et al., 2017).

$$RNH_2 + CO_2 \rightarrow RNHCOOH \quad (11.1)$$

$$CO_2 + MDEA.H_2O \Leftrightarrow MDEAH^+ + HCO_3^- \quad (11.2)$$

11.8.2 Physical adsorption

A physical phenomenon in which gaseous adsorbate is attached on solid surface adsorbent with the help of van der Waal forces is termed as adsorption. Fig. 11.3 represents the process behavior of PSA method. Adsorbents should be always microporous granulated solids because of its superior surface area per mass unit (Grande, 2012). Natural and synthetic adsorbents such as zeolite, activated carbon, silica gel, and alumina are used for physical adsorption method (Sarker et al., 2017; Chaemchuen et al., 2013). Profitably, a good adsorbent must fulfill one of the succeeding types: (1) adsorbent should have higher CO_2 selectivity and CH_4 in tiny quantity; such selectivity-based adsorbents are also called as equilibrium-dependent adsorbents (Augelletti et al., 2017); (2) kinetic diameters (KDs) vary from one gas component to another. KDs of CO_2 and CH_4 are estimated as 3.4Å and 3.8Å, respectively. The adsorbent must be selected based on the pore size of the gas molecules. The absorbents with difference in dispersion limitations of the gas molecules are called as KD-based adsorbents. One of

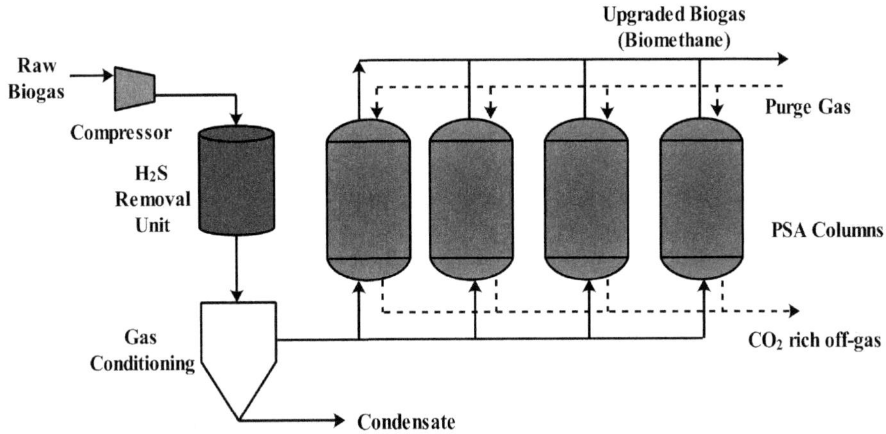

FIGURE 11.3 Flowchart for pressure swing adsorption process.

such adsorbents used to adsorb CO_2 is CMS3K carbon molecular sieve (Patterson et al., 2011). Adsorption is governed by high pressure and temperature. Hence, adsorption based on the governing parameters is termed as PSA and temperature swing adsorption (Khan et al., 2017; Grande, 2012).

11.8.3 Purification by membrane separation

The process of splitting up the gas constituents with polymeric semipermeable membranes by selective permeation is termed as membrane separation. This process works on the principle of solubilization and diffusion of gas components, which depends on concentration gradient (Scholz et al., 2013).

The process of biogas purification is illustrated in Fig. 11.4. There are two different types of membranes based on the membrane material and separation type. They are (1) glass polymeric membranes (GPMs) and (2) rubbery or glassy-substituted polyacetylenes (RPs). GPM separation based on size of gas molecule and RP is based on solubility of gas components (Tocci and Pullumbi, 2011). Polymeric membranes are made up of polysulfone, polyetherimide, polyamides, cellulose acetate, and polyimide (Basu et al., 2010; Chen et al., 2015). On comparing with PSA, membrane separation is cost-effective, less maintenance and capital cost, improved space consumption, simple operation, and so on (Miltner et al., 2009). The disadvantage with this process is replacement of expensive membrane because of pressure variation and blockage due to impurities (Cnop et al., 2007).

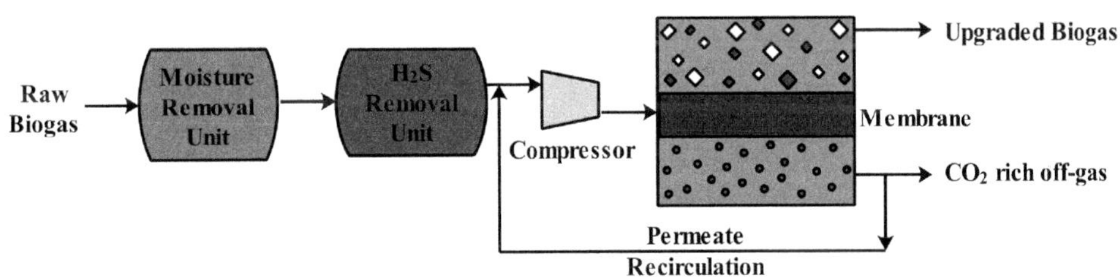

FIGURE 11.4 Flow pattern for membrane separation process.

11.8.4 Cryogenic method of separation

This method of upgradation is based on condensation and distillation at various temperatures using biogas (Goffeng, 2013). Fig. 11.5 represents the process flow diagram of cryogenic separation. By reducing the temperature in cooling device, methane and some other gases are removed. The remaining trace gases are removed in the distillation column, and pure biogas is collected (Jonsson and Westman, 2011; Yousef et al., 2016). The following are the advantages of cryogenic separation: (1) absence of special medium or any reagent for purification, (2) upgraded biogas does not require external compression, and (3) solid carbon dioxide can be produced from purified biogas (Jonsson and Westman, 2011). Disadvantages of this process are as follows: obligation of elevated pressure at very low temperature, high working cost, impurities will freeze in the devices, and proper insulation to prevent biogas from external heat.

11.8.5 Hybrid upgradation methods

Several disadvantages in the biogas upgradation methods can be overwhelmed by integrating two or more technologies, and it is called hybrid method (Sahota et al., 2018). Fig. 11.6 illustrates the development of hybrid technologies from the available methods. Membrane system integrated with high temperature and water scrubbing and cryogenic and water scrubbing process has shown a drastic reduction in processing cost and energy cost (Singhal et al., 2017). Combining adsorption with membrane technology produces 97% pure methane (Scholz et al., 2013). In future, several hybrid technologies can be achieved based on technoeconomical investigation of biogas purification process.

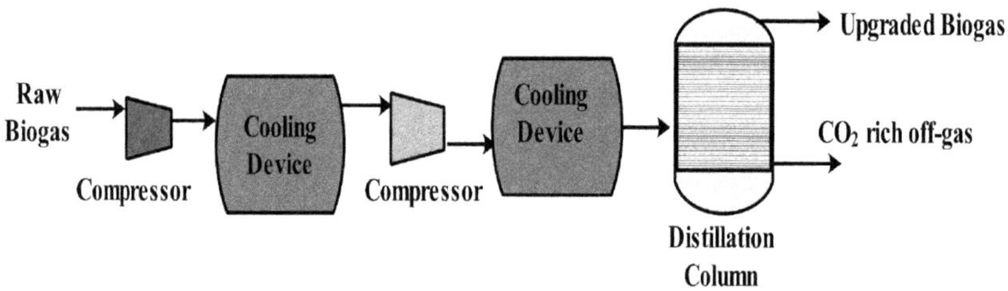

FIGURE 11.5 Flowchart for cryogenic separation method.

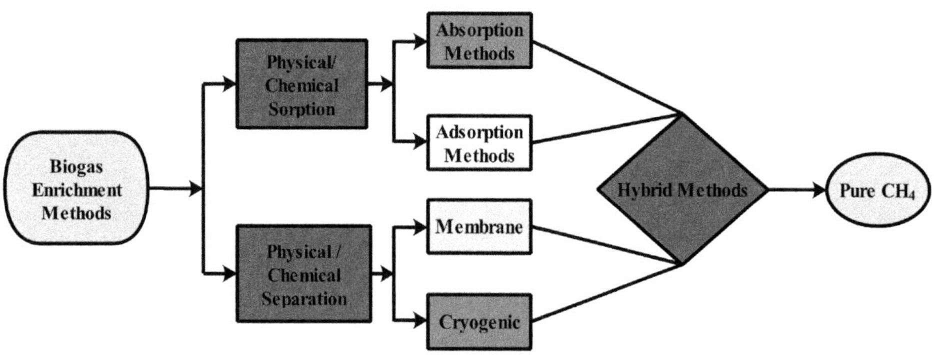

FIGURE 11.6 Flow diagram for development of hybrid methods from the available process.

11.9 Application of upgraded biogas in fuel cell

Chemical energy is transformed into electrical energy in fuel cell with the help of electrolyte layer present in between anode and cathode (Prashansa et al., 2020; Pratiksha et al., 2018). Biogas and oxygen are served into the anode and cathode, respectively, and electricity is produced due to the electrochemical reaction in the electrode (Xiang et al., 2019; Marta et al., 2019). Upgraded biogas with more methane percentage is used in polymer electrolyte membrane fuel cell since it is very sensitive toward CO_2 and sulfur impurities. When comparing solid oxide fuel cell (SOFC), molten carbonate fuel cell (MCFC), and phosphoric acid fuel cell (PAFC), SOFC is more efficient in power production (Renu et al., 2019). SOFC works on the principle of reforming of methane to hydrogen inside the cell at high temperature (750–1000°C), and it is low sensitive for sulfur impurities (Escudero and Serrano, 2019; Arnusorn et al., 2019). Hence, SOFC is the most suited fuel cell to utilize biogas as feed gas (Baldinelli et al., 2018).

Several studies have proved the importance of upgraded biogas on comparing with prereforming biogas. Reforming method revealed scoring efficiency and exergy efficiency of 37.2% and 38.6%; however, upgraded method produced 51.1% and 52.3%, respectively (Baldinelli et al., 2017). In addition to this, economic feasibility is proved to be more favorable in upgraded system than reforming system. On comparing SOFC with internal combustion engines (ICEs), extended new outcomes are needed for proper conclusion (Baldinelli et al., 2018). Pretreated biogas using chemical and thermochemical methods fits better as fuel cell fed. Based on the electrochemical performance of SOFC with feed gas composition (CH_4:CO_2:H_2; 70:25:5) at 750°C, the following are determined: (1) NH_3 in the feed gas increases the affinity to produce hydrogen, (2) presence of siloxanes reduces the degradation rate, and (3) due to corrosion in tubes and valves, chlorine in the feed gas should be detached (Escudero and Serrano, 2019; Mohammad et al., 2019). In Thailand, around 1800 biogas plants produce 1450 Mm^3 of biogas/year. Direct biogas solid oxide fuel cell (DB-SOFC) generates 45.25% higher electricity and more cost-effective when comparing with heat engine generator (Arnusorn et al., 2019; Ali Saadabadi et al., 2019).

Various stimulation-based studies proved the effectiveness of upgraded biogas in fuel cell. Based on UniSIm model, nearly 0.104–0.16 million m^3/year of CO_2 is emitted from wastewater treatment plant, and it is concluded that biogas is effective feed fed for SOFC and, hence, reduction in electrical demand and GHG emission (Jillian et al., 2017). Based on the validation of operating parameters that include steam-to-carbon ratio, temperature, and carbon deposition, system performance is analyzed for fuel processor. Temperature variation has its impact on efficiencies, for example, at 600°C–34%, at 700°C–44%, and at 800°C–46% efficient. The study concluded that pure methane and hydrogen in SOFC attain economically feasible system than using biogas (Raffaello et al., 2017).

The study conducted by Héctor et al. (2018) proved that optimized simulation results of 8000 m^3/h flow rate and molar ratio of CH_4 (2.0):H_2O (1.0):CO_2 (0.67) perform well for 14 years and supply energy less than the market price for 36,000 houses. Through water reuse system, 5.12 m^3/h of water consumption got reduced. Usage of proton exchange membrane (PEM), H_2 production increases with the increase in biogas flow rate. Biogas flow rate determines the price of energy produced via fuel cell (Héctor et al., 2018). Some of the problems in using biogas without upgradation are thermal stress, poisoning, and carbon deposition. To overcome these problems, based on literature, fibrous SOFC anode, sulfur tolerance anodes, internal dry reforming, and electrospinning are developed. Carbon deposition can be reduced based on the flow rate of CO_2, steam, or air and using nickel-based anodes. Sulfur tolerance can be achieved by nickel doped with Mo, CeO, and Cu. Improvement in catalytic behavior on CH_4 oxidation and sulfur poisoning is obtained with two-dimensional crystals of Cu, Mo, MoS, Sn, and MnO. Dynamic models in simulation predict the efficiency of SOFC and the amount of biogas needed as feed.

11.10 Conclusions

In India, biogas production from the waste is the requested technology to meet the economically feasible energy demand among poor farmers. Preliminary investment is the inhibiting parameter to implement new anaerobic digesters with the common people. Reducing global warming, solving the energy crisis, and increasing revenue find that carbon promotion is achieved with present and probable biogas plants. Biogas is well-thought-out as promising renewable energy and resourceful alternative for natural gas. Upgraded biogas results in drop-in GHG emission, sulfur, and nitrogen gases into the atmosphere and also leads to sustainable environment. Hybrid upgradation method is a proved efficient method and economically feasible technology to attain pure methane (99%) for the upcoming application. According to the literatures discussed in this chapter, claiming upgraded biogas is a proved efficient source for SOFC fuel cell to fulfill the energy demand in global market.

References

Abdeen, F.R.H., Mel, M., Jami, M.S., Ihsan, S.I., Ismail, A.F., 2016. A review of chemical absorption of carbon dioxide for biogas upgrading. Chin. J. Chem. Eng. 24, 693–702.

Ali Saadabadi, S., Aditya Thallam, T., Liyuan, F., Ralph, E.F.L., Henri, S., Aravind, P.V., 2019. Solid oxide fuel cells fuelled with biogas: potential and constraints. Renew. Energy 134, 194–214.

Amir, I.A., Mei, Y.O., Saifuddin, N., Kit, W.C., Pau, L.S., 2019. Technologies for biogas upgrading to biomethane: a review. Bioengineering 6, 92.

Andriani, D., Wresta, A., Atmaja, T.D., Saepudin, A., 2014. A review on optimization production and upgrading biogas through CO_2 removal using various techniques. Appl. Biochem. Biotechnol. 72, 1909–1928.

Angelidaki, I., Treu, L., Tsapekos, P., Luo, G., Campanaro, S., Wenzel, H., Kougias, P.G., 2018. Biogas upgrading and utilization: current status and perspectives. Biotechnol. Adv. 2, 452–466.

Arnusorn, S., Aree, K., Wutthisat, C., Jenjira, P., 2019. Prospective study of application the direct-biogas solid oxide fuel cell system to the biogas plant in Thailand. Energy Procedia 158, 978–983.

Augelletti, R., Conti, M., Annesini, M.C., 2017. Pressure swing adsorption for biogas upgrading. A new process configuration for the separation of biomethane and carbon dioxide. J. Clean. Prod. 140, 1390–1398.

Aziz, N.I.H.A., Hanafiah, M.M., Gheewala, S.H., 2019. A review on life cycle assessment of biogas production: challenges and future perspectives in Malaysia. Biomass Bioenergy 122, 361–374.

Baldinelli, L., Barelli, G.,B., 2017. Upgrading versus reforming: an energy and exergy analysis of two solid oxide fuel cell-based systems for a convenient biogas-to-electricity conversion. Energy Conver. Manage. 138, 360–374.

Baldinelli, A., Barelli, L., Bidini, G., 2018. On the feasibility of on-farm biogas-to-electricity conversion: to what extent is solid oxide fuel cells durability a threat to break even the initial investment? Int. J. Hydrog. Energy 43, 16971–16985.

Basu, S., Khan, A.L., Cano-Odena, A., Liu, C., Vankelecom, I.F., 2010. Membrane-based technologies for biogas separations. Chem. Soc. Rev. 39, 750–768.

Bhuvaneshwari, S., Hettiarachchi, H., Meegoda, J.N., 2019. Crop residue burning in India: policy challenges and potential solutions. Int. J. Environ. Res. Public Health 16, 832.

Bilgen, S., Kaygusuz, K., Sari, A., 2004. Renewable energy for a clean and sustainable future. Part A: Recovery, Utilization, and Environmental Effects Energy Sources 26 (12), 1119–1129.

Bishnoi, S., Rochelle, G.T., 2000. Absorption of carbon dioxide into aqueous piperazine: reaction kinetics, mass transfer and solubility. Chem. Eng. Sci. 55, 5531–5543.

Cebula, J., 2009. Biogas purification by sorption techniques. The Silesian University of Technology Architecture Civil Engineering Environment 2, 95–104.

Chaemchuen, S., Kabir, N.A., Zhou, K., Verpoort, F., 2013. Metal–organic frameworks for upgrading biogas via CO_2 adsorption to biogas green energy. Chem. Soc. Rev. 42, 9304–9332.

Chen, X.Y., Vinh-Thang, H., Ramirez, A.A., Rodrigue, D., Kaliaguine, S., 2015. Membrane gas separation technologies for biogas upgrading. RSC Adv. 5, 24399–24448.

Chi, C., Taejin, P., Xuhui, W., Shilong, P., Baodong, X., Rajiv, K., Chaturvedi, R.F., Victor, B., Philippe, C., Rasmus, F., Hans, T., Govindasamy, B., Zaichun, Z., Ramakrishna, R., Nemani, R., Myneni, B., 2019. China and India lead in greening of the world through land-use management. Nat. Sustain. 2, 122–129.

Cnop, T., Dortmundt, D., Schott, M., 2007. Continued Development of Gas Separation Membrane for Highly Sour Service. UOP LLC: Illinios, USA.

Debadayita, R., Pinakeswar, M., MichèleL, C., 2014. The implementation of decentralized biogasplants in Assam, NE India: the impact and effectiveness of the national biogas and manure management programme. Energy Policy 68, 80–91.

Demirbas, A., 2000. Recent advances in biomass conversion technologies. Energy Edu. Sci. Technol. 6, 19–40.

Dincer, I., 2001. Environmental issues. II. Potential solutions. Part A: Recovery, Utilization, and Environmental Effects Energy Sources 23 (1), 83–92.

Elango, D., Pulikesi, M., Baskaralingam, P., Ramamurthi, V., Sivanesan, S., 2007. Production of biogas from municipal solid waste with domestic sewage. J. Hazard. Mater. 141, 301–304.

ENVIS, 2020. Centre on Hygiene, Sanitation, Sewage Treatment Systems and Technology, International Institute of Health and Hygiene, New Delhi, Ministry of Environment, Forests & Climate Change, Govt of India. Order of the National Green Tribunal Regarding Effluent Discharge Standards for Stps, 30/04/2019. http://www.sulabhenvis.nic.in/Database/STST_wastewater_2090.aspx. (Accessed 22 January 2020).

Escudero, M.J., Serrano, J.L., 2019. Individual impact of several impurities on the performance of direct internal reforming biogas solid oxide fuel cell using W-Ni-CeO as anode. Int. J. Hyd. Energy 44, 20616–20631.

Fridleifsson, I.B., 2001. Geothermal energy for the benefit of the people. Renew. Sustain. Energy Rev. 5, 299–312.

García-Gutiérrez, P., Jacquemin, J., McCrellis, C., Dimitriou, I., Taylor, S.F.R., Hardacre, C., Allen, R.W.K., 2016. Techno-economic feasibility of selective CO_2 capture processes from biogas streams using ionic liquids as physical absorbents. Energy Fuels 30, 5052–5064.

Goffeng, B., 2013. Crynotechnology for Biogas. Dept of Chem Eng, Lund University. http://www.chemeng.lth.se/exjobb/E700.pdf.

Grande, C.A., 2012. Advances in pressure swing adsorption for gas separation. Int. Sch. Res. Netw. Chem. Engi. 13.

Héctor, N.C.S., Diego, M.P., Gilson, B.A.L., Luis, P.Z., Lisiane, V.M., 2018. A techno-economic evaluation of the energy generation by proton exchange membrane fuel cell using biogas reforming. J. Clean. Prod. 200, 598–608.

IEA bioenergy, 2000. Biogas upgrading and utilisation. In: Wheeler, P., Jaatinen, T., Lindberg, A., Lundeberg, S., Holm-Nielsen, J.B., Wellinger, A., et al. (Eds.), Task 24: Energy from Biological Conversion of Organic Waste. Paris, France. Available at: http://www.biores.eu/docs/BIOGASFUNDAMENTALS/Biogas_upgrading__utili sation.pdf.

Jillian, L., Pascale, C., Brant, P., 2017. Use of wastewater treatment plant biogas for the operation of solid oxide fuel cells (SOFCs). J. Environ. Manage. 203, 753–759.

Jonsson, S., Westman, J., 2011. Cryogenic Biogas Upgrading Using Plate Heat Exchangers. Department of Energy and Environment Division of Energy Technology. Chalmers University of Technology Göteborg, Sweden. Available online. http://publications.lib.chalmers.se/records/fulltext145544.pdf.

Kadam, R., Panwar, N.L., 2017. Recent advancement in biogas enrichment and its applications. Renew. Sustain. Energy Rev. 73, 892–903.

Khan, I.U., Othman, M.H., Hashim, H., Matsuura, T., Ismail, A.F., Rezaei-Dasht Arzhandi, M., Azelee, I.W., 2017. Biogas as a renewable energy fuel— a review of biogas upgrading, utilisation and storage. Energy. Conver. Manage. 150, 277–294.

Li, L., Voice, A.K., Li, H., Namjoshi, O., Nguyen, O., Du, Y., Rochelle, G., 2013. Amine blends using concentrated piperazine. Energy Procedia 37, 353–369.

Maile, I., Muzenda, E., Tesfagiorhis, H., 2017. Chemical absorption of carbon dioxide in biogas purification. Procedia Manuf. 7, 639–646.

Makhloufi, C., Lasseugette, E., Remigy, J.C., Belaissaoui, B., Roizard, D., Favre, E., 2014. Ammonia based CO capture process using hollow fiber membrane contactors. J. Membr. Sci. 455, 236–246.

Marta, F.S., Alfredo, E.M., Jonas Gomes dos, S., Luciana, P., Brambilla de Souza, R.F., Almir, O.N., António, G.B., Cristiane, A.O., 2019. Microbial fuel cell-induced production of fungal laccase to degradethe anthraquinone dye. Environ. Chem. Let. 17, 1413–1420. https://doi.org/10.1007/s10311-019-00876-y.

Mcleod, A., Jefferson, B., McAdam, E.J., 2014. Biogas upgrading by chemical absorption using ammonia rich absorbents derived from wastewater. Wat. Res. 67, 175–186.

Miltner, M., Makaruk, A., Bala, H., Harasek, M., 2009. Biogas Upgrading for Transportation Purposes; Operational Experience With Austria First Bio-CNG Plant. Vienna University of Technology, Vienna.

Mohammad, A.A., Waqas, H.T., Enas, T.S., El Haj Assad, M., Anis, A., Cha, S.W., 2019. On the technical challenges affecting the performance of direct internal reforming biogas solid oxide fuel cells. Renew. Sustain. Energy Rev. 101, 361–375.

Muñoz, R., Meier, L., Diaz, I., Jeison, D., 2015. A review on the state-of-the art of physical/chemical and biological technologies for biogas upgrading. Rev. Environ. Sci. Biotechnol. 14, 727–759.

Myint, M.T., Nirmalakhandan, N., 2009. Enhancing anaerobic hydrolysis of cattle manure in leach bed reactors. Bioresour. Technol. 100, 1695–1699.

Nie, H., Jiang, H., Chong, D., Wu, Q., Xu, C., Zhou, H., 2013. Comparison of water scrubbing and propylene carbonate absorption for biogas upgrading process. Energy Fuels 27, 3239–3245.

Panwara, N.L., Kaushik, S.C., Surendra, K., 2011. Role of renewable energy sources in environmental protection: a review. Renew. Sustain. Energy Rev. 15, 1513–1524.

Patterson, T., Esteves, S., Dinsdale, R., Guwy, A., 2011. An evaluation of the policy and techno-economic factors affecting the potential for biogas upgrading for transport use in UK. Energy Policy 39, 1806–1816.

Prashansa, T., Neetu, R., Asheesh, K.Y., 2020. Enhanced wastewater treatment and electricity generation using stacked constructed wetland–microbial fuel cells. Environ. Chem. Let. 18, 871–879. https://doi.org/10.1007/s10311-020-00966-2.

Pratiksha, S., Supriya, G., Vikram, G., Rouzbeh, A., Asheesh, K.Y., 2018. Up to 399 mV bioelectricity generated by a rice paddy-planted microbial fuel cell assisted with a blue-green algal cathode. Environ. Chem. Let. 17, 1045–1051. https://doi.org/10.1007/s10311-018-00824-2.

Raffaello, C., Lidia, L., Laura, T., 2017. Use of biogas from biowaste in a solid oxide fuel cell stack: application to an off-grid power plant. Renew. Energy 111, 781–791.

Rathore, N.S., Panwar, N.L., 2007. Renewable Energy Sources for Sustainable Development. New India Publishing Agency, New Delhi, India.

Rekia, B., Lyes, T., Khaled, L., Toufik, B., Mohand, T., 2020. Experimental investigation on biogas enrichment with hydrogen for improving the combustion in diesel engine operating under dual fuel mode. Int. J. Hydrog. Energy 45, 9052–9063.

Renu, B., Vaishali, G., Monoj, K.M., 2019. Improved biogas yield from organic fraction of municipal solid waste as preliminary step for fuel cell technology and hydrogen generation. Inter. Int. J. Hydrog. Energy 44, 164–173.

Rimika, K., Pooja, G., Madan, K., Virendra Kumar, V., 2019. Evaluation of biogas upgrading technologies and future perspectives: a review. Environ. Sci. Pollut. Res. 26, 11631–11661.

Rinprasertmeechai, S., Chaveda, S., Rangsunvigit, P., Kulprathipanja, S., 2012. Carbon dioxide removal from flue gas using amine-based hybrid solvent absorption. Int. J Chem. Bioeng. 6.

Sahota, S., Shah, G., Ghosh, P., Kapoor, R., Sengupta, S., Singh, P., Vijay, V., Sahay, A., Vijay, V.K., Thakur, I.S., 2018. Review of trends in biogas upgradation technologies and future perspectives. Bioresour. Technol. Rep. 1, 79–88.

Sarker, A.I., Aroonwilas, A., Veawab, A., 2017. Equilibrium and kinetic behaviour of CO_2 adsorption onto zeolites, carbon molecular sieve and activated carbons. Energy Procedia 114, 2450–2459.

Scholz, M., Melin, T., Wessling, M., 2013. Transforming biogas into biomethane using membrane technology. Renew. Sustain. Energy Rev. 1, 199–212.

Singhal, S., Agarwal, S., Arora, S., Sharma, P., Singhal, N., 2017. Upgrading techniques for transformation of biogas to bio-CNG: a review. Int. J. Energy Res. 41 (12), 1657–1669. https://doi.org/10.1002/er.3719.

Statista, 2020. Global Municipal Solid Waste Generation by Key Countries in 2017 available online. https://www.statista.com/statistics/916749/global-generation-of-municipal-solid-waste-by-country/. (Accessed August 2020).

Tippayawong, N., Thanompongchart, P., 2010. Biogas quality upgrade by simultaneous removal of CO_2 and H_2S is a packed column reactor. Energy 1–4.

Tocci, E., Pullumbi, P., 2011. Multi-scale molecular modelling approaches for designing/selecting polymers used for developing novel membrane. Gas Separation Problems with Membrane. In: Membrane Engineering for the Treatment of Gases, vol. 1. Royal Society of Chemistry, pp. 1–28.

UNDP, 2000. World Energy Assessment 2000 – Energy and the Challenge of Sustainability. UNDP, New York, ISBN 9211261260.

Urja, A., 2016. Generation of Green Energy from Paddy Straw, a Novel Initiative in Sustainable Agriculture Green Energy. Available online. https://mnre.gov.in/file-manager/akshay-urja/june-2016/30-33.pdf. (Accessed 10 January 2020).

Waste to Energy opportunities and cost in India- Energy Alternatives India (EAI), 2019. EAI.in. (Accessed 5 June 2019).

Xiang, L., Weifen, Y., Xin, L., Xiaohua, Z., 2019. Enhanced Cr reduction and bioelectricity production in microbial fuel cells using polypyrrole-coated MnO_2 on carbon cloth. Env. Chem. Let. 18, 517–525. https://doi.org/10.1007/s10311-019-00958-x.

Yousef, A.M., Eldrainy, Y.A., El-Maghlany, W.M., Attia, A., 2016. Upgrading biogas by a low-temperature CO removal technique. Alex. Eng. J. 155, 1143–1150.

Further reading

Livestock census, 20th 2020. Prepared by Ministry of Fisheries, Animal Husbandry & Dairying (MFAD), Department of Animal Husbandry & Dairying, Animal Husbandry Statistics Division, Krishi Bhawan, New Delhi, 2019. http://dahd.nic.in/about-us/divisions/statistics. (Accessed April 2020).

Monthly Renewable Energy Generation Report (MREGR), 2020. Renewable Energy Project Monitoring Division. Central Electricity Authority, Government of India. www.cea.nic.in. (Accessed 17 July 2020).

National Master Plan for Development of Waste-to-Energy in India (NMPDWEI), 2017. National Bio-Energy Board (NBB), Ministry of Non-conventional Energy Sources. (MNES), Government of India, New Delhi. http://www.seas.columbia.edu/earth/wtert/sofos/Natl_%20Master_%20Plan_of_India.pdf.

CHAPTER

12

Anaerobic digestion of agrowastes: end-of-life step of a biorefinery

Ipsita Chakravarty, Sachin A. Mandavgane

Visvesvaraya National Institute of Technology, Nagpur, Maharashtra, India

12.1 Introduction

India has an immense agrowaste-based biogas potential (Prasad et al., 2020). The inefficient utilization of agricultural residues leads to the practice of open field burning. Such irrational disposal leads to worsening in soil and air quality due to the release of greenhouse gases, particulate matter in air, and addition of inorganic salts in soils affecting community health (Singh et al., 2019). The untapped potential of agricultural waste for biogas production pertains to crucial issues such as the agrowaste supply chain, economically viable and nonstringent processing, and feasible agrobusiness models (Mittal et al., 2019). The major drawback in complete utilization of lignocellulosic agrowaste through anaerobic digestion and biogas recovery is the recalcitrant nature of lignin (Antoniou et al., 2019). Recent researchers are now pondering upon valorization of digestate of fermentation, combustion, pyrolysis, and so on for the biogas production (Bhuyan et al., 2020). Anaerobic digestion is governed by hydrolysis, acidogenesis, acetogenesis, and methanogenesis generating biogas and biofertilizer. Biogas mainly comprises methane and CO_2 (Sukhesh et al., 2018). There is a high demand for technological deployment for agrowaste valorization through anaerobic digestion. Through such efforts, the issues of natural resource recycling can be handled in an efficient manner. This chapter focuses on the significance of biogas production from agroresidues, process strategies, and modifications required to utilize the potentials of biomethane generation from agrowastes through the process of anaerobic digestion. It also highlights the role and significance of the government and industrial initiatives and responsibilities. This chapter proposes biogas production through anaerobic digestion as the final step in agrowaste biorefinery cascade processing. The ultimate outcome is biofertilizer leading to sustainable circular economy in Indian scenario.

12.2 Generation and availability of agrowastes

India generates 500 million tons of crop residues per year. The major crops of India are wheat, rice, maize, millets sugarcane, cotton,

Biofuels and Bioenergy
https://doi.org/10.1016/B978-0-323-85269-2.00005-8

jute, pulses, and oilseed crops. Rice, wheat, and sugarcane give maximum economic gain to the farmers. Current statistics of crop residue management in India is shown in Fig. 12.1. The highest agrowaste is accounted in Uttar Pradesh. Punjab holds the second place followed by Maharashtra. Cereal crops generate 70% residues followed by fibers, oilseeds, pulses, and sugarcane. Rice and sugarcane produce 50% of the crop residues. The common applications of the surplus crop residues are animal feed, domestic fuel, and shelter for animals. Apart from crop residues, 15 MT of vegetable and fruit waste are available for recycling (NPMCR, 2018).

12.3 Classification and composition of agrowastes

Agricultural wastes are classified into three major categories viz., direct field residues, semi-processed residues, and industrially processed residues as shown in Fig. 12.2. The silaging, harvesting mechanism, and time harvest of crop residues affect the composition of substrate. The biomass conversion efficiency in anaerobic digestion process is influenced by feedstock composition (De Francisci et al., 2015).

Main constituents of any lignocellulosic waste are lignin, cellulose, hemicellulose, proteins (3%–4%), and fats (1%–2%) (Das et al., 2015).

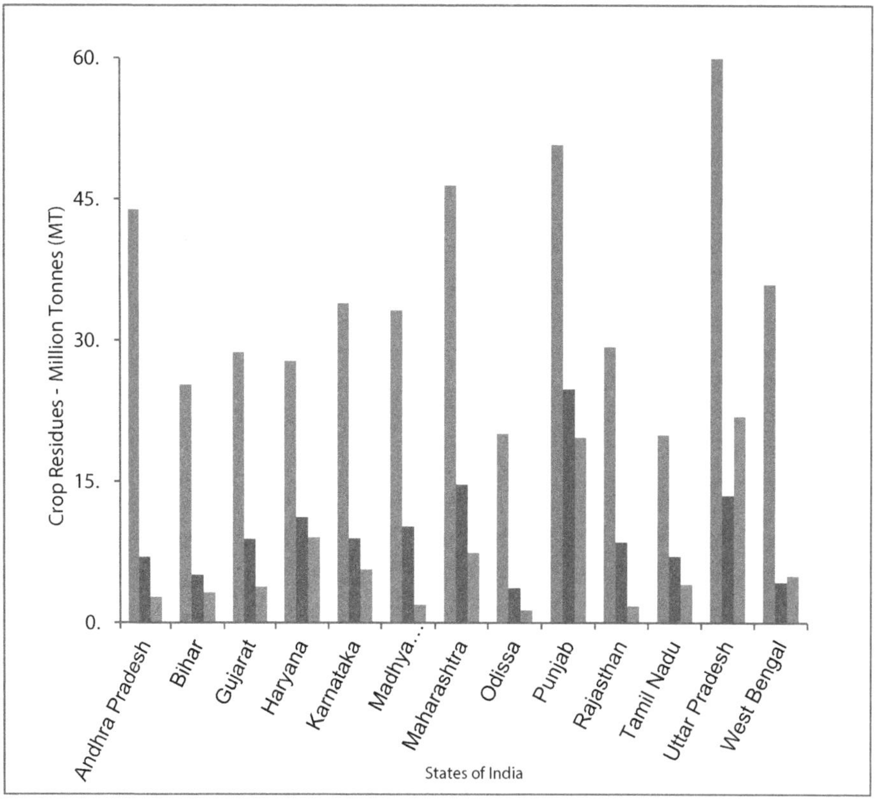

FIGURE 12.1 State-wise generation of crop residues in India (NPMCR, 2018)—■ Residue generated; ■ Residue surplus; ■ Residue burnt.

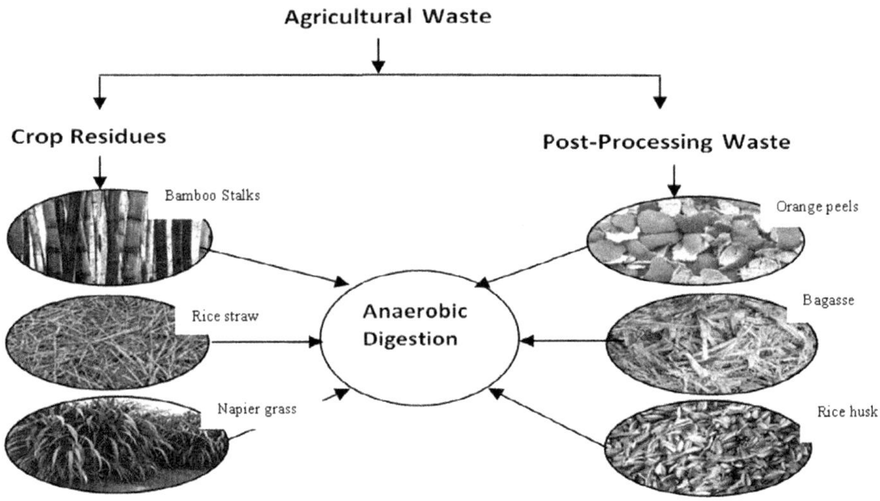

FIGURE 12.2 Types of agricultural wastes.

Cellulose is a linear polymer of cellobiose, while hemicellulose is composed of pentose and hexose. Lignin is a 3-D network of phenyl propanoid units (Dixit et al., 2020). Depolymerization and hydrolysis of lignocellulosic content is the rate-limiting step of anaerobic digestion process. The composition of various agroresidues is shown in Table 12.1. The lignocellulosic composition of the feedstock directs the pretreatment and further processing strategies to improve biomass degradation and biogas generation (Rajendran et al., 2018).

TABLE 12.1 Composition of various agricultural wastes.

Agrowaste	Cellulose	Hemicellulose	Lignin
Hardwood sstems	40–60	35–45	18–25
Softwood sstems	45–50	25–35	20–30
Nut sshells	25–30	25–30	30–40
Grasses	25–40	30–50	15–25
Cereal straws	30–40	40–50	15–20
Leaves	15–20	60–70	0

Junchen et al., 2012; Prasad et al., 2007; Saratale et al., 2008.

12.4 Process design strategies for improved biomethanation

12.4.1 Pretreatment of feedstock

The pretreatment process depends on cost of technology, ease of performance, inhibitory effect if any, energy needs, and environmental impact. The conventional methods of pretreatment are physical, chemical, and thermal. Biological methods have gained prominence in recent years due to their efficacy and eco-friendliness. Few established pretreatment processes are depicted in Table 12.2. The choice of pretreatment for any agricultural waste relies on its cost-effectiveness, toxicity, synergy with consecutive steps and the microbial strains used in the following phases, energy requirements, and efficacy.

TABLE 12.2 Pretreatment methods for various agricultural feedstocks.

Pretreatment process	Substrate	Conditions	Biogas production	References
Physical	Rice straw	Extrusion method	Methane production of the extrusion was 227.3 L/g volatile solids that is 32.5% more than that of the shredded rice straw and 72.2% more than untreated rice straw	Chen et al. (2014)
	Fruit and vegetable waste	Sonication at 20 kHz and amplitude of 80 um.	Highest methane yield of 2380 kJ/kg TS	Zeynali et al. (2017)
	Wheat straw	Ball milling to different sizes upto 0.25 mm size	43% straw biodegradability. 192 ± 25 Nm L/g VS	Motte et al. (2014)
Thermal	Wheat straw	Temperature variations given at 120, 140, 160, 1800°C	615 mL methane/g volatile solids	Rajput et al. (2018)
	Hay	Steam explosion at 160–2200°C	16% methane increased	Bauer et al. (2014)
Chemical	Cotton stalk	Alkali hydrogen peroxide treatment at 370°C	193 mL methane/g volatile solids	Zhang et al. (2018)
	Agriculture straw	3% hydrogen peroxide; 8% calcium hydroxide at 370°C	206 mL methane/g VS	Song et al. (2014)
Biological	Rice straw	*Phanerochaete chrysosporium, Coriolus versicolor, Trichoderma viride, Aspergillus niger, Gloeophyllum trabeum, Bacillus circulans, Pseudomonas aeruginosa,* and *Streptomyces badius*	Methane yield of 0.191 m^3/kg TS and the maximum methane generation rate achieved 14.85 mL/h on the 10th day	Li et al. (2020)
	Napier grass	Microbial consortium: WSD-5	49% increase in net yield of methane than untreated biomass	Wen et al. (2015)

12.4.1.1 *Physical pretreatment*

The physical pretreatment is a conventional process of breakdown of the agroresidues into fine particles, increasing the surface area. This ensures better mass transfer and contact with the microorganisms (Park et al., 2020; Al Afif et al., 2020). Processes such as ball milling cause decrystallization of cellulose and prove to be an efficient method for superior yield of simple sugars for anaerobic digestion. In a study conducted on grinding on wheat straw, 46% total carbohydrates and 72% glucose were produced after 10% reduction in cellulose crystallinity (Silva et al., 2012). The bane of physical pretreatment is its energy intensity.

12.4.1.2 *Chemical pretreatment*

Chemical pretreatment involves the use of acids, alkalis, and organic solvents for the hydrolysis of the substrates. Its potential is dominated by the type of substrate used (Solowski et al., 2020). This kind of pretreatment is not preferable for easily degradable substances. The drawback of using chemicals for treatment is their toxic effect and accumulation of volatile fatty acids, which have an inhibitory effect on the process of methanogenesis (Chen et al., 2008). Acidic pretreatment helps in hydrolysis of cellulose and hemicellulose to simple sugars and forms lignin precipitate (Chen et al., 2017) But the use of strong acids may cause the release of furfurals and hydroxymethylfurfurals (62–64) Alkali pretreatment is based on salvation and saponification phenomenon, which causes increased surface area and accessibility for microbial activities (Pellera and Gidarakos, 2018). The drawback of using chemicals for treatment is their toxic effect and accumulation of volatile fatty acids, which have an inhibitory effect on the process of methanogenesis (Fang et al., 2019).

12.4.1.3 *Biological pretreatment*

The utilization of microorganisms such as bacteria and fungi to degrade lignin, cellulose, and hemicellulose is a nontoxic and eco-friendly method. But the process takes prolonged residence time and initial investment (Li et al., 2020; Akyol et al., 2019; Sharma et al., 2019).

12.4.1.4 *Thermal pretreatment*

In this strategy, high temperature from 160 to 240°C and high pressure of >5 MPa help in the hydrolysis of complex carbohydrate structure of agricultural residues. Hot water treatment and autohydrolysis help in hydrolyzing lignocellulosic material under extremely pressurized liquid at 200°C (Veluchamy and Kalamdhad, 2017; Rajput and Visvanathan, 2018). The energy demand and operation cost are not sustainable unless the final biogas productivity is very high. Some inhibitors such as furfural and phenolic compounds formed during the process can inhibit methane generation (Scherzinger and Kaltschmitt, 2021).

Though initial investments seem to be higher than chemical, yet they are commercially useful in the long run. Basidiomycetes such as white rot fungi and brown rot fungi are significant in lignocellulosic degradation (Wan et al., 2012). White rot fungi can efficiently degrade lignin, whereas brown rot fungi can decompose cellulose and hemicellulose. Wheat straw treated with *Polyporus brumalis* enhanced biogas formation by 20% (Rouches et al., 2016). Rice straw when pretreated with a potent culture of *Pleurotus ostreatus* and *Trichoderma reesei* confirmed improved lignin degradation by 33% and biomethanation by 120% (Mustafa et al., 2016).

12.4.2 C/N balance

The agricultural wastes are predominated by lignocellulosic substrates, which have high content of cellulose and hemicellulose. Their rich carbohydrate content increases the C/N ratio. Agrowastes such as rice straw, wheat straw, bamboo stalks, and so on have C/N ratio more than 50. Fruit and vegetable wastes have C/N ratio in the range of 25–35. These substrates hold potential for anaerobic digestion, but the presence of lignin as well as low organic nitrogen need to be tackled. Cattle manure or animal waste is rich in organic nitrogen, which causes ammonia accumulation and inhibition of anaerobic digestion process because of a low C/N ratio. Codigestion involves mixing of two or more substrates to balance the C/N ratio, provide buffering capacity, and dilute toxins. Such positive synergistic combination leads to enhanced methane production as compared with monodigestion process (Hagos et al., 2017). The nutritional needs and the limiting factors for microbial activity in an anaerobic digestion system are very important (Adekunle and Okolie, 2015). The insufficient levels of media components cause variations in the predominant fermentation outcomes. The accumulation of ammonia inhibits the biochemical processes within the microbial system. C/N ratio must be balanced to avoid the accumulation of ammonia (Liu et al., 2016). On one hand, the carbon content present in the substrate acts as the energy resource for the microorganisms. On the other hand, nitrogen regulates the enzyme synthesis, amino acid, and protein formation. C/N ratio of the substrate should be adapted to the microbial metabolic activity (Yang et al., 2020; Loganath and Senophiyah-Mary, 2020). Agrowaste has huge potential due to its rich organic content to be used for energy generation via anaerobic fermentation. The potentiality of agrowaste is challenged by high carbon and low nitrogen content and also the accumulation of volatile fatty acids, which adversely affect the pH. Nitrogen supplementation and its reduced form can be beneficial to tackle the issues (Greses et al., 2020; Wainaina et al., 2020). In the course of anaerobic fermentation, several microbial interactions take place, which are sometimes affected by toxic influence of aqueous solution in the form of ammonium ion (NH_4^+) and unionized ammonia (free ammonia: NH_3). For this reason, it is important to quantify the concentration of ammoniacal nitrogen and NH_3 in the mixing liquor to ensure the stability of the AD process (Yenigün and Demirel, 2013). Most of the lignocellulosic agricultural residues have higher carbon content and lower nitrogen content. This leads to inefficient degradation. Thus, adequate nitrogen concentrations need to be maintained, and C/N ratio has to be balanced by using codigestion method. Codigestion ensures the balance of nutrients, especially in case where the substrate has a lower concentration of one of the limiting factors or essential nutrients. This helps the microorganisms to carry out the required biochemical processes, utilize the nutrients, and adapt to toxic substances in the system (Srivastava, 2020). Variation and imbalance in C/N ratio lead to insufficient production of biogas due to decrement of the specific organic loading rate. Fluctuation in specific organic loading rate resulted in accumulation and proliferation of hydrogen-consuming microorganisms. Constant specific organic loading rate stabilized the reactor for further processing. An observation has been made that at low phosphorus concentration, reduced amount of nitrogen is required by the microbial community in anaerobic fermentation process and vice versa (Mao et al., 2019). Upflow fixed-bed anaerobic reactors using urea as nitrogen source showed that the excessive cell growth caused by the addition of nutrients affected the reactor hydrodynamic pattern, hindering the liquid–gas transfer mass of hydrogen. In addition, the decrease of the hydraulic retention time increased the production of nonreduced compounds (Del Pilar Anzola-Rojas et al., 2015). Table 12.3 shows methane generation from various feedstocks through codigestion process.

12.4.3 Optimum temperature

The optimum temperature for a typical anaerobic process depends on the requirements of the microbial consortium used in the anaerobic digestion. This eventually affects the methane yield (Chae et al., 2008). The three existing categories of temperature-specific microbial communities are psychrophiles (less than 20°C), mesophiles (between 20 and 45°C), thermophiles (between 45 and 60°C), and hyperthermophiles (more than 60°C). Mesophilic systems are stable, economic, energy efficient, and sensitive to the temperature fluctuations unlike thermophilic (Liu et al., 2017). Anaerobic digestion of agricultural organic waste shows improved performance under mesophilic conditions (30–40°C).

TABLE 12.3 Codigestion of different agrowastes.

Agrowaste	Cosubstrate	Operational conditions	Yield	References
Wheat straw	Sheep manure and chicken manure	Different ratios taken as 3:7, 5:5, and 7:3 and biogas production under a mesophilic temperature (35°C)	Methane yield obtained a maximum value of 227.06 L/kg volatile solids when the manure proportion was 52%	Liu et al. (2015)
Corn stover	Chicken manure	CS:CM ratios of 3:1 and 1:1 (on volatile solid basis) maintained at 37°C	The highest methane yield of 218.8 mL/g volatile solids was achieved ratio of 3:1	Li et al. (2013)
Sugarcane bagasse	Food waste	Acid-pretreated sugarcane bagasse with food waste	Maximum biogas production of 7338 mL followed	Vats et al. (2019)
Rice straw	Kitchen waste	Single-stage batch mesophilic (37 ± 1°C) anaerobic digestion at a substrate concentration of 54 g/L based on volatile solids	Maximum biogas yield of 674.4 L/kg volatile solids	Ye et al. (2013)

Biogas formation is also significantly more (Cai et al., 2004; Cho et al., 2013). Low temperature inside the anaerobic digester leads to lower yield of biogas and unstable operational condition inside the reactor (Aryal et al., 2018). Several studies have reported the significance of initial temperature as well as the changes in temperature in the process of anaerobic fermentation. In case of low temperature inside the bioreactor, lower chemical oxygen demand (COD) removal and lower volatile fatty acids (VFA) accumulation take place causing a lower biogas production. However, there are several studies on effect of initial temperature on biogas production. But there have been fewer works on the impact of temperature changes taking place on everyday basis on the propagation of bacteria and the overall methane production. In a recent study, it has been observed that the microbial community dominant during acidogenic phase was *Acinetobacter*, *Acetitomaculum*, and *Bacillus*. *Cenarchaeum* sp. were found to flourish during the methanogenic phase. The growth and activity of the microorganisms decreased significantly at low temperature of 20°C (Wang et al., 2019). Temperature has a crucial role in the growth and metabolic activities of microbial community, thus affecting overall yield and functioning of the anaerobic digester (Alvarez and Lidén, 2008; El-Mashad et al., 2004; Xiong et al., 2012; Rademacher et al., 2012). Under mesophilic and thermophilic conditions, enzyme synthesis and release occurs during the exponential phase, which regulates the anaerobic fermentation efficiency. Changes in the operating temperature affect the microbial metabolism and the community dynamics of the anaerobic digestion system (Li et al., 2017; Meng et al., 2017; Sanz et al., 2017; Yang et al., 2015; Strauber et al., 2016). Still, the temperature requirements might vary from case to case depending on other operational conditions; especially the inoculum. Previous studies show that maximum methane was obtained at 35°C from rice straw (Yan et al., 2015), whereas at 44°C, best methane yield was obtained from corn stover (Liu et al., 2018).

12.4.4 Suitable pH

The microbial growth, its metabolic activities, and thus, the substrate utilization are affected by pH value in the anaerobic reactor. To maintain optimal pH value during various phases, anaerobic digesters are arranged in two- to three-step processes to prevent volatile fatty acid content accumulation. The optimum pH for conducting the metabolic activities and agrowaste degradation is between 6.5 and 7.5. Acidic pH of 4.5 results in reduced ammonia nitrogen (510 mg/L) but increased volatile fatty acid content (27,000 mg/L), which inhibit methanogenic activity (Shanmugam and Horan, 2009; Bolzonella et al., 2020.). Lowering of pH below 5 led to prolonged hydraulic retention time 15 days inside the acidogenic digester (Moestedt et al., 2016), whereas under alkaline conditions, volatile fatty acid generation was significantly decreased, when pH increased from 7 to 11. The pH values between 5.5 and 6.5 were often reported as an optimal range. Mao et al. (2015) revealed the effect of pH on continuous anaerobic digestion. Methanogenic activity is considered to be more, in the range of 6 and 8 with best yield at 7. There was an 88% reduced productivity when the pH fell below 6 (Latif et al., 2017). Similar findings were reported where the slightest change in alkalinity could result in accumulation of volatile fatty acids (Veluchamy et al., 2019). Previous literature reported that keeping an initial pH of 6.5 in dry anaerobic digester batch, hydrolysis and acidogenesis took place majorly giving rise to volatile fatty acid accumulation over 12,000 mg/L (Di Maria et al., 2017). But, during the methanogenesis process, there was an increase of pH up to 8, which reduced the volatile fatty acid to 1000 mg/L. The effect of pH is similar in cases of both dry and wet anaerobic digestion because of similarity of the microbial processes.

12.4.5 Hydraulic retention time

The hydraulic retention time influences the biogas generation, methane yield, pH, and volatile fatty acids. A typical example of wheat straw anaerobic digestion shows that biogas production, cellulose, and hemicellulose degradation increased with increased retention time (Cavinato et al., 2017). Biogas productions increased from 55 mL/g volatile solids in 20 days to 105.2 mL/g volatile solids in 60 days. Similarly, cellulose degradation increased from 43.8% in 20 days to 55.4% in 60 days, and hemicellulose degradation increased from 47.1% in 20 days to 76.8% in 60 days (Shi et al., 2017).

12.4.6 Adequate operational loading

Operational loading rate promises stable digestion and significant biogas yield. In the case of lower C:N ratio and easily degradable lignocellulosic agrowaste, high operational loading rate would lead to the accumulation of volatile fatty acids resulting into acidic pH and inhibition of methanogenic bacteria (Wu et al., 2019). In case of anaerobic codigestion of rice straw and pig manure, biogas obtained was 413 L/kg VS at an organic loading of 3–8 kg VS/m^{3}d. The process was inhibited by foaming and accumulation of volatile fatty acids at increased loading of 12 kg VS/m^{3}d (Li et al., 2015).

12.4.7 Solid state versus submerged state

Total solid content and moisture play a very important role in the metabolic activities that result in methane formation. Wet digestion or submerged state is run at total solids <15%, and dry digestion or solid state is carried out at total solids >15%. The conventional method of anaerobic digestion is wet digestion. Dry digestion has recently gained importance. Both have their pros and cons (Mothe et al., 2013). Wet digestion does not need any complicated mechanical equipment and removal of toxic substances. On the contrary, it requires tedious pretreatment processes, excess water, energy, and reactor capacity. Improved methane yield is obtained due to better mass transport phenomenon (Di Maria et al., 2017).

Dry digestion process is commercially beneficial, as the requirement for energy, reactor capacity, and material handling is minimal. The digestate can be used as a potential fertilizer with consistency and does not need any water recovery process. The dry digestion of wheat straw, corn stalk, and switch grass yielded five-fold higher biogas production than the wet anaerobic digestion (Ge et al., 2016; Sayara and Sánchez, 2019). The only drawback of dry digestion is poor mass transportation, prolonged retention time, and accumulation of volatile fatty acids, hydrogen, and carbon dioxide gases. Partial pressure of hydrogen gas should be limited to 4 atm for the survival of acidogenic bacteria (Wainaina et al., 2019). For practical implementation in the Indian context, mass transfer issues and lower productivity must be addressed along with economical aspects of dry digestion (Singh et al., 2020a,b). Pretreatment and multistage processes can improve the applicability and harness the potential of dry digestion (Luning et al., 2003).

12.4.8 Leachate recirculation

Leachate recirculation technique is an efficient method for anaerobic digestion. A typical Leachate recirculation technique is shown in Fig. 12.3. The digesters with Leachate recirculation hold benefits of both wet digestion and dry digestion. The recirculation improves the organic loading rate enhancing the methane production (Kusch et al., 2012). Such method was found to be useful in maintaining the buffering capacity in solid-state anaerobic digestion of rice straw with ameliorated digestate quality (Mussoline et al., 2012). The syntropy involved in anaerobic digestion steps is affected by the rate of hydrolysis and volatile fatty acids accumulation (Zhou et al., 2017). At higher operational loadings, methanogenic population is outnumbered by acidogenic microbes (Peng et al., 2018). In such cases of imbalanced microbial population, the recirculation of methanogen-

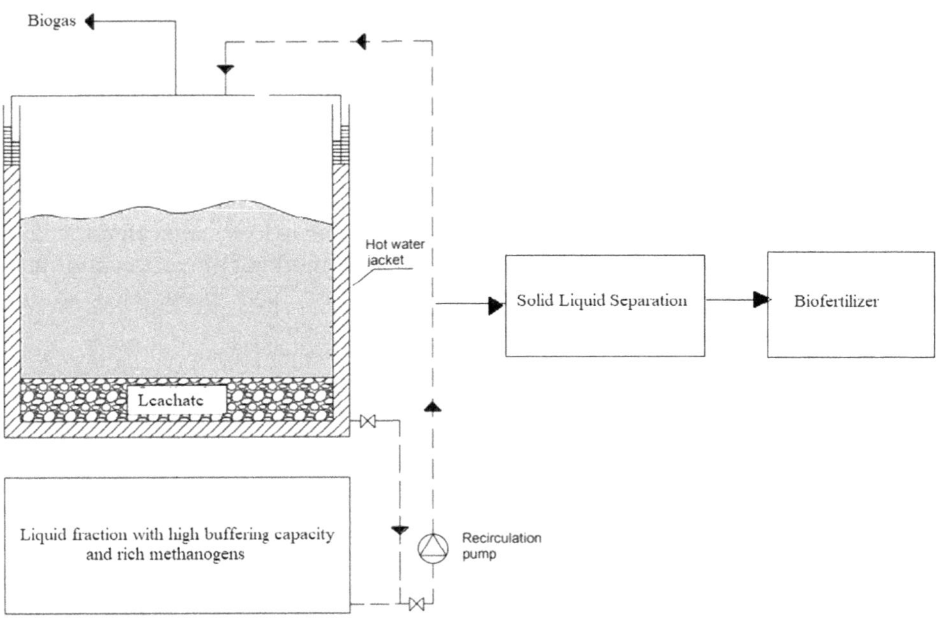

FIGURE 12.3 Leachate recirculation technique for efficient biomethanation.

rich leachate resolves the issue through improved nutritional supply, pH balance, mass transfer, and circulation of enzymes, which enhances the overall process (Pezzolla et al., 2017).

12.4.9 Influence of inoculum

An active inoculum in suitable quantity reduces the overall retention time and leads to efficient anaerobic digestion (Wang et al., 2014). In anaerobic digestion of corn stover, accumulation of volatile fatty acids was seen at low inoculum size (high substrate to inoculum ratio) (Wang et al., 2020). The peculiarity of results indicated that under mesophilic conditions, high inoculum size was preferable for methane formation, but in case of thermophilic conditions, methane formation deteriorated at high inoculum size. This could be attributed to low tolerance limit of free ammonia of thermophilic microbes (Contreas et al., 2012).

Similarly, in case of paddy straw anaerobic digestion, low inoculum led to reduced mass transfer and low methane yield (Zhou et al., 2017). Bacteria and archea are two important microbial groups in anaerobic digestion process (Ziganshin et al., 2011). The syntropy of microbial communities in anaerobic digestion process is shown in Fig. 12.4. The pretreatment process, nature of substrate, and operational conditions affect the growth, metabolism, and diversity of the microbial groups (Kondusamy and Kalamdhad, 2014). The codigestion of rice straw with pig manure resulted in increased C:N ratio, giving rise to hydrogenotrophic methanogenic *Methanothermobacter* (Zhou et al., 2016). Under thermophilic conditions in wet digestion of corn stalks, the hydrogenotrophic methanogens improved than acetotrophic methanogens (Wang et al., 2018). Similar trend was observed for high acetotrophic methanogens (methanosarcinales) in codigestion of rice straw with clay residues and pig manure (Jimenez et al., 2015). The source of methanogenic microbial strains and their adaptation to high concentration of ammonia has been an important strategy for improvement of anaerobic digestion process (Wojcieszak et al., 2017; Sayara and Sánchez, 2019; Kalamaras et al., 2020).

12.4.10 Inhibitors and toxins

Ammonium has a role in pH balance and its stabilization in the reactor. But elevated ammonium concentration may cause inhibition of the anaerobic system. Similarly, the permissible ranges of other parameters are shown in Table 12.4.

12.5 Bioreactor design aspects in anaerobic digestion

Anaerobic reactors are categorized on the basis of their mode of operation and function. The reactors that have been used so far for the

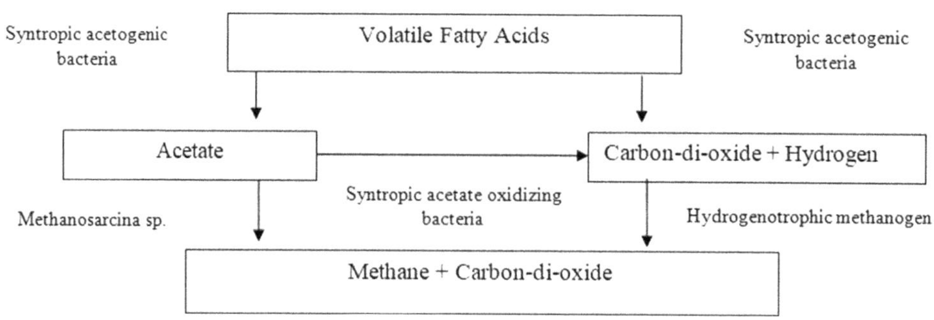

FIGURE 12.4 Syntropy involved in anaerobic digestion.

TABLE 12.4 Inhibitors in methanogenesis process.

Parameter	Range
VFA	More than 10 g L^{-1}
Oleate	More than 1.7 g L^{-1}
NH^3	More than 16 g L^{-1}
S^{2-}	More than 0.15 g L^{-1}
Ca	More than 7 g L^{-1}
Mg	More than 1 g L^{-1}
K	More than 8 g L^{-1}
Na	More than 3.5 g L^{-1}
Fe	More than 0. 005 g L^{-1}

Czatzkowska et al., 2020.

treatment of agrowastes are single-stage or two-stage batch or continuous bioreactors such as upflow anaerobic sludge blanket, sequential batch bioreactor, continuously stirred tank bioreactor, and tubular fluidized bed bioreactor (De Mes et al., 2003; Van et al., 2019; Caruso et al., 2019). Short retention time and acidic pH are inhibitory for methanogenic microbes, while these conditions favor acidogenic/acetogenic bacteria (Ghosh et al., 2016; Bassani et al., 2015). The conventional anaerobic reactors are mostly single-stage batch operation based, which fail to balance the growth of different microbial species with separate growth requirements (Vasudevan et al., 2020). During anaerobic digestion of agricultural wastes, the organic loading rate increases, which lead to extended acidogenic and short methanogenic phase. In such cases, hybrid bioreactors with two reactors operating different stages of anaerobic digestion are useful than the single-phase reactors (Xu et al., 2014). Their usage leads to improved process control and removal of toxins from acidogenic reactor with no cell washout from methanogenic reactor, resulting into enhanced methane production (Hai-Lou et al., 2002). Smooth processing of anaerobic digestion of agrowastes requires more applications of hybrid bioreactors with efficient mass transfer, maintenance of desired microbial population, and continuous operation of the process with high throughput. The characteristic features of anaerobic bioreactor improve the efficiency and performance for better biogas generation. For instance, the inclination toward developing sustainable technologies for agroresidue management by anaerobic digestion has led to the design of various anaerobic digesters, which require low energy and nutrients, and low sludge generation with improved methane production. The integration of membrane units into the anaerobic digesters has helped in complete retention of biomass. Thus, the hydraulic retention time is decoupled from solids retention time (Singh et al., 2020a,b; Dima et al., 2020).

12.5.1 Inhomogeneous system

An anaerobic bioreactor works under inhomogeneous system through solid, liquid, and gaseous phases. Solid phase consists of sludge granules, which play a vital role for successful operation of upflow—anaerobic sludge blanket reactor and expanded granular sludge bed digestion. Sludge granules contain a constant upflow hydraulic regime for considerable bacterial growth and survival. The three-phase separation phenomenon is the most characteristic feature of upflow—anaerobic sludge blanket reactor (Madaleno et al., 2020; Abdelgadir et al., 2014).

12.5.2 Time instability

Anaerobic digestion of agricultural waste consists of various physical, chemical, and physicochemical parameters governing the fermentation process. Therefore, the process design must include the factors, which can directly and indirectly impact the biogas production. A simple example is the anaerobic digestion process where the total solid content was kept 10% under the thermophilic conditions. All other factors were favorable for methane generation except the total ammoniacal nitrogen exceeded 2000 mg N/L. Which led to inhibit microbial growth and prolong digestion period. Thus, the total yield was reduced drastically (Sarker et al., 2018; Alagoz et al., 2015).

12.5.3 Space instability

The bioconversion of agricultural residue by microbial consortia involves specific functions by microorganisms at each step of the digestion process. In case of acetogenesis, *Syntrophobacter wolinii* decomposes propionate, and *Syntrophomonos wolfei* decomposes butyrate. The acid formation is done by *Clostridium* spp., *Peptococcus anaerobius, Lactobacillus,* and *Actinomyces* (Schuchmann et al., 2016). Methanogenesis is carried out in different ways. The breakdown of acetic acid to form carbon dioxide and methane and conversion of carbon dioxide by hydrogen are the basic methods. Methane generation is more in case of carbon dioxide reduction. This has been witnessed in case of submerged state fermentation of agricultural residues. Water acts as a hydrogen donor and helps in methane generation. The challenges of low methane yield and process instability hinder large-scale application of anaerobic digestion (Chow et al., 2020; Shen et al., 2020). Compartmentalized anaerobic reactors are helpful in reduction of space instability (Xing et al., 2014; Kumar and Samadder, 2020).

12.6 CO_2 mitigation and biogas upgrading

Biogas upgradation process involves methane separation from other gases. The secondary gas stream containing carbon dioxide can be treated for various applications such as algae cultivation, grain fumigation, chemical industries, and so on. The carbon dioxide separation and biogas upgradation process can be carried out using different commercial physical and chemical techniques. The presence of CO_2 in biogas reduces its calorific value and increases the expense required for compression as well as transportation. The various techniques are as follows in Table 12.5.

12.7 Anaerobic digestion as a waste valorization strategy in biorefineries

In the current Indian scenario, surplus agrowastes need alternative strategies to harness their potential for value addition in a sustainable manner (Sailwal et al., 2020). This gives rise to zero-waste concept—based biocircular economy (Pathak et al., 2018; Righetti et al., 2020; Donner et al., 2020). The schematic diagram of zero waste biorefinery is depicted in Fig. 12.5. Under this, biorefineries can be set up for different agroresidues to generate value-added products through general chemical extraction or fermentation routes. Afterward, the waste streams and the digestate generated during the processes are utilized in a sustainable way. The ultimate goal of such sequential agrowaste-based biorefineries would be circulating the resources back to the soil as biofertilizer or compost.

The upgraded biogas is a useful substitute to natural gas, and agrowaste-based digestate is a potential alternative to expensive chemical fertilizers (Singhal et al., 2017). Biogas digestate has many alternative applications such as algae production, bioethanol, and biochar production. Advanced membrane technology can be used to separate digestate into different nutrient-rich and soil-specific fertilizer classes such as nitrogen—phosphorus-rich fertilizer, nitrogen—potassium-rich fertilizer, and the like (Al Seadi and Lukehurst, 2012.). Indigenous biocultures can also help in nutrient recovery, and the residual biomass is the food for the aquatic animals (Sobhi et al., 2019).

TABLE 12.5 Upgrading biogas through carbon dioxide removal.

Technology	Application
Pressure swing adsorption	Carbon dioxide adsorbed on activated carbon
Water scrubber	Carbon dioxide adsorbed in water
Chemical adsorption	Carbon dioxide reacts with amine-based solvents
Membrane-based separation	Carbon dioxide separated through membrane
Cryoseparation	Condensation and sublimation of carbon dioxide

Sun et al., 2015.

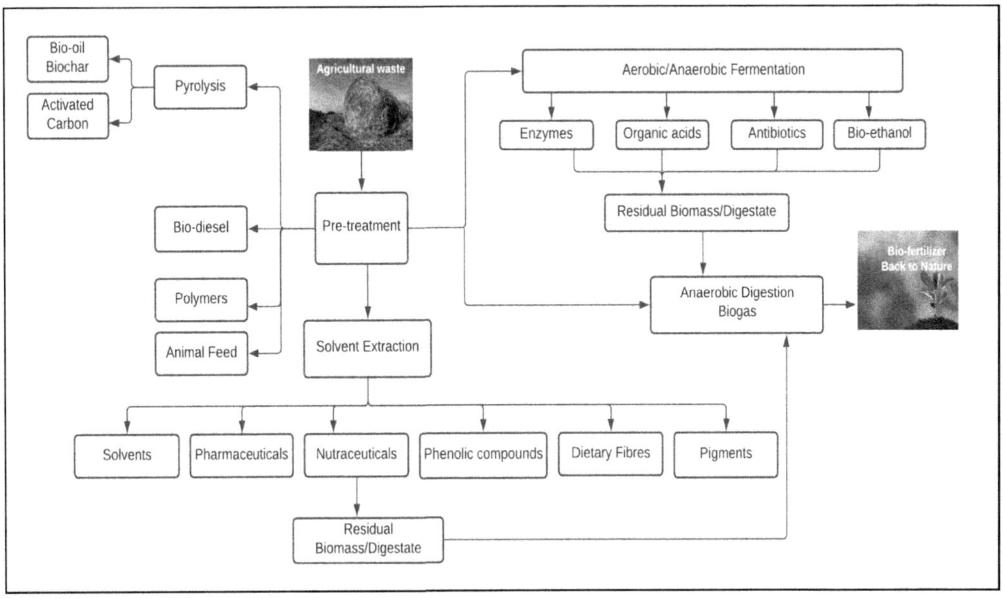

FIGURE 12.5 Agrowaste-based biorefinery.

12.8 Bottlenecks and opportunities of the anaerobic digestion process

India's agronomic policies have an important role in harnessing the potentials of agrowaste for biogas generation. Currently, very small quantity of agrowaste is being utilized for biogas production as compared with their actual potential. The policy think tank of India, NITI Aayog, envisages the solution to crop burning through technological advancement in agriwaste management. The mission of NITI Aayog and SATAT (Sustainable Alternative Towards Affordable Transportation) in India include technological implementation through enterprise-based model of bio-CNG (PIB, 2018). The New National Biogas and Organic Manure Programme (MNRE) in India and Gobardhan Scheme (DDWS) promote the significance of application of biogas in households and small-scale industries and use of biofertilizer from biogas plants. They aim to make the Indian villages self-reliant and encourage the use of clean

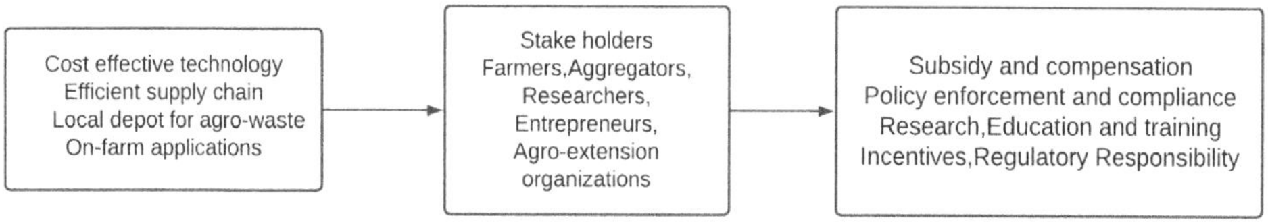

FIGURE 12.6 Roles and responsibilities of stakeholders in a futuristic approach.

fuel through energy generation from food waste, cattle manure, and paddy straw (MNRE, 2018; Mittal et al., 2018).

The responsibilities of various heads of agrowaste management are shown in Fig. 12.6. The implementation challenges of all these government measures lie in technological barriers, variations in quality and availability of agrowastes, need of storage or local depots, infant stage of supply chain infrastructure, and lack of knowledge and awareness. Thorough knowledge is required to understand the government policies, role of industries and supply chain for developing agrowaste-based biogas business models (Ravindranath and Balachandra, 2009; Bhuvaneshwari et al., 2019; Kapoor et al., 2020). Extensive efforts are required to integrate the policies and technological support for economic development in actuality.

12.9 Conclusions

The scope for agrowaste utilization through anaerobic digestion in Indian scenario is promising. Anaerobic digestion of agrowastes to produce biogas and biofertilizer holds the potential for transformation of Indian agronomy. Development of cost-effective strategies for rapid and efficient degradation and utilization of lignocellulosic wastes need to be emphasized. Biological pretreatment methods, indigenous cellulolytic and methanogenic inoculums, solid-state anaerobic digestion, and the use of hybrid reactor systems should be explored and commercialized. Sequential zero-waste biorefineries for agrowaste valorization can open up new and promising avenues for economic development. Such efforts aim to address the Government's mission of Swachh Bharat Mission and "Atmanirbhar Bharata Abhiyan" or "Self-Reliant India Mission." The biofuel policy specifically talked about banning the import and export of biofuels to stimulate the domestic market and avoid arbitrage by traders, who have been known to import cheap ethanol and sell it a higher price in domestic market for fuel blending purposes. The residual biomass from different phases of biorefineries can be a potential substrate for anaerobic digestion, leading to biogas and biofertilizer. To reduce and eliminate the banes of crop burning, there is an immediate need to generate awareness and understanding of the profitable avenues of agrowaste utilization through anaerobic digestion. Proper education would economically empower the farmers, reduce dependency on fuel imports, and ensure sustainable circular economy.

References

Abdelgadir, A., Chen, X., Liu, J., Xie, X., Zhang, J., Zhang, K., Liu, N., 2014. Characteristics, process parameters, and inner components of anaerobic bioreactors. BioMed Res. Int. 2014, 1–7.

Adekunle, K.F., Okolie, J.A., 2015. A review of biochemical process of anaerobic digestion. Adv. Biosci. Biotechnol. 6 (03), 205.

Akyol, Ç., Ince, O., Bozan, M., Ozbayram, E.G., Ince, B., 2019. Biological pretreatment with *Trametes versicolor* to enhance methane production from lignocellulosic biomass: a metagenomic approach. Ind. Crop. Prod. 140, 111659.

Al Afif, R., Wendland, M., Amon, T., Pfeifer, C., 2020. Supercritical carbon dioxide enhanced pre-treatment of cotton stalks for methane production. Energy 194, 116903.

Al Seadi, T., Lukehurst, C., 2012. Quality management of digestate from biogas plants used as fertiliser. IEA Bioenergy 37, 40.

Alagöz, B.A., Yenigün, O., Erdinçler, A., 2015. Enhancement of anaerobic digestion efficiency of wastewater sludge and olive waste: synergistic effect of co-digestion and ultrasonic/microwave sludge pre-treatment. Waste Manag. 46, 182–188.

Alvarez, R., Lidén, G., 2008. The effect of temperature variation on biomethanation at high altitude. Bioresour. Technol. 99, 7278–7284.

Antoniou, N., Monlau, F., Sambusiti, C., Ficara, E., Barakat, A., Zabaniotou, A., 2019. Contribution to circular economy options of mixed agricultural wastes management: coupling anaerobic digestion with gasification for enhanced energy and material recovery. J. Clean. Prod. 209, 505–514.

Aryal, N., Kvist, T., Ammam, F., Pant, D., Ottosen, L., 2018. An overview of microbial biogas enrichment. Bioresour. Technol. 264, 359–369.

Bassani, I., Kougias, P.G., Treu, L., Angelidaki, I., 2015. Biogas upgrading via hydrogenotrophic methanogenesis in two-stage continuous stirred tank reactors at mesophilic and thermophilic conditions. Environ. Sci. Technol. 49 (20), 12585–12593.

Bauer, A., Lizasoain, J., Theuretzbacher, F., Agger, J.W., Rincón, M., Menardo, S., Horn, S.J., 2014. Steam explosion pretreatment for enhancing biogas production of late harvested hay. Bioresour. Technol. 166, 403–410.

Bhuvaneshwari, S., Hettiarachchi, H., Meegoda, J.N., 2019. Crop residue burning in India: policy challenges and potential solutions. Int. J. Environ. Res. Public Health. 16 (5), 832.

Bhuyan, N., Narzari, R., Gogoi, L., Bordoloi, N., Hiloidhari, M., Palsaniya, D.R., Kataki, R., 2020. Valorization of agricultural wastes for multidimensional use. In: Current Developments in Biotechnol. Bioeng. Elsevier, pp. 41–78.

Bolzonella, D., Battista, F., Mattioli, A., Nicolato, C., Frison, N., Lampis, S., 2020. Biological thermophilic post hydrolysis of digestate enhances the biogas production in the anaerobic digestion of agro-waste. Renew.Sustain. Energy Rev. 134, 110174.

Cai, M., Liu, J., Wei, Y., 2004. Enhanced biohydrogen production from sewage sludge with alkaline pretreatment. Environ. Sci. Technol. 38, 3195–3202.

Caruso, M.C., Braghieri, A., Capece, A., Napolitano, F., Romano, P., Galgano, F., Genovese, F., 2019. Recent updates on the use of agro-food waste for biogas production. Appl. Sci. 9 (6), 1217.

Cavinato, C., Da Ros, C., Pavan, P., Bolzonella, D., 2017. Influence of temperature and hydraulic retention on the production of volatile fatty acids during anaerobic fermentation of cow manure and maize silage. Bioresour. Technol. 223, 59–64.

Chae, K.J., Jang, A.M., Yim, S.K., Kim, I.S., 2008. The effects of digestion temperature and temperature shock on the biogas yields from the mesophilic anaerobic digestion of swine manure. Bioresour. Technol. 99 (1), 1–6.

Chen, Y., Cheng, J.J., Creamer, K.S., 2008. Inhibition of anaerobic digestion process: a review. Bioresour. Technol. 99 (10), 4044–4064.

Chen, H., Liu, J., Chang, X., Chen, D., Xue, Y., Liu, P., Lin, H., Han, S., 2017. A review on the pretreatment of lignocellulose for high-value chemicals. Fuel Process. Technol. 160, 196–206.

Chen, X., Zhang, Y., Gu, Y., Liu, Z., Shen, Z., Chu, H., Zhou, X., 2014. Enhancing methane production from rice straw by extrusion pretreatment. Appl. Energy 122, 34–41.

Cho, S.K., Im, W.T., Kim, D.H., Kim, M.H., Shin, H.S., Oh, S.E., 2013. Dry anaerobic digestion of food waste under mesophilic conditions: performance and methanogenic community analysis. Bioresour. Technol. 131, 210–217.

Chow, W.L., Chong, S., Lim, J.W., Chan, Y.J., Chong, M.F., Tiong, T.J., Pan, G.T., 2020. Anaerobic co-digestion of wastewater sludge: a review of potential co-substrates and operating factors for improved methane yield. Processes 8 (1), 39.

Contreras, L.M., Schelle, H., Sebrango, C.R., Pereda, I., 2012. Methane potential and biodegradability of rice straw, rice husk and rice residues from the drying process. Water Sci. Technol. 65 (6), 1142–1149.

Czatzkowska, M., Harnisz, M., Korzeniewska, E., Koniuszewska, I., 2020. Inhibitors of the methane fermentation process with particular emphasis on the microbiological aspect: a review. Energy Sci. Eng. 8 (5), 1880–1897.

Das, A., Mondal, C., Roy, S., 2015. Pretreatment methods of ligno-cellulosic biomass: a review. J. Eng. Sci. Technol. Rev. 8 (5), 141–165.

De Francisci, D., Kougias, P.G., Treu, L., Campanaro, S., Angelidaki, I., 2015. Microbial diversity and dynamicity of biogas reactors due to radical changes of feedstock composition. Bioresour. Technol. 176, 56–64.

De Mes, T.Z.D., Stams, A.J.M., Reith, J.H., Zeeman, G., 2003. Methane production by anaerobic digestion of wastewater and solid wastes. Bio-methane Bio-Hydrogen 58–102.

Del Pilar Anzola-Rojas, M., da Fonseca, S.G., da Silva, C.C., de Oliveira, V.M., Zaiat, M., 2015. The use of the carbon/nitrogen ratio and specific organic loading rate as tools for improving biohydrogen production in fixed-bed reactors. Biotechnol. Rep. 5, 46–54.

Di Maria, F., Barratta, M., Bianconi, F., Placidi, P., Passeri, D., 2017. Solid anaerobic digestion batch with liquid digestate recirculation and wet anaerobic digestion of organic waste: comparison of system performances and identification of microbial guilds. Waste Manag. 59, 172–180.

Dima, A.D., Pârvulescu, O.C., Mateescu, C., Dobre, T., 2020. Optimization of substrate composition in anaerobic co-digestion of agricultural waste using central composite design. Biomass Bioenergy 138, 105602.

Dixit, S., Yadav, V.L., 2020. Comparative study of polystyrene/chemically modified wheat straw composite for green packaging application. Polym. Bull. 77 (3), 1307–1326.

Donner, M., Gohier, R., de Vries, H., 2020. A new circular business model typology for creating value from agro-waste. Sci. Total Environ. 716, 137065.

El-Mashad, H.M., Zeeman, G., van Loon, W.K., Bot, G.P., Lettinga, G., 2004. Effect of temperature and temperature fluctuation on thermophilic anaerobic digestion of cattle manure. Bioresour. Technol. 95, 191–201.

Fang, Q., Ji, S., Huang, D., Huang, Z., Huang, Z., Zeng, Y., Liu, Y., 2019. Impact of alkaline pretreatment to enhance volatile fatty acids (VFAs) production from rice husk. Biochem. Res. Int. 2019, 8489747.

Ge, X., Xu, F., Li, Y., 2016. Solid-state anaerobic digestion of lignocellulosic biomass: recent progress and perspectives. Bioresour. Technol. 205, 239–249.

Ghosh, S., Chowdhury, R., Bhattacharya, P., 2016. Mixed consortia in bioprocesses: role of microbial interactions. Appl. Microbiol. Biotech. 100 (10), 4283–4295.

Greses, S., Tomás-Pejó, E., Gónzalez-Fernández, C., 2020. Agroindustrial waste as a resource for volatile fatty acids production via anaerobic fermentation. Bioresour.Technol. 297, 122486.

Hagos, K., Zong, J., Li, D., Liu, C., Lu, X., 2017. Anaerobic co-digestion process for biogas production: progress, challenges and perspectives. Renew. Sustain. Energy Rev. 76, 1485–1496.

Hai-Lou, X., Jing-Yuan, W., Joo-Hwa, T., 2002. A hybrid anaerobic solid-liquid bioreactor for food waste digestion. Biotechnol. Lett. 24 (10), 757–761. http://agricoop.nic.in/sites/default/files/NPMCR_1.pdf.

Jiménez, J., Guardia-Puebla, Y., Cisneros-Ortiz, M.E., Morgan-Sagastume, J.M., Guerra, G., Noyola, A., 2015. Optimization of the specific methanogenic activity during the anaerobic co-digestion of pig manure and rice straw, using industrial clay residues as inorganic additive. Chem. Eng. Technol. 259, 703–714.

Junchen, L., Irfan, M., Lin, F., 2012. Bioconversion of agricultural waste to ethanol: a potential source of energy. Arch. Sci. 65 (12), 626–642.

Kalamaras, S.D., Vasileiadis, S., Karas, P., Angelidaki, I., Kotsopoulos, T.A., 2020. Microbial adaptation to high ammonia concentrations during anaerobic digestion of manure-based feedstock: biomethanation and 16S rRNA gene sequencing. J. Chem. Technol. Biotechnol. 95 (7), 1970–1979.

Kapoor, R., Ghosh, P., Kumar, M., Sengupta, S., Gupta, A., Kumar, S.S., Pant, D., 2020. Valorization of agricultural waste for biogas based circular economy in India: a research outlook. Bioresour. Technol. 304, 123036.

Kondusamy, D., Kalamdhad, A.S., 2014. Pre-treatment and anaerobic digestion of food waste for high rate methane production–A review. J. Environ. Chem. Eng. 2 (3), 1821–1830.

Kumar, A., Samadder, S.R., 2020. Performance evaluation of anaerobic digestion technology for energy recovery from organic fraction of municipal solid waste: a review. Energy 197, 117253.

Kusch, S., Oechsner, H., Jungbluth, T., 2012. Effect of various leachate recirculation strategies on batch anaerobic digestion of solid substrates. Int. J. Environ. Waste Manag. 9 (1–2), 69–88.

Latif, M.A., Mehta, C.M., Batstone, D.J., 2017. Influence of low pH on continuous anaerobic digestion of waste activated sludge. Water Res. 113, 42–49.

Li, D., Liu, S., Mi, L., Li, Z., Yuan, Y., Yan, Z., Liu, X., 2015. Effects of feedstock ratio and organic loading rate on the anaerobic mesophilic co-digestion of rice straw and pig manure. Bioresour. Technol. 187, 120–127.

Li, P., He, C., Li, G., Ding, P., Lan, M., Gao, Z., Jiao, Y., 2020. Biological pretreatment of corn straw for enhancing degradation efficiency and biogas production. Bioengineered 11 (1), 251–260.

Li, W., Guo, J., Cheng, H., Wang, W., Dong, R., 2017. Two-phase anaerobic digestion of municipal solid wastes enhanced by hydrothermal pretreatment: viability, performance and microbial community evaluation. Appl. Energy. 189, 613–622.

Li, Y., Zhang, R., Chen, C., Liu, G., He, Y., Liu, X., 2013. Biogas production from co-digestion of corn stover and chicken manure under anaerobic wet, hemi-solid, and solid state conditions. Bioresour. Technol. 149, 406–412.

Liu, C., Li, H., Zhang, Y., Liu, C., 2016. Improve biogas production from low-organic-content sludge through high-solids anaerobic co-digestion with food waste. Bioresour. Technol. 219, 252–260.

Liu, C., Wachemo, A.C., Tong, H., Shi, S., Zhang, L., Yuan, H., Li, X., 2018. Biogas production and microbial community properties during anaerobic digestion of corn stover at different temperatures. Bioresour. Technol. 261, 93–103.

Liu, L., Zhang, T., Wan, H., Chen, Y., Wang, X., Yang, G., Ren, G., 2015. Anaerobic co-digestion of animal manure and wheat straw for optimized biogas production by the addition of magnetite and zeolite. Energy Convers. Manag. 97, 132–139.

Loganath, R., Senophiyah-Mary, J., 2020. Critical review on the necessity of bioelectricity generation from slaughterhouse industry waste and wastewater using different anaerobic digestion reactors. Renew. Sustain. Energy Rev. 134, 110360.

Luning, L., Van Zundert, E.H.M., Brinkmann, A.J.F., 2003. Comparison of dry and wet digestion for solid waste. Water Sci. Technol. 48 (4), 15–20.

Madaleno, L.L., de Barros, V.G., Kesserling, M.A., Teixeira, J.R., Duda, R.M., de Oliveira, R.A., 2020. The recycling of biodigested vinasse in an upflow anaerobic sludge blanket reactor is a feasible approach for the conservation of freshwater in the biofuel ethanol industry. J. Clean. Prod. 262, 121196.

Mao, C., Feng, Y., Wang, X., Ren, G., 2015. Review on research achievements of biogas from anaerobic digestion. Renew. Sustain. Rev. 45, 540–555.

Mao, C., Wang, Y., Wang, X., Ren, G., Yuan, L., Feng, Y., 2019. Correlations between microbial community and C:N:P stoichiometry during the anaerobic digestion process. Energy 174, 687–695.

Meng, X., Yuan, X., Ren, J., Wang, X., Zhu, W., Cui, Z., 2017. Methane production and characteristics of the microbial community in a two-stage fixed-bed anaerobic reactor using molasses. Bioresour. Technol. 241, 1050–1059.

Mittal, S., Ahlgren, E.O., Shukla, P.R., 2018. Barriers to biogas dissemination in India: a review. Energy Policy 112, 361–370.

Mittal, S., Ahlgren, E.O., Shukla, P.R., 2019. Future biogas resource potential in India: a bottom-up analysis. Renew. Energy. 141, 379–389.

MNRE, 2018. Guidelines for Implementation of the Central Sector Scheme, New National Biogas and Organic Manure Programme (NNBOMP) during the Period 2017–18 to 2019–20. https://mnre.gov.in/scheme/Scheme %20of%20Biogas.

Moestedt, J., Nordell, E., Hallin, S., Schnürer, A., 2016. Two-stage anaerobic digestion for reduced hydrogen sulphide production. J. Chem. Tech. Biotechnol. 91 (4), 1055–1062.

Motte, J.C., Escudié, R., Hamelin, J., Steyer, J.P., Bernet, N., Delgenes, J.P., Dumas, C., 2014. Substrate milling pretreatment as a key parameter for solid-state anaerobic digestion optimization. Bioresour. Technol. 173, 185–192.

Motte, J.C., Trably, E., Escudié, R., Hamelin, J., Steyer, J.P., Bernet, N., Dumas, C., 2013. Total solids content: a key parameter of metabolic pathways in dry anaerobic digestion. Biotechnol. Biofuels. 6 (1), 164.

Mussoline, W., Esposito, G., Lens, P., Garuti, G., Giordano, A., 2012. Design considerations for a farm-scale biogas plant based on pilot-scale anaerobic digesters loaded with rice straw and piggery wastewater. Biomass Bioenergy 46, 469–478.

Mustafa, A.M., Poulsen, T.G., Sheng, K., 2016. Fungal pretreatment of rice straw with *Pleurotus ostreatus* and *Trichoderma reesei* to enhance methane production under solid-state anaerobic digestion. Appl.Energy. 180, 661–671.

NPMCR, 2018. NPMCR. https://agricoop.nic.in/sites/default/files/NPMCR_1.pdf.

Park, S., Han, S.K., Song, E., Kim, H., Kim, M., Lee, W., 2020. Effect of hydrothermal pre-treatment on physical properties and co-digestion from food waste and sewage sludge mixture. Waste Manag. Res. 38 (5), 546–553.

Pathak, P.D., Mandavgane, S.A., Puranik, N.M., Jambhulkar, S.J., Kulkarni, B.D., 2018. Valorization of potato peel: a biorefinery approach. Crit. Rev. Biotechnol. 38 (2), 218–230.

Pellera, F.M., Gidarakos, E., 2018. Chemical pretreatment of lignocellulosic agroindustrial waste for methane production. Waste Manag. 71, 689–703.

Peng, W., Pivato, A., Lavagnolo, M.C., Raga, R., 2018. Digestate application in landfill bioreactors to remove nitrogen of old landfill leachate. Waste Manag. 74, 335–346.

Pezzolla, D., Di Maria, F., Zadra, C., Massaccesi, L., Sordi, A., Gigliotti, G., 2017. Optimization of solid-state anaerobic digestion through the percolate recirculation. Biomass Bioenergy 96, 112–118.

PIB, 2018. Petroleum Minister to Launch SATAT Initiative to Promote Compressed Bio-Gas as an Alternative, Green Transport Fuel. https://pib.gov.in/newsite/PrintRelease.aspx?relid=183787. (Accessed 16 November 2019).

Prasad, M., Ranjan, R., Ali, A., Goyal, D., Yadav, A., Singh, T.B., Dantu, P.K., 2020. Efficient transformation of agricultural waste in India. In: Contaminants in Agriculture. Springer, Cham, pp. 271–287.

Prasad, S., Singh, A., Joshi, H.C., 2007. Ethanol as an alternative fuel from agricultural, industrial and urban residues. Resour. Conserv. Recycl. 50 (1), 1–39.

Rademacher, A., Nolte, C., Schonberg, M., Klocke, M., 2012. Temperature increases from 55 to 75°C in a two-phase biogas reactor result in fundamental alterations within the bacterial and archaeal community structure. Appl. Microbiol. Biotechnol. 96, 565–576.

Rajendran, K., Drielak, E., Varma, V.S., Muthusamy, S., Kumar, G., 2018. Updates on the pretreatment of lignocellulosic feedstocks for bioenergy production–a review. Biomass Convers. Biorefin. 8 (2), 471–483.

Rajput, A.A., Visvanathan, C., 2018. Effect of thermal pretreatment on chemical composition, physical structure and biogas production kinetics of wheat straw. J. Environ. Manage. 221, 45–52.

Ravindranath, N.H., Balachandra, P., 2009. Sustainable bioenergy for India: technical, economic and policy analysis. Energy 34 (8), 1003–1013.

Righetti, E., Nortilli, S., Fatone, F., Frison, N., Bolzonella, D., 2020. A multiproduct biorefinery approach for the production of hydrogen, methane and volatile fatty acids from agricultural waste. Waste Biomass Valorization 11, 5239–5246, 1-8.

Rouches, E., Herpoël-Gimbert, I., Steyer, J.P., Carrere, H., 2016. Improvement of anaerobic degradation by white-rot fungi pretreatment of lignocellulosic biomass: a review. Renew. Sustain. Energy. Rev. 59, 179–198.

Sailwal, M., Banerjee, A., Bhaskar, T., Ghosh, D., 2020. Integrated biorefinery concept for Indian paper and pulp industry. In: Waste Biorefinery. Elsevier, pp. 631–658.

Sanz, J.L., Rojas, P., Morato, A., Mendez, L., Ballesteros, M., González-Fernández, C., 2017. Microbial communities of biomethanization digesters fed with raw and heat pre-treated microalgae biomasses. Chemosphere 168, 1013–1021.

Saratale, G.D., Chen, S.D., Lo, Y.C., Saratale, R.G., Chang, J.S., 2008. Outlook of biohydrogen production from lignocellulosic feedstock using dark fermentation—a review. J. Sci. Ind. Res. 67 (2008), 962.

Sarker, S., Lamb, J.J., Hjelme, D.R., Lien, K.M., 2018. Overview of recent progress towards in-situ biogas upgradation techniques. Fuel 226, 686–697.

Sayara, T., Sánchez, A., 2019. A review on anaerobic digestion of lignocellulosic wastes: pretreatments and operational conditions. Appl. Sci. 9 (21), 4655.

Scherzinger, M., Kaltschmitt, M., 2021. Thermal pretreatment options to enhance anaerobic digestibility—A review. Renew. Sustain. Energy Rev. 137, 110627.

Schuchmann, K., Müller, V., 2016. Energetics and application of heterotrophy in acetogenic bacteria. Appl. Environ. Microbiol. 82 (14), 4056–4069.

Shanmugam, P., Horan, N.J., 2009. Optimising the biogas production from leather fleshing waste by co-digestion with MSW. Bioresour. Technol. 100 (18), 4117–4120.

Sharma, H.K., Xu, C., Qin, W., 2019. Biological pretreatment of lignocellulosic biomass for biofuels and bioproducts: an overview. Waste Biomass Valorization 10 (2), 235–251.

Shen, R., Jing, Y., Feng, J., Luo, J., Yu, J., Zhao, L., 2020. Performance of enhanced anaerobic digestion with different pyrolysis biochars and microbial communities. Bioresour. Technol. 296, 122354.

Shi, X.S., Dong, J.J., Yu, J.H., Yin, H., Hu, S.M., Huang, S.X., Yuan, X.Z., 2017. Effect of hydraulic retention time on anaerobic digestion of wheat straw in the semicontinuous continuous stirred-tank reactors. Biomed. Res. Int. 2017, 2457805.

Silva, G.G.D., Couturier, M., Berrin, J.G., Buléon, A., Rouau, X., 2012. Effects of grinding processes on enzymatic degradation of wheat straw. Bioresour. Technol. 103 (1), 192–200.

Singh, A., Basak, P., 2019. Economic and environmental evaluation of rice straw processing technologies for energy generation: a case study of Punjab, India. J. Clean. Prod. 212, 343–352.

Singh, B., Szamosi, Z., Siménfalvi, Z., 2020a. Impact of mixing intensity and duration on biogas production in an anaerobic digester: a review. Crit. Rev. Biotechnol. 40 (4), 508–521.

Singh, B., Szamosi, Z., Siménfalvi, Z., Rosas-Casals, M., 2020b. Decentralized biomass for biogas production. Evaluation and potential assessment in Punjab (India). Energy Rep. 6, 1702–1714.

Singhal, S., Agarwal, S., Arora, S., Sharma, P., Singhal, N., 2017. Upgrading techniques for transformation of biogas to bio-CNG: a review. Int. J. Energy Res. 41 (12), 1657–1669.

Sobhi, M., Guo, J., Cui, X., Sun, H., Li, B., Aboagye, D., Dong, R., 2019. A promising strategy for nutrient recovery using heterotrophic indigenous microflora from liquid biogas digestate. Sci. Total Environ. 690, 492–501.

Sołowski, G., Konkol, I., Cenian, A., 2020. Production of hydrogen and methane from lignocellulose waste by fermentation. A review of chemical pretreatment for enhancing the efficiency of the digestion process. J. Clean. Prod. 267, 121721.

Song, Z., Liu, X., Yan, Z., Yuan, Y., Liao, Y., 2014. Comparison of seven chemical pretreatments of corn straw for improving methane yield by anaerobic digestion. PloS One 9 (4), e93801.

Srivastava, S.K., 2020. Advancement in biogas production from the solid waste by optimizing the anaerobic digestion. WSDE 2 (2), 85–103.

Strauber, H., Lucas, R., Kleinsteuber, S., 2016. Metabolic and microbial community dynamics during the anaerobic digestion of maize silage in a two-phase process. Appl. Microbiol. Biotechnol. 100, 479–491.

Sukhesh, M.J., Rao, P.V., 2018. Anaerobic digestion of crop residues: technological developments and environmental impact in the Indian context. Biocatal. Agric. Biotechnol. 16, 513–528.

Sun, Q., Li, H., Yan, J., Liu, L., Yu, Z., Yu, X., 2015. Selection of appropriate biogas upgrading technology-a review of biogas cleaning, upgrading and utilisation. Renew. Sustain. Energy Rev. 51, 521–532.

Van, D.P., Fujiwara, T., Leu Tho, B., Toan, S., Phu, P., Hoang Minh, G., Minh, G.H., 2019. A review of anaerobic digestion systems for biodegradable waste: configurations, operating parameters, and current trends. Environ. Eng. Res. 25 (1), 1–17.

Vasudevan, Y., Govindharaj, D., Udayakumar, G.P., Ganesan, A., Sivarajasekar, N., 2020. A review on the production of biogas from biological sources. In: Sustainable Development in Energy and Environment. Springer, Singapore, pp. 1–12.

Vats, N., Khan, A.A., Ahmad, K., 2019. Observation of biogas production by sugarcane bagasse and food waste in different composition combinations. Energy 185, 1100–1105.

Veluchamy, C., Kalamdhad, A.S., 2017. Enhanced methane production and its kinetics model of thermally pretreated lignocellulose waste material. Bioresour. Technol. 241, 1–9.

Veluchamy, C., Gilroyed, B.H., Kalamdhad, A.S., 2019. Process performance and biogas production optimizing of mesophilic plug flow anaerobic digestion of corn silage. Fuel 253, 1097–1103.

Wainaina, S., Awasthi, M.K., Horváth, I.S., Taherzadeh, M.J., 2020. Anaerobic digestion of food waste to volatile fatty acids and hydrogen at high organic loading rates in immersed membrane bioreactors. Renew. Energy. 152, 1140–1148.

Wainaina, S., Lukitawesa, Kumar Awasthi, M., Taherzadeh, M.J., 2019. Bioengineering of anaerobic digestion for volatile fatty acids, hydrogen or methane production: a critical review. Bioengineered 10 (1), 437–458.

Wan, C., Li, Y., 2012. Fungal pretreatment of lignocellulosic biomass. Biotech. Adv. 30 (6), 1447–1457.

Wang, K., Yin, J., Shen, D., Li, N., 2014. Anaerobic digestion of food waste for volatile fatty acids (VFAs) production with different types of inoculum: effect of pH. Bioresour. Technol. 161, 395–401.

Wang, P., Wang, H., Qiu, Y., Ren, L., Jiang, B., 2018. Microbial characteristics in anaerobic digestion process of food waste for methane production–A review. Bioresour. Technol. 248, 29–36.

Wang, S., Ma, F., Ma, W., Wang, P., Zhao, G., Lu, X., 2019. Influence of temperature on biogas production efficiency and microbial community in a two-phase anaerobic digestion system. Water 11 (1), 133.

Wang, Y., Zhang, J., Li, Y., Jia, S., Song, Y., Sun, Y., et al., 2020. Methane production from the co-digestion of pig manure and corn stover with the addition of cucumber residue: role of the total solids content and feedstock-to-inoculum ratio. Bioresour. Technol. 306, 123172.

Wen, B., Yuan, X., Li, Q.X., Liu, J., Ren, J., Wang, X., Cui, Z., 2015. Comparison and evaluation of concurrent saccharification and anaerobic digestion of Napier grass after pretreatment by three microbial consortia. Bioresour. Technol. 175, 102–111.

Wojcieszak, M., Pyzik, A., Poszytek, K., Krawczyk, P.S., Sobczak, A., Lipinski, L., Drewniak, L., 2017. Adaptation of methanogenic inocula to anaerobic digestion of maize silage. Front. Microbiol. 8, 1881.

Wu, D., Li, L., Zhao, X., Peng, Y., Yang, P., Peng, X., 2019. Anaerobic digestion: a review on process monitoring. Renew. Sust. Energ. Rev. 103, 1–12.

Xing, Y.J., Ji, J.Y., Zheng, P., Zhang, J.Q., Ghulam, A., 2014. Microbial consortium and its spatial distribution in a compartmentalized anaerobic reactor. Appl. Microbiol. Biotech. 98 (3), 1357–1366.

Xiong, H., Chen, J., Wang, H., Shi, H., 2012. Influences of volatile solid concentration, temperature and solid retention time for the hydrolysis of waste activated sludge to recover volatile fatty acids. Bioresour. Technol. 119, 285–292.

Xu, Q., Jin, X., Ma, Z., Tao, H., Ko, J.H., 2014. Methane production in simulated hybrid bioreactor landfill. Bioresour.Technol. 168, 92–96.

Yan, Z., Song, Z., Li, D., Yuan, Y., Liu, X., Zheng, T., 2015. The effects of initial substrate concentration, C/N ratio, and temperature on solid-state anaerobic digestion from composting rice straw. Bioresour. Technol. 177, 266–273.

Yang, Z., Liu, Y., Zhang, J., Mao, K., Kurbonova, M., Liu, G., Wang, W., 2020. Improvement of biofuel recovery from food waste by integration of anaerobic digestion, digestate pyrolysis and syngas biomethanation under mesophilic and thermophilic conditions. J. Clean. Prod. 256, 120594.

Ye, J., Li, D., Sun, Y., Wang, G., Yuan, Z., Zhen, F., Wang, Y., 2013. Improved biogas production from rice straw by co-digestion with kitchen waste and pig manure. Waste Manag. 33 (12), 2653–2658.

Yenigün, O., Demirel, B., 2013. Ammonia inhibition in anaerobic digestion: a review. Process Biochem. 48 (5–6), 901–911.

Zeynali, R., Khojastehpour, M., Ebrahimi-Nik, M., 2017. Effect of ultrasonic pre-treatment on biogas yield and specific energy in anaerobic digestion of fruit and vegetable wholesale market wastes. Sustain. Environ. Res. 27 (6), 259–264.

Zhang, H., Ning, Z., Khalid, H., Zhang, R., Liu, G., Chen, C., 2018. Enhancement of methane production from Cotton Stalk using different pretreatment techniques. Sci. Rep. 8 (1), 1–9.

Zhou, J., Yang, J., Yu, Q., Yong, X., Xie, X., Zhang, L., Jia, H., 2017. Different organic loading rates on the biogas production during the anaerobic digestion of rice straw: a pilot study. Bioresour. Technol. 244, 865–871.

Zhou, S., Nikolausz, M., Zhang, J., Riya, S., Terada, A., Hosomi, M., 2016. Variation of the microbial community in thermophilic anaerobic digestion of pig manure mixed with different ratios of rice straw. J. Biosci. Bioeng. 122 (3), 334–340.

Ziganshin, A.M., Schmidt, T., Scholwin, F., Il'inskaya, O.N., Harms, H., Kleinsteuber, S., 2011. Bacteria and archaea involved in anaerobic digestion of distillers grains with solubles. Appl. Microbiol. Biotechnol. 89 (6), 2039–2052.

CHAPTER

13

Biomass gasification: a step toward cleaner fuel and chemicals

Mahendra Ram[1,2], *Monoj Kumar Mondal*[3]

[1]Indian Institute of Technology (Banaras Hindu University), Varanasi, Uttar Pradesh, India; [2]BIT Sindri, Dhanbad, Jharkhand, India; [3]Department of Chemical Engineering and Technology, Indian Institute of Technology, Banaras Hindu University, Varanasi, Uttar Pradesh, India

13.1 Introduction

One of the globally available and most wasted renewable sources of energy are wide varieties of lignocellulosic biomasses (Liew et al., 2021). It has been main energy source from the time of discovery of fire, and it is still being used for heating and cooking purposes in rural areas. Biomasses are easily available everywhere on the earth. Furthermore, increasing population of world is in demand of huge amount of food, which generates agricultural biomass. Thus, agricultural wastes are increasing day by day, which is inevitable and cannot be abolished (Ram and Mondal, 2019a,b).

During photosynthesis, the solar energy with the combination of CO_2 and water is stored in the form of chemical energy as bonds of the structural components of biomass. And thus, to harness these energies, which are stored in chemical bonds, biomass must be processed effectively. Furthermore, the biomass can be processed through thermochemical or biological processes, which may also release water vapor and CO_2 along with energy. By this way, CO_2 is available for the new plants making the process cyclic. With the production of heat and power, biomass can also be used for the production of chemicals or liquid fuels. These liquid fuels can also be used for the conveyance. During photosynthesis, the plant receives CO_2 from the atmosphere for its growth, which is released during combustion of biomass, which can be said to be CO_2 neutral emission. Also, since biomass is available all over the globe, its supply is secured than that of fossil fuels, which are concentrated in some specific places. However, other alternatives of fossil fuels such as solar PV, hydropower, wind, nuclear energy (Chai et al., 2021; Brogioli and Mantia, 2021), and so on are available for the production of heat and power; but for the production of hydrocarbon chemicals, these alternatives cannot be used, besides biomass. Nevertheless, the low energy density of the biomass requires development and improvement in conversion technologies to increase process efficiency and reduce pollution.

Gasification is a thermochemical process that converts carbonaceous material into useful and convenient gaseous fuels or chemical feedstock.

Biofuels and Bioenergy
https://doi.org/10.1016/B978-0-323-85269-2.00008-3

Biomass has many promising candidacies such as large availability, easy accessibility, lower emission of SO_X and NO_X, and carbon neutrality. There are some plants operating for the syn gas production. However, researchers and scientists are still trying to find environmentally benign and economically sustainable methods to optimize the gasification process over production cost and syn gas quality.

13.2 Biomass

Anything that is now alive or alive just a short time ago is referred as biomass. It includes living species such as plants and animals. It is formed as soon as a seed sprouts or an organism is born. Biomass reproduces, and it does not take long time to grow; thus, biomass is considered as renewable. Furthermore, animals depend on plants or on other animals for their food. Plants absorb CO_2 in the presence of sunlight for photosynthesis. And during the combustion, it liberates CO_2 that has been absorbed during photosynthesis; thus it has also been considered as carbon neutral (Basu, 2010).

Biomass includes various materials such as wood waste, bagasse, agricultural residues like sawdust, grass, food processing waste, animal waste, aquatic plants, and so on (Yaman, 2004). The major constituents of biomass are carbon, hydrogen, oxygen, and nitrogen. However, biomasses, which are not used as a food or feed, are most commonly used for the energy generation through various processes. Some of the categorizations of biomass have been listed in Table 13.1.

All living creatures need food to sustain their life. To fulfill the requirement of food, farmers need to grow crops, which generates huge amount of wastes. These wastes generate negative economic value to the farmers.

13.2.1 Generation

Biomass comprises a broad range of different kinds of biomaterials, such as wood, forest and agricultural residues, waste from wood and food industry, algae, energy grasses, straw, bagasse, sewage sludge, and so on. Rice and wheat crops produce approximately 34% and 22% of total mass as waste. It has been reported average 500 million tons biomass generated from agricultural residue per year (Bimbraw, 2019). As per the US Department of Energy, currently the United States can produce dry biomass approximately 702 million tons in a year, which will increase to 1147 million tons dry biomass by the year 2030. The 2016 Billion-Ton Report also estimates that the United States could produce 60 dollars per dry ton of dry biomass (DOE, 2017).

13.2.2 Current disposal activities

Bimbraw (2019) estimated that Punjab, a state of India, produced about 22.9 million tons and 23.1 million tons as crop residues of rice and wheat in which paddy straw (95% by volume) and wheat straw (23% by volume) were burnt in the fields (Bimbraw, 2019). At some places, crop residues left over in fields to degrade naturally, which takes time and emits CH_4, which is 100 times more toxic than CO_2, and it is also responsible for global warming (Ram and Mondal, 2018).

13.2.3 Physicochemical properties

Proximate and ultimate analyses are generally used to characterize solid fuel composition. Proximate analysis is used to determine fixed carbon, moisture, ash, and volatile matter, whereas ultimate analysis is used for the determination of main chemical elements, i.e., C, H, O, N, and S. Thermochemical reactions take place with these main chemical elements. Some mineral contents are also present in biomasses, which are converted into ash during thermochemical processes. Ash is an inert material left over after gasification as solid residue, which

TABLE 13.1 Categorizations of biomass and sources.

Sl. no.	Types of biomass	Sources
1	Woody biomass	Branches, chips, pellets, whole wood of trees, etc.
2	Herbaceous biomass	Residual part of cereal crops, pastures, oilseed crops, tubers and legumes, flowers, etc.
3	Municipal waste	Food waste, paper, cardboard, textiles, wood, parks, etc.
4	Agroindustrial wastes	Waste from the fruit canning industry, fruit mixtures, bagasse and husks, etc.

Islas, J., Manzini F., Masera, O., Vargas, V., 2019. The Role of Bioenergy in the Bioeconomy: Resources, Technologies, Sustainability and Policy. Solid Biomass to Heat and Power. 145–177 (Chapter 4(Islas et al., 2019)).

decreases the operational energy significance of a feedstock. Ash sintering and bed bridging are some technical issues, which are faced during the gasification of nonwoody biomass. It also faces tar collection problem (Guo et al., 2014; Gai and Dong, 2012).

Table 13.2 shows the characteristics of some woody and nonwoody biomass, which was taken as feedstock for gasification by researcher available in literature. Generally the higher energy content in solid fuels is found with higher carbon content. However, in reverse, high amount of moisture and ash would lead to decrease of energy content in nonwoody biomass. Beside these, the gasification is influenced by not only the carbon content but the component of moisture in elemental form (hydrogen and oxygen) also with some other oxidants present in gasification process. These are altogether responsible to form H_2, CH_4, and CO in the produced fuel gas. Furthermore, low density is generally one additional issue with nonwoody biomass particularly with herbaceous plant or the biomass initiating from grasses. This would create problems in handling mostly to control feed movement rate.

Alkalis, such as sodium and potassium, and alkaline earth minerals such as calcium and magnesium are found in biomass. It also includes some other minerals such as Fe, Si, Al, Cl, and P. Some minerals such as alkali and alkaline present in the biomass may act as catalyst for the process of thermochemical conversion. Moreover, these may form alkali silicate with silica, which would cause bridging and agglomeration inside gasifier resulting to affect gas production. Furthermore, some other minerals can possibly form ash during the gasification process (Fryda et al., 2008; Lahijani and Zainal, 2011).

13.2.4 Potential for energy and chemicals

Gasification has high process efficiency, and thus, it is the most promising technology for the conversion of biomass in fuel and chemicals (Ahrenfeldt et al., 2013). Feed of gasification can be coal, plastics, tyres, biomass, and so on or their mixtures. Thus, it is flexible in terms of feedstock and also for the products. However, the properties and composition of the produced fuel gas highly depend on the types of feedstock and operation conditions of the gasification process. In general, fuel gas can be produced from any type of biomass feedstock via gasification. The main gaseous product of gasification is mixture of hydrogen and carbon monooxide, which is a building block of synthesis of hydrocarbons. It also contains unwanted by-products such as methane and carbon dioxide. Gasification with air leads to a fuel gas with a heating value of 4–7 MJ/m^3, whereas using pure oxygen or steam as an oxidant leads to significant increase in heating value of the gas in the range of 10–18 MJ/m^3 (Yan et al., 2018). Moreover, fuel gas received from biomass gasification can

TABLE 13.2 The ultimate and proximate analysis of various lignocellulosic biomasses.

	Ultimate analysis (db, % w/w)					Proximate analysis (% w/w)					
Biomass type	C	H	O	N	S	Ash	VM	FC	M	HHV (MJ/kg)	References
Peach stone	41.0	5.7	48.4	4.9	0	1	63	29	7	18.8	Ismail et al. (2018)
Pinewood chips[a]	52.90	6.30	37.60	0.50	0.05	2.90	72.90	24.2	db	22.51	Dıaz and Martınez (2019)
Pine grass	48.20	5.56	45.19	0.36	0.07	6.90	76.60	16.5	db	21.17	Dıaz and Martınez (2019)
Miscanthus	44.5	5.2	45.0	5.3	0	2.1	64.4	22.1	11.4	18.6	Ismail et al. (2018)
Rice husk[a]	35	5.5	36	1.53	0.08	21.89	55.85	15.24	7.12	dna	Bharath et al. (2018)
Waste wood[a]	47.9	5.45	40.6	1.9	0.10	4	dna	dna	8.8	20.06	Valin et al. (2019)
Beach wood	49.6	5.69	44	0.3	410 ppm	0.7	dna	dna	8.3	20.56	Valin et al. (2019)
Pine sawdust	51.4	5.1	42.4	1.1	-	0.5	74.7	13.2	11.6	19.5	Ma et al. (2019)
Ground seed corn	48.91	5.95	41.46	1.73	0.16	1.79	86.44	11.77	db	Dna	Timmer and Brown (2019)
Waste wood pellets	47.26	6.14	44.99	0.11	dna	1.5	77.3	14.9	db	19.05	Shen et al. (2019)
Torrefied soft wood chip	47.7	5.7	45.5	1.0	0.16	0.6	76.5	22.9	db	22.3	Mafu et al. (2018)
Torrefied soft wood chip	46.6	5.5	48.0	0.1	0.01	0.6	77.3	22.1	db	22.4	Mafu et al. (2018)
Torrefied sweet sorghum bagasse	43.5	5.6	50.5	0.3	0.13	6.5	69.8	23.7	db	23.0	Mafu et al. (2018)
Palm oil empty fruit bunch	42.08	5.72	49.93	0.99	dna	3	83	9	5	18.46	Ariffin et al. (2017)
Cedar wood	51.10	5.90	42.50	0.12	0.02	0.3	80–82	18–20	db	21.72	Asadullah et al. (2004a,b)
Wood sawdust	46.2	5.1	35.4	1.5	0.06	1.3	70.4	17.9	10.4	21.27	Cao et al. (2006)
Olive oil residue[a]	50.7	5.89	36.97	1.36	0.3	4.6	76	19.4	9.5	23.66	Arvelakis et al. (2003)
Rice husk	45.8	6.0	47.9	0.3	dna	0.8	73.8	13.1	12.3	15.82	Velez et al. (2009)
Rice straw[a]	38.61	4.28	37.16	1.08	0.65	12.64	16.55	65.23	5.58	16.86	Li at al. (2009)
Pine sawdust[a]	50.54	7.08	41.11	0.15	0.57	0.55	17.16	82.29	db	23.00	Lv et al. (2004)
Spruce wood pellet[a]	49.3	5.9	44.4	0.1	dna	0.3	17.1	74.2	8.4	20.96	Miccio et al. (2009)
Coffee husk	46.8	4.9	47.1	0.6	0.6	1.0	14.3	74.3	10.4	18.9	Li at al. (2009)
Coffee ground[a]	52.97	6.51	36.62	2.8	0.05	1.0	16.7	71.8	10.5	24.46	Murkami et al. (2007)
Larch wood	44.18	6.38	49.32	0.12	dna	0.12	14.86	76.86	8.16	21.91	Weerachanchai et al. (2009)
Grapevine pruning waste[a]	46.97	5.8	44.49	0.67	0.01	2.06	19.78	78.16	db	20.37	Lapuerta et al. (2008)

TABLE 13.2 The ultimate and proximate analysis of various lignocellulosic biomasses.—cont'd

Biomass type	Ultimate analysis (db, % w/w)					Proximate analysis (% w/w)				HHV (MJ/kg)	References
	C	H	O	N	S	Ash	VM	FC	M		
Jute stick[a]	49.79	6.02	41.37	0.19	0.05	0.62	21.4−23.4	76−78	db	22.12	Asadullah et al. (2004a,b)
Sugarcane bagasse	48.58	5.97	38.94	0.2	0.05	1.26	28.74−30.74	67−70	db	21.51	Asadullah et al. (2004a,b)
Corn cob	40.22	4.11	42.56	0.39	0.04	2.97	16.11	71.21	9.71	19.11	Lu et al. (2008)
Peach stone	51.95	5.76	40.7	0.79	0.01	0.65	18.1	81.3	8.53	24.06	Arvelakis et al. (2003)
Straw[a]	36.57	4.91	40.70	0.57	0.14	8.61	17.91	64.98	8.5	17.06	Shen et al. (2008)
Camphor wood[a]	43.43	4.84	38.53	0.32	0.1	0.49	14.75	72.47	12.29	19.94	Zhou et al. (2009)
Beech wood	48.27	6.36	45.2	0.14	dna	0.8	18	81	db	21.66	Radmanesh et al. (2006)
Switchgrass[a]	47	5.3	41.4	0.5	0.1	4.6	17.1	58.4	20	21.16	Jin et al. (2009)

db, dry basis; *C*, carbon; *H*, hydrogen; *O*, oxygen; *N*, nitrogen; *S*, sulfur; *VM*, volatile matter; *FC*, fixed carbon; *M*, moisture content; *dna*, data not available.
[a] *ultimate analysis on ash-free basis.*

generate all kinds of energy or energy carriers—heat, power, biofuels, hydrogen, biomethane, and other chemicals. Syntheses of Fischer–Tropsch (FT) diesel, dimethyl ether (DME), methanol, and methane have well-established technical processes (Kuo et al., 2020).

13.3 Biomass to energy and chemicals conversion pathways

Various processes can be used to convert chemical energy stored in biomass, which can be converted into different forms of energy. A general overview of biomass conversion pathways has been given in Fig. 13.1. Factors that affect the selection of process are biomass quantity, feedstock desired energy form, project-specific factors, and environmental standards.

13.3.1 Biochemical conversion pathways

Using bacteria or enzymes and/or other living organism for the breaking down of biomass into different fuels is known as biochemical conversion process. Biochemical conversion includes aerobic conversion, anaerobic digestion, and fermentation. Organic materials are converted into gaseous mixtures directly by some selective microorganisms in the absence of O_2, which is known as anaerobic digestion. The produced gaseous mixture contains primarily CH_4, CO_2, and H_2S in a small quantity, which is collectively known as biogas. This process is suitable for organic waste having high moisture content about 70%–90%. The steps of this conversion process are hydrolysis, acidogenesis, acetogenesis, and methanogenesis as shown in Fig. 13.2. Solid residue that comes from digester is used as a biofertilizer.

As shown in Fig. 13.3, in fermentation, yeast or bacteria break down the glucose anaerobically in different organic materials. The new generation fermentation process implies the conversion of glucose or carbohydrate derived from plants into an alcohol or acid. Yeast or bacteria use plants derived sugars as a food

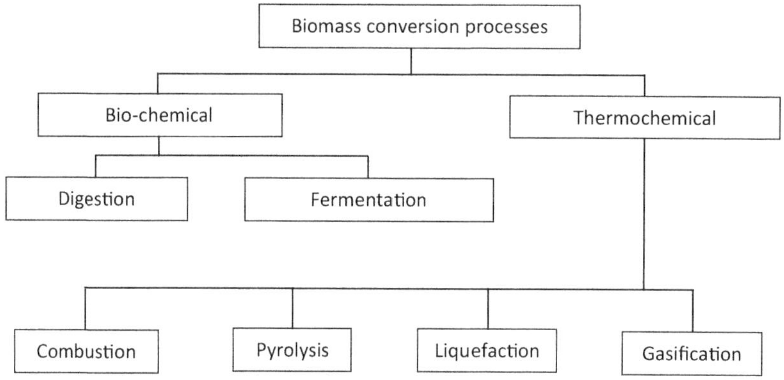

FIGURE 13.1 Overview of biomass conversion pathways.

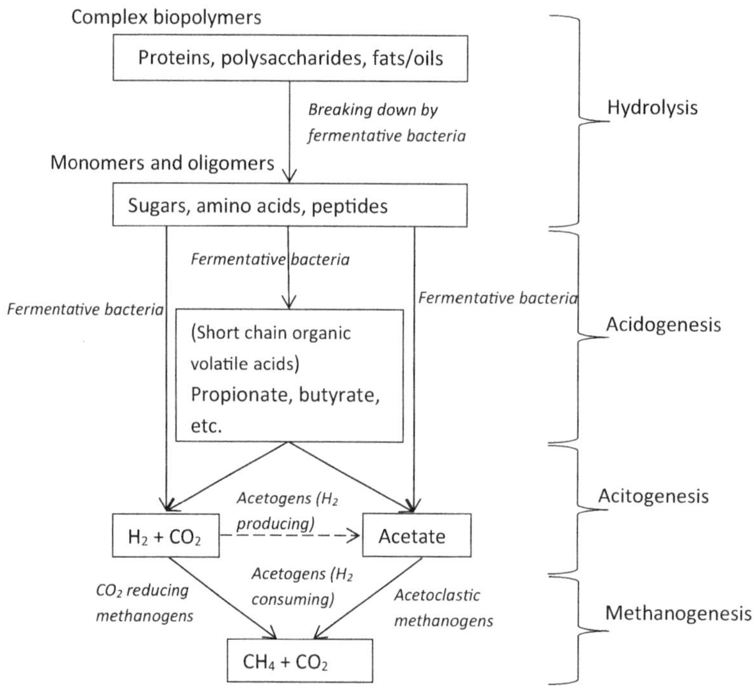

FIGURE 13.2 Anaerobic digestion process indicating four phases of conversion process.

and excrete ethanol/butanol and carbon dioxide. These can be further distilled and dehydrated to achieve higher concentration of alcohol, which can be useful for automotive fuel. The residual solid from fermentation process is used as cattle feed, or in some cases, it is used as fuel for boilers or for subsequent gasification.

13.3.2 Thermochemical conversion pathways

In thermochemical conversion pathways, bonds between adjacent carbon, hydrogen, and oxygen molecules are broken down to release their stored chemical energy that was stored during the process of photosynthesis.

Thermochemical conversion includes combustion, pyrolysis, liquefaction, and gasification. A general overview of the thermochemical process is shown in Fig. 13.4.

During all thermochemical conversion of biomass, the process changes with amount of O_2. The thermochemical conversion with excess amount of O_2 is known as combustion, whereas the conversion without O_2 supply is known as pyrolysis (Gupta et al., 2017). Furthermore, limited amount of supply of O_2 during thermochemical conversion is known as gasification. All processes have significant characteristics as listed in Table 13.3.

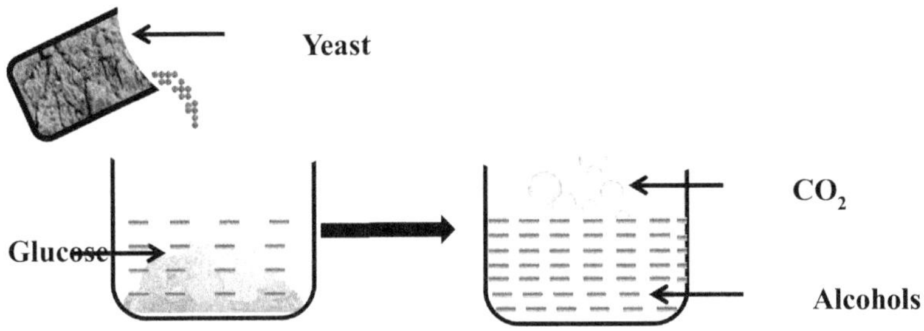

FIGURE 13.3 Fermentation process.

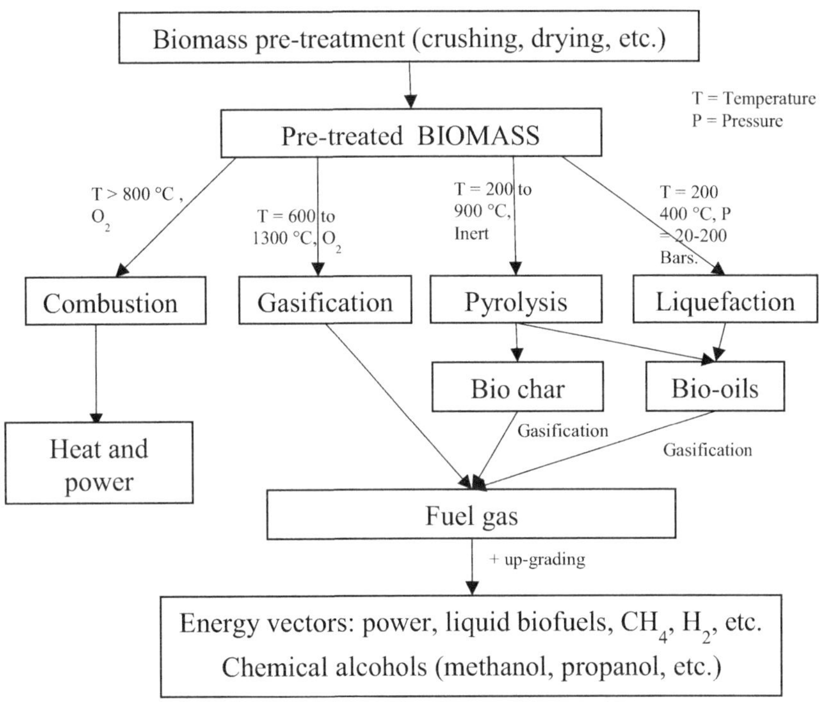

FIGURE 13.4 Overview of main thermochemical process.

TABLE 13.3 Characteristics of thermochemical process.

Sl. No.	Process	Amount of oxygen	Process temperature (°C)	Useful products
1	Combustion	Excess	Above 800	Heat
2	Gasification	Limited	700–1100	Gas
3	Pyrolysis	Absence	350–550	Char, biooil
4	Liquefaction	Limited	250–400 (High pressure; 4–22 Mpa)	Biooil

13.4 Biomass gasification overview

Biomass gasification is a physicochemical process to convert solid biomass to gaseous fuel through a chain of reactions. This physicochemical process consists of drying, pyrolysis, and combustion and char gasification. Pure oxygen, steam, air, or carbon dioxide has been used as a gasifying agent for the gasification process. The produced gas quality and its heating value depend upon the gasifying agent and other variables such as gasifying temperature, equivalence ratio, catalyst, and types of gasifying reactor. During the gasification process, gas–solid reactions and the gas–phase reactions take place (El- Shafay et al., 2020).

13.4.1 Gasification process and important chemical reactions

As gasification is the conversion of biomass to a gaseous substance considered as fuel by applying heat and/or pressure in a predefined gasifying medium such as air, CO_2, or steam with restricted amount of oxygen. The process involves the steps such as drying, pyrolysis, oxidation, and reduction (Antonopoulos et al., 2012).

Usually, biomass suitable for gasification process contains 10%–30% moisture. When biomass is heated up to about 100°C, the moisture is converted into water vapor. After drying, as heating continues, the biomass undergoes pyrolysis. Pyrolysis is heating the biomass between 300 and 550°C without supplying any oxygen. As a result, the biomass is decomposed into solids, liquids, and gases. Charcoal is solid, tar is liquid, and fuel gas is a gaseous product. Air is introduced into gasifier after the decomposition process. During oxidation, charcoal or the solid carbonized fuel reacts with the oxygen in the air at about 700–1000°C to produce carbon dioxide and heat. Furthermore, at higher temperatures under reducing condition, the following reactions take place to produce carbon dioxide, hydrogen, and methane (Cabuk et al., 2019).

$$\text{Biomass} \rightarrow H_2 + CO + CO_2 + CH_4 + HC(g) + \text{Tar}(l) + \text{Char}(g) \tag{13.1}$$

$$C + CO_2 \rightarrow 2CO \tag{13.2}$$

$$CH_4 + CO_2 \rightarrow 2CO + 2H_2 \tag{13.3}$$

$$CO + H_2O \rightarrow CO_2 + H_2 \tag{13.4}$$

$$CH_4 + H_2O(g) \rightarrow CO + 3H_2 \tag{13.5}$$

$$\text{Tar} + H_2O(g) \rightarrow CO + H_2O + CH_4 + H_2 + C_nH_m \tag{13.6}$$

$$C_nH_m + nH_2O \rightarrow nCO + \left[n + \left(\frac{m}{2}\right)H_2\right] \tag{13.7}$$

13.4.2 Biomass selection

For the gasification, biomass should have high volatile matter, high calorific value, low ash content, low moisture content, locally available, and not easily degradable (Sanchez-Silva et al., 2012).

13.4.3 Biomass pretreatment

There are varieties of biomass, which has a wide range of different properties. It has many challenging properties such as higher oxygen and moisture content, high affinity to water, high grinding energy, and low energy density (Singh et al., 2020). Thus, pretreatment becomes a crucial aspect of curtailing failure of gasification process. The main objective of pretreatment is to prepare the selected biomass a suitable feedstock for gasification. Pretreatment may include a single process or a combination of several processes such as drying, size reduction, densification, and so on.

13.4.3.1 Drying

Ideally, the moisture content should be in between 10% and 15% for the purpose of gasification (Basu, 2010). Thus, drying becomes mandatory preprocessing stage for the nonwoody biomass such as unripe coconut husk, green allege, and so on, which has high initial moisture content. Approximately 2.3 MJ is required for the vaporization of 1 kg of moisture content (Basu, 2010). Thus, moisture can reduce the overall efficiency of the process. However, sun drying may be opted for the drying in preprocessing stage to reduce the energy requirement during the process of drying. Also, heat could be recovered from the heat content of exit gas during gasification process.

13.4.3.2 Size reduction

Nonwoody biomass is found with irregular shape and size. Thus, for appropriate size of feedstock, its size reduction is often required. Uniform size is required for the other pretreatment process named densification such as pelleting. It also provides uniform size of the particles for the gasification. In general, large amount of surface area is found in smaller particles, which allow better heat and mass transfer. The feedstock size is dependent on the type and mode of operation of the gasifier used. However, the most common feedstock size varies from 1 μm to 1 cm (Souza-Santos, 2010). Additionally, for easy handling of fluidized bed and for maximizing the contact of the surface of feedstock with the oxidant during fluidization, sizing of feedstock becomes very important. Furthermore, finer particles are preferred in entrained bed-type fluidization, whereas biomass reduction takes place at a slower rate in fixed-bed gasification; thus, it requires particle size in the range of 1 cm. This size of particle is also required for delaying the rapid combustion while allowing the process of effective devolatilization.

13.4.3.3 Densification

Low bulk density, irregular size, and shape are some of the demerits of nonwoody biomass, which create problems in storing and handling. Additionally, it cannot be used in its current original form. Densification would be a best solution to eradicate the aforementioned problems. Torrefaction and pelleting are the general two methods available for the purpose of feedstock densification (Samy, 2013; Tchapda et al., 2014). The combination of these two processes can also be used for densification, which increases the density and thus improves the efficiency of the gasification process.

13.4.3.4 Torrefaction

Torrefaction is a thermal process, which converts biomass in coal-like material, which has better fuel properties than the original biomass. It is achieved by heating the biomass in inert atmosphere at a temperature range of 200–400°C. During this process, the structure of biomass changes, and it becomes more hydrophobic and brittle. It removes moisture content, volatiles materials, and reduces the level of mineral content (Sarkar et al., 2014) in the biomass. Reduction of mineral contents reduces the mineral oxides, which would be responsible of slagging or caking of ash during gasification process (Sadaka and Negi, 2009). Also, torrefied biomass has many improved properties over untreated material such as higher energy density, lower oxygen-to-carbon ratio, greater grindability, reactivity, and ignitability. Furthermore, torrefied biomass offers low transportation and storage costs (Kuo and Wu, 2016; Thanapal et al., 2014; Sarvaramini et al., 2013).

13.4.3.5 Pelleting

Another method of densification of biomass is pelleting or briquetting. Briquette particle size range of 1–2 cm is the best for fixed-bed biomass gasifier. Conventional pelletizer is typically used for the process of pelletizing. Loose biomass is fed into the pelletizer, and pressure is applied on the biomass in the presence of mold. Pelletizers are equipped with blades to cut the pellets in desired size (Chen et al., 2008). Lignin present in the biomass has the ability to bind the fibers (Kalish, 2012). Temperature of the pellets leaving pelletizer is in between 80 and 90°C; thus, cooling becomes a very important process for the pelleting. Lignin stabilization takes place during the cooling period, and the pellets become hard (Chen et al., 2008). The process of pelleting constructs a uniform-sized fuel with manyfold higher bulk density, and thus, it allures for the energy process. In addition, ash is decreased in the thermochemical conversion if the nonwoody biomass is used in pellet form. It is reported that ash was decreased by two- three fold during the combustion of cotton gin waste pellet compared with unpelleted cotton gin waste (Holt et al., 2006). Furthermore, another category of nonwoody biomass pellet has been considered for the industrial purposes (Vinterback, 2004). However, the moisture content, die temperature and configuration, pressure, and feed rate are the different properties on which the efficiency of the equipment depends. Thus, wide variation in the properties of biomass requires specific development of appropriate pelleting processes for each range of feedstocks (Uslu et al., 2008; Holt et al., 2006).

13.4.3.6 Combined torrefaction and pelleting

Densification can also be achieved by torrefaction and then pelleting. It also increases the rate of devolatilization. Sarkar et al. (2014) worked on raw, torrefied, and combination of torrefied and pelleted switchgrass, as pretreatment process. They had worked in both the inert and oxidizing atmospheres. They found the highest devolatilization rate for torrefied and pelleted switchgrass, and the lowest devolatization rate was achieved for raw-torrefied switchgrass. Thus, it could be surmised that devolatilization was increased via pelleting process. However, torrefaction and pelleting consumes extra energy and power.

13.5 Types of gasification

There are mainly two types of gasifier, which are fixed-bed and fluidized-bed gasifier. Subsections of each type of gasifier have been discussed in the following.

13.5.1 Fixed-bed gasification

Updraft and downdraft gasification are the two types of fixed-bed gasification. When the feeding is supplied from the top of the gasifier and air is supplied from the bottom into the gasification chamber, then it is known as updraft gasification process, whereas in the downdraft gasification material, air moves concurrently from top to bottom of the gasifier. Thus, the combustion zone is the lowest part of the gasifier in both types of gasifier. In downdraft gasification process, less tar content is found, as tar is thermally cracked. It is because of the exit of the produced fuel gas, which is very close to the combustion zone in downdraft gasification reactor. Low tar content in fuel gas is preferred for turbines and gas engines. A schematic diagram of updraft and downdraft gasifier is shown in Fig. 13.5.

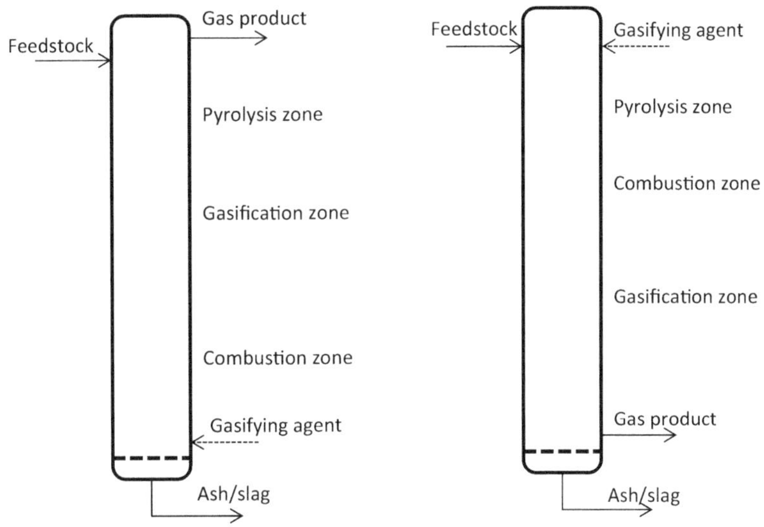

a) Updraft gasifier b) Downdraft gasifier

FIGURE 13.5 Different zones of (A) updraft gasifier and (B) downdraft gasifier.

13.5.2 Fluidized-bed gasification

Fluidized-bed gasifier involves very complex multiscale and multiphysics processes (Wang and Shen, 2020). However, fluidized-bed gasifier is widely used in recent years. There are many advantages of fluidized-bed over fixed-bed gasifier such as uniform temperature distribution, high carbon conversion, and low tar production. Furthermore, it offers a high range of biomass type, size, and feed rate. Bubbling fluidized-bed, circulating fluidized-bed, and entrained flow fluidized-bed gasifier are some of the types of fluidized-bed gasifier. In the bubbling fluidized-bed gasification system, the biomass (bed material) rests on a distributor plate, and it is fluidized with a gasifying medium. In most of the cases, air and/or nitrogen gas is used for the medium. The gas is passed through the distribution plate with a velocity approximately five times faster than that of minimum fluidization velocity. The bed temperature varies from 700 to 900°C in which biomass is pyrolyzed; thus, char and gases are produced. Then, char particles are lifted along with gasifying gas, and gasification reactions occur simultaneously. The net tar content is less than 3 gm/Nm^3 of fuel gas (Burgohain et al., 2010). For the low-grade burning biomass, circulating fluidized bed is used in which the fluidizing velocity is much higher than the terminal settling velocity of the bed material. It can be operated at high pressures. And it offers feed range from 1 to 80 tons per day (Miccio et al., 2012). Furthermore, in entrained flow gasifier, the feed material is used in pulverized form. This pulverized biomass fed pneumatically with gasifying medium (Mahinpey and Gomez, 2016), in which fuel particles dragged along with gas stream and injected to burner of the gasifier. Because of high velocity, the residence time becomes short, and thus, high temperatures and small particles are needed. Entrained flow gasifier is used under high pressure from 20 to 25 bar. The temperature at burner reaches up to 2500°C. The capacity of entrained flow gasifiers is hundreds of megawatt (Bergman et al., 2004).

13.5.3 Cogasification

When two and more than two fuels are mixed for the purpose of gasification, it is known as cogasification. It can be biomass and low or high grade coal, woody biomass with nonwoody biomass, or any other two or more than two fuels. Cogasification of biomass with low-grade coal is considered a promising technology for the production of fuel gas. During cogasification of biomass and coal, volatile component of biomass readily decomposes and forms free radicals, which react with organic matter of coal, and thus the conversion rate increases. Also, hydrogen may get attached with free radicals produced from coal as soon as their formation, and thus, it may prevent the recombination reactions to form less reactive secondary tar species. And thus hydrogen-rich fuel gases are found.

13.5.4 Catalytic gasification

When the addition of a small amount of chemical substance increases the rate of chemical equilibrium, but the substance itself does not undergo any chemical changes, it is known as catalytic reaction. Catalyst works by providing alternative mechanism, which may involve a different transition state of lower energy. Some catalysts also allow splitting the C–C and C–O ether bond of lignocellulosic biomass in its aromatic monomers (Yamaguchi et al., 2017). Catalytic agents of a catalyst generate the active sites and provide surface for the reactions to take place at the surface of catalyst. Researchers have highlighted catalytic gasification for the enhancement of hydrogen production and calorific value of the fuel gas using several catalysts based on Ni, Zn, Fe, and Cu, including other alkali metals and salt (Rapagnà et al., 2002; Rui et al., 2014; Wang et al., 2016). However, the catalyst preparation cost demotivates the researchers, which adds extra cost and needs more budget to use catalyst in the process. Another demotivation of catalyst is deactivation of catalyst by carbon deposition on the catalyst surface and fouling. Thus, some naturally occurring materials have been used as a catalyst, such as gypsum, egg shell, limestone, alumina, and so on (Fan et al., 2016; Sasaki et al., 2017; Zhao et al., 2017; Cummer and Brown, 2002). Impregnation of wastewater of pulp and paper industry on biomass surfaces was also used as a catalyst by Ram and Mondal, in both fixed- and fluidized-bed gasification (Ram and Mondal 2018, 2019a, 2019b).

13.5.5 Hydrothermal gasification

If biomass has high moisture content, then it cannot be proceeded in fixed- or fluidized-bed gasification. Thus, an interesting gasification technology has been used to convert wet biomass in hydrogen. Above critical temperature (647 K) and pressure (22.1 MPa), water behaves as gas-like properties. Thus, researchers are interested more in supercritical hydrothermal gasification (Matsumura et al., 2005; Nanda et al., 2016; Tang and Kitagawa, 2005). However, there are also a lot of researches on subcritical gasification, which takes place in the temperature between 100 and 374°C and pressure between 5 and 22 MPa (Bi et al., 2017; Su et al., 2019; Seif et al., 2015).

13.6 Effect of process parameters in gasification products yield

Amount of fuel gas produced is highly dependent on several process parameters, such as size and types of biomass feedstock, size of gasifier, gasification temperature, equivalence ratio, gasification medium, catalyst (synthesized or natural), and types of gasification.

Shen et al. (2019) used air and mixture of air and CO_2 as an oxidizing agent during gasification. At first, gasification was performed using 100% air as a gasifying medium; furthermore, 15% CO_2 was added in a gasifying medium with 85% air. Then the results of this comparative study were reported. Since CO_2 behaves as an inert gasifying medium at low temperature, it could be converted to CO at a high temperature. It was reported that CO_2 as a gasifying agent has comparable energy of produced fuel gas as that of air as a gasifying medium; also, use of CO_2 has reduced the N_2 fraction in the produced fuel gas. The energy generation of produced fuel gas per gram of wood pallet was 6.67 kJ with 15% CO_2 and 85% air as a gasifying medium, whereas 7.45 kJ of energy was found with 100% air as a gasifying medium at 800°C gasifying temperature. In addition, CO_2 air mixture with gasifying medium was also conducted in small-scale gasification experiments, and it was found that CO concentration in produced fuel gas was enhanced due to Boudouard reaction. On the other hand, CH_4 concentration was suppressed owing to CO_2 dry reforming, and the produced gas has a high $CO:H_2$ ratio due to reverse water–gas shift reaction. CO_2 addition with gasifying media also enhanced carbon conversion efficiency and cold gas efficiency by 6% and 5.8%, respectively.

Prasertcharoensuk et al. (2019) studied the behavior of lignocellulosic material after pyrolysis. Pyrolysis was performed in the temperature range of 600–900°C, which significantly influenced the char properties, and the total pore size was increased by up to 2.5–3 times. Increasing the surface area and pore size enhanced the surface area for gas–solid reaction during the gasification process. It was found that unstable compounds such as levoglucosan and its derivatives decomposed and phenolic compounds formation enhanced after pyrolysis above the temperature of 700°C. The composition of H_2 was increased from 49 to 67 mol % when gasification was performed without pyrolysis and after pyrolysis at 900 °C. Steam to carbon in biomass ratio was used 5.7 at 1000°C gasifying temperature, and the particle volume was in the range of 0.5–1 cm^3. It was attributed that amount of volatiles affected the properties of produced fuel gas; thus, pyrolysis is a crucial step for gasification.

Cogasification of two biomasses was performed by Diaz and Martinez (2019). The mixture of pinewood and pinewood grass was taken as a feedstock, and it was mixed by weight as 100%–0%, 90%–10%, 80%–20%, and 70%–30%, respectively. The experiments were conducted in a laboratory-scale updraft gasifier using air as a gasifying medium. The experiments were performed at equivalence ratios of 0.25, 0.30, and 0.34. And gasification temperature was varied from 800 to 1100°C. An increase of about 20% for H_2, 20% for CO, and 30% for CH_4 by volume was observed in the gasification of combined 80% pinewood and 20% pine grass with respect to the 100% pinewood. In addition, the mixture of 80% pinewood and 20% pine grass was found to be most promising in terms of improving carbon conversion and reducing the amount of char. It was surmised that the improvement in H_2, CO and CH_4 concentration was due to presence of K_2O in the 20% pine grass, which acted as a catalyst. However, it was prone to combustion rather than gasification when more than 20% amount of pine grass was used. Lower heating value of the produced fuel gas was found between 4.0 and 5.5 MJ/Nm^3 by the mixture of 80%–20% biomass.

Thengane et al. (2019) worked on cogasification of high ash biomass and high ash coal. Biomass taken for study contained dry litter of jackfruit, rain tree, mango, and eucalyptus tree, which showed ash content ranging from 20% to 25%. Cylindrical shape pellets are made from these garden wastes with the dimensions as length 40 mm and diameter 15 mm. High ash coal was collected from Western Coalfields Limited, Chandrapur area, which was taken in the form of chips with 10–40 mm and 5–10 mm dimensions. The highest cold gas efficiency, thermal efficiency, and lower heating value of produced fuel gas were found as 57.5%, 72.63%, and 3.05 MJ/Nm^3, respectively, for biomass ratio 0.75, whereas it was formed lowest for coal with cold gas efficiency 33.06% and thermal efficiency 49.38%. It was found that carbon conversion of feedstock to gas was increased with the increase of the amount of biomass. Furthermore, the presence of coal enhanced the rate of reaction by increasing the temperature.

Mafu et al. (2018) also used CO_2 as a gasifying agent for the gasification of biomass-derived char samples in thermogravimetric analyzer. The biomasses taken were softwood chips, hardwood chips, and sweet sorghum bagasse, and the selected temperature range was 850–950°C. Reactivity of subbituminous coal char was also compared with the reactivity of all three biomass chars under CO_2 atmosphere. The reactivity of hard wood was found lowest, whereas the reactivity of sweet sorghum bagasse was highest among all biomass char. However, the reactivity of subbituminous coal was lowest among all. Furthermore, addition of biomass from 10% to 30% by weight was also made for coal char gasification. It was found that 20% and 30% addition of biomass with coal char increased the gasification rate, whereas addition of 10% biomass had no any significant effect on the gasification reactivity of coal char.

A comparative study of steam gasification under Fe_2O_3 as catalyst and without catalyst was investigated by Shen et al. (2018) taking feedstock as a mixture of Indonesian Adaro coal and Japanese cedar in the ratio of 1:1 by weight. 10 wt% Fe_2O_3 with the mixture of Japanese cedar and coal was taken in 50% of H_2O by vol and subjected to 800°C for 1 h. The evolution of H_2 during gasification of Indonesian Adaro coal and Japanese cedar without catalyst was 100 mmol/g of feedstock, and the evolution was raised to 152 mmol/g of feedstock in the presence of Fe_2O_3 as a catalyst. It was also found that the rate of gasification process was more about 70% in the presence of Fe_2O_3 with respect to 38% the feedstock absence of Fe_2O_3.

Furuso et al. (2018) studied on alkali addition in entrained flow biomass gasification and did the thermodynamic equilibrium calculations. Four different forest wastes were taken for the study, i.e., pine stem wood, pine bark, spruce forest residue, and pine forest residues, which were further impregnated with 2%, 4%, and 8% of Na or K. It was reported that 2%–8% addition of alkali catalyst changed the biomass gasification and the cleaner fuel gas was found. And the impregnation of Na was a more promising candidate to get cleaner fuel gas.

Arifin et al. (2017) performed gasification on a medium-scale downdraft fixed-bed gasifier with a feed rate of 162 kg/h, and the feedstock was palm oil empty fruit bunch. The gasification temperature was kept in the range of 850–950°C. It was attributed that the ideal gasification temperature for palm oil empty fruit bunch was between 850 and 950°C; furthermore, the cold gas efficiency and fuel gas heating value were increased with the increase of gasification temperature. Heating value of fuel gas varied from 4.10 to 6.42 MJ/Nm^3, and cold gas efficiency (CGE) was varied from 61% to 67%. It was also attributed that the fuel gas generated from the gasification of oil palm empty fruit bunch with feed rate of 126 kg/h at gasification temperature 900°C has the capacity to generate 106 kilo Watt equivalent (kWe) electricity.

Maa et al. (2019) used olivine and dolomite to be acted as a natural catalyst in the fluidized-bed gasification. The experiments were conducted to study the effect of operating parameters in bubbling fluidized-bed cogasification of pine sawdust and brown coal. The gasification temperature varied from 700 to 1000°C, and the feed particle size varied from 2 to 3.5 mm. Furthermore, steam/fuel ratio, equivalence ratio, biomass ratio, and the catalyst loading were varied from 0.5% to 0.8%, 0.1% to 0.4%, 0% to 100%, and 3.0–12.0 wt%, respectively. H_2 yield was reached to 55.5 (g/kg-fuel) with the dolomite loading 12 wt% from 52.9 (g/kg-fuel) with the dolomite loading 3 wt%, and the yield was increased from 47.5 to 52.1 (g/kg-fuel) for olivine loading, whereas tar yield decreased drastically from 5.4 to 0.4 (g/Nm^3) and from 7.0 to 0.8 (g/Nm^3), respectively, with the presence of dolomite and olivine. It was attributed that dolomite and olivine have high potential to be used as a catalyst because of its natural occurrence and capability to enhance the H_2 production and tar upgradation.

In an another research, the effects of gasification temperature, biomass particle size, equivalence ratio, superficial velocity, and the steam addition on groundnut corn in fluidized bed were studied by Timmer and Brown (2019). During the study, the feed rate, equivalence ratio, and temperature were varied from 0.518 to 1.061 g/s, from 0.247 to 0.370, and from 715 to 805°C, respectively. Higher carbon conversion was achieved with decreasing both superficial gas velocity and particle size, and with increasing equivalence ratio. And negligible effect was recorded on carbon conversion with the change of steam concentration. The highest gas–solid chemical reaction coefficient was found as $6.32 \times 10^{-4}\ s^{-1}$, which was achieved at an equivalence ratio of 0.252, feed rate of 0.724 g/s, and gasification temperature of 797°C.

Valin et al. (2019) experimented several types of feedstocks such as bulky furniture waste, plastics, cardboards, paper, and textiles. Mixture of all these was named as solid refused fuel (SRF), which was pelletized and gasified in fluidized bed. Furthermore, gasification was done with waste wood and stem beech wood sawdust. The temperature was varied from 800 to 910°C, and the equivalence ratio was varied from 0 to 0.34. Gasification was performed with a solid feeding rate of 1–3 kg/h in a bubbling fluidized bed at 1.5 bar. With the agreement of other researchers, it was found that yield of H_2 and CO in produced gas was influenced with the temperature. Furthermore, it was found that addition of steam enhanced the yield of CO and CO_2. The optimum temperature found for the gasification of SRF was 908°C, and the yields of H_2, CO, CH_4, and CO_2 at optimum temperature were 0.2, 0.16, 0.13, and 0.36, respectively.

Tian et al. (2018) used *Enteromorpha intestinalis* biomass as a feedstock for the gasification using air and steam as an oxidizing agent in fluidized bed. During this study, calcined dolomite, limestone, and olivine were used as catalysts to crack the tar and to increase the H_2 content in produced gaseous product. To study the effect of all three natural catalysts, all the experiments were performed at equivalence ratio 0.14 and steam-to-biomass ratio at 0.5 (wt/wt). When calcined dolomite was added to the reactor, a significant amount of tar reduction was observed, and fuel gas yield was increased by more than 14%. Furthermore, the gasification temperature was varied from 800 to 1000°C, the maximum yield of H_2 was found at 1000°C, and the value was 49.1 vol %. Calcined dolomite was found most effective for the tar destruction. Increase in biomass particle size also increased the tar content; however, the variation was only for an insignificant amount. However, the char yield was almost same for all the types of catalysts; thus, it was attributed that char yield was dependent on char content in devolatilization step and the rate of reactions. At optimum temperature, the carbon conversion was found to be 60.8%, and yields of other gases were found as CO: 26.4, CH_4: 0.8, and CO_2: 23.7.

Another study was carried out by Ismail et al. (2018) in a fluidized bed using pilot plant gasifier. The biomass used as a feedstock for the study was peach stone and miscanthus. Ismail et al. (2018) also used dolomite and olivine as catalysts to enhance the tar cracking and to increase the gas yield. A study was carried out at three different gasification temperatures, i.e., 750, 800, and 850°C in air as a gasifying medium with 45 kg/h of constant feed rate. Furthermore, the operating process parameters were optimized by two-dimensional COMMENT code. Furthermore, numerical results were compared with the experimental results, and it was found in agreement with each other. Modeling was made using Eulerian approach on mass, energy, and momentum exchange. It was stated that increased gasification temperature decreased the tar production. At lower temperature, the formation of CO was high, resulting in the increased value of fuel gas heating value. It had been attributed that mathematical model can be used for biomass gasification.

Research on cogasification of Indian rice husk and Indian coal was performed by Bharath et al. (2018) using lab-scale bubbling fluidized-bed reactor. Air and steam were used as oxidizing agents at atmospheric pressure. Air-to-solid ratio was varied from 1.02 to 1.72. The high amount of fly ash was observed when the experiment was started at equivalence ratio of 0.3, and then the equivalence ratio was reduced to 0.22 for rice husk. It was found that addition of rice husk increased the carbon conversion, heating value, and cold gas efficiency of produced fuel gases. Contribution of rice husk was varied to share 0%–90% of the power. The heating value was found to be increased from the range of 2.14–2.68 MJ/Nm^3 to the range of 5.21–5.4 MJ/Nm^3 depending on the operating parameters for the case of 0% rice husk and 75% rice husk, respectively. At the share of 50%–75% of total power by rice husk, the cold gas efficiency, total carbon conversion, and heating value of produced fuel gases were found to be 78%, 89%, and 5.4 MJ/m^3 with steam. The carbon conversion was 85.7% when only air was used as a gasifying medium. There was a trace amount of methane when 8.37% of rice husk was used; however, no evidence of methane was detected when the rice husk percentage was increased to 75%.

Singh et al. (2018) used naturally available sand as bed material and groundnut shell as feedstock. This groundnut shell was gasified in bubbling fluidized-bed reactor in the presence of air as a gasifying medium. During the research, the equivalence ratio was varied from 0.29 to 0.33, feedstock feeding rate from 36 to 31.7 kg/h, and gasification temperature from 650 to 800°C. High heating value was decreased with the increase of equivalence ratio. Also, suspended particulate matters were high at higher equivalence ratio. The lowest equivalence ratio was also not favorable because of choking and burning of tar happened with the incipient of groundnut shell gasification. Therefore, the most suitable equivalence ratio was attributed as 0.31. And cold gas efficiency and carbon conversion efficiency at optimal equivalence ratio were found as 71.8% and 88%, respectively, at equivalence ratio of 0.31.

Peng at al. (2017) also worked on cogasification of wood residue and coal in fluidized bed, and catalysts were also used during these experiments of gasification. Wood residue was received from a wood product manufacturer, which was mixed with high volatile bituminous coal taken from east part of Iran. Alkali metals of $NaHCO_3$, K_2CO_3, and $KHCO_3$ prepared by one-step polymerization method were taken as catalysts, and their effects onto the products had been reported. Pure oxygen and steam were used for the gasifying medium as an oxidizing agent. With the agreement of Bharath et al. (2018), Peng et al. (2017) also reported that addition of coal to biomass had a positive impact on the gasification process. The overall conversion was increased from 81% to 94% with the increase of the coal/biomass ratio from 0.1 to 0.5. In addition, the results also showed that increasing the catalyst mass ratio was responsible for increased CO_2 and H_2 yield. As the catalyst mass ratio increased, the tar yield decreased from 13.38 to 9.27 for Al_2O_3/K_2CO_3, from 14.87 to 10.47 for $KHCO_3$, and from 14.18 to 9.62 for $Al_2O_3/NaHCO_3$. Furthermore, K_2CO_3 was found the most promising catalyst with the lowest catalyst deactivation and highest conversion to gaseous products as compared with $NaHCO_3$ and $KHCO_3$.

The summaries of works done by the various researchers on fixed- and fluidized-bed biomass gasification have been tabulated in Tables 13.4 and 13.5, respectively.

TABLE 13.4 Summary of biomass gasification in fixed bed.

Sl. No.	Feedstock	Gasification media and gasification column specification	Operating conditions	Gas composition (vol %/wt%)/gas yield (m^3/kg biomass)	Optimum condition	Efficiency/ energy recovery	Gas HHV/ LHV (MJ/ Nm^3)	Reference
1.	F: high ash biomass (garden waste) and high ash coal	GM: CO_2 GCS: H: 1.2 m OD: 0.5 m	GT: 450–650°C Biomass-to-coal ratio: 0, 0.25, 0.5, 0.75, 1	Vol % H_2: 7.2–9.6 CO: 6.8–11.7 CH_4: 1.3–1.7 CO_2: 11.2–14.8	Biomass-to-coal ratio: 0.75	dna	HHV: 5.51	Thengane et al. (2019)
2.	F: mixture of pine wood chips and pine grass	GM: air GCS: H: 0.8 m ID: 0.32 m	GT: 800–1050°C ER: 0.25, 0.30, 0.34	Vol % H_2: 4.4–8.7 CO: 15.5–23.7 CH_4: 1.6–4.7 CO_2: 13.6–9.9	Mixture of 80% pinewood and 20% pine grass	dna	HHV: 5.46 –6.46	Dıaz and Martınez (2019)
3.	F: waste wood	GM: steam GCS: H: 830 mm ID: 33 mm	GT: 1000°C After pyrolysis at: 600, 800, 900°C Steam-to-carbon in biomass ratio: 5.7	Vol % H_2: 48.8–67.2 CO: 4.5–8.8 CH_4: 7.7–2.2 CO_2: 39.0–21.8 GY: Total gas: 77.8–95.8 wt%	Pyrolysis at: 900°C	dna	dna	Prasertcharoensuk et al. (2019)
4.	F: wood sample	GM: 100% air, 15% CO_2 + 85% air GCS: H: 1.1 m ID: 46 mm	GT: 700–800°C ER: 0.24	Optimum condition H_2: 18.0 CO: 26.7 CH_4: 3.4 CO_2: 14.1 N_2: 37.8 GY: Total gas: 19.8 Nm^3/h	Air 85% + CO_2 15%, GT: 800°C	PER: 56.3% CCE: 65.2%	dna	Shen et al. (2019)
5.	F: Japanese ceder:coal: 1: 1	GM: Air, steam Catalyst: Fe_2O_3 GCS: na	GT: 800°C	H_2: 100–152 mmol/gm	Presence of Fe_2O_3	CCE 38% –70%		Shen et al. (2018)
6.	F: biomass char (softwood chips, hardwood chips, sweet	GM: CO_2 GCS: na	GT: 850, 875, 900, 925, 950°C	Biomass gasification reactivities decreased in the order: SB > SW > HW and	dna	dna	dna	Mafu et al. (2018)

Continued

TABLE 13.4 Summary of biomass gasification in fixed bed.—cont'd

Sl. No.	Feedstock	Gasification media and gasification column specification	Operating conditions	Gas composition (vol %/wt%)/gas yield (m^3/kg biomass)	Optimum condition	Efficiency/ energy recovery	Gas HHV/ LHV (MJ/ Nm^3)	Reference
	sorghum bagasse), bituminous coal			were greater than that of coal				
7.	F: palm oil empty fruit bunch	GM: air GCS: H: 2 m ID: 1.5 m	FR: 126 kg/h ER: 0.43–4.7 GT: 850–950 °C	Optimum Vol % H_2: 7.2 CO: 10.9 CH_4: 21.03 CO_2: 3.8 N_2: 57.07	T: 900°C	PER: 61% –67%	HHV: 4.02 –4.62	Ariffin et al. (2017)
8.	F: MSW refused derived fuel	GM: steam GCS: na	GT: 800°C Catalyst: Ni/SiO_2 with Ni loading 5,10, 20, and 40 wt.%	Optimum Vol% H_2: 57.9 CO: 18.4 CH_4: 2.2 CO_2: 21.5 GY Total: 68.7 wt%	20Ni/SiO_2	dna	dna	Blanco et al. (2013)

H, height; *ID*, inner diameter; *ER*, equivalence ratio; *GT*, gasification temperature; *GM*, gasifying medium; *GY*, gas yield; *F*, feed material; *dna*, data not available; *FR*, feed rate; *FS*, feed size; *SB*, steam biomass ratio; *GCS*, gasification column specification; *PER*, potential energy recovery; *CCE*, carbon conversion efficiency; *HHV*, higher heating value; *LHV*, lower heating value.

TABLE 13.5 Summary of biomass gasification in fluidized bed.

Sl. No.	Feedstock/bed material	Gasification media and gasification column specification	Operating conditions	Gas composition (vol %/wt%)/Gas yield (m^3/kg biomass)	Optimum condition	Carbon conversion efficiency/Energy recovery	Gas HHV/LHV (MJ/Nm^3)	Reference
1.	F: ground seed corn char BM: 70% silica 30% calcined limestone	GM: air, wet steam GCS: H: 81.3 + 122 cm ID: 9.5 + 15.2 cm	FR: ER: 0.24 −0.37 T: 700 −805°C	Reaction coefficients were highest at ER 0.25 and temperature 800°C	ER: 0.25 GT: 800°C	dna	dna	Timmer and Brown (2019)
2.	F: Beech wood sawdust, waste wood BM: silicon carbide powder	GM: Air, steam GCS: H: 0.95 m + 1.54 ID: 0.124 *m* + 0.20 *m*	FR: 1−3 kg/h ER: 0−0.34 GT: 800 −910°C P: 1.5 bar	At optimum Nm^3/kg H_2: 0.2 CO: 0.16 CH_4: 0.13 CO_2: 0.36	Air:steam 0.24:0.73, GT: 908°C	PER: 59%	HHV: 6.9	Valin et al. (2019)
3.	F: pine sawdust/coal BM: dolomite and olivine	GM: steam GCS: H: 55 cm ID: 8.5 cm	ER: 0.1−0.4 T: 700 −1000°C	Dolomite H_2: 52.9−55.5 g/kg-fuel Olivine H_2: 47.5−52.1 g/kg-fuel	Dolomite	For dolomite PER: 92.3 CCE: 84.2	HHV: 13.8	Ma et al. (2019)
4.	F: peach stone, miscanthus BM: dolomite, olivine	GM: air GCSS: H: 4.15 m ID: 0.5 m	FR: 45 kg/h T: 750 −850 °C	At optimum (vol %) H_2: 11.03 CO: 13.2 CH_4: 15.8 CO_2: 4.3 N_2: 55.67	T: 800°C	dna	dna	Ismail et al. (2018)
5.	F: Groundnut shell BM: Sand	GM: air GCS: H: 1600 mm ID: 200 mm	ER: 0.29, 0.31 and 0.33 GT: 650 −800°C	At optimum (vol %) H_2: 13.77 CO: 12.94 CH_4: 5.75 CO_2: 13.5 N_2: 54.04	ER: 0.31 GT: 714	CCE: 88% PER: 71.8%	dna	Singh et al. (2018)
6.	F: *Enteromorpha intestinalis* Catalyst: limestone,	GM: Air, steam GCS: H: 610 mm OD: 120 mm	ER: 0.14 GT: 800 −1000°C	Optimum (vol%) H_2: 49.1 CO: 26.4 CH_4: 0.8 CO_2: 23.7	GT: 1000°C Steam-to-biomass ratio: 1	PER: 71.5 CCE: 60.8	HHV: 10.5	Tian et al. (2018)

Continued

TABLE 13.5 Summary of biomass gasification in fluidized bed.—cont'd

Sl. No.	Feedstock/bed material	Gasification media and gasification column specification	Operating conditions	Gas composition (vol %/wt%)/Gas yield (m^3/kg biomass)	Optimum condition	Carbon conversion efficiency/Energy recovery	Gas HHV/LHV (MJ/Nm^3)	Reference
	calcined dolomite, olivine				Calcined dolomite			
7.	F: blend of Indian rice husk and coal	GM: Air, steam GCS: H: 2.5 m ID: 0.15 m	GT: 750 −850°C Rice husk: coal: 50:50, 75:25 wt%	Optimum (vol %) H_2: 8.62 CO: 13.72 CH_4: 6.68 CO_2: 11.65 N_2: 59.33	dna	PER: 78% CCE: 85%	HHV: 5.21	Bharath et al. (2018)
8.	F: Wood residue/coal BM: $NaHCO_3$, $KHCO_3$, and K_2CO_3	GM: oxygen, steam GCS: H: 1.4 m ID: 60 mm	ER: 0.4 GT: 850°C	Vol % H_2: 41.25−52.07 CO: 31.47−35.14 CH_4: 8.42:5.75 CO_2:18.85−7.04	BM: K_2CO_3	dna	HHV 12.57 −13.36 kJ/Nm^3	Peng et al. (2017)

F, feed; *FR*, feed rate; *BM*, bed material; *GM*, gasifying medium; *ER*, equivalence ratio; *GT*, gasification temperature; *GY*, gas yield; *H*, height; *ID*, inner diameter; *ER*, equivalence ratio; *F*, feed material; *dna*, data not available; *FS*, feed size; *SB*, steam biomass ratio; *GCS*, gasification column specification; *PER*, potential energy recovery; *CCE*, carbon conversion efficiency.

13.7 Outline for the production of fuels and some important chemicals from gasification products

As per Baskar et al. (2020), bio-based derived chemicals such as amino acids, propane, butanediols, ethanol, and polymers have contributed about 2% of economic growth. The prime product of gasification is fuel gas, which is a mixture of H_2 and CO. It itself is a fuel, and it is also a basic building block for many hydrocarbons. This can be used for energy feedstock, transportation fuel, and chemical feedstock such as methane, carbon monoxide, biodiesel, biogas, methanol, ammonia, fertilizer, and so on. At present, H_2 and CO are produced from natural gas, but these can be produced from biomass via gasification. Biooil, a thick viscous liquid received from pyrolysis and gasification, is a mixture of hydrocarbons similar to petrol crude. Thus, the most of the petroleum chemicals such as resins, food flavorings, agrochemicals, fertilizers, levoglucosan, adhesives, preservatives, acetic acid, and hydroxyl-acetaldehyde can also be produced from biooil (Basu, 2010).

13.8 Conclusion

Gasification is a key technology to convert low-energy-density solid fuels into high-energy-density cleaner fuel gas. Gasification converts solid biomass in small building component of carbon monooxide and hydrogen through fixed-bed and fluidized-bed operations, which can be used in multiple ways. Fuel gas produced using oxygen or steam as a gasifying medium can be used directly for electricity generation as it has higher heating value ranging from 10 to 14 kJ/Mm^3. However, when air is used as a gasifying medium, then fuel gas can be used as raw materials for the different chemicals such as dimethyl ether, biodiesel, fertilizers, and so on, as it has low higher heating value (4–6 MJ/Nm^3 nitrogen included) because of N_2 in inlet stream. Thus, it has the capabilities to reduce the use of natural gas for methane and hydrogen. However, it needs to design an efficient gasifier, which can be used for large variables of biomass feed size and various types of biomass.

References

Ahrenfeldt, J., Thomsen, T.P., Henriksen, U.L., Clausen, R., 2013. Biomass gasification cogeneration e a review of state of the art technology and near future perspectives. Appl. Therm. Eng. 50, 1407–1417.

Antonopoulos, I.S., Karagiannidis, A., Gkouletsos, A., Perkoulidis, G., 2012. Modelling of a downdraft gasifier fed by agricultural residues. Waste Manag 32, 710–718.

Ariffin, M.A., Mahmood, W.M.F.W.M., Harun, Z., Mohamed, R., 2017. Medium-scale gasification of oil palm empty fruit bunch for power generation. J. Mater. Cycles Waste Manag. 19, 1244–1252.

Arvelakis, S., Gehrmann, H., Beckmann, M., Koukios, E., 2003. Agglomeration problems during fluidized bed gasification of olive-oil residue: evaluation of fractionation and leaching as pre-treatments. Fuel 82, 1261–1270.

Asadullah, M., Miyazawa, T., Ito, S., Kunimori, K., Yamada, M., Tomishige, K., 2004a. Gasification of different biomasses in a dual-bed gasifier system combined with novel catalysts with high energy efficiency. Appl. Catal. A-Gen. 267, 95–102.

Asadullah, M., Miyazawa, T., Ito, S., Kunimori, K., Koyama, S., Tomishige, K., 2004b. A comparison of Rh/CeO_2/SiO_2 catalysts with steam reforming catalysts, dolmite and inert materials as bed materials in low throughput fluidized bed gasification systems. Biomass Bioenerg. 26, 269–279.

Baskar, G., Aiswarya, R., Kalavathy, G., Pandey, A., Gnasounou, E., Raman, J.K., Kumar, R.P., 2020. A Biorefinery approach towards development of renewable platform chemicals from sustainable biomass. Chapter 6. In: Refining Biomass Residues for Sustainable Energy and Bioproducts. Technology, Advances, Life Cycle Assessment, and Economics, pp. 135–137.

Basu, P., 2010. Biomass Gasification and Pyrolisis: Practical Design and Theory, first ed. Academic Press Massachusetts, Burlington.

Bergman, P.C.A., Bodenstaff, H., et al., May 2004. Contributions ECN biomass to "The 2nd world converence and technology exhibition on biomass for energy. Industry and climate protection". Presented in Rome, Italy 10–14.

Bharath, M., Raghavan, V., Prasad, B.V.S.S.S., Chakravarthy, S.R., 2018. Co-gasification of Indian rice husk and Indian coal with high-ash in bubbling fluidized bed gasification reactor. Appl. Therm. Eng. 137, 608–615.

Bi, Z., Zhang, J., Peterson, E., Zhu, Z., Xia, C., Liang, Y., Wiltowski, T., 2017. Biocrude from pretreated sorghum bagasse through catalytic hydrothermal liquefaction. Fuel 188, 112–120.

Bimbraw, A.S., 2019. Generation and impact of crop residue and its management. Curr. Agri. Res. 7 (3), 304–309.

Blanco, P.H., Wu, C., Onwudili, J.A., Williams, P.T., 2013. Characterization and evaluation of Ni/SiO_2 catalysts for hydrogen production and tar reduction from catalytic steam pyrolysis-reforming of refuse derived fuel. Appl. Catal. B: Environ. 134-135, 238–250.

Brogioli, D., Mantia, F.L., 2021. Innovative technologies for energy production from low temperature heat sources: critical literature review and thermodynamic analysis. Energ. Environ. Sci. 14, 1057–1082. https://doi.org/10.1039/D0EE02795B.

Buragohain, B., Mahanta, P., Moholkar, V.S., 2010. Biomass gasification for decentralized power generation: the Indian perspective. Renew. Sustain. Energ. Rev. 14, 73–92.

Cabuk, B., Duman, G., Yanik, J., Olgun, H., 2019. Effect of fuel blend composition on hydrogen yield in co-gasification of coal and on-woody biomass. Int. J. Hydrog. Energ. 45 (5), 3435–3443.

Cao, Y., Wang, Y., Riley, J., Pan, W., 2006. A novel biomass air gasification process for producing tar-free higher heating value fuel gas. Fuel Process. Technol. 87, 343–353.

Chai, Y.H., Mohamed, M., Chemg, Y.W., Chin, B.L.F., Yiin, C.L., Yusup, S., Lam, M.K., 2021. A review on potential of biohydrogen generation through waste decomposition technologies. Biomass Conv. Bioref. https://doi.org/10.1007/s13399-021-01333-z.

Chen, Q., Swithenbank, J., Sharifi, V.N., February 2008. Review of Biomass and Solid Recovered Fuel (SRF) Pelletisation Technologies, EPSRC Supergen Bioenergy Theme 4 (Heat and Power), SUWIC. Sheffield University.

Cummer, K.C., Brown, R.C., 2002. Ancillary equipment for biomass gasification. Biomass Bioenerg 23, 113–128.

Dıaz, A.P., Martınez, J.E.A., 2019. Experimental study of forestry waste gasification: pinewood chips-grass mixtures. J. Renew. Sustain. Energ. 11, 044701–044707.

DOE, Office of Energy Efficiency and renewable energy, United State, https://www.energy.gov/eere/bioenergy/bioenergy-frequently-asked-questions, Accessed on September 17, 2020.

El-Shafay, A.S., Hegazi, A.A., Zeidan, E.S.B., El-Emam, S.H., Okasha, F.M., 2020. Experimental and numerical study of sawdust air-gasification. Alex. Eng. J. 59 (5), 2859–4036.

Fan, S.M., Yuan, X.Z., Zhao, L., Xu, L., Kang, T., Kim, H., 2016. Experimental and kinetic study of catalytic steam gasification of low rank coal with an environmentally friendly, inexpensive composite K_2CO_3-eggshell derived CaO catalyst. Fuel 165, 397–404.

Fryda, L.E., Panopoulos, K.D., Kakaras, E., 2008. Agglomeration in fluidised bed gasification of biomass. Powder Technol 181, 307–320.

Furusjo, E., Ma, C., Ji, X., Carvalho, L., Lundgren, J., Wetterlund, E., 2018. Alkali enhanced biomass gasification with in situ S capture and novel syngas cleaning. Part 1: gasifier performance. Energ 157, 96–105.

Gai, C., Dong, Y., 2012. Experimental study on non-woody biomass gasification in a downdraft gasifier. Int. J. Hydrog. Energ. 37, 4935–4944.

Guo, F., Dong, Y., Dong, L., Guo, C., 2014. Effect of design and operating parameters on the gasification process of biomass in a downdraft fixed bed. Int. J. Hydrog. Energ 39, 5625–5633.

Gupta, G.K., Ram, M., Bala, R., Kapur, M., Mondal, M.K., 2017. Pyrolysis of chemically treated corncob for biochar production and its application in Cr(VI) removal. Environ. Prog. Sustain. 37 (5), 1606–1617.

Holt, G.A., Blodgett, T.L., Nakayama, F.S., 2006. Physical and combustion characteristics of pellet fuel from cotton gin by-products produced by select processing treatments. Ind. Crop. Prod. 24, 204–213.

Islas, J., Manzini, F., Masera, O., Vargas, V., 2019. The Role of Bioenergy in the Bioeconomy: Resources, Technologies, Sustainability and Policy. Chapter 4-Solid Biomass to Heat and Power, pp. 145–177.

Ismail, T.M., El-Salam, M.A., Monteiro, E., Rouboa, A., 2018. Fluid dynamics model on fluidized bed gasifier using agro-industrial biomass as fuel. Waste Manag. 73, 476–486.

Jin, H., Larson, E., Celik, F., 2009. Performance and cost analysis of future, commercially-mature gasification-based electric power generation from switchgrass. Biomass Bioenerg. 3, 142–173.

Kallis, K.X., 2012. Biomass Thermal Conversion. Pelletisation of Lignocelluloses and the Effect on the Gasification Process. Ph.D. Thesis. Centre for Energy and Resource Technology Department of Environmental Science and Technology School of Applied Sciences, Cranfield University.

Kuo, P., Illathukandy, B., Wu, W., Chang, J., 2020. Plasma gasification performance of various raw and torrefied biomass materials using different gasifying agents. Bioresour. Technol. 314, 123740.

Kuo, P., Wu, W., 2016. Thermodynamic analysis of a combined heat and power system with CO_2 utilization based on co-gasification of biomass and coal. Chem. Eng. Sci. 142, 201–214.

Lahijani, P., Zainal, Z.A., 2011. Gasification of palm empty fruit bunch in a bubbling fluidized bed: a performance and agglomeration study. Bioresour. Technol. 102, 2068–2076.

Lapuerta, M., Hernandez, J.J., Pazo, A., L'pez, J., 2008. Gasification and co-gasification of biomass wastes: effect of the biomass origin and the gasifier operating conditions. Fuel Process. Technol. 89, 828–837.

Li, J., Yin, Y., Zhang, X., Liu, J., Yan, R., 2009. Hydrogen-rich gas production by steam gasification of palm oil wastes over supported tri-metallic catalyst. Int. J. Hydrog. Energ 34 (22), 9108–9115.

Li, K., Zhang, R., Bi, J., 2009. Experimental study on syngas production by co-gasification of coal and biomass in a fluidized bed. Int. J. Hydrog. Energ. 35, 2722–2726 (a).

Liew, J.X., Minh, L.A.C., Fui, C.B.L., AlNouss, A., Shahbaz, M., Al-Ansari, T., Govindan, R., Chai, Y.H., 2021. Synergistic effects of catalytic Co-pyrolysis of corn Cob and HDPE waste mixtures using weight average global process model. Renew. Energ. 170, 948–963. https://doi.org/10.1016/j.renene.2021.02.053.

Lu, Y., Jin, H., Guo, L., Zhang, X., Cao, C., Guo, X., 2008. Hydrogen production by biomass gasification in supercritical water with a fluidized bed reactor. Int. J. Hydrog. Energ. 33, 6066–6075.

Lv, P.M., Xiong, Z.H., Chang, J., Wu, C.Z., Chen, Y., Zhu, J.X., 2004. An experimental study on biomass air–steam gasification in a fluidized bed. Bioresour. Technol. 95 (1), 95–101.

Ma, X., Zhao, X., Gu, J., Shi, J., 2019. Co-gasification of coal and biomass blends using dolomite and olivine as catalysts. Renew. Energ. 132, 509–514.

Mafu, L.D., Neomagus, W.J.P., Everson, R.C., Okolo, G.N., Strydom, C.A., Bunt, J.R., 2018. The carbon dioxide gasification characteristics of biomass char samples and their effect on coal gasification reactivity during co-gasification. Bioresour. Technol. 258, 70–78.

Mahinpey, N., Gomez, A., 2016. Review of gasification fundamentals and new findings: reactors, feedstock, and kinetic studies. Chem. Eng. Sci. 148, 14–31.

Matsumura, Y., Minowa, T., Potic, B., Kersten, S.R.A., Prins, W., van Swaaij, W.P.M., van de Beld, B., Elliott, D.C., Neuenschwander, G.G., Kruse, A., Jerry Antal Jr., M., 2005. Biomass gasification in near- and super-critical water: status and prospects. Biomass Bioenerg. 29 (4), 269–292.

Miccio, F., Piriou, B., Ruoppolo, G., Chirone, R., 2009. Biomass gasification in a catalytic fluidized reactor with beds of different materials. Chem. Eng. J. 154, 369–374.

Miccio, F., Ruoppolo, G., Kalisz, S., Andersen, L., Morgan, T.L., Baxter, D., 2012. Combined gasification of coal and biomass in internal circulating fluidized bed. Fuel Process. Technol. 95, 45–54.

Murakami, T., Xu, G., Suda, T., Matsuzawa, Y., Tani, H., Fujimori, T., 2007. Some process fundamentals of biomass gasification in dual fluidized bed. Fuel 86, 244–255.

Nanda, S., Isen, J., Dalai, A.K., Kozinski, J.A., 2016. Gasification of fruit wastes and agro-food residues in supercritical water. Energ. Conv. Manage. 110 (Suppl. C), 296–306.

Peng, W.X., Ge, S.B., Ebadi, A.G., Hisoriev, H., Esfahani, M.J., 2017. Syngas production by catalytic co-gasification of coal-biomass blends in a circulating fluidized bed gasifier. J. Clean. Prod. 168, 1513–1517.

Prasertcharoensuk, P., Bull, S.J., Phan, A.N., 2019. Gasification of waste biomass for hydrogen production: effects of pyrolysis parameters. Renew. Energ. 143, 112–120.

Radmanesh, R., Chaouki, J., Guy, C., 2006. Biomass gasification in a bubbling fluidized bed reactor: experiments and modelling. Environ. Energ. Eng. 52 (12), 4258–4272.

Ram, M., Mondal, M.K., 2018. Comparative study of native and impregnated coconut husk with pulp and paper industry waste water for fuel gas production. Energ 156, 122–131.

Ram, M., Mondal, M.K., 2019a. Conversion of unripe coconut husk into refined products using humidified air in packed bed gasification column. Biomass Convers. Biorefin. 10, 409–421.

Ram, M., Mondal, M.K., 2019b. Investigation on fuel gas production from pulp and paper waste water impregnated coconut husk in fluidized bed gasifier via humidified air and CO_2 gasification. Energ 178, 522–529.

Rapagná, S., Provendier, H., Petit, C., Kiennemann, A., Foscolo, P.U., 2002. Development of catalysts suitable for hydrogen or syn-gas production from biomass gasification. Biomass Bioenerg 22, 377–388.

Rui, Z., Feng, D., Chen, H., Ji, H., 2014. Anodic TiO_2 nanotube array supported nickel – noble metal bimetallic catalysts for activation of CH_4 and CO_2 to syngas. Int. J. Hydrog. Energ. 39, 16252–16261.

Sadaka, S., Negi, S., 2009. Improvements of biomass physical and thermochemical characteristics via torrefaction process. Environ. Prog. Sustain. 28, 427–434.

Samy, S.S., 2013. Gasification of raw and torrefied cotton gin wastes in an Auger system. Appl. Eng. Agri. 29, 405–414.

Sanchez-Silva, L., Lopez-Gonzalez, D., Villasenor, J., Sanchez, P., Valverde, J.L., 2012. Thermogravimetric-mass spectrometric analysis of lignocellulosic and marine biomass pyrolysis. Bioresour. Technol. 109, 163–172.

Sarkar, M., Kumar, A., Shankar, T.J., Patil, K.N., 2014. Thermal devolatilization kinetics of switchgrass pretreated with torrefaction and densification. Trans. ASABE 57, 1199–1210.

Sarvaramini, A., Assima, G.P., Larachi, F., 2013. Dry torrefaction of biomass-torrefied products and torrefaction kinetics using the distributed activation energy model. Chem. Eng. J. 229, 498–507.

Sasaki, T., Suzuki, T., Akasaka, Y., Takaoka, M., 2017. Generation efficiency improvement of IGCC with CO_2 capture by the application of the low temperature reactive shift catalyst. Energ 118, 60–67.

Seif, S., Tavakoli, O., Fatemi, S., Bahmanyar, H., 2015. Subcritical water gasification of beet-based distillery wastewater for hydrogen production. J. Supercrit. Fluids. 104, 212–220.

Shen, L., Nakamura, A., Murakami, K., 2018. Effect of Fe_2O_3 on steam gasification of subbituminous coal/woody biomass mixture. Energ. Sci. Eng. 6, 281–288.

Shen, Y., Li, X., Yao, Z., Cui, X., Wang, C.H., 2019. CO_2 gasification of woody biomass: experimental study from a lab-scale reactor to a small-scale autothermal gasifier. Energ 170, 497–506.

Shen, L., Gao, Y., Xiao, J., 2008. Simulation of hydrogen production from biomass gasification in interconnected fluidized beds. Biomass Bioenerg 32, 120–127.

Singh, D., Yadav, S., Rajesh, V.M., Mohanty, P., 2018. Groundnut shell gasification performance in a fluidized bed gasifier with bubbling air as gasification medium. Environ. Technol. 40 (24), 3140–3152.

Singh, S., Chakraborty, J.P., Mondal, M., 2020. Torrefaction of *Acacia nilotica*: oxygen distribution and carbon densification mechanism based on in-depth analyses of solid, liquid and gaseous products. Energ. Fuel. 34 (10), 12586–12597.

Souza-Santos, M.L., 2010. Solid Fuel Combustion and Gasification: Modeling, Simulation and Equipment Operations, second ed. CRC Press Taylor and Francis Group.

Su, H., Hantoko, D., Yan, M., Cai, Y., Kanchanatip, E., Liu, J., Zhou, X., Zhang, S., 2019. Evaluation of catalytic subcritical water gasification of food waste for hydrogen production: effect of process conditions and different types of catalyst loading. Int. J. Hydorg. Energ. 44, 21451–21463.

Tang, H., Kitagawa, K., 2005. Supercritical water gasification of biomass: thermodynamic analysis with direct Gibbs free energy minimization. Chem. Eng. J. 106 (3), 261–267.

Tchapda, A.H., Pisupati, S.V., 2014. A review of thermal coconversion of coal and biomass. Waste. Energies 7, 1098–1148.

Thanapal, S.S., Chen, W., Annamalai, K., Carlin, N., Ansley, R.J., Ranjan, D., 2014. Carbon dioxide torrefaction of woody biomass. Energ. Fuel. 28, 1147–1157.

Thengane, S.K., Gupta, A., Mahajani, S.M., 2019. Cogasification of high ash biomass and high ash coal in downdraft gasifier. Bioresour. Technol. 273, 159–168.

Tian, Y., Zhou, X., Lin, S., Ji, X., Bai, J., Xu, M., 2018. Syngas production from air-steam gasification of biomass with natural catalysts. Sci. Total Environ. 645, 518–523.

Timmer, K.J., Brown, R.C., 2019. Transformation of char carbon during bubbling fluidized bed gasification of biomass. Fuel 242, 837–845.

Uslu, A., Faaij, A.P.C., Bergman, P.C.A., 2008. Pre-treatment technologies and their effect on international bioenergy supply chain logistics. Techno-economic evaluation of torrefaction, fast pyrolysis and palletisation. Energ 33, 1206–1223.

Valin, S., Ravel, S., Vincent, P.P.D., Thiery, S., Miller, H., 2019. Fluidized bed air gasification of solid recovered fuel and woody biomass: influence of experimental conditions on product gas and pollutant release. Fuel 242, 664–672.

Vélez, J., Chejne, F., Valdés, C., Emery, E., Londoo, C., 2009. Co-gasification of Colombian coal and biomass in fluidized bed: an experimental study. Fuel 88, 424–430.

Vinterbäck, J., 2004. Pellets 2002: the first world conference on pellets. Biomass Bioenerg. 27, 513–520.

Wang, C., Liu, C., Fu, W., Bao, Z., Zhang, J., Ding, W., 2016. The water-gas shift reaction for hydrogen production from coke oven gas over $Cu/ZnO/Al_2O_3$ catalyst. Catal. Today. 263, 46–51.

Wang, S., Shen, Y., 2020. CFD-DEM study of biomass gasification in a fluidized bed reactor: effects of key operating parameters. Renew. Energ. 159, 1146–1164.

Weerachanchai, P., Horio, M., Tangsathitkulchai, C., 2009. Effects of gasifying conditions and bed materials on fluidized bed steam gasification of wood biomass. Bioresour. Technol. 100, 1419–1427.

Yamaguchi, A., Mimura, N., Shirai, M., Sato, O., 2017. Bond cleavage of lignin model compounds into aromatic monomers using supported metal catalysts in supercritical water. Sci. Rep. 7, 46172.

Yaman, S., 2004. Pyrolysis of biomass to produce fuels and chemical feedstocks. Energ. Convers. Manag. 45, 651–671.

Yan, L., Cao, Y., Zhou, H., He, B., 2018. Investigation on biomass steam gasification in a dual fluidized bed reactor with the granular kinetic theory. Bioresour. Technol. 269, 384–392.

Zhao, X., Cai, Q., Zhang, S., Luo, K., 2017. The substitution of wind power for coal-fired power to realize China's CO_2 emission reduction targets in 2020 and 2030. Energ 120, 164–178.

Zhou, J., Chen, Q., Zhao, H., Cao, X., Mei, Q., Luo, Z., Cen, K., 2009. Biomass-oxygen gasification in a high-temperature entrained-flow gasifier. Biotechnol. Adv. 27, 606–611.

Further reading

Li, C., Suzuki, K., 2009. Tar property, analysis, reforming mechanism and model for biomass gasification-An overview. Renew. Sustin. Energ. Rev. 13, 594–604.

Zhang, S., Dong, Q., Zhang, L., Xiong, Y., 2015. High quality syngas production from microwave pyrolysis of rice husk with char-supported metallic catalysts. Bioresour. Technol. 191, 17–23.

SECTION V

Bioenergy production using pyrolysis

CHAPTER

14

Pyrolysis: an alternative approach for utilization of biomass into bioenergy generation

Goutam Kishore Gupta, Monoj Kumar Mondal

Department of Chemical Engineering and Technology, Indian Institute of Technology, Banaras Hindu University, Varanasi, Uttar Pradesh, India

14.1 Introduction

The amount of energy available with any nation always plays a major role, directly or indirectly, for its growth, economical development, and social prosperity. However, the galloping world population, advancement in technology, urbanization, modernization in living style, and so on around the globe create an undesirable effect on the supply of energy, which indirectly leads to an increment in the energy demand rapidly in our day-to-day life, and it is foreseen that this energy demand will be amplified by around 50% by 2050 (Anahas and Muralitharan, 2018). In addition, the demand for petroleum-based fuels in the world is also expected to be increased to 112.2 million barrels per day by the year 2035 (Disco et al., 2017). Today, approximately 90% of the energy exploitation in developing nations is fulfilled by nonrenewable sources (coal, petroleum, and natural gas). It has been reviewed that 70% of petroleum necessities of India are met through imports and it is likely to rise in coming days. India shares merely 0.4% of the world's reserves of crude oil. According to the data provided by World Energy Outlook, by 2030, 94% of India's crude oil demand would be fulfilled through imports (Karmakar et al., 2012). According to the reports, the global energy consumption has increased from approximately 7250 M toe in 1980–13,884 M toe in 2018 (https://ourworldindata.org), and at the same time, according to IEA (2019), the change in energy demand has also increased from 49 M toe in 2015 to 328 M toe in 2018 (www.iea.org).

These energy demands are largely accomplished by exploiting the conventional energy resources, i.e., coal, petroleum, and natural gas (Singh et al., 2020a). However, their extensive use has raised concern over environmental threats such as emission of greenhouse gases, global warming, and climate change (Gupta et al., 2019a,b). Greenhouse gas emission and pollution caused with the use of these fossil fuels are deteriorating the quality and also create a

Biofuels and Bioenergy
https://doi.org/10.1016/B978-0-323-85269-2.00024-1

discrepancy in the surroundings (Kambo and Dutta, 2015). Thus, these diminishing fossil fuel supplies and growing environmental degradation have directed the concentration for the development of an alternative clean, renewable, and sustainable energy source. From the past two to three decades, extensive research has been carried out globally to find the supplement source to meet the energy demand as well as diminish the environmental concerns. Although few renewable energy sources such as wind, solar, and hydrothermal are available, they have certain limitations such as intermittent source, high cost and time, and weather dependent.

Apart from this, the utilization of biomass for bioenergy generation has gathered attention, as it is available in plenty, biodegradable, cheap, and easily accessible and has required properties to generate bioenergy. Other merits of biomass comprise carbon neutral and less amount of sulfur and nitrogen content that results into lesser amount of SO_x and NO_x in the exit stream in comparison with conventional fuels. Moreover, it is the fourth largest energy source following fossil fuels and single-handedly contributes around 15% of global energy supply that includes 38%–43% of the energy requirement in the developing countries (Ertaş and Hakkı Alma, 2010).

14.2 Biomass—an alternative

Biomass is a naturally available renewable energy source. It is an intricate organic or nonorganic solid product obtained from living or dead organisms (Tripathi et al., 2016). However, some other materials such as waste paper, manure, industrial wastes, and so on are also regarded as biomass, as these wastes are also a blend of organic and nonorganic compounds that can be treated to get bioenergy via different methods. The primary process that keeps biomass rich is photosynthesis. The radiant energy from the sun is converted to chemical energy in the form of volatiles or fixed carbon, which is released as heat or any other liquid or gaseous fuel when converted using different conversion technologies.

14.2.1 Sources of biomass

Biomass has been categorized in different ways but that depends upon certain parameters such as organic, inorganic, fluid, primary biomass, secondary biomass, and so on. Different varieties of biomass are available in our country, and they can be categorized according to their availability and amount in the surroundings. The different biomass sources include lignocellulosic wastes, energy crops, forestry wastes, food crops, municipal solid wastes, and so on (Gupta et al., 2020). Categorization of types and sources of biomass is depicted in Fig. 14.1.

Huge amount of agricultural residue are produced everywhere, which can be a potential source for energy generation. These residues include the by-products collected after harvesting the agricultural crops, and it can be primary or secondary. Wheat straw, sugarcane tops, and so on are the primary residues, whereas rice husk, bagasse, and so on are example of secondary residues. Primary residues are mostly used as animal fodder and are less available for bioenergy generation as compared with secondary residues. Forestry wastes are other sources of biomass, which basically comprise the woody biomass. This type of biomass includes stem, branches, leaves, bark, and so on from different trees such as pine, oak, redwood, and so on. Agricultural residues and woody biomass are extensively used for bioenergy generation around the world. However, along with these, municipal solid wastes and energy crops are also sources of biomass. Municipal solid wastes include the kitchen wastes, office wastes, animal and human wastes, paper, plastics, and so on, and the energy crops are the starch-producing species such as root crops, cassava, and so on.

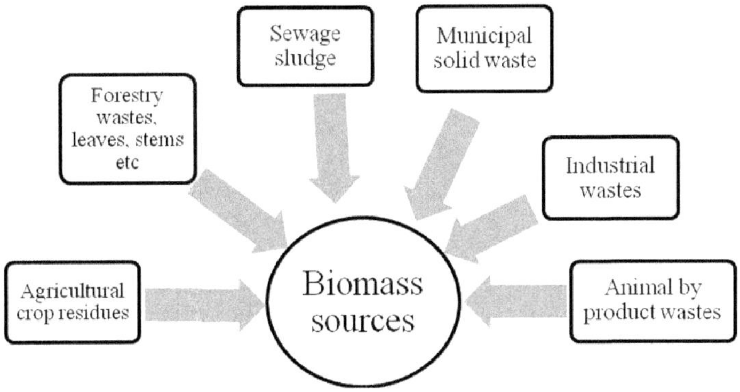

FIGURE 14.1 Different sources of biomass.

14.2.2 Physicochemical characteristics of biomass

The complete pathway (starting from transportation to final completion of product utility) for the exploitation of biomass for bioenergy generation is dependent on its nature and physicochemical characteristics. These physicochemical characteristics of the biomass have a crucial role in the designing and development of processes that are used for their conversion to biofuel. However, biomass properties vary because of the change in location, climatic condition, inherent composition, and so on.

Proximate, ultimate, compositional, and high heating value (HHV) are the most important characterization techniques that determine the utility of any raw biomass for bioenergy generation via thermochemical conversion methods. Proximate analysis comprises moisture content (MC), volatile matter (VM), ash content (AC), and fixed carbon (FC). The experiments for their determination are performed following the standard ASTM methods. MC is determined by ASTM E871, VM is determined by ASTM E872, and AC is determined by ASTM E1755. The FC is determined by the following equation:

$$FC = 100 - [MC + VM + AC]\ \text{wt.\%} \quad (14.1)$$

The ultimate analysis has widespread results in comparison with the proximate analysis. It finds out the quantity of carbon, nitrogen, hydrogen, and sulfur present in the sample using CHNS analyzer. The leftover oxygen is obtained by the following equation:

$$O = 100 - [C + H + N + S]\ \text{wt.\%} \quad (14.2)$$

HHV of biomass denotes the energy content in it, i.e., the total energy available per unit mass of the biomass and is presented in megajoule/kilogram (MJ/kg). The standard ASTM D5865-13 method is used for calculating the HHV and is determined by bomb calorimeter.

Lignocellulosic biomass comprises mostly three components, i.e., hemicelluloses, cellulose, and lignin along with some extractives. Cellulose is the natural polymer and replicate units of six-carbon ring D-glucose. The three hydroxyl groups in all the unit are bonded to each other via inter- and intramolecular hydrogen bonding and provides cellulose a crystalline and stable structure. Hemicellulose is a short diverse branched series of polymers (hexose and pentose), which environs the cellulose and provides a linking between the cellulose and lignin. Lignin is a complex, large molecular weight aromatic polymer that has cross-links of phenyl groups. Lignin is not easily degraded because of its covalent bonding. The compositional analysis of biomass can be done by following the protocols of National Renewable Energy Protocol method or Van Soest (Gupta and Mondal, 2018). Table 14.1 provides the properties of some of the biomass.

TABLE 14.1 physicochemical characteristics of biomass.

Biomass	Proximate analysis (wt.%)				Ultimate analysis (wt.%)					Compositional analysis (wt.%)				HHV (MJ/kg)	References
	MC	VM	AC	FC	C	H	N	O	S	Hem.	Cel.	Lig.	Oth.		
Saccharum munja	4.5	80.7	3.89	10.91	63.29	7.84	2.34	27.19	–	38.90	35.10	17.70	8.30	19.67	Kumar et al. (2019a,b)
Rice husk	8.70	60.21	19.70	11.39	38.50	4.79	1.01	36.10	–	21.22	39.15	13.10	26.53	15.61	Mallick et al. (2018)
Mustard stalk	9.70	70.0	7.90	12.30	43.80	5.90	0.30	43.8	0.30	18.70	39.50	22.50		17.70	Raj et al. (2015)
Wheat straw	7.67	78.33	7.0	7.0	48.24	5.64	0.56	45.55	–	36	35	18	11	15.29	Kumar et al. (2019a,b)
Pongamia pinnata	7.0	78.1	4.6	10.3	43.8	6.4	3.9	41.5	0.24	26.8	21.4	3.8		18.72	Doshi et al. (2014)
Castor seed	7.24	65.51	4.22	23.33	48.96	5.52	2.79	42.61	0.12	4.60	46.95	32.31		22.75	Santos et al. (2015)
Sagwan sawdust	3.92	76.87	4.04	15.17	48.87	5.76	1.09	44.28	–	12.9	51.1	24.7	11.3	17.97	Gupta et al. (2019a,b)
Sugar cane bagasse	7.41	80.98	4.09	7.52	46.19	6.11	0.59	47.11	–	26	43	12		18.56	Kumar et al. (2019a,b)
Mahua seed	7.88	77.05	2.08	12.63	52.48	8.69	2.62	34.86	0.34	27.33	37.92	14.20		25.88	Mishra and Mohanty (2020)

AC, Ash content; *C*, Carbon; *Cel.*, Cellulose; *FC*, Fixed carbon; *H*, Hydrogen; *Hem.*, Hemicellulose; *HHV*, High heating value; *Lig.*, Lignin; *MC*, Moisture content; *N*, Nitrogen; *O*, Oxygen; *Oth.*, Others; *S*, Sulfur; *VM*, Volatile matter.

14.3 Pretreatment of biomass

Pretreatment refers to the modification in the characteristics of biomass for its better accessibility toward final processing to biofuels or chemicals via different conversion technologies. The main objective is to disrupt the complex biomass structure to simple monomer units for enhanced output; however, the pretreatment is dependent on the type of lignocelluloses present and the pathway pursued for the conversion. Pretreatment processes basically enhance the porosity leading to higher surface area for accessibility, decrystallization of cellulose, solubilization of hemicelluloses and lignin in case of biological processes, and so on. The different pretreatment technologies can be categorized as physical, thermal, chemical, physicochemical, and biological. All these technologies have different effects on the properties of biomass; however, all these methods have major effects on the surface area.

14.3.1 Physical pretreatment

Physical pretreatment process comprises milling, chopping, grinding, extrusion, or pressure causing changes in the structure of lignocellulosic biomass that reduces biomass recalcitrance. The mechanical methods such as grinding, chipping, and so on help in reducing the particle size of the biomass and also reduce the crystallinity of the cellulose. The reduced particle size also provides enhanced surface, and it also facilitates easy feeding to the reactor for pyrolysis (Gupta and Mondal, 2019a,b). Biomass is a poor conductor, and the temperature gradient across the particles influences the pyrolysis reaction mechanism. Thus, smaller particles promote uniformity in the temperature gradient by better heat and mass transfer. However, the amount of moisture content in the biomass decides the comminution method, and the size reduction methods are quite costly that increases the overall biomass processing cost. Extrusion is a combination of multiple operations in a single unit. It leads to the abrasion of biomass and depolymerizes the biomass components, i.e., cellulose, hemicelluloses, and lignin. Extrusion improves the physical properties by reducing the particle size and increasing the surface area (Zheng et al., 2014). Under high pressure, it produces pellets of cylindrical structure with less moisture content and large volumetric energy density.

14.3.2 Thermal pretreatment

Thermal pretreatment refers to the treatment of biomass at a certain temperature before processing to products. It is performed for moisture removal, lowering of viscosity, and also for pathogens elimination in case of biological processes. Different thermal pretreatment process includes drying, steam explosion, torrefaction, liquid hot water treatment, and irradiation methods (Zheng et al., 2014). Drying of biomass refers to the elimination of inherent moisture from the surface that enhances its ability to produce biooil of better quality after pyrolysis. Torrefaction is the treatment of biomass in between 200 and 300°C, where moisture is disposed of completely along with sufficient amount of oxygen (Singh et al., 2020b). Torrefaction enhances the characteristics of biomass in terms of better fuel value, high energy density, better grindability, less susceptible to biological degradation, better feeding to the reactor, and so on (Singh et al., 2020c). Although torrefied biomass has less yield of biooil after pyrolysis, the quality of biooil produced is much better as it has lower acidity, high energy density, and so on. Since most of the oxygen is removed from the torrefied biomass, the syngas produced has more of hydrogen and methane content with lesser amount of carbon dioxide (Kan et al., 2016). In steam explosion, biomass is exposed to the high-pressure saturated steam for a short period of time, and then, it is depressurized swiftly that bursts the biomass structure, i.e., breakage of carbohydrate linkages. The operating conditions for the process are temperature 160–260°C and pressure 0.69–4.83 MPa, and time varies from few seconds to some minutes (Sun and Cheng, 2002). In liquid hot water treatment, water penetrates deeper into the biomass cell structures where solubilization of hemicelluloses, hydration of cellulose, and slight lignin removal occur. This improves the cellulose degradability for microbes and enzymes, and at the same time, partial removal of hemicellulose lessens the acetic acid content that improves the biooil quality. Some of the irradiation techniques such as microwave, ultrasonic, and gamma rays have been widely used for the pretreatment of biomass. In microwave heating, energy is produced via electromagnetic field and is transported to the biomass that provides rapid heating all through the biomass and produces hot spots. This causes disruption of lignin and cell wall structure with increased temperature and pressure. On the other hand, ultrasonic irradiation also causes rupture of cell wall, increases the specific surface, and decreases the degree of polymerization. The principle behind the process is cavitation effect, i.e., the ultrasound breaks the different linkages of lignin and leads to the formation of cavitation bubbles by splitting the polysaccharides and lignin fraction of the biomass. These bubbles grow up to a critical size, which, when collapses, increases the pressure and temperature and causes disruption (Shirkavand et al., 2016; Kumar and Sharma, 2017). Gamma ray goes deeper into the lignocelluloses and modifies lignin structure apart from decreasing the cellulose crystallinity (Kumar et al., 2020). These irradiation methods are expensive, and also it cannot hold large volumes at industrial scale.

14.3.3 Chemical pretreatment

A variety of different chemical pretreatments have been used for the fractionation of lignocellulosic biomass. These processes generally disrupt the organic fraction of the biomass, i.e., it breaks the lignin—carbohydrate linkage and the crystalline cellulose structure. The different chemicals include acid, alkali, ionic liquids, organosolvs, and so on.

Acid pretreatment is extensively studied method. This pretreatment can be accomplished either with concentrated acid at low temperatures (e.g., 30–40°C) or with dilute acids at certain higher temperatures (e.g., 120–230°C) depending upon the requirement. Different organic and inorganic acids such as nitric, maleic, hydrochloric, sulfuric, orthophosphoric, oxalic, and so on have been used for pretreating biomass. But of which, sulfuric and hydrochloric are mostly used. This pretreatment hydrolyzes hemicelluloses almost completely into its component sugar and also partially disrupts lignin. The porosity for the biomass is also improved, and it makes cellulose more accessible toward enzymes for hydrolysis (Sen et al., 2016). However, because of the toxic and corrosive nature of acids, the process requires expensive nonmetallic reactors and neutralization of pH. In case of concentrated acid, the acid must be recovered, as these are costly. Alkaline pretreatment involves the use of bases such as NaOH, KOH, Ca $(OH)_2$, $NH_3.H_2O$, and so on, which breaks the lignin structure and also cleaves the linkage between lignin and carbohydrate polymers. It generally breaks the acryl-ester, ether, and C–C bonds in the biomass (Saratale and Oh, 2015). This makes biomass more porous, swells cellulose, increases specific surface area, and decreases the degree of crystallinity. The presence of certain elements such as K, Na, Ca, and so on plays a catalytic role during pyrolysis. For example, more amount of K in the material enhances the yield of lower-molecular-weight compounds during pyrolysis (Eom et al., 2012). Ionic liquids are mainly composed of large organic cations and small inorganic anions. Ionic liquids have gathered attention, as it cian dissolve cellulose completely at gentle conditions (90–130°C and ambient pressure) with little energy input. The mechanism for cellulose dissolution involves the interaction between hydroxyl group of cellulose and ionic liquids, where hydrogen bonds are broken and breaks the chain in cellulose. Some of the ionic liquids such as 1-n-butyl-3 methylimidazolium chloride, N-methylmorpholine-*N*-oxide monohydrate, 3-methyl-*N*-butylpyridinium chloride chloride, and benzyldimethyl (tetradecyl) ammonium chloride have been used for the pretreatment of biomass. Ionic liquids are considered as green solvents because of their unique characteristics such as low toxicity, strong chemical stability, low vapor pressure, nonflammable, and so on (Kumar et al., 2020).

Organosolv pretreatment involves the use of organic solvents such as methanol, glycerol, ethanol, acetone, and so on or their mixtures. It is executed in the temperature range of 100–250°C and removes lignin and hemicellulose by tearing the bonds between them. Hence, it increases the pore volume and surface area for cellulose. But volatile nature of solvents prevents its use at industrial scale.

14.3.4 Biological pretreatment

Biological pretreatment utilizes different microbes, fungi, and enzymes that selectively disrupt hemicelluloses and lignin. These processes are slower, as it requires several days and sometimes do not need any chemicals, use less energy, and have better environment footprints (Yu et al., 2013). This type of pretreatment is mostly done in case of biochemical conversion processes and is less preferred in case of thermochemical processes. Fungal pretreatment is conducted in sterilized conditions, and it degrades hemicelluloses and lignin, which enhances cellulose digestibility. Different types of fungi such as white, brown, and soft rot have been used by different researchers for the pretreatment of lignocelluloses (Sun and Cheng, 2002). On the contrary, microbes mostly degrade cellulose and hemicelluloses and are more recalcitrant to lignin degradation. However, sterilization of biomass is not required in case of microbial pretreatment. Enzymes basically hydrolyze cellulose and hemicelluloses that enhance their activity in case of biological conversion processes. The most commonly used enzymes are cellulase and hemicellulase. However, enzymes have also been proposed for the hydrolysis of lignin prior to pyrolysis, as it enhances the production of phenols and other hydrocarbons (Lou and Wu, 2011). Chars produced are also highly porous. The application of biological pretreatment on large scale is still a challenge as it is not as efficient as other methods. The process requires long residence time, sometimes sterilization that makes the process costly, large space, and so on.

Apart from these methods, there are certain processes that combine two or three processes together for the pretreatment of lignocelluloses biomass. Fig. 14.2 depicts the different pretreatment methods for biomass.

14.4 Bioenergy conversion pathways

Energy derived from biomass is generally considered as bioenergy, and the technologies followed to extract energy out of the biomass are called bioenergy conversion technologies. However, prior to conversion, a number of stages such as harvesting, drying storage, transportation, and so on are required to convert biomass into desired form for bioenergy conversion. Biomass has the ability to be converted to different forms of energy, but this conversion is influenced by certain factors, i.e., the type, characteristics and amount of biomass, the desired final form of energy required, environmental policy and standards, economic condition, and so on (McKendry, 2002). In most of the situations, the energy conversion route is dependent on the form of energy required followed by the type and availability of biomass.

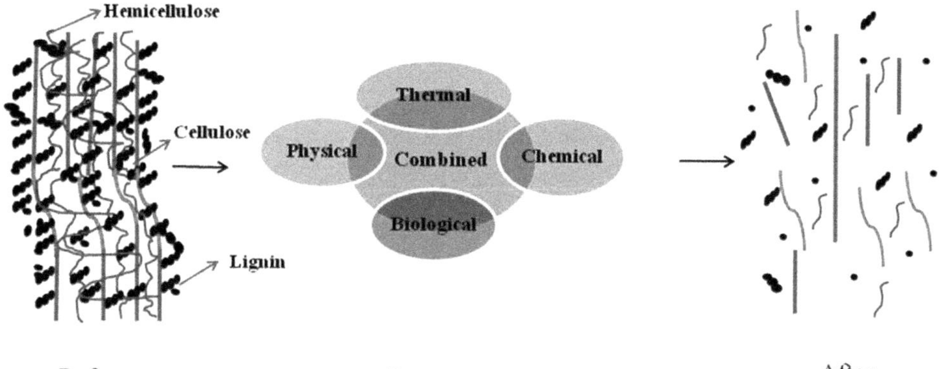

FIGURE 14.2 Effect of pretreatment methods.

Biomass can be converted to three different products, i.e., power or heat, liquid fuels, and chemicals or chemical feedstock. Today, irrespective of different technologies, biological and thermochemical techniques have congregated concentration in utilizing the biomass wastes into bioenergy generation as depicted in Fig. 14.3. In general, biological conversion processes include anaerobic digestion (AD) and fermentation that necessitate the use of microorganisms, bacteria, and so on for the biomass depolymerization; therefore, biomass with low lignin content is preferred. Thermochemical processes encompass four main process options, i.e., combustion, gasification, pyrolysis, and liquefaction (Adams et al., 2018). Low moisture is preferred for these processes, as it lessens the exploitation of additional energy in drying as well as less dilution of products occurs.

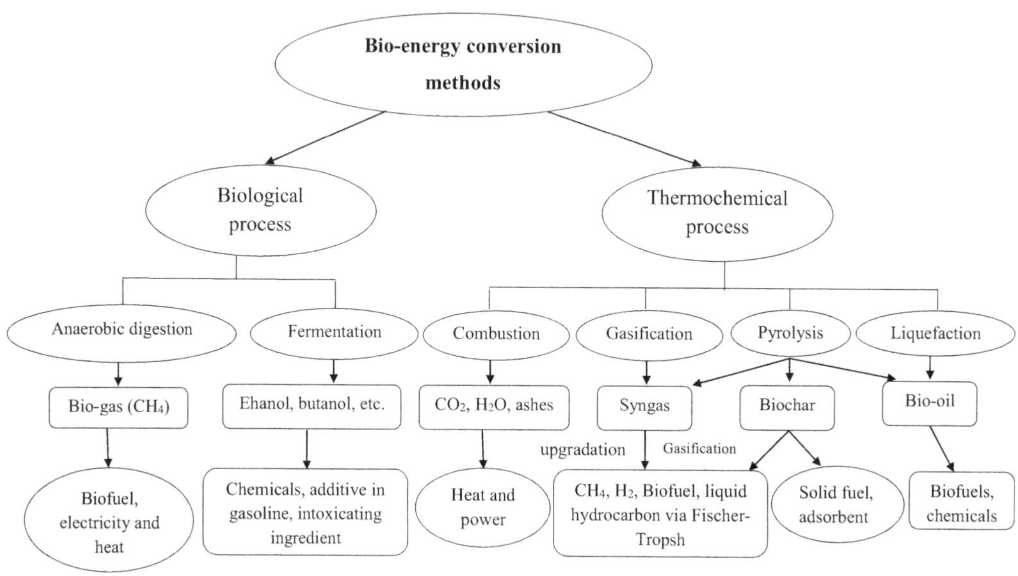

FIGURE 14.3 Bioenergy conversion technologies.

14.4.1 Biological conversion

Biological conversion of organic fractions of lignocellulosic biomass to bioenergy is a sustainable and environmentally friendly practice. These processes involve the use of different biological catalysts or biological organisms to produce bioenergy. AD and fermentation are the two mostly used biological processes.

14.4.1.1 Anaerobic digestion

AD is the conversion of organic fractions of the biomass directly to gas, which is a combination of methane, carbon dioxide, and very small fraction of hydrogen sulfide (Deublein and Steinhauser, 2008). This gas is termed as "biogas."

The process is executed in an oxygen-deficit environment using microorganisms that degrades the organic matter into gas, which has energy content around 20%–40% of the biomass feedstock. AD process is completed in four different steps, i.e., hydrolysis, acidogenesis, acetogenesis, and methanogenesis. AD is commercially established and broadly accepted technology for treating waste having high MC (80%–90%). Various feedstocks, for example, animal slurry or manure, vegetable and fruit waste, municipal solid waste, agricultural leftovers, and so on, can be used as substrate for AD. AD necessitates a satisfactory quantity of nitrogen and carbon content in the feedstock to maintain the C/N ratio. The yield and composition of the biogas is highly dependent on the inoculums and feedstock used. Biogas constitutes approximately 50%–60% methane and 40%–50% carbon dioxide and traces of some other gases. Biogas can be used in gas turbines, as a source of heat, or can be upgraded to the quality of natural gas by removing carbon dioxide. The leftover residue in the AD reactor is called digestate, which is an inert and sterile product that has the utility as biofertilizers or soil conditioner since it contains valuable plant nutrients as well as organic humus (Gupta and Mondal, 2019a,b).

14.4.1.2 Fermentation

Fermentation is also a biological process that operates in anaerobic condition and entails the exploitation of microorganisms (yeast) to process the sugar derived from biomass to mostly alcohols, acids, or blend of gases (CO and H_2S) (McKendry, 2002). The biomass is firstly degraded to small particles followed by conversion of starch to sugar and finally to alcohol. There are various substances that can be used in fermentation processes such as starch, lignocellulosic biomass, algal biomass, and so on. However, the conversion of lignocellulosic feedstocks such as wood, sawdust, and so on is complicated because of their intricate long-branched polymeric molecules and needs acid or enzymatic hydrolysis prior to fermentation of sugar to alcohol. The alcohol produced is further concentrated by the process of distillation, which is an energy-intensive process. Fermentation is a long-drawn-out time-taking process, where there is high probability of contamination from other foreign microorganisms. Biobutanol in today's world is used as blend with gasoline in many countries such as Australia, Sweden, and so on (Von Blottnitz and Curran, 2007). The leftover solid residue after fermentation process can be used as cattle feed, and if sugarcane is the feedstock for fermentation, then the leftover bagasse is used in boiler as fuel or as raw material for gasification (Coombs, 1996).

Although these processes are proficient in generating bioenergy, there are also many disadvantages such as the processes are tedious or time-consuming and they have lesser efficiency.

14.4.2 Thermochemical conversion

Thermochemical processes involve the use of heat and other chemical catalysts to generate bioenergy from the biomass. These processes can be further categorized as combustion, gasification, liquefaction, and pyrolysis (Patel et al., 2016). These processes are discussed in the following.

14.4.2.1 Combustion

Combustion is among the older and simpler thermochemical technique for the utilization of biomass. It is entirely an oxidative operation that is performed in between 800 and 1000°C where the entire chemical energy stored in the biomass is recovered as thermal energy (heat and power) together with CO_2 and water (Tripathi et al., 2016). Any type of biomass can be used for combustion, but it should have the moisture content below 50% for the better efficiency of the process (Goyal et al., 2008). The bioenergy produced from the process can be exploited in boilers, furnaces, steam turbines, and so on. However, the efficiency of the process can be improved if the biomass is cofired with coal. Certain challenges for the process are high amount of CO_2, NO_X, SO_X, and particulate matter emission along with the handling of ash.

14.4.2.2 Gasification

Gasification is a partial oxidation process, which is performed to convert carbonaceous components of the biomass to gaseous fuel. The process is operated at around 800–1000°C or even higher in the gaseous environment of O_2, CO_2, and air or with the mixture of these gases (Adams et al., 2018). Since it is a partial oxidation process, the chemical energy of the carbon gets converted to other combustible fuel gas, which has certain high efficiency than raw biomass. The composition of fuel gases includes CO, CH_4, H_2, and CO_2. Other than these gases, some liquid and char are produced as by-products that have other added values (Morrin et al., 2012). However, primarily syngas ($H_2 + CO$) can be produced if the process is operated at a certain elevated temperature, i.e., at or above 1200°C. The efficiency of the gasification process can be expressed in terms of % carbon conversion. These gases have enough calorific value that can be used as a source of heat and power in gas turbines or gas engines. The syngas has the ability to be used as the raw material in Fischer–Tropsch synthesis for liquid hydrocarbon production. Gasification is the most efficient method to produce H_2, which is regarded fuel of the coming generation.

14.4.2.3 Liquefaction

Liquefaction is a low temperature and high-pressure thermochemical technique for biomass degradation useful bioenergy. The process is performed in between 200 and 400°C and pressure of 20–200 bar. The major product of this process is biooil, but char and gases are also produced as by-products. The process has the feasibility to be operated in both batch and continuous mode; however, continuous systems need the feeding system that can be operated under pressure. The product stream of the process has high energy content and distinct characteristics of superior heat recovery as compared with other processes. The biooil obtained through this method requires additional hydrotreatment for commercial utilization, as it cannot be used directly as liquid fuel. The biooil obtained through the process has less oxygen and moisture content; thus, upgradation technique does not require complicated processes. The char obtained from the reactor can be used as fertilizer as well.

Apart from this, pyrolysis is another thermochemical technique to produce bioenergy, which is discussed in detail in the next section.

14.5 Pyrolysis

The word "pyrolysis" is a combination of two Greek words, i.e., "pyro" and "lysis." The word "pyro" refers to the fire, whereas "lysis" symbolizes breaking or cracking, respectively. Pyrolysis is an endothermic thermal depolymerization/degradation process executed in the temperature range of 300–700°C in inert atmosphere. Under the supply of continuous heat, the components of biomass are thermally cracked, which produces gases and vapors, which undergo secondary reactions and produce different products (Demirbas and Arin, 2002). It is the most efficient process to produce bioenergy in all three forms, i.e., solid (biochar), liquid (biooil), and pyrolytic gas from biomass. However, process variables such as temperature, particle size, sweeping gas flow rate, heating rate, vapor residence time, and so on play a key role in deciding the yield of products. The products of pyrolysis and their characteristics are discussed in the next section. Pyrolysis has gained much interest owing to the flexibility in operation, versatility of the technology, and adaptability to a wide range of feedstocks and products.

14.5.1 General reaction mechanism for biomass pyrolysis

Biomass pyrolysis is a complex process because of the dissimilarity in disintegration of the different components with altering reaction mechanism and reaction rate. During biomass pyrolysis, multiple reactions take place both in parallel and in series, including dehydration, decarboxylation, depolymerization, aromatization, isomerization, and charring (Vamvuka, 2011; Lange 2007; Collard and Blin, 2014). Basically, there are three stages in the pyrolysis of biomass. The first stage is the evaporation of moisture of water from the biomass. This stage occurs in the temperature range of around 100–110°C. Thereafter in the second stage, devolatilization occurs through intricate reaction mechanisms such as depolymerization, cracking, and fragmentation, which leads to the formation of condensable and noncondensable vapors. Finally, in the third stage, the charring occurs that involves the recondensation and repolymerization to form carbon-rich stable solid product (biochar).

After the moisture removal stage, the degradation of biomass components takes place at different range of temperatures. Among the lignocellulosic components, hemicellulose degrades first, i.e., in the temperature range of around 200–315°C, because of its lower thermal stability and amorphous nature. Cellulose is crystalline in nature that has long polymeric glucose units and degrades to lower monomer units from 315 to 400°C. Lignin has a heavy cross-link and complex polyaromatic molecule structure that decomposes over a wide range of temperature approximately from 180 to 900°C (Collard and Blin, 2014). In the second stage, when the devolatilization occurs, a large number of heavy organic compounds such as phenol, guaiacol, cresol, levoglucosan, furfural, levomannosan, and so on are produced, which are collected as biooil through condensation (Wang et al., 2011, 2013; Branca et al., 2013). In the final stage of charring, the volatiles liberated in the devolatilization stage either recondense or repolymerize to be a part of the biochar, or after further cracking, they are converted to noncondensable gases (H_2, CO, CO_2, and CH_4) and light organic compounds (e.g., formic acid, formaldehyde, hydroxyacetone acetic acid, acetaldehyde, etc.) that can also be condensed as biooil. Therefore, pyrolysis of biomass leads to the creation of biochar, biooil, and pyrolytic gases by the three-step mechanism discussed before.

14.5.2 Types of pyrolysis

Based on the operating conditions such as temperature, heating rate, sweeping gas flow rate, vapor residence time, and so on, pyrolysis is categorized as slow, fast, intermediate, or flash pyrolysis. Each type of pyrolysis has different main product, advantages, and certain boundaries. The product distribution is mostly dependent on the compositional contents of the biomass and interactions among them while pyrolysis.

14.5.2.1 Slow pyrolysis

It is a conventional type of pyrolysis, which is operated at lower temperature (300–500°C) and slower heating rate (0.1–0.8°C/s) and has long residence time (5–30 min or more). It mainly favors the production of biochar, but biooil and pyrolytic gases are also produced in little quantities (Demirbas and Arin, 2002). A typical slow pyrolysis produces 20%–50% biooil, 25%–35% biochar, and 20%–50% pyrolytic gases (Brownsort, 2009). The processing time can be of several hours to ensure sufficient time for the volatiles to recombine with the biochar to maximize its yield. Longer vapor residence time along with lower heating rate furnishes an appropriate environment and adequate time for the completion of different reactions. This eventually results into the creation of solid biochar.

14.5.2.2 *Fast pyrolysis*

This type of pyrolysis is carried out in the temperature range of around 450–850°C, at the heating rate of 10–200°C/s with a small vapor residence time varying from 1 to 10 s. This type of pyrolysis is performed mainly to collect biooil as the main product; however, biochar and pyrolytic gases are also produced in small amounts. Around 60%–75% of biooil, 15%–25% of biochar, and 10%–20% of pyrolytic gases are obtained from fast pyrolysis (Bridgwater, 2003). The small vapor residence time within the reactor reduces the exposure time of the volatiles with the biochar surface, and thus, it decreases the further secondary recombination of volatiles with biochar, which eventually provides more volatiles as biooil. Also, high heating rate leads to depolymerization and fragmentation of lignocellulosic components at a faster rate and concurrently accelerates the creation of lighter volatiles. This type of pyrolysis is being used in many other applications such as certain chemicals or food flavor production.

14.5.2.3 *Intermediate pyrolysis*

This process lies in between slow and fast pyrolysis, as it makes balance between biochar and biooil yields. The primary product in case of slow pyrolysis is biochar, whereas biooil is the main output for fast pyrolysis. The operating conditions for this type of pyrolysis are temperature 500–650°C, heating rate 0.1–10°C/min, and residence time 5–40 min. The product contains 40%–60% biooil, 15%–25% biochar, and 20%–30% pyrolytic gases (Kebelmann et al., 2013). Certain advantages of intermediate pyrolysis include less amount of reactive tar in the biooil that shows its direct use in engines or boilers and also the production of dry biochar, which can be used in agricultural purposes (Mahmood et al., 2013).

14.5.2.4 *Flash pyrolysis*

Flash pyrolysis is regarded as the enhanced type of fast pyrolysis process. The temperature required for this pyrolysis ranges in between 900 and 1200°C, and the heat energy provided is for a very small period of time, i.e., 0.1–1 s (Demirbas and Arin, 2002; Li et al., 2013). The product distribution is greatly affected by the heat and mass transfer, reaction kinetics, and phase alteration behavior of the biomass. The fast-heating rate along with high temperature and low vapor residence time leads to high biooil and pyrolytic gas yield, but the char yield gets decreased. Flash pyrolysis produces 25%–40% biooil, 50%–60% pyrolytic gases, and 5%–15% biochar (Foong et al., 2020).

However, the challenge in using flash pyrolysis at industrial level is in designing the reactor where biomass can stay for an infinitesimal span of time under very high heating rate. Also, the quality and stability of the biooil collected with this process is strongly exaggerated by the presence of char/ash, as it may catalyze the polymerization process within the biooil, which may increase the viscosity of biooil (Canabarro et al., 2013).

14.5.3 Effect of process variables on product yield

The process parameters play a very significant role in the pyrolysis of biomass. Some of these operating variables are temperature, heating rate, biomass particle size, sweeping gas flow rate, and so on. These parameters not only manage the pyrolysis product yield but also have influence on product characteristics. Thus, it is necessary to optimize these parameters for each feedstock to get the best yield of products as well as the better quality. These operating variables are discussed in the following.

14.5.3.1 *Effect of temperature*

Temperature has the most significant character in the yield of pyrolysis products and their characteristics. Temperature provides necessary heat required to degrade bonds in the biomass. Increase in the temperature increases the cracking reaction, thereby increasing the conversion of biomass. The increase in the temperature has a positive result on the biooil yield. It has been reported by different researchers that the optimum temperature for maximum biooil yield generally lies in the temperature range of 450–550°C (Guedes et al., 2018). However, it may vary as it also depends on the biomass type and other process parameters. But at much higher temperature (above optimum biooil yield temperature), the reverse effect has been observed, i.e., at higher temperature, secondary reactions occur, leading to cracking of larger molecular weight hydrocarbon to smaller hydrocarbons, which produces more gases and less biooil (Isahak et al., 2012). The biochar yield decreases continuously with increase in temperature because of the generation of more and more volatiles.

The pyrolysis of different feedstock showed different temperatures for maximum biooil yield, for example, rice husk (70 wt.%) at 450°C (Alvarez et al., 2014), poplar (69 wt.%) at 455°C (Makibar et al., 2015), pistachio shell (20.5 wt.%) at 550°C (Apaydin-Varol and Putun, 2007), cassava rhizome (63.23 wt.%) and cassava stalk (61.39 wt.%) at 472°C and 469°C (Pattiya and Suttibak, 2012), respectively, waste of the palm (72.4%) at 500°C (Abdullah and Gerhauser, 2008) etc. In the study by Gupta et al. (2019a,b), the effect of temperature on teak sawdust pyrolysis showed 600°C, as the maximum temperature for biooil yield and thereafter biooil yield decreased and gaseous yield increased. But the biochar decreased continuously. There was great variation in the characteristics of all three products with temperature. Besides yield of biooil, its quality also changes with pyrolysis temperature. According to Huang et al. (2014), low temperature pyrolyzed biooil contains alkenes, alkanes, long-chain fatty acids, esters, and so on; however, at higher temperature, cracking of aliphatic compounds forms aromatics, which decreases the H/C ratio and increases organonitrogen content. Similarly, Alvarez et al. (2014) found reduction in ketone and acid concentration in the biooil with increase in temperature, but increase in phenolic content was also seen. Similarly, the characteristics of biochar and pyrolytic gases are also affected with temperature.

14.5.3.2 *Effect of heating rate*

Heating rate also plays a crucial role in biomass degradation process, as it defines the type of pyrolysis, i.e., slow, fast, intermediate, or flash pyrolysis. High heating rate favors the quick degradation of biomass, which produces more volatiles and less biochar. This leads to more production of biooil because of reduction in heat and mass transfer limitations along with inhibition of secondary reactions that cause repolymerization reactions (Akhtar and Amin, 2012). However, certain secondary reactions occur that produces more and more volatiles, which on condensation enhance the biooil yield. In addition, the slower heating rate prevents the thermal cracking of biomass and thus produces more biochar. In the study by Salehi et al. (2009), the biooil yield from sawdust pyrolysis increased with the increase in heating rate from 500 to 700°C/min, but beyond this heating rate, no further increase in biooil yield was observed owing to heat and mass transfer limitations. Similarly, Putun et al. (2007) executed the pyrolysis of tobacco residues in a packed bed reactor where an increase in the biooil yield by 10 wt.% was observed as the heating rate increased from 7 to 300°C/min. Also, Aysu and Kucuhk (2013), Angin (2013), and Sensoz and Angin (2008) accounted reduction in yield of biochar from the pyrolysis of *Ferula orientalis* L., safflower seed, and *Carthamus tinctorius* L., respectively, on elevating the rate of heating from 30 to 50°C/min for the temperatures in between 400 and 600°C. Uzun et al. (2006) reported less water and oxygen in the biooil obtained at higher heating rate for the same pyrolysis temperature. This may be due to inhibition of secondary reactions that improves the biooil quality.

14.5.3.3 Effect of particle size

Since biomass particles are poor conductor of heat, it affects the heat transfer during pyrolysis and also affects the product distribution. Particle size of the biomass is one of the significant parameters that determines the rate with which heat is being provided to the biomass inner core. It has been investigated by various researchers that with increase in the biomass particle size, the biooil yield decreases and biochar yield increases. Upon increment in the particle size, the space between the outer surface and core gets enlarged, which increases the resistance for heat flow to the inner core. Also, the vapor produced has to travel longer distance, which leads to additional secondary reactions that result into the increase in the biochar yield and reduction in biooil yield (Tripathi et al., 2016). Upon pyrolysis of mallee wood, Shen et al. (2009) observed a decrease in the biooil yield with an increase in particle size. Similarly, Gupta et al. (2019a,b) studied the effect of particle size for teak sawdust pyrolysis. The study revealed a decrease in the biooil as well as pyrolytic gas yield and an increase in biochar yield when the particle size increased from 0.18 to 0.60 mm. The study from Mani et al. (2010) showed that biochar yield from wheat straw pyrolysis increased from 11.85 wt.% at 0.25 mm to 23.28 wt.% at 0.475 mm; however, further increase in particle size up to 1.35 mm showed negligible increase in biochar yield. Ertas and Alma (2010) inspected the effect of particle size on pyrolysis product yield in the range of 0.25–1.60 mm and concluded the independency of biooil and biochar yields with particle size. Onay and Kockar (2003) in their study reported reduction in the yield of biochar, as the particle size increased from 0.425 to 0.85 mm, but after exceeding 0.85 mm, the yield of biochar was found to be increasing. There are still many fluctuations in the study by different researchers, and further comprehensive investigations are needed in this study.

14.5.3.4 Effect of vapor residence time and sweeping gas flow rate

During pyrolysis, biomass is converted to significant amount of vapor that has to be condensed to get the biooil. Vapor residence time refers to the staying of vapor in the reaction zone. Long vapor residence time in the reaction chamber increases the further reactions such as thermal cracking, repolymerization, and recondensation of vapors that enhance the biochar yield and decrease the biooil yield. However, in case of short vapor residence time, there are incomplete secondary reactions, and thus, less biochar is produced (Guedes et al., 2018). Thus, to reduce these secondary reactions for better biooil yield, these vapors have to be removed from the reaction chamber to the condensation zone. The most commonly used purging gas for pyrolysis is nitrogen, as it is inert, easily available, and also very cheap. The flow rate of the purging gas decides the vapor residence time. Scot et al. (1999) performed pyrolysis of raw sorghum bagasse at 525°C and reported a decrease in the biooil yield from 75 wt.% to 57 wt.% with an increase in vapor residence time from 0.2 to 0.9 s. Onay (2007) achieved 23 wt.% more biooil from safflower seed pyrolysis when sweeping gas flow rate changed from static condition to 100 mL/min. However, a high flow rate decreases the biooil yield because of incomplete condensation of the vapors, which enhances the pyrolytic gas yield and, also, some amounts of biomass are removed out of the system before the completion of the pyrolysis process. Moreover, the product characteristics are significantly affected by vapor residence time. Wang et al. (2013a,b) investigated that longer vapor residence time for seaweed pyrolysis breaks the chain to aromatic hydrocarbon compounds and further secondary reactions enhance the number of alcohols, ketones, and other hydrocarbons. In case of biochar, long vapor residence time promotes the production of micro- and macropores, which enhances the pore size.

14.5.3.5 *Effect of pyrolysis bed height*

This process can be performed in either fixed-bed or moving-bed reactors. Moving bed reactors are those where biomass particles are in moving mode, i.e., not stationary, such as sprouted bed, rotary bed, fluidized bed, and so on, whereas fixed-bed reactors are those where there is no movement of particles. The heat transfer in case of moving-bed pyrolysis process occurs by both solid–solid (conduction) and gas–solid (convection) interaction. However, for fixed-bed system, the mode of transfer of heat is mainly conduction.

Bed height also affects the pyrolysis product yield. Gupta et al. (2019a,b) performed the teak sawdust pyrolysis by varying the bed height from 2 to 8 cm. The study reported enhancement in the yield of biooil from 40.27 wt.% to 48.80 wt.%, but both biochar and pyrolytic gas yield decreased. Similarly, Zhang et al. (2009) studied the variation of bed height in the range of 5–20 cm. The study reported enhancement in the yield of biooil from 5 to 10 cm and thereafter biooil yield decreased, whereas the biochar and pyrolytic gas yield showed reverse profile. At small bed height, the biomass bed is not dense, and the vapor residence time is also very small along with fewer amounts of vapors generated. This inhibits further secondary reactions such as repolymerization and reduces biochar yield. However, beyond optimized bed height, the amount of vapor generated is also very high. Thus, longer vapor residence time and repolymerization reactions decrease biooil yield and enhance biochar yield.

14.5.3.6 *Effect of biomass composition*

The composition of biomass influences the pyrolysis process as well. The chemical composition of biomass includes hemicelluloses, cellulose, lignin, and other extractives. The composition of biomass varies from one biomass to other; even the biomass of same origin may have dissimilar composition because of different soil or planting conditions (Omar et al., 2011). Thus, the product yield and its composition vary severely. High amount of hemicellulose and celluloses in the biomass favors more of the biooil generation, but high lignin refers to more biochar yield after pyrolysis because lignin is quite stable and very difficult to degrade. Kim et al. (2013) compared the biooil yield from three different biomasses, i.e., jatropha cake, palm kernel shell, and empty palm fruit bunches. The study revealed that biomass with more amounts of hemicelluloses and cellulose produced maximum biooil. Similarly, Casoni et al. (2015) compared the biooil yield from cellulose and biomass treated with fungus and acid. The biooil yield was maximum for cellulose pyrolysis, but fungal treatment also increased the biooil yield because of the breakdown of lignin. However, acid treatment reduced the biooil yield because acid protects the cellulose from thermal attack, which ultimately increases biochar yield. Stefanidis et al. (2014) stated that the biooil composition has variation for different components of biomass. The biooil from cellulose mainly contains sugars along with phenols, aldehydes, ketones, and alcohols in small concentrations. Similarly, the biooil from hemicellulose contains ketones, phenols, acids, and aldehydes, whereas the biooil from lignin has complex phenols having higher molecular weights.

The percentage of MC, AC, VM, and FC greatly affect the pyrolysis process. Generally, biomass with higher VM produces more biooil, whereas higher FC in the biomass yields more biochar. High AC decreases the biooil yield and enhances biochar and pyrolytic gas yield. The AC contains large percentage of inorganic compounds mainly alkali metals (e.g., K, Na, Mg, and Ca) that gets concentrated in the biochar (Kim et al., 2012). They behave as catalysts during the pyrolysis process by enhancing the cracking of tar vapors, which decreases the biooil yield and increases the acidic concentration. The MC plays a significant role in biooil characteristics. Any standard feed material for pyrolysis should have less than 10% MC because high moisture in the biomass transfers more water content to biooil, which deteriorates its properties such as calorific value.

14.6 Characteristics of pyrolysis products

Pyrolysis of biomass produces products in all three forms, i.e., biooil, biochar, and pyrolytic gas.

Biooil is the chief product collected after biomass pyrolysis, and it is a composite mixture of oxygenated organic compounds. It is generally dark brown in color having pungent smell, but its physicochemical characteristics such as density, HHV, viscosity, appearance, and so on are governed by the type of pyrolysis and also on feedstock. Biooil is a blend of numerous compounds that include acids, alcohols, aldehydes, esters, ketones, phenols, and lignin-derived oligomers (Gupta et al., 2019a,b; Gupta and Mondal, 2021). However, some of them are directly related to the unwanted properties of biooil. The advantages of biooil include its renewability along with low SO_x and NO_x emission when used as fuel, but high-water content, acidity, high AC, high moisture, and chemical instability discard its direct use (Zhang et al., 2013). It necessitates improvement processes to improve the properties so that it can be used directly in different industries as fuel.

Biochar is a black solid by-product collected after the pyrolysis of biomass. International Biochar Initiative (Lehmann and Joseph, 2015) defines biochar as a carbon-rich solid material collected after pyrolysis of biomass in oxygen-limited environment, and its characteristics depend upon the operating variables of the pyrolysis process. The solid porous structure is developed when oxygen and hydrogen present in the biomass leave at higher temperature. The physicochemical characteristic of biochar signifies a low H/C and O/C ratio that designates a higher degree of carbonization and greater stability. It also indicates an enhanced quality of fuel, as it directs to lesser energy loss and lesser smoke emission (Liu et al., 2013). The biochar is quite stable that has the life for almost 1000 years and has heating value that generally lies in the range of Indian coals.

Pyrolytic gas is the other mixture of gases collected after pyrolysis. The noncondensable gases are the combination of useful gases, such as hydrogen, methane, carbon monoxide, and so on. These gases have very high calorific values, but their composition varies with the variation in the process parameters also.

14.7 Enrichment of pyrolysis products

The products collected after the pyrolysis of biomass have impurities, and hence, it is undesirable to be used directly for different applications. Thus, these products need to be upgraded, modified, or enriched for an efficient use. The technologies that can be used for the upgradation or modification are described in the following.

14.7.1 Upgradation of biooil

The biooil produced from biomass pyrolysis has much higher oxygen content, water content, viscosity, low pH, chemical instability, and also low heating value (Xiu and Shahbazi, 2012). These demerits of the biooil make its direct use difficult, and thus, it should be upgraded for its application as liquid fuel or as resource of chemicals. Few biooil upgradation techniques are presented in Fig. 14.4.

Addition of hydrogen to the fuel products improves its quality. Hydrogenation is a widely used method that improves the stability as well as fuel quality by reducing the amount of organic acids, aldehydes, and other organic compounds (Xu et al., 2009). This saturates olefins in the biooil and also converts aromatics to naphthenes. Hydrodeoxygenation is a high-pressure upgradation method where hydrogen is used to eliminate oxygen from the biooil components such as acids, aldehydes, ketones, esters, and so on and transform it to a high-grade fuel oil. Catalysts for this process are usual hydrodesulphurization catalysts, such as $Co-MoS_2/Al_2O_3$, or metal catalysts (Pd/C) (Mortensen et al., 2011). Catalytic cracking refers to the thermal cracking under hydrogen flow, at the temperature more than 350°C and at high pressure.

The process combines hydrogenation accompanied by cracking. The dual function catalyst is used for hydrogenation and cracking. The process produces large amount of light products, but the experimental conditions are severe. Supercritical fluids such as ethanol, methanol, acetone, CO_2, and so on have been used to enhance the biooil quality in terms of decreasing the viscosity and increasing the heating value (Zhang et al., 2013). Molecular distillation is another technique to purify biooil, but the energy consumption is higher compared with conventional distillation. Biooil can also be used as combustible fuel in boiler or in transportation when emulsified with diesel with the aid of surfactant. Jiang and Ellis (2010) prepared

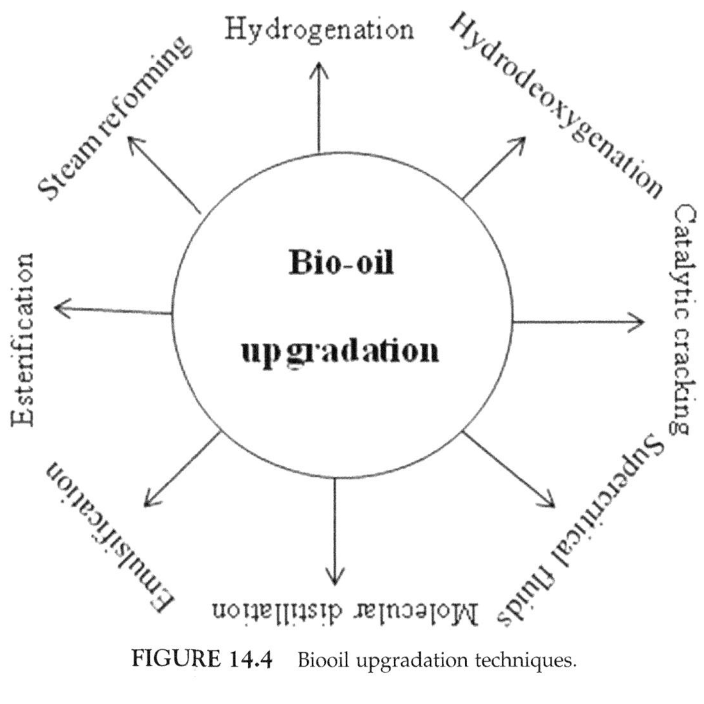

FIGURE 14.4 Biooil upgradation techniques.

the emulsion of biooil with biodiesel with octanol as emulsifier. In esterification, the organic acids present in the bio-oil are converted to their corresponding esters, which enhance the biooil property in terms of heating value, viscosity, corrosiveness, and stability (Wang et al., 2010). Steam reforming converts the biooil to more volatile components and also produces renewable hydrogen gas. But the coke deposition on the catalyst surface deactivates its properties, which is a major problem of the process.

14.7.2 Modification of biochar

To modulate the characteristics of biochar for environmental applications, different methods can be adopted as depicted in Fig. 14.5.

The introduction of acid to the biochar surface removes the impurities present and impregnates the acid functional groups. The commonly used acids are hydrochloric acid, phosphoric acid, nitric acid, and so on (Rajapaksha et al., 2016). The purpose of alkali treatment is to boost the specific surface area and incorporate the oxygen-containing functional groups in the biochar. The common alkalis are potassium hydroxide, sodium hydroxide, and so on; however, the aim of oxidizing agent is to introduce the oxygen-containing functional groups only. The most commonly used oxidizing agent is hydrogen peroxide. Metal oxides on the other hand perform different functions such as enhancement in adsorption capacity during adsorption using biochar and increase in the magnetic properties and also play a catalytic role in different applications (Wang and Wang, 2019). The role of physical technique is basically to improve the biochar structure and the surface area.

14.7.3 Enrichment of pyrolytic gases

Pyrolytic gas mainly contains CO, CH_4, CO_2, H_2, and some other impurities such as tar, H_2S, and so on. These impurities can be reduced by catalytic pyrolysis. Furthermore, cyclone separator can separate the particulate matter, and CO_2 absorption can be done via passing it through amine solutions. The tar and moisture from the gas can be reduced with the use of dichloromethane and silica gel column before gas collection.

14.8 Application of pyrolysis products

The products collected after the pyrolysis of biomass have utilities in different field described in the following.

Biooil obtained from biomass pyrolysis is a potential energy resource that has attained widespread recognition all over the globe for their

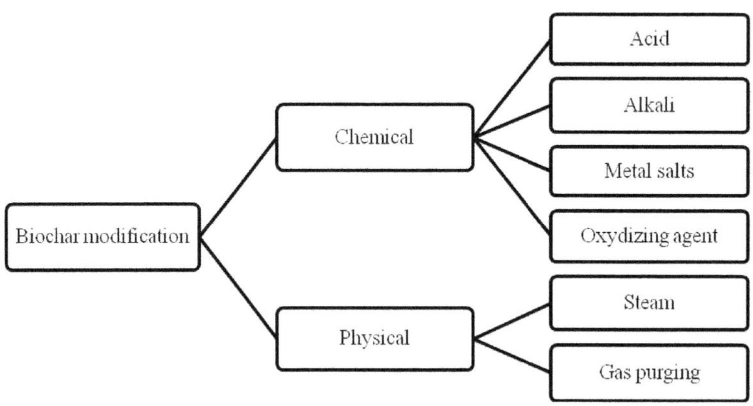

FIGURE 14.5 Biochar modification techniques.

utility as combustion fuels for boiler application, in engines or gas turbines, and also a source of valuable chemical for industries. There are several chemicals, which can be extracted from biooil, such as phenols which is used by resins industry, volatile organic acids, nitrogen cycles, and n-alkanes. The biooil is also being blended with gasoline as transportation fuel.

Biochar is a carbon-rich multifunctional substance. The different applications include its utility as solid fuel, as adsorbent for waste stream purification, and as catalyst, in soil amendment as it contains valuable nutrients required for plant growth, in carbon sequestration, and also it can be upgraded to activated carbon.

Pyrolytic gases are the blend of valuable gases such as H_2, CH_4, CO, and so on having high HHV and can be applied in the field of heat and power generation. However, the gases after cleanup can become a substitute feedstock in Fischer–Tropsch synthesis for the production of liquid hydrocarbons. Also, it can be recycled back to the system to enhance the yield and characteristics of biooil.

14.9 Conclusion

Every year enormous amounts of biomass wastes are generated from different sectors and are dumped without being utilized in open atmosphere. This open dumping of biomass not only affects the environment and human health but also affects the economical growth of the nation. Thus, these biomass wastes should be utilized in a proper way. However, there are certain drawbacks with the raw biomass such as high moisture, low HHV, different composition, and also variable particle size, but the pretreatment methods can be effective in solving such demerits. For proper utilization of biomass, different conversion technologies can be followed as described in this chapter. The enriched bioenergy products will reduce the dependency on fossil fuels. Thus, there is a need to make the local people aware regarding the utilization of biomass and different technologies available for their proper utilization. There is an urgent need to promote research and development activities for the utilization of waste biomass into value-added products for coming generation. These activities will not only reduce the amount of biomass wastes but also shorten the gap between demand and supply of energy.

Acknowledgment

The authors are highly obliged to the Department of Chemical Engineering and Technology of Indian Institute of Technology (Banaras Hindu University), Varanasi, India, for providing necessary facilities in executing the work.

References

Abdullah, N., Gerhauser, H., 2008. Bio-oil derived from empty fruit bunches. Fuel 87, 2606–2613.

Adams, P., Bridgwater, T., Lea-Langton, A., Ross, A., Watson, I., 2018. Biomass conversion technologies. Greenh. Gas Balances Bioenergy Syst. 107–139.

Akhtar, J., Amin, N.S., 2012. A review on operating parameters for optimum liquid oil yield in biomass pyrolysis. Renew. Sustain. Energy Rev. 16, 5101–5109.

Alvarez, J., Lopez, G., Amutio, M., Bilbao, J., Olazar, M., 2014. Bio-oil production from rice husk fast pyrolysis in a conical spouted bed reactor. Fuel 128, 162–169.

Anahas, A.M.P., Muralitharan, G., 2018. Characterization of heterocystous cyanobacterial strains for biodiesel production based on fatty acid content analysis and hydrocarbon production. Energy Convers. Manag. 157, 423–437.

Angın, D., 2013. Effect of pyrolysis temperature and heating rate on biochar obtained from pyrolysis of safflower seed press cake. Bioresour. Technol. 128, 593–597.

Apaydin-Varol, E., Pütün, E., Pütün, A.E., 2007. Slow pyrolysis of pistachio shell. Fuel 86, 1892–1899.

Aysu, T., Küçük, M.M., 2013. Biomass pyrolysis in a fixed-bed reactor: effects of pyrolysis parameters on product yields and characterization of products. Energy 64, 1002–1025.

Branca, C., Di Blasi, C., Mango, C., Hrablay, I., 2013. Products and kinetics of glucomannan pyrolysis. Ind. Eng. Chem. Res. 52, 5030–5039.

Bridgwater, A., 2003. Renewable fuels and chemicals by thermal processing of biomass. Chem. Eng. J. 91, 87–102.

Brownsort, P., 2009. Biomass Pyrolysis Processes: Review of Scope, Control and Variability. Biochar Research Center, Edinburgh. UK.

Canabarro, N., Soares, J.F., Anchieta, C.G., Kelling, C.S., Mazutti, M., 2013. Thermochemical processes for biofuels production from biomass. Sustain. Chem. Process 1 (1), 22.

Casoni, A.I., Bidegain, M., Cubitto, M.A., Curvetto, N., Volpe, M.A., 2015. Pyrolysis of sunflower seed hulls for obtaining bio-oils. Bioresour. Technol. 177, 406–409.

Collard, F.X., Blin, J., 2014. A review on pyrolysis of biomass constituents: mechanisms and composition of the products obtained from the conversion of cellulose, hemicelluloses and lignin. Renew. Sustain. Energy Rev. 38, 594–608.

Coombs, J., 1996. Bioconversion Assessment Study. CPL Scientific Ltd, UK (Air – CT92 – 8007).

Demirbas, A., Arin, G., 2002. An overview of Biomass pyrolysis. Energy Source 24, 471–482.

Deublein, D., Steinhauser, A., 2008. Biogas from Waste and Renewable Resources: An Introduction. Wiley VCH, Weinheim.

Disco, Y., Mahanta, P., Bora, U., 2017. Comprehensive characterization of lignocellulosic biomass through proximate, ultimate and compositional analysis for bioenergy production. Renew. Energy 103, 490–500.

Doshi, P., Srivastava, G., Pathak, G., Dikshit, M., 2014. Physicochemical and thermal characterization of nonedible oilseed residual waste as sustainable solid biofuel. Waste. Manag. 34 (10), 1836–1846.

Eom, I.Y., Kim, J.Y., Kim, T.S., Lee, S.M., Choi, D., Choi, I.G., Choi, J.W., 2012. Effect of essential inorganic metals on primary thermal degradation of lignocellulosic biomass. Bioresour. Technol. 104, 687–694.

Ertaş, M., Hakki Alma, M., 2010. Pyrolysis of laurel (*Laurus nobilis* L.) extraction residues in a fixed-bed reactor: characterization of bio-oil and bio-char. J. Anal. Appl. Pyrolysis 88 (1), 22–29.

Foong, S.Y., Liew, R.K., Yang, Y., Cheng, Y.W., Yek, P.N.Y., Mahari, W.A.W., Lee, X.Y., Han, C.S., Vo, D.V.N., Le, Q.V., Aghbashlo, M., Tabatabaei, M., Sonne, C., Peng, W., Lam, S.S., 2020. Valorization of biomass waste to engineered activated biochar by microwave pyrolysis: progress, challenges, and future directions. Chem. Eng. J. 389, 124401.

Goyal, H.B., Seal, D., Saxena, R.C., 2008. Bio-fuels from thermochemical conversion of renewable resources: a review. Renew. Sustain. Energy Rev. 12 (2), 504–517.

Guedes, R.E., Luna, A.S., Torres, A.R., 2018. Operating parameters for bio-oil production in biomass pyrolysis. A Rev. 129, 134–149.

Gupta, G.K., Mondal, M.K., 2018. Iso-conversional kinetic and thermodynamic studies of Indian sagwan sawdust pyrolysis for its bioenergy potential. Environ. Prog. Sustain. Energy 38 (4), 1–11.

Gupta, G.K., Mondal, M.K., 2019a. Bioenergy generation from agricultural wastes and enrichment of end products. Refin. Biomass Residues Sustain. Energy Bioprod. 337–356.

Gupta, G.K., Mondal, M.K., 2019b. Bio-energy generation from sagwan sawdust via pyrolysis: product distributions, characterizations and optimization using response surface methodology. Energy 170, 423–437.

Gupta, S., Mondal, P., 2021. Catalytic pyrolysis of pine needles with nickel doped gamma-alumina: reaction kinetics, mechanism, thermodynamics and products analysis. J. Clean. Prod. 286, 124930.

Gupta, G.K., Gupta, P.K., Mondal, M.K., 2019a. Experimental process parameters optimization and in-depth product characterizations for teak sawdust pyrolysis. Waste Manag. 87, 499–511.

Gupta, S., Gupta, G.K., Mondal, M.K., 2019b. Slow pyrolysis of chemically treated walnut shell for valuable products: effect of process parameters and in-depth product analysis. Energy 181, 665–676.

Gupta, S., Gupta, G.K., Mondal, M.K., 2020. Thermal degradation characteristics, kinetics,thermodynamics, and reaction mechanism analysis of pistachio shell pyrolysis for its bioenergy potential. Biomass Conv. Bioref. https://doi.org/10.1007/s13399-020-01104-2.

Huang, X., Cao, J., Shi, P., Zhao, X., Feng, X., 2014. Influences of pyrolysis conditions in the production and chemical composition of the bio-oils from fast pyrolysis of sewage sludge. J. Anal. Appl. Pyrolysis 110, 353–362.

IEA Org. https://www.iea.org/reports/global-energy-co2-status-report-2019. (Accessed on 15 May 2020).

Isahak, W.N.R.W., Hisham, M.W.M., Yarmo, M.A., Yun Hin, T.Y., 2012. A review on bio-oil production from biomass by using pyrolysis method. Renew. Sustain. Energy Rev. 16, 5910–5923.

Jiang, X., Ellis, N., 2010. Upgrading bio-oil through emulsification with biodiesel: Mixture production. Energy Fuel. 24, 1358–1364.

Kambo, H.S., Dutta, A., 2015. Comparative evaluation of torrefaction and hydrothermal carbonization of lignocellulosic biomass for the production of solid biofuel. Energy Convers. Manag. 105, 746–755.

Kan, T., Strezov, V., Evans, T.J., 2016. Lignocellulosic biomass pyrolysis: a review of product properties and effects of pyrolysis parameters. Renew. Sustain. Energy Rev. 57, 1126–1140.

Karmakar, A., Karmakar, S., Mukherjee, S., 2012. Biodiesel production from neem towards feedstock diversification: Indian perspective. Renew. Sustain. Energy Rev. 16 (1), 1050–1060.

Kebelmann, K., Hornung, A., Karsten, U., Griffiths, G., 2013. Intermediate pyrolysis and product identification by TGA and Py-GC/MS of green microalgae and their extracted protein and lipid components. Biomass Bioenergy 49, 38–48.

Kim, S.S., Ly, H.V., Choi, G.H., Kim, J., Woo, H.C., 2012. Pyrolysis characteristics and kinetics of the alga *Saccharina japonica*. Bioresour. Technol. 123, 445–451.

Kim, S.W., Koo, B.S., Ryu, J.W., Lee, J.S., Kim, C.J., Lee, D.H., Kim, G.R., Choi, S., 2013. Bio-oil from the pyrolysis of palm and Jatropha wastes in a fluidized bed. Fuel Process. Technol. 108, 118–124.

Kumar, A.K., Sharma, S., 2017. Recent updates on different methods of pretreatment of lignocellulosic feedstocks: a review. Bioresour. Bioprocess. 4, 1–19.

Kumar, M., Mishra, P.K., Upadhyay, S.N., 2019a. A comparative study of thermochemical characteristics of lignocellulosic biomasses. Bioresour. Technol. Rep. 8, 100186.

Kumar, M., Mishra, P.K., Upadhyay, S.N., 2019b. Pyrolysis of *Saccharum munja*: optimization of process parameters using response surface methodology (RSM) and evaluation of kinetic parameters. Bioresour. Technol. Rep. 8, 100332.

Kumar, B., Bhardwaj, N., Agrawal, K., Chaturvedi, V., Verma, P., 2020. Current perspective on pretreatment technologies using lignocellulosic biomass: an emerging biorefinery concept. Fuel Process. Technol. 199, 106244.

Lange, J.P., 2007. Lignocellulose conversion: an introduction to chemistry, process and economics. Biofuel Bioprod. Biorifin. 1, 39–48.

Lehmann, J., Joseph, S., 2015. Biochar for Environmental Management. Routledge, New York.

Li, L., Rowbotham, J.S., Greenwell, C.H., Dyer, P.W., 2013. An introduction to pyrolysis and catalytic pyrolysis: versatile techniques for biomass conversion. In: Suib, S.L. (Ed.), New and Future Developments in Catalysis: Catalytic Biomass Conversion. Elsevier, Amsterdam, The Netherlands, pp. 173–208B.

Liu, Z., Quek, A., Hoekman, S.K., Balasubramanian, R., 2013. Production of solid biochar fuel from waste biomass by hydrothermal carbonization. Fuel 103, 943–949.

Lou, R., Wu, S.B., 2011. Products properties from fast pyrolysis of enzymatic/mild acidolysis lignin. Appl. Energy 88, 316–322.

Mahmood, A.S.N., Brammer, J.G., Hornung, A., Steele, A., Poulston, S., 2013. The intermediate pyrolysis and catalytic steam reforming of Brewers spent grain. J. Anal. Appl. Pyrolysis 103, 328–342.

Makibar, J., Fernandez-Akarregi, A.R., Amutio, M., Lopez, G., Olazar, M., 2015. Performance of a conical spouted bed pilot plant for bio-oil production by poplar flash pyrolysis. Fuel Process. Technol. 137, 283–289.

Mallick, D., Poddar, M.K., Mahanta, P., Moholkar, V.S., 2018. Discernment of synergism in pyrolysis of biomass blends using thermogravimetric analysis. Bioresour. Technol. 261, 294–305.

Mani, T., Murugan, P., Abedi, J., Mahinpey, N., 2010. Pyrolysis of wheat straw in a thermogravimetric analyzer: effect of particle size and heating rate on devolatilization and estimation of global kinetics. Chem. Eng. Res. Des. 88 (8), 952–958.

McKendry, P., 2002. Energy production from biomass (part 2): conversion technologies. Bioresour. Technol. 83 (1), 47–54.

Mishra, R.K., Mohanty, K., 2020. Pyrolysis characteristics, fuel properties, and compositional study of *Madhuca longifolia* seeds over metal oxide catalysts. Biomass Conv. Bioref. 10, 621–637.

Morrin, S., Lettieri, P., Chapman, C., Mazzei, L., 2012. Two stage fluid bed-plasma gasification process for solid waste valorisation: technical review and preliminary thermodynamic modelling of sulphur emissions. Waste Manag. 32 (4), 676–684.

Mortensen, P.M., Grunwaldt, J.D., Jensen, P.A., Knudsen, K.G., Jensen, A.D., 2011. A review of catalytic upgrading of bio-oil to engine fuels. Appl. Catal. A General 407, 1–19.

Omar, R., Idris, A., Yunus, R., Khalid, K., Aida Isma, M.I., 2011. Characterization of empty fruit bunch for microwave-assisted pyrolysis. Fuel 90, 1536–1544.

Onay, O., 2007. Influence of pyrolysis temperature and heating rate on the production of bio-oil and char from safflower seed by pyrolysis, using a well-swept fixed-bed reactor. Fuel Process. Technol. 88, 523–531.

Onay, O., Kockar, O.M., 2003. Slow, fast and flash pyrolysis of rapeseed. Renew. Energy 28 (15), 2417–2433.

Patel, M., Zhang, X., Kumar, A., 2016. Techno-economic and life cycle assessment on lignocellulosic biomass thermochemical conversion technologies: a review. Renew. Sustain. Energy Rev. 53, 1486–1499.

Pattiya, A., Suttibak, S., 2012. Production of bio-oil via fast pyrolysis of agricultural residues from cassava plantations in a fluidised-bed reactor with a hot vapour filtration unit. J. Anal. Appl. Pyrolysis 95, 227–235.

Pütün, A.E., Önal, E., Uzun, B.B., Özbay, N., 2007. Comparison between the slow and fast pyrolysis of tobacco residue. Ind. Crops Prod. 26, 307–314.

Raj, T., Kapoor, M., Gaur, R., Christopher, J., Lamba, B., Tuli, D.K., Kumar, R., 2015. Physical and chemical characterization of various Indian agriculture residues for biofuels production. Energy Fuel 29 (5), 3111–3118.

Rajapaksha, A.U., Chen, S.S., Tsang, D.C., Zhang, M., Vithanage, M., Mandal, S., Gao, B., Bolan, N.S., Ok, Y.S., 2016. Engineered/designer biochar for contaminant removal/immobilization from soil and water: potential and implication of biochar modification. Chemosphere 148, 276–291.

Salehi, E., Abedi, J., Harding, T., 2009. Bio-oil from sawdust: pyrolysis of sawdust in a fixed-bed system. Energy Fuels 23, 3767–3772.

Santos, N.A.V., Magriotis, Z.M., Saczk, A.A., Fassio, G.T.A., Vieira, S.S., 2015. Kinetic study of pyrolysis of castor beans (*Ricinus communis* L.) presscake: an alternative use for solid waste arising from the biodiesel production. Energy Fuel 29 (4), 2351–2357.

Saratale, G.D., Oh, M.K., 2015. Improving alkaline pretreatment method for preparation of whole rice waste biomass feedstock and bioethanol production. RSC Adv. 5, 97171–97179.

Scott, D.S., Majerski, P., Piskorz, J., Radlein, D., 1999. A second look at fast pyrolysis of biomass - the RTI process. J. Anal. Appl. Pyrolysis 51, 23–37.

Sen, B., Chou, Y.P., Wu, S.Y., Liu, C.M., 2016. Pretreatment conditions of rice straw for simultaneous hydrogen and ethanol fermentation by mixed culture. Int. J. Hydrog. Energy 41, 4421–4428.

Sensöz, S., Angin, D., 2008. Pyrolysis of safflower (*Charthamus tinctorius* L.) seed press cake: part 1. The effects of pyrolysis parameters on the product yields. Bioresour. Technol. 99 (13), 5492–5497.

Shen, J., Wang, X.S., Garcia-Perez, M., Mourant, D., Rhodes, M.J., Li, C.Z., 2009. Effects of particle size on the fast pyrolysis of oil mallee woody biomass. Fuel 88, 1810–1817.

Shirkavand, E., Baroutian, S., Gapes, D.J., Young, B.R., 2016. Combination of fungal and physicochemical processes for lignocellulosic biomass pretreatment-a review. Renew. Sustain. Energy Rev. 54, 217–234.

Singh, S., Chakraborthy, J.P., Mondal, M.K., 2020a. Pyrolysis of torrefied biomass: optimization of process parameters using response surface methodology, characterization, and comparison of properties of pyrolysis oil from raw biomass. J. Clean. Prod. 272, 122517.

Singh, S., Chakraborthy, J.P., Mondal, M.K., 2020b. Torrefaction of *Acacia nilotica*: oxygen distribution and carbon densification mechanism based on in-depth analyses of solid, liquid, and gaseous products. Energy Fuel. 34 (10), 12586–12597.

Singh, S., Chakraborty, J.P., Mondal, M.K., 2020c. Intrinsic kinetics, thermodynamic parameters and reaction mechanism of non-isothermal degradation of torrefied *Acacia nilotica* using isoconversional methods. Fuel 259, 116263.

Stefanidis, S.D., Kalogiannis, K.G., Iliopoulou, E.F., Michailof, C.M., Pilavachi, P.A., Lappas, A.A., 2014. A study of lignocellulosic biomass pyrolysis via the pyrolysis of cellulose, hemicellulose and lignin. J. Anal. Appl. Pyrolysis 105, 143–150.

Sun, Y., Cheng, J.J., 2002. Hydrolysis of lignocellulosic materials for ethanol production: a review. Bioresour. Technol. 83, 1–11.

Tripathi, M., Sahu, J.N., Ganesan, P., 2016. Effect of process parameters on production of biochar from biomass waste through pyrolysis: a review. Renew. Sustain. Energy Rev. 55, 467–481.

Uzun, B.B., Pütün, A.E., Pütün, E., 2006. Fast pyrolysis of soybean cake: product yields and compositions. Bioresour. Technol. 97, 569–576.

Vamvuka, D., 2011. Bio-oil, solid and gaseous biofuels from biomass pyrolysis processes — an overview. Int. J. Energy Res. 35, 835–862.

Von Blottnitz, H., Curran, M.A., 2007. A review of assessments conducted on bio-ethanol as a transportation fuel from a net energy, greenhouse gas, and environmental life cycle perspective. J. Clean. Prod. 15 (7), 607–619.

Wang, J., Wang, S., 2019. Preparation, modification and environmental application of biochar: a review. J. Clean. Prod. 227, 1002–1022.

Wang, J., Chang, J., Fan, J., 2010. Catalytic esterification of bio-oil by ion exchange resins. J. Fuel Chem. Technol. 38 (5), 560–564.

Wang, S., Guo, X., Wang, K., Luo, Z., 2011. Influence of the interaction of components on the pyrolysis behavior of biomass. J. Anal. Appl. Pyrolysis 91, 183–189.

Wang, S., Wang, Q., Jiang, X., Han, X., Ji, H., 2013a. Compositional analysis of bio-oil derived from pyrolysis of seaweed. Energy Convers. Manage. 68, 273–280.

Wang, Z., McDonald, A.G., Westerhof, R.J., Kersten, S.R., Cuba-Torres, C.M., Ha, S., Pecha, B., Garcia-Perez, M., 2013b. Effect of cellulose crystallinity on the formation of a liquid intermediate and on product distribution during pyrolysis. J. Anal. Appl. Pyrolysis 100, 56–66.

Xiu, S., Shahbazi, A., 2012. Bio-oil production and upgrading research: a review. Renew. Sustain. Energy Rev. 16, 4406–4414.

Xu, Y., Wang, T., Ma, L., Zhang, Q., Wang, L., 2009. Upgrading of liquid fuel from the vacuum pyrolysis of biomass over the Mo–Ni/γ-Al_2O_3 catalysts. Biomass Bioenergy 33 (8), 1030–1036.

Yu, Y.Q., Zeng, Y.L., Zuo, J.E., Ma, F.Y., Yang, X.W., Zhang, X.Y., Wang, Y., 2013. Improving the conversion of biomass in catalytic fast pyrolysis via white-rot fungal pretreatment. Bioresour. Technol. 134, 198–203.

Zhang, H., Xiao, R., Huang, H., Xiao, G., 2009. Comparison of non-catalytic and catalytic fast pyrolysis of corncob in a fluidized bed reactor. Bioresour Technol 100 (3), 1428–1434.

Zhang, L., Liu, R.H., Yin, R.Z., Mei, Y.F., 2013. Upgrading of bio-oil from biomass fast pyrolysis in China: a review. Renew. Sustain. Energy Rev. 24, 66–72.

Zheng, Y., Zhao, J., Xu, F., Li, Y., 2014. Pretreatment of lignocellulosic biomass for enhanced biogas production. Prog. Energy Combust. Sci. 1–16.

SECTION VI

Bioenergy production using microbial fuel cell

CHAPTER

15

Production of biofuels in a microbial electrochemical reactor

Mamta Pal, Rajat Giri, Rakesh Kumar Sharma

Department of Biosciences, Manipal University Jaipur, Jaipur, Rajasthan, India

15.1 Introduction

Population over the globe is 6 billion, and it will be 9.4 billion in coming 30–40 years as per 2008 estimate (Conforti, 2011). Six billion people is a huge number when the world is facing energy crisis. As the population will increase, there will be upsurge in the energy requirement, which is irrespective of the countries (Logan and Regan, 2006). Energy is a basic need of life; it plays an important role in the establishment of well-developed society and human beings. There are two basic types of energy sources: renewable and nonrenewable. Firstly, fossil fuels are nonrenewable energy resources. Primary energy requirements are fulfilled by fossil fuels, e.g., petroleum, coal diesel, bitumen, and natural gases (Das and Veziroğlu, 2001). These conventional fossil fuels are continuously declining and increasing environmental pollution. Therefore, it is time to develop a better, proficient, and sustainable energy system as an alternative of nonrenewable energy source that can increase global energy supply. Secondly, solar, wind, hydro, geothermal, biomass, and additionally, biofuels such as bioethanol, biogas, and biodiesel are categorized under renewable energy resources (Naik et al., 2010). Presently, there is no single renewable energy source that has the capability to become the substitute of conventional fossil fuel–based energy. In spite of this, bioenergy seems to be a promising substitute to overcome this energy crisis (Rodionova et al., 2017; Srivastava et al., 2020). One potential bioelectrochemical technique as a substitute energy source is microbial fuel cells (MFCs). MFCs are bioelectrochemical devices and have numerous applications except bioenergy production. Nowadays, MFCs have become technology for wastewater treatment such as industrial wastewater treatment, dairy wastewater treatment, leachate treatment, agricultural wastewater treatment, wastewater treatment of paper and pulp industries, degradation of xenobiotics, degradation of azo dyes and organic pollutants, and production of hydrogen, ethanol, and methane.

https://doi.org/10.1016/B978-0-323-85269-2.00010-1

15.2 Current scenario of consumption and supply of world's energy and role of renewable energy

According to current report of world bioenergy association 2019 (https://worldbioenergy.org), the consumption of fossil fuel as primary energy has increased in 2016–17. Coal, oil, and natural gas have major contribution to primary energy supply (1.5%) than renewable energy sources (0.7%). There is similar trend to continue to 2018–19 as well. Bioenergy is the largest renewable energy source. Bioenergy has contributed for 70% of the renewable energy consumption in 2017 (Gielen et al., 2019). Traditional fossil fuel energy sources are used in very large amount because this bioenergy contribution has been reducing annually (approx. 0.5%–1%). Still renewable energy is successful to contribute a major share in the field of energy supply. In the field of renewable electricity, it contributes approximately about 25% of the electricity generation from biomass-based sources globally in 2017 and becomes third leading renewable power source after hydropower and wind (596 TWh of biopower) (Kumar and Majid, 2020). Biomass-based fuels such as bioethanol, biodiesel, and so on have the capability to replace the fossil fuel. 3% with a total contribution of biofuel was invested in transport sector in 2017. Residential, commercial, and agriculture sectors are majorly utilized biomass for cooking and heating purposes. These sectors were consumed about 40 EJ of biomass, which is approximately 95% of renewable energy in 2017. Biomass was consumed for energy purposes. Primary solid biofuels (86%), liquid biofuels (7%), and biogas (2%–3%) all were used for energy purpose for cooking and heating in 2017 (Popp et al., 2014). Liquid biofuels are mostly utilized in the transportation sector. It is used to the development of rural area, improve the fuel security, and reduce the climate change. Biofuels such as bioethanol, biodiesel, and HVO (hydrogen vegetable oil) are generated in large amount; approximately 138 billion liters of biofuel were produced in 2017. Biogas is also used as a renewable energy source; 1.33 EJ of biogas was generated globally in 2017. Agriculture sector is among the most proficient sectors for amendment of bioenergy production. Tons of agricultural waste is generated annually around the world, and it has large amount to energy in the conserved form. Agricultural residues and cereals such as maize, wheat, rice, barley, and other contribute majorly in this sector, which offers less than 3% to the bioenergy production globally (Lee et al., 2019). Municipal and industrial waste sector are also other important sector for bioenergy production. As urban areas required more energy than rural, large amount of municipal and industrial waste is used for energy in these areas (Malinauskaite et al., 2017). In 2017, 2.15 EJ bioenergy was generated from the domestic waste in which 58% was municipal waste and leftover was industrial waste.

Biomass is a unique source of bioenergy. It is a sustainable and renewable alternative for biopower or electricity generation. It will be helpful to reduce the demand of nonrenewable fuel source in the electricity sector. During the past years, biomass energy production raised approximately 4% globally; 596 TWh of bio power was produced in 2017 (Popp et al., 2021). Renewable sources contribute a major role in energy supply. Their contribution of renewable supply is constant in the past 17 years. However, the growth of renewables has been progressive during 2000–2017. 17.7% of renewable energy was consumed in 2017 globally. Bioenergy (energy from bio-based sources) contributes a major share in the energy supply so it is the largest renewable energy. It is reported that in 2017, 70% of the renewable energy was consumed (Fig. 15.1). According to the world's energy demand, very limited amount of coal is in stock, which can only fulfill the supply till

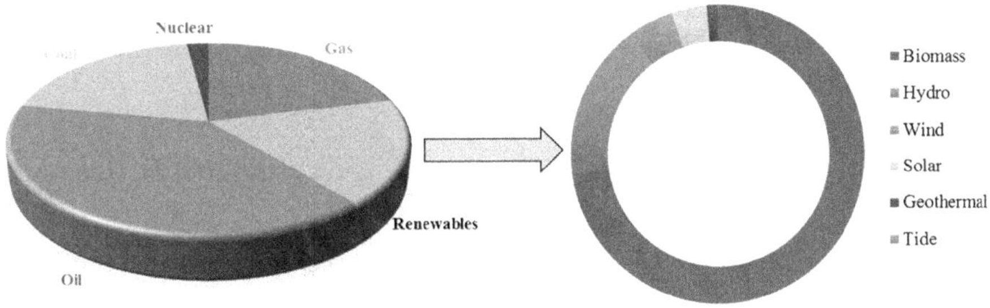

FIGURE 15.1 Gross final energy consumption in 2017.

2080. Bearing in mind both the availability and demand, it is proved that these reserved nonrenewable energy sources will not be long-lasting to complete the future energy demand. In the current situation, the greatest environmental challenge is to develop a method or techniques that supply sufficient energy to world and simultaneously to reduce the carbon footprint.

15.3 Microbial electrochemical cell

MFC is one of the efficient alternatives for renewable bioenergy technologies. In the past 20–25 years, it has attained great attention of researchers, due to the unique property of transforming organic waste directly into electricity using microbes. These are bioelectrochemical devices, in which microbes oxidize the organic matter and generate electron through their metabolism and produce electricity (Santoro et al., 2017). MFCs are sustainable system, and it does not need any kind of energy from outside. On the other hand, external energy source is a basic requirement of traditional fuel cell such as hydrogen fuel cells (Granovskii et al., 2006). MFC is not only self-sustainable and waste management device, but it also reduces the emission of greenhouse gasses. That is why, MFC is the first choice over traditional fuel cell for biopower generation. MFC technology is a modern and novel approach to generate the electricity using bacteria by the oxidation of organic matter and biomass. It has been described as "bioelectrochemical device that has ability to transform the chemical energy of organic matter into electrical energy using biocatalytic activity of microorganisms" (Logan et al., 2019). The major advantage of MFC technology is that it utilizes organic waste matter as fuel and microbes as catalytic agent.

MFC technology is a multidisciplinary approach, which utilizes the different type of alternate sources of energy. This technique indicates the convergence of basic and applied research in which chemical, physical, and life sciences come together and help in the development of a single technique. Firstly, in 1911, Potter demonstrated the generation of electricity by bacteria or yeast cell using platinum macroelectrodes in a battery type setup with sterilized medium (Potter and Waller, 1911). After this, research work in this field was increased, and many studies had been done to increase power output and attained considerable progression in MFC. Later on in 1931, Cohen confirmed the potter's experiment through a stacked MFC system and recorded the output voltage 35V at 0.2 mA of current (Cohen, 1931). All these research studies in the field of MFC technology are mostly discussed as the beginning of electromicrobiology; in 1963, the National Aeronautics and Space Administration (NASA) space program stated that in the space flights, human waste was recycled and transformed into electricity (Canfield et al., 1963; Schröder, 2011).

In 1970s, there was a little knowledge about MFCs operation and use of fuel cell as a source of electricity in the developing countries. In the early 1980s, Bennetto's work developed a better understanding of how fuel cells operated. In the early 1990s, fuel cells now become an interesting topic of research, and the development work in this field has been increased (Allen and Bennetto, 1993). After this, Habberman and Pommer firstly reported about the long-term MFC, which was utilizing municipal wastewater and continuously operated 5 years without any maintenance (Habermann and Pommer, 1991). This study first time informed about the mediated or indirect electron transfer; sulfate was used as a mediator in this study (Mokhtarian et al., 2012). However, research, which had been conducted in this time, proposed the use of chemical mediators or electron shuttles (chemicals that help in the transportation of electrons from cell to electrodes) in MFC. In 1999, Kim proposed that MFCs can be operated without chemical mediators; these mediators were not an important component in MFC (Kim and Kim 1999). A prototype of MFC was completed by the University of Queensland, Australia, with the support of Foster's Brewing in May 2007. It is a design of 10 L bioreactor, which can convert brewery wastewater into carbon dioxide, clean water, and electricity (Logan, 2010).

15.3.1 Microbial fuel cell configuration and its type

A simple MFC system configured majorly by five important tools: (1) anode, (2) cathode, (3) proton exchange membrane (PEM) or salt bridge, (4) exoelectrogens (microbes), and (5) data acquisition system (Fig. 15.2). Traditional or typical MFCs are fabricated as a double chamber system (one is anodic chamber and other one is cathodic chamber). These are two separate chambers: the anodic chamber that allows bacterial growth and subsequent oxidation of the organic matter, and the cathode chamber, where the electrons react with the catholyte and form water. Presence of oxygen in the anodic chamber will prevent the electricity generation. That is why, bioelectrochemical system must be designed such that both the growth of exoelectrogens (bacteria) and oxygen generation occur separately. Both of these chambers are connected by PEM or salt bridge that permits the transportation of protons between the electrodes. Both the electrodes are

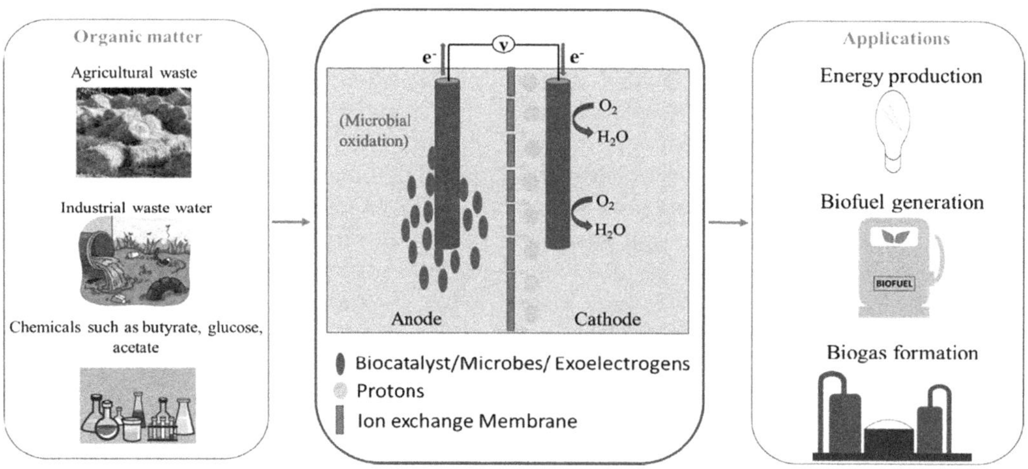

FIGURE 15.2 Bioelectrochemical system and its application.

associated by the wire containing a load or resistor. PEM is permeable to protons, which are generated in anodic chamber so that they can transfer from anode to cathode, and these protons are combined with electrons (transferred via wire from anode to cathode and form water (Goswami and Mishra, 2018). Flow of electron from anode to cathode is electricity or current. This electricity is monitored by a voltmeter or multimeter or potentiostat, which is connected with computer for essentially continuous data acquisition. MFCs may also be functioned in a continuous mode and batch mode to optimize the power output (Zhao et al., 2009).

Basically, structural designs of MFCs include double-chambered MFC, single-chambered MFC, three-chambered MFCs, stacked MFCs, constructed wetland MFCs, and sediments MFCs (Santoro et al., 2017). These different designing and pattern of compartments can enhance the power output of MFCs. Conventional rectangular-shaped MFCs, upflow mode configurations, cylindrical-shaped MFCs, miniature MFCs, and flat plate MFCs are types of dual-chambered MFC. Double-chambered MFC is commonly used and it is economical design (Logan et al., 2006).

The foremost challenge in the MFC technology is to scale up this technique due to their complex architectural configurations. To simplify the design, single-chambered MFC has been included in this field. Single-chambered MFCs are simple in design because it excludes the cathode chamber where the cathode is directly exposed to the air (Liu and Logan, 2004). First, single-chambered MFC was developed by Park and Zeikus in 2003 (Park and Zeikus, 2003). This system contained a glass bottle with centrally placed anode, porcelain layer proton exchange membrane, sewage sludge used as biocatalyst, and open air cathode. Maximum power density generated by this single-chambered MFC was 788 mW/m^2.

Miniaturization and multiplication approach are the methods that can help to scale up or industrialization of MFC. The major focus in the fuel cell research is to reduce the size of bioelectrochemical system and attain high power generation. This problem is somewhat solved by stacked MFC. It is a feasible way to increase the power output and in small setup. Stacked MFCs are multiple small bioelectrochemical setups that are linked either in series or in parallel form (Aelterman et al., 2006). When a large number of small devices connected together, they form one stacked MFC, which helps to attain the high power output. Six single continuous MFC units were connected for one stacked MFC, which produced 258 W/m^3 average power density (Dekker et al., 2009).

Sediment MFCs are set up such that one electrode is inserted into the marine sediments rich in organic matter, while the other one is overlying in oxic water. Through this setup, electricity can be generated at required levels to power the devices that are involved in marine electronics. Seawater is salt- or ion-rich water through which protons conducted into the system and generated electricity. Mostly, "bottle brush" type of cathode used in the seawater batteries helps to maintain the longevity and operation life of system because these electrodes are made up of noncorrosive material and having high surface area (Xu et al., 2015).

15.3.2 Mechanism to generate electricity in microbial fuel cells

Electrons generated via microbial metabolism in MFC system have a central role in electricity generation. Exoelectrogenic microbes facilitate electron transportation between electrodes and microbial cells. Two major mechanisms that are reported for electron transportation are (1) direct electron transfer (DET) and (2) indirect electron transfer (Fig. 15.3) (Kumar et al., 2015). DET usually occurs via physical contact between the surface of the electrodes and microbial cells (Aiyer, 2020). Microorganisms have a membrane-bound

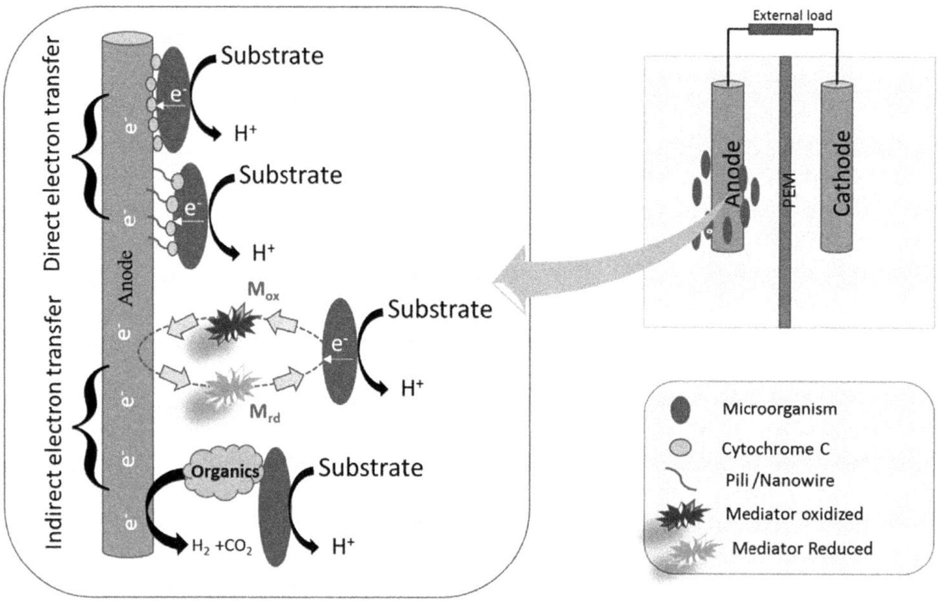

FIGURE 15.3 Electron transfer mechanism from microorganisms to anode.

electron transport protein, which helps in transportation of the electrons from cytoplasm to its surrounding environment. Anode further receives these electrons. The transportation process occurs via outer membrane cytochromes, conductive fimbriae, or electrically conductive appendages known as nanowires (Reguera, 2018). C-type cytochromes that are present on the outer surface of bacterial cell membrane act as redox protein, which facilitates the transfer of electrons to an external terminal electron acceptor, i.e., anode (Busalmen et al., 2008).

There are few microbes that are incompetent for DET. These microbes usually transport electrons in assistance of mediators. This process is known as mediated electron transfer (MET) or indirect electron transfer. Biochemical molecules such as redox-active proteins released by microbes mediate the electron transfer between electrode and cells. All these proteins and compounds are known as endogenous mediators. Redox carrier proteins involved in the process act as shuttle to transfer the electrons to terminal electron acceptor. It receives electrons from bacterial cells and gets reduced and transfers it to the anode and gets oxidized. Ieropoulos et al. stated that a good mediator shall have the following properties such as (1) nontoxicity for bacterial cells, (2) capability to cross the bacterial cell membrane, (3) soluble in enzymes, (4) a positive redox potential to enable the electron transfer, and (5) should be cheap and easily available (Ieropoulos et al., 2005).

Redox mediators are of two types: (1) endogenous mediators, which are produced by the bacteria themselves, and (2) exogenous mediators, which are added externally. Metabolites released by the microbial cells such as riboflavins, flavins, flavin mononucleotides (FMNs), phenazines, pyocyanins, and quinones act as endogenous mediators (Reguera et al., 2006). In contrast, exogenous mediators are added in the anolyte to boost the performance of MFC. In the early 1990s, the use of exogenous mediators was prominent in MFC research, as the potential of exoelectrogens was unknown. Neutral red, methylene blue, thionine, various quinone derivatives and Fe (III) EDTA, and so on are used as exogenous mediators. The use of exogenous mediators has been reduced significantly over the past few years due to toxicity and instability; though these molecules are still common to improve the MFC performances (Sekrecka-Belniak and Toczyłowska-Mamińska, 2018).

Additionally, another method of indirect electron transfer uses reduced metabolites produced by the microorganisms. This process is applicable only for those MFCs where fermentative agents are microorganisms and yeast (Schaetzle et al., 2008). This mechanics involves the secretion of reduced metabolites that may transfer the electrons to the anode (Schröder, 2018). These metabolites are often generated and secreted during fermentative production of organic acids (Moscoviz et al., 2016).

15.4 Electricity generation

MFC is an alternative for electricity generation. It has potential to convert the chemical energy of organic matter into electrical energy via microbes as biocatalysts. Chemical energy, which is released from organic matter's oxidization, is transformed into electricity rather than heat. Carnot cycle is avoided, and theoretically, it is observed that a much higher conversion efficiency is achievable (N70%) similar to other chemical fuel cells. *Rhodoferax ferrireducens* has been reported to be a potential candidate for generating electricity with an electron yield as high as of 80% (Chaudhuri and Lovley, 2003). Similarly, high electron recovery up to 89% has also been reported earlier (Rabaey et al., 2003). Recently, Ren et al. (2016) reported that highest power density over 10,000 W/m^3 can also be achieved. This group presented a miniaturized configuration of MFC with 3D graphene macroporous scaffold anode with biocompatibility along with large specific surface area, high porosity, and excellent conductivity. This resulted into a thick and dense biofilm involving *Geobacter*-enriched culture contained within the 3D scaffold. MFC demonstrated an area/volumetric current density of 15.5 A/m^2/31,040 A/m^3 and a power density of 5.6 W/m^2/11,220 W/m^3. About 3.3-fold increase was achieved in comparison with planar 2D control counterparts. MFCs are good power sources for daily life. According to Logan, power density of 0.1 to 6.9 W/m^2 is generated during the period of 1999–2008 (Logan, 2010). However, it is important to encourage large and efficient biochemical reactors with enough energy output to upscale the energy output (Slate et al., 2019). Organic substrates namely wheat straw rice, straw corn, stover orange peel, and lemon peel are used as biomass in MFCs for electricity generation (Anwar et al., 2014). Miran et al. utilized orange peel as a renewable carbon source in a double-chambered mediator-less MFC setup, which achieved a voltage generation of 0.59 ± 0.02 V at 500 Ω (Miran et al., 2016). Nevertheless, power generation through MFC is still very low, as the rate of electron abstraction is a slow process (DeLong and Chandler, 2002). A possible way to resolve this problem may be the storage of the electricity in a rechargeable device and then further distribute the stored electricity to end users (Ieropoulos et al., 2003a). Capacitors, e.g., EcoBot I are used in biologically inspired robots, which accumulate the energy produced by MFCs. MFCs can be suitable to power small elementary systems such as wireless sensors, which has low power requirement (Ieropoulos et al., 2003b; Shantaram et al., 2005). MFCs may also be served as distributed power system for underdeveloped regions. A few researchers suggested that MFCs can be a perfect energy supplying device for Gastrobots, which self-feed the biomass collected by themselves (Wilkinson, 2000). Autonomous robots may also be equipped with MFCs to utilize different fuels such as sugars, fruits, dead insects, grasses, and weed as energy supplies. EcoBot-II has MFC-enabled power supply to perform activities such as computing, communication, motion, and sensing (Ieropoulos et al., 2003a; Melhuish et al., 2006). MFCs can also be used in spaceships to generate and supply electricity by degrading biowaste generated onboard. Science is also looking at a future to design and develop a miniature MFC, which can be implanted inside a human body for powering the implantable medical device. MFC technology favors the sustainable long-term power applications. However, probable health and safety issues due to microorganisms in the MFC must be thoroughly solved before use to serve the purpose. MFC has diverse applications in waste management field due to its different designing, renewable substrate, and microbes used in the system (Table 15.1).

15.5 Hydrogen production

Hydrogen has an incredible potential as an energy source and fuel. H_2O vapors being the solely oxidation product after hydrogen burning do not contribute to atmospheric problems such as ozone depletion, acid rain, or greenhouse gas (GHG) emissions, thus showing tremendous ability as an environmentally friendly energy carrier for vehicles. Production of hydrogen from nonrenewable sources such as fuels (2%), coal (18%), petroleum-derived naphthenes, and distillates (30%) or natural gas (50%) has been practiced worldwide (www.eere.energy.gov), resulting in overextraction and depletion of

TABLE 15.1 Microbial fuel cell design, microorganism, substrate and efficiency of electricity generation, biohydrogen generation, bioethanol generation, and biogas generation.

S. No.	Fuel cell configuration	Substrate	Inoculum for MEC	Electrodes	Electrochemical performance	Biofuel efficiency	References
1	**Application: Electricity generation**						
1.1	Double chambered	Dye waste water	*Ganoderma lucidum* in cathode sludge in anode	Carbide porous ceramic	OCV: 711 mV P.D: 223 mW/m^2	–	Liu et al. (2020)
1.2	Dual cubic chambers	Acetate as feed	Mixed culture	Anode: A carbon rod and graphite fiber. Cathode electrode: Carbon rod	OCV: 0.607 V P.D: 40 W/m^3	–	Wu et al. (2020)
1.3	Single chambered	Domestic wastewater	Mixed culture	Anode: Carbon cloth Cathode: Carbon cloth	OCV: 302 ± 4 mV P.D: 538–766 mW/m^2	–	Cheng et al. (2006)
1.4	Double chambered	25-mM sodium acetate medium	*Geobacter*-enriched culture 50-mM potassium ferricyanide	Graphene	10,000 W/m^3	–	Ren et al. (2016)
1.5	Single chambered	Rice straw hydrolysate	Proteobacteria	Anode: Carbon brush anode Cathode: Gas diffusion electrode with Pt catalyst	OCV: 524.7 ± 3.2 mV P.D: 293.33 ± 7.89 mW/m^2	–	Wang et al. (2014)

TABLE 15.1 Microbial fuel cell design, microorganism, substrate and efficiency of electricity generation, biohydrogen generation, bioethanol generation, and biogas generation.—cont'd

S. No.	Fuel cell configuration	Substrate	Inoculum for MEC	Electrodes	Electrochemical performance	Biofuel efficiency	References
2	**Application: Biohydrogen production**						
2.1	Double chambered	Anaerobic sludge from wastewater	Mixed culture (anaerobic sludge)	Anode: Carbon brush Cathode: Carbon cloth coated with 0.5 mg/cm^2 Pt/C	OCV: 0.6 V P.D: 108 mW/m^2	0.25 L H_2/L/day 17%	Jain and He (2020)
2.2	Double chambered	Sugar industrial wastewater (SWW)	Mixed culture	Anode: Plain graphite plate with copper wire Cathode: Nickel foam (6 × 4 cm) coated with nickel oxide–reduced graphene oxide (NiO.rGO)	OCV: 1 V	4.38 mmol/L/day 21.16%	Jayabalan et al. (2020)
2.4	Photoanode-based MEC	Thermal power plant wastewater	Mixed culture (bacterial)	Cerium oxide–reduced graphene oxide–based carbon film	OCV: 1 V	5 L H_2/L/day 98%	Pophali et al., 2020
2.7	Single chambered	Raw food waste	Mixed culture—existing MEC effluent (acetate-fed with methanogenic inhibitors)	Anode: Sludge-modified titanium electrode Cathode: Platinum-coated titanium mesh tube	OCV: 0.8 V	4.86 L H_2/L/day 96%	Huang et al. (2020)
2.8	Double chambered	Sodium acetate	Mixed culture (salt marsh sediment)	Anode: Graphite felt anode Cathode: Stainless steel pipe	1.6 V	6.5 L H_2/L/day 66%	Rousseau et al. (2020)
2.9	Single chambered	Dairy wastewater	Mixed culture (electrogenesis bacteria) (anaerobic sludge)	Anode: Aluminum plates Cathode: Stainless steel mesh, copper and aluminum plates fabricated with polyaniline/graphene	1 V	0.81 L H_2/L/day 47%	Ghasemi et al. (2020)
3	**Application: Bioethanol production**						
3.1	Double chambered	Potassium ferricyanide	*Zymomonas mobilis*	Anode: Carbon cloth Cathode: Carbon cloth	2.0 mW/m^2	86%	Geng et al. (2020)
3.2	Double chambered	MB solution: Potassium ferricyanide	*Saccharomyces cerevisiae*	Carbon fiber brush	OCV: 655 ± 5 mV P.D: 5.2 ± 0.5 W/m^3	92.5 ± 2%	Yuan et al. (2020)

(Continued)

TABLE 15.1 Microbial fuel cell design, microorganism, substrate and efficiency of electricity generation, bio-hydrogen generation, bioethanol generation, and biogas generation.—cont'd

S. No.	Fuel cell configuration	Substrate	Inoculum for MEC	Electrodes	Electrochemical performance	Biofuel efficiency	References
3.3	Stack MFC	Potassium ferricyanide	*Saccharomyces cerevisiae* DSM 4266		OCV: 0.5 V. P.D: 14 mW/m^3	91.2%	Sugnaux et al. (2016)
3.4	Double chambered	Culture medium	*Saccharomyces cerevisiae*	Fe@Fe_2O_3/graphite electrode.	P.D:30.46 mW/m^2	11.52 g/L	Birjandi et al. (2016)
4	**Application: Biogas production**						
4.1	Double chambered	Anaerobic sludge from waste water	Extract of food waste	Anode: Graphite cathode: Carbon cloth	490 mV	4.7 L/day	Hou et al. (2020)
4.2	Single chambered	Primary sludge from waste water treatment plant	Sludge	Anode: Graphite fiber brush cathode:PFTE coated with Pt	P.D: 82 ± 18 mW/m^2	460 LN/kg	Hiegemann et al. (2016)
4.3	Dual chambered	*Azolla pinnata* biomass	Anaerobic sludge	Anode: Biochar electrode cathode:	OCV: 382 ± 4 0 mV P.D: 105 mW/m^2	215 mL/g	Hemalatha et al. (2020)
4.4	Double-chambered and stacked MFC	Raw swine wastewater	Anaerobic sludge	Anode: Graphite felt cathode: Carbon cloth	OCV: 0.77 V (parallel connections) 2.25 V P.D: 25 W/m^3	0.72 L/day	Kim et al. (2020)

MFC, microbial fuel cell.

necessary environmental resources. Thus, increasing demands of highly developed technologies for hydrogen production from renewable energy resources and biomass that reduce environmental crises are currently under high priority. However, sustainable, cost-effective, and clean hydrogen production from a broad variety of renewable biomass and wastewater including food processing wastewater (Van Ginkel et al., 2005), municipal solid waste (Lay et al., 1999; Okamoto et al., 2000), and wheat grains (Kalia et al., 1994) has been studied earlier.

A complex bioelectrochemical system (BES) sharing similarities to other fuel cell systems, i.e., microbial electrolysis cell (MEC) could show promising advancement in hydrogen production from organic matter, including renewable resources and wastewater. Use of several different fermentative dead-end products and cellulose, including acetate, propionate, and butyrate, demonstrated significant hydrogen production in an MEC. With application of external voltage, H^+ produced by the oxidation of organic material/pollutant can be reduced to H_2 gas, thus allowing production of renewable and sustainable energy. Under ideal operating conditions, high hydrogen energy recovery under lab scale has been reported earlier. However, controlling the limiting factors of MECs such as use of inoculums, material selection, voltage, and so on could significantly enhance the operating conditions and yield of innovative MEC reactor configurations.

To avoid the consumption of hydrogen by microbes and improve the purity of the produced hydrogen, addition of a membrane acts as a separator in two chamber MECs. Various membranes such as proton exchange membrane (PEM), anion-exchange membranes (AEM), and cathode-exchange membranes (CEM) have been implied for successful high hydrogen production. Thermoelectric generators (TEGs) as alternative power sources to drive hydrogen production in MEC using simulated waste heat have been studied (Jain and He, 2020). Photoanode-based MEC for hydrogen (H_2) production in wastewater under visible light irradiation had also been studied (Pophali et al., 2020). Cell cathode covered with nanoparticles such as palladium, graphene hybrid metal oxide and phosphating cobalt, and acicular nanoarray had been implicated in MECs as catalysts to investigate cost-effective hydrogen production. However, single-chamber MECs can be operated without membranes without any negative impact on efficiency of MEC, thus simplifying reactor design and reducing capital costs. Single-chambered MEC with cathode operated at high current density (90 A/m^2) and cathode fabricated with polyaniline/graphene (Rousseau et al., 2020). Ghasemi et al. proposed an efficient and cost-effective alternative for the conventional expensive cathode of MECs (Ghasemi et al., 2020). Single-chambered MEC had also been studied earlier to utilize alginate and chitosan to immobilize a *Geobacter sulfurreducens* (exoelectrogenic microbe) anode for hydrogen production (Gandu et al., 2020). Biological treatment of wastewater (MW) used as a substrate in a single-chamber MEC has been used as an appropriate technology for wastewater treatment and simultaneous hydrogen production. An integrated reactor design combining anaerobic digestion (AD) and single-chamber MEC treatment using food waste, as substrate had also been implied to achieve efficient hydrogen recovery. MEC with continuous supply of wastewater by using recycled carbon fiber mats has shown potential in producing renewable hydrogen gas and treating wastewater. Thus, use of modified reactor designs and controlled limiting factors of MEC could open the door for generating greater value and intensify the development of larger pilot systems and a more sustainable wastewater treatment with production of hydrogen gas.

Breakdown of lignocellulosic biomass present as residues in wastewater is necessary for enzymes or microbes to use accessible cellulose and hemicelluloses for metabolic activity. Hydrothermal pretreatment of the lignocellulosic biomass had been reported to produce a hemicellulose-rich liquid fraction (hydrolysate) and cellulose-rich fiber fraction. Extreme thermophilic organisms have been found to provide more than 75% of hydrogen yield due to presence of favorable thermophilic temperatures for optimal growth and less variety of soluble end products (Ivanova et al., 2009). Also, use of different substrates and reactor configurations had provided with acceptable hydrogen yields under continuous processes operated at extreme thermophilic conditions (Kongjan et al., 2010). Also, production of ethanol and hydrogen from pineapple peel by *Saccharomyces cerevisiae* had also been studied to significantly increase by immobilized cell technique. Although degradation of complex polymers by yeast *S. cerevisiae* could be quite difficult, use of mixed culture fermentation process could be adapted to convert a substrate into hydrogen with increased yield. Thus, to easily degrade complex polymers and enhance the biohydrogen production, mixed culture microbial fermentation techniques could be coupled with different MEC reactor designs providing cost-effective and sustainable biodegradation and bioenergy production (Table 15.1).

15.6 Ethanol production

Bioethanol, an eco-friendly product, is a good substitute of conventional fuel. It is largely produced by complex carbohydrates, sugar, and starch through fermentation process. Though, it can also be manufactured by chemical method. Several generation-based bioethanol has been produced using cellulosic biomass, organic wastes, crops (cotton, maize, wheat, rice, corn stover, etc.), mixed paper waste, and derived vegetable oil (Rodionova et al., 2017). This process of the conversion of lignocellulose to ethanol has been comprised at several steps such as pretreatment of biomass (physical, chemical, and biological), hydrolysis, fermentation, and finally the product recovery. Process of enzymatic hydrolysis does not require very specific environment conditions, and this process is completed in very less amount of energy.

Cellulose is a heteropolysachharide; after hydrolysis, it gets converted into glucose, while hemicellulose transforms into some pentoses and hexoses. These sugars are further consumed by specific microorganisms, which are able to produce ethanol. Several species of bacteria such as *Acetovibrio, Bacillus, Bacteriodes, Clostridium, Cellulomonas, Ruminococcus, Erwinia, Microbispora, Streptomyces,* and *Thermomonospora* have property to produce cellulase enzyme. Similarly, many fungi *Fusarium, Humicola, Penicillium, Phanerochaete, Schizophillum,* and *Trichoderma* also come into existence, which are capable to produce cellulase (Sarkar et al., 2012).

Basically, the oxidation of biomass and organic compounds using yeast and microbial metabolism favors the formation of bioethanol. Bioethanol formation occurs by the fermentation process in which 1 mole of glucose generates 2 moles of pyruvate. This pyruvate converted into 2 moles of acetaldehyde with the help of pyruvate decarboxylase as an enzyme. These 2 moles of acetaldehyde converted into 2 moles of ethanol by alcohol dehydrogenase. Alcohol dehydrogenase is an NADH-dependent enzyme, which plays a central role in the formation of ethanol. As it is an NADH-dependent enzyme, $NADH/NAD^+$ redox cycle is an important reaction, which helps in many metabolic reactions and generates electrons from the conversion of NADH to NAD^+. These electrons are used by the electrochemical system or MFC. Extracellular electron transfer mechanism (EET) occurred during ethanol production and electricity generation (Hubenova and Mitov, 2015). Recently, a cogeneration system was developed for the bioethanol and electricity production simultaneously. This system generated 755 ± 7 mV of OCV and $5.2 \pm 0.5\ W/m^3$ of power density with maximum of 92% ethanol yield (Yuan et al., 2020). Another study was conducted in which system contained a two-stage bioethanol refining system with multiliter stacked MFCs. Through this system using *S. cerevisiae* as a fermentative agent, the yield of ethanol is 96.3% (Sugnaux et al., 2016) (Table 15.1).

Nowadays, lignocellulosic biomass has become a new center of attraction in the field of MFC. Several studies are focused on the usage of lignocellulosic biomass for the production of power generation, ethanol production, biohydrogen production, and so on. This lignocellulosic biomass such as softwood, hardwood, and agricultural residues is considered as sustainable source of feedstock for biofuels. Various substrates have been explored during MFC-induced bioethanol production such as acetate, glucose, winery waste, cassava mill wastewater, food wastewater, brewery wastewater, starch processing wastewater, and so on.

Yeast is an important factor for ethanol production. It has several genes for the transportation of sugars. The genes AT5G17010 and AT5G59250 are D-xylose-proton symporter-like protein 2 and protein 3 of *Arabidopsis thaliana*, respectively, and further enhanced in *S. cerevisiae* for efficient transportation mechanism during ethanol production. Other genes are HXT2 gene (HXTs family) from *S. cerevisiae*, AtSTP2 gene (sugar transport protein 2) from *Arabidopsis thaliana*, XUT family gene from *Saccharomyces stipitis*, and others (Sharma et al., 2016). Fermentation process is a completely microbe-dependent process. There are several wild-type microorganisms such as *Candida brassicae, Candida shehatae, Escherichia coli, Mucor indicus, Pachysolen tannophilus, Pichia stipitis, Geobacter* spp., *Shewanella* spp., *S. cerevisiae*, and *Zymomonas mobilis*, which are important for fermentation process (Sarkar et al., 2012). The relation and cooperative behavior between electrogenic bacteria and fermentative bacteria needed further investigation while feeding in MFC for ethanol production. Since the control of fed substrate and bacteria can be used to optimize the biofilm, electricity generation finally will led to increase ethanol yield. There is a need to understand the relationship between the MFCs performance, power generation, and biomass generation.

15.7 Biogas production

Biogas is an energy source used for production of electricity or heat, or as car fuel in different areas of the world (Sims, 2003). Production of waste materials by several sectors such as agriculture, industries, forestry, and municipalities had stalled the sustainable development of resource recovery (biogas) and recycling of the waste materials, thus resulting in several environmental problems. Stabilization and degradation of organic materials by microbial organisms under anaerobic conditions had been widely employed in the treatment of municipal sludge and organic industrial wastewaters (Chen et al., 2008). Degradation of organic pollutants such as amoxicillin, benzoate, and phenol from anaerobic granular sludge (AGS) with functional microorganisms had been reported earlier. However, rapidly developing bioelectrochemical reactor technology could be a promising platform for simultaneous biogas production along with waste treatment. Bioelectrochemical enhancement of anaerobic digestion with a continuous single-chamber MEC under an elevated organic load and a short HRT (hydraulic residence time) had been reported (Yu et al., 2018). Anoxic/oxic (A/O) system and anaerobic digesters along with MEC combined design had also been studied to enhance the biological process and wastewater treatment. Coupling of anaerobic digestion and MEC enriched with hydrogenotrophic methanogens present in both suspended sludge and cathodic biofilm had shown significant biogas production along with enhancement in digestion of organic pollutants (Li et al., 2016).

Industrial-scale production of biogas has been carried out from some homogeneous wastes such as manures and municipal solid waste (MSW) due to their varied size, composition, structure, and properties. Other than easily degradable sugars, starches, lipids, and proteins MSW also consist of lignocelluloses and keratin, which are more difficult to degrade by microorganisms. Wood-degrading white rot fungus could play an efficient role in degradation of lignocellulosic biomass; consequently, a decrease in lignin content would increase biogas yield as reported earlier (Liew et al., 2012). To improve biogas production, pretreatment of lignocellulosic biomass is required for the accessibility of microbes and enzymes to cellulose and hemicellulose being shielded by naturally recalcitrant carbohydrate and lignin.

The potential of yeast *S. cerevisiae* (baker's yeast) to produce ethanol and its ability to tolerate the inhibitors (hydrolysates), which might be formed during pretreatment process, could be of great significance to meet the increasing need for bioenergy production. Pretreatment of lignocellulosic raw materials by wet oxidation also enhanced its biodegradability by *S. cerevisiae*, thus resulting in significant increase of ethanol and biogas production (Srivastava et al., 2020). Yeast *S. cerevisiae* had also been studied to show increased biogas produce under optimal pH (8 pH) conditions from tofu wastewater in Indonesia (Rusdi, 2016). However, pretreated hydrolysate of lignocelluloses biomass could be used as direct effluent in anaerobic digesters without removing the ethanol fermentation inhibitors (furfural and hydroxymethylfurfural), as anaerobic bacteria are tolerant to these inhibitors as compared with the ethanol fermenting yeasts. A gram-negative, facultative anaerobic bacterium *Z. mobilis* had been demonstrated to produce bioethanol, which might be due to presence of important characteristics for its special Entner–Doudoroff (ED) pathway making it an ideal ethanologenic strain (Yu and Zhang, 2003). Bioconversion of the sugar to ethanol by mixed fermentation of *S. cerevisiae* and *Z. mobilis* by alkaline (NaOH)-pretreated rice straw had also been reported earlier (Das et al., 2013). Thus, the use of both *Saccharomyces* and *Zymomonas* for fermentation of pretreated substrate along with MEC reactor designs could significantly enhance the production of biogas along with bioethanol production (Table 15.1).

Moreover, unwanted impurities present in biogas produced by anaerobic digestion may contribute negatively on biogas compression and reduce the quality of biogas. Electromethanogens in MEC have been found to utilize hydrogen or electrons formed at cathode to capture and convert unwanted CO_2 (residual) into methane, which could be used as renewable natural gas (low carbon fuel substituting) for natural gas vehicle transportation and electricity generation. Thus, coupling the microbial digestion of pretreated complex polymer (present in waste water) and MEC reactor designs under optimized limiting factors would result in increased production of valuable energy resources that could reach their threshold with utmost purity and cost-effectiveness.

15.8 Future applications and prospects

MFCs have potential to working in different fields, and they have various applications. Nowadays, scientific community becomes more curious to know about fuel cell working and to upscale the reactor and system. There is extensive scope for the development and advancement of MFC technology. Initially, MFC technology was known only for power generation from the waste product, without using any external or additional energy, but now MFC emerges as a technology for the production of ethanol, hydrogen, biogas, enzymes, and bioremediation of wastewater. Still MFC technology is a budding or growing technology; it needs continuous and impactful research, which can be incorporated as method or tool for human or social welfare. In the MFC, various kinds of substrates are used for the electricity production and waste management. Recent days, some unconventional substrates are used in MFC, which has potential to operate the MFC in terms of power generation and waste management. In the near future, it is expected that more research work will be focused on the development of cost-effective fuel cell, more varieties of substrate will be used, and MFC will be a sustainable and economical way of bioenergy. In the next few years, MFC will become a technique of next generations, which will be proven a boon to find the solutions to our environmental issues. It depends on us that how it will be possible in a promising way.

References

Aelterman, P., Rabaey, K., Pham, H.T., Boon, N., Verstraete, W., 2006. Continuous electricity generation at high voltages and currents using stacked microbial fuel cells. Environ. Sci. Technol. 40, 3388–3394.

Aiyer, K.S., 2020. How does electron transfer occur in microbial fuel cells? World J. Microbiol. Biotechnol. 36, 19.

Allen, R.M., Bennetto, H.P., 1993. Microbial fuel-cells. Appl. Biochem. Biotechnol. 39, 27–40.

Anwar, Z., Gulfraz, M., Irshad, M., 2014. Agro-industrial lignocellulosic biomass a key to unlock the future bio-energy: a brief review. J. Radiat. Res. Appl. Sci. 7, 163–173.

Birjandi, N., Younesi, H., Ghoreyshi, A.A., Rahimnejad, M., 2016. Electricity generation, ethanol fermentation and enhanced glucose degradation in a bio-electro-Fenton system driven by a microbial fuel cell. J. Chem. Technol. Biotechnol. 91, 1868–1876.

Busalmen, J.P., Esteve-Núñez, A., Berná, A., Feliu, J.M., 2008. C-type cytochromes wire electricity-producing bacteria to electrodes. Angew. Chemie Int. Ed. 47, 4874–4877.

Canfield, J.H., Goldner, B.H., Lutwack, R., 1963. NASA Technical Report, vol. 63. Magna Corp, Anaheim, CA.

Chaudhuri, S.K., Lovley, D.R., 2003. Electricity generation by direct oxidation of glucose in mediatorless microbial fuel cells. Nat. Biotechnol. 21, 1229–1232.

Chen, Y., Cheng, J.J., Creamer, K.S., 2008. Inhibition of anaerobic digestion process: a review. Bioresour. Technol. 99, 4044–4064.

Cheng, S., Liu, H., Logan, B.E., 2006. Increased performance of single-chamber microbial fuel cells using an improved cathode structure. Electrochem. Commun. 8, 489–494.

Cohen, B., 1931. The bacterial culture as an electrical half-cell. J. Bacteriol. 21, 18–19.

Conforti, P., 2011. Looking Ahead in World Food and Agriculture: Perspectives to 2050. Food and Agriculture Organization of the United Nations (FAO).

Das, A., Paul, T., Jana, A., Halder, S.K., Ghosh, K., Maity, C., Mohapatra, P.K., Das, Pati, B.R., Mondal, K.C., 2013. Bioconversion of rice straw to sugar using multizyme complex of fungal origin and subsequent production of bioethanol by mixed fermentation of *Saccharomyces cerevisiae* MTCC 173 and Zymomonas mobilis MTCC 2428. Ind. Crops Prod. 46, 217–225.

Das, D., Veziroğlu, T.N., 2001. Hydrogen production by biological processes: a survey of literature. Int. J. Hydrogen Energy 26, 13–28.

Dekker, A., Heijne, A.T., Saakes, M., Hamelers, H.V.M., Buisman, C.J.N., 2009. Analysis and improvement of a scaled-up and stacked microbial fuel cell. Environ. Sci. Technol. 43, 9038–9042.

DeLong, E.F., Chandler, P., 2002. Power from the deep. Nat. Biotechnol. 20, 788–789.

Gandu, B., Rozenfeld, S., Hirsch, L.O., Schechter, A., Cahan, R., 2020. Immobilization of bacterial cells on carbon-cloth anode using alginate for hydrogen generation in a microbial electrolysis cell. J. Power Sources 455, 227986.

Geng, B.-Y., Cao, L.-Y., Li, F., Song, H., Liu, C.-G., Zhao, X.-Q., Bai, F.-W., 2020. Potential of *Zymomonas mobilis* as an electricity producer in ethanol production. Biotechnol. Biofuels 13, 1–11.

Ghasemi, B., Yaghmaei, S., Abdi, K., Mardanpour, M.M., Haddadi, S.A., 2020. Introducing an affordable catalyst for biohydrogen production in microbial electrolysis cells. J. Biosci. Bioeng. 129, 67–76.

Gielen, D., Boshell, F., Saygin, D., Bazilian, M.D., Wagner, N., Gorini, R., 2019. The role of renewable energy in the global energy transformation. Energy Strateg. Rev. 24, 38–50.

Goswami, R., Mishra, V.K., 2018. A review of design, operational conditions and applications of microbial fuel cells. Biofuels 9, 203–220.

Granovskii, M., Dincer, I., Rosen, M.A., 2006. Environmental and economic aspects of hydrogen production and utilization in fuel cell vehicles. J. Power Sources 157, 411–421.

Habermann, W., Pommer, E.H., 1991. Biological fuel cells with sulphide storage capacity. Appl. Microbiol. Biotechnol. 35, 128–133.

Hemalatha, M., Sravan, J.S., Min, B., Mohan, S.V., 2020. Concomitant use of *Azolla* derived bioelectrode as anode and hydrolysate as substrate for microbial fuel cell and electro-fermentation applications. Sci. Total Environ. 707, 135851.

Hiegemann, H., Herzer, D., Nettmann, E., Lübken, M., Schulte, P., Schmelz, K.-G., Gredigk-Hoffmann, S., Wichern, M., 2016. An integrated 45 L pilot microbial fuel cell system at a full-scale wastewater treatment plant. Bioresour. Technol. 218, 115–122.

Hou, Q., Yang, Z., Chen, S., Pei, H., 2020. Using an anaerobic digestion tank as the anodic chamber of an algae-assisted microbial fuel cell to improve energy production from food waste. Water Res. 170, 115305.

Huang, J., Feng, H., Huang, L., Ying, X., Shen, D., Chen, T., Xu, Y., 2020. Continuous hydrogen production from food waste by anaerobic digestion (AD) coupled single-chamber microbial electrolysis cell (MEC) under negative pressure. J. Waste Manag. 103, 61–66.

Hubenova, Y., Mitov, M., 2015. Extracellular electron transfer in yeast-based biofuel cells: a review. Bioelectrochemistry 106, 177–185.

Ieropoulos, I., Greenman, J., Melhuish, C., 2003a. Imitating metabolism: energy autonomy in biologically inspired robots. In: Proceedings of the AISB, pp. 191–194.

Ieropoulos, I., Melhuish, C., Greenman, J., 2003b. Artificial metabolism: towards true energetic autonomy in artificial life. In: European Conference on Artificial Life. Springer, pp. 792–799.

Ieropoulos, I.A., Greenman, J., Melhuish, C., Hart, J., 2005. Comparative study of three types of microbial fuel cell. Enzyme Microb. Technol. 37, 238–245.

Ivanova, G., Rákhely, G., Kovács, K.L., 2009. Thermophilic biohydrogen production from energy plants by *Caldicellulosiruptor saccharolyticus* and comparison with related studies. Int. J. Hydrogen Energy 34, 3659–3670.

Jain, A., He, Z., 2020. Improving hydrogen production in microbial electrolysis cells through hydraulic connection with thermoelectric generators. Process Biochem. 94, 51–57.

Jayabalan, T., Matheswaran, M., Preethi, V., Mohamed, S.N., 2020. Enhancing biohydrogen production from sugar industry wastewater using metal oxide/graphene nanocomposite catalysts in microbial electrolysis cell. Int. J. Hydrog. Energy 45 (13), 7647–7655.

Kalia, V.C., Jain, S.R., Kumar, A., Joshi, A.P., 1994. Fermentation of biowaste to H_2 by *Bacillus licheniformis*. World J. Microbiol. Biotechnol. 10, 224–227.

Kim, B., Kim, H., H, M., P, D., 1999. Direct electrode reaction of Fe(III)-reducing bacterium, *Shewanella putrefaciens*. J. Microbiol. Biotechnol. 9.

Kim, T., An, J., Jang, J.K., Chang, I.S., 2020. Determination of optimum electrical connection mode for multi-electrode-embedded microbial fuel cells coupled with anaerobic digester for enhancement of swine wastewater treatment efficiency and energy recovery. Bioresour. Technol. 297, 122464.

Kongjan, P., O-Thong, S., Kotay, M., Min, B., Angelidaki, I., 2010. Biohydrogen production from wheat straw hydrolysate by dark fermentation using extreme thermophilic mixed culture. Biotechnol. Bioeng. 105, 899–908.

Kumar, J.,C.R., Majid, M.A., 2020. Renewable energy for sustainable development in India: current status, future prospects, challenges, employment, and investment opportunities. Energy. Sustain. Soc. 10, 2.

Kumar, R., Singh, L., Wahid, Z.A., Din, M.F.M., 2015. Exoelectrogens in microbial fuel cells toward bioelectricity generation: a review. Int. J. Energy Res. 39, 1048–1067.

Lay, J.-J., Lee, Y.-J., Noike, T., 1999. Feasibility of biological hydrogen production from organic fraction of municipal solid waste. Water Res. 33, 2579–2586.

Lee, S.Y., Sankaran, R., Chew, K.W., Tan, C.H., Krishnamoorthy, R., Chu, D.-T., Show, P.-L., 2019. Waste to bioenergy: a review on the recent conversion technologies. BMC Energy 1, 4.

Li, Y., Zhang, Y., Liu, Y., Zhao, Z., Zhao, Z., Liu, S., Zhao, H., Quan, X., 2016. Enhancement of anaerobic methanogenesis at a short hydraulic retention time via bioelectrochemical enrichment of hydrogenotrophic methanogens. Bioresour. Technol. 218, 505–511.

Liew, L.N., Shi, J., Li, Y., 2012. Methane production from solid-state anaerobic digestion of lignocellulosic biomass. Biomass Bioenergy 46, 125–132.

Liu, H., Logan, B.E., 2004. Electricity generation using an air-cathode single chamber microbial fuel cell in the presence and absence of a proton exchange membrane. Environ. Sci. Technol. 38.

Liu, S.-H., Tsai, S.-L., Guo, P.-Y., Lin, C.-W., 2020. Inducing laccase activity in white rot fungi using copper ions and improving the efficiency of azo dye treatment with electricity generation using microbial fuel cells. Chemosphere 243, 125304.

Logan, B.E., 2010. Scaling up microbial fuel cells and other bioelectrochemical systems. Appl. Microbiol. Biotechnol. 85, 1665–1671.

Logan, B.E., Hamelers, B., Rozendal, R., Schröder, U., Keller, J., Freguia, S., Aelterman, P., Verstraete, W., Rabaey, K., 2006. Critical review microbial fuel cells: methodology and technology. Am. Chem. Soc. 40, 5181–5192.

Logan, B.E., Regan, J.M., 2006. Microbial fuel cells—challenges and applications. Environ. Sci. Technol. 40, 5172–5180.

Logan, B.E., Rossi, R., Ragab, A., Saikaly, P.E., 2019. Electroactive microorganisms in bioelectrochemical systems. Nat. Rev. Microbiol. 17, 307–319.

Malinauskaite, J., Jouhara, H., Czajczyńska, D., Stanchev, P., Katsou, E., Rostkowski, P., Thorne, R.J., Colón, J., Ponsá, S., Al-Mansour, F., Anguilano, L., Krzyżyńska, R., López, I.C., Vlasopoulos, A., Spencer, N., 2017. Municipal solid waste management and waste-to-energy in the context of a circular economy and energy recycling in Europe. Energy 141, 2013–2044.

Melhuish, C., Ieropoulos, I., Greenman, J., 2006. Artificial gills for robots: underwater energy autonomy with MFCs. In: Proceedings of the TAROS'06 Conference. University of Surrey Guildford, UK, pp. 118–125.

Miran, W., Nawaz, M., Jang, J., Lee, D.S., 2016. Conversion of orange peel waste biomass to bioelectricity using a mediator-less microbial fuel cell. Sci. Total Environ. 547, 197–205.

Mokhtarian, N., Wan Ramli, W.D., Rahimnejad, M., Najafpour, G.D., 2012. Bioelectricity generation in biological fuel cell with and without mediators. World Appl. Sci. J. 18, 559–567.

Moscoviz, R., Toledo-Alarcón, J., Trably, E., Bernet, N., 2016. Electro-fermentation: how to drive fermentation using electrochemical systems. Trends Biotechnol. 34, 856–865.

Naik, S.N., Goud, V.V., Rout, P.K., Dalai, A.K., 2010. Production of first and second generation biofuels: a comprehensive review. Renew. Sustain. Energy Rev. 14, 578–597.

Okamoto, M., Miyahara, T., Mizuno, O., Noike, T., 2000. Biological hydrogen potential of materials characteristic of the organic fraction of municipal solid wastes. Water Sci. Technol. 41, 25–32.

Park, D.H., Zeikus, J.G., 2003. Improved fuel cell and electrode designs for producing electricity from microbial degradation. Biotechnol. Bioeng. 81, 348–355.

Pophali, A., Singh, S., Verma, N., 2020. Simultaneous hydrogen generation and COD reduction in a photoanode-based microbial electrolysis cell. Int. J. Hydrogen Energy 45, 25985–25995.

Popp, J., Kovács, S., Oláh, J., Divéki, Z., Balázs, E., 2021. Bioeconomy: biomass and biomass-based energy supply and demand. N. Biotechnol. 60, 76–84.

Popp, J., Lakner, Z., Harangi-Rákos, M., Fári, M., 2014. The effect of bioenergy expansion: food, energy, and environment. Renew. Sustain. Energy Rev. 32, 559–578.

Potter, M.C., Waller, A.D., 1911. Electrical effects accompanying the decomposition of organic compounds. Proc. R. Soc. Lond. Ser. B Contain. Pap. Biol. Charact. 84, 260–276.

Rabaey, K., Lissens, G., Siciliano, S.D., Verstraete, W., 2003. A microbial fuel cell capable of converting glucose to electricity at high rate and efficiency. Biotechnol. Lett. 25, 1531–1535.

Reguera, G., 2018. Microbial nanowires and electroactive biofilms. FEMS Microbiol. Ecol. 94.

Reguera, G., Nevin, K.P., Nicoll, J.S., Covalla, S.F., Woodard, T.L., Lovley, D.R., 2006. Biofilm and nanowire production leads to increased current in Geobacter sulfurreducens fuel cells. Appl. Environ. Microbiol. 72, 7345–7348.

Ren, H., Tian, H., Gardner, C.L., Ren, T.-L., Chae, J., 2016. A miniaturized microbial fuel cell with three-dimensional graphene macroporous scaffold anode demonstrating a record power density of over 10000 W m^{-3}. Nanoscale 8, 3539–3547.

Rodionova, M.V., Poudyal, R.S., Tiwari, I., Voloshin, R.A., Zharmukhamedov, S.K., Nam, H.G., Zayadan, B.K., Bruce, B.D., Hou, H.J.M., Allakhverdiev, S.I., 2017. Biofuel production: challenges and opportunities. Int. J. Hydrogen Energy 42, 8450–8461.

Rousseau, R., Ketep, S.F., Etcheverry, L., Délia, M.-L., Bergel, A., 2020. Microbial electrolysis cell (MEC): a step ahead towards hydrogen-evolving cathode operated at high current density. Bioresour. Technol. Rep. 9, 100399.

Rusdi, R., 2016. Kinetics studies impact of initial pH and addition of yeast *Saccharomyces cerevisiae* on biogas production from tofu wastewater in Indonesia. Int. J. Eng. 29, 1037–1046.

Santoro, C., Arbizzani, C., Erable, B., Ieropoulos, I., 2017. Microbial fuel cells: from fundamentals to applications. A review. J. Power Sources 356, 225–244.

Sarkar, N., Ghosh, S.K., Bannerjee, S., Aikat, K., 2012. Bioethanol production from agricultural wastes: an overview. Renew. Energy 37, 19–27.

Schaetzle, O., Barrière, F., Baronian, K., 2008. Bacteria and yeasts as catalysts in microbial fuel cells: electron transfer from micro-organisms to electrodes for green electricity. Energy Environ. Sci. 1, 607–620.

Schröder, U., 2018. A basic introduction into microbial fuel cells and microbial electrocatalysis. ChemTexts 4, 19.

Schröder, U., 2011. Discover the possibilities: microbial bioelectrochemical systems and the revival of a 100-year–old discovery. J. Solid State Electrochem. 15, 1481–1486.

Sekrecka-Belniak, A., Toczyłowska-Mamińska, R., 2018. Fungi-based microbial fuel cells. Energies 11, 2827.

Shantaram, A., Beyenal, H., Veluchamy, R.R.A., Lewandowski, Z., 2005. Wireless sensors powered by microbial fuel cells. Environ. Sci. Technol. 39, 5037–5042.

Sharma, N.K., Behera, S., Arora, R., Kumar, S., 2016. Potential role of xylose transporters in industrial yeast for bioethanol production: a perspective review. In: Proceedings of the First International Conference on Recent Advances in Bioenergy Research. Springer, pp. 81–93.

Sims, R.E.H., 2003. Bioenergy Options for a Cleaner Environment: In Developed and Developing Countries. Elsevier.

Slate, A.J., Whitehead, K.A., Brownson, D.A.C., Banks, C.E., 2019. Microbial fuel cells: an overview of current technology. Renew. Sustain. Energy Rev. 101, 60–81.

Srivastava, R.K., Shetti, N.P., Reddy, K.R., Aminabhavi, T.M., 2020. Biofuels, biodiesel and biohydrogen production using bioprocesses. A review. Environ. Chem. Lett. 18, 1049–1072.

Sugnaux, M., Happe, M., Cachelin, C.P., Gloriod, O., Huguenin, G., Blatter, M., Fischer, F., 2016. Two stage bioethanol refining with multi litre stacked microbial fuel cell and microbial electrolysis cell. Bioresour. Technol. 221, 61–69.

Van Ginkel, S.W., Oh, S.-E., Logan, B.E., 2005. Biohydrogen gas production from food processing and domestic wastewaters. Int. J. Hydrogen Energy 30, 1535–1542.

Wang, Z., Lee, T., Lim, B., Choi, C., Park, J., 2014. Microbial community structures differentiated in a single-chamber air-cathode microbial fuel cell fueled with rice straw hydrolysate. Biotechnol. Biofuels 7, 9.

Wilkinson, S., 2000. "Gastrobots"—benefits and challenges of microbial fuel cells in food powered robot applications. Auton. Robots 9, 99–111.

Wu, D., Sun, F., Chua, F.J.D., Zhou, Y., 2020. Enhanced power generation in microbial fuel cell by an agonist of electroactive biofilm – Sulfamethoxazole. Chem. Eng. J. 384, 123238.

Xu, B., Ge, Z., He, Z., 2015. Sediment microbial fuel cells for wastewater treatment: challenges and opportunities. Environ. Sci. Water Res. Technol. 1, 279–284.

Yu, Z., Leng, X., Zhao, S., Ji, J., Zhou, T., Khan, A., Kakde, A., Liu, P., Li, X., 2018. A review on the applications of microbial electrolysis cells in anaerobic digestion. Bioresour. Technol. 255, 340–348.

Yu, Z., Zhang, H., 2003. Pretreatments of cellulose pyrolysate for ethanol production by *Saccharomyces cerevisiae*, *Pichia* sp. YZ-1 and *Zymomonas mobilis*. Biomass Bioenergy 24, 257–262.

Yuan, J., Liu, S., Jia, L., Ji, A., Chatterjee, S.G., 2020. Cogeneration system of bioethanol and electricity with microbial fuel cell technology. Energy Fuels 34, 6414–6422.

Zhao, F., Slade, R.C.T., Varcoe, J.R., 2009. Techniques for the study and development of microbial fuel cells: an electrochemical perspective. Chem. Soc. Rev. 38, 1926–1939.

CHAPTER

16

Advances in microbial fuel cell technology for zero carbon emission energy generation from waste

Mpumelelo Thomas Matsena, Evans Martin Nkhalambayausi Chirwa

Water Utilization and Environmental Engineering Division, Department of Chemical Engineering, University of Pretoria, Pretoria, Gauteng, South Africa

List of abbreviations

ACNQ 2-amino-3-carboxy-1,4-naphthoquinone
ADP Adenosine diphosphate
AEM Anion exchange membrane
AFM Atomic force microscopy
AHQDS Anthrahydroquinone-2,6-disulfonate
ANME Anaerobic methanotrophic archaea
AOM Anaerobic oxidation of methane
AQDS Anthraquinone-2,6-disulfonate
ATP Adenosine triphosphate
Bio-PdNPs Biogenic palladium nanoparticles
BOD Biological oxygen demand
BPM Bipolar membrane
CCS Carbon capture and storage
CEM Cation exchange membrane
CNT Carbon nanotubes
COD Chemical oxygen demand
DET Direct electron transfer
DIET Direct interspecies electron transfer
DO Dissolved oxygen
EAB Electrochemically active bacteria
EET Extracellular electron transfer
FO Forward osmosis
FO-MFC Forward osmosis microbial fuel cell
GTL Gas-to-liquid
HA Humic acid
MDC Microbial desalination cell
MEC Microbial electrolysis cell
MEC-MFC Microbial electrolysis cell-Microbial fuel cell
MET Mediated electron transfer
MFC Microbial fuel cell
MWCNT Multiwalled carbon nanotubes
$NADP^+$ Nicotinamide adenine dinucleotide phosphate
NADPH Reduced nicotinamide adenine dinucleotide phosphate
NASA National Aeronautics and Space Administration
P-ETC Photosynthetic electron transport chain
PAH Polycyclic aromatic hydrocarbons
PCB Polychlorinated biphenyl
PdNPs Palladium nanoparticles
PEM Proton exchange membrane
PSI Photosystem I
PSII Photosystem II
SWCNT Single walled carbon nanotubes
UMFC Upflow microbial fuel cell
UPBR Upflow packed bed reactor
UPBR-UMFC Upflow packed bed reactor-Upflow microbial fuel cell
WWTP Wastewater treatment plant

Biofuels and Bioenergy
https://doi.org/10.1016/B978-0-323-85269-2.00013-7

16.1 Introduction

16.1.1 The history and emergence of microbial fuel cells

An increase in the demand for energy has resulted in the global availability of fossil fuel resources declining. In addition, the added pressure on the environment due to the emission of greenhouse gases has disturbed the ecological balance. This has led to the need for more cleaner and renewable energy sources. One such potential technology that can provide a clean energy source is a fuel cell. The operational principles of fuel cells date as far back as the 18th century. In 1791, a physician and a professor at the University of Bologna named Luigi Galvani presented his results in which he observed the leg of a frog twitching when he was investigating the nature and effect of electricity in the tissue of an animal (Kipnis, 1987). By passing a spark of an electrical current through the isolated frog, he observed that the muscles of the frog can move by exposing them to an electrical current and coined the term "animal electricity." From which he concluded from further observations that the tissue of a frog also had an innate and vital "animal electricity" that activated the nerve and muscle when spread across with metal probes or touched by another frog's nerve and muscle. This was one of the first reported observations of bioelectricity. Today, Galvani is known as the first electrochemist and considered a pioneer of bioelectricity and even the famous Galvanic cell is named after him.

After the discovery that Galvani made, a professor of experimental physics at Pavia University named Alessandro Volta looked into the experiments that Galvani conducted and argued that the observations that Galvani made were not due to an innate animal electricity but were caused by the metal probes that Galvani used to connect the nerves and muscles in his experiments. This led to the Galvani–Volta controversy at the end of the 18th century which also gave rise to the invention of an early battery resulting from Volta experiments (Volta, 1800). Volta had began conducting experiments where he constructed a pile of alternating discs of copper with silver or zinc with tin that were separated by a pasteboard disc or a spongy material soaked in brine. He connected the pile to an electrical wire and observed a steady flow of electricity to which he coined the terms "electrical current" and "electromotive force" from his observations. This led to the development of a battery in which in a letter dated March 20, 1800, he conveyed his findings to Joseph Banks who was at the time the president of Royal Society (Dicks and Rand, 2018). He called the pile "Voltaic pile" which was the first nonrechargeable power source. With Humphry Davy, who was part of the Royal Institution in London realising that Volta had actually produced electricity via the reactions of chemicals at the metal–solution interface in which hydrogen was produced at the "positive" copper disc, and zinc was consumed at the "negative" disc, this led to the recognition of the relationship between chemical and electrical effects which prompted Davy to coin the word "electrochemical" (Dicks and Rand, 2018). This led to the development of the science of "electrochemistry."

During that time, the debate between Galvani and Volta was still taking place. However, in the later years, the debate was resolved. Galvani's observations that the nerves and muscle generated their own innate or internal electricity were confirmed in the mid-19th century when the electrical impulses in nerves were observed through indirect methods with the first primitive amplifiers. Also, when the vacuum tube amplifier was invented in the early 20th century, it was unequivocally concluded that Galvani was correct (Adrian, 1926). The vacuum tube amplifier successfully showed the presence of individual electric oscillations in response to physiological stimulation of frogs. The observations which Volta made that the frog's legs were twitching due to the electricity generated by two metals connected to a frog were also confirmed by Humphry Davy (Dicks and Rand, 2018). Hence today, the Galvanic cell is also known as the Voltaic cell named after Volta.

The principle of a fuel cell effect as it is known today was first demonstrated by a professor of chemistry and physics at Basel University, Christian Friedrich Schönbein. By using dilute sulfuric acid as an electrolyte and platinum (Pt) wire as an electrode, he was able to generate electricity by having one compartment filled with hydrogen and the other one exposed to air. The results were published in The London and Edinburgh Philosophical Magazine and Journal of Science in January 1839 (Schönbein, 1839). In the same journal, in February 1839, a Welsh judge and a physical scientist named William Robert Grove demonstrated a more efficient voltaic cell. The difference between his and the one demonstrated by Schönbein was that the "Grove cell" as it was called at the time was separated by a porous wall and contained only a zinc electrode in dilute sulfuric acid, the other compartment only contained a Pt electrode in concentrated nitric acid (Wisniak, 2015).

In February 1839, Grove also gave an account of an experiment in which he saw the galvanometer being permanently deflected when he connected two strips of platina in tubes containing oxygen and hydrogen. In 1842, he decided to return to his experiment in which he gave an account on in February 1839 and decided to expand on the work including the work done by Schönbein (Grove, 1842). He developed the first working prototype of a fuel cell in which he arranged the two Pt electrodes in tubes submerged in dilute sulfuric acid. He partially filled the first tube with oxygen and the second one with hydrogen and after that, he combined several of these electrodes in a series circuit and created a "gas battery" in which today it is known as the first fuel cell prototype (Grove, 1842). Furthermore, the interconnections between several components of the fuel cell including cations, anions, electrodes, electrolytes, oxidizing and reducing agent were experimentally determined in 1893 and related to the physical properties and chemical reactions by one of the founders of the field of physical chemistry, Friedrich Wilhelm Ostwald (Flimban et al., 2019). His contribution specifically in the chemistry of fuel cells laid the foundation for researchers to better understand the fuel cell workings.

One of the researches who also expanded on the knowledge of fuel cells was a professor of botany at the University of Durham, Michael Cresse Potter. In 1910, when he was researching on how microbes degrade organic compounds, he observed that microorganisms actually have the ability to generate a potential difference and deliver current (Potter, 1910). In 1911, he continued on the work and used Pt wires as electrodes and microbial cultures of *Escherichia coli* and *Saccharomyces cerevisiae* to further demonstrate the ability of microbes to generate electricity (Potter, 1911). This work led to the discovery of the first microbial fuel cell (MFC) which was further explored by Barnett Cohen, a biochemist at Cambridge University. Cohen connected these MFCs in series and demonstrated in 1931 how this type of connection can improve energy generation (Cohen, 1931).

It was not until the 1960s that MFCs took part of a program that was launched by the National Aeronautics and Space Administration as an opportunity to recycle waste into electrical energy during a space flight (Liengen et al., 2014). This never attracted much attention up until 1983 when Bennetto et al. (1983) and Stirling et al. (1983) at Kings College put emphasis on the usage of redox mediators in MFCs to improve energy generation. In addition, the interest increased in 1999 when it was demonstrated by the Byung Hong Kim group at the Korean Institute of Science and Technology that electrochemically active bacterial (EAB) species in MFCs did not require mediators for electron transport (Kim et al., 1999). Since the beginning of the 21st century, the research in MFCs has exponentially increased with notable advances being made in the improvement of energy recovery. Table 16.1 shows the summary of the history and emergence of MFCs.

16.1.2 Advances and limitations

From the time when Luigi Galvani reported the first observations of bioelectricity in the 18th century to the time when Michael Cresse Potter discovered the first MFC in the 20th century, the technology has grown and has attracted an increasing number of researchers in recent years as shown in Fig. 16.1.

The first major research outputs on MFCs was published in 1910 to 1931, which includes research by Potter (1910, 1911) and Cohen (1931) on the demonstration of MFC and how series connection improves performance. Between 1962 and 1999, only 29 papers on the research based on MFCs were published according to Scopus. These include the work done by Bennetto et al. (1983) and Stirling et al. (1983) on the mediator and Kim et al. (1999) on the mediatorless MFC. It is not up until the years between 2000 and 2009 that the number began to increase. During this time, 965 papers where published (Fig. 16.1). The most prominent publications that were most cited based on Scopus were the work done by Chaudhuri and Lovley (2003)

TABLE 16.1 Summary of the history and the emergence of microbial fuel cells.

Scientist	Year	Contribution	References
Galvani, L. (University of Bologna)	1791	Reported the first observations of bioelectricity.	Kipnis (1987)
Volta, A. (Pavia University)	1800	Discovered a "voltaic pile" which was the first nonrechargeable power source.	Volta (1800)
Davy, H. (Royal Institution)	1800	Recognized the relationship between chemical and electrical effects.	Dicks and Rand (2018)
Schönbein, C.F. (Basel University)	1839	The first demonstration of the principle of a fuel cell effect.	Schönbein (1839)
Grove, W.R. (Royal Institution)	1839	Demonstrated a more efficient voltaic cell called the "Grove cell."	Wisniak (2015)
Grove, W.R. (Royal Institution)	1842	Developed the first working prototype of a fuel cell.	Grove (1842)
Ostwald, F.W. (Leipzig University)	1893	Described the interconnections between several components of the fuel cell.	Flimban et al. (2019)
Potter, M.C. (University of Durham)	1910	Observed that microorganisms actually have an ability to generate a potential difference.	Potter (1910)
Potter, M.C. (University of Durham)	1911	Used *Escherichia coli* and *Saccharomyces cerevisiae* for electricity generation.	Potter (1911)
Cohen, B. (Cambridge University)	1931	Connected microbial fuel cells in series to improve energy generation.	Cohen (1931)
Bernetto et al. (Kings College)	1983	Put emphasis on the usage of redox mediators in MFCs to improve energy generation.	Bennetto et al. (1983)
Kim et al. (Korean Institute of Science and Technology)	1999	Demonstrated that no mediators for electron transport are required for electrogens.	Kim et al. (1999)

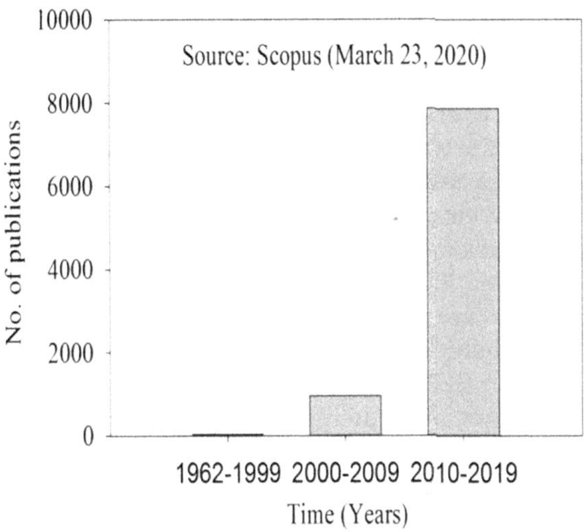

FIGURE 16.1 The total number of publications in the years of 1962–99, 2000–09, and 2010–19. *The data are based on the number of articles mentioning "microbial fuel cell" in Scopus till March 23, 2020.*

and Rabaey et al. (2004) which confirmed a mediatorless MFC, Liu and Logan (2004) which presented a single chamber MFC, Gorby et al. (2006) which expanded on the knowledge of electrically conductive bacterial nanowires, and Logan et al. (2007) on anode modification using graphite fiber brush.

A spike in the number of publications was experienced in the years between 2010 and 2019 with 7866 publications. This is because MFCs do not only provide solutions to today's environmental problems but they can also be applied over a wide range of applications. These applications presented in the years 2010–2019 include doping carbon nanotubes (CNTs) with nitrogen as cathodic catalysts for oxygen reduction (Feng et al., 2011); the simultaneous nitrification, denitrification, and carbon removal (Virdis et al., 2010); salt removal in a continuously operated upflow microbial desalination cell (MDC) with an air cathode (Jacobson et al., 2011); and self-powered supercapacitive MFC that uses high-power operation of an internal supercapacitor (Santoro et al., 2016). From Fig. 16.2, it can also be seen that in the last 10 years, the MFC research has been growing meaning in future, there exists a potential for new applications on MFCs to be discovered including the introduction of MFCs in large-scale applications.

It should be mentioned that although MFC research has made significant advances to date, there are various limitations that are present especially in the operation of the fuel cell itself. These include activation, ohmic and concentration losses. Activation losses are related to the kinetic performance of the MFC overall electrochemical reaction, ohmic losses are related to the internal resistance which is mostly dominated by ionic conductivity but can include electrical conductivity, and concentration losses are related to the mass transport limitations at the electrolyte–electrode interface (Rismani-Yazdi et al., 2008). These cause the fuel cell to operate at an output potential difference lower than the theoretical one. Therefore, better solutions have to be discovered to reduce the effect of these limitations on the performance of the MFC.

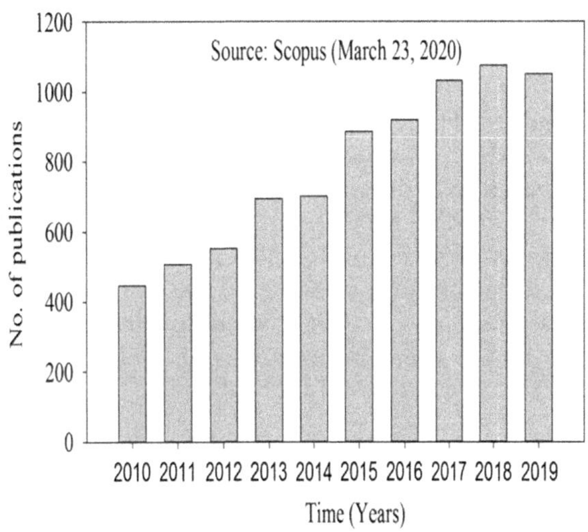

FIGURE 16.2 The total number of publications in each year from 2010 to 2019. *The data are based on the number of articles mentioning "microbial fuel cell" in Scopus till March 23, 2020.*

16.1.3 Versatility of application

The advantageous characteristic of MFCs is in the usage of bacteria as biocatalysts at the anode chambers. This characteristic harnesses the ability of the microorganisms to mediate electrons which are generated from the oxidation of organic matter and transferred from the outer cell membrane to the surface of the anode. Various different methods have been explored in order to increase the electric output of the MFCs. These methods include anode modification using nanomaterials such as copper-doped iron oxide nanoparticles (Sekar et al., 2019), or using a synthetic flavin pathway from *Bacillus subtilis* expressed in *Shewanella oneidensis* MR-1 to secrete a very high amount of flavins and increase the power output (Yang et al., 2015). Upon trying to optimize or increase electrical output, MFCs have also been applied in various applications such as the degradation of organic matter (Ieropoulos et al., 2005; Rabaey et al., 2003). Due to the reason that bacteria are capable to degrading organic matter, MFCs have also been used to remove organic pollutants in wastewater such as phenol (Luo et al., 2009) and PAHs (Gambino et al., 2017). MFCs are also modified with microbial electrolysis cells (MECs) which unlike MFCs use energy to within the range of 0.2–0.8 V to produce biohydrogen. Previous studies have demonstrated the successful operation of MEC-MFC in which an MFC is used to supply power to the MEC to produce biohydrogen (Sun et al., 2008, 2009). The additional applications of MFCs include desalination, bioremediation, and usage as biosensors.

16.1.4 Implication to the energy industry—water/energy nexus

The lack of access to portable water is a cause for concern in the world. Two billion people have no access to safely managed drinking water services and about 4.5 billion people lack safely managed sanitation services (World Health Organization, 2019). Millions of people die every year with almost 297,000 children under five dying every year from diarrheal diseases due to poor sanitation and/or unsafe drinking water (Hutton and Chase, 2016). Also, lack of access to clean water also affects food production. Due to about 3% of the water on earth being fresh and with 68.7% of the fresh water locked in glaciers and icecaps, it is therefore essential that the fresh water which we have be utilized in an efficient manner (IEEE Spectrum, 2010). The global distribution of the water resource is summarized in Fig. 16.3.

Due to the linkages that exist between the energy and water resource, improving the efficiency of water-use should be looked at holistically in conjunction with reducing electricity consumption. This is because the linkage is set to rise due to the energy and water resources experiencing an increase in the demands and constraints as a result of climate change, the growth in population and the economy. Fig. 16.4 shows the rise in global electricity consumption from 1999 to 2019 which should be reduced at the same time when water is used efficiently (Enerdata, 2020).

It is projected that energy-related water consumption is set to increase by 60% in the year 2040 due to several energy technologies which can have high water demands such as biofuels production, nuclear power, and carbon capture and storage (IEA, 2017). In addition, although it is estimated that the global electricity consumption by the supply, treatment, and desalination of water will remain at about 4%, the increase in electricity consumption due to water-related issues in the Middle East is set to increase from 9% in 2015 to 16% by 2040 (IEA, 2017). This is actually something that will mostly be experienced by countries which rely on nuclear and coal-fired power plants such as China, India, and South Africa, many of which are located in areas that are experiencing water shortages. As such, making the efforts to find

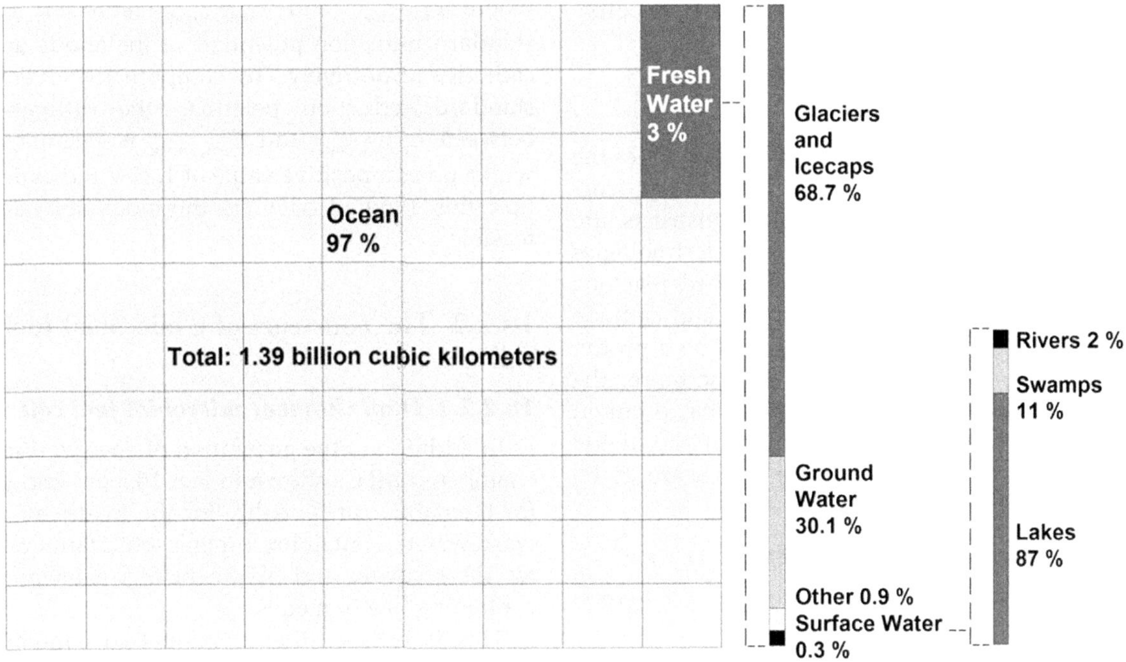

FIGURE 16.3 Global distribution of the water resource.

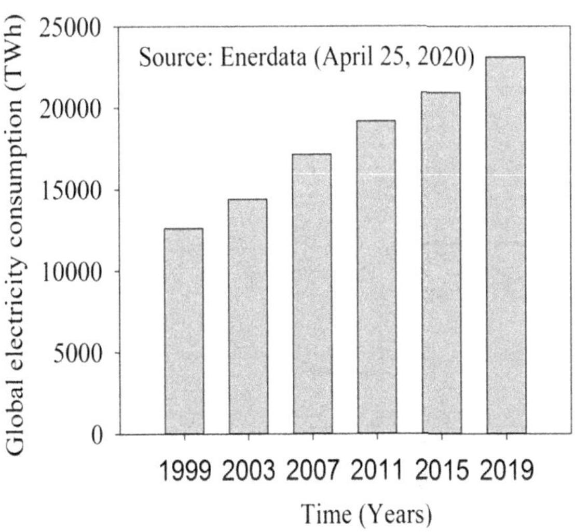

FIGURE 16.4 Global electricity consumption.

new technologies which can reduce both water-use and energy consumption necessary. A technology that currently provides such capability is MFC. Although it is still in the laboratory stages, it is a promising technology that can address the challenges of treating waste while generating renewable energy. It functions independent of the sun or the wind, and it provides additional ways in which electricity can be generated by converting organic matter to energy by the use of microorganisms. It should be noted that MFCs should never be thought of as a technology that can be a major energy supplier, however, it should be thought of as a new approach in solving global water-energy problems which can be integrated into already existing clean and renewable energy solutions such as wind and solar power.

16.2 The science of microbial fuel cells

16.2.1 Electrochemistry and thermodynamic feasibility

Intense amount of research has gone into the successful implementation of different constructions of MFCs. Most of these constructs are adapted from the already existing technologies such as the fuel cell and packed bed reactors. MFCs are comprised of the anode and cathode compartments which are separated by a proton exchange membrane (PEM). Furthermore, the general electrochemical reactions that occur in each compartment are oxidation at the anode and reduction at the cathode which are given by Eq. (16.1) and Eq. (16.2):

$$\text{Substrate} \leftrightarrow \text{Oxidized product} + H^+ + e^- \quad (16.1)$$

$$\text{Oxidant} + H^+ + e^- \leftrightarrow \text{Reduced oxidant} \quad (16.2)$$

The electrons produced by oxidizing the substrate or organic matter using the EAB move from the anode to the cathode to reduce the oxidant. In addition, the charge balance is maintained by the movement of hydrogen ions (H^+) through the PEM from the cathode to the anode. For the electrochemical reactions in MFC to be thermodynamically feasible, an overall positive standard reduction potential should be obtained. This will be illustrated by using glucose ($C_6H_{12}O_6$) as a substrate and oxygen (O_2) as an oxidant. This is the most widely used MFC. The redox half-reactions are given in Eq. (16.3) and Eq. (16.4) (Sharma and Li, 2010):

$$C_6H_{12}O_6 + 6H_2O \leftrightarrow 6CO_2 + 24H^+ + 24e^- \; E^o_{\text{red, anode}} = -0.42\text{ V} \quad (16.3)$$

$$6O_2 + 24H^+ + 24e^- \leftrightarrow 12H_2O \; E^o_{\text{red, cathode}} = 0.82\text{ V} \quad (16.4)$$

where $E^o_{\text{red, anode}}$ and $E^o_{\text{red, cathode}}$ represent the standard reduction potentials of the anode and cathode, respectively. To compute the overall standard reduction potential, the difference between $E^o_{\text{red, cathode}}$ and $E^o_{\text{red, anode}}$ is calculated, which gives a positive value of 1.24 V indicating that this type of MFC is thermodynamically feasible.

16.2.2 The construct of a microbial fuel cell

16.2.2.1 Dual chamber microbial fuel cell

In addition to the generation of energy, dual-chambered MFCs shown in Fig. 16.5 are known for their dual functionality for the treatment of wastewaters containing organic pollutants and for the recovery and treatment of wastewaters containing heavy metals.

This is because they contain two compartments, namely the anode with the ability to breakdown organic waste, and the cathode with the ability for metal recovery or reduction to less toxic forms. In addition, this is the reason why this construct is highly favored in the

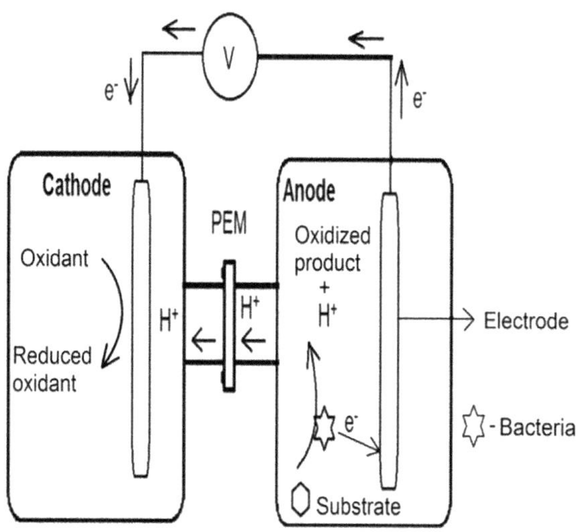

FIGURE 16.5 Schematic of a dual-chambered microbial fuel cell.

treatment of waste in general due to its versatility in its applications. The organic pollutants that have been treated in previous studies include phenol (Luo et al., 2009), PAHs (Gambino et al., 2017), and polychlorinated biphenyl (Xu et al., 2015b). Also, the heavy metals that have been recovered using this construct include vanadium (Zhang et al., 2012), copper (Heijne et al., 2010), silver (Choi and Cui, 2012), and chromium (Sophia and Sai, 2016).

16.2.2.2 *Single chamber microbial fuel cell*

Single-chambered MFCs can only be implemented in the treatment of organic waste without the functionality of metal recovery. This is because the only oxidant that they can utilize is the available oxygen in air and hence they are also known as air-cathode MFCs. The introduction of single-chambered MFCs was due to the problems that were experienced in the mass transfer of dissolved oxygen (DO) to the cathode electrode in dual-chambered MFCs. This led to the removal of the cathode compartment and the implementation of air-cathodes which can be easily modified to improve the mass transfer of oxygen and performance of MFC. Some of the modification of air-cathodes to improve performance of MFC which have been implemented include using nitrogenous mesoporous carbon air cathode coated with cobalt and copper nanoparticles (Liang et al., 2020), diffusion-layer-free air cathode based on ionic conductive hydrogel (Li et al., 2020), activated carbon-supported Fe–Ag–N multi-doped graphene (Lv et al., 2019), and manganese dioxide on carbon Vulcan air-cathode (Majidi et al., 2019). The schematic of the single-chambered MFC is shown in Fig. 16.6.

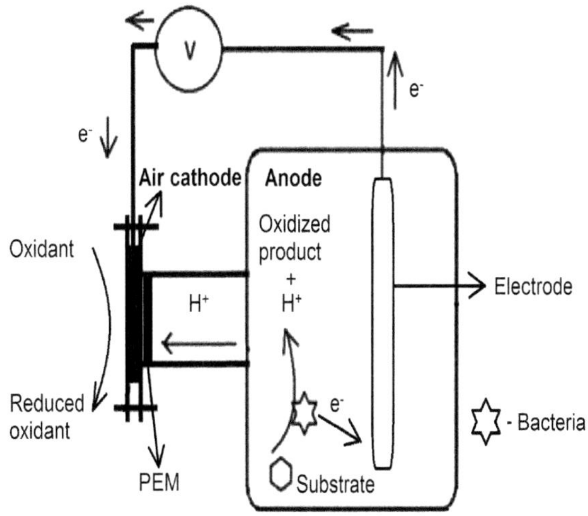

FIGURE 16.6 Schematic of a single-chambered microbial fuel cell.

16.2.2.3 *Upflow microbial fuel cell*

In the treatment processes, biofilm systems are known to offer an opportunity for the bacterial community to survive under high concentrations of either organic or metallic pollutants. This is achieved by presenting an environment with a variable survivability range due to the biofilm thickness forming a dynamic concentration gradient (Matsena et al., 2018). Therefore, under high loading concentrations of pollutants, the mass transport resistance leads to zones in the biofilm where the concentration is low enough to allow for the survival of detoxifying microbial agents. An example of such a biofilm system is an upflow packed bed reactor (UPBR). To reduce the energy requirements of such systems, He et al. (2005, 2006a) developed and demonstrated that a UPBR can be integrated with an upflow MFC (UMFC) to realize a continuously operated MFC with improved performance (Fig. 16.7). This also led to the combined advantages of both dual-chambered MFCs and UPBR. Since then, UPBR-UMFC systems have been used in studies such as thermophilic microbial communities fed with acetate (Dessì et al., 2019), an anaerobic aged sludge fueled with a synthetic acetate-based and actual domestic wastewater (Ismail and Jaeel, 2016),

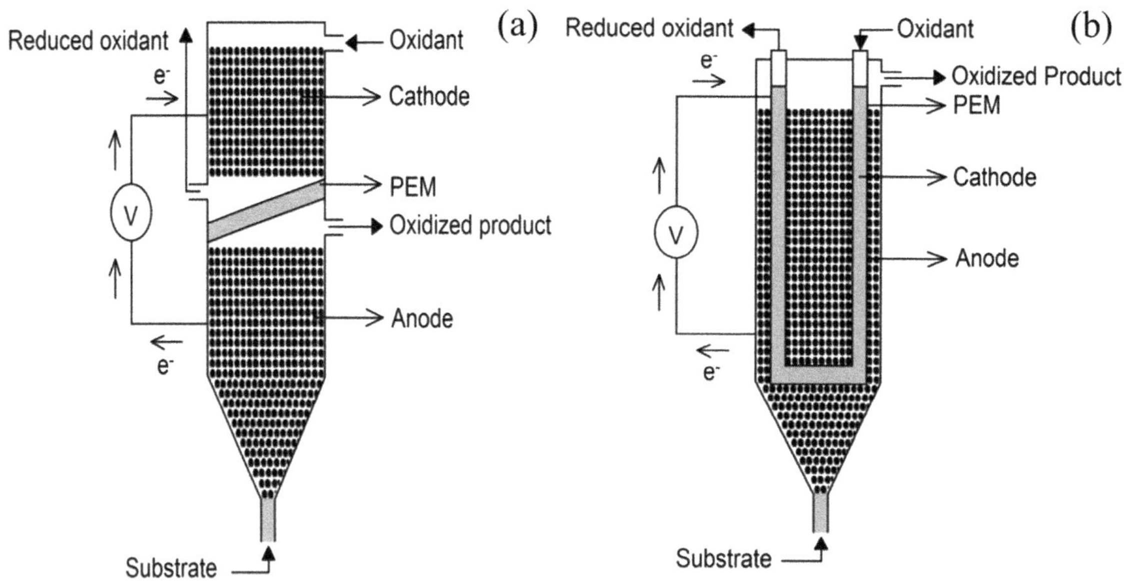

FIGURE 16.7 Schematic of (A) an upflow microbial fuel cell and (B) an upflow microbial fuel cell with interior proton exchange membrane.

development of double layer capacitor materials modified anodes (Chen et al., 2018), and evaluation of electrochemical performance of a photosynthetic UMFC (Zheng et al., 2017).

16.2.2.4 *Stacked microbial fuel cell*

Due to the fact that the output potential difference of an individually operated construct of MFC can never be above a theoretical value, even when neglecting losses and limitations, this leads to finding other means in order to enhance the performance of MFCs. One of the solutions is to use a stacked MFC which can be classified into three groups, namely, series, parallel, and hybrid (Fig. 16.8).

The usage of stacked MFCs was demonstrated by Aelterman et al. (2006) in which it was shown that stacked MFCs of either series or parallel connections have the ability to produce high-power densities. Zhang et al. (2017) then showed that these two work in different ways in order to achieve maximum power densities. A series connection improves operational output potential difference and on the other hand, a parallel connection improves operational current output. Due to this trade-off between operational output potential difference and current output, Zhang et al. (2017) demonstrated that a hybrid configuration can be used for stacked MFCs in order to improve both current output and output potential difference, although the power density will be lower as compared to parallel and series connections.

16.2.3 Factors affecting performance

The performance of MFC is affected by various losses which limit the attainment of theoretical output potential difference under open circuit conditions. This is due to irreversible losses that occur in the MFC which are also known as overpotentials. These three main irreversible losses are identified as activation losses, ohmic losses, and concentration

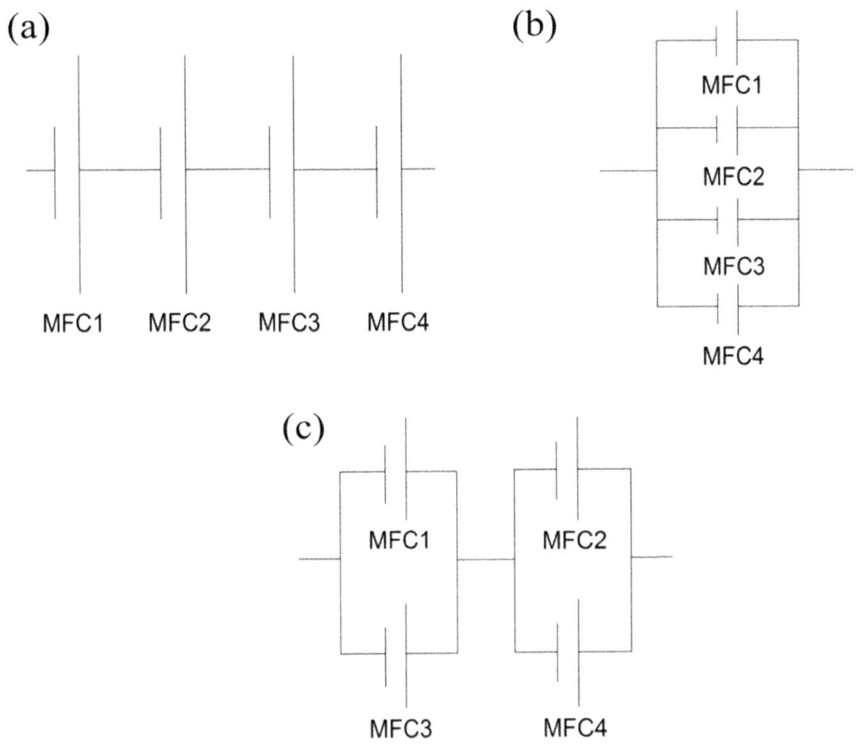

FIGURE 16.8 Schematic of different stacked microbial fuel cell (MFC) configurations such as (A) series MFC connection, (B) parallel MFC connection, and (C) hybrid MFC connection.

losses as shown in Fig. 16.9A. These losses occur due to the thermodynamic requirement to activate the electrochemical reactions, resistance to charge flow, and mass transfer limitations. Other losses which affect the performance of MFC are parasitic losses which are mainly caused by the crossover of substrate and unwanted side reactions. The level of magnitude of these losses varies from system to system. A typical power density curve is shown in Fig. 16.9B.

16.2.3.1 Activation losses

The performance of the MFC depends on the kinetics of the electrochemical reactions which take place at the cathode and anode. The kinetics are limited by the activation energy of each reaction and are mostly dominant at low current densities/output as shown in Fig. 16.9A (Fu et al., 2010). Various steps can be taken to reduce the activation losses. These include increasing the porosity and surface area of the anode, introduction of soluble mediators to improve the anode–bacteria interaction or using microbes that use conducting pili to attach to the electrode, increasing the operating temperature of the anode, and establishing an enriched biofilm on the anode electrode (Kundu and Dutta, 2018).

16.2.3.2 Ohmic losses

Performance of MFC is also restricted by the ohmic losses. These are the losses caused by the resistance to the flow of electrons through the interconnections, electrodes, PEM, catholyte,

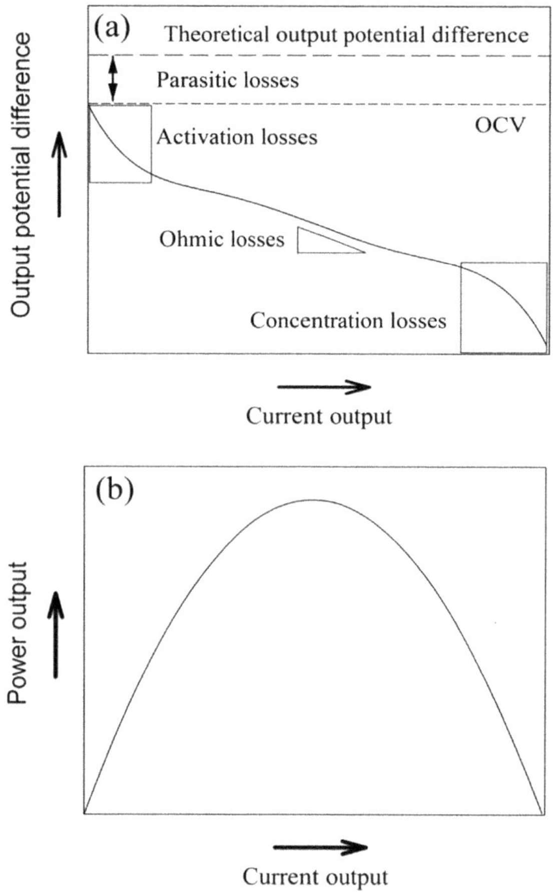

FIGURE 16.9 (A) A typical polarization curve and output potential difference losses in microbial fuel cell and (B) a typical power output curve.

and anolyte. This loss is located on the linear region of the polarization curve as shown in Fig. 16.9A and can be calculated using Ohm's law in which the magnitude of the gradient represents the internal resistance (Kundu and Dutta, 2018). There are several ways which can be employed to reduce this type of loss, namely by minimizing the spacing of the cathode and anode electrode, reducing the electron travel distance within the electrodes, and using low resistive electrodes. It should be noted that the internal resistance is dominated by ionic resistance due to the electrolyte, and in addition, the electronic resistance is mostly negligible.

16.2.3.3 *Concentration losses*

The process of supplying substrates and oxidants at the electrode and the removal of the oxidized products and reduced oxidants is governed by mass transfer. When there is insufficient mass transfer, this leads to low oxidant or substrate concentration at the surface of the electrode. This in turn results in a loss of output potential difference which is caused by a decrease in reaction rate at the electrode surface. These losses normally occur at high current densities and increase as the current density increases as shown in Fig. 16.9A. In order to reduce this type of loss, one can modify the electrode design such as porosity, composition, thickness, and specific surface area (Rismani-Yazdi et al., 2008). Ideally this should improve hydrodynamic flow to improve mass transfer and prevent product accumulation. Employing continuous flow MFCs such as UMFC can improve the flux of substrate or oxidant to the electrode. This has been reported in a previous study by Jong et al. (2006) in which the reduction of retention time from 2.7 to 0.7 min increased the performance of the MFC.

16.2.3.4 *Parasitic losses*

Additional losses to the three main losses are known as parasitic losses. These losses occur due to substrate crossover and unwanted reactions which have been reported in previous studies (Elangovan and Dharmalingam, 2016, 2017; Kim et al., 2016). Ideally, the substrate should not pass through the membrane but due to molecular diffusion and electroosmosis (Jiang and Chu, 2003), there is some form of substrate crossover that occurs. This results in the lowering of performance of MFC due to the decrease in coulombic efficiency since less substrate is consumed at the anode. Furthermore, the substrate and its oxidized products may lead to the poisoning of the cathode surface. Therefore, this means that a careful consideration on both the material and thickness of the membrane needs to be looked at so that parasitic losses are reduced.

16.3 Electrochemically active bacteria

16.3.1 Classification of electrochemically active bacteria

EAB can be classified as either electrogenic bacteria or electrotrophic bacteria. Electrogenic bacteria are microorganisms that have the ability to transfer electrons outside of the bacterial cell to extracellular electron acceptors. They are used in the anode as biocatalysts in order to generate electrons from the substrate and improve performance of MFC. The examples of electrogens that have been identified in previous studies include *Geobacter sulfurreducens* (Bond and Lovley, 2003), *S. oneidensis* MR-1 (Bretschger et al., 2007), and *Shewanella putrefaciens* (Yang et al., 2017). In addition, weak electrogens which produce low current densities that have been identified include *Bacillus subtilis* and *Klebsiella aerogenes* (Logan et al., 2019). The other type of EAB includes electrotrophic bacteria which are microorganisms that have the ability to transfer electrons into the bacterial cell from extracellular electron donors. They can be used in cathode as biocatalysts in order to consume electrons and enhance the performance of the MFC. Microorganisms that have been identified as electrotrophs include *Geobacter metallireducens* (Gregory et al., 2004), *Klebsiella pneumoniae* (Rhoads et al., 2005), and *Acidithiobacillus ferrooxidans* (Ishii et al., 2015).

Recently, it has been shown that EAB are not limited to mesophilic conditions or typical ranges of pH and salinity. This is because several EAB can be classified as extremophiles, which are EAB that can survive extreme environmental conditions. These include acidophiles, alkaliphiles, halophiles, and thermophiles. An example of an acidophile EAB which survived an acidic MFC environment includes *Acidiphilium* sp. (Malki et al., 2008), and an alkaliphile EAB which survived an alkaline MFC environment includes *Geoalkalibacter* spp. (Badalamenti et al., 2013). Both also improved performance of MFC. Furthermore, halophilic and thermophilic EAB which improved performance of MFC are *Geoalkalibacter* (Luo et al., 2017) and *Calditerrivibrio nitroreducens* (Fu et al., 2013). The usage of extremophilic EAB demonstrates that there are no limitations in the generation of current under extreme conditions. Also, the usage of extreme conditions such as high temperatures can provide advantages such as reduced contamination and fast electrochemical reaction rates in the MFC.

16.3.2 Types of electron transfer mechanisms

Electron transfer to and from the EAB is one of the most important factor which plays a role in the performance of the MFC. Also, numerous studies have been performed in order to understand the mechanisms involved in the transfer of electrons. This is due to the fact that the knowledge acquired from these studies will be able to help researchers to better select the microorganisms suitable for MFC application through correct classification of the EAB and identification of the mechanisms involved in the electron transfer. The extracellular electron transfer mechanisms can be classified into two main categories, namely direct electron transfer (DET) and mediated electron transfer (MET). Furthermore, electron transfer through conductive pili of EAB can be used to represent an extension of DET. Fig. 16.10 illustrates the schematic representation of the mechanisms.

16.3.2.1 Direct electron transfer

DET is related to the transfer of electrons through the adherence of bacterial cells on the electrode without exogenous mediators (Fig. 16.10A). Most of the bacterial cells that

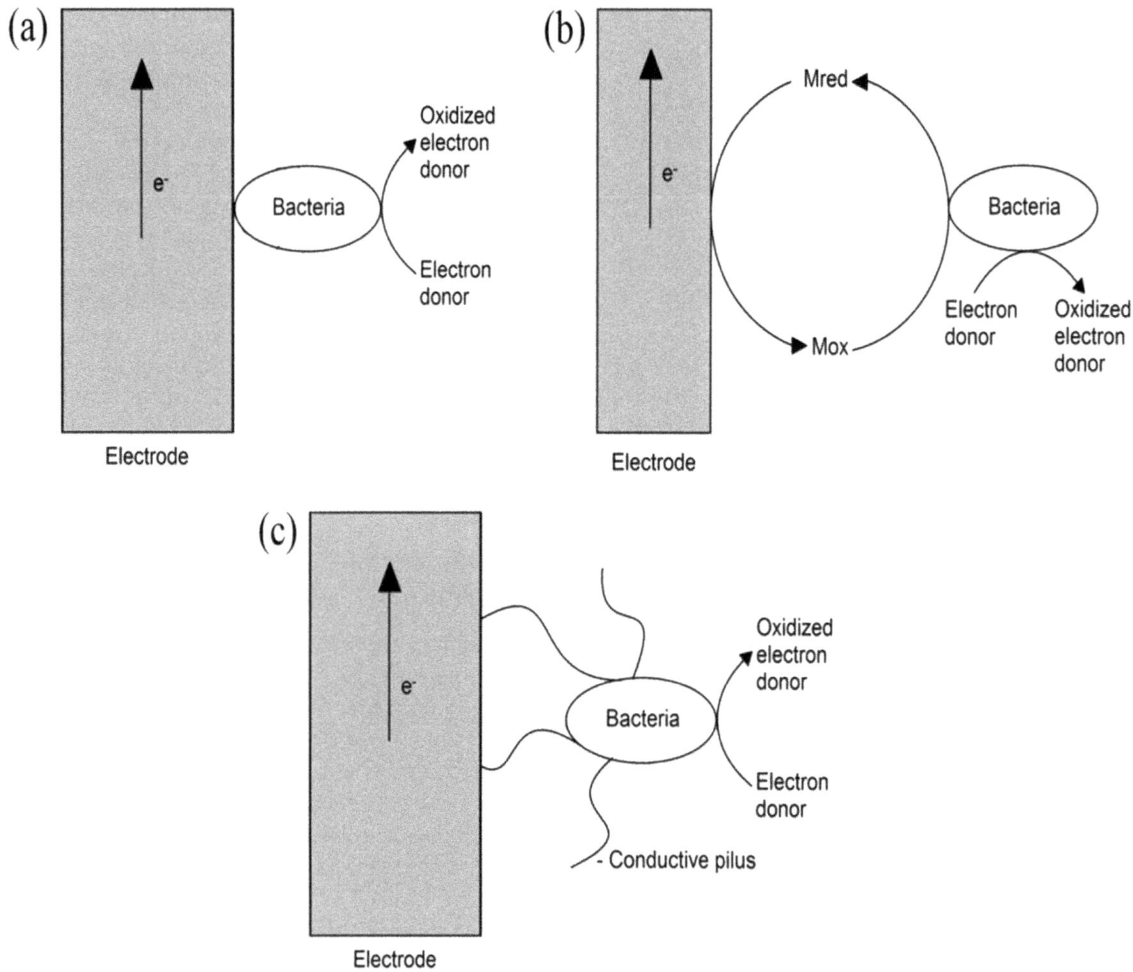

FIGURE 16.10 Electron transfer mechanisms that exist between anode and the electrochemically active bacteria: (A) direct electron transfer, (B) mediated electron transfer, and (C) via conductive pili.

use this type of mechanism belong to the *Geobacteraceae* family. Although it must be mentioned that the first genus to demonstrate the transfer of electrons without exogenous mediators through possibly DET was *Shewanella* (Kim et al., 1999). Furthermore, *Geobacter* sp. was confirmed to transfer electrons through possibly DET (Basséguy et al., 2014). DET is known to use membrane-bound redox proteins which require a direct contact between the surface of the bacteria and the electrode. This was demonstrated in a study by Holmes et al. (2006) who showed that these redox proteins are c-cytochromes. C-cytochromes were oxidized and reduced by losing and gaining electrons from the iron atom located in the heme of the bacteria. Some of the c-type cytochromes used for DET that exist on the surface of the bacterial cells include OmcS, OmcE, and OmcB.

16.3.2.2 *Mediated electron transfer*

MET can occur via two types of mediators: exogenous mediator and endogenous mediator. MET via exogenous mediator involves an electron transfer between bacteria and the electrode using an external soluble mediator (Fig. 16.10B). The oxidized form of this mediator enters the bacteria and gets reduced, and upon its release into the electrolyte solution, it gets reoxidized and transfers electrons to the electrode. Various compounds have been tested in the past as exogenous mediators. Quinine, phenoxazine, phenozine, and phenothiazine are one of the compounds tested and with thionine being the most widely used artificial mediator (Basséguy et al., 2014). However, due to the fact that the bacteria are dependent on the external mediator to transfer electrons, most of these MFCs which use exogenous mediators contain bacteria which are not intrinsically electroactive. Also, the mediators produce low current densities and coulombic efficiencies, i.e., 54% of the consumed glucose leads to energy generation in MFC (Schröder, 2007). Additional disadvantages include constant addition of the mediator when using a continuous MFC and the mediators are toxic to humans and cannot be used in an environment that is open. Therefore, in order to avoid having to use exogenous mediators, one has to consider possibly using EAB which employ MET via endogenous mediator.

Previous studies have shown that bacteria have the ability to transfer electrons using endogenous mediators (Xin et al., 2020; Zou et al., 2017). The endogenous mediators are naturally derived from the bacterial cell and include flavinic compounds from *S. oneidensis* MR-1 (Yang et al., 2015) and 2-amino-3-carboxy-1,4-naphthoquinone secreted by *Shewanella putrefaciens* (Freguia et al., 2010). They are excreted from the cells in reduced forms and are oxidized to transfer electrons at the electrode. Most bacteria that generate electricity in MFC without exogenous mediators are electroactive. In addition, the bacteria produce their own mediators and eliminate the requirement of a constant mediator addition during continuous operation of the MFC. Since most of the mediators naturally produced by bacteria are nontoxic, this enables the MFC to be easily handled by humans without being detrimental to their health.

16.3.2.3 *Electron transfer via conductive pili*

From the earliest to present studies, the investigations on DET have focused on the c-type cytochromes which are located in the periplasm and outer membrane (Shi et al., 2016). However, in order to explain the improvement of MFC due to thicker biofilms, a hypothesis was presented which described the long-distance electron transfer in mediatorless MFCs. This involved a pilus acting as a conductive nanowire produced by the bacterial cells as a possible explanation (Fig. 16.10C). Many studies used *Geobacter* as a model species. One of the first studies to report on conductive nanowire was a study done by Childers et al. (2002). This study found that when *Geobacter metallireducens* was grown on insoluble Fe(III) oxides as an electron acceptor, it expressed pili and led to the detection of gene transcripts for the putative pilin monomer, PilA. A similar study done by Reguera et al. (2005) also observed the same pili expression during the growth of *G. sulfurreducens* on Fe(III) oxide, but not with soluble Fe(III) and determined the diameter of the conductive nanowires to be 3 nm. In addition, there was no filament development and reduction in Fe(III) oxide observed when the PilA pilin monomer gene was deleted from the wild-type *G. sulfurreducens*. This meant that *G. sulfurreducens* depended on PilA pilin monomer for the development of pili and reduction of Fe(III) oxide.

Reguera et al. (2005) found that in addition to using Fe(III) oxide for pili growth in *G. sulfurreducens*, fumarate can also be used as an electron acceptor substitute. Moreover, they observed that when adding poorly crystalline Fe(III) oxide to *G. sulfurreducens* grown using fumarate, the nanoparticles attached to the pili. To determine if the pili were conductive, Reguera et al. (2005) used conductive-tip atomic force microscopy which revealed that the pili were conductive. This finding coupled with the observation that pilA-deficient *G. sulfurreducens* could not reduce Fe(III) oxide, led to the discovery that there is a long-distance electron transfer mechanism involved in addition to the DET mechanism through direct membrane attachment. This was further demonstrated for bioelectrochemical systems, which suggested that *G. sulfurreducens* nanowires played an important role in the long-distance electron transfer to the electrode (Reguera et al., 2006). To avoid confusion with other conductive structures such as flagella, pilin-based microbial nanowires are now referred to as e-pili (Lovley, 2017).

It should be noted that there is still more work that needs to be done. This is because it has been demonstrated that the *G. sulfurreducens* outer surface *c*-type cytochrome known as OmcS can form filaments (Wang et al., 2019). Which introduces the question that was the e-pili detected not OmcS filaments? One recent study even makes such an interesting claim that previous studies on *G. sulfurreducens* e-pili had made a mistake, and were in fact studying filaments of OmcS (Wang et al., 2019). However, due to the evidence that exists in some studies, that the filaments were retaining the diameter of an e-pili, and differing greatly in conductivity depending on the density of aromatic amino acids in the pilin monomer (Lovley and Walker, 2019), it is difficult to accept such a claim. Provided that the filaments are OmcS, how could the aromatic density of the PilA monomer influence the OmcS filament conductivity? Maybe the two coexisted. That is still a question that needs to be answered which makes the study of conductive nanowires intriguing.

16.4 Other organisms used in microbial fuel cells

16.4.1 Algal microbial fuel cell

An algae is simply a photosynthetic eukaryotic organism. In an algae, there is a process called photosynthesis which takes place in the organelles named chloroplast. This process allows the organism to simply convert light energy to chemical energy which can be used by the organism to fuel its activities. In the chloroplast, there are thylakoids organized in a structure similar to a stack of coins where light-dependent photosynthetic reactions take place (Sekar and Ramasamy, 2015). Light-independent photosynthetic reactions take place in the inner matrix of the chloroplast named stroma. These two photosynthetic reactions are responsible for the net-generation of sugars and oxygen from both water and carbon dioxide (CO_2). A light-dependent reaction is responsible for the generation of oxygen, reduced nicotinamide adenine dinucleotide phosphate (NADPH), and energy in the form of adenosine triphosphate (ATP) through photosynthetic electron transport chain (P-ETC) by using water, nicotinamide adenine dinucleotide phosphate ($NADP^+$), and adenosine diphosphate (ADP). Then the light-independent reaction converts the ATP and NADPH from the light-dependent reaction including CO_2 to form ADP, $NADP^+$, and sugars such as glucose using the Calvin cycle. The schematic is shown in Fig. 16.11.

As mentioned, the light-dependent reaction encompasses P-ETC which is essential in the production of ATP and NADPH. P-ETC contains various protein complexes such as photosystem II (PSII), photosystem I (PSI), ATP synthase, soluble proteins such as ferredoxin and NADP reductase that are located in the thylakoid membrane (Sekar and Ramasamy, 2015). Protein complexes PSI and PSII contain chlorophylls. These chlorophylls in PSII use light to generate

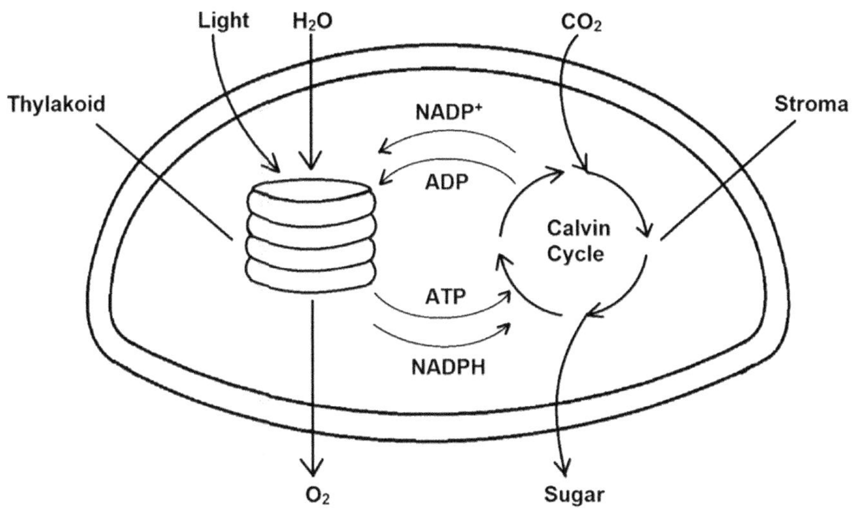

FIGURE 16.11 Photosynthetic light-dependent and independent reactions in the chloroplast.

photo-excited high energy electrons and oxygen. The electrons are then transported through the chain from PSII to ferredoxin to PSI (Shukla and Kumar, 2018). In PSI, they are then used to reduce the $NADP^+$ to NADPH and create an energy gradient which is used to pump protons outside the thylakoid membrane to generate ATP by using ATP synthase. The ATP is then used to maintain and fuel the algae activities. The NADPH is used in the light-independent reaction by CO_2. However, as of recent, researchers have found a way to either use oxygen or NADPH generated from algae in MFCs in both the anode and cathode compartments.

16.4.1.1 Anodic algae

Algae can be used in the anode of MFCs. This has been demonstrated by Xu et al. (2015a) who used *Chlorella pyrenoidosa* under different light intensities, starved the anode with CO_2 or any substrate and continuously fed nitrogen. This led to the generation of energy by the MFC. This is because by not feeding CO_2, the Calvin cycle was not activated and this allowed the NADPH to be oxidized at the anode electrode and generate electrons instead of being utilized by the Calvin cycle. Xu et al. (2015a) also showed that the continuous sparging of nitrogen is essential since the electrons generated by algae in the anode which are meant to be transferred to the anode electrode can be consumed by the by-product of algae, namely oxygen. Further demonstration that algae can be used in the anode of the MFC was done by Ramanathan et al. (2011) who used algal species of *Tetraselmis gracilis*, *Chlorella* sp., and *Mougeotia*.

16.4.1.2 Cathodic algae

Due to the fact that algae can generate oxygen, some researchers have tried to use it in the cathode in order to enhance the performance of a dual-chambered MFC. Successful improved energy generation in MFC by using algae in the cathode was shown by previous studies done by Xu et al. (2015a) and Xiao et al. (2012). Xu et al. (2015a) also demonstrated that integrating both the anodic algae and cathodic algae in a dual-chambered MFC enhances performance instead of only using anodic algae. In addition, the usage of cathodic algae is innovative since it allows algae to be used as a biocatalyst in enhancing performance of MFC instead of using expensive and toxic cathode electrocatalysts such as Pt.

16.4.2 Fungal microbial fuel cell

Fungi are eukaryotic organisms which can be classified into four subdivisions: Ascomycota (yeasts and sac fungi), Zygomycota (bread molds), Chytridiomycota (chytrids), and the Basidiomycota (club fungi). The internal structure and the shape of the sporangium which produces spores is useful in the identification of these various major groups. In previous years, the usage of fungi in MFCs was less investigated. This was attributed to the lack of evidence which could suggest the existence of fungal electrogens, and of those experiments that used fungi, low power densities were observed in comparison to bacteria (Sekrecka-Belniak and Toczyłowska-Mamińska, 2018). However, in recent years, the investigation of fungal MFCs has increased since there is evidence of electrogenic fungi which utilizes DET mechanism similar to bacteria by using enzymes in the cell membrane such as lactate dehydrogenase and ferricyanide reductase (Prasad et al., 2007). Ascomycota (yeast) is the most widely used fungi in the study of fungal MFCs. Yeast-based fungal MFCs have been proven to use DET via c-cytochrome (Wilkinson et al., 2006). Other fungal species from different subdivisions have also been proven to be electrogenic. This includes wood-degrading white-rot fungi which belong to the Basidiomycota (Pontié et al., 2019). White-rot can be utilized as a biocathode (Wu et al., 2012). Various other species of fungi which have been used include *Saccharomyces cerevisiae* and *Candida* sp. in the anode, and *Trametes versicolor* in the cathode (Sekrecka-Belniak and Toczyłowska-Mamińska, 2018).

16.4.3 Humus powered fuel cell

The soil has several layers which are named horizons in which the arrangement of these layers is called soil profile. These horizons can be easily identified due to their different textures, structure, and color (Upton and de Groot, 2008). The nature of the horizons can then be used to classify different soil types to help provide information about the ground that is most suitable for a specific application such as farming and construction. A soil profile generally contains the following horizons: humus, topsoil, subsoil, weathered rock fragments, and bedrock. The schematic is shown in Fig. 16.12.

Humus, which contains a variety of organic compounds that are chemically heterogeneous, is distributed on both land and aquatic environments. Furthermore, humus is also a stable

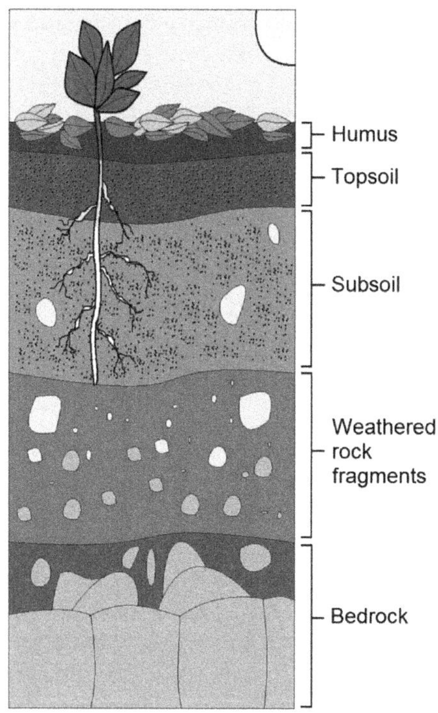

FIGURE 16.12 Schematic showing the soil profile.

organic material that accumulates the soil when plant and animal matter decay. It is considered to be a product of phenol polymerization, and the phenols are derived from both fungal breakdown of lignin and as secondary metabolites in plants and bacteria. In addition, the polymerization occurs due to oxidative coupling which is catalyzed by oxidative enzymes such as phenol oxidase and laccases (Martinez et al., 2013). During oxidative coupling, amino acids are also copolymerized into the polymers (Field, 2001). Although humus is not easily biodegradable, it is very useful in providing humic substances which can be utilized as redox mediators, electron donors, and electron acceptors.

Humic substances contain several redox-active moieties such as (hydro)quinone, sulfhydyl, carboxylic, and phenolic functional groups which can be utilized by bacteria (Lipczynska-Kochany, 2018). They are utilized by bacteria by functioning as electron sinks, and can also function as electron sources in the form of reduced-humic substances. The first demonstration of humic substances as electron sinks or acceptors was done by Lovley et al. (1996). They showed that *Geobacter metallireducens* and *Shewanella alga* can use humic acid (HA) as a terminal electron acceptor and oxidize acetate. On the other hand, it was shown by Coates et al. (2002) that reduced humic substances can act as electron sources or donors for the reduction of NO_3^- using anaerobes. Since humus contains quinone moieties, it has also been shown that they have the potential of acting as redox mediators. This was done by using humic hydroquinone model compound, anthrahydroquinone-2,6-disulfonate, and humic quinone model compound, anthraquinone-2,6-disulfonate (Lovley et al., 1999; Scott et al., 1998).

Due to the reason that humus can act as an electron donor, electron acceptor, and redox mediator, its use was expanded to MFCs. One of the studies which showed that the introduction of humic substances in the anode can promote the reduction of azo dye and enhance the performance of MFC was done by Sun et al. (2013). Also, it was shown by another study that humus can act as a sole carbon source for an enriched microbial inoculum in the anode and effectively improve the performance of the MFC (Adekunle et al., 2017).

16.4.4 Biogas powered fuel cell

Biogas is a combination of various gases mainly consisting of CO_2 and methane (CH_4) which are produced from the breakdown of organic matter under anaerobic conditions. The organic matter used to produce biogas includes agricultural organic waste, municipal organic solid waste, and food organic waste. Also, the microorganisms used to produce biogas are normally methanogens or anaerobes. Due to the fact that biogas production requires organic matter, the growth of the organic matter such as plants and algae can be used to reduce CO_2 emissions. In addition, CH_4 can be used as a source of fuel for electricity generation. Various technologies have been used to convert methane to fuels such as gas-to-liquids (GTLs) technologies. However, GTL plants operate at large scales and are capital-intensive (Ren, 2017). Direct fuel cells have also been used for methane conversion to fuels but the process is highly energy-intensive and experiences problems such as catalytic instability (O'Hayre et al., 2016). Therefore, this makes it essential to consider other processes for the conversion of methane to generate not only fuels but also electricity and chemicals.

One of the processes to consider is the use of MFCs. This is due to both flexibility in scale and the ability to integrate MFC with catalytic processes to allow for selection of desired products. However, previous studies had problems in achieving anaerobic oxidation of methane (AOM) due to difficulties in microbial culturing and the complexity of the process. For example, anaerobic methanotrophic archaea

(ANME) which are single-celled organisms that perform AOM through reverse methanogenesis are difficult to isolate as pure cultures under laboratory conditions because of syntrophic association with other bacteria. In addition, due to a complex step of methane activation without oxygen-derived radicals, it can take up to months to culture ANME. Therefore, researchers had to find ways to work around these problems. One solution was to take advantage of the syntrophic association between an engineered ANME strain of *Methanosarcina acetivorans* which is widely used, and an electrogen of *Geobacter sulfurreducens*, as well as sludge inoculum (Ren, 2017). This helped due to the fact that the engineered *M. acetivorans* is able to produce methyl-coenzyme M reductase and convert methane and hydrogen carbonate (HCO_3^-) to acetate. Then the acetate can be used as a substrate by *G. sulfurreducens* to generate electrons at the anode and reduce the oxidant at the cathode. The reason for the addition of sludge inoculum is to provide electron shuttles. The schematic in Fig. 16.13 illustrates the process.

Various studies have successfully used *M. acetivorans* and *Geobacter* sp. species such as *Geobacter sulfurreducens* (McAnulty et al., 2017; Yamasaki et al., 2018). This process is still at its developing stage, however, it introduces an interesting application into the generation of clean energy by using biogas. This is beneficial due to the process being less energy-intensive and also cost-effective.

16.5 Physical/structural parameters

16.5.1 H^+ transport systems

One of the most important steps in MFC operation is the transfer of protons, mainly hydrogen ions, from the anode to the cathode chamber. This step neutralizes the build up of hydrogen ions generated in the anode. In addition, it

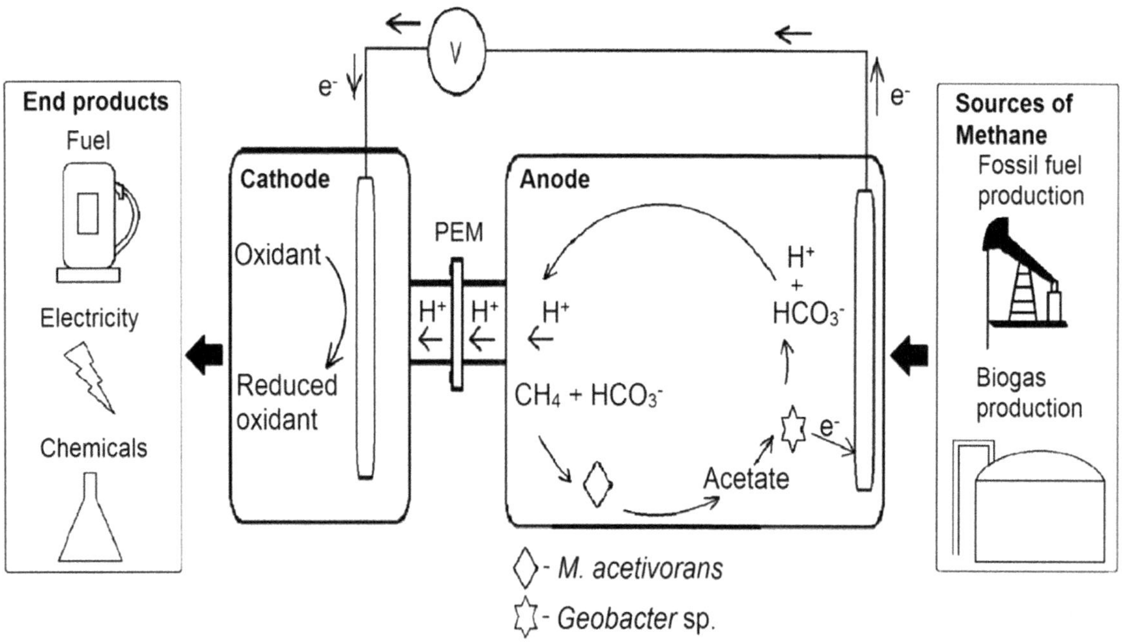

FIGURE 16.13 Anaerobic oxidation of methane in a dual-chambered microbial fuel cell.

ensures that the electrons generated from the anode are consumed at the cathode in order to complete the cell. Several ways can be employed to improve the transfer of hydrogen ions to the cathode such as the usage of salt bridge, ion selecting membranes, and forward osmosis (FO) membranes.

16.5.1.1 Salt bridges

During the early stages of MFC research, salt bridges were widely used ensure a complete MFC connection and neutralization of hydrogen build up in the anode. Most of the studies used agar salt bridges (Khan et al., 2012; Kumar et al., 2012; Sevda and Sreekrishnan, 2012). Agar would come in the form of a powder, which when dissolved in water and heated, it would form a gelatinous substance upon cooling. By adding agar and various salts in water, an agar salt bridge could then be prepared. In addition, the salts would mainly consist of potassium chloride, sodium chloride, and potassium nitrate. However, almost all of the studies reported low power densities. This was attributed to high internal resistance and substrate cross-over caused by the salt bridge during operation (Min et al., 2005). Recently, various researchers are trying to minimize the internal resistance and substrate cross-over caused by salt bridge. Sarma et al. (2019) used the Taguchi method in the optimization of salt bridge parameters. Hernández-Flores et al. (2019) synthesized and implemented characterization methods for both proton transfer and oxygen permeability which gave detailed information on how various changes affect performance. Therefore, although salt bridge is an old concept, continuous research dedicated to successful utilization of salt bridges in MFCs is still relevant today because they are less expensive, and their success in improving performance of MFC can help in the development of large-scale MFCs.

16.5.1.2 Ion-selective membranes

A useful characteristic that an MFC chamber separator should have is the ability to select the type of ions which permeate through it in addition to not allowing substrate or oxygen cross-over. There are various ion-selective separators that are available which can be used in MFCs. These include cation exchange membranes (CEMs) which allow positively charged ions to permeate and anion exchange membranes (AEMs) which allow negatively charged ions to permeate. Other ion-selective separators that can be used include bipolar membranes (BPMs) which allow the permeation of both the anions and cations. The schematic of the ion-selective membranes is shown in Fig. 16.14.

Membranes which allow the permeation of positively charged ions, mainly protons, are also known as PEMs. PEMs which are commonly used in MFCs are normally made from a Nafion. Nafion is a perfluorosulphuric acid with a backbone of hydrophobic fluorocarbons ($-CF_2-CF_2-$) to which the negatively charged sulphonated groups (SO_3^-) are attached. The presence of SO_3^- is what allows the Nafion to bring into effect proton conductivity or permeation. However, pH splitting is a problem when operating MFCs with Nafion. The splitting of pH occurs due to a decrease in the anode pH as hydrogen ions are generated and an increase in the cathode pH as hydrogen ions are consumed. This leads to an increase in internal resistance and decreased performance of MFC (Rahimnejad et al., 2015). Kim et al. (2007) found that this can be solved by using an AEM. Reason being that since phosphate is normally used as a buffer in MFCs, orthophosphate anions (HPO_4^{2-} and $H_2PO_4^-$) could help buffer the pH drop in the anode chamber while maintaining charge balance in both chambers. Therefore, Kim et al. (2007) used an AEM and showed an enhanced performance of MFC as compared

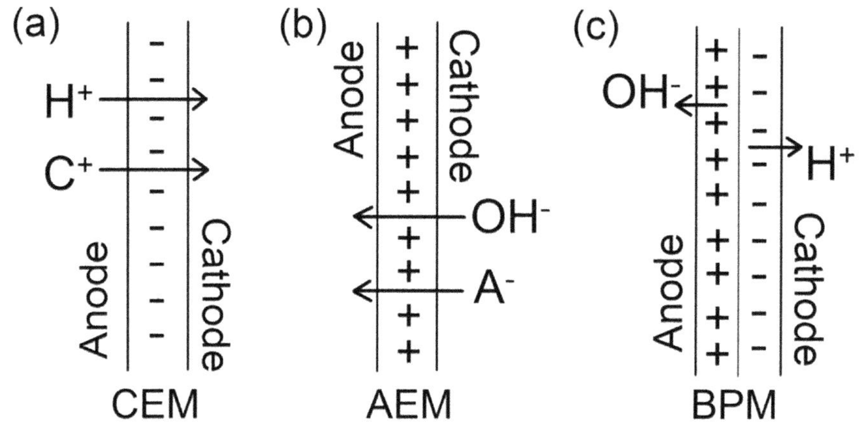

FIGURE 16.14 Charge transport in ion-selective membranes, namely (A) cation exchange membrane (CEM), (B) anion exchange membrane (AEM), and (C) bipolar membrane (BPM). Cations (C^+) and anions (A^-).

to using a Nafion. Koók et al. (2020) also demonstrated a better performance when using an AEM. The usage of BPM was shown by a previous study (ter Heijne et al., 2006), however, in comparison to both PEM and AEM, it leads to the lowest performance of MFC (Harnisch et al., 2008).

It has to be noted that ion-selective membranes do have disadvantages. In addition to them being expensive, during operation they can undergo fouling. Igboamalu et al. (2019) demonstrated that PEM is susceptible to fouling which can affect both proton transfer and performance. Therefore, finding ways to reduce membrane fouling will be beneficial when considering ion-selective membranes.

16.5.1.3 *Ceramic membranes*

In simple terms, ceramic materials are made of pottery that is then hardened by heat and can be explained by a more scientific definition which describes them as inorganic, nonmetallic solids prepared by the action of heat and subsequent cooling (Yousefi et al., 2017). Additionally, these materials can be used as membranes in order to improve the performance of MFC. Ceramic membranes have recently been used as separators in MFCs due to their good stability, availability, low production cost, and high structural strength (Yousefi et al., 2017).

The usage of ceramic membranes in MFCs has been demonstrated by Chatterjee and Ghangrekar (2014) who compared the air-cathode MFC which was coated with carbon on both sides of the ceramic membrane. This air-cathode performed better by producing 5.2 times higher power density as compared to a dual-chambered MFC with aqueous cathode. In addition, Bagchi and Behera (2021) demonstrated that proton exchange across the ceramic membrane can be improved by anolyte recirculation in a continuously operated MFC.

Due to their low cost, innovative designs that utilize ceramic membranes which can achieve acceptable power efficiencies using even large-scale MFCs is possible. Ghadge and Ghangrekar (2015) presented novel scalable 26 L air-cathode clayware-MFC with multiple electrodes. Three of which were embedded pairs of anode and cathode electrodes in this large-scale MFC. The different types of electrical configurations were examined, in which parallel configuration presented the best performance with low internal resistance.

16.5.1.4 Forward osmosis membrane systems

FO has been used to draw clean water from wastewater. FO is used to promote the flow of water from a high water potential to a low water potential area. The driving force in the process is the concentrated solution called a draw solution on the permeate side of the membrane, which has high osmotic efficiency and can be easily separated from the solution to produce portable water (Zhang et al., 2011). The advantage of using FO is that it leads to low fouling, has high rejection for a large range of contaminants, and uses low hydraulic pressure.

FO can be integrated in the operation of MFCs. In FO-microbial fuel cells (FO-MFCs), the construction of a dual-chambered MFC is coupled with an installation of FO membrane instead of an ion-selective membrane (Kim et al., 2015). The wastewater and microbes are in the anode, and an oxidant and the draw solution (NaCl solution or seawater) are used in the cathode. This enables both water permeation through the FO-membrane due to the difference in osmotic pressure between the cathode and the anode chamber, and promotes energy generation. FO-MFC operation was demonstrated by Zhang et al. (2011) who successfully achieved wastewater treatment, water extraction, and seawater desalination. The same study also showed that FO membranes can improve proton/ion transport.

FO membranes integrated in MFCs can perform better than AEMs or CEMs and improve MFCs power output. This is attributed to the FO membranes having a low internal resistance. One study even showed that the FO membrane can achieve an internal resistance as low as 54 Ω when compared to the 63 and 73 Ω of AEM and CEM, respectively (Werner et al., 2013). In the same study, the FO membrane's low internal resistance was consistent with ease of transport of protons and water permeation through the membrane.

16.5.2 Electrode compositions

16.5.2.1 Material

The role played by an electrode material on performance of MFC is important. This is attributed to the fact that different materials have different characteristics which can affect performance of MFC, and of the most important one between them all being electrical conductivity. This explains why most research in MFCs is conducted using carbon as the prime material for the electrode. Carbon is not only chemically inert under ambient conditions, but it is also highly conductive. Different types of carbon materials have been used to enhance performance of MFC such as graphite (Li et al., 2019), carbon cloth (Tran et al., 2019), and carbon paper (Quaglio et al., 2019). However, research is still underway in developing novel materials for application to MFCs such as CNTs. CNTs are cylindrical molecules which are made of rolled-up sheets of single-layer carbon atoms (graphene). A diameter between 0.4 and 6 nm represents a single-walled CNT (SWCNT) (Shi et al., 2015). A multiwalled CNT (MWCNT) consists of concentric interlinked nanotubes with diameters reaching more than 100 nm. In comparison to the widely used carbon cloth, paper, or graphite, CNTs are recently receiving attention in their possible applications to MFCs. Erbay et al. (2015) reported that electrogens grown over CNTs result in excellent electron transfer due to $\pi-\pi$ stacking between the carbon atoms and the pili of electrogenic bacteria. Also, it was reported in the same study that the spaces between CNTs provide openings for microbial growth. Furthermore, forming a highly conductive anode with a large surface area by coating CNTs over carbon cloth led to an increase in maximum power density by 250% when compared to non-modified carbon cloth (Tsai et al., 2009). Therefore, CNTs provide an alternative to widely used electrode materials in order to improve both conductivity and surface area.

16.5.2.2 *Surface area*

The surface area of an electrode has an influence on the performance of MFC. When increased, more electrogenic bacteria are able to attach which improves electron transfer by DET. Also, more reaction sites become available for oxidant reduction on the cathode electrode. Therefore, surface area optimization of the electrode provides an approach to enhance performance of MFC. This was demonstrated by Oh and Logan (2006) who increased power output by only changing the cathode surface area from 2 to 22.5 cm^2. Similar results have been reported by previous studies (Bian et al., 2018; Penteado et al., 2018). UMFCs which provide high surface area to volume ratios have also been reported (Dessì et al., 2019; Subha et al., 2019). In addition, the UMFC scalable characteristic makes them suitable for developing large-scale MFCs.

16.5.3 Electrode spacing

The distance between the anode and cathode electrodes has an impact on the ohmic losses of the MFC. If the distance is reduced, the proton transfer is enhanced which leads to low ohmic losses/internal resistance. Several studies have therefore investigated and tried to optimize the spacing between the electrodes. Liu et al. (2005) showed that the decrease in the spacing from 4 to 2 cm reduced the ohmic loss and resulted in an increase in power output. This observation has also been confirmed by several studies (Ghangrekar and Shinde, 2007; Kim et al., 2007). However, making the electrodes too close, increased oxidant or substrate cross-over (Fan et al., 2007). Therefore, electrode spacing optimization is critical for MFCs.

16.5.4 Culture composition (pure vs. mixed culture)

Different culture compositions have been used in various MFC studies, mainly pure and mixed cultures. Mixed cultures are normally obtained in the form of activated sludge or wastewater from various industries such as food, beverage, and dairy. Then once a bioelectrochemical study is conducted, potential electrogenic pure cultures are isolated from the mixed culture and tested. However, different studies have shown that using only a pure culture in performance of MFC might not be desirable. This is because pure cultures are associated with high internal resistance due to lack of diversity in the microbial community (Tkach et al., 2017). Most electrogenic sludge can actually contain a variety of electrogens that can play a role in the generation of electricity. In addition, it has been shown that sludge can also act as a mediator through a recently discovered electron transfer mechanism known as direct interspecies electron transfer (Logan et al., 2019). Therefore, although it is important to isolate pure cultures to study and understand MFCs, it is important to understand that the end goal is to create a consortium or mixed culture of electrogens capable of providing a diverse microbial community for MFC applications. Table 16.2 shows some of the studies that have compared the performance of MFC of pure and mixed cultures.

16.5.5 Anode temperature

An optimal MFC anode temperature depends on the type of electrogens used in the cell. This is because electrogens can be classified based on their optimal growth temperature range such as: psychrophiles (−5 to 20°C), mesophiles (20–45°C), and thermophiles (above 45°C). Most studies performed in MFC research optimized temperatures in the mesophilic range (Li et al., 2013; Song et al., 2017), however, the limits are endless. A study done Larrosa-Guerrero et al. (2010) obtained the highest coulombic efficiency using a temperature as low as 4°C, and a study by Fu et al. (2015) optimized performance of MFC to a temperature as high as 95°C. This

TABLE 16.2 Comparison of pure and mixed culture performance of microbial fuel cells for selected studies done by Tkach et al. (2017), Varanasi et al. (2016) and Mohamed et al. (2018).

Culture composition	Carbon source	Construct	Anode electrode	Maximum power density	References
Pure	Sodium acetate	Dual chamber	Carbon fiber brush	83.7 mW/m^2	Tkach et al. (2017)
Mixed	Sodium acetate	Dual chamber	Carbon fiber brush	465.3 mW/m^2	Tkach et al. (2017)
Pure	Sodium acetate	Single chamber	Platinum-tungsten oxide	1.16 W/m^2	Varanasi et al. (2016)
Mixed	Sodium acetate	Single chamber	Platinum-tungsten oxide	1.47 W/m^2	Varanasi et al. (2016)
Pure	Glucose	Single chamber	Carbon paper	57.6 mW/m^2	Mohamed et al. (2018)
Mixed	Glucose	Single chamber	Carbon paper	78.5 mW/m^2	Mohamed et al. (2018)

means that MFCs can be operated at a wide range of temperatures. It all depends on the electrogens used. It must be mentioned that although performance of MFC depends on the temperature where the optimal growth for the electrogens is achieved, high optimal temperatures are favored due to high MFC power densities as compared to low optimal temperatures. This is due to an increased rate of bioelectrochemical reaction when the temperature is increased (Zhang et al., 2019).

16.6 Chemical and biochemical parameters

16.6.1 Carbon sources

A carbon source simply refers to any molecule which consists of a carbon that a microorganism uses to build biomass. It can be inorganic or organic. Microorganisms which can produce their own carbon source from carbon fixation in the Calvin cycle are known as autotrophs. Those that cannot are known as heterotrophs. Heterotrophs can directly utilize organic carbon sources or use inorganic compounds as sources of energy, and autotrophs only need inorganic carbon (CO_2) as shown in Table 16.3.

Examples of electrogenic heterotrophs and autotrophs utilized in MFC applications are shown in Table 16.4. Heterotrophs, specifically chemoorganoheterotrophs, are the most widely used.

16.6.2 Aeration requirements

Dual-chambered MFCs which use oxygen as their oxidant require DO levels to be optimized for them to achieve better performance. This is because when DO levels are too low, a decrease in performance will result due to low amount of oxygen available to accept the electrons from the anode. In addition, an increase in DO at concentrations above 6 mg/L can begin not to have an impact on performance of MFC (Juang et al., 2012; Rodrigo et al., 2010). However, it should be noted that during aeration, an increase in DO level is limited by the diffusion of oxygen molecules or bubbles in water. Also, it is important to avoid large bubble formation during

TABLE 16.3 Types of heterotrophs and autotrophs.

Description	Type	Source	Examples
Heterotrophs	Chemoorganoheterotroph	Organic carbon	Fungi, nonphotosynthetic bacteria
	Chemolithoheterotroph	Inorganic compound	Nitrifying bacteria
	Photoheterotroph	Organic carbon	Nonsulphur bacteria
Autotrophs	Chemoautotroph	Carbon dioxide	Methanogens
	Photoautotroph	Carbon dioxide	Green algae, cyanobacteria

TABLE 16.4 Electrogenic heterotrophs and the carbon sources utilized in microbial fuel cell applications.

Name	Type	Carbon source	Construct	Maximum power density (mW/m^2)	References
Rhodobacter sphaeroides	Photoheterotroph	Sodium acetate	Dual chamber	408.06	Cadirci (2018)
Shewanella putrefaciens	Chemoorganoheterotroph	Lactate	Dual chamber	928	Zhao et al. (2020)
Geobacter sp.	Chemoorganoheterotroph	Sodium acetate	Single chamber	329	Kondaveeti et al. (2020)
Chlorella vulgaris	Photoautotroph	CO_2	Dual chamber	248	Aiyer (2021)

aeration since it can hinder oxygen mass-transfer to both the liquid and the cathode electrode, and result in a poor oxygen reduction and a decrease in power density (Commault et al., 2014).

16.6.3 Redox potential and anode pH

Redox potential measures the reducing or oxidizing potential of a substrate or oxidant, respectively. It can be used to determine a suitable electron acceptor or donor. A more positive redox potential indicates a better electron acceptor, and a more negative redox potential indicates a better electron donor. The operational pH of the anode is specific to the intended application and the electrogens utilized. For example, some electrogens are acidophilic such as *Acidiphilium* sp. (Malki et al., 2008), and some are alkaliphilic such as *Geoalkalibacter* spp. (Badalamenti et al., 2013). Hence what is normally suggested is to conduct lab experiments to optimize the pH provided the electrogen utilized is not known. For example, Vélez-Pérez et al. (2020) conducted MFC experiments under acidic anode conditions when performing the co-treatment of industrial acid mine drainage and municipal wastewater. Also, Geng et al. (2020) conducted experiments under alkaline anode conditions to improve performance of MFC. Provided that there are methanogens which hinder the performance of MFC, the pH can be optimized in order to reduce their impact and improve performance (Igboamalu et al., 2019).

16.7 Innovations/integration of emerging science

16.7.1 Carbon nanotubes advanced membrane

In Section 16.5.2.1, it was mentioned that CNTs can enhance the performance of an anode. However, not only are they suitable for anode enhancement but they can also be used in the desalination process. As stated, CNTs are graphite sheets which are rolled into a tubular-like structure, and their appearance is of a lattice work fence. Also, those with a diameter between 0.4 and 6 nm are known as SWCNTs, and those that can reach more than 100 nm are MWCNTs (Shi et al., 2015). They find applicability in desalination due to providing a near-frictionless flow of water through them and retention of wide variety of pollutants in water (Das et al., 2014). In addition, the smoothness of the hydrophobic walls, high aspect ratios, and nanotubular structure allow for ultra-efficient permeation of water molecules. Most CNTs in length are greater than 1 μm but can reach up to 550 mm (Shi et al., 2015; Zhang et al., 2013). The advantage of using CNT membranes integrated in FO-MFCs is that they provide better water permeability and salt rejection (Ahn et al., 2012). Also, due to the pressure drop required being negligible, CNT membranes are a low energy alternative which can improve the cost function in many membrane processes such as desalination. Some of the wastewater components that are rejected by the CNT membrane include bacteria, viruses, proteins, sugars, oil emulsions, and algae.

16.7.2 Biogenic palladium nanoparticles anode

The anode material properties can greatly affect the transfer of electrons between electrogens and the anode electrode, as a result, the enhancement of anode material properties is a crucial step in improving the efficiency of MFC. Palladium (Pd) which is a metal closely related to Pt has been previously used to modify the anode and improve its electrocatalytic activity. However, due to the chemical synthesis of Pd nanoparticles (PdNPs) involving toxic chemicals under severe conditions, a biological synthesis method has been proposed as an alternative synthesis method. This is due to the utilization of less chemical agents under less harsh conditions during biological synthesis. Previous studies which have reported on improving the anode electrocatalytic activity using biogenic palladium nanoparticles (Bio-PdNPs) include studies by both Matsena et al. (2020) and Quan et al. (2015). They utilized the ability of Pd-reducing bacteria to reduce Pd(II) to elemental Pd (Pd(0)) to synthesize Bio-PdNPs. This is achieved by Pd(II) crossing the outer membrane, reduction taking place via the periplasmic hydrogenase, and redeposition of Pd(0) on the cells outer surface. Further applications of Bio-PdNPs in MFC include evans blue removal (Quan et al., 2018a) and iohexol degradation (Quan et al., 2018b).

16.7.3 In situ biogenic biodeposited catalytic cathode

To increase the electrocatalytic performance of the cathode electrode, in-situ biodeposition of catalytic metals on the surface of the electrode can be considered. Most metal-reducing bacteria have the capability of reducing dissolved metals and redepositing them onto the surface of their outer cell walls, which can help in the in-situ bio-deposition of catalytic metals. This can be done by the usage of biocathodes to which metal-reducing bacteria will attach, and by feeding an effluent from mining wastewater containing electrocatalytic heavy metals such as platinum (Pt), the biocathode can reduce and redeposit the Pt onto the cathode surface. This process has the potential of not only treating wastewater from mining sites but it also provides means to improve the electrocatalytic activity of the cathode and enhancing energy generation by MFCs. Although most MFC research on cathodic electrodes has been done on the replacement of chemically synthesized Pt with biocathodes (Commault et al., 2014; He and Angenent, 2006b; Lefebvre et al., 2008), the in-situ biodeposition of biologically synthesized Pt on the surface of the cathode can also be studied in order to improve performance of MFC.

16.8 Applications of microbial fuel cells

16.8.1 Wastewater treatment plants

Wastewater refers to any water that is contaminated by the domestic, commercial, and industrial use. Wastewater treatment plants (WWTPs) which are used for the decontamination and waste removal can be classified based on the type of wastewater that needs to be treated, i.e., agricultural, municipal, and industrial wastewater. All of these wastewater types have different treatment requirements and thus, designs may vary. However, a generic design shown in Figs. 16.15 and 16.16 can be utilized to explain the impact that MFCs will have if

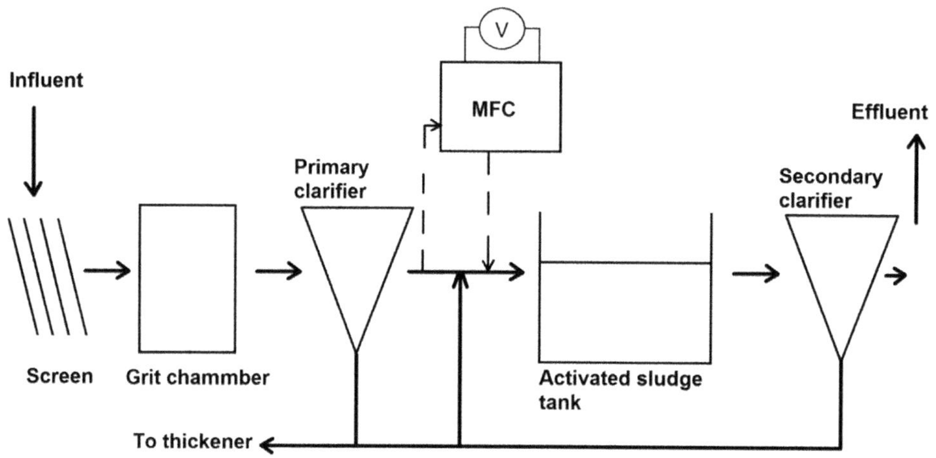

FIGURE 16.15 Typical wastewater treatment process from wastewater treatment plant integrated with microbial fuel cell.

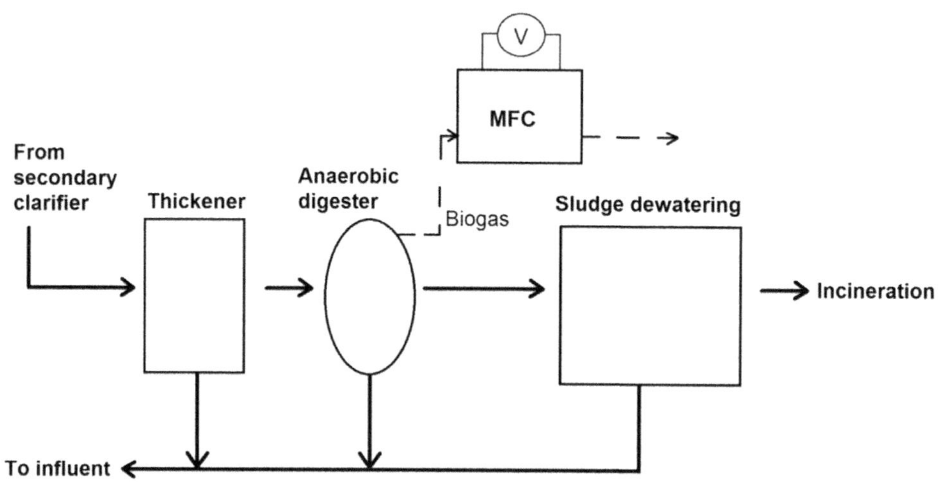

FIGURE 16.16 Typical sludge treatment process from wastewater treatment plant integrated with microbial fuel cell.

implemented in the large-scale WWTPs. One of the first options that provide an opportunity for energy production is the stream that enters the activated sludge tank (Fig. 16.15). It can be directed to the MFC to not only generate energy but also for COD removal. A second option is to use the biogas generated from the anaerobic digester as a substrate to the MFC for energy generation (Fig. 16.16). The proof of concept of this process has already been done by Hiegemann et al. (2016), who showed that by integrating MFC technology in WWTPs, energy savings can be achieved on the overall plant. Also, it was shown from the same study that stable power output can be achieved at COD concentrations even lower that 130 mg/L.

16.8.2 Landfill site remediation

A landfill is a site used for the disposal of waste materials. Normally at a landfill site, waste naturally degrades and produces both gases which contain methane and CO_2, and leachate which is comprised of COD (Fig. 16.17) (Li et al., 2010; Themelis and Ulloa, 2007). Due to methane, CO_2 and COD being possible substrates that can be used in MFC for energy generation, this provides an opportunity for energy recovery at a landfill site. In addition, since methane and CO_2 are greenhouse gases, this enables the landfill to be environmentally friendly by reducing greenhouse gas emissions. Examples of MFCs that can be used at landfill sites include algal MFC (Nguyen et al., 2017) and methane/biogas-powered MFC (Ding et al., 2017).

16.8.3 Desalination plants

Due to population growth and climate change leading to water shortage, the increase in freshwater demand has increased. Several technologies exist that implement solutions into solving water shortage such as seawater desalination by the use of membranes. However, due to the energy required to operate such technologies, especially the most widely used reverse osmosis membranes, it becomes costly to meet the

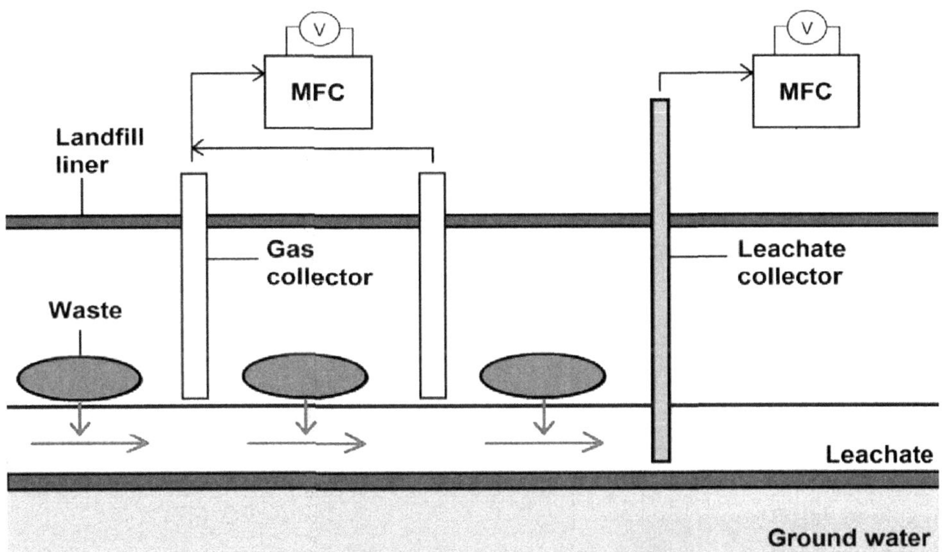

FIGURE 16.17 A landfill site integrated with microbial fuel cells.

demand for freshwater. Therefore, solutions to implement energy efficient desalination processes are needed. One such solution is to use an FO-MFC integrated with MDC. This is because MDC integrated with FO-MFC simultaneously desalinates water and generates electricity. The FO-MFC firstly separates water from viruses, bacteria, suspended solids and oil emulsions using CNT membrane (Ahn et al., 2012), then the MDC contains both CEM and BPM which separates water from the concentrated salt solution obtained from FO-MFC (Tawalbeh et al., 2020). The operation of MDC integrated with FO-MFC is shown in Fig. 16.18.

16.8.4 Soil remediation systems

In Section 16.4.3, it was mentioned that soil contains a layer with humus which can be used to power MFCs. This is because humus contains humic substances which are capable of acting as carbon sources (Lipczynska-Kochany, 2018), and therefore provides an added application in which soils that are polluted with heavy metals can be remediated. Normally, mining of metals leaves a lot of negative environmental impact which leads to contamination of the soil (Liao et al., 2008). Several methods have been used for soil remediation which includes biological reduction, however, this process is limited by the level of metal toxicity to the microbes. To solve this problem, a possible solution can be the usage of humus powered MFCs. This is done by submerging the anode in the humus layer and the cathode in the topsoil layer (Fig. 16.19). This is because metals have been found to accumulate in the topsoil (Afolayan, 2018). This process of using humus and topsoil to generate electricity has already been proven to be successful (Goto et al., 2015), and presents a solution where mining processes which are associated with soil pollution can be merged with MFCs to remediate the soil while generating energy.

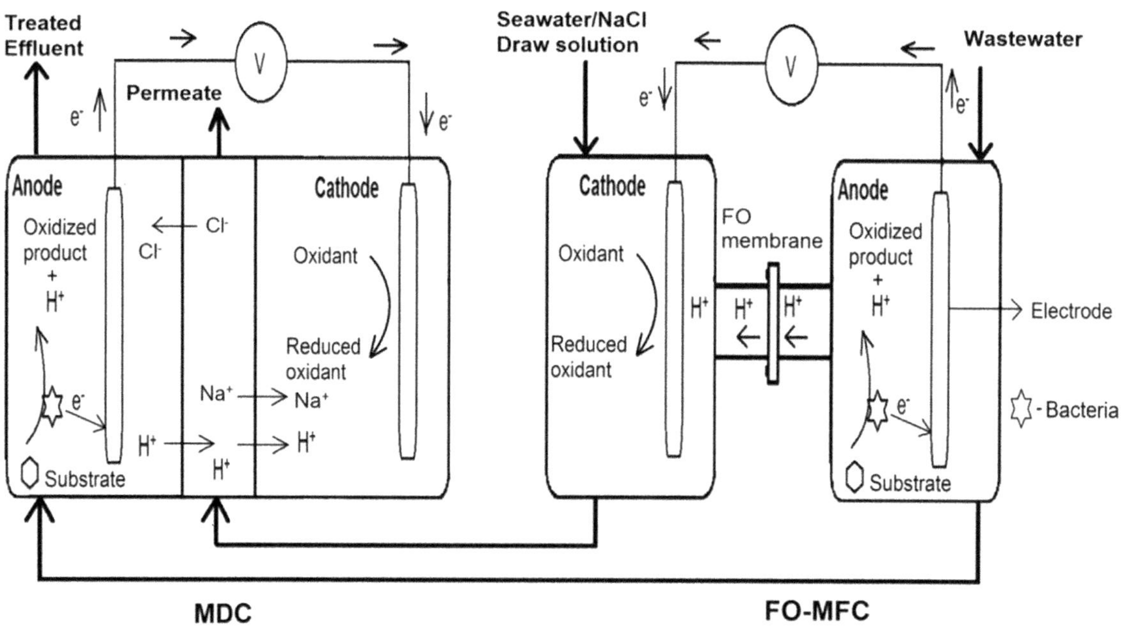

FIGURE 16.18 A microbial desalination cell integrated with forward osmosis microbial fuel cell.

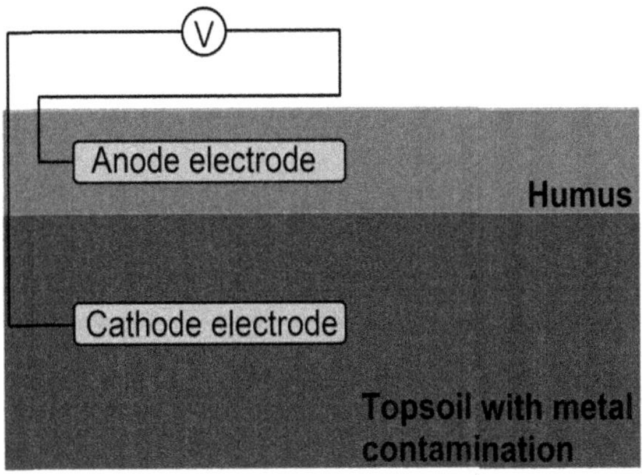

FIGURE 16.19 Soil remediation by using microbial fuel cells.

16.9 Conclusions

The research community is supporting the development of environmentally friendly energy generation from waste by the use of MFCs. A considerable amount of research has gone into the development of novel materials, the identification and also the studying of a wide range of MFC applications. These include the development of CNT membranes, ceramic membranes, FO membrane, and biogenic PdNPs to improve the energy recovery of MFCs. The MFCs have also been studied in various applications, namely in wastewater treatment plants, landfill sites, desalination plants, and soil remediation. Additionally, various organisms have been used to develop a variety of MFCs such as algal MFCs, fungal MFCs, microorganisms that utilize humic substances for energy generation, and the usage of ANME to generate electricity. Although there are a wide variety of configurations that have been developed for MFCs (such as dual chamber, single chamber, upflow, and stacked arrangement), and numerous efforts made to reduce losses (activation, ohmic, concentration, and parasitic losses), there are still areas for continued development to make MFCs more feasible with improved energy generation.

Funding sources

Funding was received from National Research Foundation (NRF) for Rated Researchers (IFRR) Grant No. IFR170214222643 awarded to Professor Evans M. N. Chirwa of the University of Pretoria.

Acknowledgments

Authorship and contributions
Mpumelelo Thomas Matsena: Writing—original draft, Writing—review and editing. **Evans Martin Nkhalambayausi Chirwa**: Funding acquisition, Conceptualization, Writing—review and editing.

References

Adekunle, A., Raghavan, V., Tartakovsky, B., 2017. Carbon source and energy harvesting optimization in solid anolyte microbial fuel cells. J. Power Sources 356, 324–330.

Adrian, E.D., 1926. The impulses produced by sensory nerve endings. J. Physiol. 61 (1), 49–72.

Aelterman, P., Rabaey, K., Pham, H.T., Boon, N., Verstraete, W., 2006. Continuous electricity generation at high voltages and currents using stacked microbial fuel cells. Environ. Sci. Technol. 40 (10), 3388–3394.

Afolayan, A.O., 2018. Accumulation of heavy metals from battery waste in topsoil, surface water, and garden grown maize at Omilende area, Olodo, Nigeria. Glob. Chall. 2 (3), 1700090.

Ahn, C.H., Baek, Y., Lee, C., Kim, S.O., Kim, S., Lee, S., et al., 2012. Carbon nanotube-based membranes: fabrication and application to desalination. J. Ind. Eng. Chem. 18 (5), 1551–1559.

Aiyer, K.S., 2021. Synergistic effects in a microbial fuel cell between co-cultures and a photosynthetic alga Chlorella vulgaris improve performance. Heliyon 7 (1), e05935.

Badalamenti, J.P., Krajmalnik-Brown, R., Torres, C.I., 2013. Generation of high current densities by pure cultures of anode-respiring Geoalkalibacter spp. under alkaline and saline conditions in microbial electrochemical cells. mBio 4 (3) e00144–00113.

Bagchi, S., Behera, M., 2021. Evaluation of the effect of anolyte recirculation and anolyte pH on the performance of a microbial fuel cell employing ceramic separator. Process Biochem. 102, 207–212.

Basséguy, R., Délia, M.L., Erable, B., Bergel, A., 2014. Electroactive biofilms. In: Liengen, T., Féron, D., Basséguy, R., Beech, I.B. (Eds.), Understanding Biocorrosion. Woodhead Publishing, Oxford, pp. 107–143.

Bennetto, H.P., Stirling, J.L., Tanaka, K., Vega, C.A., 1983. Anodic reactions in microbial fuel cells. Biotechnol. Bioeng. 25 (2), 559–568.

Bian, B., Shi, D., Cai, X., Hu, M., Guo, Q., Zhang, C., et al., 2018. 3D printed porous carbon anode for enhanced power generation in microbial fuel cell. Nano Energy 44, 174–180.

Bond, D.R., Lovley, D.R., 2003. Electricity production by Geobacter sulfurreducens attached to electrodes. Appl. Environ. Microbiol. 69 (3), 1548–1555.

Bretschger, O., Obraztsova, A., Sturm, C.A., Chang, I.S., Gorby, Y.A., Reed, S.B., et al., 2007. Current production and metal oxide reduction by *Shewanella oneidensis* MR-1 wild type and mutants. Appl. Environ. Microbiol. 73 (21), 7003–7012.

Cadirci, B.H., 2018. An electricity production study by *Rhodobacter sphaeroides*. Int. J. Hydrog. Energy 43 (38), 18001–18006.

Chatterjee, P., Ghangrekar, M.M., 2014. Design of clayware separator-electrode assembly for treatment of wastewater in microbial fuel cells. Appl. Biochem. Biotechnol. 173 (2), 378–390.

Chaudhuri, S.K., Lovley, D.R., 2003. Electricity generation by direct oxidation of glucose in mediatorless microbial fuel cells. Nat. Biotechnol. 21 (10), 1229–1232.

Chen, W., Liu, Z., Hou, J., Zhou, Y., Lou, X., Li, Y., 2018. Enhancing performance of microbial fuel cells by using novel double-layer-capacitor-materials modified anodes. Int. J. Hydrog. Energy 43 (3), 1816–1823.

Childers, S.E., Ciufo, S., Lovley, D.R., 2002. Geobacter metallireducens accesses insoluble Fe(III) oxide by chemotaxis. Nature 416 (6882), 767–769.

Choi, C., Cui, Y., 2012. Recovery of silver from wastewater coupled with power generation using a microbial fuel cell. Bioresour. Technol. 107, 522–525.

Coates, J.D., Cole, K.A., Chakraborty, R., O'Connor, S.M., Achenbach, L.A., 2002. Diversity and ubiquity of bacteria capable of utilizing humic substances as electron donors for anaerobic respiration. Appl. Environ. Microbiol. 68 (5), 2445–2452.

Cohen, B., 1931. The bacterial culture as an electrical half-cell. J. Bacteriol. 21, 18–19.

Commault, A.S., Lear, G., Novis, P., Weld, R.J., 2014. Photosynthetic biocathode enhances the power output of a sediment-type microbial fuel cell. N. Z. J. Bot. 52 (1), 48–59.

Das, R., Ali, M.E., Hamid, S.B.A., Ramakrishna, S., Chowdhury, Z.Z., 2014. Carbon nanotube membranes for water purification: a bright future in water desalination. Desalination 336, 97–109.

Dessì, P., Chatterjee, P., Mills, S., Kokko, M., Lakaniemi, A.-M., Collins, G., et al., 2019. Power production and microbial community composition in thermophilic acetate-fed up-flow and flow-through microbial fuel cells. Bioresour. Technol. 294, 122115.

Dicks, A.L., Rand, D.A.J., 2018. Introducing fuel cells. In: Dicks, A.L., Rand, D.A.J. (Eds.), Fuel Cell Systems Explained. John Wiley & Sons Ltd., New Jersey, pp. 1–26.

Ding, J., Lu, Y.-Z., Fu, L., Ding, Z.-W., Mu, Y., Cheng, S.H., et al., 2017. Decoupling of DAMO archaea from DAMO bacteria in a methane-driven microbial fuel cell. Water Res. 110, 112–119.

Elangovan, M., Dharmalingam, S., 2017. Application of polysulphone based anion exchange membrane electrolyte for improved electricity generation in microbial fuel cell. Mater. Chem. Phys. 199, 528–536.

Elangovan, M., Dharmalingam, S., 2016. Preparation and performance evaluation of poly (ether-imide) based anion exchange polymer membrane electrolyte for microbial fuel cell. Int. J. Hydrog. Energy 41 (20), 8595–8605.

Enerdata, 2020. Global Energy Statistical Yearbook 2020. https://yearbook.enerdata.net/electricity/electricity-domestic-consumption-data.html. (Accessed 24 April 2020).

Erbay, C., Pu, X., Choi, W., Choi, M.-J., Ryu, Y., Hou, H., et al., 2015. Control of geometrical properties of carbon nanotube electrodes towards high-performance microbial fuel cells. J. Power Sources 280, 347–354.

Fan, Y., Hu, H., Liu, H., 2007. Enhanced Coulombic efficiency and power density of air-cathode microbial fuel cells with an improved cell configuration. J. Power Sources 171 (2), 348–354.

Feng, L., Yan, Y., Chen, Y., Wang, L., 2011. Nitrogen-doped carbon nanotubes as efficient and durable metal-free cathodic catalysts for oxygen reduction in microbial fuel cells. Energy Environ. Sci. 4 (5), 1892–1899.

Field, J.A., 2001. Recalcitrance as a catalyst for new developments. Water Sci. Technol. 44 (8), 33–40.

Flimban, S.G.A., Ismail, I.M.I., Kim, T., Oh, S.-E., 2019. Overview of recent advancements in the microbial fuel cell from fundamentals to applications: design, major elements, and scalability. Energies 12 (17).

Freguia, S., Tsujimura, S., Kano, K., 2010. Electron transfer pathways in microbial oxygen biocathodes. Electrochim. Acta 55 (3), 813–818.

Fu, L., You, S.-J., Yang, F.-l., Gao, M.-m., Fang, X.-h., Zhang, G.-q., 2010. Synthesis of hydrogen peroxide in microbial fuel cell. J. Chem. Technol. Biotechnol. 85 (5), 715–719.

Fu, Q., Fukushima, N., Maeda, H., Sato, K., Kobayashi, H., 2015. Bioelectrochemical analysis of a hyperthermophilic microbial fuel cell generating electricity at temperatures above 80 °C. Biosci. Biotechnol. Biochem. 79 (7), 1200–1206.

Fu, Q., Kobayashi, H., Kawaguchi, H., Wakayama, T., Maeda, H., Sato, K., 2013. A thermophilic gram-negative nitrate-reducing bacterium, calditerrivibrio nitroreducens, exhibiting electricity generation capability. Environ. Sci. Technol. 47 (21), 12583–12590.

Gambino, E., Toscanesi, M., Del Prete, F., Flagiello, F., Falcucci, G., Minutillo, M., et al., 2017. Polycyclic aromatic hydrocarbons (PAHs) degradation and detoxification of water environment in single-chamber air-cathode microbial fuel cells (MFCs). Fuel Cells 17 (5), 618–626.

Geng, Y.-K., Yuan, L., Liu, T., Li, Z.-H., Zheng, X., Sheng, G.-P., 2020. Thermal/alkaline pretreatment of waste activated sludge combined with a microbial fuel cell operated at alkaline pH for efficient energy recovery. Appl. Energy 275, 115291.

Ghadge, A.N., Ghangrekar, M.M., 2015. Performance of low cost scalable air–cathode microbial fuel cell made from clayware separator using multiple electrodes. Bioresour. Technol. 182, 373–377.

Ghangrekar, M.M., Shinde, V.B., 2007. Performance of membrane-less microbial fuel cell treating wastewater and effect of electrode distance and area on electricity production. Bioresour. Technol. 98 (15), 2879–2885.

Gorby, Y.A., Yanina, S., McLean, J.S., Rosso, K.M., Moyles, D., Dohnalkova, A., et al., 2006. Electrically conductive bacterial nanowires produced by Shewanella oneidensis strain MR-1 and other microorganisms. Proc. Natl. Acad. Sci. U.S.A. 103 (30), 11358–11363.

Goto, Y., Yoshida, N., Umeyama, Y., Yamada, T., Tero, R., Hiraishi, A., 2015. Enhancement of electricity production by graphene oxide in soil microbial fuel cells and plant microbial fuel cells. Front. Bioeng. Biotechnol. 3 (42).

Gregory, K.B., Bond, D.R., Lovley, D.R., 2004. Graphite electrodes as electron donors for anaerobic respiration. Environ. Microbiol. 6 (6), 596–604.

Grove, W.R., 1842. LXXII. On a gaseous voltaic battery. Lond. Edinb. Dubl. Phil. Mag. J. Sci. 21 (140), 417–420.

Harnisch, F., Schröder, U., Scholz, F., 2008. The suitability of monopolar and bipolar ion exchange membranes as separators for biological fuel cells. Environ. Sci. Technol. 42 (5), 1740–1746.

He, Z., Angenent, L.T., 2006. Application of bacterial biocathodes in microbial fuel cells. Electroanalysis 18 (19-20), 2009–2015.

He, Z., Wagner, N., Minteer, S.D., Angenent, L.T., 2006. An upflow microbial fuel cell with an interior cathode: Assessment of the internal resistance by impedance spectroscopy. Environ. Sci. Technol. 40 (17), 5212–5217.

He, Z., Minteer, S.D., Angenent, L.T., 2005. Electricity generation from artificial wastewater using an upflow microbial fuel cell. Environ. Sci. Technol. 39 (14), 5262–5267.

Heijne, A.T., Liu, F., Weijden, R.v.d., Weijma, J., Buisman, C.J.N., Hamelers, H.V.M., 2010. Copper recovery combined with electricity production in a microbial fuel cell. Environ. Sci. Technol. 44 (11), 4376–4381.

Hernández-Flores, G., Andrio, A., Compañ, V., Solorza-Feria, O., Poggi-Varaldo, H.M., 2019. Synthesis and characterization of organic agar-based membranes for microbial fuel cells. J. Power Sources 435, 226772.

Hiegemann, H., Herzer, D., Nettmann, E., Lübken, M., Schulte, P., Schmelz, K.-G., et al., 2016. An integrated 45L pilot microbial fuel cell system at a full-scale wastewater treatment plant. Bioresour. Technol. 218, 115–122.

Holmes, D.E., Chaudhuri, S.K., Nevin, K.P., Mehta, T., Methé, B.A., Liu, A., et al., 2006. Microarray and genetic analysis of electron transfer to electrodes in Geobacter sulfurreducens. Environ. Microbiol. 8 (10), 1805–1815.

Hutton, G., Chase, C., 2016. The knowledge base for achieving the sustainable development goal targets on water supply, sanitation and hygiene. Int. J. Environ. Res. Public Health 13 (6), 536.

IEA, 2017. Water-energy Nexus. https://www.iea.org/reports/water-energy-nexus. (Accessed 23 March 2020).

IEEE Spectrum, 2010. IEEE Spectrum Special Report: Water vs Energy. https://spectrum.ieee.org/static/special-report-water-vs-energy. (Accessed 20 April 2020).

Ieropoulos, I.A., Greenman, J., Melhuish, C., Hart, J., 2005. Comparative study of three types of microbial fuel cell. Enzyme Microb. Technol. 37 (2), 238–245.

Igboamalu, T.E., Chirwa, E.M.N., Bezuidenhout, N., Matsena, M.T., 2019. Microbial fuel cell power output and growth: effect of substrate pH on anaerobic microbe consortium. Chem. Eng. Trans. 76, 1381–1386.

Ishii, T., Kawaichi, S., Nakagawa, H., Hashimoto, K., Nakamura, R., 2015. From chemolitho autotrophs to electrolitho autotrophs: CO_2 fixation by Fe(II)-oxidizing bacteria coupled with direct uptake of electrons from solid electron sources. Front. Microbiol. 6, 994.

Ismail, Z.Z., Jaeel, A.J., 2016. Modelling study of an upflow microbial fuel cell catalysed with anaerobic aged sludge. Int. J. Ambient. Energy 37 (1), 85–93.

Jacobson, K.S., Drew, D.M., He, Z., 2011. Efficient salt removal in a continuously operated upflow microbial desalination cell with an air cathode. Bioresour. Technol. 102 (1), 376–380.

Jiang, R., Chu, D., 2003. Comparative studies of methanol crossover and cell performance for a DMFC. J. Electrochem. Soc. 151 (1), A69.

Jong, B.C., Kim, B.H., Chang, I.S., Liew, P.W.Y., Choo, Y.F., Kang, G.S., 2006. Enrichment, performance, and microbial diversity of a thermophilic mediatorless microbial fuel cell. Environ. Sci. Technol. 40 (20), 6449–6454.

Juang, D.-F., Lee, C.-H., Hsueh, S.-C., Chou, H.-Y., 2012. Power generation capabilities of microbial fuel cells with different oxygen supplies in the cathodic chamber. Appl. Biochem. Biotechnol. 167 (4), 714–731.

Khan, M., Bhattacharjee, R., Amin, M.S.A., 2012. Performance of the salt bridge based microbial fuel cell. Int. J. Eng. Technol. 1, 115–123.

Kim, B., Hyun, M.S., Park, D.H., 1999. Direct electrode reaction of Fe(III)-reducing bacterium, *Shewanella putrefaciens*. J. Microbiol. Biotechnol. 9, 127–131.

Kim, I.S., Kim, L.H., Kim, S.-J., Kim, K.-Y., 2015. Biofouling in osmotic membrane bioreactor. In: Shon, H.K., Phuntsho, S., Zhang, T.C., Surampalli, R.Y. (Eds.), Forward Osmosis: Fundamentals and Applications. American Society of Civil Engineers, Reston, pp. 241–275.

Kim, J., Kim, B., An, J., Lee, Y.S., Chang, I.S., 2016. Development of anode zone using dual-anode system to reduce organic matter crossover in membraneless microbial fuel cells. Bioresour. Technol. 213, 140–145.

Kim, J.R., Cheng, S., Oh, S.-E., Logan, B.E., 2007. Power generation using different cation, anion, and ultrafiltration membranes in microbial fuel cells. Environ. Sci. Technol. 41 (3), 1004–1009.

Kipnis, N., 1987. Luigi Galvani and the debate on animal electricity, 1791–1800. Ann. Sci. 44 (2), 107–142.

Kondaveeti, S., Lee, S.-H., Park, H.-D., Min, B., 2020. Specific enrichment of different *Geobacter* sp. in anode biofilm by varying interspatial distance of electrodes in air-cathode microbial fuel cell (MFC). Electrochim. Acta 331, 135388.

Koók, L., Žitka, J., Bakonyi, P., Takács, P., Pavlovec, L., Otmar, M., et al., 2020. Electrochemical and microbiological insights into the use of 1,4-diazabicyclo[2.2.2]octane-functionalized anion exchange membrane in microbial fuel cell: a benchmarking study with Nafion. Sep. Purif. Technol. 237, 116478.

Kumar, S., Kumar, H.D., Gireesh Babu, K., 2012. A study on the electricity generation from the cow dung using microbial fuel cell. J. Biochem. Technol. 3 (4).

Kundu, P.P., Dutta, K., 2018. Progress and Recent Trends in Microbial Fuel Cells. Elsevier Science, Amsterdam.

Larrosa-Guerrero, A., Scott, K., Head, I.M., Mateo, F., Ginesta, A., Godinez, C., 2010. Effect of temperature on the performance of microbial fuel cells. Fuel 89 (12), 3985–3994.

Lefebvre, O., Al-Mamun, A., Ng, H.Y., 2008. A microbial fuel cell equipped with a biocathode for organic removal and denitrification. Water Sci. Technol. 58 (4), 881–885.

Li, H., Xu, H., Yang, Y.-L., Yang, X.-L., Wu, Y., Zhang, S., et al., 2019. Effects of graphite and Mn ore media on electro-active bacteria enrichment and fate of antibiotic and corresponding resistance gene in up flow microbial fuel cell constructed wetland. Water Res. 165, 114988.

Li, L., Sun, Y., Yuan, Z.-H., Kong, X., Li, Y., 2013. Effect of temperature change on power generation of microbial fuel cell. Environ. Technol. 34, 1929–1934.

Li, W., Zhou, Q., Hua, T., 2010. Removal of organic matter from landfill leachate by advanced oxidation processes: a review. Int. J. Chem. Eng 2010, 270532.

Li, Y., Yang, W., Liu, X., Guan, W., Zhang, E., Shi, X., et al., 2020. Diffusion-layer-free air cathode based on ionic conductive hydrogel for microbial fuel cells. Sci. Total Environ. 743, 140836.

Liang, B., Ren, C., Zhao, Y., Li, K., Lv, C., 2020. Nitrogenous mesoporous carbon coated with Co/Cu nanoparticles modified activated carbon as air cathode catalyst for microbial fuel cell. J. Electroanal. Chem. 860, 113904.

Liao, G.-l., Liao, D.-x., Li, Q.-m., 2008. Heavy metals contamination characteristics in soil of different mining activity zones. Trans. Nonferrous Met. Soc. China 18 (1), 207–211.

Liengen, T., Basseguy, R., Feron, D., Beech, I., Birrien, V., 2014. Understanding Biocorrosion: Fundamentals and Applications. Elsevier Science, Amsterdam.

Lipczynska-Kochany, E., 2018. Humic substances, their microbial interactions and effects on biological transformations of organic pollutants in water and soil: a review. Chemosphere 202, 420–437.

Liu, H., Cheng, S., Logan, B.E., 2005. Power generation in fed-batch microbial fuel cells as a function of ionic strength, temperature, and reactor configuration. Environ. Sci. Technol. 39 (14), 5488–5493.

Liu, H., Logan, B.E., 2004. Electricity generation using an air-cathode single chamber microbial fuel cell in the presence and absence of a proton exchange membrane. Environ. Sci. Technol. 38 (14), 4040–4046.

Logan, B., Cheng, S., Watson, V., Estadt, G., 2007. Graphite fiber brush anodes for increased power production in air-cathode microbial fuel cells. Environ. Sci. Technol. 41 (9), 3341–3346.

Logan, B.E., Rossi, R., Ragab, A.a., Saikaly, P.E., 2019. Electroactive microorganisms in bioelectrochemical systems. Nat. Rev. Microbiol. 17 (5), 307–319.

Lovley, D.R., Walker, D.J.F., 2019. Geobacter protein nanowires. Front. Microbiol. 10 (2078).

Lovley, D.R., 2017. Electrically conductive pili: biological function and potential applications in electronics. Curr. Opin. Electrochem. 4 (1), 190–198.

Lovley, D.R., Fraga, J.L., Coates, J.D., Blunt-Harris, E.L., 1999. Humics as an electron donor for anaerobic respiration. Environ. Microbiol. 1 (1), 89–98.

Lovley, D.R., Coates, J.D., Blunt-Harris, E.L., Phillips, E.J.P., Woodward, J.C., 1996. Humic substances as electron acceptors for microbial respiration. Nature 382 (6590), 445–448.

Luo, H., Xu, G., Lu, Y., Liu, G., Zhang, R., Li, X., et al., 2017. Electricity generation in a microbial fuel cell using yogurt wastewater under alkaline conditions. RSC Adv. 7 (52), 32826–32832.

Luo, H., Liu, G., Zhang, R., Jin, S., 2009. Phenol degradation in microbial fuel cells. Chem. Eng. J. 147, 259–264.

Lv, C., Liang, B., Zhong, M., Li, K., Qi, Y., 2019. Activated carbon-supported multi-doped graphene as high-efficient catalyst to modify air cathode in microbial fuel cells. Electrochim. Acta 304, 360–369.

Majidi, M.R., Shahbazi Farahani, F., Hosseini, M., Ahadzadeh, I., 2019. Low-cost nano wired α-MnO_2/C as an ORR catalyst in air-cathode microbial fuel cell. Bioelectrochemistry 125, 38–45.

Malki, M., De Lacey, A.L., Rodríguez, N., Amils, R., Fernandez, V.M., 2008. Preferential use of an anode as an electron acceptor by an acidophilic bacterium in the presence of oxygen. Appl. Environ. Microbiol. 74 (14), 4472–4476.

Martinez, C.M., Alvarez, L.H., Celis, L.B., Cervantes, F.J., 2013. Humus-reducing microorganisms and their valuable contribution in environmental processes. Appl. Microbiol. Biotechnol. 97 (24), 10293–10308.

Matsena, M.T., Tichapondwa, S.M., Chirwa, E.M.N., 2020. Synthesis of biogenic palladium nanoparticles using *Citrobacter* sp. for application as anode electrocatalyst in a microbial fuel cell. Catalysts 10 (8), 838.

Matsena, M.T., Igboamalu, T.E., Chirwa, E.M.N., 2018. Phenol degrading and chromium (VI) reducing biofilm system: effect of shock-loading. Chem. Eng. Trans. 70, 207.

McAnulty, M.J., Poosarla, V.G., Kim, K.-Y., Jasso-Chávez, R., Logan, B.E., Wood, T.K., 2017. Electricity from methane by reversing methanogenesis. Nat. Commun. 8 (1), 15419.

Min, B., Cheng, S., Logan, B.E., 2005. Electricity generation using membrane and salt bridge microbial fuel cells. Water Res. 39 (9), 1675–1686.

Mohamed, H.O., Sayed, E.T., Obaid, M., Choi, Y.-J., Park, S.-G., Al-Qaradawi, S., et al., 2018. Transition metal nanoparticles doped carbon paper as a cost-effective anode in a microbial fuel cell powered by pure and mixed biocatalyst cultures. Int. J. Hydrog. Energy 43 (46), 21560–21571.

Nguyen, H.T.H., Kakarla, R., Min, B., 2017. Algae cathode microbial fuel cells for electricity generation and nutrient removal from landfill leachate wastewater. Int. J. Hydrog. Energy 42 (49), 29433–29442.

O'Hayre, R., Cha, S.W., Colella, W., Prinz, F.B., 2016. Fuel Cell Fundamentals. Wiley, New Jersey.

Oh, S.-E., Logan, B.E., 2006. Proton exchange membrane and electrode surface areas as factors that affect power generation in microbial fuel cells. Appl. Microbiol. Biotechnol. 70 (2), 162–169.

Penteado, E.D., Fernandez-Marchante, C.M., Zaiat, M., Gonzalez, E.R., Rodrigo, M.A., 2018. Optimization of the performance of a microbial fuel cell using the ratio electrode-surface area/anode-compartment volume. Brazilian J. Chem. Eng. 35 (1), 141–146.

Pontié, M., Jaspard, E., Friant, C., Kilani, J., Fix-Tailler, A., Innocent, C., et al., 2019. A sustainable fungal microbial fuel cell (FMFC) for the bioremediation of acetaminophen (APAP) and its main by-product (PAP) and energy production from biomass. Biocatal. Agric. Biotechnol. 22, 101376.

Potter, M.C., 1911. Electrical effects accompanying the decomposition of organic compounds. Proc. R. Soc. Lond. B. Biol. Sci. 84 (571), 260–276.

Potter, M.C., 1910. On the difference of potential due to the vital activity of microorganisms. Proc. Univ. Durham Phil. Soc. 3, 245–249.

Prasad, D., Arun, S., Murugesan, M., Padmanaban, S., Satyanarayanan, R.S., Berchmans, S., et al., 2007. Direct electron transfer with yeast cells and construction of a mediatorless microbial fuel cell. Biosens. Bioelectron. 22 (11), 2604–2610.

Quaglio, M., Massaglia, G., Vasile, N., Margaria, V., Chiodoni, A., Salvador, G.P., et al., 2019. A fluid dynamics perspective on material selection in microbial fuel cell-based biosensors. Int. J. Hydrog. Energy 44 (9), 4533–4542.

Quan, X., Zhang, X., Sun, Y., Zhao, J., 2018b. Iohexol degradation by biogenic palladium nanoparticles hosted in anaerobic granular sludge. Front. Microbiol. 9 (1980).

Quan, X., Xu, H., Sun, B., Xiao, Z., 2018a. Anode modification with palladium nanoparticles enhanced Evans Blue removal and power generation in microbial fuel cells. Int. Biodeterior. Biodegrad. 132, 94–101.

Quan, X., Sun, B., Xu, H., 2015. Anode decoration with biogenic Pd nanoparticles improved power generation in microbial fuel cells. Electrochim. Acta 182, 815–820.

Rabaey, K., Boon, N., Siciliano, S.D., Verhaege, M., Verstraete, W., 2004. Biofuel cells select for microbial consortia that self-mediate electron transfer. Appl. Environ. Microbiol. 70 (9), 5373–5382.

Rabaey, K., Lissens, G., Siciliano, S.D., Verstraete, W., 2003. A microbial fuel cell capable of converting glucose to electricity at high rate and efficiency. Biotechnol. Lett. 25 (18), 1531–1535.

Rahimnejad, M., Adhami, A., Darvari, S., Zirepour, A., Oh, S.-E., 2015. Microbial fuel cell as new technology for bioelectricity generation: a review. Alexandria. Eng. J. 54 (3), 745–756.

Ramanathan, G., Birthous, R.S., Abirami, D., Highcourt, D., 2011. Efficacy of marine microalgae as exoelectrogen in microbial fuel cell system for bioelectricity generation. World J. Fish Marine Sci. 3 (1), 79–87.

Reguera, G., Nevin, K.P., Nicoll, J.S., Covalla, S.F., Woodard, T.L., Lovley, D.R., 2006. Biofilm and nanowire production leads to increased current in *Geobacter sulfurreducens* fuel cells. Appl. Environ. Microbiol. 72 (11), 7345–7348.

Reguera, G., McCarthy, K.D., Mehta, T., Nicoll, J.S., Tuominen, M.T., Lovley, D.R., 2005. Extracellular electron transfer via microbial nanowires. Nature 435 (7045), 1098–1101.

Ren, Z.J., 2017. Microbial fuel cells: running on gas. Nat. Energy 2 (6), 17093.

Rhoads, A., Beyenal, H., Lewandowski, Z., 2005. Microbial fuel cell using anaerobic respiration as an anodic reaction and biomineralized manganese as a cathodic reactant. Environ. Sci. Technol. 39 (12), 4666–4671.

Rismani-Yazdi, H., Carver, S., Christy, A., Tuovinen, O., 2008. Cathodic limitations in microbial fuel cells: an overview. J. Power Sources 180, 683–694.

Rodrigo, M.A., Cañizares, P., Lobato, J., 2010. Effect of the electron-acceptors on the performance of a MFC. Bioresour. Technol. 101 (18), 7014–7018.

Santoro, C., Soavi, F., Serov, A., Arbizzani, C., Atanassov, P., 2016. Self-powered supercapacitive microbial fuel cell: the ultimate way of boosting and harvesting power. Biosens. Bioelectron. 78, 229–235.

Sarma, D., Barua, P.B., Dey, N., Nath, S., Thakuria, M., Mallick, S., 2019. Investigation and taguchi optimization of microbial fuel cell salt bridge dimensional parameters. J. Inst. Eng. India Ser. C. 100 (1), 103–112.

Schönbein, C.F., 1839. X. On the voltaic polarization of certain solid and fluid substances. Lond. Edinb. Dubl. Phil. Mag. J. Sci. 14 (85), 43–45.

Schröder, U., 2007. Anodic electron transfer mechanisms in microbial fuel cells and their energy efficiency. Phys. Chem. Chem. Phys. 9 (21), 2619–2629.

Scott, D.T., McKnight, D.M., Blunt-Harris, E.L., Kolesar, S.E., Lovley, D.R., 1998. Quinone moieties act as electron acceptors in the reduction of humic substances by humics-reducing microorganisms. Environ. Sci. Technol. 32 (19), 2984–2989.

Sekar, A.D., Jayabalan, T., Muthukumar, H., Chandrasekaran, N.I., Mohamed, S.N., Matheswaran, M., 2019. Enhancing power generation and treatment of dairy waste water in microbial fuel cell using Cu-doped iron oxide nanoparticles decorated anode. Energy 172, 173–180.

Sekar, N., Ramasamy, R.P., 2015. Photosynthetic energy conversion: recent advances and future perspective. Interface Mag. 24 (3), 67–73.

Sekrecka-Belniak, A., Toczyłowska-Mamińska, R., 2018. Fungi-based microbial fuel cells. Energies 11 (10), 2827.

Sevda, S., Sreekrishnan, T.R., 2012. Effect of salt concentration and mediators in salt bridge microbial fuel cell for electricity generation from synthetic wastewater. J. Environ. Sci. Health A 47 (6), 878–886.

Sharma, Y., Li, B., 2010. The variation of power generation with organic substrates in single-chamber microbial fuel cells (SCMFCs). Bioresour. Technol. 101 (6), 1844–1850.

Shi, D., Guo, Z., Bedford, N., 2015. Carbon nanotubes. In: Shi, D., Guo, Z., Bedford, N. (Eds.), Nanomaterials and Devices. William Andrew Publishing, Oxford, pp. 49–82.

Shi, L., Dong, H., Reguera, G., Beyenal, H., Lu, A., Liu, J., et al., 2016. Extracellular electron transfer mechanisms between microorganisms and minerals. Nat. Rev. Microbiol. 14 (10), 651–662.

Shukla, M., Kumar, S., 2018. Algal growth in photosynthetic algal microbial fuel cell and its subsequent utilization for biofuels. Ren. Sustain. Energy Rev. 82, 402–414.

Song, Y., An, J., Chae, K.-J., 2017. Effect of temperature variation on the performance of microbial fuel cells. Energy Technol. 5 (12), 2163–2167.

Sophia, A.C., Sai, S., 2016. Modified microbial fuel cell for Cr(VI) reduction and simultaneous bio-electricity production. J. Environ. Chem. Eng. 4 (2), 2402–2409.

Stirling, J.L., Bennetto, H.P., Delaney, G.M., Mason, J.R., Roller, S.D., Tanaka, K., et al., 1983. Microbial fuel cells. Biochem. Soc. Trans. 11 (4), 451–453.

Subha, C., Kavitha, S., Abisheka, S., Tamilarasan, K., Arulazhagan, P., Banu, J.R., 2019. Bioelectricity generation and effect studies from organic rich chocolaterie wastewater using continuous upflow anaerobic microbial fuel cell. Fuel 251, 224–232.

Sun, J., Li, W., Li, Y., Hu, Y., Zhang, Y., 2013. Redox mediator enhanced simultaneous decolorization of azo dye and bioelectricity generation in air-cathode microbial fuel cell. Bioresour. Technol. 142, 407–414.

Sun, M., Sheng, G.-P., Mu, Z.-X., Liu, X.-W., Chen, Y.-Z., Wang, H.-L., et al., 2009. Manipulating the hydrogen production from acetate in a microbial electrolysis cell–microbial fuel cell-coupled system. J. Power Sources 191 (2), 338–343.

Sun, M., Sheng, G.-P., Zhang, L., Xia, C.-R., Mu, Z.-X., Liu, X.-W., et al., 2008. An MEC-MFC-coupled system for biohydrogen production from acetate. Environ. Sci. Technol. 42 (21), 8095–8100.

Tawalbeh, M., Al-Othman, A., Singh, K., Douba, I., Kabakebji, D., Alkasrawi, M., 2020. Microbial desalination cells for water purification and power generation: a critical review. Energy 209, 118493.

ter Heijne, A., Hamelers, H.V.M., de Wilde, V., Rozendal, R.A., Buisman, C.J.N., 2006. A bipolar membrane combined with ferric iron reduction as an efficient cathode system in microbial fuel cells. Environ. Sci. Technol. 40 (17), 5200–5205.

Themelis, N.J., Ulloa, P.A., 2007. Methane generation in landfills. Renew. Energy 32 (7), 1243–1257.

Tkach, O., Sangeetha, T., Maria, S., Wang, A., 2017. Performance of low temperature microbial fuel cells (MFCs) catalyzed by mixed bacterial consortia. J. Environ. Sci. 52, 284–292.

Tran, T.V., Lee, I.C., Kim, K., 2019. Electricity production characterization of a sediment microbial fuel cell using different thermo-treated flat carbon cloth electrodes. Int. J. Hydrog. Energy 44 (60), 32192–32200.

Tsai, H.-Y., Wu, C.-C., Lee, C.-Y., Shih, E.P., 2009. Microbial fuel cell performance of multiwall carbon nanotubes on carbon cloth as electrodes. J. Power Sources 194 (1), 199–205.

Upton, D., de Groot, P., 2008. Planting and Establishment of Tropical Trees. Commonwealth Secretariat, London.

Varanasi, J.L., Nayak, A.K., Sohn, Y., Pradhan, D., Das, D., 2016. Improvement of power generation of microbial fuel cell by integrating tungsten oxide electrocatalyst with pure or mixed culture biocatalysts. Electrochim. Acta 199, 154–163.

Vélez-Pérez, L.S., Ramirez-Nava, J., Hernández-Flores, G., Talavera-Mendoza, O., Escamilla-Alvarado, C., Poggi-Varaldo, H.M., et al., 2020. Industrial acid mine drainage and municipal wastewater co-treatment by dual-chamber microbial fuel cells. Int. J. Hydrog. Energy 45 (26), 13757–13766.

Virdis, B., Rabaey, K., Rozendal, R.A., Yuan, Z., Keller, J., 2010. Simultaneous nitrification, denitrification and carbon removal in microbial fuel cells. Water Res. 44 (9), 2970–2980.

Volta, A., 1800. XVII. On the electricity excited by the mere contact of conducting substances of different kinds. In a letter from Mr. Alexander Volta, F. R. S. Professor of Natural Philosophy in the University of Pavia, to the Rt. Hon. Sir Joseph Banks, Bart. K.B. P. R. S. Phil. Trans. Royal Soc. Lond. 90, 403–431.

Wang, F., Gu, Y., O'Brien, J.P., Yi, S.M., Yalcin, S.E., Srikanth, V., et al., 2019. Structure of microbial nanowires reveals stacked hemes that transport electrons over micrometers. Cell 177 (2), 361–369.e310.

Werner, C.M., Logan, B.E., Saikaly, P.E., Amy, G.L., 2013. Wastewater treatment, energy recovery and desalination using a forward osmosis membrane in an air-cathode microbial osmotic fuel cell. J. Membr. Sci. 428, 116–122.

Wilkinson, S., Klar, J., Applegarth, S., 2006. Optimizing biofuel cell performance using a targeted mixed mediator combination. Electroanalysis 18 (19-20), 2001–2007.

Wisniak, J., 2015. Historical notes: electrochemistry and fuel cells: the contribution of William Robert Grove. Indian J. Hist. Sci. 50, 476–490.

World Health Organization, 2019. WHO Global Water, Sanitation and Hygiene: Annual Report 2018. https://apps.who.int/iris/bitstream/handle/10665/327118/WHO-CED-PHE-WSH-19.147-eng.pdf?ua=1. (Accessed 13 April 2020).

Wu, C., Liu, X.-W., Li, W.-W., Sheng, G.-P., Zang, G.-L., Cheng, Y.-Y., et al., 2012. A white-rot fungus is used as a biocathode to improve electricity production of a microbial fuel cell. Appl. Energy 98, 594–596.

Xiao, L., Young, E.B., Berges, J.A., He, Z., 2012. Integrated photo-bioelectrochemical system for contaminants removal and bioenergy production. Environ. Sci. Technol. 46 (20), 11459–11466.

Xin, X., Pang, H., She, Y., Hong, J., 2020. Insights into redox mediators-resource harvest/application with power production from waste activated sludge through freezing/thawing-assisted anaerobic acidogenesis coupling microbial fuel cells. Bioresour. Technol. 311, 123469.

Xu, C., Poon, K., Choi, M.M.F., Wang, R., 2015a. Using live algae at the anode of a microbial fuel cell to generate electricity. Environ. Sci. Pollut. Res. 22 (20), 15621–15635.

Xu, X., Zhao, Q.L., Wu, M.S., 2015b. Improved biodegradation of total organic carbon and polychlorinated biphenyls for electricity generation by sediment microbial fuel cell and surfactant addition. RSC Adv. 5 (77), 62534–62538.

Yamasaki, R., Maeda, T., Wood, T.K., 2018. Electron carriers increase electricity production in methane microbial fuel cells that reverse methanogenesis. Biotechnol. Biofuels 11 (1), 211.

Yang, L., Deng, W., Zhang, Y., Tan, Y., Ma, M., Xie, Q., 2017. Boosting current generation in microbial fuel cells by an order of magnitude by coating an ionic liquid polymer on carbon anodes. Biosens. Bioelectron. 91, 644–649.

Yang, Y., Ding, Y., Hu, Y., Cao, B., Rice, S.A., Kjelleberg, S., et al., 2015. Enhancing bidirectional electron transfer of Shewanella oneidensis by a synthetic flavin pathway. ACS Synth. Biol. 4 (7), 815–823.

Yousefi, V., Mohebbi-Kalhori, D., Samimi, A., 2017. Ceramic-based microbial fuel cells (MFCs): a review. Int. J. Hydrog. Energy 42 (3), 1672–1690.

Zhang, B., Feng, C., Ni, J., Zhang, J., Huang, W., 2012. Simultaneous reduction of vanadium (V) and chromium (VI) with enhanced energy recovery based on microbial fuel cell technology. J. Power Sources 204, 34–39.

Zhang, F., Brastad, K.S., He, Z., 2011. Integrating forward osmosis into microbial fuel cells for wastewater treatment, water extraction and bioelectricity generation. Environ. Sci. Technol. 45 (15), 6690–6696.

Zhang, L., Li, J., Zhu, X., Ye, D.-d., Fu, Q., Liao, Q., 2017. Response of stacked microbial fuel cells with serpentine flow fields to variable operating conditions. Int. J. Hydrog. Energy 42 (45), 27641–27648.

Zhang, R., Zhang, Y., Zhang, Q., Xie, H., Qian, W., Wei, F., 2013. Growth of half-meter long carbon nanotubes based on Schulz–Flory distribution. ACS Nano 7 (7), 6156–6161.

Zhang, X., Li, X., Zhao, X., Li, Y., 2019. Factors affecting the efficiency of a bioelectrochemical system: a review. RSC Adv. 9 (34), 19748–19761.

Zhao, X., Deng, W., Tan, Y., Xie, Q., 2020. Promoting electricity generation of shewanella putrefaciens in a microbial fuel cell by modification of porous poly(3-aminophenylboronic acid) film on carbon anode. Electrochim. Acta 354, 136715.

Zheng, W., Cai, T., Huang, M., Chen, D., 2017. Comparison of electrochemical performances and microbial community structures of two photosynthetic microbial fuel cells. J. Biosci. Bioeng. 124 (5), 551–558.

Zou, L., Qiao, Y., Zhong, C., Li, C.M., 2017. Enabling fast electron transfer through both bacterial outer-membrane redox centers and endogenous electron mediators by polyaniline hybridized large-mesoporous carbon anode for high-performance microbial fuel cells. Electrochim. Acta 229, 31–38.

CHAPTER

17

Concomitant bioenergy production and wastewater treatment employing microbial electrochemical technologies

M.M. Ghangrekar[1,2], *Santosh Kumar*[2], *Azhan Ahmad*[1], *Sovik Das*[1]

[1]Department of Civil Engineering, Indian Institute of Technology Kharagpur, Kharagpur, West Bengal, India; [2]PK Sinha Centre for Bioenergy and Renewables, Indian Institute of Technology Kharagpur, Kharagpur, West Bengal, India

17.1 Introduction

Fossil fuels have become a fundamental part of daily life of human beings due to the rapid advancement in the transport system, tremendous growth in the industrial sector, and other domestic uses. Generally, fossil fuels are employed to generate electrical energy required for transportation and industries, and these sectors have a crucial role to play in the improvement of human living standards as well as in accelerating technological developments. However, fossil fuels are limited and cannot be replenished once exhausted. Moreover, burning of fossil fuels causes numerous adverse environmental impacts, such as emission of greenhouse gases (GHGs), which is strongly accountable for global warming (Brennan and Owende, 2010). The energy crisis has started hitting all over the world due to exorbitant population growth as well as rapid industrialization. Therefore, renewable energy sources are the only hope for present as well as future generations toward the attainment of sustainable development. Several renewable energy sources are there in practice such as geothermal, hydrothermal, solar, and wind. Recent advancements in biotechnology and related disciplines have opened a scope for energy recovery from waste biomass using microbial electrochemical technologies (METs), which have been successfully developed and employed by different researchers as an alternative to fossil fuels, which are also serving the purpose of pollution mitigation (Owusu and Asumadu-Sarkodie, 2016).

Rapid industrialization and accelerated population growth are the two major contributors of wastewater generation and environmental pollution. Treatment of generated wastewater is imperative due to its adverse effect on natural water bodies, thus having an adverse impact on the public health (Ghangrekar et al., 2020).

Biofuels and Bioenergy
https://doi.org/10.1016/B978-0-323-85269-2.00015-0

The basic idea behind wastewater remediation is the removal of both suspended and dissolved impurities, chemical and organic impurities, and pathogens (Ahmad et al., 2020). Moreover, clean water, energy, nutrient, and other value-added by-product recovery with simultaneous wastewater treatment can be accomplished employing METs (Bose et al., 2020). The METs represent a recent trend in emerging wastewater treatment technologies that treat wastewater with concurrent bioenergy recovery, biomass production, and CO_2 sequestration (Pant et al., 2012).

The METs comprise anodic and cathodic chambers that are disconnected by a proton exchange membrane (PEM), and mechanisms of METs rely on the metabolic activity of electrogenic bacteria present in the anodic chamber (Das, 2017). Oxidation of decomposable organic matter occurs with the help of microbes living in the anodic chamber, which leads to the synthesis of chemicals, bioelectricity, biofuels, and CO_2 sequestration (Ramírez-Vargas et al., 2018). The anode material affects not only the performance of METs but also their fabrication cost (Das and Ghangrekar, 2020). The material selected for anode should be noncorrosive, electrochemically active toward oxidation reaction of interest, and physically and chemically stable, and it should be compatible to the electrogenic bacteria for the formation of biofilm. The efficiency of METs also strongly depends on the cathode material used for the reduction of terminal electron acceptor (TEA) by the electrons coming from an exterior circuit from the anode (Das et al., 2020f). The cathode material used in METs depends on the application of MET; though the common cathode materials used in METs are carbon-based materials, such as graphite plates, rods, carbon foam, carbon cloth, carbon felt, carbon paper, and carbon nanotube (CNT)–based composites, metals such as titanium mesh and stainless steel are also being used (Ghorai et al., 2020; Roy and Pandit, 2018).

Microbial fuel cell (MFC) and microbial carbon capture (MCC) are the most investigated MET configurations employed for wastewater treatment with concomitant bioenergy recovery (Das et al., 2019c). Moreover, microbial electrosynthesis (MES) is also an emerging MET, where bioenergy recovery or synthesis of higher chain carbon compounds is possible through carbon sequestration (Das et al., 2020e; Das and Ghangrekar, 2018). However, typically, wastewater treatment is not being targeted so far in MES research (Das et al., 2018); hence, discussion on MES has not been presented in this chapter.

The MFCs are the first set of bioelectrochemical technologies in which electricity is produced by using the electrons derived from bacterial-catalyzed biochemical reactions with concomitant bioelectricity recovery (Das et al., 2020d). The working methodology of MFC is comparable with that of a conventional fuel cell with an alteration that bacteria act as biocatalyst in anaerobic anodic chamber and organic load in wastewater acts as the fuel (Logan et al., 2006). In an MFC, cathodic and anodic chambers are divided by a separator also known as proton exchange membrane (PEM). Microbes oxidize the biodegradable matter present in the anodic chamber to generate protons and electrons. Electrons ultimately reach the cathode through the exterior circuit, whereas the protons transport through the PEM to the cathodic chamber, where these substances react generally with dissolved oxygen to generate water and thus producing electricity in an MFC (Das et al., 2020g).

An MCC is a modified version of an MFC, in which algal cells are cultivated in the cathodic chamber. In MCCs, bacterial catalysis in anodic chamber transforms the chemical energy available in wastewater into electrical energy, and in cathodic chamber, the oxygen produced by algal cells functions as TEA (Varanasi and Das, 2018). An MCC is the most suitable, effective, and innovative technology, where oxygen produced by algal biomass in the cathodic chamber is used as TEA for the reduction of oxygen through two- or four-electron pathways that accomplish electricity generation as well as concurrent CO_2 sequestration, production of algal biomass, and wastewater remediation (Nagendranatha Reddy et al., 2019).

17.2 Microbial electrochemical technologies

In comparison with other wastewater remediation technologies, METs are one of the most sustainable, environmentally friendly, and leading technologies to produce bioenergy, biofuel, and other valuable products. The METs are basically considered as a hybrid and interdisciplinary technology having applications in diverse areas, i.e., bioenergy production, CO_2 sequestration, waste remediation, resource recovery, bioelectronics, desalination, and so on (Das et al., 2019c). It has applications in multidisciplines, which include electrochemistry, microbiology, environmental engineering, material sciences, and biochemistry. The primary advantage of this technology is that the waste can be used in any physical state (solid, liquid, or gaseous) as raw material for numerous value-added products recovery (Mohan et al., 2019). The MET possesses a lot of potential toward the attainment of cleaner future; however, numerous bottlenecks and challenges associated with the scaling up of these technology need to be circumnavigated, leading to the commercialization of this innovative technology.

17.2.1 Factors affecting the performance of microbial electrochemical technologies

The METs are considered as an emerging technology due to their well-documented applications in multidisciplines; however, METs are mostly operated on a small scale to demonstrate the practical value of this technology, although there are a few large-scale METs, e.g., world's first pilot-scale MFC implemented in Yatala Brewery and the Ugold technology, which converted human urine into clean liquid fertilizer (Freguia et al., 2019; Gajda et al., 2018) and pilot-scale MFC installed at IIT Kharagpur to treat real human waste and lighting toilet premises for the nocturnal hours (Das et al., 2020b). The main limitations associated with the pilot-scale METs are their efficiency and the cost of their implementation. The performance of METs is governed by anodic and cathodic reaction kinetics, electrode materials, operating conditions, and the physical architecture of the setup as well as the amount, type, and activity of microorganisms employed in them (Gil et al., 2003).

To overcome the anodic limitation, appropriate choice of anodic material is required, and the chosen anode material should be biocompatible, nontoxic, noncorrosive, and chemically stable and should have high surface area with high conductivity and economical and locally available (Das and Ghangrekar, 2020). On the other hand, the significant limitation associated with cathode is the poor oxygen reduction reaction (ORR) kinetics caused due to high activation energy required for the conversion of the oxidant into a reduced form (Roy and Pandit, 2018). To overcome this limitation, different cathode catalysts are being used, which increase the interactive surface area and decrease the activation energy to accelerate the ORR rate and product output (Das et al., 2020f). Performance of METs also depends on bacterial inoculum, electrocatalyst, membrane characteristics, overall configuration of the system, complexity of the wastewater used as anolyte, and the type of catholyte used (Chakraborty et al., 2020).

The structural configurations of METs, such as the shape and type of METs; electrode material used that may cause high internal resistance and electrochemical energy losses; interelectrode distance; and specific surface area of electrode, also play a remarkable role in dictating the efficacy of METs. In this regard, single-chamber MFC has been proposed in which air cathode is used, instead of submersed cathode in cathodic chamber as used in dual-chambered MFC, to reduce the internal resistance (Das et al., 2020d). The membrane separating the two chambers also has a significant part to play in the working of an MET by preventing the physical mixing of the two electrolytes and allowing only protons to transfer from the anodic to the cathodic chamber (Das et al., 2020a). Furthermore, the power generation for a single MFC is very less, and thus to overcome this, multiple sets of MFCs need to be joined either in series or in parallel connection, leading to stacking, which would improve the overall electrical output of the setup (Rinaldi et al., 2008).

Biological activity of the microorganisms, e.g., ability to form biofilm and electron transfer, substrate utilization rate, favorable growth conditions, and so on, also has an imperative role on the performance of METs. Therefore, to overcome the biological limitations associated with METs, there is a need of detailed understanding of extracellular electron transfer mechanism. When working with real wastewater, selection of suitable microorganisms and adaptation in the MET is considered as biological optimization. Effective optimization of biological parameters can offer an improvement in current density rendering the METs as a sustainable and applicable technology for real-life implementations (Veerubhotla et al., 2018).

Physiochemical parameters such as temperature, pH, substrate composition, degree of mixing, retention time, external resistance, and substrate loading rate can affect the microbial activity by affecting the electroactive biofilm (EAB) formation, thus governing electricity generation and treatment efficiency of MFCs. Biofilm formed by electroactive microorganisms on the electrode surface, which is responsible for the extracellular electron transfer in METs, also significantly affects the power generation of an MET. Furthermore, anodic electrolyte pH can hinder the growth of specific microbial species responsible for EAB formation. The temperature at which the system is operated also dictates the metabolic activity of microorganisms responsible for EAB formation. The external resistance is another parameter that greatly influences the METs operation, as it not only regulates the electron movement in circuit but it can also alter the start-up time, metabolic activity, and EAB formation (Li et al., 2010; Veerubhotla et al., 2018).

Substrate composition is one of the chief parameters of MET that greatly affects its performance. Substrate loading rate, mixing, and hydraulic retention time can affect the microbial metabolism by providing mechanical shear stress that results in either cell aggregation or cell death, leading to the alteration in EAB formation (Varanasi et al., 2015). Optimization of all these diverse challenges associated with METs is only possible by adopting the interdisciplinary approach by including expertise from biotechnology, biochemistry, material science, electrochemistry, and physics as well as mathematical modeling. As discussed previously, this technology is too diverse and has numerous real-life applications, and lots of research studies are ongoing on this topic to overcome the limitations associated with the technology to make the METs more sustainable and cost-effective for real-life applications.

17.2.2 Electrode materials used in microbial electrochemical technologies

Electrode materials used as anode or cathode highly influence the performance and power generation of METs. The conductivity and resistance of the anode material used in MET should be high enough to generate significant electricity. High biocompatibility of electrode material provides better biofilm formation, resulting in increased electrical output. Anticorrosion properties, hence chemical stability, of anode material provide longevity to the electrode. Physical properties that include larger surface area, toughness, and mechanical strength also affect the choice of an anode material. Most frequently employed anode materials are carbon-based materials. Recently, CNTs are reported to be comparatively best anode material because of higher specific surface area, high enough mechanical strength, and superior conductivity (Chen et al., 2019). More recently, CNTs/conductive polymer composites have captured significant interest of the scientific community, as incorporation of CNTs in conductive polymers can lead to a synergistic effect between adhesion and electrocatalytic properties (Table 17.1). Though, CNTs are better anode material; however this material is comparatively expensive (Nourbakhsh et al., 2017). Few other frequently used non–carbon-based anode materials are titanium and gold; however, these materials are costly, and similar power can be obtained using low-cost carbon-based material in comparison with these expensive metals as anode (Richter et al., 2008; Ter Heijne et al., 2008).

TABLE 17.1 Different electrode material used in microbial electrochemical technologies.

Wastewater/Anolyte	Anodic inoculum	Electrode material	Used as	Power density (mW/m^2)	References
Phosphate-buffered basal medium	*Klebsiella pneumoniae*	Ni_3Mo_3C	Anode	250	Zeng et al. (2015)
Artificial wastewater	Mixed culture	Mo	Anode	1296	Yamashita and Yokoyama (2018)
Sugarcane bagasse effluent	Bacterial consortium isolated form sugarcane bagasse effluent	Iron/titanium Ni–P composite	Anode	2100	Chandraserkharan Meenu et al. (2018)
Phosphate-buffered saline	*Escherichia coli*	rGO/PANI/Pt	Anode	2059	Gnana kumar et al. (2014)
LB medium	*Shewanella oneidensis*	PANI-TiO_2-GN	Anode	79.3	Han et al. (2018)
Synthetic wastewater	*E. coli*	MWCNTs/SnO_2	Anode	1421	Mehdinia et al. (2014)
Synthetic wastewater	Mixed anaerobic sewage sludge	WO_3	Anode	143.67	Das and Ghangrekar (2020)
			Cathode	126.94	
Modified PBS	Anaerobic sludge	V_2O_5	Cathode	1073	Ayyaru et al. (2019)
Synthetic wastewater	–	$FeMn_2$	Cathode	1940	Guo et al. (2019)
Synthetic wastewater	Anaerobic digester sludge	Co/N-CNT	Cathode	1260	Yang et al. (2019)
Modified PBS	Bacterial mixture	Biofilm covered Pt/C cathode	Cathode	970	Jin et al. (2020)
Synthetic wastewater	Mixed anaerobic sludge	CuZn	Cathode	75	Das et al. (2020b)
		Pd	Cathode	63.94	Das et al. (2020d)
Sucrose-based synthetic wastewater	Mixed anaerobic sludge	Rh/Ac	Cathode	65.52	Bhowmick et al. (2019a)

Cathode surface has a very vital role on the reduction kinetics of TEA; therefore, efficient cathode material depicts the performance of METs (Bhowmick et al., 2019a). Hence, increasing the specific cathode surface area might be a successful way to improve the performance of an MET. Carbon-based cathode materials are generally used in METs; however, they demonstrate very poor ORR kinetics that limits the performance of METs, which can be improved by using effective cathode catalysts (Bhowmick et al., 2019b). The potential properties of cathode catalyst are that it should exhibit high catalytic activity and high specific surface area. Moreover, electrical conductivity should be high enough to minimize the resistive losses. Additionally, it should be stable at high pH, nontoxic, easily synthesizable at a large scale, and commercially available. Therefore, the selection of a suitable electrode material is imperative that governs the efficacy of an MET.

17.3 Microbial fuel cell

An MFC is an escalating bioelectrochemical equipment that transforms organic impurities present in the wastewater into electrical energy by the electrochemical catalytic activity of microbes (Das et al., 2020d). In an MFC, microbes residing in the anodic chamber break the bonds of organic pollutants present in wastewater to produce CO_2, protons, and free electrons (Das et al., 2020a). Free electrons produced in the anodic chamber, in the biofilm made of the electroactive microbes on the anodic surface, reduce the anode. The electrons pass from anode to the cathode assisted by an exterior circuit, and simultaneously, protons are allowed to transfer through the PEM from the anodic to the cathodic chamber. Thus, on cathode, reduction of oxidizing agent such as O_2 occurs to produce electricity as well as clean H_2O (Fig. 17.1).

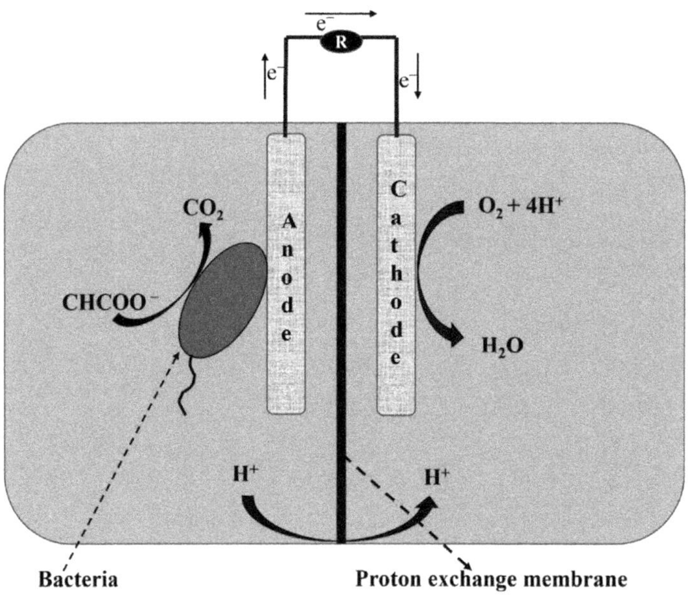

FIGURE 17.1 Schematics of a microbial fuel cell.

The overall biochemical reactions that take place in the anodic and cathodic chambers of an MFC are presented in Eq. (17.1) through Eq. (17.3).

At anode:

$$CH_3COO^- + 2H_2O \rightarrow 2CO_2 + 7H^+ + 8e^- \quad (17.1)$$

At cathode:

$$2O_2 + 8H^+ + 8e^- \rightarrow 4H_2O \quad (17.2)$$

$$O_2 + 2H^+ + 2e^- \rightarrow H_2O_2 \quad (17.3)$$

17.3.1 Application and challenges associated with microbial fuel cell

The potential application of MFCs is wastewater remediation (including organic matter, sulfate, and nitrate removal) as well as electricity generation (Das et al., 2020c). However, it has many other applications such as production of biogas and H_2O_2 and applications as biosensors to detect the existence of volatile fatty acids, biochemical oxygen demand (BOD), dissolved oxygen (DO), and also harmful substances in water streams (Gupta et al., 2020). Recently, biocathode-based MFC has captured a lot of attention due to diverse applications, such as wastewater treatment with concurrent electric power generation, CO_2 sequestration, biofuel generation, and other valuable products recovery.

The generation of wastewater is cumulating day by day due to the global population and industrial bloom. Wastewater is one of the primary components of environmental contamination, and hence, its treatment is very essential. The traditional wastewater treatment methods are expensive as well as require huge energy and time. However, wastewater treatment using MFC technology is considered as one of the effective and sustainable methods. In an investigation on wastewater treatment using MFC; chemical oxygen demand (COD), total nitrogen, and ammonia nitrogen removal efficiencies were reported to be 75.59%, 76.15%, and 83.23%, respectively (Xia et al., 2019). In another investigation, when constructed wetland-microbial fuel cell (CW-MFC) was used for wastewater remediation, 82.2% of decolorization and 70% reduction in COD were reported (Rathour et al., 2019). In a pilot-scale CW-MFC of 30 L capacity, 91.7% of COD and 97.3% of ammoniacal nitrogen removal were reported (Tang et al., 2019). Using palm oil industry effluent as substrate in MFC, power generation was reported to be 14.8 W/m^3, when *Pseudomonas aeruginosa* and *Klebsiella variicola* were cocultured in the anodic chamber (Islam et al., 2020). In a polyaniline-coated anode of an MFC, the maximum power density of 28 W/m^3 was reported. Again in a report of sewage sludge stabilization using MFC, a power generation of 27.65 W/m^3 was observed (Raad et al., 2020). Thus, simultaneous power recovery using MFC during wastewater treatment makes the technology relatively viable.

Though, MFC is known as the oldest MET; however, it has become prevalent in the past few years only (Das et al., 2019b). Moreover, the yield of bioelectricity from an MFC is still far away for its practical utilization and hence applications (Das et al., 2019a). The main challenge in the construction of field-scale MFCs is the higher construction cost, especially because of involvement of the costly electrode and PEM (Logan and Regan, 2006). Another considerable challenge is the storage of generated electricity; therefore, development of appropriate electrical storage device and circuit is imperative. Thus, for the real-life application and scaling up of MFCs, more fundamental research on electrode materials and PEM is required (Jadhav et al., 2018). If wastewater remediation and bioelectricity production is to be considered, among all bioelectrochemical system, MFC still has shown some level of realistic practical applications. However, to make it applicable on a larger scale, there is the need to investigate each and every parameter, including design of MFC, selection of appropriate microbes, and physicochemical parameters affecting performance.

17.4 Microbial carbon-capture cell

The MCC has drawn considerable attention in the field of wastewater remediation, clean electricity generation, capture of CO_2 discharged from the anodic chamber and its fixation in the cathodic algal biomass synthesis, and concurrent algal biomass harvesting to be converted into produces, such as biodiesel, ethanol, biofertilizers, hydrogen, and amino acid (Hu et al., 2015; Pandit et al., 2012). In MCCs, the oxidation of biodegradable matter present in the anodic chamber takes place by the bacterial catalysts action and generates CO_2 as well as electrical energy (Fig. 17.2). Carbon dioxide generated in the anodic chamber by the degradation of biodegradable matter can be used for growth of algal biomass through the photosynthetic pathway (Jadhav et al., 2019). The overall biochemical reactions that take place in both the chambers of an MCC are presented in Eqs. (17.4) through (17.6), respectively.

At the cathode: (algae as biocatholyte)
Light reaction: (in the presence of light)

$$nCO_2 + nH_2O \rightarrow (CH_2O)_n + nO_2 \quad (17.4)$$

Dark reaction: (in the absent of light)

$$C_2H_4O_2 + 2O_2 \rightarrow 2CO_2 + 2H_2O \quad (17.5)$$

Net cathodic reaction:

$$O_2 + 4H^+ + 4e^- \rightarrow 2H_2O \quad (17.6)$$

17.4.1 Application of microbial carbon capture

17.4.1.1 Biofuel production and CO_2 sequestration

An MCC is a modified MFC system in which cultivation of microalgae is facilitated in the cathodic chamber of an MFC. In this system, CO_2 sequestration and its fixation occur, where CO_2 produced in the anodic chamber is supplied to algal culture present in the cathodic chamber.

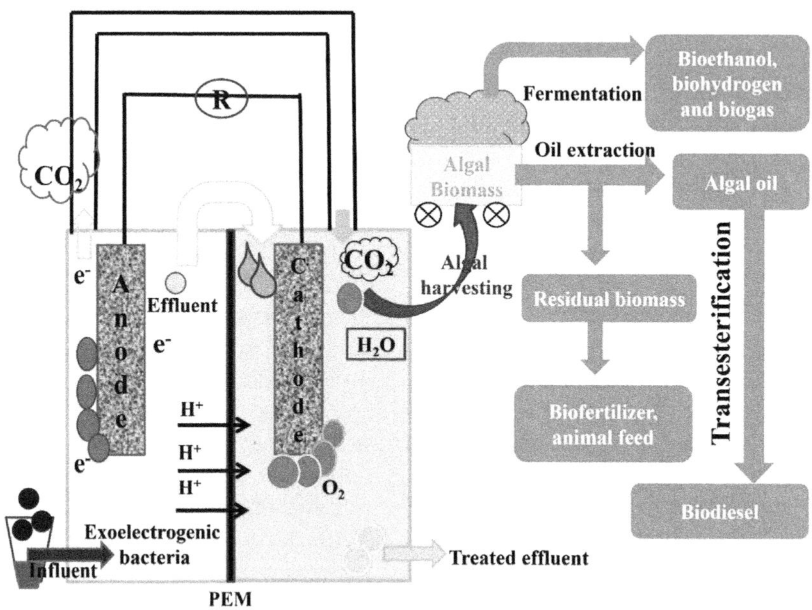

FIGURE 17.2 Schematic representation of application and working mechanism of a microbial carbon-capture cell and further downstream processing of harvested algae for bioenergy production.

Due to the algal cultivation and O_2 production, cathodic ORR is supported to increase power generation. Additionally, harvested algal biomass cultivated in the cathodic chamber can be used for biofuel and other products recovery. Biofuels produced from microalgae are termed as the third-generation biofuels. Microalgae are regarded as one of the oldest living microorganisms on the earth (Song et al., 2008). Microalgae reproduce at an exponentially faster rate, which is about 100 times quicker than the plants, and can also increase their biomass by twofold in less than a day (Lam and Lee, 2012). Moreover, some microalgal strains are also able to store huge quantity of lipids in their cells, and this lipid can be further converted into biodiesel (Chisti, 2008; Lam and Lee, 2012).

Microalgae are photosynthetic and free-living organisms with the size of few microns, which can grow in varied aquatic environments, which include freshwater, wastewater, brackish water (<3.5% salt), marine water (~3.5% salt), and hypersaline (>3.5% salt) environment with a widespread pH and temperature range. Lipid accumulation in algal biomass depends on different cultivation conditions as well as on the algal species. Microalgal lipids are generally classified into two categories—storage lipids (nonpolar) and structural lipids (polar). The storage lipids are commonly in the form of triacylglycerols (TAG), which can be both saturated and unsaturated fatty acids and can be further transesterified to generate biodiesel. In adverse conditions, many microalgae modify their lipid synthesis pathways and accumulate lipids (20% −50% dry cell weight) primarily in the form of TAG, allowing them to tolerate hostile condition (Sharma et al., 2012).

Microalgae have numerous attractive characteristics for biodiesel production, such as costs related to their transportation and harvesting are much lesser in comparison with other biomass resources and microalgae are also capable of CO_2 sequestration and fixation, thus alleviating increasing atmospheric CO_2 levels, which is one of the culprits of global warming. Moreover, microalgae are readily decomposable, and they can also be chemically treated; hence, algal cultivation and biodiesel extraction from algal biomass is not a complex process, and algae can grow almost everywhere, where there is sufficient sunlight (Demirbas, 2011). Algal biomass concentration and lipid productivity suggest that microalgae present in the cathodic chamber of MCCs might be a promising application of biofuel production using MCCs (Table 17.2).

TABLE 17.2 Application of microbial carbon capture for the remediation of different pollutants and simultaneous power recovery.

Wastewater treatment				Algal biomass and lipid (concentration/productivity)		
COD removal (%)	Nitrate removal (%)	Phosphate removal (%)	Power generation (W/m^3)	Biomass (g/L)	Lipid (mg/L.day)	References
92.0	98.0	82.0	2.20	0.13	–	Xiao et al. (2012)
86.7	70.5	69.2	0.97	–	109.44	Hu et al. (2015)
87.3	–	–	6.40	–	66.40	Jadhav et al. (2017)
72.1	–	–	3.20	2.45	–	Neethu et al. (2018)
95.0	–	–	2.32	0.80	–	Varanasi et al. (2020)
–	–	–	5.94	–	234.3	Li et al. (2019)
70.0	–	–	1.78	1.34	–	Jadhav et al. (2019)

17.4.1.2 Wastewater treatment

Remediation of decomposable contaminants present in wastewater takes place in the anodic chamber of an MCC, where microbes oxidize this organic matter to generate electrical energy as well as CO_2; however, some fraction of organic matter is utilized by microbes as carbon source to perform different metabolic functions. The CO_2 generated in anodic chamber is further utilized for microalgal culture cultivation in the cathodic chamber of an MCC. Removal of inorganic pollutants, such as P and N, occurs in the anodic as well as in the cathodic chamber of an MCC. Microalgae have gained much attention because of its capability to eliminate N, P, and organic matter from wastewater as well as it generates O_2, which acts as TEA in the cathodic chamber of MCCs.

In a work by Xiao et al., decrease in phosphorus (82% of P as phosphate), nitrogen (98% of N as ammonium), and organic matter (92% in terms of COD) was observed from municipal wastewater, when *Chlamydomonas reinhardtii* and *Pseudokirchneriella subcapitata* were cultivated in an integrated photobiochemical system (installing MFCs inside an algal bioreactor) (Xiao et al., 2012). In the same research, it was also suggested that 27% of phosphate reduction took place in the anodic chamber by bacterial activity and 55% of phosphate reduction in cathodic chamber due to algal activity. In an airlift-type MCC, 86.7%, 70.5%, and 69.2% reductions in COD, nitrate, and phosphate, respectively, were reported when *Chlorella vulgaris* was cultivated in cathodic chamber (Hu et al., 2015). In an another investigation, when *C. vulgaris* was cultivated in photosynthetic algal MFC, removal efficiencies of 85.6% for ammonium nitrate, 70.2% for total nitrogen, and 93.9% for total organic carbon were reported (Zhang et al., 2019). Thus, considering the aforementioned literature, MCCs can be considered as one of the leading and sustainable techniques for wastewater treatment to demonstrate high removal efficiency.

17.4.1.3 Bioelectricity production

The electricity generation in MCCs depends on numerous physicochemical parameters including pH, temperature, light intensity, and DO concentration of the catholyte and electrode material, which favor growth of microbes on the anode as well as in the cathodic chamber in the form of microalgae. Power generation in MCCs also depends on the microbial and algal species, and it has been reported that when *Chlorella pyrenoidosa* was cultivated, instead of *Anabaena ambigua*, there was 1.5-fold higher power generation. This higher power generation of 6.4 W/m^3 might have been due to the enhanced growth rate of *C. pyrenoidosa* (Jadhav et al., 2017). In an experiment, increase in power density from 4.1 to 5.6 W/m^3 was reported, when the optical density (OD) of cathodic *C. vulgaris* culture improved from 0.21 to 0.85 (OD at 659 nm) (Wang et al., 2010).

In an another investigation, 3.72 W/m^3 of power generation was reported when *C. vulgaris* was cultivated in photosynthetic algal MFC, which was 1.6 times more in comparison with the control MFC, and the increase voltage indicated the importance of algal cultivation in MCCs (Zhang et al., 2019). Output voltage of MCCs varies with the fluctuation in the O_2 concentration in the cathodic chamber, which is again governed by the growth of algal cells in the chamber. Power production of MCCs can be amplified significantly by optimizing the numerous parameters, leading toward the real-life application of the technology in the near future.

17.4.1.4 Algal biomass production

Algal cultivation in the cathodic chamber of MCCs provides additional advantage as it supplies O_2 that increases ORR, and CO_2 produced in the anodic chamber is used in cathodic chamber to increase the algal growth rate. The basic contents of algal biomass are lipid, carbohydrate, and protein. Due to the diverse applications of algal biomass in the food manufacturing industry, aquaculture and bioenergy applications, especially for biodiesel generation and mass production of algae, are attractive and highly beneficial (Pulz, 2001). However, the practical application of algal biomass derived from MCCs for biodiesel is still costly, in comparison with other renewable energy sources and fossil fuels (Nigam and Singh, 2011). Investigation on MCCs generally demonstrates very low biomass productivity due to less CO_2 sequestration; however, in a recent work, 2.45 g/L of algal biomass was cultivated in MCCs (Table 17.2). Again in an another investigation, 1.07 g/L of biomass was obtained after 21 days of incubation period, when microalgal strain named *Tetraselmis indica* was cultivated in MCC (Amit et al., 2020). In a recent research, an increase in microalgal biomass concentration from 1.2 g/L in control to 2.1 g/L was reported, when the algal culture was dosed with quorum-sensing molecules (Das et al., 2019d). Considering these investigations, biofuel production using MCCs might come up as a cost-effective and sustainable approach with the selection of optimum algal strain and growth conditions.

17.4.1.5 Biofuel production

Algal biomass is regarded as a promising substrate for biofuel generation. There are many kinds of biofuels, which include biodiesel, bioethanol, biohydrogen, and biomethane. Depending on the type of biofuel intended to synthesize, different processes of biofuel synthesis from algal biomass are employed (Fig. 17.3). Pyrolysis is one of the biofuel synthesis processes, where algal biomass can be used to synthesis biochar as well as gaseous products such as CH_4 and H_2. Pyrolysis is an endothermic thermal putrefaction process of algal biomass at elevated temperature (over 300°C) in a closed inert atmosphere to produce biochar and gaseous product. After thermal treatment, when the formed biochar is washed with diluted acid or water, the solution contains sufficient carbon source for microbial growth to produce bioethanol (Özçimen and İnan, 2015). Biomass pyrolysis is a very complex reaction mechanism, and thus to understand the process of pyrolysis, the entire pyrolysis reaction mechanism should be divided into three basic steps. Initially, conversion of biomass leads to the formation of unreacted residues and water; then in the second step, conversion of unreacted residues into primary biochar, volatile, and gaseous products takes place; finally, in the third and last step, primary biochar is converted into volatile and gaseous products (Pourkarimi et al., 2019). Condensation of volatile products obtained from pyrolysis produces biofuel (Pourkarimi et al., 2019). Pyrolysis of algal biomass provides biofuel, syngas, and biochar, which has wide practical applications.

Gasification is the thermochemical transformation of algal biomass into H_2, CO, CH_4, and other wide range of hydrocarbons. Gasification is one of the clean, reliable, and flexible processes to produce syngas (H_2+CO) in which low-value substrate is converted into high-value products. The desired yield of the gaseous product can be obtained by altering the operating conditions, such as temperature, rate of heating, catalyst used, and reaction time, as well as selecting the optimized algal strain (Raheem et al., 2018).

Hydrothermal liquefaction (HTL) is the procedure of thermochemical transformation of biomass into liquid biofuel in a hot, pressurized water atmosphere for an adequate time, leading to the breakdown of solid biopolymer mainly present in liquid components. In HTL, high moisture-containing biomass can be converted directly into liquid biofuel. Biofuel production using algal biomass by the method of HTL is considered as technically and economically viable option for biofuel production.

17.4.1.6 Biodiesel

Biodiesel from algal biomass has captured a considerable attention in the past few years, as it is considered as the third-generation biofuel.

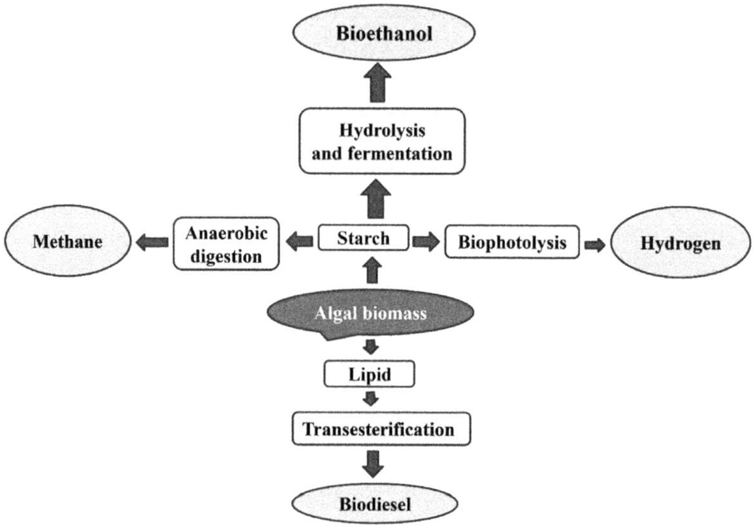

FIGURE 17.3 Process of biofuel synthesis from algal biomass.

Compared with the commercial diesel, combustion of biodiesel generates no sulfur oxides, no net emission of CO_2, less CO emission, less particulate matter, smoke, and hydrocarbon emissions. The obstacle associated with algal biodiesel commercialization is the inferior lipid content of algae. Although there are algal strains such as *Chlorella*, *Scenedesmus*, *Neochloris*, and *Nanochloropsis* in which an average lipid content of 40%–60% in dry cell mass has been reported (Gouveia and Oliveira, 2009).

Algal culture with elevated biomass yield and high lipid content is in high demand; however, cultivation of such culture in field-scale cultivation systems is a major challenge. Moreover, to improve lipid productivity, two-step culture system can be applied: (1) focus on cultivation of algal biomass with high productivity, and (2) provide optimal stress to improve lipid accumulation (Rodolfi et al., 2009). Hence, the influence of culture environment on the fatty acid composition and stress is the key factor for both biodiesel quality and quantity (Praveenkumar et al., 2012).

Pyrolysis of algal biomass obtained from MCCs can be considered as a sustainable and clean process, as generated CO_2 can be used for algal cultivation. The dry pyrolysis of algal biomass produces a light-colored liquid composed of different hydrocarbons. In an investigation, 55% algal oil (liquid phase), 35% residue, and 10% gaseous products (noncombustible gas) were obtained at 150°C; however, in the same investigation, it was found that the liquid phase yield is governed by different operating parameters such as temperature, residence time, rate of heating, algal biomass, and additives used (Kaneria and Khunt, 2018).

Biodiesel can be explained as the monoalkyl ester of fatty acids produced from algal oil, vegetable oils, animal fats, and alcohols through a catalyst by the process of transesterification (Ahmad et al., 2011). Transesterification is the chemical procedure of biodiesel synthesis, in which triacylglycerol (TAG) reacts with alcohol in the presence of catalyst to form alkayl ester and glycerol. In biodiesel synthesis, the most frequently employed alcohols are methanol and ethanol, especially methanol due to its cost-effectiveness and chemical as well as physical advantages (chemically polar and short chain). The stoichiometry of transesterification indicates that 1 mole of TAG reacts with 3 moles of alcohol and generates 3 moles of methyl ester and 1 mole of glycerol. Literature suggests that alkali-catalyzed transesterification reaction is faster as compared with acid-catalyzed; therefore, alkali catalysis is commercially preferred. However, selection of acid-catalyzed or alkali-catalyzed transesterification depends on free fatty acid concentration in TAG. If free fatty acid content is low, alkali catalysis is preferred, whereas if free fatty acid and water contents are more, acid catalysis is preferred (Özçimen and İnan, 2015).

17.4.1.7 Bioethanol

Bioethanol is a renewable fuel that is used as additive in gasoline to improve its quality. Ethanol is soluble in gasoline, and therefore, the mixture of ethanol in gasoline in all proportion provides a homogeneous mixture that can be used as fuel. Algal bioethanol has a high octane number, which is generated from the fermentation of algal polysaccharides (starch, sugar and cellulose). The carbohydrate content of microalgae differs from species to species, and in green algae, it is about 25%–50%, in red algae, 30%–60%, and in brown algae, 30%–50%. Some of the algal strains contain the highest polysaccharides, such as *Porphyra* (40%–76%), *Palmaria* (38%–74%), and *Ascophyllum* (42%–70%) (Özçimen and İnan, 2015). Fermentation of carbohydrates present in algal biomass by the action of microbes results in bioethanol production. Therefore, biomass having more carbohydrates is desirable for bioethanol production, and under specific condition, the carbohydrate content of microalgae can be increased up to 70% (Brányiková et al., 2011).

Algal fermentation in the chemical breakdown of algal biomass, due to the action of yeast, results in the generation of ethanol. The transformation of glucose to ethanol and CO_2 proceeds as stated in Eq. (17.7).

$$C_6H_{12}O_6 \rightarrow 2CH_3CH_2OH + 2CO_2 \qquad (17.7)$$

The stoichiometry of the equation indicates that 1 mole of glucose converts into 2 moles of ethanol as well as 2 moles of CO_2; however, in practice, microbes used in the fermentation reaction also utilize carbon source for their metabolic activity; therefore, 100% substrate conversion is not possible. The bioethanol yield depends on numerous other parameters such as carbohydrate content of algal cell. This carbohydrate content also depends on type of algal strain, which again under specific growth condition carbohydrate content can be increased in microalgal cell as discussed earlier. Ethanol yield also depends on the process of ethanol production employed from algal biomass. Therefore, to increase the ethanol yield, selection of proper algal strain and optimization of process for algal growth and fermentation are essential.

17.4.1.8 Biohydrogen

Hydrogen is a clean gaseous fuel, which is synthesized from algal starch by biophotolysis. There are two types of biophotolysis (direct biophotolysis and indirect biophotolysis) that is used for biohydrogen synthesis. In direct biophotolysis, the breakdown of H_2O molecule due to algal photosynthesis occurs (Sharma and Arya, 2017). This is a very fast process, and amount of biohydrogen produced is higher. The limitation associated with this process is that O_2 present in water can inhibit the activity of enzyme hydrogenase. Direct biophotolysis follows the chemical reaction as stated in Eq. (17.8).

$$H_2O \rightarrow H_2 + 1/2O_2 \qquad (17.8)$$

In indirect biophotolysis of biohydrogen production, the electrons are derived from starch molecules of microalgae. This process is a two-phase reaction, wherein in the first phase, storage of photosynthesis and carbohydrate occurs, whereas in the second phase, carbohydrate fermentation results in the generation of biohydrogen. The entire process is shown in Eqs. (17.9) and (17.10):

$$12H_2O + 6CO_2 + Light \rightarrow C_6H_{12}O_6 + O_2 \quad (17.9)$$

$$C_6H_{12}O_6 + 12H_2O + Light \rightarrow 12H_2 + 6CO_2 \quad (17.10)$$

Biohydrogen production at the rate of 12.5 mL H_2/h.g dry cell weight was stated in biophotolysis process using cyanobacterium *Anabaena variabilis* (Markov, 1997). In other investigation under biophotolysis using cyanobacterium *A. variabilis*, when pH was maintained in the range of 6.8–8.3, optimum hydrogen generation was reported (Troshina, 2002). Temperature optimization in the range of 30–40°C caused twofold escalation in the hydrogen generation (Show et al., 2018). In another research, 48.95 mol% (18.77 mol/kg biomass) of H_2 was reported in catalytic gasification of *C. vulgaris* at 851°C temperature with 16.4% (wt.%) loading of ZnO–Ni–CaO catalyst for 28.8 min of reaction time (Raheem et al., 2018). Therefore, high algal biomass production and process development for algae to H_2 generation can lead to the increase in yield.

17.4.1.9 Biomethane

Biomethane also known as natural gas is produced from anaerobic digestion of algal waste biomass after lipid extraction. Anaerobic digestion is governed by the nature of biomass, and if the biomass contains larger fractions of cellulose and hemicellulose, it may resist anaerobic digestion, and then pretreatment is required. Algal biomass could be a potential substrate for anaerobic digestion for biomethane production due to its high productivity as well as low ash content. Some algal strains do not contain cellulose and hemicellulose in their cell wall, and their cell wall is protein based. Also, some algal strains possess no cell wall and, therefore, algal strain selection in this regard is also important, as the initiation of the process will depend on the cell wall composition. For faster and efficient conversion of biomass to methane, algal strain with high productivity and low ash content is necessary. In anaerobic digestion of algal biomass by the action of microbes, the CH_4 and CO_2 produced vary in the range of 50%–70% and 30%–45%, respectively. Generally, before anaerobic digestion of biomass, pretreatment is required. The frequently applicable pretreatment types are mechanical, thermal, chemical, and biological (Milledge et al., 2019).

In a batch fermentation of algal biomass for 12–15 days of retention time under addition of volatile solids (VS_{added}), the methane yield of 350–480 mL/g VS_{added} was reported (Hansson, 1983). In a different work, methane yield was found to be 9.27 L/kg VS_{added} at 40°C temperature and 16 h of retention time (Santos et al., 2014). Therefore, considering the aforementioned outcomes of methane yield, it can be concluded that process optimization is important as methane production depends on numerous factors, such as temperature, pH, hydraulic retention time, and organic loading rate.

17.4.1.10 Other value-added by-products recovery

After lipid extraction from algal biomass, the residual part can be used for the generation of biofertilizers, bioplastic, medicinal valued product, animal feed, etc. Algal biomass from MCCs can also be used for the synthesis of biochar by thermochemical processes, such as pyrolysis. Biochar has a widespread applications, and thus, it can be employed for the development of cost-effective green electrode material that may be used as electrode in MCCs to make the system self-dependent (Chakraborty et al., 2020). Algal biomass has methanogen suppression properties; therefore, it can be used in the anodic chamber of an MCC for suppression of activity of methanogens to make the MCC systems more efficient.

Microalgae are considered as potential source of medicinal products related to health sector, and some of these products include cholesterol reducing medicines, antioxidants, and omega-3 fatty acid used as supplements as well as anti-cancerous products (Kothari et al., 2017). Some microalgae are rich in organic metabolites that are used as skin care products and protect from sun burn, sporopollenin, which is chemically inert polymer that provides skin protection from ultraviolet radiation. Scytonemin is another algal-derived secondary metabolite, yellowish brown in color that protects skin from sunburn, and it is also commercially used in beauty products (Zhang et al., 2014). Considering all these benefits and diverse applications of microalgae, it can be concluded that microalgae cultivated in MCCs could become a sustainable and cost-effective approach for the generation of multiple valuables with concurrent wastewater treatment.

17.4.2 Advantages of microbial carbon captures over other sequestration technologies

An MCC is a new and emerging carbon sequestration technology, which offers several environmental (wastewater treatment, CO_2 sequestration, lower sludge generation for disposal, heavy metal recovery), economical (low operational cost, eliminate downstream processes in wastewater treatment plants, extra revenue from simultaneous bioelectricity, biofuel production and other value-added product recovery), energy (biofuel and bioelectricity), and operational benefits (self-generation of microbes in anodic chamber as well as microalgae in cathode chamber) (Li et al, 2009, 2013). Moreover, MCCs can be operated at room temperature and provide clean electricity with concomitant CO_2 fixation in the cathodic chamber; therefore, it is considered as an environmentally friendly technology (Gude, 2016).

In MCCs, the conversion of biomass to electricity occurs in only one step with comparatively high efficiency. Sludge is generated in anodic chamber of MCCs by an anaerobic process, and thus, the volume of the sludge generated is much less as compared with the aerobic oxidation. Additionally, MCCs provide economical and operational benefits, as it can be operated in a stable way for extended time without any extra expense. Therefore, MCC can be considered as a superior technology, where, apart from organic matter oxidation, reduction of inorganic impurities present in the wastewater occurs with simultaneous bioelectricity recovery.

Moreover, it has been reported that MFC having microalgae as biocathode generates 47% of more electricity as compared with control (without biocathode) (Yadav et al., 2020). Output voltage of MCC is also affected by the concentration of the algal cells, as increase in power generation from 4.1 to 5.6 W/m^3 has been reported, when the OD of algal suspension is improved from 0.21 to 0.85 at a wavelength 658 nm (Wang et al., 2010). Therefore, the expectation for MCCs as a futuristic wastewater treatment technology with simultaneous bioelectricity and biofuel production is high; however, there is a need for further development and investigation on algal and bacterial strain, electrode material, PEMs, and different physicochemical operational parameters for achieving higher efficiency of this technology.

17.4.3 Challenges in scaling-up of microbial carbon capture and future perspective

The power generation in MCC is very low with respect to other renewable energy sources such as solar and wind energy because of its thermodynamic limitations. For the commercialization of MCCs, there is a need of optimization of numerous parameters, such as development of efficient electrode material, selection of potential microbes for the formation of appropriate anodic biofilm, selection of efficient microalgae from the perspective of downstream biofuel production, process development, and development of low-cost membranes. In a recent work, a low-cost PEM was developed using biochar originating from food waste by pyrolyzing it at the temperature of 600°C followed by the deposition of biochar on polyvinyl alcohol–based matrix after sulfonating the biochar. This fabricated PEM demonstrated ion transport number, proton conductivity, and oxygen diffusion coefficient of 0.891, 0.07 S/cm, and 6.46×10^{-9} m^2/s, respectively, and 32 times higher proton conductivity per unit cost and 7% higher COD removal efficiency were reported in MFC as compared with Nafion-based MFC (Chakraborty et al., 2020).

The thermodynamic limitation of MCCs, also known as cathodic limitations, includes ohmic loss, activation loss, mass transfer losses, and substrate diffusion through the PEM from the anodic to cathodic chamber. Power reduction (ohmic loss) caused due to the internal resistance that is unavoidable due to resistance of electrode and other interconnections, and resistance due to electrolyte as well as PEM used also restricts MCC's performance. Ohmic losses can be reduced by using electrodes with higher conductivity, by minimizing the space between anode and cathode, and by amplifying the conductivity of electrolyte up to the optimum level (Moseley, 2001).

In a chemical reaction, there is a need of activation energy, and in MCCs, the current produced largely depends on the cathodic reduction kinetics, and a share of the cathode potential is spent to overcome this activation obstacle (Moseley, 2001; Singh et al., 2010). The activation barrier can be minimized by electrode modifications with catalysts that reduce resistance and provide larger surface area for ORR, as well as optimization of operational conditions (Singh et al., 2010).

Mass transfer losses in the anodic chamber are considered as negligible, and it is the cathodic mass transfer losses that are responsible for the inferior performance of MCCs (Rismani-Yazdi et al., 2008). In the cathodic chamber of an MCC, the mass transfer is governed by the supply of oxidant (O_2) and withdrawal of the product formed (H_2O). Reactant depletion and product accumulation at the cathode surface limit mass transfer that lessens the performance of an MCC in terms of electrical output. Theoretically, PEM must permit only protons to pass through from anodic chamber to cathodic chamber; however, practically, substrate crossover, due to diffusion and electroosmosis, occurs that is responsible for potential drop in the cathodic chamber (Jiang and Chu, 2004).

The installation cost of MCCs is also one of the significant factors that hinders their field-scale application. Cost of the MCCs includes material cost (electrode, membrane, catalyst, capacitor and circuit, current collector, and reactor material) and labor cost. For an effective field-scale application of MCCs, there is a need of cost reduction, which can be only possible when low-cost materials are employed for their fabrication. Moreover, for higher revenue generation, there is the need to integrate different systems in conjugation, such that the by-product of the first system might be feedstock for the other system, and in this way, resource recovery from waste is possible in a systematic manner.

Algal growth rate in the cathodic chamber as well as biofilm development on the anode regulates the performance of an MCC; therefore, selection of algal and microbe species is a very important factor and should be decided wisely. Algal growth is affected by numerous physiological parameters, such as temperature, light intensity, DO concentration, pH, CO_2 concentration, and cation concentration, which can hamper the photosynthesis; hence, optimization of these parameters is imperative. Microalgae harvested from MCCs provide additional benefits in the form of biofuel, biochar, and other valuable products recovery. Though limitations are associated with MCC system, these can be reduced by following appropriate scientific approach as discussed earlier. Due to diverse applications of MCC, it has sustainability and environmental benefits, and thus, can be considered as a futuristic technology.

17.5 Technoeconomic analysis, environmental impact, and life cycle assessment of microbial electrochemical technology

17.5.1 Technoeconomic assessment

Technoeconomic (TE) analysis is carried out to explore the economic and technical efficacy of any process or product, and it is an essential parameter elucidating the practical applicability of the process. The process of TE analysis includes the cost of feedstock collection, manufacturing and equipment cost, and operational cost (Kumar et al., 2020). Moreover, the major steps involved in TE analysis are market study, creation of process flow diagram, energy balance, economic evaluation, and sensitivity analysis (Fig. 17.4), which portraits a vibrant picture of the technoeconomic feasibility of the method (Mu et al., 2020).

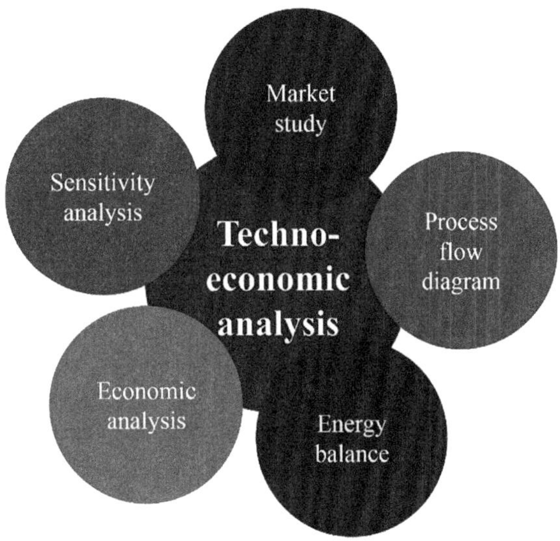

FIGURE 17.4 Methodology followed during a technoeconomic analysis.

The TE analysis on MFC and MCC is not available in the literature, and therefore, the same could not be discussed; however, to understand the TE feasibility of METs, the limited TE analyses available on MES and MEC are presented here. In an exploration, TE analysis for the synthesis of five compounds, namely acetic acid, propionic acid, formic acid, methanol, and ethanol, was explored, and it was observed that the synthesis of methanol, propionic acid, and acetic acid demonstrated no profit in 15 years of operation due to higher production cost and inferior revenue generation (Li et al., 2019). For instance, the production cost of acetic acid employing renewable energy is 3.84 £/kg, and its market price is 0.48 £/kg; thereby, no net positive revenue is generated, and the capital cost could not be paid back (Li et al., 2019). However, ethanol and formic acid have an estimated payback period of 15 years because the synthesis costs using renewable energy and market price of ethanol are 0.88 and 1.09 £/kg, respectively, whereas the synthesis costs using renewable energy and market price of formic acid are 0.30 and 0.38 £/kg, respectively. Hence, in case of formic acid and ethanol, positive revenue was generated; as a result, a payback period of 15 years was projected. Furthermore, the discounted cash flow rates of return of the MES setups for the production of formic acid and ethanol were 21% and 14%, respectively. This indicates that both of these are more economically attractive products compared with the 11.6% return for industrial-scale production of these compounds from different chemicals abiotically.

Cusick et al. (2010) performed an economic analysis for the synthesis of hydrogen gas employing MEC, and estimated \$4.51/kg of H_2

and \$3.01/kg of H_2 are the hydrogen production cost from winery and domestic wastewater, respectively (Cusick et al., 2010). In addition, this cost of hydrogen production from wastewater was less compared with the market cost of hydrogen gas (\$6 kg/of H_2). Thus, MEC can be envisioned as a profitable technology for the production of hydrogen. In a recent investigation, economic analysis of ammonia recovery from urine via MET was performed, demonstrating the application of MET for urine treatment was less energy intensive, when compared with other recovery technologies such as freeze thawing, stripping, and electrodialysis (Rodríguez Arredondo et al., 2015; Tice and Kim, 2014).

In an investigation, TE analysis of a dual-chambered MEC for concomitant wastewater remediation in anodic chamber and the production of formic acid (HCOOH) in cathodic chamber were explored (Shemfe et al., 2018). Moreover, the influence of change in COD concentration and applied potential on the production cost was also evaluated. It was observed that the production cost of HCOOH was € 0.015−€ 0.005 per g for the COD removal of 0.038−0.106 kg/yr and HCOOH production of 0.094−0.26 kg/yr (Shemfe et al., 2018). Moreover, when the initial COD of the anolyte was amplified from 0.25 to 1 g/L, production cost of HCOOH decreased from € 0.08 to € 0.015 per g, whereas while changing applied cell potential to 0.75 from 0.90 V, diminution in the production cost from € 0.46 to € 0.02 per g was observed due to the increase in the productivity of HCOOH. Thus, applied voltage has more influence on the production cost than anolyte COD. However, both applied voltage and initial COD should be simultaneously optimized to reduce the production cost. In addition, the limitations of the METs, such as MFC, are high cost of membrane and electrodes. However, this can be circumnavigated by hybridizing conventional wastewater treatment technologies with MFC fabricated by using low cost materials, which would assist in providing feasible solution in terms of economic sustainability to the technology. Thus, to overcome the economic barriers, the construction cost of the microbial electrochemical technology systems must be kept as low as possible.

17.5.2 Impact of microbial electrochemical technology on the environment

The main features of MET are the use of electrodes as electron donor or acceptor for the degradation of organic matter; as a result, very small external energy and no extra chemical addition are required. In addition to this, energy is harvested from these systems for onsite usage. This is beneficial compared with conventional physical, chemical, or even biological treatment methods because these processes require addition of external chemicals and consume a large amount of nonrenewable energy (Wang et al., 2015). Thereby, the use of METs can save a lot of energy, and the process could last for decades for the remediation of pollutants from the wastewater. On the other hand, MET can capture CO_2 via algal sequestration or bioelectrochemical reduction at the cathode, and this is beneficial because CO_2 is primarily a greenhouse gas and majorly contributes to climate change. Hence, MET can use CO_2 as feedstock and simultaneously convert it into valuable products such as biofuels, thereby, decreasing the potential menace of global warming.

Additionally, nutrients such as N and P need to be eliminated from wastewater, as these can cause pollution in streams and the condition of streams can result in eutrophication. The recovery of nutrients in the form of struvite can help in converting these pollutants into value-added products, which can be utilized as an alternative to chemical fertilizers for agriculture practices (Ahmad and Gupta, 2019). Traditional nitrification−denitrification for nitrogen removal has disadvantages of requiring energy for aeration in the first step and requiring external organic substrate for denitrification in the second step (Wang et al., 2015). The mechanism involved in MET for nitrogen recovery is to convert ammonium to ammonia gas (95%) at an elevated pH of about 10.5 and then strip out the ammonia for its recovery in acidic solution (Wu and Modin, 2013). Hence, nitrogen recovery using MET can save energy for excess nutrient recovery from contaminated water, and as a result, it can minimize the chances of eutrophication. Thus, it can be concluded that using MET for wastewater treatment could help in overcoming the limitations of conventional systems along with minimizing the impacts on the environment.

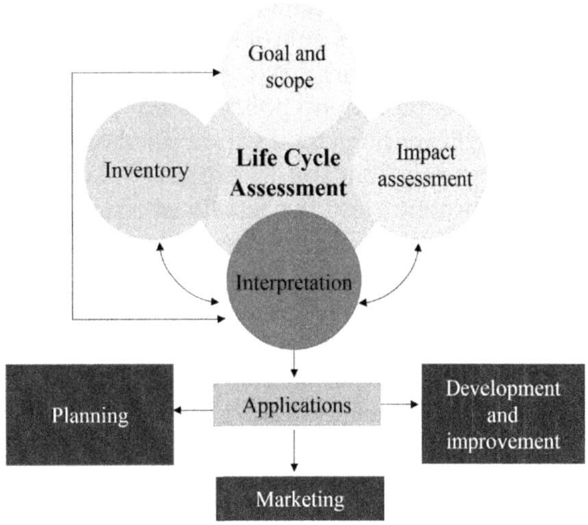

FIGURE 17.5 Methodology of life cycle analysis.

17.5.3 Life cycle analysis

The scientific technique of life cycle assessment (LCA) evaluates numerous implications of a process or product on the environment, human health, and material utilization related to the products' life, operation, and other associated activities (Morales et al., 2015). The LCA is a cradle-to-grave methodology employed to assemble and analyze the inputs, outputs, and possible environmental effects through all the stages of a system, and it is the most preferred tool to convince the end users for the potential applications of METs for wastewater remediation, bioenergy generation, and CO_2 capture (Jadhav et al., 2018). The LCA includes defining the goal and scope of the investigation, life cycle inventory analysis, evaluation of different impacts associated with the process, and the interpretation of results (Fig. 17.5). In METs, major input includes wastewater from sewage collection system or industrial effluents, electricity used in pumping, mixing, and other chemicals, if used. The major outputs are bioenergy, electricity, and various gas emissions. For the LCA of METs, the environmental effects of all the inflows and outflows are estimated, and based on it, the overall environmental impact of MET as a process is determined.

The comparative LCA of MFC and microbial electrolysis cell (MEC) showed that MFC alone does not deliver noteworthy environmental advantage as compared with other conventional anaerobic wastewater treatment options; however, MEC offers substantial environmental benefits, such as the reduction in greenhouse gas releases as well as other ecologically toxic emissions (e.g., aromatic hydrocarbons) (Foley et al., 2010). The comparative LCA analysis was carried out for conventional constructed wetlands (CW) system (without MFC implementation) and a CW system attached with an MFC fabricated with gravel-based anode and another CW system joined with MFC comprising a graphite-based anode. In this investigation, the CW system attached with graphite-based anode MFC emerged as the utmost ecofriendly option, which can substitute traditional CWs due to the 20% reduction in the land requirements (Pant et al., 2011). However, the arrangement was 1.5 times more costly than the traditional CW system (Corbella et al., 2017). Therefore, to make

MFC feasible, investigation on cheaper materials for different components of MFC is necessary for implementation with CW system.

In an exploration, policy-based LCA with Thai Eco Scarcity method was applied on biofuel production in Thailand to explore the environmental benefits, and the results were compared with conventional energy sources (Lecksiwilai and Gheewala, 2020). When the system boundaries had only three impact categories, which included freshwater consumption, land use pattern, and GHG emissions, the environmental profile for biofuels exhibited better results than fossil fuels. This can be attributed to the involvement of very few impact categories for the production of biofuel. When 16 impact categories were included, a significant variation in the result was observed. It was noticed that for B100s (pure biodiesel), negative impact score on environment was 15% higher than diesel (Lecksiwilai and Gheewala, 2020). The increase in the impacts of biofuel is due to the significant contribution of land use pattern, pesticides in soil, and particulate matter discharges compared with the fossil fuel. Hence, a detailed investigation of the ecological impacts can provide clear picture on the effects of biodiesel production in comparison with its fossil fuels. However, when only GHG impact was evaluated for B100, only 10% contribution was observed for B100 as compared with 76% for fossil fuels, thus suggesting, when only GHGs emission was considered, B100 is clearly better than diesel (Lecksiwilai and Gheewala, 2020).

In another demonstration, comparative LCA of microalgae, *Scenedesmus dimorphus*, through autotrophic cultivation in raceway pond and its transformation to biogas and biofuel was carried out (Mediboyina et al., 2020). Moreover, the influence of nutrient addition and reusing it with respect to energy requirement and environmental impacts was also explored. The relative percentage contribution of three impact categories, acidification potential (AP in kg SO_2 eq.), eutrophication potential (EP in kg phosphate eq.), and freshwater aquatic ecotoxicity potential (FAETP in kg DCB eq.) was calculated by the standardization of each impact on different situations. It was observed that in all the scenarios, raceway pond is the chief offender in AP and FAETP; thereby, open raceway ponds have a higher negative impact on the surroundings and dominated the process steps. However, the downstream process steps associated with biodiesel production scheme demanded about 91% of the primary energy, which was higher than the biogas production scheme. The significant decrease in principal energy demand and ecological impacts is because of recycling the nutrient-laden water into the open raceway pond. Moreover, it was also observed from LCA that when microalgae were used for biogas and biofuel production, the biogas conversion pathway was more sustainable and energy efficient than the biodiesel pathway. Furthermore, the overall exploration showed that the generation of biogas and biodiesel through microalgae-mediated CO_2 sequestration pathway can be energy efficient, when biomass productivity was higher ($>10\ g/m^2$.day) and energy efficient systems were installed for the mixing the cultures (specific electrical energy consumption of $<2\ W/m^3$) (Mediboyina et al., 2020). When cultivation of algae is practiced in cathodic chamber of MFCs, which provides the extra benefits as providing biomass that can be used for biofuel generation as well as other valuable recovery that results in rendering more sustainability to the technology. Hence, the METs are efficient in producing sustainable energy and numerous other valuable products from wastewater supplemented with the simultaneous pollutant removal. Thus, METs can become a better alternative to replace energy-intensive wastewater treatment processes.

17.6 Challenges and future perspectives

Providing secure, cheap, renewable, and clean energy is one of the major obstacles of the 21st century, as due to the proliferation in the world population and rapid development in the economic sector, it has significantly led to the rise in the energy demand (Tsita et al., 2020). Hence, converting waste into clean and renewable energy such as bioenergy can alleviate the global environmental problems. However, there are challenges to such conversion, such as commercialization of bioenergy and biofuel production, inferior yield of these processes, and economic acceptance of the process (Moustakas et al., 2020). On the other hand, METs such as MFC and MCCs are efficient for harvesting green electricity from wastewater along with the synthesis of a wide range of valuables such as biofuels, whereas the traditional treatment technologies such as activated sludge process require mechanical aeration, which consume nonrenewable energy and volatize some percentage of COD into atmosphere, which as a result contributes to GHG emission and global warming (Pant et al., 2012). Thus, METs possess enormous potential to replace conventional technologies for concurrent wastewater remediation and bioenergy generation.

The electrodes and catalysts employed for the fabrication of METs require high conductivity, porosity, biochemical stability, and superior electrochemical properties. To achieve these properties, electrocatalysts, such as platinum, are required over carbon-based electrode, which are expensive and thus limit the real-life application of this innovative technology. Furthermore, the membranes used in METs face numerous challenges, such as loss of substrates, biofouling, and reduced permeability of protons; as a result, periodic replacement of membranes is required for efficient contaminant removal and energy production (Ivase et al., 2020). To overcome this, low-cost membranes can be used as PEM instead of expensive polymer-based membranes (Sophia Ayyappan et al., 2018). Hence, more research should be carried out for developing low-cost membranes and electrocatalysts for their application in METs, which can be further employed for the scaling up. The rise in the waste to energy sector has been affected by numerous economic, technological, and political obstacles such as the dearth of national rules and regulations, limited funds, and collection of insufficient data and related assessment. Hence, analyzing and discussing these limitations can aid the nation for the further introduction of such technologies, especially when the development of this sector is essential for the growth of a nation (Khajuria et al., 2010).

The practice of producing biodiesel from microalgae has been investigated for about past 60 years. Using algae for bioenergy production helps in reducing production time and demands minimum freshwater requirement. However, only limited investigations have demonstrated pilot-scale application of this technology, thus prohibiting its commercialization (Su et al., 2017). The foremost drawback pertaining to biodiesel generation from microalgae is the insufficient biomass concentration and yield with low volatile solids, which is linked with the high cost of harvesting and dewatering. Moreover, the distortion of algal cells is a challenge for the successful generation of valuables from algal biomass via the biorefinery technique (Günerken et al., 2015). These limitations were existing since the first exploration; however, only a few researchers have focused on these engineering bottlenecks (Chen et al., 2011; Zabed et al., 2020). Hence, more investigations on improving the energy harvesting techniques and hybridizing biomass production technologies with algal biorefineries or wastewater treatment should be carried out (Elshobary et al., 2019).

17.7 Conclusion

Wastewater treatment and bioelectricity production are considered as the chief potential applications of METs; however, concurrent CO_2 sequestration, biofuel, and other valuable production can also be accomplished though METs. Technologies such as MFCs and MCCs can be installed as additional bioenergy source to overcome the energy associated problems in the rural sector. However, the power generation by individual MFCs or METs is very low for practical application, and therefore for its successful application, yield of METs needs to be increased considerably. Though METs are among the upcoming technologies for wastewater treatment due to well-documented applications in multiple domains, numerous bottlenecks associated with this technology need to be overcome prior to its successful commercialization. Therefore, vast research scope on METs is available, which would be a major stride toward the efficacious real-life implementations of this innovative technology.

References

Ahmad, A., Das, S., Ghangrekar, M.M., 2020. Removal of xenobiotics from wastewater by electrocoagulation: a mini-review. J. Indian Chem. Soc. 97, 493–500.

Ahmad, A., Gupta, A., 2019. Urine a source for struvite: nutrient recovery – a review. J. Indian Chem. Soc. 96, 507–514.

Ahmad, A.L., Yasin, N.H.M., Derek, C.J.C., Lim, J.K., 2011. Microalgae as a sustainable energy source for biodiesel production: a review. Renew. Sustain. Energy Rev. 15, 584–593.

Amit, Nayak, J.K., Ghosh, U.K., 2020. Microalgal remediation of anaerobic pretreated pharmaceutical wastewater for sustainable biodiesel production and electricity generation. J. Water Process Eng. 35, 101192.

Ayyaru, S., Mahalingam, S., Ahn, Y.-H., 2019. A non-noble V_2O_5 nanorods as an alternative cathode catalyst for microbial fuel cell applications. Int. J. Hydrogen Energy 44, 4974–4984.

Bhowmick, G.D., Das, S., Adhikary, K., Ghangrekar, M.M., Mitra, A., 2019a. Using rhodium as a cathode catalyst for enhancing performance of microbial fuel cell. Int. J. Hydrogen Energy 44, 22218–22222.

Bhowmick, G.D., Das, S., Verma, H.K., Neethu, B., Ghangrekar, M.M., 2019b. Improved performance of microbial fuel cell by using conductive ink printed cathode containing Co_3O_4 or Fe_3O_4. Electrochim. Acta 310, 173–183.

Bose, D., Rawat, R., Bahuguna, R., Vijay, P., Gopinath, M., 2020. Sustainable approach to wastewater treatment and bioelectricity generation using microbial fuel cells. Curr. Dev. Biotechnol. Bioeng. 37–50.

Brányiková, I., Maršálková, B., Doucha, J., Brányik, T., Bišová, K., Zachleder, V., Vítová, M., 2011. Microalgae-novel highly efficient starch producers. Biotechnol. Bioeng. 108, 766–776.

Brennan, L., Owende, P., 2010. Biofuels from microalgae-a review of technologies for production, processing, and extractions of biofuels and co-products. Renew. Sustain. Energy Rev. 14, 557–577.

Chakraborty, I., Das, S., Dubey, B.K., Ghangrekar, M.M., 2020. Novel low cost proton exchange membrane made from sulphonated biochar for application in microbial fuel cells. Mater. Chem. Phys. 239, 122025.

Chandraserkharan Meenu, P., Sreelekshmy, B.R., Basheer, R., Sadasivan, S.M., Vijayakumari Ramakrishnan, R.M., Shibli, S.M.A., 2018. Development of a high-performance mediatorless microbial fuel cell comprising a catalytic steel anode. ACS Appl. Bio Mater. 1, 1124–1133.

Chen, C.-Y., Yeh, K.-L., Aisyah, R., Lee, D.-J., Chang, J.-S., 2011. Cultivation, photobioreactor design and harvesting of microalgae for biodiesel production: a critical review. Bioresour. Technol. 102, 71–81.

Chen, S., Patil, S.A., Brown, R.K., Schröder, U., 2019. Strategies for optimizing the power output of microbial fuel cells: transitioning from fundamental studies to practical implementation. Appl. Energy 233–234, 15–28.

Chisti, Y., 2008. Biodiesel from microalgae beats bioethanol. Trends Biotechnol. 26, 126–131.

Corbella, C., Puigagut, J., Garfí, M., 2017. Life cycle assessment of constructed wetland systems for wastewater treatment coupled with microbial fuel cells. Sci. Total Environ. 584–585, 355–362.

Cusick, R.D., Kiely, P.D., Logan, B.E., 2010. A monetary comparison of energy recovered from microbial fuel cells and microbial electrolysis cells fed winery or domestic wastewaters. Int. J. Hydrogen Energy 35, 8855–8861.

Das, I., Das, S., Agrawal, A., Ghangrekar, M.M., 2019a. Synthesis and application of novel low cost $Cu_{0.5}Mn_{0.5}Fe_2O_4$ spinel as cathode catalyst in microbial fuel cell. Abstr. Pap. Am. Chem. Soc. 258, 1155.

Das, I., Das, S., Chakraborty, I., Ghangrekar, M.M., 2019b. Bio-refractory pollutant removal using microbial electrochemical technologies: a short review. J. Indian Chem. Soc. 96, 493–497.

Das, I., Das, S., Dixit, R., Ghangrekar, M.M., 2020a. Goethite supplemented natural clay ceramic as an alternative proton exchange membrane and its application in microbial fuel cell. Ionics (Kiel). 26, 3061–3072.

Das, I., Das, S., Ghangrekar, M.M., 2020b. Application of bimetallic low-cost CuZn as oxygen reduction cathode catalyst in lab-scale and field-scale microbial fuel cell. Chem. Phys. Lett. 751, 137536.

Das, I., Das, S., Sharma, S., Ghangrekar, M.M., 2020c. Ameliorated performance of a microbial fuel cell operated with an alkali pre-treated clayware ceramic membrane. Int. J. Hydrogen Energy 45, 16787–16798.

Das, S., 2017. Resource Recovery from Wastewater Using Bioelectrochemical System.

Das, S., Chakraborty, I., Rajesh, P.P., Ghangrekar, M.M., 2020d. Performance evaluation of microbial fuel cell operated with Pd or MnO_2 as cathode catalyst and *Chaetoceros* pretreated anodic inoculum. J. Hazardous, Toxic Radioact. Waste 24, 04020009.

Das, S., Chatterjee, P., Ghangrekar, M.M., 2018. Increasing methane content in biogas and simultaneous value added product recovery using microbial electrosynthesis. Water Sci. Technol. 77, 1293–1302.

Das, S., Das, I., Ghangrekar, M.M., 2020e. Role of applied potential on microbial electrosynthesis of organic compounds through carbon dioxide sequestration. J. Environ. Chem. Eng. 8, 104028.

Das, S., Das, S., Das, I., Ghangrekar, M.M., 2019c. Application of bioelectrochemical systems for carbon dioxide sequestration and concomitant valuable recovery: a review. Mater. Sci. Energy Technol. 2, 687–696.

Das, S., Das, S., Ghangrekar, M.M., 2019d. Quorum-sensing mediated signals: a promising multi-functional modulators for separately enhancing algal yield and power generation in microbial fuel cell. Bioresour. Technol. 294, 122138.

Das, S., Ghangrekar, M.M., 2020. Tungsten oxide as electrocatalyst for improved power generation and wastewater treatment in microbial fuel cell. Environ. Technol. 41, 2546–2553.

Das, S., Ghangrekar, M.M., 2018. Value added product recovery and carbon dioxide sequestration from biogas using microbial electrosynthesis. Indian J. Exp. Biol. 56, 470–478.

Das, S., Mishra, A., Ghangrekar, M.M., 2020f. Production of hydrogen peroxide using various metal-based catalysts in electrochemical and bioelectrochemical systems: mini review. J. Hazardous Toxic Radioact. Waste 24, 06020001.

Das, S., Mishra, A., Ghangrekar, M.M., 2020g. Concomitant production of bioelectricity and hydrogen peroxide leading to the holistic treatment of wastewater in microbial fuel cell. Chem. Phys. Lett. 759, 137986.

Demirbas, A., 2011. Waste management, waste resource facilities and waste conversion processes. Energy Convers. Manag. 52, 1280–1287.

Elshobary, M.E., Abo-Shady, A.M., Khairy, H.M., Essa, D., Zabed, H.M., Qi, X., Abomohra, A.E.F., 2019. Influence of nutrient supplementation and starvation conditions on the biomass and lipid productivities of Micractinium reisseri grown in wastewater for biodiesel production. J. Environ. Manage. 250, 109529.

Foley, J.M., Rozendal, R.A., Hertle, C.K., Lant, P.A., Rabaey, K., 2010. Life cycle assessment of high-rate anaerobic treatment, microbial fuel cells, and microbial electrolysis cells. Environ. Sci. Technol. 44, 3629–3637.

Freguia, S., Logrieco, M.E., Monetti, J., Ledezma, P., Virdis, B., Tsujimura, S., 2019. Self-powered bioelectrochemical nutrient recovery for fertilizer generation from human urine. Sustain 11, 5490.

Gajda, I., Greenman, J., Ieropoulos, I.A., 2018. Recent advancements in real-world microbial fuel cell applications. Curr. Opin. Electrochem. 11, 78–83.

Ghangrekar, M.M., Das, S., Tiwari, B.R., 2020. Integration of bioelectrochemical systems with other existing wastewater treatment processes. Integr. Microb. Fuel Cells Wastewater Treat. 229–248.

Ghorai, A., Roy, S., Das, S., Komber, H., Ghangrekar, M.M., Voit, B., Banerjee, S., 2020. Chemically stable sulfonated polytriazoles containing trifluoromethyl and phosphine oxide moieties for proton exchange membranes. ACS Appl. Polym. Mater. 2, 2967–2979.

Gil, G.C., Chang, I.S., Kim, B.H., Kim, M., Jang, J.K., Park, H.S., Kim, H.J., 2003. Operational parameters affecting the performance of a mediator-less microbial fuel cell. Biosens. Bioelectron. 18, 327–334.

Gnana kumar, G., Kirubaharan, C.J., Udhayakumar, S., Karthikeyan, C., Nahm, K.S., 2014. Conductive polymer/graphene supported platinum nanoparticles as anode catalysts for the extended power generation of microbial fuel cells. Ind. Eng. Chem. Res. 53, 16883–16893.

Gouveia, L., Oliveira, A.C., 2009. Microalgae as a raw material for biofuels production. J. Ind. Microbiol. Biotechnol. 36, 269–274.

Gude, V.G., 2016. Wastewater treatment in microbial fuel cells – an overview. J. Clean. Prod. 122, 287–307.

Günerken, E., D'Hondt, E., Eppink, M.H.M., Garcia-Gonzalez, L., Elst, K., Wijffels, R.H., 2015. Cell disruption for microalgae biorefineries. Biotechnol. Adv. 33, 243–260.

Guo, X., Jia, J., Dong, H., Wang, Q., Xu, T., Fu, B., Ran, R., Liang, P., Huang, X., Zhang, X., 2019. Hydrothermal synthesis of Fe Mn bimetallic nanocatalysts as high-efficiency cathode catalysts for microbial fuel cells. J. Power Sources 414, 444–452.

Gupta, A., Das, S., Ghangrekar, M.M., 2020. Optimal cathodic imposed potential and appropriate catalyst for the synthesis of hydrogen peroxide in microbial electrolysis cell. Chem. Phys. Lett. 754, 137690.

Han, T.H., Parveen, N., Shim, J.H., Nguyen, A.T.N., Mahato, N., Cho, M.H., 2018. Ternary composite of polyaniline graphene and TiO_2 as a bifunctional catalyst to enhance the performance of both the bioanode and cathode of a microbial fuel cell. Ind. Eng. Chem. Res. 57, 6705–6713.

Hansson, G., 1983. Methane production from marine, green macro-algae. Resour. Conserv. 8, 185–194.

Hu, X., Liu, B., Zhou, J., Jin, R., Qiao, S., Liu, G., 2015. CO_2 fixation, lipid production, and power generation by a novel air-lift-type microbial carbon capture cell system. Environ. Sci. Technol. 49, 10710–10717.

Islam, M.A., Karim, A., Mishra, P., Dubowski, J.J., Yousuf, A., Sarmin, S., Khan, M.M.R., 2020. Microbial synergistic interactions enhanced power generation in co-culture driven microbial fuel cell. Sci. Total Environ. 738, 140138.

Ivase, T.J.P., Nyakuma, B.B., Oladokun, O., Abu, P.T., Hassan, M.N., 2020. Review of the principal mechanisms, prospects, and challenges of bioelectrochemical systems. Environ. Prog. Sustain. Energy 39, 13298.

Jadhav, D.A., Ghangrekar, M.M., Duteanu, N., 2018. Recent progress towards scaling up of MFCs. Microb. Fuel Cell 443–457.

Jadhav, D.A., Jain, S.C., Ghangrekar, M.M., 2017. Simultaneous wastewater treatment, algal biomass production and electricity generation in clayware microbial carbon capture cells. Appl. Biochem. Biotechnol. 183, 1076–1092.

Jadhav, D.A., Neethu, B., Ghangrekar, M.M., 2019. Microbial carbon capture cell: advanced bio-electrochemical system for wastewater treatment, electricity generation and algal biomass production. Appl. Microalgae Wastewater Treat. 317–338.

Jiang, R., Chu, D., 2004. Comparative studies of methanol crossover and cell performance for a DMFC. J. Electrochem. Soc. 151, A69.

Jin, X., Yang, N., Liu, Y., Guo, F., Liu, H., 2020. Bifunctional cathode using a biofilm and Pt/C catalyst for simultaneous electricity generation and nitrification in microbial fuel cells. Bioresour. Technol. 306, 123120.

Kaneria, K., Khunt, R., 2018. Production of algal oil from algae by dry pyrolysis. Int. J. Pet. Sci. Technol. 12, 1–4.

Khajuria, A., Matsui, T., Machimura, T., Morioka, T., 2010. Assessment of the challenge of sustainable recycling of municipal solid waste management in India. Int. J. Environ. Technol. Manag. 13, 171–187.

Kothari, R., Pandey, A., Ahmad, S., Kumar, A., Pathak, V.V., Tyagi, V.V., 2017. Microalgal cultivation for value-added products: a critical enviro-economical assessment. 3 Biotech 7, 243.

Kumar, A., Saini, K., Bhaskar, T., 2020. Hydochar and biochar: production, physicochemical properties and techno-economic analysis. Bioresour. Technol. 310, 123442.

Lam, M.K., Lee, K.T., 2012. Microalgae biofuels: a critical review of issues, problems and the way forward. Biotechnol. Adv. 30, 673–690.

Lecksiwilai, N., Gheewala, S.H., 2020. Life cycle assessment of biofuels in Thailand: implications of environmental trade-offs for policy decisions. Sustain. Prod. Consum. 22, 177–185.

Li, F., Sharma, Y., Lei, Y., Li, B., Zhou, Q., 2010. Microbial fuel cells: the effects of configurations, electrolyte solutions, and electrode materials on power generation. Appl. Biochem. Biotechnol. 160, 168–181.

Li, M., Zhou, M., Tan, C., Tian, X., 2019. Enhancement of CO2 biofixation and bioenergy generation using a novel airlift type photosynthetic microbial fuel cell. Bioresour. Technol. 272, 501–509.

Li, W.-W., Yu, H.-Q., He, Z., 2013. Towards sustainable wastewater treatment by using microbial fuel cells-centered technologies. Energy Environ. Sci. 7, 911–924.

Li, Y., Lu, A., Ding, H., Jin, S., Yan, Y., Wang, C., Zen, C., Wang, X., 2009. Cr(VI) reduction at rutile-catalyzed cathode in microbial fuel cells. Electrochem. Commun. 11, 1496–1499.

Logan, B.E., Hamelers, B., Rozendal, R., Schröder, U., Keller, J., Freguia, S., Aelterman, P., Verstraete, W., Rabaey, K., 2006. Microbial fuel cells: methodology and technology. Environ. Sci. Technol. 40, 5181–5192.

Logan, B.E., Regan, J.M., 2006. Microbial fuel cells—challenges and applications. Environ. Sci. Technol. 40, 5172–5180.

Markov, S., 1997. Photoproduction of hydrogen by cyanobacteria under partial vacuum in batch culture or in a photobioreactor. Int. J. Hydrogen Energy 22, 521–524.

Mediboyina, M.K., Banuvalli, B.K., Chauhan, V.S., Mudliar, S.N., 2020. Comparative life cycle assessment of autotrophic cultivation of Scenedesmus dimorphus in raceway pond coupled to biodiesel and biogas production. Bioprocess Biosyst. Eng. 43, 233–247.

Mehdinia, A., Ziaei, E., Jabbari, A., 2014. Multi-walled carbon nanotube/SnO2 nanocomposite: a novel anode material for microbial fuel cells. Electrochim. Acta 130, 512–518.

Milledge, J., Nielsen, B., Maneein, S., Harvey, P., 2019. A brief review of anaerobic digestion of algae for bioenergy. Energies 12, 1166.

Mohan, S.V., Sravan, J.S., Butti, S.K., Krishna, K.V., Modestra, J.A., Velvizhi, G., Kumar, A.N., Varjani, S., Pandey, A., 2019. Sustainable platform. Biomass, Biofuels Biochem. 3–18.

Morales, M., Quintero, J., Conejeros, R., Aroca, G., 2015. Life cycle assessment of lignocellulosic bioethanol: environmental impacts and energy balance. Renew. Sustain. Energy Rev. 42, 1349–1361.

Moseley, P.T., 2001. Fuel cell systems explained. J. Power Sources 93, 285.

Moustakas, K., Loizidou, M., Rehan, M., Nizami, A.S., 2020. A review of recent developments in renewable and sustainable energy systems: key challenges and future perspective. Renew. Sustain. Energy Rev. 119.

Mu, D., Xin, C., Zhou, W., 2020. Life cycle assessment and techno-economic analysis of algal biofuel production. Microalgae Cultiv. Biofuels Prod. 281–292.

Nagendranatha Reddy, C., Nguyen, H.T.H., Noori, M.T., Min, B., 2019. Potential applications of algae in the cathode of microbial fuel cells for enhanced electricity generation with simultaneous nutrient removal and algae biorefinery: current status and future perspectives. Bioresour. Technol. 292, 122010.

Neethu, B., Bhowmick, G.D., Ghangrekar, M.M., 2018. Enhancement of bioelectricity generation and algal productivity in microbial carbon-capture cell using low cost coconut shell as membrane separator. Biochem. Eng. J. 133, 205–213.

Nigam, P.S., Singh, A., 2011. Production of liquid biofuels from renewable resources. Prog. Energy Combust. Sci. 37, 52–68.

Nourbakhsh, F., Pazouki, M., Mohsennia, M., 2017. Impact of modified electrodes on boosting power density of microbial fuel cell for effective domestic wastewater treatment: a case study of Tehran. J. Fuel Chem. Technol. 45, 871–879.

Owusu, P.A., Asumadu-Sarkodie, S., 2016. A review of renewable energy sources, sustainability issues and climate change mitigation. Cogent. Eng. 3, 1–14.

Özçimen, D., İnan, B., 2015. An Overview of Bioethanol Production From Algae. Biofuels - Status and Perspective. IntechOpen, London, UK.

Pandit, S., Nayak, B.K., Das, D., 2012. Microbial carbon capture cell using cyanobacteria for simultaneous power generation, carbon dioxide sequestration and wastewater treatment. Bioresour. Technol. 107, 97–102.

Pant, D., Singh, A., Van Bogaert, G., Gallego, Y.A., Diels, L., Vanbroekhoven, K., 2011. An introduction to the life cycle assessment (LCA) of bioelectrochemical systems (BES) for sustainable energy and product generation: relevance and key aspects. Renew. Sustain. Energy Rev. 15, 1305–1313.

Pant, D., Singh, A., Van Bogaert, G., Irving Olsen, S., Singh Nigam, P., Diels, L., Vanbroekhoven, K., 2012. Bioelectrochemical systems (BES) for sustainable energy production and product recovery from organic wastes and industrial wastewaters. RSC Adv. 2, 1248–1263.

Pourkarimi, S., Hallajisani, A., Alizadehdakhel, A., Nouralishahi, A., 2019. Biofuel production through micro- and macroalgae pyrolysis – a review of pyrolysis methods and process parameters. J. Anal. Appl. Pyrolysis 142, 104599.

Praveenkumar, R., Shameera, K., Mahalakshmi, G., Akbarsha, M.A., Thajuddin, N., 2012. Influence of nutrient deprivations on lipid accumulation in a dominant indigenous microalga Chlorella sp., BUM11008: evaluation for biodiesel production. Biomass Bioenergy 37, 60–66.

Pulz, O., 2001. Photobioreactors: production systems for phototrophic microorganisms. Appl. Microbiol. Biotechnol. 57, 287–293.

Raad, N.K., Farrokhi, F., Mousavi, S.A., Darvishi, P., Mahmoudi, A., 2020. Simultaneous power generation and sewage sludge stabilization using an air cathode-MFCs. Biomass Bioenergy 140, 105642.

Raheem, A., Ji, G., Memon, A., Sivasangar, S., Wang, W., Zhao, M., Taufiq-Yap, Y.H., 2018. Catalytic gasification of algal biomass for hydrogen-rich gas production: parametric optimization via central composite design. Energy Convers. Manag. 158, 235–245.

Ramírez-Vargas, C., Prado, A., Arias, C., Carvalho, P., Esteve-Núñez, A., Brix, H., 2018. Microbial electrochemical technologies for wastewater treatment: principles and evolution from microbial fuel cells to bioelectrochemical-based constructed wetlands. Water 10, 1128.

Rathour, R., Patel, D., Shaikh, S., Desai, C., 2019. Eco-electrogenic treatment of dyestuff wastewater using constructed wetland-microbial fuel cell system with an evaluation of electrode-enriched microbial community structures. Bioresour. Technol. 285, 121349.

Richter, H., McCarthy, K., Nevin, K.P., Johnson, J.P., Rotello, V.M., Lovley, D.R., 2008. Electricity generation by Geobacter sulfurreducens attached to gold electrodes. Langmuir 24, 4376–4379.

Rinaldi, A., Mecheri, B., Garavaglia, V., Licoccia, S., Di Nardo, P., Traversa, E., 2008. Engineering materials and biology to boost performance of microbial fuel cells: a critical review. Energy Environ. Sci. 1, 417–429.

Rismani-Yazdi, H., Carver, S.M., Christy, A.D., Tuovinen, O.H., 2008. Cathodic limitations in microbial fuel cells: an overview. J. Power Sources 180, 683–694.

Rodolfi, L., Zittelli, G.C., Bassi, N., Padovani, G., Biondi, N., Bonini, G., Tredici, M.R., 2009. Microalgae for oil: strain selection, induction of lipid synthesis and outdoor mass cultivation in a low-cost photobioreactor. Biotechnol. Bioeng. 102, 100–112.

Rodríguez Arredondo, M., Kuntke, P., Jeremiasse, A.W., Sleutels, T.H.J.A., Buisman, C.J.N., ter Heijne, A., 2015. Bioelectrochemical systems for nitrogen removal and recovery from wastewater. Environ. Sci. Water Res. Technol. 1, 22–33.

Roy, S., Pandit, S., 2018. Microbial electrochemical system: principles and application. Biomass Biofuels Biochem. Microb. Electrochem. Technol. Sustain. Platf. Fuels Chem. Remediat. 19–48.

Santos, N.O., Oliveira, S.M., Alves, L.C., Cammarota, M.C., 2014. Methane production from marine microalgae *Isochrysis galbana*. Bioresour. Technol. 157, 60–67.

Sharma, A., Arya, S.K., 2017. Hydrogen from algal biomass: a review of production process. Biotechnol. Rep. 15, 63–69.

Sharma, K.K., Schuhmann, H., Schenk, P.M., 2012. High lipid induction in microalgae for biodiesel production. Energies 5, 1532–1553.

Shemfe, M., Gadkari, S., Yu, E., Rasul, S., Scott, K., Head, I.M., Gu, S., Sadhukhan, J., 2018. Life cycle, techno-economic and dynamic simulation assessment of bioelectrochemical systems: a case of formic acid synthesis. Bioresour. Technol. 255, 39–49.

Show, K.-Y., Yan, Y., Ling, M., Ye, G., Li, T., Lee, D.-J., 2018. Hydrogen production from algal biomass – advances, challenges and prospects. Bioresour. Technol. 257, 290–300.

Singh, D., Pratap, D., Baranwal, Y., Kumar, B., Chaudhary, R.K., 2010. Microbial fuel cells: a green technology for power generation. Ann. Biol. Res. 1, 128–138.

Song, D., Fu, J., Shi, D., 2008. Exploitation of oil-bearing microalgae for biodiesel. Chin. J. Biotechnol. 24, 341–348.

Sophia Ayyappan, C., Bhalambaal, V.M., Kumar, S., 2018. Effect of biochar on bio-electrochemical dye degradation and energy production. Bioresour. Technol. 251, 165–170.

Su, Y., Song, K., Zhang, P., Su, Y., Cheng, J., Chen, X., 2017. Progress of microalgae biofuel's commercialization. Renew. Sustain. Energy Rev. 74, 402–411.

Tang, C., Zhao, Y., Kang, C., Yang, Y., Morgan, D., Xu, L., 2019. Towards concurrent pollutants removal and high energy harvesting in a pilot-scale CW-MFC: insight into the cathode conditions and electrodes connection. Chem. Eng. J. 373, 150–160.

Ter Heijne, A., Hamelers, H.V.M., Saakes, M., Buisman, C.J.N., 2008. Performance of non-porous graphite and titanium-based anodes in microbial fuel cells. Electrochim. Acta 53, 5697–5703.

Tice, R.C., Kim, Y., 2014. Energy efficient reconcentration of diluted human urine using ion exchange membranes in bioelectrochemical systems. Water Res. 64, 61–72.

Troshina, O., 2002. Production of H_2 by the unicellular cyanobacterium *Gloeocapsa alpicola* CALU 743 during fermentation. Int. J. Hydrogen Energy 27, 1283–1289.

Tsita, K.G., Kiartzis, S.J., Ntavos, N.K., Pilavachi, P.A., 2020. Next generation biofuels derived from thermal and chemical conversion of the Greek transport sector. Therm. Sci. Eng. Prog. 17, 100387.

Varanasi, J.L., Das, D., 2018. Bioremediation and power generation from organic wastes using microbial fuel cell. Microb. Fuel Cell 285–306.

Varanasi, J.L., Prasad, S., Singh, H., Das, D., 2020. Improvement of bioelectricity generation and microalgal productivity with concomitant wastewater treatment in flat-plate microbial carbon capture cell. Fuel 263, 116696.

Varanasi, J.L., Roy, S., Pandit, S., Das, D., 2015. Improvement of energy recovery from cellobiose by thermophillic dark fermentative hydrogen production followed by microbial fuel cell. Int. J. Hydrogen Energy 40, 8311–8321.

Veerubhotla, R., Varanasi, J.L., Das, D., 2018. Biofilm formation within microbial fuel cells. Prog. Recent Trends Microb. Fuel Cells 231–242.

Wang, H., Luo, H., Fallgren, P.H., Jin, S., Ren, Z.J., 2015. Bioelectrochemical system platform for sustainable environmental remediation and energy generation. Biotechnol. Adv. 33, 317–334.

Wang, X., Feng, Y., Liu, J., Lee, H., Li, C., Li, N., Ren, N., 2010. Sequestration of CO_2 discharged from anode by algal cathode in microbial carbon capture cells (MCCs). Biosens. Bioelectron. 25, 2639–2643.

Wu, X., Modin, O., 2013. Ammonium recovery from reject water combined with hydrogen production in a bioelectrochemical reactor. Bioresour. Technol. 146, 530–536.

Xia, T., Zhang, X., Wang, H., Zhang, Y., Gao, Y., Bian, C., Wang, X., Xu, P., 2019. Power generation and microbial community analysis in microbial fuel cells: a promising system to treat organic acid fermentation wastewater. Bioresour. Technol. 284, 72–79.

Xiao, L., Young, E.B., Berges, J.A., He, Z., 2012. Integrated photo-bioelectrochemical system for contaminants removal and bioenergy production. Environ. Sci. Technol. 46, 11459–11466.

Yadav, G., Sharma, I., Ghangrekar, M., Sen, R., 2020. A live bio-cathode to enhance power output steered by bacteria-microalgae synergistic metabolism in microbial fuel cell. J. Power Sources 449, 227560.

Yamashita, T., Yokoyama, H., 2018. Molybdenum anode: a novel electrode for enhanced power generation in microbial fuel cells, identified via extensive screening of metal electrodes. Biotechnol. Biofuels 11, 39.

Yang, W., Lu, J.E., Zhang, Y., Peng, Y., Mercado, R., Li, J., Zhu, X., Chen, S., 2019. Cobalt oxides nanoparticles supported on nitrogen-doped carbon nanotubes as high-efficiency cathode catalysts for microbial fuel cells. Inorg. Chem. Commun. 105, 69–75.

Zabed, H.M., Akter, S., Yun, J., Zhang, G., Zhang, Y., Qi, X., 2020. Biogas from microalgae: technologies, challenges and opportunities. Renew. Sustain. Energy Rev. 117, 109503.

Zeng, L.-Z., Zhao, S.-F., Li, W.-S., 2015. Ni_3Mo_3C as anode catalyst for high-performance microbial fuel cells. Appl. Biochem. Biotechnol. 175, 2637–2646.

Zhang, X., Rong, J., Chen, H., He, C., Wang, Q., 2014. Current status and outlook in the application of microalgae in biodiesel production and environmental protection. Front. Energy Res. 2, 1–15.

Zhang, Y., Zhao, Y., Zhou, M., 2019. A photosynthetic algal microbial fuel cell for treating swine wastewater. Environ. Sci. Pollut. Res. 26, 6182–6190.

SECTION VII

Microalgae as source of biofuels and bioenergy

CHAPTER

18

Microalgae—the ideal source of biofuel

C.V. Vidhya

Department of Microbiology, Ethiraj College for Women, Chennai, Tamil Nadu, India

18.1 Microalgae

Microalgae are unicellular organisms that usually occur as individual cells or in chains or groups. They are found in both freshwater and marine water. Like plants, microalgae also perform photosynthesis. They utilize the greenhouse gas (carbon dioxide) for their photoautotrophic growth. They along with bacteria, form the basic trophic level in the food web, as they are the primary source of feed energy for the trophic levels above them.

Microalgae are the large community that are untapped for their potential products. As per the estimation, approximately, 200,000–800,000 species are present, but only 50,000 species have been discovered and studied (Starckx, 2012). About 15,000 compounds such as enzymes, peptides, sterols, fatty acids, toxins, carotenoids, antioxidants, various polymers were found to be produced from microalgae (Cardozo et al., 2009). Microalgae are ideal hosts to accumulate a large variety of desired products by changing physical and physiological sources such as light source, pH, temperature, nutrients, salts and supply of carbon dioxide. Also, they can be genetically manipulated to achieve the same.

18.2 Biofuel production from microalgae

With the rich history of production of various products from microalgae, they are also identified as the richest and safest source for the biofuel production. Microalgae possess a higher concentration of lipids that are extractable. Later, they are easily converted to biofuel. In general, among various microalgae, *Chlorella, Spirulina, Chlamydomonas, Scenedesmus* and *Botryococcus brauni* are more suitable for biofuel production. Algae, when modified genetically, can produce and accumulate extremely high lipid content, which could be converted to biofuel with energy yield up to 40%. The chemical formula of biofuel is $C_{106}H_{263}O_{110}N_{16}$.

Along with biofuel production, remediation of wastewater and value-added products production can also be attained. Microalgal strains are usually selected based on various factors such as rapid growth, oil content, downstream processing and production yield. Apart from this, the selection criteria involve the ability of microalgae to grow in the presence of highly variable temperature, oxygen concentration, and chemical nature of water (Thomas et al., 2018).

The merits of microalgae in biofuel production include the usage of natural sources such

Biofuels and Bioenergy
https://doi.org/10.1016/B978-0-323-85269-2.00007-1

as sunlight, CO_2, water, minimal mineral nutrients and small-scale land. They efficiently utilize potassium, phosphorous, and nitrogen present in the domestic, agricultural and industrial wastewater along with phycoremediation. The ability of the microalgae acumen to adapt to a variety of habitats can be exploited so that transportation costs can be reduced. Unlike plants, the cultivation of microalgae can be done in areas with extreme conditions such as deserts, seashore and so on, and their cultivation is independent of seasons (round the year). High yields can be attained by manipulating microalgae by synthetic biology, genetic engineering and metabolic engineering. Biofuels from microalgae are nontoxic and highly biodegradable (Thomas et al., 2015). Apart from biofuel production, the remnant biomass can be used for various other applications such as aquaculture and extraction of high-value products such as protein supplements, eicosapentaenoic acid (EPA), fertilizers, docosahexaenoic acid (DHA), animal feed, biocontrol agents, therapeutics and other nutraceuticals. Microalgal oil (biooil) can either be directly used as a fuel or converted into biodiesel by transesterification. Table 18.1 displays a few important microalgal species and their biofuel products.

TABLE 18.1 Microalgae and their biofuel products.

S. No	Name of the Microalgae	Name of the biofuel
1.	*Botryococcus* sp.	Biodiesel
2.	*Chlamydomonas* sp.	Biogas, biohydrogen
3.	*Chlorella* sp.	Biodiesel, biogas
4.	*Dunaliella* sp.	Bioethanol, biodiesel, biogas
5.	*Nannochloropsis* sp	Biodiesel, biogas
6.	*Phaeodactylum* sp	Biodiesel, biogas
7.	*Saccharomyces* sp	Bioethanol
8.	*Spirogyra* sp	Biohydrogen
9.	*Scenedesmus* sp	Bioethanol

18.2.1 Biofuel and its types

A biofuel is a fuel from biological resources and not by geological processes that lead to production of fossil fuels, such as coal, petroleum and so on. Biofuel, in general, is produced from plants, animals, microbes such as microalgae and also from wastes generated from agricultural, industrial and domestic processes. Microalgal biofuel is generally light green in color.

There are four generations of biofuels. Each one differs in their source of production and also source cultivation systems. The biofuel produced from food crops is known as first-generation biofuel. It is otherwise called as conventional biofuel. Transesterification process converts the vegetable oil, starch, or sugar in the crops into bioethanol or biodiesel. Transesterification can be replaced by yeast-based fermentation process. Biofuels belonging to second generation are obtained from biomass rich in lignocellulose. This includes plant and animal materials, agricultural residues and so on. The only advantage of second-generation biofuels over first generation is that no long stretches of land are utilized, but the disadvantage is that the recovery process is little difficult. Biofuels produced from oil rich algae are the third-generation biofuels. These biofuels are advantageous because small-scale wastewater system or bioreactors are used. And the recovery process is quicker and easier. Fourth-generation biofuels include electrofuels and photobiological solar fuels, wherein arable land or biomass destruction is not necessary (Lee et al., 2013).

The most common types of biofuel obtained from microalgal biomass are bioethanol and biodiesel. The other biofuel types include biobutanol, biogasoline, biogas, methane, green diesel and jet biofuel. In this chapter, production of bioethanol and biodiesel is discussed.

18.2.1.1 Bioethanol

Bioethanol, in general, is a fermented product of carbohydrates from crops such as sugarcane, corn, and so on. It can also be produced from microalgal biomass. Ethanol at its purest form is used as a fuel for vehicles. Presently, it is slowly getting replaced by bioethanol. The United States and Brazil largely use bioethanol as a vehicle fuel. One of the important research institutions called as National Corn-to-Ethanol Research Center (NCERC), Southern Illinois University Edwardsville, had come up with ethanol-based biofuel research projects (Acien et al., 2012).

18.2.1.2 Biodiesel

Biodiesel can be produced by transesterification of oil and/or fats from animals, plants such as jatropha, soy, pennycress, mahua, rapeseed, flax, mustard, palm oil, sunflower, *Pongamia pinnata,* hemp, and also from microalgae. Currently, it is the commonly used fuel in Europe, as it is safe to handle and transport, nontoxic, and biodegradable.

Biodiesel has a very high flash point (300°F (148°C)) when compared with diesel and petroleum, which have a flash point of 125°F (52°C). The calorific value of biodiesel is about 37.27 MJ/kg. It can be used in its purest form or in combination with mineral diesel. The advantage of biodiesel is the better lubricity and complete combustion that leads to increased engine energy output. But the disadvantage is that, during winter season, maintenance and performance problems occur, as it becomes highly viscous at lower temperatures. The viscosity depends on the feedstock resource used. Various steps in production of bioethanol and biodiesel from microalgae are detailed in Fig. 18.1.

18.2.1.3 Biobutanol

Microalgae's another product is biobutanol from the green wastes that remain during the extraction of oil. *Palmaria palmate, Laminaria saccharina, Alaria esculenta* and sea weeds such as *Ulva lactuca, Chaetomorpha linum and Enteromorpha compressa* are few important species used for the production of biobutanol (Milledge et al., 2014). Its energy density is 10% less than gasoline and greater than that of ethanol. Apart from that, performance is better, and it resists corrosion, when it is in combination with gasoline.

18.2.1.4 Miscellaneous bioproducts from microalgae

Biogasoline produced from microalgae is used for internal combustion engines. Biogas and its constituent methane can be produced from microalgal wastes and are used for generating electricity. Green diesel, otherwise known as renewable diesel or hydrotreated oil–derived or hydrogen-derived renewable diesel, is derived from microalgae. In 2008, trials with jet biofuels produced from microalgae were carried out by Virgin Atlantic and Lufthansa. Since 2015, production of fatty acid methyl esters and alkenones from the microalgae, *Isochrysis* and using the same as jet biofuel feedstock is under research (Reddy et al., 2015). The production of jet fuel from microalgae improved more by 2017.

18.3 Massive production of microalgae

The massive production of microalgae includes nutritional requirements and cultivation systems. The nutritional requirements involve simple resources such as sunlight, water or wastewater and mineral nutrients (micro and macro). Sunlight is the primary resource for the growth of microalgae, and as it is a limiting factor, many of the industries started to use all possible technologies to develop artificial lighting system. OriginOil is a technocompany that had developed Helix BioReactorTM. This equipment had a rotating vertical shaft with low-energy light that was arranged in a helix pattern.

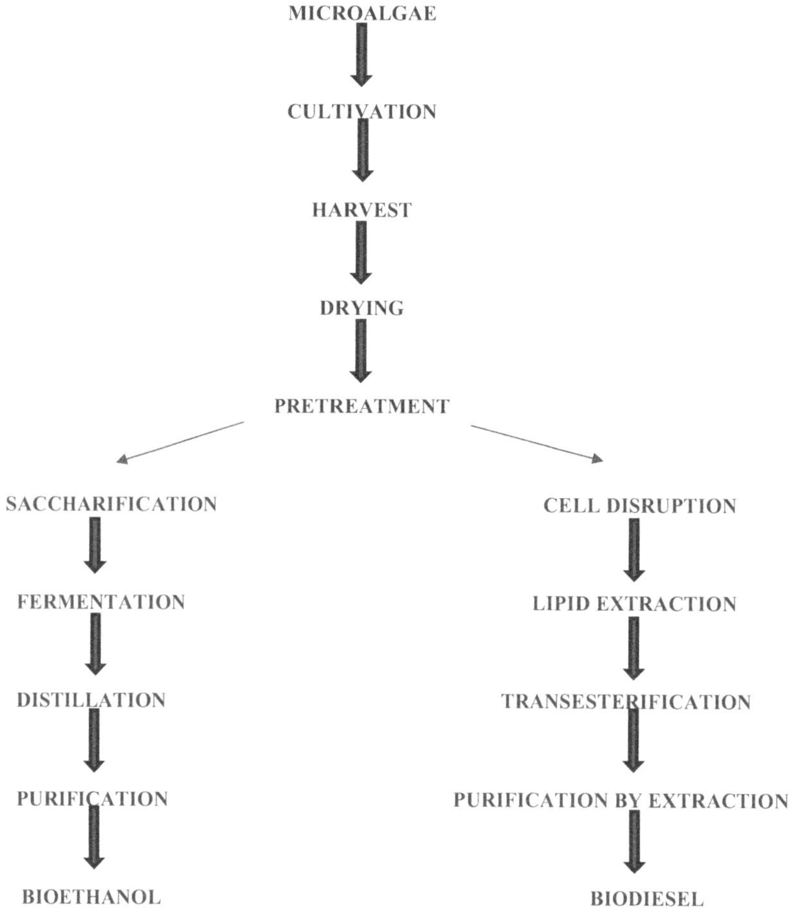

FIGURE 18.1 Production of bioethanol and biodiesel from microalgae.

Water and its temperature has their own effect on the metabolic rate of microalgae. The increase in water current velocity has a greater effect on algae growth and metabolic rate, by increasing nutrient uptake. Wastewater is an easily available nutrient source and is highly recommended instead of freshwater, because of its recyclable feature. Wastewater can be obtained from effluent of treated sewage, agricultural and flood plain run-off. As this wastewater would be highly contaminated, it has to be treated before supplied as a source for microalgal growth. Organic fertilizer obtained after sludge digestion is much suitable for microalgal growth, which is filtered and sterilized for its cultivation. (Pittman et al., 2011). In open pond systems, few algal strains resist high concentrations of heavy metals. (Schenk et al., 2008). In some instances, it was proved that microalgal strains can remove approximately 90% of zinc and nickel from industrial effluent, in a very short time periods (Chong et al., 2000).

Mineral nutrients include both micro and macronutrients. Macronutrients such as nitrogen, phosphorous, and potassium and micronutrients such as carbon dioxide, silica and iron, and trace elements such as cobalt, zinc, manganese, molybdenum, magnesium, calcium, and sulfur are essential for growth and metabolism of microalgae. Nitrogen is an essential substrate for algal growth. Among the various sources of nitrogen, nitrate was found to be the most preferred nitrogen source, followed by urea that is a readily available source and the best economical substitute for nitrate with respect to large-scale cultivation of microalgae (Arumugam et al., 2013). Phosphorus stands as the second most essential nutrient for microalgal growth, as it is involved in almost all the metabolic pathways and cycles. Carbon dioxide also helps in increasing productivity and yield of microalgal biomass. Utilization of about 1.8 tons of carbon dioxide/ton of dry microalgal biomass is ideal. This ratio may vary with different species. In Perthshire, the Glenturret Distillery, during whisky distillation, discovered that each ton of microalgae absorbed two tons of CO_2 (Aylott, 2010).

Although there are two commercially available enrichment media, Guillard's F_2 medium and Walne medium, which can be used to grow most of the microalgal species, they are usually not recommended for the large-scale production, due to their high complexity and cost. Hence, enrichment media formulated for the mass production of algae contain mineral nutrients with agriculture-grade fertilizers.

18.4 Types of cultivation systems

Microalgae grow faster as the harvesting cycle takes only 1–10 days. Their cultivation is cost-effective when compared with any sources of biofuel. This is because their cultivation can be repeated many times in shorter time period on smaller land area. The types of cultivation systems include open pond system, closed-loop system, photobioreactors (PBRs) and algal turf scrubber (ATS).

18.4.1 Open pond system

Open pond systems are simply ground ponds with fitted paddle wheel. In this system, nutritional and power requirements, operating costs, and capital costs are comparatively lower. In general, almost all the commercial microalgae producers follow open pond systems. The only demerit is that they are open to the environment and hence open to contamination (Benemann and Oswald, 1996).

18.4.2 Closed-loop system

On the contrary, closed-loop system is not exposed to environment and hence avoid the risk of contamination. The commercial problem in this method is the quest for a cheaper source for sterile carbon dioxide. This problem was overcome by the use of smoke stacks (Johnson et al., 2018).

18.4.3 Photobioreactors

Photobioreactor (PBR) system is a controlled environment system, which completely rules out the chance of contamination and continuously monitors the feeding of the source material and the output. This system is fitted with light guides and sparger for the efficient process (Johnson et al., 2018). As an added advantage, the closed-loop cogeneration system can be integrated with the photobioreactor, to make it as an advanced conventional method.

18.4.4 Algal turf scrubber

A modern cultivation system, ATS, was designed for clearing up the nutrient wastes and pollutants, from water bodies, using algal turfs. This method was primarily used for water bodies and systems, but now it is used for microalgal biomass production. ATS is as exact as the algal turfs of a naturally developed coral reef, in drawing in nutrient-rich natural water sources and waste streams and pulsed over a sloppy surface. The surface had been coated with a screen or a rough plastic membrane. This coating allows native strains of microalgae to colonize on the surface. Once the microalgae biomass had been established itself on the coated surface, it could be regularly harvested, every 5–15 days. This leads to the production of 18 metric tons of microalgal biomass/hectare/year (Adey et al., 2011).

On the contrary to the methods that use only single high yielding strain, this method permits the biomass formation of polycultures of natural microalgae. The advantage in this method is lower lipid content of the microalgae that is suitable for fuels such as methane, ethanol and butanol. The remnant microalgal biomass is subjected to further treatment with a process namely hydrothermal liquefaction. This leads to secondary production of biogasoline, biodiesel, jet fuel and so on.

Merits of ATS includes:

- Higher yield
- Operation and production cost is low
- Rules out contamination issues (polyculture of natural microalgae)

18.5 Production of bioethanol

The primary process involves the microalgal biomass production on a small scale or large scale. The cultivation methods were discussed in the previous section. The recovery of bioethanol from the microalgal biomass follows the pretreatment of biomass that speeds up the recovery process.

18.5.1 Harvest of microalgal biomass

Harvesting the microalgal biomass is the fundamental step in the recovery of bioethanol. The method involves filtration, manual sieving, press filtration technique and centrifugation.

Filtration is the classical method followed in harvesting microalgal biomass. Manual sieving, one among the conventional methods, is still followed in open pond cultivation. (Thein, 2011). Press filtration technique is followed in dewatering the sludge. The disadvantage of compressible filter cakes is the demand for high maintenance and low specific permeability. This problem can be resolved by two methods. In the first method, the shear force is applied tangential to the filtration membrane, so that this force will prevent the formation of a filter cake. The second method is the belt filtration technique in which there is continuous removal of the cake and belt drying is done by pressing. One of the most effective and reliable methods with respect to large-scale operation is centrifugation. Disk stack separators, a novel centrifugation instrument especially for microalgae (e.g., Evodos), are operated at 3000–4000 g. This is to lower the damages caused to cells The efficiency is obtained even below 1 kWh/m^3 (Pascal et al., 2012).

Prior to the harvesting process, a preconcentration technique can be employed. Flocculation is one for the preconcentration technique that can be used to make the dewatering step more efficient. Mineral salts such as ferric chloride and organic polymers such as chitosan and aluminum sulfate act as flocculants. But this method has a disadvantage of being expensive as well as harmful to the environment as chemicals are involved (Uduman et al., 2010). To overcome this, autoflocculation is recommended. This is attained by halting the aeration by pH shift.

18.5.2 Drying of microalgal biomass

The drying of microalgal biomass is required to form a dry cake, which is later subjected to cell disruption and product extraction. The drying methods include dehydration in the presence of solvent, drying and hydrothermal liquefaction. Dehydration is one of the simple and cost-effective methods in which the biomass is extracted in the presence of solvents such as hexane (Chisti, 2007). Simple drying procedure involves several methods such as spray drying, drum drying, convectional drying, refractance, freeze drying, fluidized-bed drying, window dehydration method, low-pressure shelf drying, and simple sun drying. This procedure increases shelf life of the biomass cake. Hydrothermal liquefaction, an alternate method, is a continuous process wherein wet microalgae is subjected to high temperatures of 350°C and pressures up to 3000 pounds/sq inch (21,000 kPa) respectively (Elliott et al., 2013).

18.5.3 Pretreatment of biomass

The pretreatment procedures are usually four types that include physical, chemical, physicochemical and biological. The selection of the pretreatment procedure is one of the important factors that decide the cost of the bioethanol produced. In this procedure, sugars are obtained either directly or indirectly by hydrolysis without the loss of the same. In general, there are no special techniques considered as the best option, as each process has their own merits and demerits (Sarkar et al., 2012).

18.5.3.1 Physical pretreatment

The physical pretreatment includes mechanical and microwave oven pretreatments. Mechanical treatment involves the simple grinding and milling processes that lead to communition that improves the process by decreasing the polymerization of particles. The efficiency of this method depends on the raw material, moisture content and dimensions of the particles. An ideal example is the production of higher yield when the rice straw was pretreated with wet milling than the dry one (Hideno et al., 2009).

Microwave oven treatment is a simple and rapid method with low energy utilization with higher heating capacity. The heat generated during the process leads to polar bond vibration. This directly causes explosion of particles and degradation of lignocellulosic structure. The acid substance released from lignocellulosic material leads to further hydrolysis. Nowadays, this treatment is performed in combination with chemicals such as NaOH, Na_2CO_3, $Ca(OH)_2$, switchgrass, and coastal bermudagrass (Keshwani et al., 2010).

18.5.3.2 Physicochemical pretreatment

The physicochemical pretreatment includes steam explosion methods such as, liquid hot water method, ammonia fiber explosion (AFEX), CO_2 explosion and wet oxidation methods.

In steam explosion method, the biomass is heated under high pressure steam (20–50 bar, 160–270°C) for a few minutes, followed by arriving at room temperature. The steam diffusion into biomass disperses the fibers. The products formed after this process include alcohols, xylitol and levulinic acid. This process is catalyst-independent process (Balat and Balat, 2008).

Liquid hot water method is a process in which the biomass is subjected to water with high temperature and pressure, for 15 min, without the addition of chemical or catalyst. On the contrary, this method does not require a drop in pressure and temperature and also does not require a quick drop in pressure or expansion such as steam explosion method. Pressure applied is for the prevention of evaporation and for the stabilization of water (Ballesteros et al., 2006). It is chemical and catalyst-free process making it more advantageous. The demerit is the formation of lower amounts of undesirable compounds such as carboxylic acid and furfural that inhibit microalgal growth (Mabee et al., 2006).

AFEX method is a simple and rapid method. In this method, liquid ammonia with steam explosion is performed. The standardization involves the following conditions: biomass with 15%–30% moisture content requires liquid ammonia at a loading ratio of 1–2 kg NH_3/kg dry biomass with pressure above 12 atm to reach the required high temperature. Ammonia is very effective on shredding biomass fibers but not on mass with high lignin content. The crystallization of sugars is partial, and hence, enzymatic degradation of hemicellulose and cellulose is required as an additional process (Chundawat et al., 2007; Taherzadeh and Karimi, 2008). The ammonia involved in the process can be recovered and reused, thus making the process economical. About 90% conversion of cellulose and hemicellulose can be obtained by this technique (Prasad et al., 2007).

CO_2 explosion is a simple and rapid method similar to AFEX method but needs a low temperature. The merits of this technique are the absence of the formation of inhibitors and very high yield, when compared with the steam explosion method (Hill et al., 2006).

Wet oxidation method is basically a simple and cost-effective method in which biomasses are treated with water and the catalyst used is air or oxygen with the temperature around 120°C. The sugar molecules released in this process are oligomers. There is increased solubility of hemicellulose and lignin (Martin et al., 2007).

18.5.3.3 Chemical pretreatment

The Pretreatment methods involve chemicals such as dilute acids, organic solvents, alkaline solutions and ammonia. This method is rapid and yields are high (Carvalheiro et al., 2008).

Acid pretreatment involves two methods where biomass is being subjected to either concentrated acids or diluted acids. The former method is not preferred due to the production of inhibiting compounds and corrosion of equipments (Vincent et al., 2014). In general, phosphoric acid, nitric acid, hydrochloric acid and sulfuric acid are used. Acids convert hemicelluloses to xylose, which is further converted to furfural. Furfural, a toxic component, is removed by distillation during the recovery of bioethanol.

Alkaline pretreatment is carried out at low temperature, and lignins are easily removed, when compared with acid treatment. This process involves ammonia, sodium hydroxide, calcium hydroxide and potassium hydroxide as catalysts. Still, the production of unrecoverable salts and inhibitory by-products such as formic acid, furfural and hydroxymethylfurfural makes this process limited (Tabil et al., 2011; Miranda et al., 2012).

Pyrolysis is a common method that requires high temperature of 300°C for the degradation of biomass. It is an endothermic process. The process requires low energy input. The cellulose is degraded to char and gaseous products such as hydrogen and carbon monooxide. Later, the char is washed with diluted acids or simply with water. The remaining solution possesses sufficient amount of carbon source to support microalgal growth for bioethanol production. It is indeed an efficient method.

Organosolv pretreatment involves treatment with organic solvents such as acetone, methanol, ethylene glycol, ethanol, and catalysts such as sodium hydroxide, ammonia, sulfuric acid and hydrochloric acid. After the pretreatment, solid lignin, cellulosic fibers, and hemicellulose sugars (aqueous) are obtained. The disadvantages of this method are volatilization and oxidation. Also, there is a high risk conditions when the process is performed under high pressure. After the process, the solvents must be recovered to avoid accumulation of inhibitory substances such as furfural and soluble phenols.

18.5.3.4 Biological pretreatment

Biological methods are much preferable because of no chemicals, less energy input and production of less inhibitory substances. The rate of hydrolysis of the microalgal biomass depends on the polymerization and crystal formation of cellulose. Enzymatic hydrolysis of cellulose is by specific cellulase enzymes. Cellulase enzymes include β-glucosidase, exoglucanase and endoglucanase. Endoglucanase attacks cellulose fibers with low crystallinity. The enzyme exoglucanase aids the removal of cellulose, along with endoglucanase, and then the molecule is degraded. Reducing sugars released in the process is mainly glucose, which is produced due to hydrolysis of cellulose by β-glucosidase. This method is very cost-effective and is carried out at low pH (4.8) and temperature of (45–50°C) (Ramesh et al., 2018).

Cellulase enzymes used in hydrolysis can be produced from bacteria and fungi. These microorganisms can be anaerobic, aerobic, mesophilic, or thermophilic. Few cellulose-producing bacteria are *Bacillus, Clostridium, Thermomonospora, Cellulomonas, Ruminococcus, Erwinia, Bacteriodes, Microbispora, Acetovibrio* and *Streptomyces*; cellulose-producing fungi include *Penicillium, Trichoderma, Phanerochaete, Fusarium, Schizophillum sp.*, and *Humicola* (Carvalheiro et al., 2008).

There are two factors that greatly affect hydrolysis, the presence of lignin and hemicellulose, and pore size. Lignin and hemicellulose are physical barriers that prevent the digestible parts of cellulose to hydrolyze. The most important limiting factor in enzymatic hydrolysis process is the substrate's pore size. But enhancement of the pore size can be achieved by removal of hemicellulose and the hydrolysis becomes efficient.

18.5.4 Downstream processing in recovery of bioethanol

After the microalgal biomass is harvested, pretreated and dried into biomass cake, the downstream process of bioethanol starts. Similar to the pretreatment procedure, vigorous hydrolysis can be done by saccharification and fermentation. This is followed by the extraction methods. The extraction method involves two important processes, namely, dehydration and hydrothermal liquefaction. The extraction is followed by distillation, purification and transportation.

18.5.4.1 Saccharification and fermentation

Saccharification is basically a process by which complex carbohydrates are broken down into ethanol. This process includes chemicals and/or enzymes. Each method has its own merits and demerits. Chemical saccharification is done by sulfuric acid, hydrochloric acid and nitric acid. The standardized condition involves temperature of 120–140°C, sustained for around 15 and 30 min of reaction time. In one study, acid hydrolysis of *Scenedesmus obliquus* at 120°C with 2–3 N sulfuric acid for 30 min, on the solid concentration between 20 and 500 g/L completely hydrolyzed the carbohydrate content (71%–97% of carbohydrate content), which was made of 65% glucose (Kim et al., 2014; Gomis et al., 2015).

Enzymatic hydrolysis involves the usage of very low temperatures and decreased degradation risks. Enzymes such as cellulases, pectinases and amylases are used for the saccharification of microalgal biomass. But this procedure may require cell disruption by autoclave/bead-beating/sonication. Among the enzymes used, pectinase is a more effective enzyme. In a study (Limayem and Ricke, 2012), *Chlamydomonas reinhardtii* biomass was subjected to enzymatic hydrolysis by amylase from *Bacillus licheniformis* l. The yield was 50 g/L of biomass concentration, which had 59.7% of carbohydrates. At pH 4.5 and temperature of 90°C for 30 min, saccharification was performed. This attained 94% hydrolysis of carbohydrates in microalgae. Furthermore, approximately 60% of yield was obtained in the fermentation using *Saccharomyces cerevisiae* S288C (Kumar et al., 2010). More details regarding different methods of hydrolysis are discussed in the previous section.

18.5.4.2 Distillation and purification

Following the saccharification, fermentation and removal of microalgal biomass, the remaining solution consists of crude bioethanol. This crude form is then distilled and purified. Removal of water is the important step in distillation, to obtain pure bioethanol, as the mixture contains bioethanol, gasoline and water. The separation of a bioethanol and water mixture (azeotropic mixture) cannot be attained in a single distillation step. It involves many processes, either alone or in combinations.

A distillation process involves many methods in combinations such as azeotropic distillation and heteroazeotropic distillation using a gasoline additive as entrainer, liquid–liquid extraction, a heat pump–assisted extractive distillation, desorption on molecular sieves, evaporation, extractive batch distillation, hybrid processes such as combinations of distillation/vapor permeation/adsorption, pressure swing absorption, pressure swing distillation and extractive distillation with ionic liquids (Kumar et al., 2010).

Distillation process is performed with simple procedures wherein ethanol–water mixture is subjected to boiling. The water's boiling point is 100°C, which is usually higher than that of ethanol, which is 78°C. Hence, ethanol gets vaporized before water (Mesa et al., 2011). Since it is an azeotropic mixture, the process requires higher energy for distillation. Azeotropic mixtures can be separated by adding specific agents such as pentane, hexane, benzene, cyclohexane, diethyl ether and acetone that change the azeotropic structure. These agents change the volatility of the azeotropic mixture by triggering strong attractions among the molecules in the mixture (Bargos et al., 2016).

The distillation process is performed in the large-sized distillation column unit that has top and bottom chambers. Bioethanol is extracted in the top chamber, and the water is collected in the bottom chamber. 37% bioethanol is then concentrated in rectifying column to approach the concentration of 95%. (Middleberg, 1995). The remaining product in the bottom chamber is fed to stripping column for the removal of the excess water. Anhydrous bioethanol is generally more efficient than the hydrous one, but the latter is commonly used as the fuel.

18.5.4.3 Transportation

A bioethanol distillation unit location has to be best defined on the basis of raw materials, availability of water, energy, supply chain, taxes and other miscellaneous factors. Both entry and exit point of the distillation unit should be easily accessible. Hence, transportation is also very important factor and needs to be optimized. This will reduce the loss of ethanol (volatility) and costs of carbon emissions (Mendes et al., 2001).

18.6 Production of biodiesel

Biodiesel is a diesel fuel obtained from biological sources such as microorganisms, plants, and animals. It can be used in its purest form (B100) or can be mixed with petroleum diesel in the form of B2 (98% petroleum diesel and 2% biodiesel), B5 (95% petroleum diesel and 5% biodiesel), B20 (80% petroleum diesel and 20% biodiesel). Biodiesel not only prevents pollution but also aids a country's economy by not depending on other countries for its fuel demands.

Comparatively, biodiesel is a clean burning fuel, as it neither emits carcinogenic pollutants nor causes global warming. It has a higher flash point than petrol and diesel. The by-product glycerin is formed during the process and can be used by other industries.

18.6.1 Cultivation, harvest and downstream processing of microalgal biodiesel

The cultivation systems for microalgal biomass production, pretreatment of biomass and drying methods are similar to bioethanol production method, and the same is discussed in the previous section. The downstream processing of biodiesel includes the cell disruption processes, conversion of algal oil to biodiesel by transesterification and purification of biodiesel by various extraction methods and the same are described in this section.

18.6.1.1 Cell disruption methods

Microalgal biomass drying process is followed by the disruption of the microalgae cell so as to release intracellular metabolites formed. Several methods are used for cell disruption that depends on the nature of cell wall and the product to be obtained. The methods involved are on the basis of mechanical methods such as autoclave, cell homogenizers, bead mills, spray drying; nonmechanical methods such as osmotic shock, ultrasounds, organic solvents, freezing, usage of acids and bases and enzyme-based reactions. These methods are usually used singly or in combinations to attain better results (Lee et al., 2010).

Ultrasound-assisted extraction (UAE) is a simple technique that allows the absorption of ultrasound waves by microalgae that creates cavitation bubbles in a solvent material. This absorption differs with different species. Microwave-assisted extraction (MAE) allows the microalgae to absorb microwave energy which amount varies with each species. In both cases, the waves propagate into liquid media creating alternating high and low pressure (compression and rarefaction) cycle. In low pressure cycle, vacuum bubbles are created near the microalgal cells, and during the high pressure cycles (cavitation), bubbles collapse near the cells, with high energy that creates shear forces disrupting the cell membranes. Ultrasonication in combination with grinding method was found to be successful in a study, during the extraction of lipids from diatom *Chaetoceros gracilis*. Ultrasound and solvent extraction method with solvents (methanol: chloroform) was found very effective in extracting fatty acids from *Thraustochytrium* sp. Mechanical homogenizer is a method in which, when microalgal biomass was made to pass through high pressure homogenizer, the cell walls were easily disrupted. These mechanical methods were found to be successful, but the percentage of success differs with species.

Nonmechanical methods such as, enzymatic hydrolysis, is a biological process in which cellulase is employed in the conversion of cell wall cellulose to glucose and phospholipases in converting the membrane phospholipids to fatty acids and glycerol.

18.6.1.2 Combinations of methods

In a study, the combination of methods, such as autoclaving, enzymatic hydrolysis and mechanical homogenizer, were very successful. From *Haematococcus pluvialis*, carotenoid astaxanthin, a pigment, was extracted (Sommerfeld et al., 2010). The procedure involved the following conditions: autoclaving for 30 min at 121°C and 1 atm; acidification with HCl 0.1 M for 15 and 30 min; alkalinization with NaOH 0.1 M for 15 and 30 min; enzymatic treatment with a mixture of 0.1% protease K and 0.5% Driselase in a phosphate buffer at pH 5.8 and 30°C, for 1 hour; spray drying, inlet 180°C, outlet 115°C; and mechanical disruption with a cell homogenizer. Combination of autoclaving, bead-beating, microwaves and sonication was found to be the most effective method for lipids extraction from *Scenedesmus* sp, *Botryococcus* sp., and *Chlorella vulgaris* (Demirbas and Demirbas, 2010). Quartz sand, solvents addition and ultrasound extraction methods were successful and efficient in the extraction of pigment from fatty acid–resistant algae such as *S. obliquus*. In this procedure, freeze-dried microalgal biomass was subjected to quartz sand and solvent mixture, followed by the extraction in an ultrasound bath for 90 min at 4°C. The success rate was 90%.

Electroporation or electropermeabilization (Sobczuk and Chisti, 2010) alters the cell wall andmembranes, when subjected to electric field, releasing the lipid molecules. This method is simple, efficient and safe as it is a one-step process with no addition of any chemical or biological materials. In a study, the combinations of mechanical homogenizer and three-phase continuous centrifugation aided in concentrating the sample by centrifugation, and it is separated into the lipid stream, the organic and protein-rich water stream, and the biomass cake. This is followed by subjecting the biomass cake to homogenizers that disrupt the cell membranes and release the lipid content.

18.6.1.3 Conversion of algal oil to biodiesel

Microalgal oil molecules are basically polar (phospholipids and glycolipids) and nonpolar (mono-, di-, and triglycerides, certain pigments, vitamins, cyclic and acyclic hydrocarbons, wax esters, and sterols) lipids. 30%–40% of the total lipids and the free fatty acids (FFAs) are saponifiable and are present at about 1%–5% of the total lipid fraction. Once the biomass cells are disrupted and the lipid fraction is extracted, the conversion of lipid to biodiesel process by transesterification follows. After oil extraction from microalgae, the remaining biomass, which is rich in protein, can be used as feed for livestock. Hence, this is a value-added process.

Transesterification is a reaction that converts the reactant fatty acids and alcohol to fatty acid methylesters—biodiesel. It is a reversible reaction and involves catalysts such as strong base or strong acid to speed up the process. In large-scale production, sodium or potassium methanolate is used as a catalyst. The end products are raw biodiesel and raw glycerol. Sometimes, to increase the production, methanol is added, and later, it is recovered and reused. Transesterification is followed by distillation. The glycerol, which is obtained along with the biodiesel, can be used in the food industry, cosmetic industries and oleochemical industry (Ohlrogge and Jaworski, 1997). The physical characteristics of biodiesel contained form microalgae are no less than of fossil diesel.

18.6.1.4 Extraction of biodiesel

The extraction of biodiesel usually uses organic solvents such as chloroform or methanol. As they impose toxic effect over biodiesel, less volatile and less toxic solvents such as ethanol, 1-butanol, or isopropanol are currently being used. A novel method, expeller press solvent extraction method which is supercritical fluid extraction (SFE) with hexane, used as the solvent, was performed to extract biodiesel from microalgae at its best. The former method extracts 70%–75% of the oils out of algae, while the latter (otherwise known as accelerated solvent extraction [ASE]) extracts particularly labile and light-sensitive chemicals such as nutraceuticals. SFE is more efficient than solvent separation methods, as supercritical fluids are selective, highly pure, high concentration of product and of 100% success (Radakovits et al., 2010).

In pressurized fluid extraction (PFE),dried or ASE-dried biomass is subjected to nonaqueous solvents enclosed in a cartridge; the extraction is performed at elevated temperature and pressure. Compressed inert gas is then purged to extract oil from the cell and collected in a vessel. PFE is similar to Soxhlet extraction, except that in the former method, the solvents are used near their supercritical region where they have very high extraction properties. SFE is another technique that is based on the enhanced solvating power of fluids, for example, CO_2, water, nitrous oxide, ethane, n-butane, pentane, methanol and sulfur hexafluoride, with diffusion coefficients greater than those of a liquid. In this process, carbon dioxide is subjected to a very high pressure that liquefies it. It is further heated to the point where it has a dual properties of both liquid and gas. This liquefied fluid is then used as the solvent in oil extraction.

18.7 Merits and challenges of biofuel from microalgae

Microalgal-based biofuel has the following advantages:

- Microalgae has rapid growth rate.
- It needs no special land space for its growth. Its per-acre yield is 7–31 times more than palm (palm oil), which is the second best choice.

- It can utilize wastewater and salt water bodies. This saves the usage of freshwater bodies.
- High volumes of biomass production lead to higher and better yield when compared with other sources.
- Biofuel is nontoxic and highly biodegradable and reduces carbon emissions.
- The presence of high levels of polyunsaturates makes this biofuel suitable for cold climate.
- Innovations in algal biomass production and genetic manipulation of microalgal strains make this field more productive and support country's economy.

The challenges in efficient biofuel production from microalgae are as follows:

- Direct sunlight for a longer term can kill microalgae.
- Constant temperature maintenance is required.
- Overcrowding will inhibit microalgal growth.
- The excess oxygen should be continuously removed from the water.
- Open algal ponds are exposed to sunlight and rainfall, which causes fluctuations in pH and salinity.
- Sometimes, native species of microalgae often overgrow the desired strain.

18.8 Role of genetic engineering in biofuel production

More knowledge about the metabolic pathways involved in biofuel production can be useful to manipulate microalgae into improved strain. In addition, controlling the environmental factors such as efficient illumination, cell quiescence and sequestration of carbon can also improve the strains.

The lipid biosynthesis is directly proportional to the quantity of fatty acid available (Hsieh et al., 2012). Hence, research studies were towards the heightened expression of genes that code few enzymes such as acetyl-CoA carboxylase enzyme that converts carboxylates acetyl-CoA to malonyl-CoA (the gene involved is ACCase) in biosynthesis of fatty acids.

When genes that code for enzymes involved in synthesis of starch are knocked, the accumulation of lipid is increased. This may, sometimes, result in reduced biomass production (Ajjawi et al., 2017). This problem can be managed by silencing the gene by RNAi along with inducible promoters.

In a study with *Chlorella minutissima* (Wei et al., 2017), expression of few acyltransferases such as GPAT, LPAAT, phosphatidic acid phosphatase, glycerol-3-phosphate dehydrogenase and DGAT directly doubled up the lipid accumulation. This controls metabolic flux toward the overproduction of lipid. Manipulating the regulation of transcription of the promoter genes, in the lipid biosynthesis operon, can display the same metabolic flux (for example: PSR1). In a study, the knocking down of the single transcription regulator ZnCys in *Nannochloropsis gaditana* ended up with 98% increase in lipid content (approximately 5 g of yield/m^2/day) (Szyjka et al., 2017).

Rubisco activase (RCA) enzyme, an important regulatory enzyme in Calvin cycle that regulates Rubisco. This enzyme has a direct effect over the rate of carbon fixation, when high proportion of catalytically competent active sites of Rubisco are made available. In a study, RCA was made to get overexpressed in *Nannochloropsis oceanica*. This resulted in elevation of photosynthesis, which led to the accumulation of a very large content of biomass and lipid accumulation needful for biofuel production (Milledge et al., 2014).

Abiotic factors can also be manipulated so as to improve the microalgal strains. One of the methods was to improve the photosynthetic yield by controlled reduction of light-capturing antenna system size in microalgae. This will minimize the energy loss due to NPQ. Research studies have developed mutant microalgal strains with truncated antenna systems.

This truncation downregulated the expression of genes that code for LHC pigment-binding proteins and thereby reduced energy loss.

Multiple gene activation or gene silencing can be experimented by simply introducing guide RNAs for every target genes into the dCas9/sgRNA variant of the CRISPR/Cas9 system. The application of the CRISPR/Cas9 system is a versatile and powerful method in metabolic pathway engineering. CRISPR technology is a rising opportunity in strain improvement especially in microalgae. Various advancements in CRISPR/Cas and other gene editing tools had led way for many triumphant attempts in improving microalgal strains, for effective generation of targeted mutants. This offers easy multiplexing on comparison with the traditional mutagenesis methods and RNAi-mediated knockdown and knockout approaches. This method led way for less complications and highly programmed protocol that manipulated metabolic pathways, which were involved in lipid biosynthesis.

Another approach is simultaneous blocking of the metabolic routes that are competitive to lipid production such as β-oxidation, starch production and degradation of lipid. Another method includes utilizing dCas9 variant that activates stress-responsive elements of the lipid synthesis pathway, under nonstress conditions. This bypasses the inhibitory effects on the biomass produced.

However, the success rate and sustainability of these genetic methods and tools in microalgal strain improvement largely depend on type of strain, cultivation system and cost of the production methods. Open pond cultivation system is much suitable for genetically modified strains, but in one study, when *Acutodesmus dimorphus* was cultured in an open pond, both the wild type and the mutant strains of the same could not outgrow the native strains (LaMonica, 2008). Hence a lot of research study is recommended to draw a conclusion regarding biofuel production from genetically modified strains.

18.9 Applications of biofuels

In 2018, the amount of biofuel produced was about 152 billion liters. The International Energy Agency anticipates that the level of biofuels production will definitely meet the world's demand for transportation fuels, by 2050. It will also reduce dependency on other countries for petroleum and diesel. However, the production and consumption of biofuels could not meet the demand of IEA's sustainable development scenario. It is assumed that between the periods, 2020–30, production of biofuel can have a hike by 10% per every year so as to meet the demand of IEA.

The disadvantage is the high cost of biofuels when compared with fossil fuels. If renewable biofuels are subsidized, then production, transportation and marketing costs could be lowered.

Bioethanol apart from being renewable bioresources, reduces the use of cancer-causing gasoline compounds such as toluene, xylene, benzene and ethylbenzene. It also reduces the emission of greenhouse gases, small particulates and soot from motor fuels. Biodiesel, nowadays, is used in motor engines that were designed for diesel. Biodiesel has a reduced sulfur production during the combustion process. But it delivers comparatively less energy than petroleum-based diesel. This fuel involves only minimal engine modifications that involve installation of synthetic rubber hoses and seals in the fuel system, as natural rubbers ere easily degraded by the biofuel.

Nowadays, in few of the countries, the manufacturers provide warranty for the usage of B100 in their diesel engines. But, Volkswagen, a German company, advices the consumers to check with the environmental services department of the company, before using B100. Since 1994, biodiesel was found to be compatible with diesel engines that use "Viton," the synthetic rubber in their mechanical fuel injection systems. However, before 2014, the vehicles were not certified to use pure biodiesel. This is because there were no emission control protocols available for biodiesel in that period.

Biobutanol is an alcoholic biofuel that can be used as a direct gasoline substitute in internal combustion engines without any modifications. Biobutanol consumption is as similar as that of gasoline. Biobutanol mixed with gasoline rendered a better performance and corrosion resistance than that of ethanol or E85. Biogasoline or BG100, or 100% biogasoline, can be used as a drop-in substitute for petroleum gasoline in any conventional gasoline engine without modification. Green diesel is almost similar to traditional diesel fuel. It can easily replace or can be much useful blended stock, which will improve the quality of the existing diesel pool. Jet biofuel is considered as a valid replacement for a fuel that reduces carbon footprint. After a prolonged and continuous multiyear technical review from oil companies, aircraft makers and motor engine manufacturers, biofuels were approved for commercial use in July 2011. Since then, some airlines have experimented the usage of biofuels in commercial flights.

18.10 Future prospects

The main goals of the new and emerging processes for biofuel production are reduction in the reaction time and development of a continuously operating model, which would aid the increased production capacity to meet the demand for biofuel and reduction in operation costs. Research studies, based on in-depth knowledge about metabolic pathways and genes that control the metabolic machinery of the microalgae, will definitely favor the production of various biofuels, efficiently. On the other side, focus should be laid on construction of cost-effective cultivation systems. All these efforts will positively lead to the production of cost-effective, pollution-free biofuel in future.

18.11 Conclusion

There are various technical, environmental and socioeconomic issues with the production and usage of biofuels. Yet with improved and/or genetically tailored strains and effective downstream processing, high-quality microalgal biofuels can be produced. Apart from these, the biofuel should be cost-effective. As a concern about nature, the biofuel should be free of toxic materials and pollutants.

References

Acien Fernandez, F.G., González-López, C.V., Fernández Sevilla, J.M., Molina Grima, E., 2012. Conversion of CO_2 into biomass by microalgae: how realistic a contribution may it be to significant CO_2 removal? Appl. Microbiol. Biotechnol. 96 (3), 577–586.

Adey, W.H., Kangas Patrick, C., Mulbry, W., 2011. Algal turf scrubbing: cleaning surface waters with solar energy while producing a biofuel. BioScience 61 (6), 434–441.

Ajjawi, I., Verruto, J., Aqui, M., Soriaga, L.B., Coppersmith, J., Kwok, K., 2017. Lipid production in nannochloropsis gaditana is doubled by decreasing expression of a single transcriptional regulator. Nat. Biotechnol. 35, 647–652.

Arumugam, M., Agarwal, A., Arya, M.C., Ahmed, Z., 2013. Influence of nitrogen sources on biomass productivity of microalgae Scenedesmus bijugatus. Bioresour. Technol. 131, 246–249.

Aylott, M., 2010. Forget palm oil and soya, microalgae is the next big biofuel source. Ecologist - informed by Nature. Newsletter. 24th Sptember 2010 issue.

Balat, M., Balat, H., 2008. Progress in bioethanol processing. Prog. Energy Combust. Sci. 34 (5), 551–573.

Ballesteros, I., Negro, M., Oliva, J.M., Caban, A., Manzanares, P., Ballesteros, M., 2006. Ethanol production from steam-explosion pretreated wheat straw. Appl. Biochem. Biotechnol. 130, 496–508.

Bargos, F.F., Lamos, W.Q., Bargos, D.C., Neto, M.B., Pardal, P.C.P.M., 2016. Location problem method applied to sugar and ethanol mills location optimization. Renew. Sustian. Energy Rev. 65, 274–282.

Benemann, J.R., Oswald, W.J., 1996. Systems and Economic Analysis of Microalgae Ponds for Conversion of CO_2 to Biomass. Final Report to the Pittsburgh Energy Technology Center. DOE/PE/93204-T5.

Cardozo, K.H.M., Thais, G., Marcelo, P., Barros, Vanessa, R., Falcao, Angela, P., Tonon, Norberto, P., Lopes, Sara, C., Moacir, A., Torres, Anderson, O., Courchesne, N.M.D., Parisien, A., Wang, B., Lan, C.Q., 2009. Enhancement of lipid production using biochemical, genetic and transcription factor engineering approaches. J. Biotechnol. 141, 31–41.

Carvalheiro, F., Duarte, L.C., Gírio, F.M., 2008. Hemicellulose biorefineries: a review on biomass pretreatments. J. Sci. Ind. Res. 67, 849–864.

Chisti, Y., 2007. Biodiesel from microalgae. Biotechnol. Adv. 25 (3), 294–306.

Chong, A.M.Y., Wong, Y.S., Tam, N.F.Y., 2000. Performance of different microalgal species in removing nickel and zinc from industrial wastewater. Chemosphere 41, 251–257.

Chundawat, S.P., Venkatesh, B., Dale, B.E., 2007. Effect of particle size based separation of milled corn stover on AFEX pretreatment and enzymatic digestibility. Biotechnol. Bioeng. 96 (2), 219–231.

Demirbas, A., Demirbas, M.F., 2010. Algae Energy: Algae as a New Source of Biodiesel. Springer-Verlag, London.

Elliott, D.C., Hart, T.R., Schmidt, A.J., Neuenschwander, G.G., Rotness, L.J., Olarte, M.V., Zacher, A.H., Albrecht, K.O., Hallen, R.T., Holladay, J.E., 2013. Process development for hydrothermal liquefaction of algae feedstocks in a continuous-flow reactor. Algal Res. 2 (4), 445–454.

Gomis, V., Pedraza, R., Saquete, M.A., Font, A., Garcia Cano, J., 2015. Ethanol dehydration via azeotropic distillation with gasoline fraction mixtures as entrainers: a pilot-scale study with industrially produced bioethanol and naptha. Fuel Process. Technol. 140, 198–204.

Hideno, A., Inoue, H., Tsukahara, K., Fujimoto, S., Minowa, T., Inoue, S., 2009. Wet disk milling pretreatment without sulfuric acid for enzymatic hydrolysis of rice straw. Bioresour. Technol. 100, 2706–2711.

Hill, J., Nelson, E., Tilman, D., Polasky, S., Tiffany, D., 2006. Environmental, economic, and energetic costs and benefits of biodiesel and ethanol biofuels. Proc. Natl. Acad. Sci. 103 (30), 11206–11210.

Hsieh, H.J., Su, C.H., Chien, L.J., 2012. Accumulation of lipid production in Chlorella minutissima by triacylglycerol biosynthesis-related genes cloned from *Saccharomyces cerevisiae* and *Yarrowia lipolytica*. J. Microbiol. 50, 526–534.

Johnson, T.J., Katuwal, S., Anderson, G.A., Gu, L., Ruanbao, Z., Liping, G., Gibbons, W.R., 2018. Photobioreactor cultivation strategies for microalgae and cyanobacteria. Biotechnol. Prog. 34, 811–827.

Kim, K.H., Choi, I.S., Kim, H.M., Wi, S.G., Bae, H., 2014. Bioethanol production from the nutrient stress-induced microalga *Chlorella vulgaris* by enzymatic hydrolysis and immobilized yeast fermentation. Bioresour. Technol. 153, 47–54.

Keshwani, D.R., Cheng, J.J., 2010. Microwave-based alkali pretreatment of switchgrass and coastal bermudagrass for bioethanol production. Biotechnol. Prog. 26, 644–652.

Kumar, S., Singh, N., Prasad, R., 2010. Anhydrous ethanol: a renewable source of energy. Renew. Sustian. Energy Rev. 14 (7), 1830–1844.

LaMonica, M., 2008. New Energy Act to Fuel Flow of Biogasoline. CNET. CBS Interactive Inc.

Lee, J.-Y., Yoo, C., Jun, C.Y., Ahn, C.Y., Oh, H.M., 2010. Comparison of several methods for effective lipid extraction from microalgae. Bioresour. Technol. 101 (1), S75–S77.

Lee, R.A., Lavoie, J.-M., 2013. From first- to third-generation biofuels: challenges of producing a commodity from a biomass of increasing complexity. Anim. Front. 3 (2), 6–8.

Limayem, A., Ricke, S.C., 2012. Lignocellulosic biomass for bioethanol production: current perspectives, potential issues and future prospects. Prog. Energy Combust. Sci. 38 (4), 449–467.

Mabee, W.E., Gregg, D.J., Arato, C., Berlin, A., Bura, R., Gilkes, N., Saddler, J.N., 2006. Updates on softwood-to-ethanol process development. Appl. Biochem. Biotechnol. 129 (1–3), 55–70.

Martin, C., Klinke, H.B., Thomsen, A.B., 2007. Wet oxidation as a pretreatment method for enhancing the enzymatic convertibility of sugarcane bagasse. Enzyme Microb. Technol. 40 (3), 426–432.

Mendes Pinto, M.M., Raposo, M.F.J., Bowen, J., Young, A.J., Morais, R., 2001. Evaluation of different cell disruption processes on encysted cells of *Haematococcus pluvialis*: effects on astaxanthin recovery and implications for bioavailability. J. Appl. Phycol. 13 (1), 19–24.

Mesa, L., González, E., Cara, C., González, M., Castro, E., Mussatto, S.I., 2011. The effect of organosolv pretreatment variables on enzymatic hydrolysis of sugarcane bagasse. Chem. Eng. J. 168 (3), 1157–1162.

Middleberg, A.P.J., 1995. Process-scale disruption of microorganisms. Biotechnol. Adv. 13 (3), 491–551.

Milledge, J., Smith, B., Dyer, P., Harvey, P., 2014. Macroalgae-derived biofuel: a review of methods of energy extraction from seaweed biomass. Energies 7 (11), 7194–7222.

Miranda, J.R., Passarinho, P.C., Gouveia, L., 2012. Pretreatment optimization of Scenedesmus obliquus microalga for bioethanol production. Bioresour. Technol. 104, 342–348.

Ohlrogge, J.B., Jaworski, J.G., 1997. Regulation of fatty acid synthesis. Annu. Rev. Plant Biol. 48, 109–136.

Pascal, S., Gerold, G., Robert, D., Rosa, R.S., Clemens, P., 2012. Composition of algal oil and its potential as biofuel. J. Combust. 285185.

Pittman, J.K., Dean, A.P., Osundeko, O., 2011. The potential of sustainable algal biofuel production using wastewater resources. Bioresour. Technol. 102 (1), 17–25.

Prasad, S., Singh, A., Joshi, H.C.B., 2007. Ethanol as an alternative fuel from agricultural, industrial and urban residues. Resour. Conserv. Recycl. 50 (1), 1–39.

Radakovits, R., Jinkerson, R.E., Darzins, A., Posewitz, M.C., 2010. Genetic engineering of algae for enhanced biofuel production. Eukaryot. Cell. 9, 486–501.

Ramesh, C.R., Ramachandran, S., 2018. Bioethanol Production from Food Crops: Sustainable Sources, Interventions, and Challenges. Academic Press.

Reddy, C., O'Neil, G., 2015. Jet Fuel from Algae? Scientists probe fuel potential in common ocean plant.

Sarkar, N., Ghosh, S.K., Bannerjee, S., Aikat, K., 2012. Bioethanol production from agricultural wastes: an overview. Renew. Energy 37 (1), 19–27.

Schenk, P.M., Thomas-Hall, S.R., Stephens, E., Marx, U.C., Mussgnug, J.H., Posten, C., Kruse, O., Hankamer, B., 2008. Second generation biofuels: high-efficiency microalgae for biodiesel production. BioEnergy Res. 1, 20–43.

Sobczuk, T.M., Chisti, Y., 2010. Potential fuel oils from the microalga Choricystis minor. J. Chem. Technol. Biotechnol. 85 (1), 100–108.

Sommerfeld, M., Chen, W., Hu, Q., Giorgi, D., Navapanich, T., 2010. Application of Electroporation for Lipid Extraction from Microalgae. http://www.optiswitch.com/products/pdfs/Application%20of%20.

Starckx, S., 2012. A Place in the Sun - Algae Is the Crop of the Future, According to Researchers in Geel Flanders Today.

Szyjka, S.J., Mandal, S., Schoepp, N.G., Tyler, B.M., Yohn, C.B., Poon, Y.S., 2017. Evaluation of phenotype stability and ecological risk of a genetically engineered alga in open pond production. Algal Res. 24, 378–386.

Tabil, L., Adapa, P., Kashaninejad, M., 2011. Biomass Feedstock Pre-processing—part 1: Pre-treatment. Biofuels Engineering Process Technology.

Taherzadeh, M.J., Karimi, K., 2008. Pretreatment of lignocellulosic wastes to improve ethanol and biogas production: a review. Int. J. Mol. Sci. 9 (9), 1621–1651.

Thein, M., 2011. The use of natural resource for sustainable production of *Spirulina* in Myanmar crater lakes. In: Proceedings of the 5th International Algae Congress, Microalgae & Biomass, Berlin, Germany.

Thomas, K.M., Archana, T., Bhaskar, M.V., 2015. A new novel solution to grow diatom algae in large natural water bodies and its impact on CO_2 capture and nutrient removal. J. Algal Biomass Util. 6 (2), 22–27.

Thomas, K.M., Narasimha Reddy, P., Archana, T., 2018. Potential of diatom consortium developed by nutrient enrichment for biodiesel production and simultaneous nutrient removal from waste water. Saudi J. Biol. Sci. 25, 704–709.

Uduman, N.,Y., Qi, M.K., Danquah, G.M., Forde, A.H., 2010. Dewatering of microalgal cultures: a major bottleneck to algae-based fuels. J. Renew. Sustain. Energy 2 (1), 012701.

Vincent, M., Pometto III, A.L., van Leeuwen, J., 2014. Ethanol production via simultaneous saccharification and fermentation of sodium hydroxide treated corn stover using *Phanerochaete chrysosporium* and *Gloeophyllum trabeum*. Bioresour. Technol. 158, 1–6.

Wei, L., Wang, Q., Xin, Y., Lu, Y., Xu, J., 2017. Enhancing photosynthetic biomass productivity of industrial oleaginous microalgae by overexpression of RuBisCO activase. Algal Res. 27, 366–375.

CHAPTER

19

Principles and technology advances in wastewater treatment through microalgae for bioenergy production

Saptaswa Biswas, Rahulkumar Maurya, Kaustubha Mohanty

School of Energy Science and Engineering, Indian Institute of Technology Guwahati, Guwahati, India

19.1 Introduction

19.1.1 General wastewater sources and characteristics

The word "wastewater" is something that evokes all kinds of distaste in our mental image, vivid with all things bad, dark, murky, gooey, unhygienic, insanitary, and whatnot. Every sane human being avoids looking at that image whether it be in front of his eyes or the back of his mind. Given so much repugnance toward this particular entity called wastewater, one would expect a strong course of action toward its eradication, but what we see is a general apathy toward this concern. While technological advances and research prowess have led to all things great, there has been a major gap in addressing this problem that has been staring at our faces for too long. Even in current times, a large proportion approximating 80% of all wastewater is being directly dumped in the environment globally without treatment causing widespread water pollution, and this menace is seriously impacting environmental sustainability, water management, human health, and even the economy (WWAP - United Nations World Water Assessment Programme, 2017). A dire need to mitigate this looming problem is needed on a global scale. But before delving into the finer aspects of this problem—wastewater, our perspective of looking at wastewater has to change, and we need to consider it as a valuable resource, a relatively untapped one. The good thing is that wastewater is literally 99% water and the remaining 1% consists of suspended, colloidal, and dissolved solids. But how do we define wastewater? Wastewater has been defined in different ways over the years such as "used water," "sick waster" or just "effluent," but the currently accepted one is according to the UN-Water convention which is itself adapted from a 2008 report on wastewater (UN-Water, 2015; Raschid-Sally and Jayakody, 2008):

Wastewater can be considered as a combination of one or more of these categories: domestic effluent comprising blackwater (urine, excreta

Biofuels and Bioenergy
https://doi.org/10.1016/B978-0-323-85269-2.00026-5

and related sludge) and graywater (used water from kitchen and bathroom); water from commercial establishments and institutions, including hospitals; industrial effluent, storm water, and miscellaneous urban runoff; and agricultural, horticultural, and aquaculture runoff.

To find the remedy to this ever-looming problem of wastewater mitigation, we need to take a closer look at the sources and identify the technological gaps in the production of this wastewater. Wastewater based on its origin can be broadly divided into (1) domestic or municipal wastewater; (2) industrial wastewater, and (3) agricultural wastewater.

19.1.1.1 *Domestic or municipal wastewater*

Domestic wastewater mainly originates from household activities, sanitation, and so on in places of or around human settlements and is mostly in urban areas. Sewerage systems and septic tanks are the major sources around urban areas, and open sanitation dumps around slums and rural areas. The composition is mostly biodegradable (human excreta) with minor traces of heavy metals and other pollutants. Domestic wastewater is potentially less hazardous as compared with industrial wastewater, although recently with the rapidly increasing drug, cosmetics, and medication usage and very low awareness of proper disposal methods, cause for concern is high as even minute concentrations of these substances can have substantial long-term impacts (WWAP - United Nations World Water Assessment Program, 2017). Highly dysfunctional treatment and effluent management systems in most urban areas and direct release of wastewater into mainstream water bodies are the major reasons for the dreadful consequences of domestic wastewater.

19.1.1.2 *Industrial wastewater*

Globally, 5%–20% of all water used goes in industries. Because of different governances, enforcement, and local policies in different areas, there is a huge gap in wastewater control on industries, ranging from almost nonexistent in underdeveloped or developing nations, to highly functional control in some developed nations. Almost 70% of all industrial discharge in developing nations is estimated to be dumped untreated (UN-Water, 2009). Different industrial sectors have different wastewater constituents and thus require separate treatment systems. Wastewater from the pulp and paper industry is characterized by a high amount of chlorinated matter and other colored components, whereas wastewater from dairy industries mostly contains sugars, proteins, fats, and additive residues. Similarly, other sectors such as energy, textiles, iron and steel, the food industry, and mines and quarries all have specific wastewater characteristics. This poses a serious technological setback in most of the underdeveloped and developing nations, leading to an onslaught of poorly setup industrial covers with minimum attention to waste generation and treatment.

19.1.1.3 *Agricultural wastewater*

The unique aspect of the agricultural sector is that it is both the producer and consumer of wastewater; thus, essentially it is capable of creating this menace as well as faces its devastating consequences. Agricultural wastewater encompasses wastewater originated from aquaculture and livestock activities along with the conventional plant and crop production sector (Raschid-Sally and Jayakody, 2008). The wastewater pathways are mainly by surface runoff and flow to rivers and other water streams and seepage into groundwater, and thus this type of wastewater is not obnoxious when compared with urban and industrial wastewater. Major pollutants here are nutrients mainly nitrates and phosphates present in fertilizers and animal excreta, pesticides, organic matter, suspended matter (TSS), salts, and minute amounts of metals.

19.1.2 Problems regarding disposal and reuse

In most cases, wastewater should not be allowed to be disposed of directly, owing to the presence of dangerous chemicals and other pollutants, but the case is exactly the opposite. In terms of municipal wastewater, sewer systems and septic tanks are useful, but for industrial wastewater, specific disposal regulations for different industries need on-site treatment and disposal management systems. Because all of these sectors require proper treatment before disposal, various methods have been developed such as the use of stabilization ponds or wetland generation, or treatment by anaerobic digestion (AD) or activated sludge, or the use of membranes, or chemical treatment such as UV radiation, ozonation, or advanced oxidation. All these methods have their own benefits and disadvantages. These methods all suffer from incomplete treatment, and some of them, such as a wetland, require a high land use and are time-taking (Li et al., 2019). Chemical methods are unsafe for plant and animal life, including humans in contact with them. Because of these problems, biological methods are being preferred nowadays so that disposed water after treatment can be reused in some way or the other, without harming living beings or being uneconomical.

19.1.3 Wastewater components as algal nutrient sources

The best thing about algal wastewater remediation is that nitrates and phosphates, the primary pollutants in almost all kinds of wastewater, are the major nutrients for algal growth. A simplified molecular formula of microalgae ($C_{106}H_{263}O_{110}N_{16}P$) suggests that it would take 0.063 g of nitrogen and 0.009 g of phosphorus for 1 g biomass of algae and carbon would contribute more than half of the mass (Xue et al., 1988; Li et al., 2019), and all of these components are aplenty in wastewaters. Such is the reason that unsightly algal blooms are so prevalent in lakes, sewers, effluent ponds, and the like. Agricultural wastewater and wastewater from allied industries such as dairy, piggeries, and so on are more suited for algal remediation since they contain high amounts of nitrates and phosphates and the presence of carbon content also enhances algal biomass productivity. Heavy metals such as zinc and copper present in municipal and industrial wastewater serve as micronutrients for algal growth. Many other contaminants, both organic and inorganic, are also found to similarly assimilate in different metabolic pathways in microalgae.

This basic concept lays forward the foundation for this chapter, where we look at the different aspects of wastewater treatment (WWT), and how algae can be an integral part of it. Modern ways of looking at the wastewater problem and its subsequent remediation by microalgae lie in assessing the environmental and economic sustainability, using state-of-the-art technologies/methods such as consortia engineering, building-integrated systems, and the incorporation of automation and artificial engineering for a better optimized and fast system. These pioneering attempts at WWT have been meticulously detailed in this chapter, which has not been explored in previous literature.

19.2 Microalgae as a wonderful tool in wastewater remediation

19.2.1 Microalgae: strains and wastewater remediation principles

19.2.1.1 Strain selection

The basic idea of wastewater remediation using algae arose due to the abundance of algal growth in wastewater contaminated areas. Microalgae are microscopic photosynthetic organisms capable of living in a variety of water environments, freshwater, marine, waste streams, and water columns, to even partially submerged areas such as rocks. Due to the sheer variety of characteristics found in different microalgal strains, they have been selectively exploited for different purposes. For WWT, certain algal strains are more better in remediating than others. This is inherent in their capacity for nutrient uptake such as nitrates and phosphates. Due to this, screening of algal strains is a prerequisite for any WWT system. The main considerations are that the strain should have fast growth and sufficiently high biomass productivity, a high nutrient removal rate, and robustness in its capability to adapt to different wastewater types and weather conditions. Low lipid content of a strain can be compensated by higher productivity, thereby improving overall performance in terms of treatment capability. Microalgal strains that have stood up to these benchmarks are most commonly from *Chlorella* species like *Chlorella vulgaris*, *Chlorella sorokiniana*, *Chlorella pyrenoidosa*, *Chlorella reinhardtii*, and so on. Other algal species such as *Scenedesmus* (*Scenedesmus obliquus*, *Scenedesmus rubescens*, *Scenedesmus dimorphus*), *Botryococcus braunii*, *Haematococcus* sp., *Nannochloropsis* sp., and so on have also been commonly used for WWT (Li et al., 2019). Only a few strains from *Scenedesmus* sp. and *Chlorella* sp. are suited for agricultural and industrial wastewater, while most algal strains could be used for municipal wastewater.

19.2.1.2 Remediation mechanism

For understanding microalgal remediation mechanisms, understanding of the basic metabolic pathways of microalgae is necessary. Although predominantly autotrophic in nature, sufficing on water, light, and an inorganic carbon source such as CO_2, some algal strains can be heterotrophic, relying on organic carbon sources such as glucose and acetate, and can grow without any light, while others can use both inorganic and organic sources of carbon in presence of light, which is called mixotrophic (Gupta et al., 2019). Microalgae in wastewaters employ heterotrophic or mixotrophic mode to metabolize organic matter. The three basic metabolic pathways, carbon, nitrogen, and phosphorus, govern most of the WWT mechanisms of microalgae.

19.2.1.2.1 Carbon

The main carbon intake by microalgae is inorganic carbon in the form of atmospheric CO_2 in autotrophic mode or using soluble carbonates, which can be directly utilized or converted to free carbon dioxide by carbonic anhydrase activity (Gupta et al., 2019). When considering inorganic sources of carbon, one way of fixing carbon is by oxidation of organic carbon to CO_2 and utilizing that in photosynthesis through the Calvin cycle. The other way is by heterotrophic or mixotrophic metabolism where different organic carbon sources follow different pathways. The organic load of wastewater is quantified by its BOD (biological oxygen demand) or COD (chemical oxygen demand) and can be characterized by the presence of organic carbon sources such as glucose, acetate, and volatile fatty acids (VFAs). In microalgae, glucose is metabolized in the glycolysis pathway, forming pyruvate, which is then converted to acetyl-CoA in mitochondria. The acetyl-CoA is a determining molecule, which goes in the TCA cycle to produce citrate and energy as ATP. In stress conditions, the accumulated citrate in the mitochondrial matrix is transported out to the cytosol where it is converted to acetyl-CoA, and this goes to the chloroplast to take part in the fatty acid synthesis process (Perez-Garcia et al., 2011). For VFAs, it starts with acetylation to their respective coenzymes such as propionic acid to propionyl CoA, which is then converted to acetyl-CoA, which can take part in the TCA cycle or form malate in the glyoxylate cycle, which is again used in the fatty acid synthesis process. Acetate metabolism begins with acetylation to acetyl-CoA, and the rest is similar to the VFA pathway (Gupta et al., 2019).

19.2.1.2.2 Nitrogen

Nitrogen is a key element for any organism, in this case, microalgae, and is prevalent in various biological systems such as proteins, enzymes, chlorophylls, energy molecules such as ATP, and genetic materials (DNA and RNA) in organic form. Nitrogen in wastewater sources, however, is found mainly in inorganic forms such as nitrates, nitrites, or ammonia. To utilize this, these inorganic forms of nitrogen are converted by microalgae to organic forms in a process called assimilation. Ammonium is the preferred source of nitrogen because of the absence of any redox reaction, thus minimizing energy requirement (Maestrini et al., 1986). Nitrates and nitrates are first converted to ammonia with the help of reductase enzymes, and ammonia is incorporated into amino acids by GS/GOGAT (glutamine synthetase/glutamate synthase) and GDH (glutamate dehydrogenase) pathways (Perez-Garcia et al., 2011). Wastewater containing a higher amount of ammonia than nitrates or nitrates is beneficial to microalgae; however, excess ammonia has also been noted to repress algal growth (Morris and Syrett, 1963).

19.2.1.2.3 Phosphorus

Phosphorus is another common contaminant in wastewaters and an essential component in microalgal energy metabolism and biological substances such as nucleic acids, proteins, lipids, and so on. Abundant forms of phosphorus in wastewater are polyphosphates, orthophosphates (HPO_4, H_2PO_4), and organic phosphates, and microalgae are known to readily assimilate orthophosphates, which are further assimilated to organic components such as phospholipids, proteins, and nucleic acids by phosphorylation (Gupta et al., 2019). Microalgae also fix phosphorus when in high concentration by storing the excess phosphate in polyphosphate granules of the cell, which act as a reservoir to be used during nutrient deplete conditions(Cai et al., 2013). Different studies have utilized a varied range of wastewater sources, targeting nutrient remediation, such as nitrates and phosphates using different microalgal strains. Table 19.1 gives an idea of the recent studies concerning algal culture systems and nutrient uptake.

19.2.2 Microalgae consortia for wastewater treatment—an approach better than axenic strain

Biological WWT is mostly done by using a consortium of microbes rather than pure or axenic cultures. This is because wastewater carries with itself several microbes, known or unknown, and even during treatment, it is exposed to outside contamination; thus essentially, achieving axenic condition is difficult to near to impossible. However, this predicament is not necessarily a handicap, since mixed cultures and microbial consortia have been found to be more effective at wastewater remediating than pure cultures. The most apparent reasons are that complex waste degradation processes are better done by a consortium of microbes with proper division of labor, and the waste remediating system becomes more robust and resilient to environmental changes, and less susceptible to outside contamination. A microalgal consortium consists of different microalgal species or strains, whereas a microalgal–bacterial consortium consists of mutually compatible algal and bacterial species. There are two methods of using microalgal consortia, one is by testing and mixing different axenic strains to determine their mutualistic capability with each other (artificial consortium), and the other method is to use microalgal consortia already found in nature, or in and around wastewater streams (natural consortium). The advantage in the first method is that we can model the treatment system as we know the definite strains and can identify their interactions, but the problem with the first method is that it is a time-taking process,

TABLE 19.1 Microalgae nutrient removal and productivity in different wastewater types and culture systems.

Wastewater type	Source	Strains used	Culture system	Nutrient removal (%)			Productivity (mg/L d)		References
				TN	TP	COD	Biomass	Lipid	
Municipal wastewater	Tertiary wastewater	*Chlorella vulgaris*	Membrane photobioreactor (PBR)	86–99	100	NA	2 g/L	NA	Praveen et al. (2016)
	Sanitation effluent	*C. vulgaris*	Alginate beads in bioreactor	99	60	NA	NA	NA	Filippino et al. (2015)
	Sewer wastewater	Consortium of *Chlorella* sp., *Nitzschia* sp., *Navicula* sp., *Stigeoclonium* sp.	High-rate algal pond	99	NA	80	NA	NA	Gutiérrez et al. (2016)
	Digested sludge wastewater	Consortium of *Coelastrum microporum* and wastewater bacteria	Duran bottle PBR	35.2–87.8	43.3–88.7	59.5–80.3	NA	NA	Lee et al. (2015)
Agricultural wastewater	Cattle manure leachate	*Coelastrum* sp.	Flasks	80	100	41	281	11	Mousavi et al. (2018)
	Livestock wastewater	*Chlorella* and *Phormidium* consortium	Slanted algal biofilm reactor	95	93	89	4 g/m^2 d	NA	Choudhary et al. (2017)
	Domestic graywater			95	90	70	3.6 g/m^2 d		
	Anaerobically digested slurry			93	88	80	3.1 g/m^2 d		
	Agrowaste digestate	*Parachlorella kessleri*	Flask	98	59	39	62	NA	Koutra et al. (2018)
Industrial wastewater	Raw dairy wastewater	*Ascochloris* sp.	Column and flat-plate PBR	82	97	96	284–292	94–98	Kumar et al. (2019)
	Petroleum refining wastewater	Polyculture biofilm microalgae	Rotating algae biofilm reactor	70.8	61.3	Negligible	4.1 g/m^2 d	NA	Hodges et al. (2017)

	Brewery effluent	*Scenedesmus obliquus*	Cylindrical bubble column PBR	88	30	71	NA	NA	Ferreira et al. (2019)
	Soybean processing wastewater	*Chlorella* sp.	Conical bottle	84.7	97.3	70.5	NA	7.22	Qiu et al. (2019)
	Pulp and paper industry biosludge digestate	*C. vulgaris* and *Scenedesmus acuminatus*	Glass bottle PBR	100	97	NA	685	NA	Tao et al. (2017)
Mixed wastewater medium	20% kitchen waste, 10 g/L of poultry litter waste and 5% flue gas	*Chlorella minutissima*	Raceway pond	100	91	85	6.2 g/m^2 d	1.93	De Bhowmick et al. (2019)
	20:80 ratio of piggery and winery wastewater	*Chlorella* sp.	Flasks	89.3	56.5	NA	NA	NA	Ganeshkumar et al. (2018)

without any surety of a definite result. This is where the second method becomes useful as it is already an optimized system, although modeling the system and establishing purity becomes difficult. In a 2010 report, microalgal strains were isolated from carpet wastewater, and a consortium of 15 selected strains were prepared, which showed a >96% nutrient removal in WWT containing 85%–90% carpet industry effluents with 10%–15% municipal sewage, which was much higher than the individual strains (Chinnasamy et al., 2010). The settleable algal community in consortia was selectively enriched according to a 2017 report, by manipulating settling time (ST) and volume exchange ratio (VER) in photobioreactors (PBRs), thus generating a highly efficient wastewater removal system with phosphorus removal greater than 99% and easily harvestable biomass (Hu et al., 2017). Many other types of microalgal consortia such as microalgae–bacteria (Le et al., 2019), microalgae–yeast (Zeng et al., 2018), and microalgae–fungi (Zhou et al., 2012; Barnharst et al., 2018) apart from microalgal–microalgal consortia and their efficacies in WWT have also been reported.

19.2.3 Microalgae and bacteria consortia for wastewater treatment

When WWT using microalgae is concerned, there are almost always some bacteria involved, whether intentionally or unintentionally. This is because these two work well with each other in a mutualistic relationship and exchange nutrients and effectively cycle carbon, nitrogen, phosphorus, sulfur complementing each other's growth, and productivity (Gonçalves et al., 2017). Although it is general practice to use consortia in WWT, there is little technical knowledge about the different algal and bacterial populations in a consortium, and the mechanisms and interactions involved. The basic suggested mechanism follows that bacteria contribute to the oxidation of organic matter to inorganic components (NH_4^+, PO_4^{-3}, CO_2) and microalgae utilize these inorganic compounds, along with CO_2 and light, which drives photosynthesis producing algal biomass and releasing O_2, which is utilized by the bacteria, thus closing the loop (Fig. 19.1). The actual mechanism is far too complex, and a valid model has not yet been (Le et al., 2019).

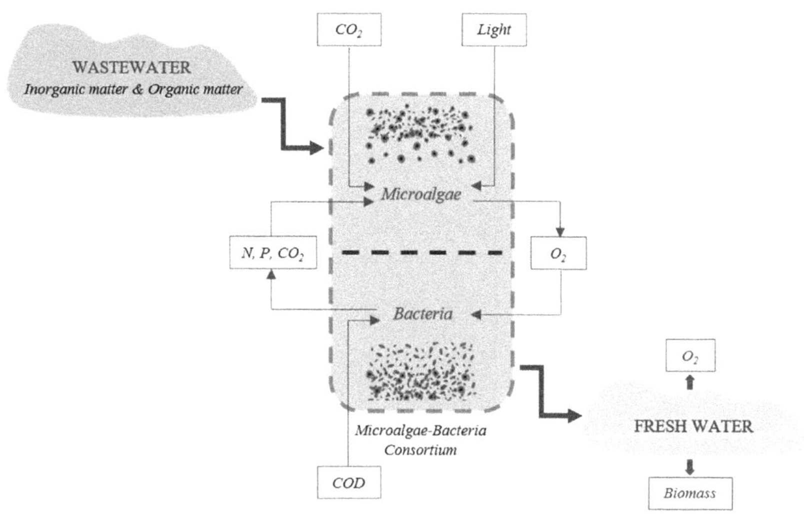

FIGURE 19.1 Mechanism of microalgae–bacteria consortium in wastewater treatment.

The main growth mode of microalgae is phototrophic, which releases hydroxyl ions during nitrate assimilation and conversion to ammonia, mixotrophic, and heterotrophic growth modes, which, if possible, have a very low impact on the overall metabolism. In the case of bacteria, heterotrophic growth modes could be aerobic or anoxic, which helps in denitrification, and autotrophic mode in aerobic conditions helps in nitrification (cyanobacteria). The basic reactions governing the mutualistic relationship between microalgae and bacteria are given in Table 19.2.

Although theoretically bacteria and microalgae should contribute equally to achieve equilibrium in a proper consortium, in experimental studies, it has been seen that microalgae are the major driving force in WWT, with minimal contribution from bacteria and decreased growth after the initial few days (Le et al., 2019). This is due to the faster metabolism of bacteria compared with microalgae. For proper treatment efficiency, the wastewater compositions and culture conditions need to be varied and optimized. Different bacterial and cyanobacterial consortia with microalgae have been used and studied for wastewater remediation. The communities are also dependent on the wastewater used.

19.3 Microalgae-based wastewater treatment technologies

19.3.1 Algae culture systems

Microalgae culturing has a long history and a similarly long list of culture systems. The culturing systems can be broadly divided into suspended and attached (nonsuspended) systems. Suspended systems have been at the forefront for a long time, and only recently attached systems are being researched vigorously (Li et al., 2019). Even microalgal strains have their own preferences with some preferring suspended growth, while others in the attached mode and some are known to exist in both modes simultaneously. Both systems have their own share of advantages and shortcomings and have also been used together so as to mitigate each other's shortcomings.

19.3.1.1 Suspended culture system: high-rate algal pond (open) and photobioreactor (close)

Suspended culture systems are further divided into open and closed systems. Traditionally, algal culture systems were all open systems, mainly raceway ponds. The ease of open systems lies in the design, construction, and functioning

TABLE 19.2 Basic microalgal–bacterial consortium metabolic reactions in a wastewater treatment process.

Microalgae-based reactions	
$CO_2 + H_2O + NH_4^+ + PO_4^{-3} \rightarrow$ microalgal biomass $+ O_2$	Photosynthesis
$NO_3^- + 4H_2O \rightarrow NH_4^+ + 7OH^-$	Nitrate reduction
Bacteria-based reactions	
$COD + O_2 +$ nutrients $\rightarrow$ bacterial biomass $+ CO_2$	Heterotrophic metabolism
$NH_4^+ + 2O_2 \rightarrow NO_3^- + 2H^+ + H_2O$	Nitrification
$8NO_3^- + 5CH_3COOH \rightarrow 8HCO_3^- + 6H_2O + 2CO_2 + 4\ N_2$	Denitrification

and comes at a very low price as compared with closed systems (Gonçalves et al., 2017). The most common version is the high-rate algal pond (HRAP) where the pond is built in a raceway fashion and paddle wheels are installed for circulating the culture for mixing and preventing sedimentation, the height not exceeding 0.2–0.4 m for efficient light penetration. Being able to be placed in any outdoor scenario with unlimited scale-up potential, open systems feel like the obvious choice, but it suffers from some serious problems. Evaporation losses, inadequate biomass productivity due to low surface to volume ratio, limited parameter adjustment capability, and high contamination probability hinder the efficiency of open ponds. Recently, another open system, the thin-layer cascade system (TLC) is gaining popularity, which addresses some of the drawbacks of HRAP. It has a very thin layer of circulating culture in a sloping cascade-like configuration, enabling very high biomass productivity (Grivalský et al., 2019).

Closed systems or PBRs are highly controllable systems with much higher productivity than traditional open systems. Parameters such as pH, temperature, CO_2 and O_2 concentration, mixing, and substrate feeding can be optimized accordingly. Evaporation losses and outside contamination are much lower in closed systems (Zhou et al., 2020). But high costs in setup and operation, and low robustness in outdoor conditions and scaling up, make the closed system a relatively less sought-out system on an industrial scale. The different configurations that are commonly used are flat plate reactors, column reactors, and tubular reactors, tubular being the most common in industrial scale. In the case of WWT, open systems are preferred rather than closed systems, due to cost constraints, and low requirement for contamination prevention.

19.3.1.2 Nonsuspended culture system: biofilm (open) and matrix immobilization (close)

Microalgae tend to be planktonic in nature and favor suspended growth but can exist as biofilms (attached growth), some species being more inclined toward this type of growth. A biofilm is defined as a layer of organized microbial communities accumulating at a solid–liquid interface (Flemming and Wingender, 2010). With proper nutrient and light availability and a moist environment, biofilm can form at any surface. A biofilm is mainly composed of water, microscopic cells, as well as a mixture of polymers, nutrients, metabolites, cell lysis products, and attachment material components (EPS, extracellular polymeric substances). A biofilm is characterized by the microorganism diversity it contains, which could be only bacterial cells, or a mixture of bacteria, cyanobacteria, microalgae, protozoa, and fungi (Singh et al., 2006). Various factors such as nutrient availability, temperature, pH, and age of the culture determine the EPS chemical composition. This EPS attachment matrix provides the structure of the biofilm and provides stability, protection from grazers, desiccation, and shear stress (Mantzorou and Ververidis, 2019). The present algal biofilm system borrow its form from its predecessor, the first biofilm system established back in 1971 by Torpey et al. for WWT using a mixed consortium of algae and bacteria biomass attached to rotating aluminum disks partially submerged in wastewater (Torpey et al., 1971). Bacterial biofilms–removed carbonaceous organic matter and filamentous algae helped in removing nutrients under fluorescent light. A different kind of algal biofilm system was developed by Johnson and Wen in 2010, which consisted of a rocking shaker where the biofilm support alternated between the culture medium (liquid phase) and open air (gaseous phase), thus helping firm growth of algal biofilm and effective nutrient uptake and CO_2 intake (Johnson and Wen, 2010). Biomass was scrapped off from attachment and showed comparative water content as from a centrifuged biomass from suspended culture, which makes the system cost- and energy-efficient. A permeated biofilm system was made by Ozkan et al., in 2012, where culture media were dripped using nozzles on a thick concrete surface where the microalgae *B. braunii* was immobilized (Ozkan et al., 2012). The aerial biomass productivity at 0.71 g/m^2 d can be considered low with respect to other biofilm systems, but the system was found to be water- and energy-efficient with a 99.7% lower energy consumption and 45% lower water consumption than current raceway ponds and PBRs.

Christenson and Sims developed a system of rotating algal biofilm reactor in 2012, where cotton cords are wrapped around a drum on which the biofilm is formed, and the drum remains partially submerged in wastewater (Christenson and Sims, 2012). They developed a spool harvester, which can scrape off the biomass from the cords. Applaudable aerial productivity of up to 31 g/m^2 d and a net positive energy balance was reported in this study at outdoor conditions. Another kind of rotating algal biofilm reactor was developed by Gross et al., in 2013, which had a triangular configuration of drive shafts on which the attachment material (cotton duct) was placed and the lowest-most shaft was submerged in the culture medium, and it was continuously rotated to make the biofilm alternate between CO_2 and nutrients (Gross et al., 2013). The system was scaled up to pilot scale in 2014 along with a vertical model like a conveyor belt system, which has higher aerial productivity than triangular systems (Gross and Wen, 2014). The whole pilot model included an 8.5 m^2 raceway pond retrofitted with two triangular and six vertical RAB units and showed a significantly high aerial productivity of 21.5 g/m^2 d. The optimized system with conveyer belt RAB was tested with municipal wastewater in 2018 and showed high capabilities in wastewater remediation as compared with raceway ponds (Zhao et al., 2018).

Blanken et al., in 2014, developed a system of series of vertical rotating stainless steel woven mesh disks, called Algadisks that remain partially suspended in a nutrient medium, and it showed good biomass productivity of 20.1 g/m^2 d (Blanken et al., 2014). The system was scaled up to pilot scale and subjected to a 6-month study by Sebestyén et al. as reported in 2016; however, the productivity was considerably lower at 8.4 g/m^2 d owing to various mechanical problems during the experiments. The research boasted a long-term operation of more than 20 weeks without the need for reinoculation, which makes this system worth considering (Sebestyén et al., 2016).

Mainly these biofilm systems are open systems since they are directly exposed to the air. In a study in 2007, a type of closed biofilm system was developed known as a twin-layer system, which used immobilized algal biofilm on the substrate layer and culture medium/wastewater circulated in a source layer (Shi et al., 2007). This system was enhanced in 2017, which saw six times increase in EPS and 10 times increase in cell dry weight when compared with similar suspended culture (Ekelhof and Melkonian, 2017). With high WWT capability, low contamination probability, and easy harvesting, it turned out to be a successful model.

19.3.2 Recent advances on wastewater treatment technologies

WWT technologies using microalgae have advanced over the years, but the basic mechanisms almost remained the same. The advances have largely been in the designing of bioreactors and other culture systems, and methods of system modeling, which recently has been dominated by artificial intelligence (AI)–based models. In terms of culture system design changes, there has been an increase in the use of nonconventional systems like hybrid PBRs, thin-layer cascade systems, attached systems, and so on. A quite different approach to bioreactors called membrane PBRs was developed in which the internal membrane filtration system is used to separate the algal biomass from the discharge and run in continuous mode, thus effectively decreasing the need for harvesting (Sheng et al., 2017). It also enables higher nutrient removal efficiencies and lower footprint than a conventional PBR, and due to its ability to decouple HRT and SRT, a low amount of sludge is produced (Gao et al., 2016). Microalgal community analysis has also recently attracted much attention in identifying efficient microalgal strains for WWT and determined their metabolic functioning, and genome sequencing has been the cornerstone in this regard (Nguyen et al., 2020). Microalgae have been integrated with aquaponics system, and fish-generated waste has been used as a nutrient source, thereby allowing efficient production of vegetables and algal biomass, integrated with wastewater remediation (Addy et al., 2017). Nowadays, microalgal cultures are being used to remediate a diverse range of materials from wastewater, some of which could pave the way to a much cleaner and greener environment. In a 2020 research paper, the authors suggested the use of microalgal-based biopolymer as a biocompatible solution to plastic pollution, where EPS produced by the strain *Cyanothece* sp. was used to aggregate nanoplastics and microplastics, thus facilitating a better remediation approach (Cunha et al., 2020). Recent research has suggested the use of microalgae as a potent approach for removal of antibiotics from wastewater, which could work by processes such as adsorption, hydrolysis, and photodegradation and have also been integrated with other processes such as ultraviolet irradiation (photocatalysis) and advanced oxidation, and microalgal genetic engineering has also been considered as a viable method in this regard (Leng et al., 2020). In a very futuristic approach, architects and biotechnologists have collaborated in making building-integrated PBRs and WWT systems. One such example is the BIQ building consisting of "bioadaptive facades," where microalgae are grown in glass louvres to generate energy and provide shade at the same time. When the biomass becomes too concentrated, it is harvested and sent to a biogas plant (Elrayies, 2018). Another state-of-the-art system has been developed where large-scale printing of immobilized algae-laden hydrogels is done using robotic extrusions, which form biocompatible scaffolds of algal cells, which can be used to create hybrid structures and incorporated in building walls (Malik et al., 2020). This method was used to print algal gels into modular leaf vein–shaped clay tiles termed "Indus" by Bio-ID lab, where wastewater is passed and the algal hydrogel can filter out the toxic chemicals and heavy metals from the wastewater (Bio-ID Lab, 2020). Collaboration between researchers from different fields such as biology, chemical engineering, electrical engineering, computer science, architecture, and so on is the only way forward to realize the full capability of microalgae in WWT.

19.4 Use and processing of microalgae grown on wastewater

The algal culture system and biomass quality can be characterized based on the productivity and composition of the biomass generated. When grown in wastewater instead of a traditional vulture medium, the biomass quality and the lipid content can vary greatly. After the wastewater has been treated, the algal biomass is harvested, and this biomass has a lot of potential based on its quantity and quality (Mishra et al., 2019). Harvesting is a cost- and energy-intensive process, and various harvesting techniques have been developed, which comprise either or a combination of centrifugation, sedimentation, floatation, filtration, and flocculation. Preference of techniques is based on the algal strain or consortia used and the cost or energy trade-off. The algal biomass can be processed in various manners to yield different products such as biofuels, pigment and vitamin production, bioplastics, or as an animal or fish feed.

19.4.1 Biomass to bioenergy

The most common use of microalgae after harvesting is the production of biofuels. There are many biofuel conversion routes, which can be classified into biochemical, thermochemical, and chemical routes. Factors such as energy efficiency, feed quality and quantity, microalgal biomass composition, and setup costs play a major role in determining the pathway to be chosen. Table 19.3 summarizes some of the different conversion routes applied to different microalgal strains.

19.4.1.1 Biochemical conversion

Fermentation to produce bioethanol and biobutanol is a common biochemical conversion method carried out in anaerobic conditions and has a simple mechanism. In alcoholic fermentation, yeast or bacteria are used to hydrolyze and ferment the algal polysaccharides (starch, cellulose) to ethanol after an initial pretreatment stage (Mishra et al., 2019). For butanol production, *Clostridium* sp. is used to ferment the polysaccharides in ABE fermentation process to produce acetone, butanol, and ethanol in the ratio of 3:6:1 (Mishra et al., 2019). Although these products are very valuable, research is limited due to concerns such as price and energy efficiency.

AD is another conversion route where the biomass is broken down into biogas or methane in anaerobic conditions following a series of reactions. In WWT systems, AD is a preferred

TABLE 19.3 Different conversion routes for microalgal biomass.

Types of conversion	Conversion method	Algae	Wastewater source	Pretreatment	Output	References
Biochemical	Anaerobic digestion	*Chlorella pyrenoidosa*	Sewage treatment plant wastewater and synthetic wastewater	Thermal (900°C, 8 h), hydrothermal (121°C, 15 psi, 15 min), ultrasonic (100 W, 15 min), and Fenton (140 rpm, 3 h)	Untreated: Total biogas yield of 479 mL and methane yield of 147 mL volatile solid^{-1} (g) Ultrasonic, thermal, hydrothermal, and Fenton-pretreated microalgae biomass show 15.5%, 8.2%, 18.4%, and 10.9% enhancement in methane production	Kumari et al. (2021)
	Fermentation	*Nannochloropsis oculata Tetraselmis suecica*	Seawater and sewerage wastewater	Alkaline pretreatment	Bioethanol yield: 0.41%–7.26%	Reyimu and Özçimen (2017)
	Anaerobic digestion	Microalgae consortium	Domestic wastewater	Microwave pretreatment	Methane yield: 0.17–0.27 L CH_4/g VS	Passos et al. (2014)
	Anaerobic digestion	Consortium of *Nostoc, Phormidium, Geitlerinema*	Municipal wastewater	Extracted and unextrated phycobiliproteins	Methane yield: 152–209 mL CH_4/g VS	Arashiro et al. (2020)

Continued

TABLE 19.3 Different conversion routes for microalgal biomass.—cont'd

Types of conversion	Conversion method	Algae	Wastewater source	Pretreatment	Output	References
	Anaerobic digestion	*C. vulgaris*, *S. obliquus*, and *C. reindhardtii*	Swine manure slurry	NA	Methane yields: 106 −146 CH_4 g/COD (batch 171 mL CH_4/gCOD (semicontinuous)	Molinuevo-Salces et al. (2016)
	Fermentation	*Scenedesmus obliquus*	Municipal wastewater	NA	Highest ethanol concentration: 2490/mgL, corresponding to an ethanol yield of 25% by weight	Walls et al. (2019)
	Anaerobic digestion	*Chlorella vulgaris*, *Scenedesmus obliquus*, and *Scenedesmus arvernensis*	Domestic, brewery, olive mill wastewater	NA	0.10–0.16 m^3 CH_4/kg volatile solids	Assemany et al. (2020)
	Anaerobic digestion	*Nannochloropsis limnetica*	Municipal wastewater	NA	Methane yield 408 N mL/g VS	Tsapekos et al. (2018)
Thermal	Anaerobic digestion/hydrothermal liquefaction	Consortium of *C. pyrenoidosa*, *Phormidium*	Municipal wastewater	NA	Cumulative biomethane production: 346.59 ± 5 mL/g VS Biocrude yield: 43 ± 2%	Naaz et al. (2019)
	Pyrolysis	Consortium of *Coelastrum* sp., *Actinastrum* sp., diatom sp., *Myopordon pulchellum*	Wastewater treatment HRAP	NA	Overall, < 15% of the WWT HRAP biomass energy (19.7 MJ/kg) was recovered in the pyrolysis biooil at 500°C	Mehrabadi et al. (2017)
	Hydrothermal liquefaction	*Monoraphidium* sp.	Sewage treatment plant effluent	Co-HTL with domestic sewage sludge	Biocrude yield: 39.38%	Mishra and Mohanty (2020)
	Hydrothermal liquefaction	*Galdieria sulphuraria* (monoculture and polyculture)	Municipal wastewater	NA	Biocrude yield: 28.1%	Cheng et al. (2019)

TABLE 19.3 Different conversion routes for microalgal biomass.—cont'd

Types of conversion	Conversion method	Algae	Wastewater source	Pretreatment	Output	References
	Hydrothermal liquefaction	*Chlorella vulgaris*	Municipal WTP	NA	Biocrude yield: 29.37%	Arun et al. (2018)
Chemical	Transesterification	*Chlamydomonas reinhardtii*	Starchy wastewater	Solvent extraction	Biodiesel: 90.34–119.61 g/kg algal biomass	Radhakrishnan et al. (2021)
	Transesterification	*Monoraphidium* sp.	Domestic sewage wastewater	Solvent extraction	30.07% lipid having 55.32% total saturated and 37.07% unsaturated fatty acid	Mishra and Mohanty (2019)
	Immobilized lipase for transesterification	*Acutodesmus obliquus*	Domestic wastewater	Solvent extraction	Highest FAME conversion: 95.36% lipid productivity: 57.03/mg Ld	Guldhe et al. (2019)
	Transesterification	*Desmodesmus* sp. and mixed culture	Domestic wastewater	Solvent extraction	Lipid content: 26.3% 2.5%–3.0% FAME conversion/dry algal biomass	Komolafe et al. (2014)

HRAP, high-rate algal pond; *HTL*, hydrothermal liquefaction; *WTP*, water treatment plant; *WWT*, wastewater treatment.

mode of conversion, and the digested liquid can be used again for microalgal growth, thus closing the loop (Pancha et al., 2019).

A relatively less explored route is biohydrogen production using microalgae. Certain algal strains such as *Chlamydomonas reinhardtii*, *Platymonas subcordiformis*, *Chlorella fusca*, *Scenedesmus obliquus*, and *Chlorococcum littorale* due to the presence of hydrogenase enzyme in their cells are capable of producing hydrogen. Any of the three mechanisms (1) direct biophotolysis, (2) indirect biophotolysis, and (3) two-stage dark/photo fermentation can be used to produce hydrogen; however, these routes have not been commercialized yet due to technical difficulties and low yields (Nagarajan et al., 2017).

19.4.1.2 Thermochemical conversion

Hydrothermal liquefaction (HTL) is a commonly used thermochemical conversion method operating in a near supercritical state, which is used to convert wet microalgal biomass into biocrude or biooil. Although a costly process by itself, HTL enables the whole of algal biomass to be processed to a single entity without any need of dewatering, thus minimizing other post-processing problems (Chiaramonti et al., 2015). The biooil can then be processed in a refinery to produce different biofuels such as jet fuel. Pyrolysis is similar to HTL and is a high temperature (400 and 600°C) process carried out in an inert atmosphere, which converts the algal biomass to biocrude, char, and other gases (Chiaramonti et al., 2017).

19.4.1.3 Chemical conversion

The chemical conversion route, which is transesterification, is the most used and most researched route for algal biomass due to the presence of high concentration of lipids in algal cells. In this process, triglycerides (lipids/fatty acids) in algal biomass are converted to fatty acid alkyl esters (FAAEs), which are commonly called biodiesel and glycerol in the presence of alcohol and a catalyst. Extraction of lipids from biomass and transesterification or direct transesterification of biomass are two of the many routes to form biodiesel (Hemalatha et al., 2019).

19.4.2 Biomass to bioactives (a biorefinery approach)

Microalgae and cyanobacteria have long been known for their nutritional properties, the most popular being Spirulina. Different microalgal strains are known to produce different amounts of bioactive compounds such as vitamins (riboflavin, thiamine), pigments (b-carotene; chlorophyll, etc.), protein (C-phycocyanin, A-phycocyanin), and fatty acids (palmitic acid, linoleic acid, stearic acid, oleic acid, etc.), which can replace their chemically synthesized counterparts (Cuellar-Bermudez et al., 2017). Algal lipids are a good way of producing docosahexaenoic acid (DHA) and eicosapentaenoic acid (EPA), which are used for both food and pharmaceutical industries. Polyunsaturated fatty acids (PUFAs), mostly omega-3 fatty acids, have been in great demand as essential nutrients and as a cure for cardiovascular diseases (Chandra et al., 2019; Liang et al., 2015). Extraction of these compounds poses technical difficulties, and the cost of production is essentially high, and thus integration in what is called a microalgal biorefinery is an efficient way to bring the process to a feasible one. A biorefinery is similar to that of a petroleum refinery that converts an initial raw material, in this case, biomass, into products of value and biofuel. This is a cascading system, where each step of fractionation (cascading phase) leads to various intermediates, which are further converted to multifarious value-added products (Cherubini, 2010). For processes that are uneconomical but are of utmost industrial importance, integrating them with a refinery approach is the only viable way to keep the processes running. The same is the case for algal biofuel production. Research till now has not been able to make the algal biofuel production cost-efficient, so it has been incorporated in a biorefinery where wastewater and CO_2 from exhausts can be used as raw materials, and algal biomass is fractionated into different products starting from food, nutraceuticals, cosmetics, bioplastics, and pharmaceutically active compounds. Apart from biodiesel, which is obtained from the lipid part of algal biomass after chemical treatment, the deoiled biomass can be converted into bioethanol by fermentation, biomethane by AD, biohydrogen, and so on (Chandra et al., 2019).

Algae and cyanobacteria have recently been used for bioplastic production, which is mainly Polyhydroxyalkanoates (PHAs). In nutrient-deprived conditions or excess organic carbon sources, PHAs are accumulated, which act as a store of carbon and energy. Polyhydroxybutyrate (PHB), a kind of PHA, produced from microalgae has similar properties to the common plastic, polypropylene, and is even better as it is nontoxic, biocompatible, and biodegradable (Poirier et al., 1995). The cost and energy requirements of bioplastic production from algae are high, and integration with biorefinery seems to be the only viable option.

An ideal integrated biorefinery approach aims to utilize the algal biomass to its fullest, aiding in biofuel production, with bioactive production, and coupling with WWT and CO_2 sequestration, taking a closed-loop approach, an example of circular bioeconomy, a schematic of which is shown in Fig. 19.2.

19.5 Artificial intelligence and modeling of wastewater treatment systems

Gone are the days when every biological system used to be handled and monitored by

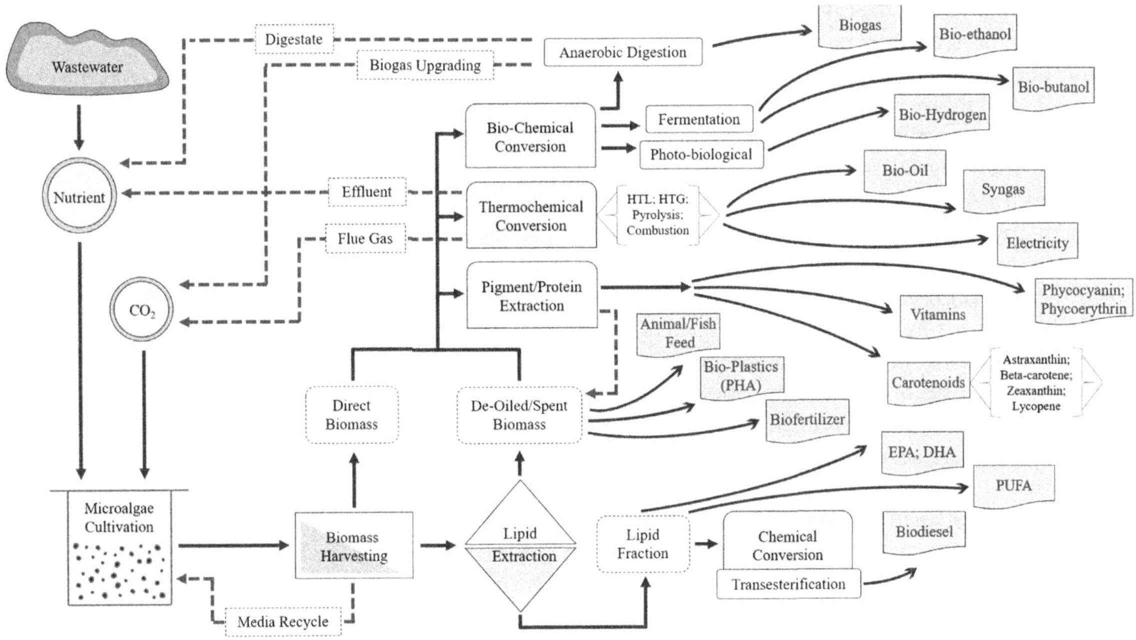

FIGURE 19.2 Schematic of an integrated microalgal biorefinery.

human hands. Every technological field nowadays makes use of data and AI in the maximum possible manner. Although biological systems started using AI very late in the game, it is now widely being researched and implemented upon. Modeling of complex biological systems and using these in a technological aspect, automation of routine tasks, and monumental data analysis are some of the paths where AI plays a big role. Now, these approaches are being used in the WWT scenario because of obvious advantages such as minimal experimental work, less chemical consumption, and effective treatment results. In the case of WWT plants, there are various cascading processes, and all these processes must meet industrial guidelines so that it can be considered safe, and thus, a wholesome model is desirable. Also, wastewater is a dynamic entity, and measurement of the incoming wastewater parameters every time is not only a hassle, but sometimes is not at all possible, such as BOD which takes 5 days to obtain. Drawing a correlation between easily obtainable parameters with complex parameters is a time-consuming process and would require an immense amount of experimental data. Scenarios such as these are exactly where AI comes into play, and in this case, researchers use what is called an artificial neural network (ANN). ANN is one of the AI tools; others such as fuzzy logic, genetic algorithms (GA), particle swarm optimization, and Monte Carlo simulation are also used extensively for the optimization of plants and modeling of process parameters (Ye et al., 2020). Nowadays the most commonly used AI tools in WWT systems are ANN and GA, or a combination of both.

19.5.1 Artificial neural network

ANN derives its name and functioning from biological neural networks in the human brain. Like the brain, it has the ability to learn from existing information to solve problems. The basic unit of ANN is a neuron, which contains information either from external input or from other neurons, and many of these neurons are grouped in layers. These artificial neurons can receive and transmit their signals to other neurons just like synapses present in the brain. There are (1) input layer or the studied variable, which in a WWT scenario could be pH, pollutant concentration or flow rate, (2) the output layer or the response variable, e.g., removal efficiency or rate of pollutant removal; and (3) hidden layer, which could be just one or more than one (Fig. 19.3). The hidden layer is the mainframe of an ANN, and by providing several example data of inputs and outputs and an algorithm, the ANN learns or is trained to form the connections of hidden layers, where the task is to predict the output as close to the real values as possible (minimum error) and thus is subjected to repeated calculations.

ANNs for their capability of self-learning and adapting have been used widely for solving nonlinear processes and design experiments, and for process simulation, prediction, and optimization. For wastewater systems, ANNs have been trained to predict BOD and also removal efficiency of BOD in treatment plants (Heddam et al., 2016; Ahmadi et al., 2018). Nutrient, heavy metal, and synthetic dye removal models have also been developed using ANNs (da Silva Ribeiro et al., 2019; Sekulić et al., 2017; Ebrahimpoor et al., 2019). The models generated by ANN have been used to predict effluent quality and quantity and in real-time monitoring of plants (Lee et al., 2011; Najafzadeh and Zeinolabedini, 2019). In the case of microalgal WWT systems, ANNs have been used to design whole experimental setup and parameters, or estimate microalgal growth and for efficient treatment in a biorefinery system (Nayak et al., 2018; Yazdani et al., 2020). In a 2016 study, ANN (BP-LM training algorithm) was used to model the removal efficiency of toxic element arsenic (As(III) and As(V)) in synthetic wastewater by the microalgae *Botryococcus braunii* (Podder and Majumder, 2016). A high correlation coefficient of $R^2 = 0.9998$ was achieved between the

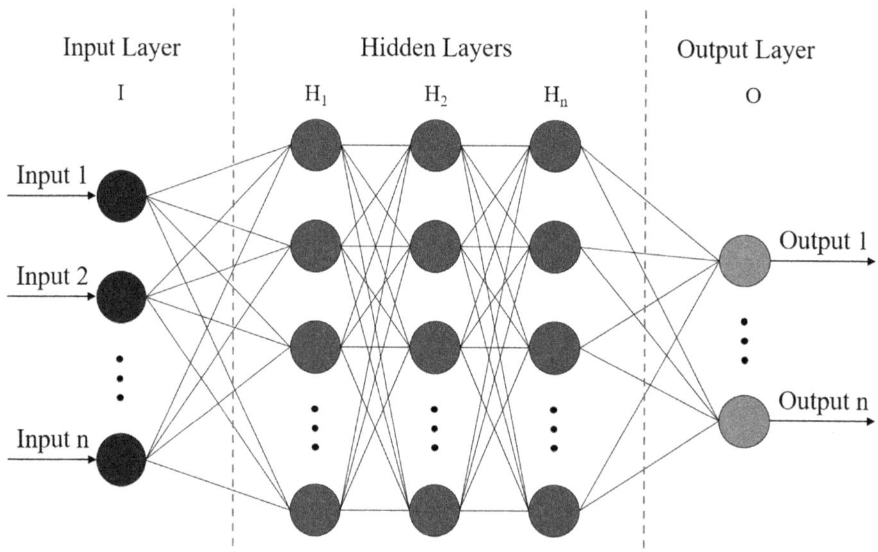

FIGURE 19.3 Basic structure of a layered artificial neural network.

experimental and modeled outputs, thus exhibiting the efficacy of the ANN model. Another study, although not directly related to WWT, but could be useful in designing WWT systems, trained an ANN model to discriminate between algal monocultures and mixed cultures by using just the data of their spectral signatures (Franco et al., 2019). In many systems, GA has been used in conjunction with ANN for even better process optimization (Dineshkumar et al., 2015; Nayak et al., 2018).

19.5.2 Genetic algorithm

GA, a subset of evolutionary algorithms (EA), specializes in generating superior-level solutions in search and optimization problems (Grefenstette, 1986). These algorithms are several notches above a random search, with its capability to direct solutions in the region of better performance. GA follows the rules of Darwinian natural selection and genetics and contains similarly named operators such as selection, crossover, and mutation, which are applied to possible solutions called chromosomes. Simply put, GAs contain randomized data called chromosomes, and these are made up of characters or bits of information called genes with each chromosome assigned a fitness score, a higher value means a more optimal solution. In each iteration, high fitness value chromosomes are selected and allowed to reproduce with each other by crossing over of genes, thus producing even better offspring having higher fitness scores. Every next generation of results is better than the previous, and sometimes random mutations (random addition, deletion, or exchange of genes) are done to prevent premature convergence and maintain diversity (Fig. 19.4). After several iterations, the optimal solution will be achieved by the GA (Mulligan and Brown, 1998).

GAs have been used in desalination plants and in determining flocculant doses in WWT (Abdulrahim and Alasfour, 2010; Dawood and Li, 2013). In a 2016 paper, GA was used alongside water quality models to determine minimum required efficiencies of sewage treatment for watershed, and the optimized model was used to design the treatment system at the minimum cost possible (Louzadavalory et al., 2016). GA-ANN models have been found to be popular than just using GA alone. In comparison studies, using both GA and ANN was found to be more robust and superior than just GA and even other models (Mohan et al., 2015). Like in a 2018 study, GA-ANN model was developed for the removal of Cd(II) from aqueous solution using natural waste substances (Nag et al., 2018). In a report from 2020, the effectiveness of an ANN-GA model was demonstrated in optimizing the operational conditions for treating dye wastewater using an electrooxidation (EO) process (Picos-Benítssez et al., 2020). In microalgae wastewater systems, GA has been used to determine parameters for efficient lead removal by a microalgae–bacteria consortium (Hernández-Melchor et al., 2018). An ANN-GA model

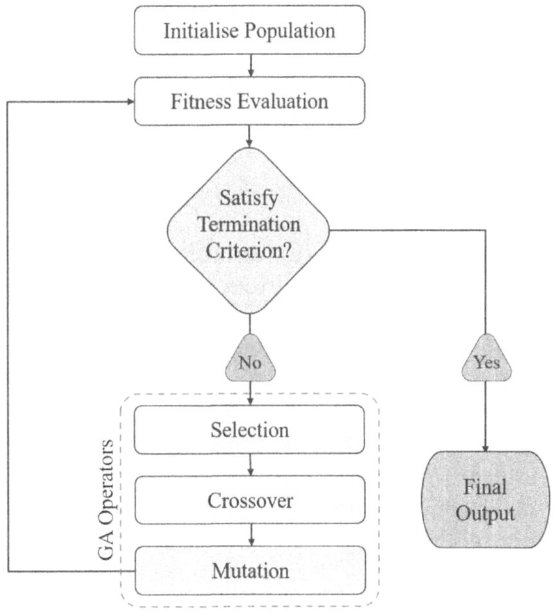

FIGURE 19.4 Basic structure of a genetic algorithm (GA).

was developed to optimize process parameters in a microalgal (*Scenedesmus* sp.) biorefinery in a 2018 study, for treating domestic wastewater along with CO_2 sequestration from flue gas (Nayak et al., 2018).

19.6 Assessing sustainability (life cycle assessment and technoeconomic analysis)

No process should be developed which is not sustainable, whether it be economical or environmental; this statement seems like a no-brainer, but regrettably, most of today's industrial processes and systems are not sustainable to the environment, thus jeopardizing the earth's future. Even if they are made to be environmentally sustainable, it is hard to make them economic. Earlier due to lack of public or scientific outcry over environmental degradation, most systems used to be pretty harsh to Mother Nature, and there were almost zero laws and policies, which would keep control over the amount of leeway given toward such companies. As our environment went lush to grim over the course of years, laws and policies by governments and international bodies mandating checks on how much a system is allowed to pollute became not only desirable but also utterly necessary. So nowadays, every product or commodity reaching the market or so has to go through a series of checks, which determine its holistic value in relation to people, market, and the environment. Two of the most sought-after devices in this respect are life cycle assessment (LCA) and technoeconomic analysis (TEA). LCA is the core for any sustainability assessment system and describes the environmental interactions and impact, whereas TEA is a relatively less used device used to describe the feasibility of a system considering the technology and economic impact (Garlapati et al., 2019).

19.6.1 Life cycle assessment

LCA, as its name suggests, is a tool that analyses a product or process's whole life cycle in terms of sustainability toward energy and environment. Usually called a cradle-to-grave analysis, LCA estimates the environmental impact at each stage of a product/process starting from acquiring raw material (cradle) and processing to manufacturing, distribution to its use, and finally disposal (grave). The LCA process is divided into four steps: (1) aim and scope; (2) inventory; (3) impact assessment; and (4) interpretation. In simple terms, the first step defines what the study will do and determines boundaries, assumptions, and impact categories (e.g., toxicity, global warming). The second step, "inventory analysis" develops the flow model of processes considering all inputs (raw materials, energy) and outputs (products, pollutants). In "impact assessment" (third step), conclusions can be drawn as to which parameters are to be considered for the model to be in our liking. Environmental impacts are evaluated based on their priorities, for example, whether to decrease air pollution but compromising on land use, thus making it more environmentally sustainable. Comparing models to find the most sustainable approach is a part of the third step. The last step or "interpretation" evaluates the whole model and information gained from the inventory and assessment stages and gauges whether it is well substantiated. LCEA (life cycle energy analysis) is an accompaniment to LCA and concerns energy sustainability. Different software packages have been specifically made for LCA studies, the most common being GaBi, GREET Model, and SimaPro (Shah et al., 2016). Considering research on microalgae and WWT, there have been many LCA studies, mostly focusing on biodiesel production (de Souza et al., 2019). LCA analyses suggested that microalgae cultivation was more feasible economically when cultivated in raceway ponds rather than PBRs, but PBRs tended to be more environmentally sustainable (Jorquera et al., 2010; Zaimes and Khanna, 2013; Silva et al., 2015). Another LCA study showed that the use of saltwater or wastewater was more sustainable eliminating the need of adding most nutrients (Yang et al., 2011). Two terminologies GHG (greenhouse gas emissions) and NER (net energy ratios) have been commonly used to quantify sustainability in microalgal systems. In a 2019 comparative study between different algal biofuel conversion routes, LCA showed that biogas production using AD with pretreatment (NER: 0.72; GHG: 60.84 g $CO_{2\text{-eq}}$ (MJ biogas)$^{-1}$) was more industrially and environmentally sustainable than biodiesel/biooil production using either transesterification, hydrothermal liquefaction, or pyrolysis (Sun et al., 2019). A quite insightful research on algal WWT was carried out in a 2014 study where GHG emissions from algal jet fuel production were estimated using LCA in two scenarios (Fortier et al., 2014). One scenario was where HTL is performed at microalgae WWT site and biocrude transported to refinery to produce algal jet fuel, and the other was to transport dewatered algae to refinery and perform HTL there; the first scenario (35.2 kg CO_{2eq}/GJ) was found to be much lower in GHG emissions than the second (86.5 kg CO_{2eq}/GJ). A somewhat ironic LCA study from 2017 found that the potential environmental impact due to discharge of untreated effluents from sewage wastewater was quite less than one, which was microalgae-treated and suggested the correlation with the amount of electric consumption needed for microalgae cultivation (Diniz et al., 2017). Although there have been many LCA studies over the years regarding microalgae in WWT, the consensus regarding its environmental and energy sustainability remains highly divisive. Nowadays, LCA is mostly used alongside TEA for a better understanding of the processes and their viability in real life (Kern et al., 2017, 2019; DeRose et al., 2019; Barlow et al., 2016).

19.6.2 Technoeconomic Analysis

Before a product or process is brought to the public or market, it has to go through several models and designs each with its own set of pros and cons, and this is where TEA comes into play. By assessing each model in terms of its technical practicability and economic feasibility, TEA can determine the best possible model. A major aim of TEA is to calculate the cost of production. The TEA process can be broadly divided into three steps: (1) defining the technology and system boundaries; (2) process designing (mass and energy balance, workflow); and (3) economic analysis (capital and production costs). TEA has been widely used to model microalgal WWT processes and subsequently to determine the selling price of algal biofuels produced (Ranganathan and Savithri, 2019; Ansari et al., 2019). Some studies have estimated the selling price of algal biofuels from as low as \$2.23/gallon (Xin et al., 2016), while some claiming it to be high as \$33.16/gallon (Richardson et al., 2012). In a 2020 study, TEA was used to evaluate a dairy WWT system using microalgae in a V-shaped pond and was found to have an IRR of 118%, payback period of 1.9 years for 1 MLD plant capacity, and the biomass production price to be \$482/ton (Kumar et al., 2020). An integrated LCA-TEA was performed in a 2019 study to compare two processing pathways of Algal Tuft Scrubber, a highly successful WWT system: the biochemical pathway, where the harvested biomass is subjected to fermentation and then HTL, and the thermochemical pathway, where the whole biomass is directly subjected to HTL (DeRose et al., 2019). TEA estimates showed a higher fuel selling price of \$12.85 GGE^{-1} for the biochemical pathway than the thermochemical one (\$10.41 GGE^{-1}). LCA estimates corroborated that the thermochemical pathway was more sustainable as it had a net negative GHG emissions of −2g CO_{2eq} MJ/fuel, whereas biochemical pathway had GHG emissions of 111.2 g CO_{2eq} MJ/fuel. However, sensitivity analysis considering lower ash content and feedstock cost projected a lower MFSP of \$3.85 GGE^{-1} for the biochemical pathway than thermochemical pathway, as the coproduct value is much higher in the first pathway. All things considered, with so many sustainability studies being done on algal use, WWT, and biofuel production; there has been almost no system that can be considered sustainable on all levels, whether it be technology, environment, energy, or economy.

19.7 Challenges and future prospects

WWT using microalgae is a highly sought-after method in recent times, but this process also suffers from some liabilities and limitations in its current state. Different wastewater sources have different compositions with varying N:P ratios, which is a determining factor in microalgal growth (Salama et al., 2017). So, a single system or a specific type of algal strain or culture does not suffice for every kind of wastewater, each system requiring its own modeling, thereby decreasing its potential for effective commerciality. Another problem in the biological treatment system is the presence of toxic compounds in wastewater, and due to the dynamic nature of wastewater, it cannot be predicted efficiently, which can result in sudden death of microalgae, thus halting the whole system (Lavrinovičs and Juhna, 2018). Filtration systems have been designed for the removal of competing microbes and suspended solids, which could hamper light penetration, but it adds to the overall operation costs making the system economically less attractive (Salama et al., 2017). Selection of algal strains tolerant to a specific wastewater type is a time-consuming process, and when added to the subsequent optimization steps, microalgal WWT seems like a very slow process. LCA and TEA studies have suggested that most of these wastewater systems are not sustainable on a large or industrial scale, and the biorefinery approach has been suggested as a savior option.

With increasing technologies and the advent of powerful AI, the future is progressing to a cleaner and greener earth. Modeling of wastewater parameters, treatment systems, and automation is going to hugely decrease the overall costs while increasing energy efficiency. Algal culture systems have also been developed, which seem almost futuristic, such as growing algae on the walls of buildings, algae tiles that extract toxic dyes from water, and some architects have even envisioned whole cities with self-sustaining structures that produce their own electricity and remediate the wastes produced (Elrayies, 2018; Malik et al., 2020). Biorefinery approach as a holistic process encompassing a zero-waste system has immense potential to be the future of WWT across the globe. Many commercial enterprises have started working on algal biofuel production and WWT on a large scale. Companies such as Algenol Biofuels Inc., BP Amoco plc., ExxonMobil Corp., and Martek Biosciences Corp. have done considerable research and development in terms of algal biorefinery (Salama et al., 2017). In Japan, the microalgal company Euglena has announced plans to generate commercial-scale jet fuel from microalgae. Already a billion-dollar industry, which is mainly due to nutraceuticals and microalgae commercialization, can soon become one of the most attractive when proper WWT systems are developed and incorporated in biorefinery (Sukla, 2019).

19.8 Conclusion

Microalgae have been deemed as a savior in terms of a zero-fossil fuel future for many decades now, but the rate of progress in research has been very slow. Sustainability in terms of both energy and environment has not been very satisfactory, but now interest has again resurfaced with advances in automation, modeling, and newer culture techniques. As a wastewater remediating tool, microalgae are impeccable in their efficiency and ease of operation. Certain design modifications are necessary to harness their true possible, where other systems can be incorporated because as of now, very few large-scale algal WWT systems have been designed, which act as a true biorefinery. To make the process economic, and environmentally sustainable, closed-loop systems or cradle-to-cradle systems are needed.

References

Abdulrahim, H.K., Alasfour, F.N., 2010. Multi-objective optimisation of hybrid MSF-RO desalination system using genetic algorithm. Int. J. Exergy 7 (3), 387–424.

Addy, M.M., Kabir, F., Zhang, R., Lu, Q., Deng, X., Current, D., et al., June 2017. Co-cultivation of microalgae in aquaponic systems. Bioresour. Technol. 245, 27–34.

Ahmadi, A., Fatemi, Z., Nazari, S., 2018. Assessment of input data selection methods for BOD simulation using data-driven models: a case study. Environ. Monit. Assess 190 (4).

Ansari, F.A., Ravindran, B., Gupta, S.K., Nasr, M., Rawat, I., Bux, F., 2019. Techno-economic estimation of wastewater phycoremediation and environmental benefits using Scenedesmus obliquus microalgae [Internet] J. Environ. Manag. 240, 293–302. https://doi.org/10.1016/j.jenvman.2019.03.123 (November 2018), Available from:

Arashiro, L.T., Ferrer, I., Pániker, C.C., Gómez-Pinchetti, J.L., Rousseau, D.P.L., Van Hulle, S.W.H., et al., 2020. Natural pigments and biogas recovery from microalgae grown in wastewater. ACS Sustain. Chem. Eng. 8 (29), 10691–10701.

Arun, J., Varshini, P., Prithvinath, P.K., Priyadarshini, V., Gopinath, K.P., 2018. Enrichment of bio-oil after hydrothermal liquefaction (HTL) of microalgae *C. vulgaris* grown in wastewater: Bio-char and post HTL wastewater utilization studies. Bioresour. Technol. 261 (March), 182–187. https://doi.org/10.1016/j.biortech.2018.04.029.

Assemany, P., de Paula Marques, I., Calijuri, M.L., Reis, A., 2020. Complementarity of substrates in anaerobic digestion of wastewater grown algal biomass. Waste Biomass Valorization 11 (11), 5759–5770. https://doi.org/10.1007/s12649-019-00875-8.

Barlow, J., Sims, R.C., Quinn, J.C., 2016. Techno-economic and life-cycle assessment of an attached growth algal biorefinery [Internet] Bioresour. Technol. 220, 360–368. https://doi.org/10.1016/j.biortech.2016.08.091. Available from:

Barnharst, T., Rajendran, A., Hu, B., 2018. Bioremediation of synthetic intensive aquaculture wastewater by a novel feed-grade composite biofilm. Int. Biodeterior. Biodegrad. 126, 131–142 (October 2017).

De Bhowmick, G., Sarmah, A.K., Sen, R., January 2019. Performance evaluation of an outdoor algal biorefinery for sustainable production of biomass, lipid and lutein valorizing flue-gas carbon dioxide and wastewater cocktail [Internet] Bioresour. Technol. 283, 198–206. https://doi.org/10.1016/j.biortech.2019.03.075. Available from:

Bio-ID Lab, 2020. Indus I Saving Water, Saving Lives [Internet] [cited 2020 Sep 30]. Available from: https://www.induswater.co/.

Blanken, W., Janssen, M., Cuaresma, M., Libor, Z., Bhaiji, T., Wijffels, R.H., 2014. Biofilm growth of *Chlorella sorokiniana* in a rotating biological contactor based photobioreactor. Biotechnol. Bioeng. 111 (12), 2436–2445.

Cai, T., Park, S.Y., Li, Y., 2013. Nutrient recovery from wastewater streams by microalgae: status and prospects [Internet] Renew Sustain. Energy Rev. 19, 360–369. https://doi.org/10.1016/j.rser.2012.11.030. Available from:

Chandra, R., Iqbal, H.M.N., Vishal, G., Lee, H.S., Nagra, S., 2019. Algal biorefinery: a sustainable approach to valorize algal-based biomass towards multiple product recovery. Bioresour. Technol. 278, 346–359 (November 2018).

Cheng, F., Jarvis, J.M., Yu, J., Jena, U., Nirmalakhandan, N., Schaub, T.M., et al., 2019. Bio-crude oil from hydrothermal liquefaction of wastewater microalgae in a pilot-scale continuous flow reactor. Bioresour. Technol. 294 (September), 122184. https://doi.org/10.1016/j.biortech.2019.122184.

Cherubini, F., 2010. The biorefinery concept: using biomass instead of oil for producing energy and chemicals [Internet] Energy Convers. Manag. 51 (7), 1412–1421. https://doi.org/10.1016/j.enconman.2010.01.015. Available from:

Chiaramonti, D., Prussi, M., Buffi, M., Casini, D., Rizzo, A.M., 2015. Thermochemical conversion of microalgae: challenges and opportunities [Internet] Energy Procedia 75, 819–826. https://doi.org/10.1016/j.egypro.2015.07.142. Available from:

Chiaramonti, D., Prussi, M., Buffi, M., Rizzo, A.M., Pari, L., 2017. Review and experimental study on pyrolysis and hydrothermal liquefaction of microalgae for biofuel production [Internet] Appl. Energy 185, 963–972. https://doi.org/10.1016/j.apenergy.2015.12.001. Available from:

Chinnasamy, S., Bhatnagar, A., Hunt, R.W., Das, K.C., 2010. Microalgae cultivation in a wastewater dominated by carpet mill effluents for biofuel applications [Internet] Bioresour. Technol. 101 (9), 3097–3105. https://doi.org/10.1016/j.biortech.2009.12.026. Available from:

Choudhary, P., Prajapati, S.K., Kumar, P., Malik, A., Pant, K.K., 2017. Development and performance evaluation of an algal biofilm reactor for treatment of multiple wastewaters and characterization of biomass for diverse applications [Internet] Bioresour. Technol. 224, 276–284. https://doi.org/10.1016/j.biortech.2016.10.078. Available from:

Christenson, L.B., Sims, R.C., 2012. Rotating algal biofilm reactor and spool harvester for wastewater treatment with biofuels by-products. Biotechnol. Bioeng. 109 (7), 1674–1684.

Cuellar-Bermudez, S.P., Aleman-Nava, G.S., Chandra, R., Garcia-Perez, J.S., Contreras-Angulo, J.R., Markou, G., et al., 2017. Nutrients utilization and contaminants removal. A review of two approaches of algae and cyanobacteria in wastewater [Internet] Algal. Res. 24, 438–449. https://doi.org/10.1016/j.algal.2016.08.018. Available from:

Cunha, C., Silva, L., Paulo, J., Faria, M., Nogueira, N., Cordeiro, N., 2020. Microalgal-based biopolymer for nano- and microplastic removal: a possible biosolution for wastewater treatment. Environ. Pollut. 263.

Dawood, A.S., Li, Y., 2013. Modeling and optimization of new flocculant dosage and pH for flocculation: removal of pollutants from wastewater. Water (Switzerland) 5 (2), 342–355.

DeRose, K., DeMill, C., Davis, R.W., Quinn, J.C., 2019. Integrated techno economic and life cycle assessment of the conversion of high productivity, low lipid algae to renewable fuels [Internet] Algal. Res. 38, 101412. https://doi.org/10.1016/j.algal.2019.101412 (July 2018), Available from:

Dineshkumar, R., Dash, S.K., Sen, R., 2015. Process integration for microalgal lutein and biodiesel production with concomitant flue gas CO_2 sequestration: a biorefinery model for healthcare, energy and environment [Internet] RSC Adv. 5 (90), 73381–73394. Available from: http://xlink.rsc.org/?DOI=C5RA09306F.

Diniz, G.S., Tourinho, T.C.O., Silva, A.F., Chaloub, R.M., 2017. Environmental impact of microalgal biomass production using wastewater resources. Clean Technol. Environ. Pol. 19 (10), 2521–2529.

Ebrahimpoor, S., Kiarostami, V., Khosravi, M., Davallo, M., Ghaedi, A., 2019. Bees metaheuristic algorithm with the aid of artificial neural networks for optimization of acid red 27 dye adsorption onto novel polypyrrole/$SrFe_{12}O_{19}$/graphene oxide nanocomposite. Polym. Bull. 76 (12), 6529–6553.

Ekelhof, A., Melkonian, M., November 2017. Enhanced extracellular polysaccharide production and growth by microalga *Netrium digitus* in a porous substrate bioreactor [Internet] Algal Res. 28, 184–191. https://doi.org/10.1016/j.algal.2017.11.003. Available from:

Elrayies, G.M., January 2018. Microalgae: prospects for greener future buildings [Internet] Renew Sustain. Energy Rev. 81, 1175–1191 (August 2016), Available from: https://linkinghub.elsevier.com/retrieve/pii/S1364032117311747.

Ferreira, A., Ribeiro, B., Ferreira, A.F., Tavares, M.L.A., Vladic, J., Vidović, S., et al., September 2, 2019. Scenedesmus obliquus microalga-based biorefinery – from brewery effluent to bioactive compounds, biofuels and biofertilizers – aiming at a circular bioeconomy [Internet] Biofuels Bioprod. Bioref. 13 (5), 1169–1186. Available from: https://onlinelibrary.wiley.com/doi/abs/10.1002/bbb.2032.

Filippino, K.C., Mulholland, M.R., Bott, C.B., 2015. Phycoremediation strategies for rapid tertiary nutrient removal in a waste stream [Internet] Algal. Res. 11, 125–133. https://doi.org/10.1016/j.algal.2015.06.011. Available from:

Flemming, H.C., Wingender, J., 2010. The biofilm matrix. Nat. Rev. Microbiol. 8 (9), 623–633.

Fortier, M.-O.P., Roberts, G.W., Stagg-Williams, S.M., Sturm, B.S.M., June 2014. Life cycle assessment of bio-jet fuel from hydrothermal liquefaction of microalgae [Internet] Appl. Energy 122, 73–82. https://doi.org/10.1016/j.apenergy.2014.01.077 (July 2011), Available from:

Franco, B.M., Navas, L.M., Gómez, C., Sepúlveda, C., Acién, F.G., 2019. Monoalgal and mixed algal cultures discrimination by using an artificial neural network [Internet] Algal Res. 38, 101419. https://doi.org/10.1016/j.algal.2019.101419 (August 2018), Available from:

Ganeshkumar, V., Subashchandrabose, S.R., Dharmarajan, R., Venkateswarlu, K., Naidu, R., Megharaj, M., 2018. Use of mixed wastewaters from piggery and winery for nutrient removal and lipid production by *Chlorella* sp. MM3 [Internet] Bioresour. Technol. 256, 254–258. https://doi.org/10.1016/j.biortech.2018.02.025 (December 2017), Available from:

Gao, F., Li, C., Yang, Z.H., Zeng, G.M., Feng, L.J., Liu, J.Z., et al., 2016. Continuous microalgae cultivation in aquaculture wastewater by a membrane photobioreactor for biomass production and nutrients removal [Internet] Ecol. Eng. 92, 55–61. https://doi.org/10.1016/j.ecoleng.2016.03.046. Available from:

Garlapati, V.K., Tewari, S., Ganguly, R., 2019. Life cycle assessment of first-, second-generation, and microalgae biofuels. In: Advances in Feedstock Conversion Technologies for Alternative Fuels and Bioproducts [Internet]. Elsevier, pp. 355–371. https://doi.org/10.1016/B978-0-12-817937-6.00019-9. Available from:

Gonçalves, A.L., Pires, J.C.M., Simões, M., 2017. A review on the use of microalgal consortia for wastewater treatment. Algal Res. 24, 403–415.

Grefenstette, J., January 1986. Optimization of control parameters for genetic algorithms [Internet] IEEE Trans. Syst. Man. Cybern. 16 (1), 122–128. Available from: http://ieeexplore.ieee.org/document/4075583/.

Grivalský, T., Ranglová, K., da Câmara Manoel, J.A., Lakatos, G.E., Lhotský, R., Masojídek, J., 2019. Development of thin-layer cascades for microalgae cultivation: milestones (review). Folia Microbiol. 64 (5), 603–614.

Gross, M., Henry, W., Michael, C., Wen, Z., 2013. Development of a rotating algal biofilm growth system for attached microalgae growth with in situ biomass harvest [Internet] Bioresour. Technol. 150, 195–201. https://doi.org/10.1016/j.biortech.2013.10.016. Available from:

Gross, M., Wen, Z., 2014. Yearlong evaluation of performance and durability of a pilot-scale Revolving Algal Biofilm (RAB) cultivation system [Internet] Bioresour. Technol. 171, 50–58. https://doi.org/10.1016/j.biortech.2014.08.052. Available from:

Guldhe, A., Singh, P., Renuka, N., Bux, F., 2019. Biodiesel synthesis from wastewater grown microalgal feedstock using enzymatic conversion: A greener approach. Fuel 237 (September 2018), 1112–1118. https://doi.org/10.1016/j.fuel.2018.10.033.

Gupta, S., Pawar, S.B., Pandey, R.A., 2019. Current practices and challenges in using microalgae for treatment of nutrient rich wastewater from agro-based industries [Internet] Sci. Total Environ. 687, 1107–1126. https://doi.org/10.1016/j.scitotenv.2019.06.115. Available from:

Gutiérrez, R., Ferrer, I., González-Molina, A., Salvadó, H., García, J., Uggetti, E., 2016. Microalgae recycling improves biomass recovery from wastewater treatment high rate algal ponds. Water Res. 106, 539–549.

Heddam, S., Lamda, H., Filali, S., 2016. Predicting effluent biochemical oxygen demand in a wastewater treatment plant using generalized regression neural network based approach: a comparative study. Environ. Process 3 (1), 153–165.

Hemalatha, M., Sravan, J.S., Min, B., Venkata Mohan, S., March 2019. Microalgae-biorefinery with cascading resource recovery design associated to dairy wastewater treatment [Internet] Bioresour. Technol. 284, 424–429. https://doi.org/10.1016/j.biortech.2019.03.106. Available from:

Hernández-Melchor, D.J., López-Pérez, P.A., Carrillo-Vargas, S., Alberto-Murrieta, A., González-Gómez, E., Camacho-Pérez, B., 2018. Experimental and kinetic study for lead removal via photosynthetic consortia using genetic algorithms to parameter estimation. Environ. Sci. Pollut. Res. 25 (22), 21286–21295.

Hodges, A., Fica, Z., Wanlass, J., VanDarlin, J., Sims, R., 2017. Nutrient and suspended solids removal from petrochemical wastewater via microalgal biofilm cultivation [Internet] Chemosphere 174, 46–48. https://doi.org/10.1016/j.chemosphere.2017.01.107. Available from:

Hu, Y., Hao, X., van Loosdrecht, M., Chen, H., 2017. Enrichment of highly settleable microalgal consortia in mixed cultures for effluent polishing and low-cost biomass production [Internet] Water Res. 125, 11–22. https://doi.org/10.1016/j.watres.2017.08.034. Available from:

Johnson, M.B., Wen, Z., 2010. Development of an attached microalgal growth system for biofuel production. Appl. Microbiol. Biotechnol. 525–534.

Jorquera, O., Kiperstok, A., Sales, E.A., Embiruçu, M., Ghirardi, M.L., February 2010. Comparative energy life-cycle analyses of microalgal biomass production in open ponds and photobioreactors [Internet] Bioresour. Technol. 101 (4), 1406–1413. https://doi.org/10.1016/j.biortech.2009.09.038. Available from:

Kern, J.D., Gorelick, D.E., Characklis, G.W., Macklin, C.M., 2019. Multiobjective optimal siting of algal biofuel production with municipal wastewater treatment in watersheds with nutrient trading markets. J. Water. Resour. Plan Manag. 145 (2), 1–13.

Kern, J.D., Hise, A.M., Characklis, G.W., Gerlach, R., Viamajala, S., Gardner, R.D., 2017. Using life cycle assessment and techno-economic analysis in a real options framework to inform the design of algal biofuel production facilities [Internet] Bioresour. Technol. 225, 418–428. https://doi.org/10.1016/j.biortech.2016.11.116. Available from:

Komolafe, O., Velasquez Orta, S.B., Monje-Ramirez, I., Noguez, I.Y., Harvey, A.P., Orta Ledesma, M.T., 2014. Biodiesel production from indigenous microalgae grown in wastewater. Bioresour Technol 154, 297–304. https://doi.org/10.1016/j.biortech.2013.12.048.

Koutra, E., Grammatikopoulos, G., Kornaros, M., 2018. Selection of microalgae intended for valorization of digestate from agro-waste mixtures [Internet] Waste Manag. 73, 123–129. https://doi.org/10.1016/j.wasman.2017.12.030. Available from:

Kumar, A.K., Sharma, S., Dixit, G., Shah, E., Patel, A., 2020. Techno-economic analysis of microalgae production with simultaneous dairy effluent treatment using a pilot-scale high volume V-shape pond system [Internet] Renew. Energy 145, 1620–1632. https://doi.org/10.1016/j.renene.2019.07.087. Available from:

Kumar, A.K., Sharma, S., Patel, A., Dixit, G., Shah, E., May 12, 2019. Comprehensive evaluation of microalgal based dairy effluent treatment process for clean water generation and other value added products [Internet] Int. J. Phytoremed. 21 (6), 519–530. https://doi.org/10.1080/15226514.2018.1537248. Available from:

Kumari, P., Varma, A.K., Shankar, R., Thakur, L.S., Mondal, P., 2021. Phycoremediation of wastewater by Chlorella pyrenoidosa and utilization of its biomass for biogas production. J. Environ. Chem. Eng. 9 (1), 104974. https://doi.org/10.1016/j.jece.2020.104974.

Lavrinovičs, A., Juhna, T., 2018. Review on challenges and limitations for algae-based wastewater treatment. Constr. Sci. 20 (1), 17–25.

Le, T.G., Tran, D.T., van Do, T.C., Van Nguyen, T., 2019. Design considerations of microalgal culture ponds and photobioreactors for wastewater treatment and biomass cogeneration. In: Microalgae Biotechnology for Development of Biofuel and Wastewater Treatment, pp. 535–567.

Lee, C.S., Lee, S.A., Ko, S.R., Oh, H.M., Ahn, C.Y., 2015. Effects of photoperiod on nutrient removal, biomass production, and algal-bacterial population dynamics in lab-scale photobioreactors treating municipal wastewater. Water. Res. 68, 680–691.

Lee, J.W., Suh, C., Hong, Y.S.T., Shin, H.S., 2011. Sequential modelling of a full-scale wastewater treatment plant using an artificial neural network. Bioprocess Biosyst. Eng. 34 (8), 963–973.

Leng, L., Wei, L., Xiong, Q., Xu, S., Li, W., Lv, S., et al., 2020. Use of microalgae based technology for the removal of antibiotics from wastewater: a review [Internet] Chemosphere 238, 124680. https://doi.org/10.1016/j.chemosphere.2019.124680. Available from:

Li, K., Liu, Q., Fang, F., Luo, R., Lu, Q., Zhou, W., et al., July 2019. Microalgae-based wastewater treatment for nutrients recovery: a review [Internet] Bioresour. Technol. 291, 121934. https://doi.org/10.1016/j.biortech.2019.121934. Available from:

Liang, Y., Kashdan, T., Sterner, C., Dombrowski, L., Petrick, I., Kröger, M., et al., 2015. Algal biorefineries. In: Industrial Biorefineries and White Biotechnology.

Louzadavalory, J.P., Reis, J.A.T.D., Mendonça, A.S.F., 2016. Combining genetic algorithms with a water quality model to determine efficiencies of sewage treatment systems in watersheds. J. Environ. Eng. (United States) 142 (3), 1–9.

Maestrini, S.Y., Robert, J.M., Leftley, J.W., Collos, Y., 1986. Ammonium thresholds for simultaneous uptake of ammonium and nitrate by oyster-pond algae. J. Exp. Mar. Bio. Ecol. 102 (1), 75–98.

Malik, S., Hagopian, J., Mohite, S., Lintong, C., Stoffels, L., Giannakopoulos, S., et al., January 11, 2020. Robotic extrusion of algae-laden hydrogels for large-scale applications [Internet] Glob. Challeng. 4 (1), 1900064. Available from: https://onlinelibrary.wiley.com/doi/abs/10.1002/gch2.201900064.

Mantzorou, A., Ververidis, F., 2019. Microalgal biofilms : a further step over current microalgal cultivation techniques [Internet] Sci. Total Environ. 651, 3187–3201. https://doi.org/10.1016/j.scitotenv.2018.09.355. Available from:

Mehrabadi, A., Craggs, R., Farid, M.M., 2017. Pyrolysis of wastewater treatment high rate algal pond (WWT HRAP) biomass. Algal Res. 24, 509–519. https://doi.org/10.1016/j.algal.2016.08.007.

Mishra, S., Mohanty, K., 2019. Comprehensive characterization of microalgal isolates and lipid-extracted biomass as zero-waste bioenergy feedstock: An integrated bioremediation and biorefinery approach. Bioresour. Technol. 273 (November 2018), 177–184. https://doi.org/10.1016/j.biortech.2018.11.012.

Mishra, S., Mohanty, K., 2020. Co-HTL of domestic sewage sludge and wastewater treatment derived microalgal biomass – An integrated biorefinery approach for sustainable biocrude production. Energy Convers. Manag. 204 (November 2019), 112312. https://doi.org/10.1016/j.enconman.2019.112312.

Mishra, S., Roy, M., Mohanty, K., August 2019. Microalgal bioenergy production under zero-waste biorefinery approach: recent advances and future perspectives [Internet] Bioresour. Technol. 292, 122008. https://doi.org/10.1016/j.biortech.2019.122008. Available from:

Mohan, S., Singh, Y., Verma, D.K., Hasan, S.H., 2015. Synthesis of CuO nanoparticles through green route using Citrus limon juice and its application as nanosorbent for Cr(VI) remediation: process optimization with RSM and ANN-GA based model [Internet] Process Saf. Environ. Prot. 96, 156–166. https://doi.org/10.1016/j.psep.2015.05.005. Available from:

Molinuevo-Salces, B., Mahdy, A., Ballesteros, M., González-Fernández, C., 2016. From piggery wastewater nutrients to biogas: Microalgae biomass revalorization through anaerobic digestion. Renew. Energy 96, 1103–1110.

Morris, I., Syrett, P.J., 1963. The development of nitrate reductase in *Chlorella* and its repression by ammonium. Arch. Mikrobiol. 47 (1), 32–41.

Mousavi, S., Najafpour, G.D., Mohammadi, M., Seifi, M.H., April 3, 2018. Cultivation of newly isolated microalgae *Coelastrum* sp. in wastewater for simultaneous CO_2 fixation, lipid production and wastewater treatment [Internet] Bioprocess Biosyst. Eng. 41 (4), 519–530. https://doi.org/10.1007/s00449-017-1887-7. Available from:

Mulligan, A.E., Brown, L.C., 1998. Genetic algorithms for calibrating water quality models. J. Environ. Eng. 124 (3), 202–211.

Naaz, F., Bhattacharya, A., Pant, K.K., Malik, A., 2019. Investigations on energy efficiency of biomethane/biocrude production from pilot scale wastewater grown algal biomass. Appl. Energy 254, 113656. https://doi.org/10.1016/j.apenergy.2019.113656.

Nag, S., Mondal, A., Roy, D.N., Bar, N., Das, S.K., 2018. Sustainable bioremediation of Cd(II) from aqueous solution using natural waste materials: kinetics, equilibrium, thermodynamics, toxicity studies and GA-ANN hybrid modelling. Environ. Technol. Innov. 11, 83–104.

Nagarajan, D., Lee, D.J., Kondo, A., Chang, J.S., 2017. Recent insights into biohydrogen production by microalgae – from biophotolysis to dark fermentation [Internet] Bioresour. Technol. 227 (1), 373–387. https://doi.org/10.1016/j.biortech.2016.12.104. Available from:

Najafzadeh, M., Zeinolabedini, M., 2019. Prognostication of waste water treatment plant performance using efficient soft computing models: an environmental evaluation [Internet] Meas. J. Int. Meas. Confed. 138, 690–701. https://doi.org/10.1016/j.measurement.2019.02.014. Available from:

Nayak, M., Dhanarajan, G., Dineshkumar, R., Sen, R., 2018. Artificial intelligence driven process optimization for cleaner production of biomass with co-valorization of wastewater and flue gas in an algal biorefinery [Internet] J. Clean. Prod. 201, 1092–1100. https://doi.org/10.1016/j.jclepro.2018.08.048. Available from:

Nguyen, L.N., Commault, A.S., Kahlke, T., Ralph, P.J., Semblante, G.U., Johir, M.A.H., et al., 2020. Genome sequencing as a new window into the microbial community of membrane bioreactors – a critical review [Internet] Sci. Total Environ. 704, 135279. https://doi.org/10.1016/j.scitotenv.2019.135279. Available from:

Ozkan, A., Kinney, K., Katz, L., Berberoglu, H., 2012. Reduction of water and energy requirement of algae cultivation using an algae biofilm photobioreactor [Internet] Bioresour. Technol. 114, 542–548. https://doi.org/10.1016/j.biortech.2012.03.055. Available from:

Pancha, I., Chokshi, K., Mishra, S., 2019. Application of microalgae in wastewater treatment. Appl. Microalgae Wastewater Treat. 173–193.

Passos, F., Hernández-Mariné, M., García, J., Ferrer, I., 2014. Long-term anaerobic digestion of microalgae grown in HRAP for wastewater treatment. Effect of microwave pretreatment. Water Res. 49.

Perez-Garcia, O., Escalante, F.M.E., de-Bashan, L.E., Bashan, Y., 2011. Heterotrophic cultures of microalgae: metabolism and potential products [Internet] Water Res. 45 (1), 11–36. https://doi.org/10.1016/j.watres.2010.08.037. Available from:

Picos-Benítez, A.R., Martínez-Vargas, B.L., Duron-Torres, S.M., Brillas, E., Peralta-Hernández, J.M., 2020. The use of artificial intelligence models in the prediction of optimum operational conditions for the treatment of dye wastewaters with similar structural characteristics. Process Saf. Environ. Pro. 143, 36–44.

Podder, M.S., Majumder, C.B., 2016. The use of artificial neural network for modelling of phycoremediation of toxic elements As(III) and As(V) from wastewater using *Botryococcus braunii* [Internet] Spectrochim. Acta - Part A Mol. Biomol. Spectrosc. 155, 130–145. https://doi.org/10.1016/j.saa.2015.11.011. Available from:

Poirier, Y., Nawrath, C., Somerville, C., February 1995. Production of Polyhydroxyalkanoates, a family of biodegradable plastics and elastomers, in bacteria and plants. Nat. Biotechnol. 13, 142–150.

Praveen, P., Heng, J.Y.P., Loh, K.C., 2016. Tertiary wastewater treatment in membrane photobioreactor using microalgae: comparison of forward osmosis & microfiltration [Internet] Bioresour. Technol. 222, 448–457. https://doi.org/10.1016/j.biortech.2016.09.124. Available from:

Qiu, Y., Zu, Y., Song, C., Xie, M., Qi, Y., Kansha, Y., et al., April 2019. Soybean processing wastewater purification via *Chlorella* L166 and L38 with potential value-added ingredients production [Internet] Bioresour. Technol. Rep. 7, 100195. https://doi.org/10.1016/j.biteb.2019.100195. Available from:

Radhakrishnan, R., Banerjee, S., Banerjee, S., Singh, V., Das, D., 2021. Sustainable approach for the treatment of poultry manure and starchy wastewater by integrating dark fermentation and microalgal cultivation. J. Mater. Cycles Waste Manag. 23, 790–803. https://doi.org/10.1007/s10163-021-01173-z.

Ranganathan, P., Savithri, S., March 2019. Techno-economic analysis of microalgae-based liquid fuels production from wastewater via hydrothermal liquefaction and hydroprocessing [Internet] Bioresour. Technol. 284, 256–265. https://doi.org/10.1016/j.biortech.2019.03.087. Available from:

Raschid-Sally, L., Jayakody, P., 2008. Drivers and characteristics of wastewater agriculture in developing countries: results from a global assessment. IWMI Res. Rep. 127.

Reyimu, Z., Özçimen, D., 2017. Batch cultivation of marine microalgae *Nannochloropsis oculata* and *Tetraselmis suecica* in treated municipal wastewater toward bioethanol production. J. Clean Prod. 150, 40–46. https://doi.org/10.1016/j.algal.2012.04.001.

Richardson, J.W., Johnson, M.D., Outlaw, J.L., 2012. Economic comparison of open pond raceways to photo bio-reactors for profitable production of algae for transportation fuels in the Southwest [Internet] Algal Res. 1 (1), 93–100. https://doi.org/10.1016/j.algal.2012.04.001. Available from:

Salama, E., Kurade, M.B., Abou-shanab, R.A.I., El-dalatony, M.M., May 2017. Recent progress in microalgal biomass production coupled with wastewater treatment for biofuel generation [Internet] Renew. Sustain. Energy Rev. 79, 1189–1211. https://doi.org/10.1016/j.rser.2017.05.091. Available from:

Sebestyén, P., Blanken, W., Bacsa, I., Tóth, G., Martin, A., Bhaiji, T., et al., 2016. Upscale of a laboratory rotating disk biofilm reactor and evaluation of its performance over a half-year operation period in outdoor conditions. Algal Res. 18, 266–272.

Sekulić, Z., Antanasijević, D., Stevanović, S., Trivunac, K., 2017. Application of artificial neural networks for estimating Cd, Zn, Pb removal efficiency from wastewater using complexation-microfiltration process. Int. J. Environ. Sci. Technol. 14 (7), 1383–1396.

Shah, A., Baral, N.R., Manandhar, A., 2016. Technoeconomic analysis and life cycle assessment of bioenergy systems [Internet]. In: Advances in Bioenergy, Vol. 1. Elsevier. https://doi.org/10.1016/bs.aibe.2016.09.004. Available from:

Sheng, A.L.K., Bilad, M.R., Osman, N.B., Arahman, N., 2017. Sequencing batch membrane photobioreactor for real secondary effluent polishing using native microalgae: process performance and full-scale projection [Internet] J. Clean. Prod. 168, 708–715. https://doi.org/10.1016/j.jclepro.2017.09.083. Available from:

Shi, J., Podola, B., Melkonian, M., 2007. Removal of nitrogen and phosphorus from wastewater using microalgae immobilized on twin layers: an experimental study. J. Appl. Phycol. 417–423.

Silva, A.G., Carter, R., Merss, F.L.M., Corrêa, D.O., Vargas, J.V.C., Mariano, A.B., et al., March 2015. Life cycle assessment of biomass production in microalgae compact photobioreactors [Internet] GCB Bioenergy 7 (2), 184–194. Available from: http://doi.wiley.com/10.1111/gcbb.12120.

da Silva Ribeiro, T., Grossi, C.D., Merma, A.G., dos Santos, B.F., Torem, M.L., 2019. Removal of boron from mining wastewaters by electrocoagulation method: modelling experimental data using artificial neural networks [Internet] Miner. Eng. 131, 8–13. https://doi.org/10.1016/j.mineng.2018.10.016 (August 2018), Available from:

Singh, R., Paul, D., Jain, R.K., 2006. Biofilms: implications in bioremediation. Trends Microbiol. 14 (9), 389–397.

de Souza, M.P., Hoeltz, M., Brittes Benitez, L., Machado, Ê.L., de Souza Schneider, R. de C., 2019. Microalgae and clean technologies: a review. Clean - Soil Air Water 47 (11), 1–18.

Sukla, L.B., 2019. Artificial intelligence and virtual environment for microalgal source for production of nutraceuticals. Biomed. J. Sci. Tech. Res. 13 (5).

Sun, C.H., Fu, Q., Liao, Q., Xia, A., Huang, Y., Zhu, X., et al., 2019. Life-cycle assessment of biofuel production from microalgae via various bioenergy conversion systems [Internet] Energy 171, 1033–1045. https://doi.org/10.1016/j.energy.2019.01.074. Available from:

Tao, R., Kinnunen, V., Praveenkumar, R., Lakaniemi, A.M., Rintala, J.A., 2017. Comparison of Scenedesmus acuminatus and *Chlorella vulgaris* cultivation in liquid digestates from anaerobic digestion of pulp and paper industry and municipal wastewater treatment sludge. J. Appl. Phycol. 29 (6), 2845–2856.

Torpey, W.N., Heukelekian, H., Kaplovsky, A.J., Epstei, R., 1971. Rotating disks with biological growths prepare wastewater for disposal or reuse. J. Water Pollut. Control. Fed. 43 (11), 2181–2188.

Tsapekos, P., Kougias, P.G., Alvarado-Morales, M., Kovalovszki, A., Corbière, M., Angelidaki, I., 2018. Energy recovery from wastewater microalgae through anaerobic digestion process: Methane potential, continuous reactor operation and modelling aspects. Biochem. Eng. J. 139 (March), 1–7. https://doi.org/10.1016/j.bej.2018.08.004.

UN-Water, 2009. The United Nations World Water Development Report 3: Water in a Changing World. UNESCO, Paris.

UN-Water, 2015. Wastewater Management: A UN-water Analytical Brief. UN-Water.

Walls, L.E., Velasquez-Orta, S.B., Romero-Frasca, E., Leary, P., Yáñez Noguez, I., Orta Ledesma, M.T., 2019. Non-sterile heterotrophic cultivation of native wastewater yeast and microalgae for integrated municipal wastewater treatment and bioethanol production. Biochem. Eng. J. 151, 107319. https://doi.org/10.1016/j.bej.2019.107319.

WWAP - United Nations World Water Assessment Programme, 2017. The United Nations World Water Development Report 2017 - Wastewater, the Untapped Resource. UNESCO, Paris.

Xin, C., Addy, M.M., Zhao, J., Cheng, Y., Cheng, S., Mu, D., et al., 2016. Comprehensive techno-economic analysis of wastewater-based algal biofuel production: a case study [Internet] Bioresour. Technol. 211, 584–593. https://doi.org/10.1016/j.biortech.2016.03.102. Available from:

Xue, H.B., Stumm, W., Sigg, L., 1988. The binding of heavy metals to algal surfaces. Water. Res. 22 (7), 917–926.

Yang, J., Xu, M., Zhang, X., Hu, Q., Sommerfeld, M., Chen, Y., January 2011. Life-cycle analysis on biodiesel production from microalgae: water footprint and nutrients balance [Internet] Bioresour. Technol. 102 (1), 159–165. https://doi.org/10.1016/j.biortech.2010.07.017. Available from:

Yazdani, H., Khoshhal, A., Mousavi, N.S., April 2020. Evaluating the performance of a sequencing batch reactor for sanitary wastewater treatment using artificial neural network. Environ. Prog. Sustain. Energy.

Ye, Z., Yang, J., Zhong, N., Tu, X., Jia, J., Wang, J., 2020. Tackling environmental challenges in pollution controls using artificial intelligence: a review [Internet] Sci. Total Environ. 699, 134279. https://doi.org/10.1016/j.scitotenv.2019.134279. Available from:

Zaimes, G.G., Khanna, V., 2013. Microalgal biomass production pathways: evaluation of life cycle environmental impacts [Internet] Biotechnol. Biofuels 6 (1), 88. Available from: http://biotechnologyforbiofuels.biomedcentral.com/articles/10.1186/1754-6834-6-88.

Zeng, Y., Xie, T., Li, P., Jian, B., Li, X., Xie, Y., et al., 2018. Enhanced lipid production and nutrient utilization of food waste hydrolysate by mixed culture of oleaginous yeast *Rhodosporidium toruloides* and oleaginous microalgae *Chlorella vulgaris* [Internet] Renew. Energy 126, 915–923. https://doi.org/10.1016/j.renene.2018.04.020. Available from:

Zhao, X., Kumar, K., Gross, M.A., Kunetz, T.E., Wen, Z., 2018. Evaluation of revolving algae biofilm reactors for nutrients and metals removal from sludge thickening supernatant in a municipal wastewater treatment facility [Internet] Water Res. 143, 467–478. https://doi.org/10.1016/j.watres.2018.07.001. Available from:

Zhou, W., Cheng, Y., Li, Y., Wan, Y., Liu, Y., Lin, X., et al., 2012. Novel fungal pelletization-assisted technology for algae harvesting and wastewater treatment. Appl. Biochem. Biotechnol. 167 (2), 214–228.

Zhou, W., Lu, Q., Han, P., Li, J., 2020. Microalgae cultivation and photobioreactor design [Internet]. In: Microalgae Cultivation for Biofuels Production. Elsevier Inc. https://doi.org/10.1016/B978-0-12-817536-1.00003-5. Available from:

CHAPTER

20

Weeds as a renewable bioresource: prospects for bioconversion to biofuels and biomaterials through a cascade of approaches

Rupam Kataki[1], *Manika D. Kataki*[2]

[1]Department of Energy, Tezpur University, Tezpur, Assam, India; [2]Department of Biotechnology, Darrang College, Tezpur, Assam, India

20.1 Introduction

Weeds are the unwanted plants that grow with cultivated crops in agricultural fields, swamps, and wastelands. As it is not possible to fully eradicate the weeds in the fields, they can be used as a source of energy. In the absence of an effective weed control mechanism, the invasion of weeds has led to huge losses in the field of agriculture, health, the environment, and economics (Lerdau and Wickham, 2011). Bhan et al. (1999) assessed that weeds India can lower crop production by up to 31.5% (in winter up to 22.7% and while in up to 36.5%). In a few separate investigations, it was reported that weeds caused a substantial decline in agricultural production along with its quality (DWSR, 2013). In the survey by Weed Scientists in India, it was recorded that weeds-related losses may range from as low as 10%–100%. However, for bigger economies such as India, even a docile estimate 10% loss due to weed may result in huge losses of food grains of approximately $ 13 billion in value (Yaduraju, 2012). Additionally, weeds also cause many indirect effects on human and animal health, loss of biodiversity, depletion of nutrients, harmful effects on forest and aquatic ecosystem, and so on. To mitigate the damage from weed on the agro/forest/aquatic ecosystem, several approaches, viz. manual, mechanical, chemical, and biological ones, have been devised (Rao and Chauhan, 2015). A probable and more realistic approach for reducing/recovering energy intensity of weed management practices would be toward utilization of these weeds for some useful purpose. In this regard, weeds can be utilized as fodder, food, vermicompost, mulch, traditional and herbal medicine, biopesticide, and technological applications such as for making paper, particleboard,

Biofuels and Bioenergy
https://doi.org/10.1016/B978-0-323-85269-2.00021-6

furniture, and so on (Kumar and Vishwakarma, 2005; Varshney and Prasad, 2008; Sharma and Pant, 2020). In addition to all these uses, weeds have also been seen as a potential source of energy. Weeds as a bioresource can be utilized for energy production by employing common biomass conversion routes such as biochemical and thermochemical. Both biochemical and thermochemical conversion, e.g., biomethanation and pyrolysis of weed biomass to biogas have been studied for several aquatic and perennial weeds (Baruah, 1995; Konwer et al., 2007). Among different aquatic weeds, water hyacinth has established itself as a highly feasible candidate for conversion to bioenergy. Biomethanation of water hyacinth to biogas has been studied by several researchers all over the world. Besides, water hyacinth has been found suitable for thermochemical conversion and fermentation also (NASA, 1978; Thomas and Eden, 1990; Gunnarssona and Petersen, 2007). Weeds as a resource of bioenergy and biofuel have twofold advantages. First of all, weeds have to be anyway disposed of from the field. So, utilization of this otherwise wasted resource for energy production helps in easy and safe disposal. Secondly and more importantly, the present resources for biofuels and bioenergy have been under intense criticism for their competition with food crops for agricultural land. This has gone to such a dimension that a global debate over food versus fuel is triggered. This has necessitated looking out for new alternate resources of biofuels and bioenergy, not in competition with food crops. In this regard, the utilization of weeds as a resource for bioenergy and biofuel fulfills the purpose.

The present work involves a review of the present state of utilization of weeds for bioenergy and biofuels. The different conversion pathways involved in the conversion of weed biomass to bioenergy are discussed.

20.2 Energy utilization for weed management

Energy use on weed management can be annulled by its utilization for energy production so that a part of the energy used in weed management is recovered. Researchers from throughout the globe have reported the energy use efficiency in weed management in different crops, viz. soybean (Goette et al., 1980), corn (Clements et al., 1995; Pimentel and Burgess, 1980; Peart, 1999), fodder oats (Sharma and Subhash, 2006), and grain corn (Lague and Khelifi, 2001). Based on the aforementioned studies, it can be safely concluded that weed management is an energy-intensive process depending upon crop and field requirement, weeding strategies adopted, and level of weeding desired. Energy use is an important parameter in calculating the efficiency of production. Energy conservation and enhanced energy efficiency for higher energy productivity of agriculture are prime movers in this direction.

20.3 Weed as a source of bioenergy and biofuel

Weed biomass utilization for energy production is being considered as an area of importance for obvious reasons such as increasing needs for energy supply from raw materials, which are renewable and sustainable in supply to produce energy, which is environmentally benign and socioeconomically sustainable. More and more interest has been paid for technological upgradation that uses novel and renewable energy sources such as biomass, geothermal, solar power, wind, and hydropower (Demirbas et al., 2006). Biomass is considered as significant contributor to the world economy. It has the capability to solve various burning energy-related issues such as "greenhouse" effect (Miao et al., 2004). Other than heat and power production, different fuels can be produced from the biomass sources (Yaman, 2004).

In addition to this energy production from weed biomass, it possesses other environmental and socioeconomic benefits, which includes the following:

- ➢ Higher energy availability at farm level and reduction of net energy requirement in agriculture
- ➢ Management of streams and water bodies
- ➢ Streams of several products are obtained from biomass that includes biofertilizers, activated carbon, chemicals and fuels, and so on
- ➢ Fire hazard reduction
- ➢ Link to biotechnology
- ➢ Related with carbon sequestration as well as management
- ➢ Electricity can be dispatched
- ➢ Distributed generation: It can be helpful in the reduction of energy losses in the transmission as well as distributed networks and also defer capital expenditure on transmission works
- ➢ Response to government programs
- ➢ Local environmental benefits including odor reduction and local air purification and enhancement of water quality

Considering the aforementioned point, Table 20.1 enlists some of the latest research on the utilization of weed biomass for bioenergy applications and its associated environmental or socioeconomic benefits.

20.4 Biomass conversion technologies

Renewable feedstock can be used in various forms, such as solid, liquid, or gaseous forms. It is also used for the production of electric power, heat, various chemicals, and gaseous and liquid fuels. Various biochemical and thermochemical conversion technologies have been used to convert weed biomass resources into energy. Thermochemical conversion methods include three subcategories, which are pyrolysis, gasification, and liquefaction. Combustion alone contributes significantly (over 97%) to world's bioenergy production. Liquefaction can be carried out directly or indirectly. Liquid tars and oils can be produced by direct liquefaction that includes hydrothermal liquefaction and rapid pyrolysis. Noncondensable gaseous by-products of pyrolysis can be transformed into liquid products via indirect liquefaction. Fast pyrolysis of biomass is used to generate a product that can be used both as a platform chemical and as an energy source. The liquid portion of these pyrolysis products contains two phases, which are a low molecular weight–containing aqueous phase and a nonaqueous phase that contains high-molecular-weight insoluble organic products. Biomass gasification has emerged as the latest biomass conversion and energy generation process. Improved efficacy and overall reduction in bioelectricity generation are the two main advantages of this process. Thus the thermochemical conversion methods offer greater flexibility in terms of usability of almost all types of biomass as feedstock and end product for further conversion to fuels and chemicals. Figs. 20.1 and 20.2 demonstrate the various types of biomass that exits together with energy conversion technologies.

20.4.1 Thermochemical conversion of biomass

20.4.1.1 Combustion

Direct combustion is the most prevalent method for transforming biomass to heat at a large or commercial scale. Extremely hot combustion gases are obtained on burning solid feedstock under controlled conditions. These gases are mostly utilized for product drying.

TABLE 20.1 Weeds used for various end uses.

Weed biomass	Bioenergy & biofuel products	Environmental/socioeconomic benefit	References
Water hyacinth (*Pontederia crassipes* Mart.)	Biomethane	Mitigation of eutrophic waters Reduced weed management cost	Castro and Agblevor (2020)
Seaweeds (macroalgae)	Biofuels, biochar	CO_2 sequestration, mitigation of CO_2 emissions, generation of useful chemical products such as hydrocolloids/phycocolloids, alginate, agar, and carrageen	Sondak et al. (2017)
Weedy biomass and other crop residues	Bioethanol	Development of enzymatic cocktails for saccharification	Pandiyan et al. (2019)
Water hyacinth (*Eichhornia crassipes*)	Biogas, biofuel, biofertilizer, etc.	Vermicompost for organic farming of groundnut (*Arachis hypogaea*)	Snehalata and Rao (2018)
Aquatic weeds (*Salvinia molesta* and *Pistia stratiotes*)	Biogas, biofertilizer	Phytoremediation of wastewater	Mustafa and Hayder (2021)
Aquatic, terrestrial, and invasive weeds	Biofuels	Production of sustainable fuels and chemicals Rural employment generation Response to government policies	Bhuyan et al. (2019)
Aquatic weeds	Biofuels	Integrated biofuel production with phytoremediation Net energy gain, economically feasible	Kaur et al. (2018)

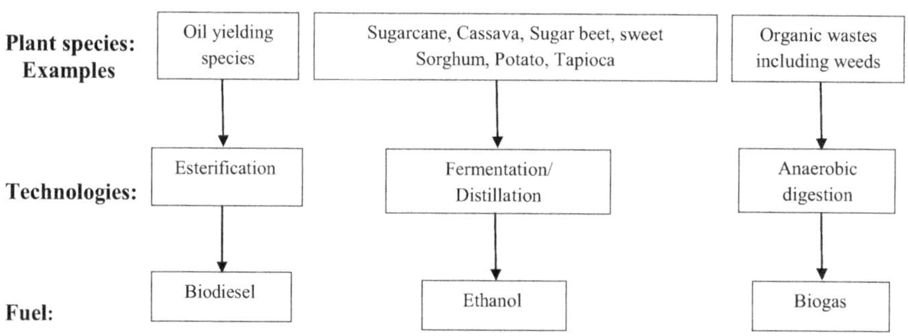

FIGURE 20.1 Conversion possibility for biomass abundant in oils, sugars, and starch (El Bassam, 1998).

However, recently, these gases are used for production of hot air, hot water, and stream using heat exchanger. The layout and the design of the combustor system vary with grates (used for supporting bed of fuel and to mix combustion air) to make room for varied quality of solid biomass with high to low ash content.

Efficient combustion of feedstocks emits flue gases that comprise mostly carbon dioxide and water vapor along with small traces of other products. These gases are purified using flue gas scrubbers, bag filters, and other methods such as electrostatic precipitators. Further chemical processing can also be done if needed for minimization of NOx as well as other pollutants.

Stream electricity can be generated from biomass combustion as mentioned earlier. This method involves production of steam and

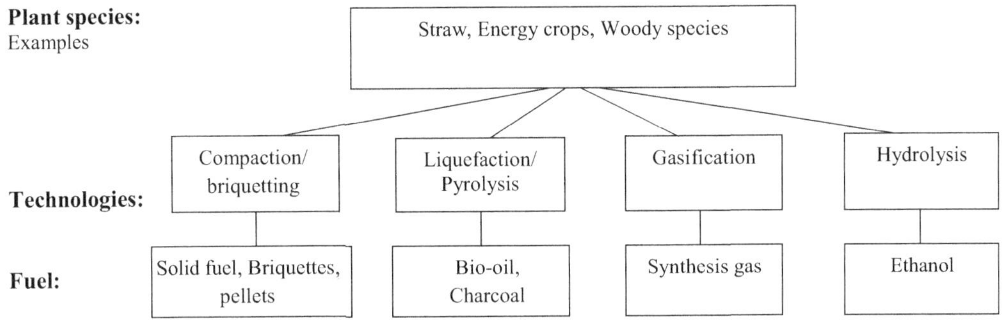

FIGURE 20.2 Conversion possibility for lignocellulosic biomass (El Bassam, 1998).

utilizing the generated steam to power an engine or turbine for producing electricity. The conversion efficiency was in the range of 10%–18%.

Biomass fuels can be used together with fossil fuels in preexisting power stations. This usually involves serving relatively low amount of biomass (commonly below 5% of the total fuel) with the conventional fuel, without any obstruction to the operation of the plant.

Bioelectricity as combined heat and power (CHP) is at the helm of current technological trends. Biomass as agroindustrial residues and municipal solid wastes is currently utilized for production of electricity with conventional steam turbine power systems. The cost-competitiveness of electricity generation from biomass can be improved if issues like increase in conversion efficiencies, lower/no cost of biomass collection, and decrease in capital cost, areresolved favourably (El Bassam, 1998; Schuck, 2006). In this context, weed biomass, which is available almost at a zero cost, can be an effective contributor to the success of the previously discussed processes.

20.4.1.2 *Gasification*

Gasification is a process used for transformation of carbon-rich waste materials into combustible gas at elevated temperatures in nonavailability of air. The gaseous product obtained from biomass consists of carbon monoxide, carbon dioxide, hydrogen, methane, water, and small amounts of heavy hydrocarbons. The gas produced from gasification in the presence of air is termed as "producer gas." It has reduced the quantity and quality of atmospheric nitrogen and carbon dioxide. Producer gases also have a relatively low calorific value (CV), often in the range of 4–6 MJ/Nm^3. Fig. 20.3 shows the flowchart of a gasification process.

In some newer applications, air is supplied for gasification, and the gas obtained is referred as "syngas" with a CV in the range of 10–15 MJ/Nm^3 (Demirbas, 2007).

20.4.1.3 *Pyrolysis*

Pyrolysis is the process of thermal conversion of materials performed in an inert atmosphere, generally at temperatures ranging from 400 to 800°C. It results in a mixture of gases, vapors, liquids, oils, solid char, and ash. The composition and proportions of these products depend on input composition, pretreatment, temperatures, residence time, and reaction rates. At high temperatures, in the range of 500°C and short reaction times (under 2 s), pyrolysis oils are produced. Conversion efficiency to biooil can be up to 85% by weight. The solid product of pyrolysis, i.e., charcoal/biochar has been traditionally used as fuel for domestic applications. Additionally, its (biochar) nonfuel use including soil application as soil amendment, C-sequestration, heavy metal adsorption, and numerous industrial applications has

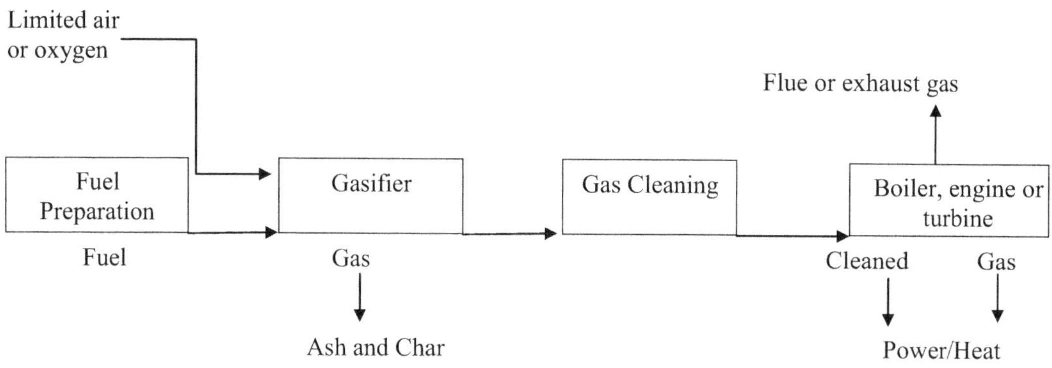

FIGURE 20.3 Flowchart of a gasification process.

recently attracted attention of researchers and policy-makers throughout the globe. An overall yield of 35% by weight can be achieved by manipulating process parameters through application of long reaction times (hours to days) and low temperatures (~350°C), and the heat required in traditional kilns can be obtained by burning the gas and liquid by-products. Indigenous traditional kilns in northeast India have a much lower efficiency of c. 20% by weight with no provision for utilization of by-products as a source of heat (Kataki, 2005). Fig. 20.4 shows the flowchart of pyrolysis process.

The oil produced from biomass feedstock using pyrolysis has many advantages. The biooil can be transported from the production site to the location of its use. Upgrading of pyrolysis oil to more familiar petroleum fuels is another advantage. Employment of low temperature in the process hinders ash melting reactions and softening of ash and thus retains the ash in the char. This also makes it a favored technology for conversion of weed biomass and other crop residues (El Bassam, 1998).

20.4.1.4 Liquefaction

Liquefaction is the process of heating biomass in the absence of air at temperatures ranging from 525 to 600K in water under pressure of 5–20 MPa. Some solvents or catalysts can be added during the liquefaction process whenever it is needed. The disadvantages of biomass liquefaction are (1) difficulty to reach the operation conditions and (2) lower hydrogen production.

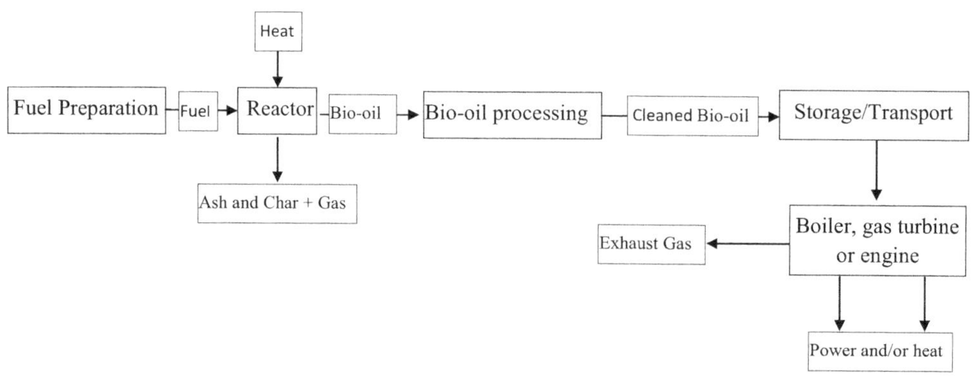

FIGURE 20.4 Flowchart of pyrolysis process.

Therefore, liquefaction is not considered a favorable process for the production of hydrogen. Nowadays, other thermochemical processes such as pyrolysis and gasification are reported as feasible methods and receiving much attention in the production of hydrogen.

20.4.1.4 *Briquetting*

Briquetting is a technique aiming at densification of loose, low-energy-density biomass resources. Two main high-pressure technologies are widely used for briquettes. They are ram or piston press and screw extrusion machines. The briquettes formed by piston press are completely solid, whereas screw press briquettes have a concentric hole that offers better combustion properties because of higher specific area. These briquettes are also homogenous and do not disintegrate easily, which have high combustion rate, and that is why can substitute coal for many applications. Briquettes possess density of 1.2 g/cm^3 from loose biomass of bulk density of 0.1–0.2 g/cm^3.

20.5 Biochemical conversion of biomass

20.5.1 Esterification for biodiesel

Worldwide, there are a large number of plant species with fairly considerable oil contents in the seeds, fruits, tubers, and/or roots. The oil can be extracted by compressing the seeds mechanically with screw presses and expellers with or without preheating. Pressing with preheating allows for the removal of up to 95% of the oil, while extraction by use of organic solvents eliminates almost 99% of the oil (El Bassam, 1998). Vegetable oils can be modified to suit the requirements of the diesel engine through the process of transesterification.

In India, more than 300 different species of oil-bearing trees exist. Traditional uses of these seeds for lightning have been for decades, and only a small quantity of these seeds are used for different commercial purposes. According to the National Oilseeds and Vegetable Oil development Board of the Indian Ministry of Agriculture (NOVOD), there are about 10 species with economic potential for biodiesel production, viz. *Jatropha, Pongammia, Azadirachta, Madhuca indica,* and *Simarouba glauca* (GTZ/TERI, 2005; Altenburg, 2008). Indigenous biodiesel production in India can bring about much-needed reduced energy dependency on oil imports and reduction of CO_2 emissions; contribute to the rural economy, employment, and income; safeguard rural energy security; and maintain ecological balance and rejuvenation of wasteland through the protection of natural resource. Lately, several weed plants feedstock such as seeds of common purslane (*Portulaca*), *Cephalaria syriaca* weed (Turkish Pelemir), *Argemone Mexicana,* and dried *Salvinia molesta* (aquatic weed) were utilized as novel nonedible feedstocks for biodiesel production using transesterification reaction of extracted oil (Hoseini et al., 2019; Mubarak et al., 2016; Karabas, 2018; Anjum et al., 2019). Research is also focused on optimization of transesterification conditions such as methanol to oil molar ratio, catalyst loading, reaction temperature and time, stirring speed, and so on for maximum biodiesel yield from weedy biomass (Hoseini et al., 2019).

20.5.2 Fermentation/distillation for ethanol

Ethanol is produced by the fermentation of sugars, generally from sugarcane and sugar beet. Starch hydrolysis of cellulose degradation is also used in some cases followed by sugar fermentation and then by subsequent distillation. Fermentation of sugar from crops using yeast produces alcohol. Alcohol can also be derived from crops with starch such as wheat, cassava, sweet sorghum, and so on. The glucose derived by hydrolysis of starch can also be fermented to produce alcohol. Sweet sorghum has emerged as a candidate tree for conversion to ethanol, which can be grown in temperate and tropical regions, and has a very high yield in terms of grains, sugar, and lignocellulosic biomass with less requirement of seed rate than other starch crops such as corn and wheat. Ethanol production from cellulosic biomass holds great promise, as bioethanol from lingocellulosic biomass appears to have near-term economic potential. There exists four methodologies for bioethanol production from cellulosic biomass, viz. concentrated acid hydrolysis, dilute acid hydrolysis, enzymatic acid hydrolysis, and biomass gasification and fermentation (Planning Commission, 2003).

Feedstocks for bioethanol production can be quite large—agricultural residues and wastes, municipal and industrial wastes, and many more. The development of microbes was another concern for cellulosic feedstocks' conversion to ethanol, as the hemicellulose part of biomass was not fully utilized for fermentation. Of late, some new strains of yeast as well as bacteria were produced having the capability of fermenting the full spectrum of available sugars in feedstocks to ethanol. Recently, bioethanol was produced by fermentation of mixed sugars (pentose and hexose) in the hydrolysate obtained after acid–alkali pretreatment of the mixed feedstock of eight different species of invasive weeds (Borah et al., 2019). Two different yeast species viz. *Saccharomyces cerevisiae* and *Candida shehatae* were utilized for fermentation of C_6 and C_5 sugars separately. Similarly, *Pachysolen tannophilus* was reported as a potent xylose-fermenting yeast strain for ethanol production from cellulose-degrading enzymes pretreated water hyacinth (*Eichhornia crassipes*) (Manivannan and Narendhirakannan, 2015). In another study, prior acclimatization of yeast strain *Candida lusitaniae* to a high concentration of galactose was found as a suitable strategy to enhance ethanol yield from pretreated seaweed *Gracilaria verrucosa*. The acclimated yeast to a higher concentration of specific sugars was able to ferment mixed sugars present in the seaweed hydrolysate medium (Ra et al., 2015).

20.5.3 Anaerobic digestion to fuel and fertilizer

Anaerobic digestion (AD) refers to the decomposition of organic wastes into gaseous fuel by bacteria in an environment that is devoid of oxygen. The process proceeds in stages to successfully break down the organic matter into simpler organic compounds. The final product, known as biogas, is a mixture of methane (CH_4) and carbon dioxide (CO_2), and some trace gas. Hence, biogas can be produced by fermenting organic materials in the absence of air (or oxygen) with the help of bacteria to break down materials into intermediate such as alcohols and fatty acids and finally to methane, carbon dioxide, and water. AD involves three phases namely hydrolysis covering breakdown of large molecules into smaller ones decomposable by bacteria; the acid phase when complicated molecules such as proteins, fats, and carbohydrates are decomposed by acid-forming bacteria into organic acids (short-chain fatty acids), CO_2, H_2, ammonia, and some impurities; and lastly methane phase involving methane forming bacteria convert fatty acids into methane.

The important factors concerning biogas generation are the pH of the slurry and its C:N ratio. A pH of 7 or close to it is most suitable for action of microorganisms. A correct C:N ratio is important for the efficient utilization of slurry. Total biogas production rate depends upon organic material being digested, digester loading rate, and environmental conditions. Cow dung is the most widely used raw material for biogas generation and is usually mixed in the proportion of 1:1 by weight to bring the total solid content to around 8%–10%. The untreated cow dung contains roughly 80%–85% moisture, and the rest is totally solid. The adjustment of total solid content helps in the digestion of the materials at a faster rate and also in deciding the mixing of the various crop residues, weeds, and plants, with cow dung as feedstocks in a biogas digester.

The advantages of biogas in a country like India are numerous. It is a cooking fuel that helps to maintain the health and hygiene of the women and cooking place. It contributes to environmental improvement by reducing the fuel wood requirement for cooking, rendering more time for the women for other productive works, and supplying manure for field application. It can also be utilized for decentralized power generation by using a dual fuel engine (80% biogas and 20% diesel). Biogas, through a specially designed mantle, can be used for lightning and can partially replace diesel to run IC engines for water pumping and small industries such as oil mill, flour mill, sawmill, and so on. Recently, AD and vermicomposting of terrestrial weeds like *Parthenium hysterophorus, Ageratum conyzoides, Saccharum spontaneum,* and *Lantana camara* have been suggested as a viable alternative weed management practice for the production of biogas and valuable fertilizer for application in agriculture (Saha et al., 2018). Similarly, hydrothermal pretreatment and AD of pelagic *Sargassum* (brown seaweed) were reported as efficient valorization technologies for biogas and nutrient-rich biofertilizer (Thompson et al., 2020). In another study, AD of seaweed *Laminaria digitata* and reed *Phragmites australis* was reported for potential large-scale biogas production with high methane yield and nutrient recycling (by absorbing nutrients and oxygen during growth) for reduced eutrophication of water bodies (Ohlsson et al., 2020).

20.6 Weeds used for energy generation

The approach of weed utilization for energy generation is a comparatively recent one. For energetic utilization, weeds as biomass require specific characteristics, which are highly variable and are also dependent on postharvest treatment. Low moisture herbaceous weeds are most suitable for thermochemical decomposition, whereas biological processes such as microbic digestion and fermentation are most suitable for high moisture aquatic weeds.

20.6.1 Weed characteristics

Weeds are considered to grow in harsh environments where soil or natural vegetative cover has been altered or damaged very frequently. These disturbances give the weeds advantages over other crops, pastures, or ornamental plants. Generally, it contains different quantities of cellulose, hemicellulose, lignin, and a small number of other extractives. Weed species are generally characterized as perennial, average growth rate, with loosely held fibers, indicating a lower proportion of lignin that binds together the cellulosic fibers. The relative proportions of cellulose and lignin are one of the determining factors in identifying the suitability of plant species for subsequent processing as energy crops. Cellulose is a long linear chain polymer of β-(1−4)-D-glucopyranose units. Hemicellulose is a combination of polysaccharides, composed almost completely of sugars such as glucose, mannose, xylose, arabinose, and methyl glucuronic. The average molecular weight is less than 30,000. This is in stark contrast to cellulose.

20.6.2 Fuel properties of weed biomass

The fuel property of biomass is dependent on its various physical as well as chemical characteristics, which also affect the entire thermal or chemical process. The major characteristics of the biomass during subsequent processing as an energy source are related to the following properties:

20.6.2.1 Moisture content

Moisture content is the main physical fuel property of weed biomass. Moisture content is influenced by subsequent contamination of the harvested biomass from various foreign sources, which give a very negative effect on other "material" properties of the biomass during the following treatment or processing. Biomass moisture content and suitable bioconversion technique are directly related to each other where thermal conversion requires low moisture content feedstock (typically <50%), and bioconversion can utilize high moisture content feedstocks. Therefore, woody weeds such as *Ipomea, Lantana,* and so on are suitable for thermal conversion to liquid fuels and value-added chemicals, whereas aquatic weeds such as water hyacinth, duckweed, and so on are suitable for biochemical conversions such as fermentation and AD to produce ethanol or methane.

20.6.2.2 Calorific value

The CV of any material is the representation of its energy content or heat value that is released when burned in presence of air. CV is usually measured in terms of the energy stored per unit mass or volume. The calorific value of weed biomass depends on moisture content, decreasing linearly with rising moisture content (Jenkins et al., 1998; McKendry, 2002). Furthermore, CVs have been correlated with ash content and are expected to follow an inverse trend concerning ash (Jenkins et al., 1996).

20.6.2.3 Proportions of fixed carbon and volatile matter

Fuel analysis is done based on VM content, ash, and moisture. Fixed carbon (FC) is determined by difference. The chemical energy of biomass is stored in two forms fixed carbon and volatile matter (VM):

- VM of biomass is that portion driven off as a gas (including moisture) by heating (to 900°C for 2 min).
- The FC content is the mass remaining after the releases of volatiles, excluding the ash and moisture contents.

20.6.2.4 Ash/residue content

Ash is the general term used to describe the inorganic matter present in the fuel. In the case of biomass, the ash content may originate from the biomass itself, e.g., materials that the plant absorbed from the water or the soil during its growth, or the supply chain, e.g., soil collected along with biomass. Generally, the ash content of herbaceous biomass is higher than that of woody biomass.

20.6.2.5 Mineral content

There are major differences in the concentrations of these elements between woody and herbaceous crops, and herbaceous crops generally have more N, Cl, and K, but less Ca than woody crops (Vassilev et al., 2010; Tao et al., 2012a). Though not essential for survival, Si is accumulated to high levels in many types of grass, up to 10% dry weight (Epstein, 1999).

Fuel properties of weed biomass vary in a wide range as the type and nature of weed vary in different ecosystems. Weeds are classified botanically as monocot or dicot, and there exist several other classification systems based on different criteria. Weeds for energy conversion can be viewed as either herbaceous or woody. Few pieces of literature had been reported on the chemical composition of different weeds. However, Obernberger (1998) and Lu et al. (2009) studied the fuel properties of several weeds biomass (Table 20.2). Table 20.2 shows a comparison of elemental composition and ash value between weeds and wood fuel.

The chemical and physical characteristics of biofuel are affected by the stage of vegetation. If the weed feedstock is harvested late, the quality of the biofuel was improved due to a decline in the undesirable elements such as chlorine, sulfur, nitrogen, and potassium as established by several systematic investigations (Lewandowski and Kicherer, 1997).

TABLE 20.2 A comparison of the chemical composition of weeds.

Biomass type	C	H	O	N	S	Ash
Grass	49 (wt% daf)	6.3 (wt% daf)	43 (wt% daf)	1.4 (wt% daf)	0.2 (wt% daf)	7.0 (wt% db)
Water hyacinth	81 (w/w, %)	11.3 (w/w, %)	4.95 (w/w, %)	2.78 (w/w, %)	0.0	–
Wood fuel	49 (wt% daf)	6.2 (wt% daf)	44 (wt% daf)	0.5 (wt% daf)	0.05 (wt% daf)	2.0 (wt% db)

20.7 Weeds for energy

The traditional and current feedstock used for "bioenergy and biofuel" has come under severe criticism for several reasons. Forest denudation, habitat destruction, biodiversity loss, and above all their competition for agricultural land giving rise to the "food versus fuel" debate have made the researchers, policy-makers, and others concerned to look for alternative feedstocks. Weeds having the same botanical and biochemical features can be a suitable alternative to biomass feedstocks for energy conversion in use today. In this regard, however, the availability of weeds for energy conversion is a matter of concern, albeit, whatever weed biomass produced/wasted is worth trying for converting to energy.

The information on the quantity of weed available from the agricultural ecosystem is not much available in the literature. Osadolor (2009) studied the availability and productivity of weeds as a possible source of sustainable, highly renewable replacement for wood as an energy source in Nigeria. Several types of regenerative grass, weeds, and leaves were identified as an alternative for renewable energy. The viability of a feedstock is judged by its abundance and the turnover rate ranging from 26 g/m^2 in the savanna to 400 g/m^2 in the forest zone. In another work, the availability of woody weeds such as *Ipomoea carnea* was estimated to be about 100 tons from an area of 1 ha of land (Konwer et al., 2007).

Aquatic weeds exhibit very high growth rate and overall higher productivity levels and therefore have received widespread attention of scientists and engineers for biomass production for energy use. Various forms of this type of biomass include freshwater aquatic plants, marine plants, and microalgae. The harvested biomass may be fed to a digester to produce methane or convert to electricity. Varma and Behera (2003) reported the growth rates of some aquatic plants in Table 20.3.

20.7.1 Weeds as feedstocks for biomethanation

The chemical composition of weed may vary depending upon the species, thereby influencing

TABLE 20.3 Growth rates of some aquatic plants.

Aquatic plants	Growth
Berula/Ranunculus sp.	8.5 dt/ha/yr
Ceratophyllum demersum	19.0
Phragmites sp.	7.5–13 dt/ha/yr
Typha sp.	10.7–25 dt/ha/yr
Sasitana sp.	27 dt/ha/yr
Eichhornia sp.	15–44 dt/ha/yr
Kelp	4.5 m per week
Gracilara	9–10 dt/ha/yr
Euchina	13 dt/ha/yr
Cladophora prolifer	24 dt/ha/yr
Chlorella and *Spirulina*	80 dt/ha/yr
Lemna monor	3.8 g/dry wt. m. day
Spirodela polyrhiza	3.4 g/dry wt. m. day
Hydrilla verticillata	21.7 g/dry wt. m. day

the yield of methane. Prochnow et al. (2009) reviewed the progress in the area of weed/grass biomethanation for its suitability and sustainability for AD. Comparative study of several grass species revealed minor differences in biogas yields and decomposition rates of cocksfoot, timothy, and reed canary grass, which ranged from 490 to 540 L/kg_{ODM}. Similarly, Mähnert et al. (2002) also conducted batch experiments with eight grass species and reported the biogas yield in the range 554–641/kg_{ODM}.

Biogas production from aquatic weeds has been a matter of interest the world over. Baruah (2005) studied fuel and fertilizer production from three aquatic weed species viz. *Euryale ferox* Salisb, *Pistia stratiotes* L., and *Echinochloa crusgalli* L. *E. ferox was* found to contain the highest amount of carbohydrate and exhibited maximum biogas yield among the three. pH value was found to be stable for 5–6 weeks in the slurry of *P. stratiotes*. The highest methane content (64%) was found in the biogas produced from *E. crusgalli*.

Water hyacinth (*E. crassipes*) has received worldwide attention in this regard. It is a nitrogen-rich (approximately 32% of dry matter) freshwater fern from the Pontederiaceae family with a C/N ratio of about 15 (Gunnarssona and Petersen, 2007). Owing to its rapid growth and robust nature, it can become twice within a number of days under high temperature and humidity (Tag EI-Din, 1992) and produce up to 140 tons of dry matter/ha/yr (Abdelhamid and Gabr, 1991). However, there are certain problems associated such as its dense mats block navigation and disrupt irrigation, recreation, fishing, and power generation. Also, the mats prevent penetration of sunlight and aeration, which is responsible for oxygen deficiency in the water, which reduces biological diversity. Biogas production by utilization of water hyacinth and some other aquatic species is given in Table 20.4.

TABLE 20.4 Biogas production from aquatic and marine plants/weeds (Varma and Behera, 2003).

Biomass	Biogas production
1 *Eichhornia crassipes*	7.045 L/kg. wet wt.
2 *Thalassia testudinum* shoot (marine seagrass)	10.4 L/kg. wet wt.
3 *T. testudinum* root (marine seagrass)	7.4 L/kg. wet wt.
4 *Syringodium filiforme* (marine seagrass)	7.377 L/kg. wet wt.
5 *Gracilaria tikrahiae* (algae)	22.2–32.4 L/kg. wet wt.
6 *Hypnea musciformis* (algae)	7.6 L/kg. wet wt.
7 *Sargassum fluitans* (algae)	8.5 L/kg. wet wt.
8 *Ulva* sp. (algae)	32–43 wet wt.
9 *Ipomoea carnea*	387–429 L/kg. vs. added

Biogas production from *E. crassipes* has been also widely studied in India. The Central Institute of Agricultural Engineering (CIAE), Bhopal, has installed a biphasic biogas plant at farmers' fields near Dharwad with the substrate as common weeds. The biphasic plant had the advantage of separate acid and methane reactors where optimum parameters can be maintained easily. It prevents the problems of pretreatments, scum breaking, and slurry flow. More recently, research is oriented toward optimization of various process parameters for enhanced biomethane yield from weeds using response surface methodology or artificial neural networks. For instance, anaerobic conversion of thermoalkaline pretreated Siam weeds (*Chromolaena odorata*) led to an enhanced biogas yield of 49.20% as compared with untreated samples (Dahunsi et al., 2018). Influence of various parameters (pretreatment temperature, time) on biogas production was optimized using statistical tools before digestion. Moreover, a comprehensive study related to the effect of harvest month on the biochemical properties of alligator weed for the coproduction of biohydrogen and biomethane was evaluated by Li et al. (2020). It was reported that the biomethane yield is strongly influenced by the chemical oxygen demand (COD) content of the hydrogenogenic effluent in the AD process. Biomethanation of invasive weed water hyacinth (*Pontederia crassipes* Mart.) from eutrophic water bodies was reported as a sustainable postweed management practice for developing countries by Castro and Agblevor (2020). The estimated energy generated by biomethanation was 846.5 MJ per ton of fresh biomass as compared with the energy consumption of 57.9 MJ associated with mechanical harvesting. Similarly, thermally pretreated *Lemna gibba* (duckweeds; DW) were used for biomethanation along with waste-activated sludge (WAS) to obtain a maximum methane yield of 468 mL CH_4/g volatile solids in DW:WAS (50: 20) (Gaur et al., 2017).

20.7.2 Weeds for briquetting/pelleting

In the recent decade, North America, Europe, and Asia have shown their attention toward utilization of weed biomass as solid biofuels (Lewandowski et al., 2003b; Xiong et al., 2008; Singh Bisen et al., 2020). High moisture content, low bulk density, and irregular shapes and sizes are few challenges linked to biomass handling, transportation, storage, and utilization in its original form. In this regard, densification of biomass into solid fuel via pelleting and briquetting is one measure to overcome this issue. Mani et al. (2006) reported that physical homogeneity, energy, and bulk density were improved via pellets/briquettes requiring pretreatment such as drying, grinding, and mixing with additives if required. The reason behind the compression of loose and low energy density biomass is to increase its energy density, durability, and strength. The compression force, particle size, moisture, and binders decide the success of the compaction (Kaliyan and Morey, 2009; Mani et al., 2006). The economic feasibility of pelleting and briquetting is necessary for its commercialization (Werther et al., 2000).

20.7.2.1 Soda weeds

Yumak et al. (2010) investigated the briquetting conditions of Russian tumbleweed, *Salsola tragus*, which grows in salty soils. Soda weeds were first chopped coarsely in a local thresher and then chopped finely in a hammer mill. Soda weed (*Salsola tragus* L) can be densified into briquettes having density of 600–950 kg/m^3 without any binding materials. Cylindrical and conical shapes can be used to make stable briquettes. However, conical (truncated cone) form can be preferred to reduce resistance of friction between the material and the die in addition to the advantage of easy removal.

20.7.2.2 Water hyacinth

Munjeri et al., (2016) investigated the use of the harvested water hyacinth weed as a source of energy in the form of briquettes and also compared with other briquettes made from leaf samples. A molasses binder utilized the strength of briquette. Briquetting was performed with hydraulic press machine. The gross calorific value (GCV = 14.55 MJ/kg) reveals that water hyacinth briquette is a probable energy source for heating applications. The water hyacinth briquette has higher ash content and a lower percentage of VM compared with other briquettes. The water hyacinth briquette's VM content was found to be 65.7%. The briquette sample with the highest VM was observed to have the highest GCV.

20.7.3 Weeds for carbonization and briquetting

20.7.3.1 Ipomoea carnea

Carbonization of the fresh woody stem of *I. carnea* was reported by Konwar et al. (2007b) in a lab-scale pyrolyzer with a temperature range of 350–600°C and heating rate of 5°C/min. *I. carnea* is considered among one of the noxious weeds available in the wetlands of tropical regions. Charcoal yield ranged from 24.23% to

37.89 wt%, and CV varied from 17.29 to 33.47 MJ/kg. Solid fuel was prepared from charcoal obtained at a maximum pyrolyzing temperature, i.e., 600°C. Although the density of charcoal obtained was found to be lower than commercial-grade wood charcoal, after briquetting the density increased from 0.210 to 0.689, which also improved combustion properties when compared with firewood. Fig. 20.5 shows the *I. carnea* (A) growing on roadside and (B) blocking water in the irrigation canal. Fig. 20.6 shows the *I. carnea* (A) charcoal and (B) handmade charcoal briquette.

20.7.3.2 Lantana camera *and* Chromolaena odorata

Goswami et al. (2016), Mandal et al. (2013), Saikia et al. (2015) reported the availability of weed biomasses in northeast India and its possibility of making briquettes from the biomass. It had been reported that annually about 20 t/ha of weed biomass is produced in NE India. Both *Lantana camera* and *C. odorata* were assessed for their biochar yield and quality. The beehive briquette properties were also evaluated for these biochars. Biochar obtained after sieving with 5 mm sieve was mixed with mud to obtain

FIGURE 20.5 *Ipomoea carnea* (A) growing on roadside and (B) blocking water in the irrigation canal.

FIGURE 20.6 *Ipomoea carnea* (A) charcoal and (B) hand-made charcoal briquette.

beehive briquettes. The mud acts both as a binder and as a burning controller as it decreases the burning rate. Biochar production efficiency of *Lantana and Chromolaena* was found to be 27.72% and 18.34%, respectively. The carbon content of *Lantana* was found to be 65.99%, whereas *Chromolaena* was 61.22%. The CVs of beehive briquettes made from *Lantana* and *Chromolaena* biochar were 18.6 and 18.1 MJ/kg, respectively. The burning duration of *Lantana* biochar briquette was 133 min that was 10 and 4 min lesser than pinewood and *Chromolaena* biochar briquette, respectively. The burning duration of all briquettes is sufficient for cooking three items in an everyday meal.

20.7.4 Weeds for gasification

Gasification is a form of thermochemical conversion of biomass into combustible gases in presence of air, producing synthesis gas consisting of H_2, CO, CO_2, CH_4, and light hydrocarbons. Production of a gaseous product with low heating value and high tar content are some of the disadvantages of low-temperature gasification (Ali et al., 2020). Hence, high temperature or catalytic substances are employed to increase gas yields and reduce tar content.

Wongsiriamnuay and Tippayawong (2012) used Mimosa to assess its suitability for the production of gaseous fuel via gasification using a bench-scale, fixed-bed gasifier under atmospheric pressure. The investigation aimed to study the consequence of process parameters such as temperature, collection time, and catalyst to biomass ratio on its product yields and composition. It was observed that temperature rise results in enhancement of production of H_2, CO, and CO_2 (11, 18, and 24 mol %). However, in the case of CH_4, it was a reverse scenario. It can be stated that the rates of production for H_2 and CO are high during the volatile releasing step while the rates are low during carbonization. It is also believed that the utilization of catalysts can improve product distribution during gasification. It was also reported that with a 1:1 catalyst-to-biomass ratio, the highest concentrations of H_2 and CO, i.e., 27 and 29 mol%, respectively, were achieved.

Wongsiriamnuay et al. (2008) performed thermogravimetric analysis and gasification of Mimosa under atmospheric pressure in fixed-bed reactor to estimate mass loss rate, composition, and yield of gaseous product. The gas produced was analyzed by gas chromatography. From the results obtained, Mimosa was found to be potentially suitable as biofuel. It contains a high proportion of holocellulose and was rich in carbon and VM, and low in ash content. Its heating value, in comparison with other biomass, was higher than most agricultural residues. The product gas from gasification contains high H_2 and CO, which is responsible for the lower heating value of gaseous fuel. Therefore, it can be concluded that weeds can act as a renewable fuel source.

20.7.5 Weeds for bioethanol

Bioethanol production is renewable, as the CO_2 emitted from burning is recycled by plants during their photosynthesis. According to Huber et al. (2006), bioethanol production from weed biomass is promising and cheaper, as their cellulosic substrates do not need extra additional economic effort to grow on waste land or water bodies. Due to the difference in their structure and cell wall components from other plant species, the digestion in the bioconversion process is affected during bioethanol production (Sarkar et al., 2012). The most potential fraction of the cell wall for the production of bioethanol is cellulose and hemicelluloses (Chandel et al., 2007; Kumar and John Thomas, 2002), whereas through acidic and enzymatic (hemicellulase/cellulase) reactions, carbohydrate is possible to be transformed to fermentable monomeric sugars. These sugars through microbial fermentation processes are converted into 2,3-butanediol, xylitol, and ethanol (Chandel et al.,

2010). According to Chandel and Singh (2011), the complex composition of the weedy substrate can be a limiting factor for its potential utilization. Therefore, exploration of efficient and economical approaches to digest complex chemicals is critical.

Yanagisawa et al. (2011) worked with agar, chigaiso, and sea lettuce, which represents three different types of seed weeds namely red, brown, and green, respectively. It reported methods to achieve high ethanol concentrations. In comparison with lignocellulosic biomass, seaweeds were reported to be low in cellulose and consisted of β-1,3-glucan, starch, and cellulose as glucans. The absence of lignin makes the polysaccharides in these seaweeds readily available for hydrolysis. The only hydrolyzable polysaccharides present in sea lettuce and chigaiso are glucans. It was observed that for these seaweeds, about 3% more ethanol yield was obtained. However, agar consists of both galactan and glucan, which is responsible for a higher concentration of ethanol obtained via combined saccharification. Combined saccharification means acid hydrolysis of galactan to produce galactose, which is then converted to glucose via enzymatic hydrolysis of glucan. Finally, *Saccharomyces cerevisiae* IAM 4178 were employed to ferment sugars to obtain high concentration viz. 5.5% of ethanol, which was higher than the declared economically feasible concentration for distillation. Hence, it can be concluded that utilization of new saccharification methods and seaweeds was an effective pathway for ethanol production.

Lemna minor, another nonfood aquatic crop, is gaining attention due to its ability to convert pollutants to biofuel production system (Ge et al., 2012). Lack of lignin and hemicellulose as compared with terrestrial plants and duckweed biomass could be directly broken down to glucose with the help of enzymes without any pretreatment. Hence, apart from its potential utilization for phytoremediation, *L. minor* may also serve as an alternative source of bioethanol.

About 12.4 g dry biomass m^2/day of duckweed can be generated from a pilot-scale anaerobically treated swine wastewater culture pond (Xu et al., 2011). The study reported that after 10 days the starch content tripled in duckweed under nutrient-deprived conditions. The technology developed with duckweed in this study was found to generate 50% more ethanol compared with maize-to-ethanol conversion.

Chen et al. (2012) reported the abundant availability and novelty behind the use of *L. punctata* for ethanol preparation. Pectinase was utilized for enhancement in the glucose yield, which results in increased ethanol concentration as well as productivity. The study reported that pretreated *L. punctata* mash can produce about 30.8 g/L of ethanol for 14 h optimal conditions such as temperature = 45°C, time for pretreatment = 300 min, and 26.54 PECTU/g mash of pectinase. Recent research is focused on enhanced bioethanol production from a variety of weeds keeping in mind the sustainability and economic feasibility of the production process. Table 20.5 enlists some of the recent studies regarding bioethanol production from various weed species.

20.7.6 Weeds for biooil

The application of biomass for production of chemicals in a biorefinery should be encouraged more. The present state is very much similar to the early days of chemical industry from coal and coke/ or the late petrochemical industry. In an attempt to identify the basic components of weed biomass, researchers have reported >300 compounds in biooil; however, only a small fraction of the biooil (40%–50%) has been structurally characterized. Mohan et al. (2006) reported that a variety of functional groups such as alcohols, acids, ketones, sugars, aldehydes, phenols, furans, and so on are present in pyrolysis oil obtained from wood/biomass. Characterization of biooil acquired from various weeds has been discussed in the following.

TABLE 20.5 Conversion technologies and ethanol yield of weeds.

Weed feedstock	Conversion technology			Ethanol yield	References
	Pretreatment method	Saccharification	Fermentation		
Aquatic weeds (*Eichhornia crassipes*, *Lemna minor*, and *Azolla microphylla*)	Hydrothermal coupled with anaerobic digestion	Commercial cellulase from *Trichoderma reesei*	*Saccharomyces cerevisiae*	0.167–0.231 g/g	Kaur et al. (2019)
Parthenium hysterophorus	Acid/alkali	Cold and hot hydrolysis using commercial cellulases	*Torulaspora delbrueckii*, *Schizosaccharomyces pombe*, and *Saccharomyces cerevisiae*	0.24–0.27 g/g	Tavva et al. (2016)
Water hyacinth (*E. crassipes*)	Microbial pretreatment using lignolytic fungus (*Phanerochaete chrysosporium*) followed by dilute acid/alkali treatment	–	*S. cerevisiae*	1.4 g/L (ethanol conc.)	Zhang et al. (2018)
Kariba weed (*Salvinia molesta*)	Thermal acid hydrolysis followed by sonication	Commercial cellulases (Celluclast, Viscozyme, and Cellic CTec2)	*S. cerevisiae*, *Kluyveromyces marxianus*, and *Pichia stipitis*	15.9 g/L, 0.45 (*P. stipitis*), 14.7 g/L, 0.41 (*K. marxianus*), 13.2 g/L, 0.37 (*S. cerevisiae*)	Kityo et al. (2020)
Seaweed *Ulva rigida*	–	Thermal acid hydrolysis in presence of acid catalyst; 4% (v/v) H_2SO_4	*Pachysolen tannophilus*	0.12 g/g	M. El Harchi et al. (2018)
Brachiaria mutica (para grass)	Dilute alkali pretreatment	Commercial cellulase	*S. cerevisiae*	3.86 g/L	Sahoo et al. (2017)
Gooseweed (*Sphenoclea zeylanica* Gaertn.)	1% NaOH, followed by 1% H_2O_2	Cellulase (β-glucosidase)	*Saccharomyces cerevisiae*	11.84 g/L	Vu et al. (2018)
Prosopis juliflora	3% NaOH	CTec2 cellulase enzyme	*S. cerevisiae*	21.84 g/L, 0.27 g/g	Palled et al. (2018)
Agarophyte red seaweed, *Gelidium elegans*	Thermal acid hydrolysis		*S. cerevisiae*	13.27 g/L	Hessami et al. (2019)
Napier grass (*Pennisetum purpureum* Schumach)	Acid/alkali	Commercial cellulase (Cellic CTec2)	*S. cerevisiae* and *Scheffersomyces shehatae*	187.4 g/kg of Napier grass at 44.7 g/L ethanol conc.	Kongkeitkajorn et al. (2020)

Saikia et al. (2015) reported that at 550°C the maximum biooil, i.e., 41% was obtained of which 11.5% was oil phase from *I. carnea* through pyrolysis. The presence of macromolecules containing oxygen and water (sugars, hydroxy aldehydes, carboxylic acids, hydroxy ketones, and oligomeric phenolic compounds) is responsible for the higher biooil (1142 kg/m^3) density compared with commercial diesel (820–860 kg/m^3) (Oasmaa and Czernik, 1999). Gas chromatography–mass spectroscopy (GC-MS) analysis revealed that biooil is a complex mixture of various organic compounds. It mostly consists of branched/long chains, phenol and its derivatives, mixed acids, esters, alcohols, and various aromatics, and the rest are furans and ketones.

Muradov et al. (2010) studied thermal characteristic of pyrolysis oil obtained from *L. minor* performing around 400–700°C applying different flow rate (36–150 mL/min) for Ar. It was observed that the maximum biooil yield of about 40 wt.% was obtained at 500°C. The GC-MS analysis of biooil obtained at 400, 500, and 700°C revealed that the product state was minutely influenced by pyrolysis temperature, whereas temperature greatly influenced its relative quantity. The higher temperature resulted in increased production of lighter components such as 4-methyl phenol, phenol, and 2-furan methanol-H-imidazole-4-methanol.

Aysu and Durak (2015) experimented with *Datura stramonium* L. using supercritical liquefaction and slow pyrolysis both with and without catalysts (zinc oxide, calcium hydroxide). A cylindrical reactor was used with organic solvents (isopropanol, acetone) under supercritical conditions with varying temperatures (275, 300, and 325°C) for liquefaction, whereas pyrolysis was performed in a fixed-bed tubular reactor at the temperature range of 400–600°C. The liquids obtained from liquefaction and pyrolysis at 300 and 500°C were characterized and analyzed. It was observed that both biooils were low in oxygen and had higher carbon contents. Higher heating value (HHV) of liquefaction biooils (29.73 in acetone and 28.02 MJ/kg in isopropanol) was higher compared with raw biomass 14.39 MJ/kg. The low HHV of pyrolytic oil, i.e.,19.40 MJ/kg compared with liquefaction biooil may be attributed to higher content aromatic and ester compounds in liquefaction biooils than pyrolysis biooils. Apart from this, high content of oxygenated compounds in pyrolysis oil is responsible for low heating value.

Miranda et al. (2014) reported the ability of aquatic plant *Elodea canadensis* and *Marsilea quadrifolia* to treat selenium-rich mining wastewater and to produce petrochemicals. GC-MS analysis reveals that straight chain of C_{14}–C_{20} alkanes in the biooil can be used as a supplement in diesel fuel, or glycerine-free component of biodiesel.

In recent times, the effects of various operation parameters such as pyrolysis time, temperature, nitrogen flow, and heating rate on the biooil yield and quality obtained from Crofton weed were evaluated by Cheng et al. (2019). It was observed that biooil yield and quality are strongly influenced by pyrolysis temperature, time, and heating rate, respectively. High-quality biooil containing aldehyde (28.09%), ketone (15.39%), and phenol (13.04%) was obtained at optimized parameters of pyrolysis temperature (600°C), time (50 min), and heating rate (20°C/min). Furthermore, large amounts of combustible matter rich in biogas were also produced from the pyrolysis of Crofton weed. In another study, biooil rich in phenolic compounds, ketones, and furans was obtained by the pyrolysis of phumdi, paragrass, and a mixture of both the biomass (Awasthi et al., 2019). Biochar with high organic carbon content was also obtained as a copyrolysis product suggesting the mixture of phumdi and paragrass as a potential feedstock for bioenergy production.

20.7.7 Weeds for biochar

The biochars obtained from weed are high in C and N content and low in H compared with the original. In comparison with the original feedstock, the heating values for chars were much high and comparable with some coals strengthening its candidature for potential utilization as fuel with HHV ranging from 19.4 to 22.9 MJ/kg (Yu et al., 2011). According to Bird and Benson (1987), biochar obtained from seaweed could be a potential soil ameliorant, fertilizer, and an additional revenue stream. However, further investigation is essential to explore its commercial and economical feasibility of seaweed as feedstock for fuel and chemicals, where ash content obtained was 52.4% (Ross et al., 2008). HHV of the seaweed biochar was found to be 33 MJ/kg close to the graphite of 33 MJ/kg (Plummer, 1930) and that of wood charcoal of 32 MJ/kg (Misginna and Rajabu, 2012; Milledge et al., 2015). The effect of temperature on the properties of biochar obtained from Crofton weed (*Eupatorium adenophora*) was reported by (Fan et al., 2019). It was observed that high pyrolysis temperature–produced biochar had high metal adsorption properties, higher aromaticity, alkalinity, and stability, whereas lower pyrolysis temperature yields porous biochar with a higher proportion of water-soluble nitrogen, phosphorus, magnesium, and calcium. Recently, mixed weed biomass such as *Carthamus oxyacantha*, *Asphodelus tenuifolius*, and *Chenopodium album* was used to produce biofuels such as desirable quality bio-oil (57.5%), biogas (23.75%), and biochar (18.75%) at a lower temperature (400°C) using nanocatalyst-assisted gasification process (Ali et al., 2020). The produced biochar from weed feedstock was rich in organic matter, nitrogen, phosphorous, and potassium and could be used as a soil amendment. Similarly, biochar obtained by the slow pyrolysis of invasive weed species *Prosopis juliflora* was used as a stabilizing material to amend expansive black cotton soil (GuhaRay et al., 2019). Furthermore, soil–biochar composite obtained from water hyacinth (*E. crassipes*) was used as a soil amendment material to improve soil fertility and metal adsorption by increasing the water retention capacity and decreasing desiccation cracks in soils (Bordoloi et al., 2018).

20.8 Future prospects

Weeds have been considered as a menace in traditional agriculture because of its adverse effects in agricultural production. Ill effects of weeds are also visible in animal husbandry, human health, aquatic ecosystem, forests, and pastures, and therefore, various mechanisms such as physical, mechanical, chemical, and biological have been devised, widely used, and integrated into modern agricultural practices. However, weeds can be of great economic importance, as they are a renewable bioresource and a suitable alternative to conventional bioresources for bioconversion to various purposes including biofuels and biomaterials through different conversion pathways. Weeds as a renewable bioresource also enjoy several distinct advantages such as their ubiquitous nature and ability to grow and proliferate in diverse and stressed agroecosystems. As a result of this, effective weed control is always a challenge and a costly proposition. In view of the aforementioned, it has become necessary to view weeds not as an unnecessary biomass but as an opportunity to utilize it in a gainful manner to recover the chemical components of weeds for conversion to useful substances. This will also help in achieving the twin objectives of circular bioeconomy and overall sustainability.

An important area of future perspectives of weeds utilization will be to explore its chemical makeup for possible development of materials of pharmaceutical significance. A number of weeds exhibit pharmacological attributes, which indicates their possible role as a chemotherapeutic agent. The lignocellulosic groups abundant in weed biomass could be the starting point for conversion to various materials for phytoextraction and biosorption of pollutants, biochar for c-sequestration, mosquito repellents, and so on.

Quite a few industrial houses have been using weed biomass for pharmaceutical purposes, while R & D Team at TATA International have developed a weed (Parthenium) biomass-derived preservative, which is environmentally benign and cheaper than its nonrenewable resource-derived chemical alternative.

Many nongovernmental organizations (NGOs), working in close association with rural people, are trying to reconnect the use of weeds for fulfilling a number of needs of the rural population. It is in this context that weeds may serve as an excellent low-cost feedstock for various industries. There are lots of logistics and technical issues, which might require important consideration, and these are collection, transportation to a common point for further transportation, high moisture content causing perishability, varied compositions of different weeds, and so on.

20.9 Conclusion

In India, manual weeding is the traditionally adopted strategy to manage weeds, which is not only labor-intensive but expensive too. Therefore, there has been a need to search for other alternatives such as herbicides, which are blooming with an annual rate of 15%. Besides using different techniques to weeding this biomass, it can be more profitable to use it as a renewable energy source to produce bioenergy biofuel. Use of weed biomass for biochemical and thermochemical conversion leads to a variety of liquid and solid fuels, which have potential value in the energy sector. Various biomass types including woody and herbaceous and aquatic weed are garnering much attention as a substitute for energy production with suitable conversion technologies. Currently, weed biomass utilization for energy production uses only a few select technologies, which have their limitations and prospects. The production of synthetic natural gas, lignocellulosic bioethanol, and synthetic biofuels may be a part of future technology. Weed biomass from agricultural and nonagricultural fields may also be a contribution toward green biorefineries. Production of biogas is a sustainable technology to utilize weed/grass biomass to fit within green biorefinery concepts. Briquettes' successful commercialization will have a constructive effect on both environment and economics of society. It will create new job opportunities and provide a sustainable weed management alternative that will reduce not only its prevalence but also the intensive use of wood as a fuel.

References

Abdelhamid, A.M., Gabr, A.A., 1991. Evaluation of water hyacinth as a feed for ruminants. Archiv. Für. Tierernaehrung. 41 (7–8), 745–756.

Ali, S., Shafique, O., Mahmood, S., Mahmood, T., Khan, B.A., Ahmad, I., 2020. Biofuels production from weed biomass using nanocatalyst technology. Biomass Bioenergy 139, 105595.

Altenburg, T., Dietz, H., Hahl, M., Nikolidakis, N., Rosendahl, C., Seelige, K., 2008. Biodiesel Policies for Rural Development in India. German Development Institute.

Anjum, S.S., Prakash, O., Pal, A., 2019. Conversion of non-edible *Argemone Mexicana* seed oil into biodiesel through the transesterification process. Energy Sources Part A: Recov. Utiliz. Environ. Effect 41 (19), 2356–2363.

Awasthi, A., Singh, G., Dhyani, V., Kumar, J., Reddy, Y.S., Adarsh, V.P., et al. Bhaskar, T., 2019. Co-pyrolysis of phumdi and para grass biomass from Loktak lake. Bioresour. Technol. 285, 121308.

Aysu, T., Durak, H., 2015. Thermochemical conversion of *Datura stramonium* L. by supercritical liquefaction and pyrolysis processes. J. Supercrit. Fluids. 102, 98–114.

Baruah, U., 1995. Production of Fuel and Fertilizer by Anaerobic Digestion of Some Aquatic Weeds of North-East India Unpblished MSc (Agri) Thesis Submitted to AAU, Jorhat-13, pp. 10–25.

Bhan, V.M., Kumar, S., Raghuwanshi, M.S., 1999. Weed management in India. Indian J. Plant Prot. 27 (1/2), 171–202.

Bhuyan, N., Sut, D., Gogoi, L., Kataki, R., Goud, V.V., 2019. Rural biorefinery: a viable solution for production of fuel and chemicals in rural India. In: Sustainable Bioenergy. Elsevier, pp. 21–47.

Bird, K.T., Benson, P.H., 1987. Seaweed Cultivation for Renewable Resources. Elsevier, Bergen, Norway.

Borah, A.J., Agarwal, M., Goyal, A., Moholkar, V.S., 2019. Physical insights of ultrasound-assisted ethanol production from composite feedstock of invasive weeds. Ultrason. Sonochem. 51, 378–385.

Bordoloi, S., Garg, A., Sreedeep, S., Lin, P., Mei, G., 2018. Investigation of cracking and water availability of soil-biochar composite synthesized from invasive weed water hyacinth. Bioresour. Technol. 263, 665–677.

Castro, Y.A., Agblevor, F.A., 2020. Biomethanation of invasive water hyacinth from eutrophic waters as a post weed management practice in the Dominican Republic: a developing country. Environ. Sci. Pol. Resear. 1–12.

Chandel, A.K., Kapoor, R.K., Singh, A., Kuhad, R.C., 2007. Detoxification of sugarcane bagasse hydrolysate improves ethanol production by Candida shehatae NCIM 3501. Bioresour. Technol. 98 (10), 1947–1950.

Chandel, A.K., Singh, O.V., Chandrasekhar, G., Rao, L.V., Narasu, M.L., 2010. Key drivers influencing the commercialization of ethanol-based biorefineries. J. Commer. Biotechnol. 16 (3), 239–257.

Chandel, A.K., Singh, O.V., 2011. Weedy lignocellulosic feedstock and microbial metabolic engineering: advancing the generation of 'Biofuel'. Appl. Microbiol. Biotechnol. 89 (5), 1289–1303.

Chen, Q., Jin, Y., Zhang, G., Fang, Y., Xiao, Y., Zhao, H., 2012. Improving production of bioethanol from duckweed (*Landoltia punctata*) by pectinase pretreatment. Energies 5, 3019–3032.

Cheng, S., Shu, J., Xia, H., Wang, S., Zhang, L., Peng, J., et al. Zhang, Q., 2019. Pyrolysis of Crofton weed for the production of aldehyde rich bio-oil and combustible matter rich bio-gas. Appl. Therm. Eng. 148, 1164–1170.

Clements, D.R., Weise, S.F., Brown, R., Stonehouse, D.P., Hume, D.J., Swanton, C.J., 1995. Energy analysis of tillage and herbicide inputs in alternative weed management systems. Agri. Ecosys. Environ. 52 (2–3), 119–128.

Dahunsi, S.O., Oranusi, S., Efeovbokhan, V.E., Olayanju, A., Zahedi, S., Ojediran, J.O., et al. Aladegboye, O.J., 2018. Anaerobic conversion of *Chromolaena odorata* (Siam weed) to biogas. Energy Rep. 4, 691–700.

Demirbas, A., 2007. The influence of temperature on the yields of compounds existing in bio-oils obtained from biomass samples via pyrolysis. Fuel Process. Technol. 88, 591–597.

Demirbas, A., Pehlivan, E., Altun, T., 2006. Potential evolution of Turkish agricultural residues as bio-gas, bio-char and bio-oil sources. Int. J. Hydrog. Energy 31, 613–620.

DWSR/, 2013. Vision 2050. Directorate of Weed Science Research, Jabalpur, India.

El Bassam, N., Graef, M., Jakob, K., 1998. Sustainable energy supply for communities from biomass. Sust. Agri. Food Energy Ind. 837–843.

El Harchi, M., Kachkach, F.F., El Mtili, N., 2018. Optimization of thermal acid hydrolysis for bioethanol production from *Ulva rigida* with yeast *Pachysolen tannophilus*. S. Afri. J. Bot. 115, 161–169.

Epstein, E., 1999. SILICON. Annu. Rev. Plant Physiol. Plant Mol. Biol. 50 (1), 641–664.

Fan, L., Zhou, X., Liu, Q., Wan, Y., Cai, J., Chen, W., et al. Luo, H., 2019. Properties of *Eupatorium adenophora* Spreng (Crofton weed) biochar produced at different pyrolysis temperatures. Environ. Eng. Sci. 36 (8), 937–946.

Gaur, R.Z., Khan, A.A., Suthar, S., 2017. Effect of thermal pretreatment on co-digestion of duckweed (*Lemna gibba*) and waste activated sludge on biogas production. Chemosphere 174, 754–763.

Ge, X., Zhang, N., Phillips, G.C., Xu, J., 2012. Growing Lemna minor in agricultural wastewater and converting the duckweed biomass to ethanol. Bioresour. Technol. 124, 485–488.

Gesellschaft für Technische Zusammenarbeit (GTZ)/The Energy Research Institute (TERI), 2005. Liquid Biofuels for Transportation: India Country Study on Potential and Implications for Sustainable Agriculture and Energy, New Delhi.

Goette, J.M., Currey, W.L., Brecke, B.J., Green, M.B., Fluck, R.C., 1980. Comparisons of energy requirements in conventional and no-tillage soybeans. In: Proceedings of the 3rd Annual Southern Conservation Tillage Conference for Sustainable Agriculture 19 June, 1980, Gainesville, Florida.

Goswami, R., Shim, J., Deka, S., Kumari, D., Kataki, R., Kumar, M., 2016. Characterization of Cadmium removal from aqueous solution by biochar produced from Ipomoea fistulosa at different pyrolytic temperatures. Ecol. Engg. 97, 444–451.

GuhaRay, A., Guoxiong, M., Sarkar, A., Bordoloi, S., Garg, A., Pattanayak, S., 2019. Geotechnical and chemical characterization of expansive clayey soil amended by biochar derived from invasive weed species *Prosopis juliflora*. Innov. Infrastruct. Solut. 4 (1), 44.

Gunnarssona, C.C., Petersen, C.M., 2007. Water hyacinths as a resource in agriculture and energy production: a literature review. Waste Manag. 27, 117–129.

Hessami, M.J., Cheng, S.F., Ambati, R.R., Yin, Y.H., Phang, S.M., 2019. Bioethanol production from agarophyte red seaweed, Gelidium elegans, using a novel sample preparation method for analysing bioethanol content by gas chromatography. 3 Biotech 9 (1), 25.

Hoseini, S.S., Najafi, G., Sadeghi, A., 2019. Chemical characterization of oil and biodiesel from common Purslane (Portulaca) seed as novel weed plant feedstock. Indus. Crop. Prod. 140, 111582.

Huber, G.W., Iborra, S., Corma, A., 2006. Synthesis of transportation fuels from biomass: chemistry, catalysts, and engineering. Chem. Rev. 106, 4044.

Jenkins, B.M., Bakker, R.R., Wei, J.B., 1996. On the properties of washed straw. Biomass Bioenergy 10, 177–200.

Jenkins, B.M., Baxter, L.L., Miles, T.R., Miles, T.R., 1998. Combustion properties of biomass. Fuel Process. Technol. 54, 17–46.

Kaliyan, N., Morey, V.R., 2009. Factors affecting strength and durability of densified biomass products. Biomass Bioenergy 33, 337–359.

Karabas, H., 2018. Investigation of the use of Cephalaria syriaca weed seeds (Turkish Pelemir) as feedstock in biodiesel production. Fresenius Environ. Bullet. 27 (11), 7265–7269.

Kataki, R., 2005. Studies on Carbonization of Some Indigenous Wood Species of North-East India. PhD thesis submitted to Tezpur University, Tezpur.

Kaur, M., Kumar, M., Sachdeva, S., Puri, S.K., 2018. Aquatic weeds as the next generation feedstock for sustainable bioenergy production. Bioresour. Technol. 251, 390–402.

Kaur, M., Kumar, M., Singh, D., Sachdeva, S., Puri, S.K., 2019. A sustainable biorefinery approach for efficient conversion of aquatic weeds into bioethanol and biomethane. Energy Conver. Manag. 187, 133–147.

Kityo, M.K., Sunwoo, I., Kim, S.H., Park, Y.R., Jeong, G.T., Kim, S.K., 2020. Enhanced bioethanol fermentation by sonication using three yeasts species and Kariba weed (*Salvinia molesta*) as biomass collected from lake Victoria, Uganda. Appl. Biochem. Biotechnol. 192 (1), 180–195.

Kongkeitkajorn, M.B., Sae-Kuay, C., Reungsang, A., 2020. Evaluation of Napier grass for bioethanol production through a fermentation process. Process 8 (5), 567.

Konwar, D., Kataki, R., Saikia, P., 2007a. Pyrolysis of some indigenous tree species of northeast India: effect of pyrolysis temperature and heating rate on the products yield and char quality. Energy Source Part A. 29 (15), 1433–1442.

Konwar, D., Kataki, R., Saikia, M., 2007b. Production of solid fuel from Ipomoea carnea wood. Energy Source Part A. 29 (9), 817–822.

Kumar, S., Vishwakarma, K., 2005. Nutritional value of alligator weed [*Alternanthera philoxeroides* (Mart.) Griseb.] and its possible utility as a fodder in India. Indian J. Weed Sci. 37 (1&2), 152.

Kumar, S.M.G., John Thomas, K., 2002. Macrophytes of the Muriyad wetlands: ecological role in relation to rice cultivation and fishing. In: Paper Presented in the Symposium on Conservation, Restoration, Management of Aquatic Ecosystems, Indian Institute of Science, Bangalore, India, December 9–13.

Laguë, C., Khelifi, M., 2001. Energy use and time requirements for different weeding strategies in grain corn. Canadian Biosyst. Eng. 43, 2–13.

Lerdau, M., Wickham, J.D., 2011. Non-natives: four risk factors. Nature 475, 36–37.

Lewandowski, I., Kicherer, A., 1997. Combustion quality of biomass: practical relevance and experiments to modify the biomass quality of *Miscanthus* × *giganteus*. Eur. J. Agr. 6, 163–177.

Lewandowski, I., Scurlock, J.M.O., Lindvall, E., Christou, M., 2003. The development and current status of perennial rhizomatous grasses as energy crops in the US and Europe. Biomass Bioenergy 25, 335–361.

Li, H., Song, W., Cheng, J., Ding, L., Zhou, J., Li, Y.Y., 2020. Effects of harvest month on biochemical composition of alligator weed for biohydrogen and biomethane cogeneration: identifying critical variations in microbial communities. Int. J. Hydrog. Energy 45 (7), 4161–4173.

Lu, W., Wang, C., Yang, Z., 2009. The preparation of high caloric fuel (HCF) from water hyacinth by deoxy-liquefaction. Bioresour. Technol. 100, 6451–6456.

Mähnert, P., Heiermann, M., Pöchl, M., Schelle, H., Linke, B., 2002. Alternative use for grassland cuts – forage grasses as biogas co-substrates. Landtechnik 57, 260–261.

Mandal, S., Singh, R.K., Kumar, A., Verma, B.C., Ngachan, S.V., 2013. Characteristics of weed biomass-derived biocharand their effect on properties of beehive briquettes. Indian J. Hill Farm. 26 (1), 8–12.

Mani, S., Tabil, L.G., Sokhansanj, S., 2006. Effects of compressive force, particle size and moisture content on mechanical properties of biomass pellets from grasses. Biomass Bioenergy 30, 648–654.

Manivannan, A., Narendhirakannan, R.T., 2015. Bioethanol production from aquatic weed water hyacinth (*Eichhornia crassipes*) by yeast fermentation. Waste Biomass Valoriz. 6 (2), 209–216.

McKendry, P., 2002. Energy production from biomass (Part 1): overview of biomass. Bioresour. Technol. 83, 37–46.

Miao, X.L., Wua, Q.Y., 2004. High yield bio-oil production from fast pyrolysis by metabolic controlling of *Chlorella protothecoides*. J. Biotechnol. 110, 85–93.

Milledge, J.J., Staple, A., Harvey, P.J., 2015. Slow pyrolysis as a method for the destruction of Japanese wireweed, *Sargassum muticum*. Environ. Nat. Resour. Res. 5 (1), 28–37.

Miranda, A.F., Muradov, N., Gujar, A., Stevenson, T., Nugegoda, D., Ball, A.S., Mouradov, A., 2014. Application of aquatic plants for the treatment of selenium-rich mining wastewater and production of renewable fuels and petrochemicals. J. Sust. Bioenergy Sys. 4, 97–112.

Misginna, M.T., Rajabu, H.M., 2012. Yield and chemical characteristics of charcoal produced by TLUDND gasifier cookstove using eucalyptus wood as feedstock. In: Paper presented at the Second International Conference on Advances in Engineering and Technology, Nagapattinam, India.

Mohan, D., Pittman, C.U., Steele, P.H., 2006. Pyrolysis of wood/biomass for bio-oil: a critical review. Energy Fuels 20, 848.

Mubarak, M., Shaija, A., Suchithra, T.V., 2016. Ultrasonication: an effective pre-treatment method for extracting lipid from *Salvinia molesta* for biodiesel production. Resour. Effi. Technol. 2 (3), 126–132.

Munjeri, K., Ziuku, S., Maganga, H., Siachingoma, B., Ndlovu, S., 2016. On the potential of water hyacinth as a biomass briquette for heating applications. Int. J. Energy Environ. Eng. 7, 37–43.

Muradov, N., Fidalgo, B., Gujar, A.C., Raissi, A.T., 2010. Pyrolysis of fast-growing aquatic biomass – Lemna minor (duckweed): characterization of pyrolysis products. Bioresour. Technol. 101, 8424–8428.

Mustafa, H.M., Hayder, G., 2021. Recent studies on applications of aquatic weed plants in phytoremediation of wastewater: a review article. Ain Shams Eng. J. 12 (1), 355–365.

NASA, 1978. Bioconversion Study - JPL Publication 79-9, Jet Propulsion Laboratory. California Institute of Technology, Pasadena, California. https://ntrs.nasa.gov/api/citations/19790011279/downloads/19790011279.pdf.

Oasmaa, A., Czernik, S., 1999. Fuel oil quality of biomass pyrolysis oils state of the art for the end users. Energy Fuel. 13 (4), 914–921.

Obernberger, I., 1998. Decentralized biomass combustion: state-of-the art and future development. Biomass Bioenergy 14 (l), 33–56.

Ohlsson, L.O., Karlsson, S., Rupar-Gadd, K., Albers, E., Welander, U., 2020. Evaluation of *Laminaria digitata* and *Phragmites australis* for biogas production and nutrient recycling. Biomass Bioenergy 140, 105670. Online: http://planningcommission.nic.in/reports/genrep/cmtt_bio.pdf.

Osadolor, O.O., 2009. Availability of grasses, weeds and leaves as energy resources. Renew. Energy 34, 486–491.

Palled, V., Anantachar, M., Veerangouda, M., Prakash, K.V., Ramachandra, C.T., et al. Ramesh, B., 2018. Prosopis juliflora—a potential problematic weed for lignocellulosic ethanol production. In: Energy and Environment. Springer, Singapore, pp. 191–206.

Pandiyan, K., Singh, A., Singh, S., Saxena, A.K., Nain, L., 2019. Technological interventions for utilization of crop residues and weedy biomass for second generation bioethanol production. Renew. Energ. 132, 723–741.

Peart, R.M., 1999. Energy and the environment. In: Kitani, O., Jungbluth, T., Peart, R.M., Ramdani, A. (Eds.), CIGR Handbook of Agricultural Engineering, Energy & Biomass Engineering, vol. V. ASAE, St. Joseph, MI, pp. 42–52.

Pimentel, D., Burgess, M., 1980. Energy inputs in corn production. In: Pimentel, D. (Ed.), CRC Handbook of Energy Utilization in Agriculture. CRC Press, Inc, Boca Raton, FL, pp. 67–84.

Planning Commission, 2003. Report of the Commission on Development of Bio-Fuel, Planning Commission. Government of India, New Delhi.

Plummer, W.B., 1930. Heat of combustion of Carbon1. Indus. Eng. Chem. 22 (6), 630–632.

Prochnow, A., Heiermann, M., Plöchl, M., Linke, B., Idler, C., Amon, T., Hobbs, P.J., 2009. Bioenergy from permanent grassland – a review 1. Biogas. Bioresour. Technol. 100, 4931–4944.

Ra, C.H., Choi, J.G., Kang, C.H., Sunwoo, I.Y., Jeong, G.T., Kim, S.K., 2015. Thermal acid hydrolysis pretreatment, enzymatic saccharification and ethanol fermentation from red seaweed, Gracilaria verrucosa. Microbiol. Biotechnol. Lett. 41 (1), 9–15.

Rao, A.N., Chauhan, B.S., 2015. Weeds and weed management in India - a review. Weed. Sci. Anthe. Asian-Pacific Reg. 87–118.

Ross, A.B., Jones, J.M., Kubacki, M.L., Bridgeman, T., 2008. Classification of macroalgae as fuel and its thermochemical behaviour. Bioresour. Technol. 99 (14), 6494–6504.

Saha, B., Devi, C., Khwairakpam, M., Kalamdhad, A.S., 2018. Vermicomposting and anaerobic digestion—viable alternative options for terrestrial weed management—a review. Biotechnol. Rep. 17, 70–76.

Sahoo, D., Ummalyma, S.B., Okram, A.K., Sukumaran, R.K., George, E., Pandey, A., 2017. Potential of Brachiaria mutica (Para grass) for bioethanol production from Loktak Lake. Bioresour. Technol. 242, 133–138.

Saikia, P., Gupta, U.N., Barman, R.S., Kataki, R., Chutia, R.S., Baruah, B.P., 2015. Production and characterization of bio-oil produced from Ipomoea carnea bio-weed. Bio-Energy Res. 8 (3), 1212–1223.

Sarkar, N., Ghosh, S.K., Bannerjee, S., Aikat, K., 2012. Bioethanol production from agricultural wastes: an overview. Renew. Energy 37 (1), 19–27.

Schuck, S., 2006. Biomass as an energy source. Int. J. Environ. Stud. 63 (6), 823–835.

Sharma, V., Pant, S., 2020. Weeds as a underutilized bioresource and management tool: a comprehensive tool. Waste Biomass Valoriz. https://doi.org/10.1007/s12649-018-0212-2.

Sharma, K.C., Subhas, C., 2006. Effect of cultural and chemical weed control on productivity of fodder oat (*Avena sativa*) in arid Rajasthan. Indian J. Agro. 51 (4), 327–330.

Singh Bisen, K., Sharma, P., Gupta, B., Baredar, P., 2020. Development and experimental characterization of energy efficient poultry litter & plant weeds based briquettes (PLPWBB) by comparing with rice husk briquettes. Mater. Today Proc. https://doi.org/10.1016/j.matpr.2020.09.130 (in Press).

Snehalata, A., Rao, K.R., 2018. Bioenergy conversion from aquatic weed water hyacinth into agronomically valuable vermicompost. In: Biosynthetic Technology and Environmental Challenges. Springer, Singapore, pp. 245–269.

Sondak, C.F., Ang, P.O., Beardall, J., Bellgrove, A., Boo, S.M., Gerung, G.S., et al. Largo, D., 2017. Carbon dioxide mitigation potential of seaweed aquaculture beds (SABs). J. Appl. Phycol. 29 (5), 2363–2373.

Tag EI-Din, A.R., 1992. Utilization of water-hyacinth hay in feeding of growing sheep. Indian J. Ani. Sci. 62 (10), 989–992.

Tao, G., Geladi, P., Lestander, T.A., Xiong, S., 2012. Biomass properties in association with plant species and assortments. II: a synthesis based on literature data for ash elements. Renew. Sustain. Energy Rev. 16, 3507–3522.

Tavva, S.M.D., Deshpande, A., Durbha, S.R., Palakollu, V.A.R., Goparaju, A.U., Yechuri, V.R., et al. Muktinutalapati, V.S.R., 2016. Bioethanol production through separate hydrolysis and fermentation of *Parthenium hysterophorus* biomass. Renew. Energy 86, 1317–1323.

Thomas, T.H., Eden, R.D., 1990. Water hyacinth – a major neglected resource. Mater. Sci. Wind Energy Biomass Technol. 3, 2092–2096.

Thompson, T.M., Young, B.R., Baroutian, S., 2020. Efficiency of hydrothermal pretreatment on the anaerobic digestion of pelagic Sargassum for biogas and fertiliser recovery. Fuel 279, 118527.

Varma, A., Behera, B., 2003. Green Energy. Concept publishing Company, New Delhi.

Varshney, J.G., Prasad, B.M., 2008. Future scenario of weed management in India. Indian J. Weed Sci. 40 (1&2), 01–09.

Vassilev, S.V., Baxter, D., Andersen, L.K., Vassileva, C.G., 2010. An overview of the chemical composition of biomass. Fuel 89, 913–933.

Vu, P.T., Unpaprom, Y., Ramaraj, R., 2018. Impact and significance of alkaline-oxidant pretreatment on the enzymatic digestibility of *Sphenoclea zeylanica* for bioethanol production. Bioresour. Technol. 247, 125–130.

Werther, J., Saenger, M., Hartge, E.U., Ogada, T., Siagi, Z., 2000. Combustion of agricultural residues. Progress Energy Comb. Sci. 26 (1), 1–27.

Wongsiriamnuay, T., Tippayawong, N., 2012. Product gas distribution and composition from catalyzed gasification of mimosa. Int. J. Renew. Energy Res. 2 (3), 363–368.

Wongsiriamnuay, T., Phengpom, T., Panthong, P., Tippayawong, N., 2008. Renewable energy from thermal gasification of a giant sensitive plant (Mimosa pigra L.). 5th International Conference on Combustion. Incineration/Pyrolysis and Emission Control, Chiang Mai, Thailand.

Xiong, S., Zhang, Q.G., Zhang, D.Y., Olsson, R., 2008. Influence of harvest time on fuel characteristics of five potential energy crops in northern China. Bioresour. Technol. 99, 479–485.

Xu, J., Cui, W., Cheng, J.J., Stomp, A.M., 2011. Production of high-starch duckweed and its conversion to bioethanol. Biosyst. Eng. 110, 67–72.

Yaduraju, N.T., 2012. Weed management perspectives for India in the changing agriculture scenario in 'the'- country. Pak. J. Weed Sci. Res. 18, 703–710.

Yaman, S., 2004. Pyrolysis of biomass to produce fuels and chemical feedstocks. Energy Conver. Manag. 45, 651–671.

Yanagisawa, M., Nakamura, K., Ariga, O., Nakasaki, K., 2011. Production of high concentrations of bioethanol from seaweeds that contain easily hydrolyzable polysaccharides. Process Biochem. 46, 2111–2116.

Yu, J., Wang, C., Wang, Y., Yang, Z., 2011. Deoxy-liquefaction products obtained from Crofton weed at different temperatures. J. Anal. Appl. Pyrol. 92 (1), 68–75.

Yumak, H., Ucar, T., Seyidbekiroglu, N., 2010. Briquetting soda weed (*Salsola tragus*) to be used as a rural fuel source. Biomass Bioenergy 34, 630–636.

Zhang, Q., Wei, Y., Han, H., Weng, C., 2018. Enhancing bioethanol production from water hyacinth by new combined pretreatment methods. Bioresour. Technol. 251, 358–363.

Further reading

Ahmad, M., Moon, D.H., Vithanage, M., Koutsospyros, A., Lee, S.S., Yang, J.E., et al. Ok, Y.S., 2014. Production and use of biochar from buffalo-weed (*Ambrosia trifida* L.) for trichloroethylene removal from water. J. Chem. Technol. Biotechnol. 89 (1), 150–157.

Anonymous, 1976. Making Aquatic Weeds Useful-Some Perspectives for Developing Countries. National Academy of Sciences (US), Washington D.C.

Bauer, A., Hrbek, R., Amon, B., Kryvoruchko, V., Bodiroza, V., Wagentristl, H., et al. Amon, T., 2007. Potential of biogas production in sustainable biorefinery concepts. In: Proceedings of the Fifth Research and Development Conference of Central- and Eastern European Institutes of Agricultural Engineering. Part 2, pp. 20–31.

Bhattacharjee, N., Biswas, A.B., 2018. Pyrolysis of *Alternanthera philoxeroides* (alligator weed): effect of pyrolysis parameter on product yield and characterization of liquid product and bio char. J. Energy Instit. 91 (4), 605–618.

Emanuel, E., 1999. Annu. Rev. Plant Physiol. Plant Mol. Biol. 50, 641–664.

Fahmi, R., Bridgwater, A.V., Darvell, L.I., Jones, J.M., Yates, N., Thain, S., Donnison, I.S., 2007. The effect of alkali metals on combustion and pyrolysis of Lolium and Festuca grasses, switchgrass and willow. Fuel 86 (10–11), 1560–1569.

Fahmi, R., Bridgwater, A.V., Donnison, I., Yates, N., Jones, J.M., 2008. The effect of lignin and inorganic species in biomass on pyrolysis oil yields, quality and stability. Fuel 87 (7), 1230–1240.

Fraud-Williams, R.J., 2002. Weed competition. In: Nayler, R.E.L. (Ed.), Weed Management Handbook, ninth ed. Blackwell Science, UK.

Fuwape, J.A., 1996. Effects of carbonization temperature on charcoal from some tropical trees. Bioresour. Technol. 56, 1–4.

Hirata, T., Kawamoto, S., Nishimoto, T., 1991. Thermogravimetry of wood treated with water insoluble retardants and a proposal for development of fire-retardant wood materials. Fire Mater. 15, 27–36.

International Labour Organization, 1985. Fuelwood and Charcoal Preparation. Geneva, 48pp.

Kataki, R., Konwar, D., 2007. Carbonization of some traditionally preferred indigenous tree species of north-east India. Energy Source 2, 203–211.

Kataki, R., Konwer, D., 2001. Fuelwood characteristics of some indigenous woody species of north-east India. Biomass Bioenergy 20, 17–23.

Kataki, R., Konwer, D., 2002. Fuelwood characteristics of indigenous tree species of north-east India. Biomass Bioenergy 22, 433–437.

Kromus, S., Kamm, B., Kamm, M., Fowler, P., Narodoslawsky, M., 2006. The green biorefinery concept – fundamentals and potential. Biorefineries Indus. Process. Prod. 12, 235–285.

Muradov, N., Taha, M., Miranda, A.F., Kadali, K., Gujar, A., Rochfort, S., Stevenson, T., Ball, A.S., Mouradov, A., 2014. Dual application of duckweed and azolla plants for wastewater treatment and renewable fuels and petrochemicals production. Biotechnol. Biofuel 7, 1–17.

Pahkala, K., Aalto, M., Isolahti, M., Poikola, J., Jauhiainen, L., 2008. Large-scale energy grass farming for power plants – a case study for Osthrobotnia. Finland. Biomass Bioenergy 32, 1009–1015.

Pimentel, D., Zuniga, R., Morrison, D., 2005. Update on the environmental and economic costs associated with alien-invasive species in the United States. Ecol. Econ. 52, 273–288.

Schwarz, A., Gorniak, S.L., Bernardi, M.M., Dagli, M.L.Z., Spinosa, H.S., 2003. Effects of Ipomoea carnea aqueous fraction intake by dams during pregnancy on the physical and neurobiological development of rate offspring. Neurotoxi. Teratol. 25, 615–626.

Sensöz, S., Yorgun, S., Angin, D., Culcuoglu, D., Ozcimen, D., Karaosmanoglu, F., 2001. Fixed-bed pyrolysis of rapeseed cake. Energy Source 23, 873–877.

Thorp, J.R., 2000. Weeds of National Significance. LANTANA (*Lantana camara*) Strategic Plan. National Weed Strategy Executive Committee, Launceston, Tas., Australia. https://www.weeds.org.au/docs/lanstrat.pdf.

Tillman, D.A., 2000. Biomass cofiring: the technology, the experience, the combustion consequences. Biomass Bioenergy 19, 365–384.

Tonn, B., Thumm, U., Claupein, W., 2007. Grassland biomass for combustion: quality optimisation by choice of grassland community and delayed harvest. In: Proceedings of the 15th European Biomass Conference and Exhibition, Berlin, pp. 189–195.

Yaman, S., Sahan, M., Haykiri-Acma, H., Sasen, K., Kucukbayrak, S., 2000. Production of fuel briquettes from olive refuse and paper mill waste. Fuel Process. Technol. 68, 23–31.

CHAPTER

21

Critical parameters affecting large-scale production of microalgal biomass in outdoor open raceway ponds

E. Baldev[1], *D. Mubarak Ali*[2], *R. Sathya*[2], *N. Thajuddin*[3]

[1]Sreedhareeyam Farmherbs India Pvt. Ltd., Ernakulam, Kerala, India; [2]School of Life Sciences, B.S. Abdur Rahman Crescent Institute of Science and Technology, Chennai, Tamil Nadu, India; [3]National Repository for Microalgae and Cyanobacteria - Freshwater (DBT Sponsored), Bharathidasan University, Tiruchirappalli, Tamil Nadu, India

21.1 Introduction

Algal studies are mainly focused on controlling the major challenges facing by the world such as global warming, energy demand, and food security. Microalgae utilize and mitigate CO_2 during their photosynthesis. The importance of this microalgal biomass includes that it provides high valuable products such as cosmetics, pharmaceuticals, and nutraceuticals and it contains protein, lipids, and carbohydrates (Thajuddin, 2010; Mubarak Ali and Thajuddin, 2009; Mubarak Ali et al., 2011a, 2011b, 2012, 2013; Mendoza et al., 2013; Morais Junior et al., 2020). The application of microalgae in the field of biofuels has been widely studied by many researchers and algae has been declared as the superior feedstock of different bioenergies, including bioethanol, biohydrogen, and biodiesel (Baldev et al., 2018; Praveenkumar et al., 2012; Kumar et al., 2015). The applications of microalgae in various green processes and product developments are shown in Fig. 21.1.

Microalgae, one of the oldest life forms, have been known to exist on Earth for billions of years. Microalgae are either unicellular or multicellular photosynthetic organisms which are very efficient at utilizing light energy and developing biomass. Microalgae are a rich source of a huge number of bio- and other derived compounds, which has made algae a primary sustainable source. Microalgae possess a simple cell organization with a huge surface area, which facilitates a huge quantity of nutrients (Demirbas and Demirbas, 2010). Microalgae are very efficient in translating light energy into biomass

Biofuels and Bioenergy
https://doi.org/10.1016/B978-0-323-85269-2.00006-X

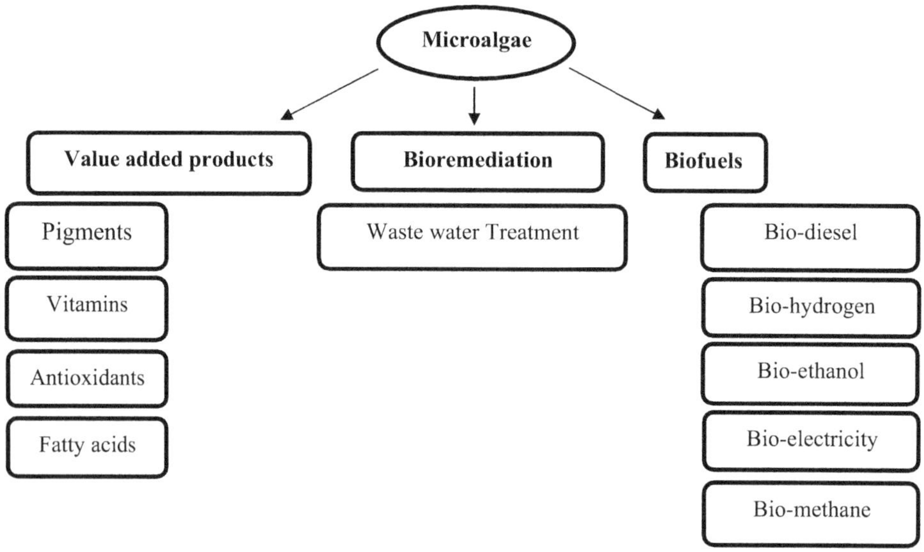

FIGURE 21.1 Microalgal applications in various green processes and product development.

and are a rich source of lipids. The high lipid content makes microalgae a supreme candidate for biofuel production, specifically biodiesel production. In addition to biodiesel, microalgae have been reported as a feasible candidate for the production of biohydrogen, biomethanol, bioethanol, biofertilizer, biogas, bioplastics, animal feed, cosmetics, and medicinal added-value products (Afzal et al., 2017; Tong et al., 2014; Gebreslassie et al., 2013; Thajuddin et al., 2002, 2007, 2010, 2015; Baldev et al., 2021; Sathya et al., 2021). Microalgae are a storehouse for a variety of compounds such as proteins, pigments, lipids, fatty acids, enzymes, polysaccharides, sterols, and vitamins, having high nutritional and pharmaceutical value. Around 350,000 microalgal species have been documented so far (Shahid et al., 2017). From various studies it is generalized that the biochemical composition of algae is comprised of, as high as, carbohydrates (60%), proteins (65%), and lipids (70%), and other essential amino acids (Afzal et al., 2017; Thajuddin, 2010). Algae have the capacity to accumulate large amounts of nonpolar lipids in the form of triacylglycerides or hydrocarbons up to a maximum level of 20%−50% of dry cell weight (Borowitzka and Moheimani, 2012). Microalgae are a promising biomass feedstock for many useful products because of their advantages including a short growth cycle, which gives greater yield of biomass productivity, harvesting index, carbon fixation capacity, and no requirement for a large amount of arable land for cultivation, as wastewater or seawater can be used as the medium (Khan et al., 2018; Miranda et al., 2017; Posadas et al., 2017). It is reported that 1 kg of microalgal biomass can sequestrate 1.83 kg of CO_2 and that it accounts for about 40% of global CO_2 sequestration (Chisti 2007; Ng et al., 2017). Across the world, many researchers and industrialists are currently undertaking the cultivation and processing of algae with the aim of discovering different species, their growth conditions and processes, and the development of competitive economic methods for producing algal biomass and other associated products. However, further work is needed to fully exploit the potential of microalgae for application in large-scale production of biofuels, pharmaceuticals, and nutraceuticals.

The development of new algal biomass production techniques will increase the chances of discovering new value-added products with high commercial value (Bechet et al., 2013).

Algal biomass production mainly uses open raceway ponds or closed photobioreactors for large-scale processes with their own advantages and disadvantages. Photobioreactors have the advantage of having significantly higher volumetric algal productivities (0.2–3.8 g/L/d) than raceway ponds (0.12–0.48 g/L/d) (Brennan and Owende, 2010; Ketheesan and Nirmalakhandan, 2012). At the same time, disadvantages of these photobioreactors are that they are expensive and difficult to maintain. Large-scale biomass production had its beginning in the 1950s with the development of high-rate algal ponds by Oswald (Oswald, 1960). Since then, large-scale algal cultivations are mainly on raceway ponds. It is reported that 95% of total worldwide algal production is based on raceway pond systems (Mendoza et al., 2013). To maintain highly productive large-scale raceway ponds for a longer period, a detailed analysis of various parameters is necessary. Here, an analysis of the various challenges in large-scale cultivation of microalgae in raceway ponds is discussed in detail. Timely pointing out and overcoming these issues are a must for scaling up of algal biomass production to commercial-scale production.

21.2 Microalgal biomass production

The production of sufficient biomass is the primary and necessary step for any applications using microalgae where light, CO_2, H_2O, and inorganic salts are the essential components for photosynthetic activity. The production of microalgal biomass is much more expensive than other crops. The medium is enriched with essential elements such as nitrogen, phosphorus, iron, and, rarely, silicon, along with inorganic elements to enable the growth of algal cells. The minimal nutritional requirements of microalgal biomass can be calculated using the Grobbelaar method (2004), that is, $CO_{0.48}H_{1.83}N_{0.11}P_{0.01}$. Certain nutrients may not form the metal ion complex, hence, in such cases more nutrients will need to be added. The required supplements for the growth of marine microalgae are commercially used nitrate and phosphate fertilizers (Grima et al., 2009; Thajuddin et al., 2016). Approximately, the dry weight of this microalgal biomass contains 50% carbon (Miron et al., 2003). A fixed amount of algal biomass production contains 100 kg and provides 183 kg of CO_2 (Chisti, 2007). The microalgal biomass can be cultivated by four different methods: photoautotrophic, heterotrophic, mixotrophic, and photoheterotrophic (Chojnacka and Marquez-Rocha, 2004).

21.3 Microalgal cultivation systems

21.3.1 Raceway ponds

Raceway ponds are manmade systems commonly used for the production of microalgal biomass. A raceway pond resembles an oval shape with a recirculation channel of depth 1 ft. A paddle wheel provides proper mixing and rotation. The flow is guided by bends and baffles (Fig. 21.2a). Raceway channels are either built using concrete or earth components, or lined with white plastic. During photosynthesis, the culture needs to feed in front of the paddlewheel where the flow begins. Once the circulation is completed the paddlewheel prevents the sedimentation and harvesting of broth, and the microalgal biomass culture has been used since the 1950s (Chisti, 2007).

The area with the largest biomass-producing raceway ponds is of 440,000 m^2 (Spolaore et al., 2006). This setup, focused on the production of cyanobacterial biomass for food, is owned by Earthrise Nutritionals (www.earthrise.com). During the evaporation process the cooling effect

FIGURE 21.2 (A) Raceway pond with capacity of 35,000 L operating at Bharathidasan University Tiruchirappalli, and (B) 1500 L operating at Centre for Microalgal and Nano Biotechnology (CeMANT), CIIC, B.S. Abdur Rahman Crescent Institute of Science and Technology, Chennai, Tamilnadu, India (12°52′17.4″N 80°05′02.8″E). *(A) Thajuddin, N., Baldev, E., Mubarak Ali, D., 2016. Microalgal biomass production: lab-to-land practices at Bharathidasan university, India. ENVIS Newsletter 14 (3), 2–4.; Baldev, E., Mubarak Ali, D., Saravanakumar, K., Arutselvan, C., Naiyf, S.A., Ali Alharbi, S., Sivasubramanian, V., Thajuddin, N., 2018. Unveiling algal cultivation using raceway ponds for biodiesel production and its quality assessment. Renew. Energy, 123, 486–498.*

is achieved in the raceway pond. Significant water loss can occur due to temperature fluctuations within a diurnal cycle and also losses to the Earth's atmosphere, raceways' use of CO_2 is much less efficient than by photobioreactors. The microorganisms in the waste water mainly affect the biomass production. Due to the improper mixing of raceways the biomass concentration remains low and cannot be sustained in an optically dark zone. The production of biodiesel using microalgal biomass from the raceways has been extensively evaluated (Sheehan et al., 1998). While comparing these photobioreactors, raceways are cost effective, and easily to construct and operate; however the drawback of these photobioreactors is that they provide low biomass production.

21.3.2 Photobioreactors (PBRs)

PBRs can be optimized according to the physiological and biological characteristics of the cultivated microalgal species (Mata et al., 2000). They are mainly used for the cultivation of a single algal species for a prolonged duration. Large-scale production of microalgal biomass using photobioreactors has been successful (Grima et al., 2009; Tredici, 1999; Pulz, 2001; Carvalho and Malcata, 2001). The closed configurations of this system are helpful for preventing contamination because the strains are more sensitive. By using these PBRs, the cell mass is increased and, significantly, there is a reduction in the harvesting costs (Carvalho et al., 2006). se tubular PBRs are composed of an array of linear transparent tubes that are made from glass or plastic and the size is 0.1 m or less in diameter, they can be aligned either horizontally, vertically, inclined, or as a helix (Grima et al., 2001; Miron et al., 1999; Ugwu et al., 2002; Watanabe and Saiki, 1997; Chisti, 2007). Earlier the flat-plate PBR was in use, and received a great deal of attention due to its large surface area (Ugwu et al., 2008). The low accumulation of dissolved oxygen and the high photosynthetic efficiency make flat-plate PBRs more suitable for algal biomass production than the tubular versions (Richmond et al., 2000).

A column PBR is very compact, cost effective, and can easily transfer a large volume of biomass, providing efficient mixing and showing better growth conditions. These columns are illuminated via optical walls and aerated from the bottom (Eriksen, 2008; Suh and Sun, 2003). Natural illumination is less expensive than artificial illumination of tubular PBRs and provides valuable products, however the necessity for natural illumination requires large-scale biomass production (Pulz, 2001). Turbulent flow can operate either using an airlift pump or mechanical pump which prevents sedimentation of the microalgal biomass and also maintaining turbulent flow speed. Airlift pumps have been used more successfully than the mechanical version (Chisti, 2007; Acién Fernández et al., 2001). The production of biodiesel can be prepared either by raceways or tubular PBR using a suitable method, along with the adaptability of the selected strains (Table 21.1).

21.4 Factors affecting large-scale raceway pond cultivation

The profitable scale production of algal biomass has been carried out mainly in raceway ponds under different environmental conditions, such as light, temperature, and nutrients. Some of the important factors affecting open large-scale cultivation are addressed in this chapter.

21.4.1 Strain selection

The selection of a fast-growing, highly productive strain, easily acclimatizing to the local climatic conditions is the basis for an outdoor algae biomass production unit. Among the freshwater and marine water organisms, the latter enjoy an added advantage of noncompetition with different crops from the usage of water. In any microalgal biomass systems, high biomass productivity, along with a high content of the desired product is aimed at, which reduces the extraction and purification costs per unit product. For example, the production of biomass and higher lipid content are important criteria for microalgae species producing biodiesel. There are many species of biotechnology-important microalgae yet to be unexplored. Strain selection can be highly improved and this is technically feasible with the help of well-maintained microalgal germplasms. It is believed that around 10,000 microalgal species exist with much fewer amounts having been biotechnologically explored, mainly because of the lack of availability of culture collections (Spolaore et al., 2006). Genetic engineering is another important area in the development of modified strains. The constraints and risks in the usage of engineered strains also hinders the selection of improved strains for open cultivation.

21.4.2 Inoculum development and scale up

The selected strain has to be isolated in unialgal form and sufficient mass has to be developed for further scaling up. In general, subsequent scale up from the preceding step was followed by a factor of 10, that is, from 10 to 100 mL and so on. Depending on the nature of growth of algae the scale up volume and the time interval between each step also change, where the extended cultivation time required not only increases costs, but also the chance of contamination (Borowitzka, 2013). In certain cases, the adaptation of lab cultures to the outdoor environment is a crucial factor which determines the successful initiation of large-scale production. In general, keeping a decreased culture depth of inoculum ponds and also having some coverings which provide control over the amount of direct sunlight may help in successful inoculum development and also speed up the subsequent steps.

TABLE 21.1 Summary of earlier reports on raceway pond cultivation of microalgae and their lipid content.

Microalgal species	Cultivation mode	Culture volume (L)	Lipid content (%) in DW	References
Nannochloropsis salina	Continuous	300	16.5–21	Boussiba et al. (1987)
Platymonos sp.	–	–	20–40	Sheehan et al. (1998)
Dunaliella salina	Semicontinuous	110	35	Moheimani and Borowitzka (2006)
Pleurochrysis casterae	Semicontinuous	110	33	Moheimani and Borowitzka (2006)
Haematococcus pluvialis	Batch	50,000	25	Huntley and Redalje (2007)
Muriellopsis sp.	Semicontinuous	100	23.5–25	Blanco et al. (2007)
Botryococcus braunii	Batch	1800	19	Ashokkumar and Rengasamy (2012)
Scenedesmus sp.	Continuous	23	20–22	Ketheesan and Nirmalakhandan (2012)
Botryococcus braunii	Batch	80	24	Rao et al. (2012)
Chlorella variabilis	Continuous	150	10	De Bhowmick et al. (2014)
Coelastrella sp.	Continuous	–	12–16.2	Hong et al. (2016)
Chroococcus turgidus	Batch	1000	28	Karthikeyan et al. (2016)
Desmodesmus sp.	Continuous	1000	20	Nagappan and Verma (2016)
Graesiella sp.	Batch	40,000	33.4	Wen et al. (2016)
Chlorella sp.	Continuous	35,000	26	Baldev et al. (2018)
Nannochloropsis gaditana	Continuous	10,000	14–15	Rodríguez López et al. (2019)
Botryococcus braunii	Batch	3675	69.43	Ruangsomboon et al. (2020)

21.4.3 Aeration

Aeration is very important in an open pond system for good productivity. It increases the photosynthetic efficiency of microalgae by facilitating better access to O_2 and light penetration. The mass transfer coefficient of any system depends on aeration, which in turn is based on the depth and impeller speed of the pond (Sreekumar et al., 2016). Aeration serves the dual purpose of dissolving the oxygen into the system as well as removing excess air from the system. The dissolved oxygen (DO) level is a critical factor which affects biomass productivity. A very low or very high DO level is harmful to algal growth. Therefore maintaining an optimal DO is crucial in algal cultivation. In open pond cultivation, simple aeration solves this but the case is different in closed reactors where O_2 has nowhere to escape to.

21.4.4 Agitator speed

The prevention of settling of culture in algal cultivation ponds is very important for better productivity and healthy growth of algal cultures. Establishing satisfactory mixing is a key concern in open raceway ponds. The problem of settling as well as a reduction of the shading effect of the culture is controlled by a paddlewheel or agitator. As it has to run for a long time, the paddle wheel is a significant energy consumer for cultivation in a large-scale process (Rogers et al., 2014). About 69% of the total utility cost is due to the mixing purpose (Hreiz et al., 2014). It is reported that nearly a 10-fold increment in algal productivity was achieved by optimizing the mixing (Hreiz et al., 2014). The speed of the agitator should be well controlled in order to achieve maximum biomass production (Norsker et al., 2011; Moazami et al., 2012). Agitation is a very important factor, being an economic deciding factor for scalability and also affecting the economic feasibility of biomass production. An unwanted high paddle wheel results in high turbulence in the pond, preventing proper data obtaining and thus resulting in energy loss (Rogers et al., 2014). Mixing in a raceway pond was found to occur mainly near the sump and paddle wheels bend with a maximum 0.07 m^{-2} s^{-1} dispersion coefficient (Mendoza et al., 2013). In general, the long part of the raceway pond always receives poor mixing, with mainly mixing in the vertical order (Prussi et al., 2014). Baffles in raceway ponds increase the vertical mixing, usually at the cost of energy dissipated in the axial direction (Mendoza et al., 2013). The cells in the dark part at the bottom of the culture pond are brought up to the light surface, and this cyclical process continues mainly by means of vertical mixing and thus is more important than axial mixing.

Mixing also serves various other purposes such as periodic exposure of cells to sunlight, maintaining cells in suspension, freeing the availability of nutrients to algal cells, removal of excess photosynthetic oxygen, etc. (Prussi et al., 2014). Apart from this mixing, it enhances the light utilization efficiency by avoiding photolimitation and photo-inhibition (Chiaramonti et al., 2013). At the same time, mixing should be optimal and unnecessary mixing should be avoided. Roger et al. reported a 25% biomass loss during the nighttime due to mixing. Therefore, it is better to reduce or avoid mixing at night. Depending upon the size of raceway ponds, the culture velocity is generally kept at between 5 and 30 cm s^{-1}. As the size of the pond increases the culture velocity is also found to be higher. The Reynolds number (Re) is used for measuring mixing, with a high Re number indicating optimal mixing.

21.4.5 Light

Light is the major factor which affects the production of algal biomass. The growth rate of algae will be at its maximum at the optimum intensity of light. Light utilization efficiency can be determined by the light intensity at which the culture growth becomes saturated. Algae overcome of light limitations by denaturation of chloroplast membranes (Mock and Kroon et al., 2002). A higher light intensity above the saturation limits results in chloroplast disruption and the inactivation of enzymes in CO_2 fixation (Iqbal and Zafar, 1993). A high light intensity can cause deleterious stress to cells by distressing their PS I and II, and eventually the formation of ROS and in turn a huge reduction in photosynthetic efficiency (Lee, 1999).

In open raceway pond cultivation, the light intensity is an uncontrollable variable. At the same time, open ponds have a huge effective exposure area, which is an added advantage for algal productivity. In closed systems light intensity can be controlled, but the area of exposure will be minimum. The biomass concentration, depth, and turbulence regime mainly determine the quantity of light absorbed by each algal cell (Sutherland et al., 2014). The degree of light attenuation is determined by the depth of the pond and the concentration of biomass. The turbulence in the pond regime regulates the place to which a cell travels into and out of the favorable light condition (Sutherland et al., 2014). This is the reason for the growth rate of shallower open ponds being high compared to deeper ones. At the same time, a decrease in pond depth negatively affects the areal algal productivity. Pigmentation of algal cells is another factor which blocks light significantly. In an earlier study, it was found that a biomass concentration up to 0.66 g L^{-1} does not restrict light penetration (Kumar et al., 2013; Tredici 2004). On the other hand, keeping a low concentration of biomass in ponds is not important as it decreases the total biomass productivity and enhances the downstream costs, making the process uneconomical. Torzillo et al. reported that microalgal growth saturated about one-tenth of the maximum light intensity available outdoors during summer (Torzillo and Vonshak, 2013). Sunlight is the cheapest source of light for open pond cultivation but has some disadvantages such as changing the environmental conditions (weather, day and night cycles, and seasonal changes). Thus, by guaranteeing sufficient mixing and a proper light/dark cycle, efficient production of biomass can be achieved in an economically feasible way.

21.4.6 Pond depth

The raceway pond depth is another notable parameter which has a vital role in productivity. It determines the rate of photosynthesis and thereby biomass productivity. The adequate penetration of sunlight is ensured by constructing the raceway pond in a shallow manner and the paddle wheel in the raceway ponds is limited to a depth of 20–30 cm. Apart from influencing the light penetration, a deeper pond will promote growth, which is unfavorable to algal cultivation systems. The total volume of the pond is directly proportional to the depth of the pond. As the depth of the pond increases, the productivity of the surface area increases (Rogers et al., 2014). The depth of raceway ponds influences factors such as temperature, efficiency of light utilization, mixing, and power consumption by the paddlewheel (Rogers et al., 2014). In a particular site, the pond temperature varies depending on the depth.

21.4.7 Temperature

The most important environmental factor that influences algal growth rate, cell size, biochemical composition, and nutrient requirements is temperature. At optimum temperature the utilization of carbon and nitrogen is efficient. It has been reported that at varying temperatures the viscosity of cytoplasm changes, which results in poor utilization of carbon and nitrogen, and thereby affects productivity (Hope and Walker 1975). Microalgae are able to grow under a wide range of temperatures. Photosynthesis and cellular division are carried out by microalgae at a temperature range of 15–30°C. A temperature range between 20 and 25°C, an increase up to 40°C, and a decrease down to 17°C provide maximum growth rates for mesophilic, thermophilic, and psychrophilic strains, respectively (Sandnes et al., 2005). Most algae have shown a capacity to withstand temperature variations of about 5–10°C in open pond systems. For example, *Nannochloropsis oceanica* has an optimum temperature of 26.7°C with a standing higher and lower temperature variation of around 7°C (Sandnes et al., 2005). In open ponds, the temperature is also an uncontrollable factor which depends on many other factors such as the intensity of irradiation, depth of pond, and other climatic conditions. Naturally, evaporation occurs as a response to environmental conditions and helps to manage the temperature of ponds. Although temperature variations affect the growth rate of microalgae, implementing temperature control strategies in large-scale open ponds is highly energy intensive and uneconomical.

21.4.8 pH

The pH is a crucial parameter influencing the growth and metabolism of algae. The saturated and available CO_2 and micronutrients are mainly determined by the pH (Hansen 2002). Changing the pH in media may affect algal metabolism and will limit algal growth (Goldman et al., 1992). The life cycle completion is directly proportional to the alkaline pH and increases the suppleness of mother cell walls, preventing autospore release (Guckert and Cooksey, 1990). A pH of 78 is suitable to most algae (González et al., 2012). Usually, the pH will tend to increase with algal growth and can have a negative effect on algal growth (Hulatt and Thomas, 2011). In order to avoid this, CO_2 is added to the culture system, thus ensuring that biomass productivity is not affected. Therefore, pH is monitored carefully and as it crosses the optimal level CO_2 is input. At alkaline pH, CO_2 absorption is accelerated as hydroxyl ions and CO_2 react to produce bicarbonate. There are some cultures which can grow at much higher pH without compromising the productivity. For example, *Spirulina* sp. can grow affectively at pH greater than 10 and this has a great benefit of minimizing the chances of contamination. pH also plays a vital role in the flocculation of culture. Flocculation is one of the appropriate processes for harvesting large quantities of algal biomass. The harvesting of algal biomass from a medium is a critical step in large-scale production, and consumes ~30% of the total capital cost (Oh et al., 2001). The pH increment promotes the flocculation significantly by increasing the flocculants (Mc Causland et al., 1999). In effect, pH is an easy method for cost-effective, less laborious, eco-friendly method for microalgal flocculation.

21.4.9 CO_2

Microalgae utilize CO_2 as the main source of their carbon requirement, leading to increases in the concentration of CO_2, and use of fossil fuels which are responsible for global warming. Many researchers have suggested and utilized algae as a tool which is cost effective for reducing the carbon footprint. Moreover, the algal CO_2 mitigation strategy ends up with different high-value products at the end of the process (Uday and Ahluwalia, 2013; Anguselvi 2019; Baldev et al., 2021). A decrease in pH can be seen because carbonic acid forms when CO_2 is sparged to the growth medium. Thus, the rate of utilization of CO_2 can be used as a tool for measuring the algal growth; at the same time it controls the pH of the system. Li et al. (2013) highlighted that 8%–27% of the production cost is required for the carbon source. A CO_2 concentration of 65 μmol/L at pH 8.5 was found to be optimal for most algal growth (Weissman et al., 1988). Based on algal species, the CO_2 utilization efficiency was also found to vary widely. Chaudhary et al. (2018) reported that *C. vulgaris* and *S. obliquus* exhibited a CO_2 fixation rate and efficiency of converting CO_2 of 140.91 and 129.82 mg/L/day and 14.9% and 13.85%, respectively. The poor residence time of gas bubbles is a major problem in maximum utilization of CO_2 by algae. Some researchers have suggested bicarbonate salts as a better alternative to CO_2 gas for algal cultivation as it has some advantages such as high solubility, efficient mass transfer, and no dripping out of gas bubbles (Chi et al., 2011; Jennifer et al., 2012).The need for CO_2 management, specifically, its capture and storage, is a global issue and will continue to be so until the emergence of an alternative energy source. Microalgae growing in an area of about 4000 m^2 are reported to capture 2.7 tons/day of CO_2 (Benemann et al., 2003). Overall, large-scale cultivation of microalgae serves a major role in the reduction of greenhouse gas (GHG) emissions at the same time as producing a sustainable feedstock for various industries, ranging from biofuels to nutraceuticals.

21.4.10 Salinity

Based on their salinity requirement, microalgae can be broadly classified as freshwater and marine cultures. Variation in the concentration of salinity is a common problem in open ponds mainly due to rain and evaporation. The fluctuating nature of salinity greatly affects the growth and cell composition of microalgae. Based on the tolerance to salinity, microalgae can be divided into oligohaline, mesohaline, and polyhaline (Venkata Mohan and Devi, 2014). Normally, marine algal cultures prefer a salinity range of 30–35 ppt. Some of the commercially important microalgal species, *Dunalila salina* and *Tetraselmis* species are able to grow in hypersaline conditions. Alteration of the biochemical composition of algal cells happens through an important factor called salinity. In response to the environmental salinity the lipid composition of microalgae is reported to change (Takagi and Yoshida, 2006). In *Dunaliella* the saturated and monosaturated fatty acids are increased as there is an increase in salinity (Xu and Beardall, 1997). *Dunaliella tertiolecta* (oleaginous species) cells have high tolerance for salinity, which is appropriate for large-scale outdoor cultivation (Elenkov et al., 1996). The increase in salinity of *Botryococcus braunii* (freshwater alga) showed an increase in growth rate, carbohydrate, and lipid content, suggesting salinity as a stress which can aid in the selective accumulation of some of the biocomponents. Lipid production was stimulated in the freshwater alga *Scenedesmus* sp., through the addition of NaCl (Salama et al., 2013). However, a reduction in biomass and lipid productivity results when salinity stress is excessive as it will affect photosynthesis.

21.4.11 Algal predators

Algal cultivation in open ponds is exposed to various predators as it is in an open condition. Usually birds, insects, and frogs are found to contaminate open ponds. Maintaining unialgal cultures with different species in nearby ponds is a difficult task due to these unavoidable carriers or predators. Apart from these, a range of microorganisms such as bacteria, fungi, virus, rotifers, ameba, and ciliates are serious threats to algal ponds as these organisms feed on microalgae, causing severe biomass loss (Carney and Lane, 2014; Kumar et al., 2015). Many researchers suggest the introduction of extremophilic algal species as one of the best ways to reduce the risk of contamination and predation. Some species like *D. salina* which requires high salinity, *Arthrospira* requiring huge alkalinity, or *Chlorella* requiring huge amounts of nutrients can be grown on a large scale with minimum contamination due to their highly selective growth parameters (Borowitzka and Vonshak, 2017). The key factor is to reduce the risk from the contaminants and predators where the selected species are exposed, to maintain the optimum growth conditions from the different environment (Flynn et al., 2017).

21.5 Biomass productivity in raceway ponds

It is reported that algal species like *Chlorella* sp. have a very short doubling time of 3–4 h and thus have an exponential growth rate (Chisti 2007). Lewis and Nocera (2006) examined that the fastest growing terrestrial crop, switch grass (*Panicum virgatum* L.), where less than 0.5% of the incident solar energy was used. At the same time, selected microalgal biomass can be produced at more than 50 times the rate of switchgrass. Photosynthesis and respiration are the two competitive factors which determine the productivity of algal biomass during algal cultivation. The rate of algal photosynthesis depends on many factors, as discussed earlier in this chapter, such as light intensity, nutrient level, pH, temperature, etc. Therefore, keeping these factors at their optimal levels is the key to a successful algal production system. In the case of outdoor raceway pond cultivation, some factors like temperature and light intensity are not controllable. In outdoor ponds, microalgal biomass productivity was high during a shorter period of time with appropriate conditions. Moheimani and Borowitzka (2006) reported a biomass productivity of 20–22 g dry wt/m^2/day for about a year. Photosynthetic efficiency is one of the important parameters for determining the rate of biomass production. It has a photoautotrophic growth mainly defined for energy storage as new biomass per unit of incident light energy (Akkerman et al., 2002). The quantitatively estimated biomass, oil productivity, and photosynthetic efficiency of microalgae were 11.42%, 6%, and 3%, respectively (Sudhakar and Premalatha, 2012). However, the magnitude effects of photosynthetic efficiency were decreased in the field, due to many factors such as poor absorption, transmission, reflection, respiration, and photoinhibition (Miyamoto, 2009).

21.6 Open raceway ponds for microalgal biofuel production

The need for new biofuels is now a global demand in view of the energy crisis and depleting oil resources. The utmost care should be taken to reduce the competitiveness between the production of food products and biofuel and also to address difficulties due to the use of fresh water, land, and fertilizers. Microalgal biomass has been studied as a suitable candidate for biofuel production because of the various features described earlier in this chapter. The key to a successful algal biofuel technology is to produce increased biomass and to find alternative substrates that can be used in metabolic pathways; this will helpful in reducing the costs and making the industrialization of these fuels viable. The biorefinery concept is an important area which has real potential to take algal biofuel to a commercial level. Microalgae grown in raceway ponds provide higher biofuel production, lipid production, and carbohydrate contents with few species than other crops(Brennan and Owende, 2010; Ishika et al., 2017; Baldev et al., 2018). In an open raceway pond the microalgal biomass requires less energy, is easy to operate, and produces lower biomass with lower maintenance costs than closed reactors. The main disadvantages in the closed system are the requirements for sterilization stages and oxygen oversaturation which affect biofuel production more than in open ponds (Chisti, 2007; Brennan and Owende, 2010; Ishika et al., 2017; Baldev et al., 2018).

21.7 Future outlook and prospects

In this chapter, the major factors affecting the large-scale cultivation of microalgae in raceway ponds are discussed. Researchers from all over the world have made significant progress in this field and have paved the way for an understanding of the various factors such as availability of nutrients, environmental conditions, harvesting, growth impact, and quality of biomass. The studies on the molecular basis of these factors affecting algal growth will give an added advantage for the betterment of pond management, thereby increasing algal productivity. With the significant advancements in biotechnology, the full genome data of many algal species are available, which makes genetic manipulation to meet specific requirements simpler. Continued research in the area of open pond systems and the factors affecting them will help to develop more precise models for algal biomass production. In addition to these factors, energy balance is a challenging one for some cases such as biofuel production from algae. To minimize those high costs, a refinery approach with the integration of different processes and industries would be the best option. The integration of technologies for the development of biofuel and algal products would be economically feasible and sustainable when compared to the standalone biofuel production processes.

21.8 Conclusion

Microalgae have many advantages over traditional energy crops, mainly because of their high biomass production capacity in fresh, marine, or even waste waters. Microalgae are biofactories, being the source of renewable, sustainable, and economic bioproducts. Based on several factors, the biomass production of algae can be carried out in outdoor systems. A better understanding of the growth requirements and factors affecting growth will help to achieve a higher degree of reliable and successful commercial production. This chapter has described the major challenges and factors affecting large-scale outdoor biomass production, the associated challenges and limitations, and how to surmount these to make algal cultivation viable for commercial production. A basic understanding of the factors influencing algal growth is crucial for large-scale processes and different methods of sustainable and successful biomass production and product development.

Acknowledgment

The authors are thankful to B.S. Abdur Rahman Crescent Institute of Science and Technology for providing the CSM and CIIC-BioNest Program.

References

Acién Fernández, F.G., Fernández Sevilla, J.M., Sánchez Pérez, J.A., Molina Grima, E., Chisti, Y., 2001. Airlift-driven external-loop tubular photobioreactors for outdoor production of microalgae: assessment of design and performance. Chem. Eng. Sci. 56 (8), 2721–2732.

Afzal, I., Shahid, A., Ibrahim, M., Liu, T., Nawaz, M., Muhammad, A., Mehmood, M.A., 2017. Chapter 3 microalgae: a promising feedstock for energy and high-value products. In: Khalid, M.Z., Mohammad, Z., Muhammad, A. (Eds.), Algae Based Polymers, Blends, and Composites. Elsevier, pp. 55–75.

Akkerman, I., Janssen, M., Rocha, J., Wijffels, R.H., 2002. Photobiological hydrogen production: photochemical efficiency and bioreactor design. Int. J. Hydrog. Energ. 27 (11), 1195–1208.

Anguselvi, V., Ebhin Masto, R., Mukherjee, A., Pradeep, K.S., 2019. CO_2 capture for industries by algae. In: Algae, Yee Keung Wong. IntechOpen.

Ashok kumar, V., Rengasamy, R., 2012. Mass culture of *Botryococcus braunii* Kutz. under open raceway pond for biofuel production. Bioresour. Technol 104, 394–399.

Baldev, E., MubarakAli, D., Saravanakumar, K., Arutselvan, C., Naiyf, S.A., Ali Alharbi, S., Sivasubramanian, V., Thajuddin, N., 2018. Unveiling algal cultivation using raceway ponds for biodiesel production and its quality assessment. Renew. Energy 123, 486–498.

Baldev, E., Pugazhendhi, A., MubarakAli, D., Thajuddin, N., 2021. Wastewater as an economical and eco-friendly green medium for microalgal biofuel production. Fuel 294, 120484.

Bechet, Q., Shilton, A., Guieysse, B., 2013. Modeling the effects of light and temperature on algae growth: state of the art and critical assessment for productivity prediction during outdoor cultivation. Biotechnol. Adv. 31 (8), 1648–1663.

Benemann, J.R., Van Olst, J.C., Massingill, M.J., Weissman, J.C., Brune, D.E., 2003. The controlled eutrophication process: using microalgae for CO_2 utilization and agricultural fertilizer recycling. In: Proceedings of Conference on Greenhouse Gas Technologies. Kyoto, Japan.

Blanco, A.M., Moreno, J., Del Campo, J.A., Rivas, J., Guerrero, M.G., 2007. Outdoor cultivation of lutein-rich cells of *Muriellopsis* sp. in open ponds. Appl. Microbiol. Biotechnol. 73, 1259–1266.

Borowitzka, M.A., 2013. High-value products from microalgae—their development and commercialisation. J. Appl. Phycol. 25 (3), 743–756.

Borowitzka, M.A., Moheimani, N.R., 2012. Open Pond Culture Systems In: Algae for Biofuels and Energy. Springer, pp. 133–152.

Borowitzka, M.A., Vonshak, A., 2017. Scaling up microalgal cultures to commercial scale. Eur. J. Phycol. 52, 407–418.

Boussiba, S., Vonshak, A., Cohen, Z., Avissar, Y., Richmond, A., 1987. Lipid and biomass production by the halotolerant microalga *Nannochloropsis salina*. Biomass 12, 37–47.

Brennan, L., Owende, P., 2010. Biofuels from microalgae – a review of technologies for production, processing, and extractions of biofuels and co-products. Renew. Sustain. Energy Rev. 14, 217–232.

Carney, L.T., Lane, T.W., 2014. Parasites in algae mass culture. Front. Microbiol. 5, 278.

Carvalho, A., Malcata, F., 2001. Transfer of carbon dioxide with in culture of microalgae: plain bubbling versus hollow-fiber modules. Biotechnol. Prog. 17, 265–272.

Chaudhary, R., Dikshit, A.K., Tong, Y.W., 2018. Carbon-dioxide biofixation and phycoremediation of municipal wastewater using *Chlorella vulgaris* and *Scenedesmus obliquus*. Environ. Sci. Pollut. Res. 25 (21), 20399–20406.

Chi, Z., O'Fallon, J.V., Chen, S., 2011. Bicarbonate produced from carbon capture for algae culture. Trends Biotechnol. 29 (11), 537–541.

Chiaramonti, D., Prussi, M., Casini, D., Tredici, M.R., Rodolfi, L., Bassi, N., 2013. Review of energy balance in race way ponds for micro algae cultivation:re-thinking a traditional system is possible. Appl. Energy 102, 101–111.

Chisti, Y., 2007. Biodiesel from microalgae. Biotechnol. Adv. 25 (3), 294–306.

Chojnacka, K., Marquez-Rocha, F.J., 2004. Kinetic and stoichiometric relationships of the energy and carbon metabolism in the culture of microalgae. Biotechnology 3 (1), 21–34.

De Bhowmick, G., Subramanian, G., Mishra, S., Ramkrishna, S., 2014. Raceway pond cultivation of a marine microalga of Indian origin for biomass and lipid production: a case study. Algal Res. 6, 201–209.

Demirbas, A., Demirbas, M.F., 2010. Algae Energy: Algae as a New Source of Biodiesel. Springer Science & Business Media.

Elenkov, I., Stefanov, K., Dimitrova-Konaklieva, S., Popov, S., 1996. Effect of salinity on lipid composition of *Cladophora vagabunda*. Phytochemistry 42 (1), 39–44.

Eriksen, N.T., 2008. Production of phycocyanin–a pigment with applications in biology, biotechnology, foods and medicine. Appl. Microbiol. Biotechnol. 80 (1), 1–14.

Flynn, K.J., Kenny, P., Mitra, A., 2017. Minimising losses to predation during microalgae cultivation. J. Appl. Phycol. 29, 1829–1840.

Gebreslassie, B.H., Waymire, R., You, F., 2013. Sustainable design and synthesis of algae-based biorefinery for simultaneous hydrocarbon biofuel production and carbon sequestration. AIChE J 59 (5), 1599–1621.

Goldman, J.C., Azov, Y., Riley, C.B., Dennett, M.R., 1992. The effect of pH in intensive microalgal cultures. I. biomass regulation. J. Exp. Mar. Biol. Ecol. 57, 1–13.

González-López, C.V., Acién Fernández, F.G., Fernández-Sevilla, J.M., Sánchez Fernández, J.F., Molina Grima, E., 2012. Development of a process for efficient use of CO_2 from flue gases in the production of photosynthetic microorganisms. Biotechnol. Bioeng. 109 (7), 1637–1650.

Grima, E.M., Belarbi, E.H., Acien Fernandez, F.G., Robles Medina, A., Chisti, Y., 2001. Tubular photobioreactor design for algal cultures. J. Biotechnol. 92 (2), 113–131.

Grima, E.M., Fernández Sevilla, J.M., Acién Fernández, F.G., Flickinger, M.C., 2009. Microalgae, mass culture methods. In a Book on Encyclopedia of Industrial Biotechnology: Bioprocess, Bioseparation, and Cell Technology. 1–24.

Guckert, J.B., Cooksey, K.E., 1990. Triglyceride accumulation and fatty acid profile changes in *Chlorella* (Chlorophyta) during high pH induced cell cycle inhibition. J. Phycol. 26, 72–79.

Hansen, P.J., 2002. Effect of high pH on the growth and survival of marine phytoplankton: implications for species succession. Aquat. Microb. Ecol. 28, 279–288.

Hong, J.W., Kim, O.H., Kim, H., Jo, S.W., Cho, H.W., et al., 2016. Mass cultivation from a Korean raceway pond system of indigenous microalgae as potential biofuel feedstock. Oil Gas Res. 2, 108.

Hope, A.B., Walker, N.A., 1975. The Physiology of Giant Algal Cells. Cambridge University Press, London, UK.

Hreiz, R., Sialve, B., Morchain, J., Escudié, R., Steyer, J.-P., Guiraud, P., 2014. Experimental and numerical investigation of hydro dynamics in raceway reactors used for alga culture. Chem. Eng. J. 250, 230–239.

Hulatt, C.J., Thomas, D.N., 2011. Productivity, carbon dioxide uptake and net energy return of microalgal bubble column photobioreactors. Bioresour. Technol. 102 (10), 5775–5787.

Huntley, M.E., Redalje, D.G., 2007. CO_2 mitigation and renewable oil from photosynthetic microbes: a new appraisal. Mitig. Adapt. Strateg. Glob. Chang. 12, 573–608.

Iqbal, M., Zafar, S., 1993. Effects of photon flux density, CO_2, aeration rate, and inoculum density on growth and extracellular polysaccharide production by *Porphyridium cruentum*. Folia Microbiol. 38, 509–514.

Ishika, T., Moheimani, N.R., Bahri, P.A., 2017. Sustainable saline microalgae co-cultivation for biofuel production: a critical review. Rev. Environ. Sci. Biotechnol. 78, 356–368.

Jennifer, L.M., Jeffrey, C.C., Lawrence, E.P., Sally, M.B., Himadri, B.P., 2012. Perspectives on biofuels: potential benefits and possible pitfalls. Am. Chem. Soc. 1116, 107–141.

Karthikeyan, D., Muthukumaran, M., 2016. Mass Cultivation of Microalgae in Open Raceway Pond for Biomass and Biochemicals Production. Int. J. Adv. Res. Biol. Sci. 3 (2), 247–260.

Ketheesan, B., Nirmalakhandan, N., 2012. Feasibility of microalgal cultivation in a pilot- scale airlift-driven raceway reactor. Bioresour. Technol. 108, 196–202.

Kumar, K., Sirasale, A., Das, D., 2013. Use of image analysis tool for the development of light distribution pattern inside the photobioreactor for the algal cultivation. Bioresour. Technol. 143, 88–95.

Khan, M.I., Shin, J.H., Kim, J.D., 2018. The promising future of microalgae: current status, challenges, and optimization of a sustainable and renewable industry for biofuels, feed, and other products. Microb. Cell Fact 17, 36.

Kumar, K., Mishra, S.K., Shrivastav, A., Park, M.S., Yang, J.W., 2015. Recent trends in the mass of algae in raceway ponds. Renew. Sustain. Energy Rev. 51, 875–885.

Lee, C.-G., 1999. Calculation of light penetration depth in photobioreactors. Biotechnol. Bioprocess Eng. 4 (1), 78–81.

Lewis, N.S., Nocera, D.G., 2006. Powering the planet: chemical challenges in solar energy utilization. Proc. Natl. Acad. Sci. USA 103, 15729–15735.

Li, S., Luo, S., Guo, R., 2013. Efficiency of CO_2 fixation by microalgae in a closed raceway pond. Bioresour. Technol. 136, 267–272.

Mata, T.M., Martins, A.A., Caetano, N.S., 2000. Microalgae for biodiesel production and other applications: a review. Renew. Sustain. Energy Rev. 14 (1), 217–232.

McCausland, M.A., Brown, M.R., Barrett, S.M., Diemar, J.A., Heasman, M.P., 1999. Evaluation of live microalgae and microalgal pastes as supplementary food for juvenile Pacific oysters (*Crassostrea gigas*). Aquaculture 174 (3), 323–342.

Mendoza, J.L., Granados, M.R., de Godos, I., Acién, F.G., Molina, E., Heaven, S., 2013. Oxygen transfer and evolution in microalgal culture in open raceways. Bioresour. Technol. 137, 188–195.

Miron, A.S., Gomez, A.C., Camacho, F.G., Grima, E.M., Chisti, Y., 1999. Comparative evaluation of compact photobioreactors for large-scale monoculture of microalgae. J. Biotechnol. 70 (1–3), 249–270.

Miranda, A.F., Ramkumar, N., Andriotis, C., Höltkemeier, T., Yasmin, A., Rochfort, S., Wlodkowic, D., Morrison, P., Roddick, F., Spangenberg, G., Lal, B., Subudhi, S., Mouradov, A., 2017. Applications of microalgal biofilms for wastewater treatment and bioenergy production. Biotechnol. Biofuels. 10, 120.

Miron, A.S., Garcia, M.C.C., Gomez, A.C., Camacho, F.G., Grima, E.M., Chisti, Y., 2003. Shear stress tolerance and biochemical characterization of *Phaeodactylum tricornutum* in quasi steady-state continuous culture in outdoor photobioreactors. Biochem. Eng. J. 16, 287–297.

Miyamoto, K. (Ed.), 2009. Biological Energy Production, Renewable Biological Systems for Alternative Sustainable Energy Production. FAO Food and Agriculture Organization of the United Nations, Agricultural Consumer Protection.

Moazami, N., Ashori, A., Ranjbar, R., Tangestani, M., Eghtesadi, R., Nejad, A.S., 2012. Large-scale biodiesel production using microalgae biomass of *Nannochloropsis*. Biomass Bioenergy 39, 449–453.

Mock, T., Kroon, B.M.A., 2002. Photosynthetic energy conversion under extreme conditions—II: the significance of lipids under light limited growth in Antarctic sea ice diatoms. Phytochemistry 61, 53–60.

Moheimani, N.R., Borowitzka, M.A., 2006. The long term culture of the coccolithophore Pleurochrysis carterae (Haptophyta) in outdoor raceway ponds. J. Appl. Phycol. 18, 703–712.

Morais Junior, W.G., Gorgicha, M., Corrêa, P.S., Martins, A.A., Mata, T.M., Caetano, N.S., 2020. Microalgae for biotechnological applications: cultivation, harvesting and biomass processing. Aquaculture 528, 735562.

Mubarak Ali, D., Thajuddin, N., 2009. Biosynthesis of colloidal silver nanoparticles using fresh water cyanobacterium *Oscillatoria earlei* NTDM03. Seaweed Res. Utiln. 31, 219–226.

Mubarak Ali, D., Thajuddin, N., Jeganathan, K., Gunasekaran, M., 2011a. Plant extract mediated synthesis of silver and gold nanoparticles and its antibacterial activity against clinically isolated pathogens. Colloids Surf. B 85, 360–365.

Mubarak Ali, D., Sasikala, M., Gunasekaran, M., Thajuddin, N., 2011b. Biosynthesis and characterization of Silver nanoparticles using marine cyanobacterium *Oscillatoria willei* NTDM01. Dig. J. Nanomater. Biostruct. 6 (2), 415–420.

Mubarak Ali, D., Arunkumar, J., Suriya, K.R., Ishack, K.A.S.S., Thajuddin, N., 2012. Molecular modelling and phylogenetic analysis of C-phycocyanin gene sequence from marine cyanobacterium *Phormidium tenue* NTDM05. Seaweed Res. Utiln. 34, 35–44.

MubarakAli, D., Arunkumar, J., Nag, K.H., Ishack, K.A.S.S., Baldev, E., Pandiaraj, D., Thajuddin, N., 2013. Gold nanoparticles from Pro and eukaryotic photosynthetic microorganisms - comparative studies on synthesis and its application on bio labelling. Colloids Surf. B 103, 166–173.

Nagappan, S., Verma, S.K., 2016. Growth model for raceway pond cultivation of *Desmodesmus* sp. MCC34 isolated from a local water body. Eng. Life Sci.16 45–52.

Norsker, N.H., Barbosa, M.J., Vermuë Marian, W.R., 2011. Microalgal production - a close look at the economics. Biotechnol. Adv. 29, 24–27.

Oh, H.-M., Lee, S.J., Park, M.-H., Kim, H.-S., Kim, H.-C., Yoon, J.-H., Kwon, G.-S., Yoon, B.-D., 2001. Harvesting of *Chlorella vulgaris* using a bioflocculant from *Paenibacillus* sp. AM49. Biotechnol. Lett. 23 (15), 1229–1234.

Oswald, W.J., Golueke, C.G., 1960. Biological transformation of solar energy. Adv. Appl. Microbiol. 11, 223–242.

Posadas, E., Marín, D., Blanco, S., Lebrero, R., Muñoz, R., 2017. Simultaneous biogas upgrading and centrate treatment in an outdoors pilot scale high-rate algal pond. Bioresour. Technol. 232, 133–141.

Praveenkumar, R., Shameera, K., Mahalakshmi, G., Akbarsha, M.A., Thajuddin, N., 2012. Influence of nutrient deprivations on lipid accumulation in a dominant indigenous microalga *Chlorella* sp., BUM11008: Evaluation for biodiesel production. Biomass Bioenergy 37, 60–66.

Prussi, M., Buffi, M., Casini, D., Chiaramonti, D., Martelli, F., Carnevale, M., 2014. Experimental and numerical investigations of mixing in raceway ponds for algae cultivation. Biomass Bioenergy 67, 390–400.

Pulz, O., 2001. Photobioreactors: production systems for phototrophic microorganisms. Appl. Microbiol. Biotechnol. 57, 287–293.

Rao, A.R., Ravishankar, G.A., Sarada, R., 2012. Cultivation of green alga *Botryococcus braunii* in raceway, circular ponds under out door conditions and its growth, hydrocarbon production. Bioresour. Technol. 123, 528–533.

Richmond, A., 2000. Microalgal biotechnology at the turn of the millennium: a personal view. J. Appl. Phycol. 12, 441–451.

Rodríguez López, A., Silvia, B.R., Vallejo, R.A., García, P.G., Macías-Sánchez, M.D., Gutiérrez Díaz, M., González Librán, R., Fernández Acero, F.J., 2019. Sustainable cultivation of *Nannochloropsis gaditana* microalgae in outdoor raceways using flue gases for a complete 2-year cycle: a circular economy challenge. J. Appl. Phycol. 31, 1515–1523.

Rogers, J.N., Rosenberg, J.N., Guzman, B.J., Oh, V.H., Mimbela, L.E., Ghassemi, A., 2014. A critical analysis of paddle wheel-driven raceway ponds for algal biofuel production at commercial scales. Algal Res. 4, 76–78.

Ruangsomboon, S., Dimak, J., Jongput, B., Wiwatanaratanabutr, I., Kanyawongha, P., 2020. Outdoor open pond batch production of green microalga *Botryococcus braunii* for high hydrocarbon production: enhanced production with salinity. Sci. Rep. 10, 2731.

Salama, el-S., Kim, H.C., Abou-Shanab, R.I., Ji, M.K., Oh, Y.K., Kim, S.H., 2013. Biomass, lipid content, and fatty acid composition of freshwater *Chlamydomonas mexicana* and *Scenedesmus obliquus* grown under salt stress. Bioprocess Biosyst. Eng. 36, 827–833.

Sandnes, J., Kallqvist, T., Wenner, D., Gislerød, H.R., 2005. Combined influence of light and temperature on growth rates of *Nannochloropsis oceanica*: linking cellular responses to large-scale biomass production. J. Appl. Phycol. 17 (6), 515–525.

Sathya, R., MubarakAli, D., Mohamed Saalis, J., Kim, J.-W., 2021. A systemic review on microalgal peptides: bioprocess and sustainable applications. Sustainability 13 (6), 3262. https://doi.org/10.3390/su13063262.

Shahid, A., Khan, AZ., Liu, T., Malik, S., Afzal, I., Mehmood, MA., 2017. Production and processing of algal biomass. In: Algae based polymers, blends, and composites. Elsevier, San Diego, pp. 273–299.

Sheehan, J., Dunahay, T., Benemann, J., Roessler, P., 1998. Look back at the U.S. Department of energy's aquatic species Program. Biodiesel from Algae; Close-Out Report. 1–328. NREL Report No. TP-580-24190. https://www.nrel.gov/docs/legosti/fy98/24190.pdf.

Spolaore, P., Joannis-Cassan, C., Duran, E., Isambert, A., 2006. Commercial applications of microalgae. J. Biosci. Bioeng. 101, 87–96.

Sreekumar, N., GiriNandagopal, M.S., Aneesh, V., Rahul, A., Selvaraju, N., 2016. Marine microalgal culturing in open pond systems for biodiesel production—critical Parameters. J. Renew. Sustain. Energy 8, 023105.

Sudhakar, K., Premalatha, M., 2012. Theoretical assessment of algal biomass potential for photobioreactors. AIChE. J. 40 (6), 1055–1066.

Suh, S., Sun, B.L.A., 2003. Light distribution model for an internally radiating photobioreactor. Biotechnol. Bioengineer. 82, 2–20.

Sutherland, D.L., Turnbull, M.H., Craggs, R.J., 2014. Increased pond depth improves algal productivity and nutrient removal in wastewater treatment high- rate algal ponds. Water Res. 53, 271–281.

Takagi, M., Yoshida, T., 2006. Effect of salt concentration on intracellular accumulation of lipids and triacylglyceride in marine microalgae *Dunaliella* cells. J. Biosci. Bioeng. 101 (3), 223–226.

Thajuddin, N., 2010. Biotechnological potential of cyanobacteria and their industrial applications. In: Industrial Exploitation of Microorganism. I K International, pp. 70–85.

Thajuddin, N., Nagasathya, A., Challadevi, R., Saravanan, L., 2002. Biodiversity of cyanobacteria in different salt pan of Pudukottai district, Tamilnadu. Seaweed Res. Util. 24, 1–11.

Thajuddin, N., Sandra, A., Nierzwicki-Bauer, Subramanian, G., 2007. Cyanobacterial biodiversity and recent trends in their taxonomy. BDU J. Sci. Tech. 1, 1–14 (in Tamil); 15-28 (in English).

Thajuddin, N., Muralitharan, G., Sundaramoorthy, M., Ramamoorthy, R., Ramachandran, S., Akbarsha, M.A., Gunasekaran, M., 2010. Morphological and genetic diversity of symbiotic cyanobacteria from cycads. J. Basic. Microbiol. 50, 254–265.

Thajuddin, N., Ilavarasi, A., Baldev, E., MubarakAli, D., Alharbi, N.S., Arunachalam, C., Alharbi, S.A., 2015. Stress Induced lipids accumulation in Naviculoid marine diatoms for bioenergy application. Int. J. Biotechnol. Wellness Ind. 4, 18–24.

Thajuddin, N., Baldev, E., Mubarak Ali, D., 2016. Microalgal biomass production: lab-to-land practices at Bharathidasan university, India. ENVIS Newslett. 14 (3), 2–4.

Tong, K.L., You, F.Q., Rong, G., 2014. Robust design and operations of hydrocarbon biofuel supply chain integrating with existing petroleum refineries considering unit cost objective. Comput. Chem. Eng. 68, 128–139.

Torzillo, G., Vonshak, A., 2013. Environmental stress physiology with reference to mass cultures. In: Richmond, A., Hu, Q. (Eds.), Handbook of Microalgal Culture: Applied Phycology and Biotechnology. John Wiley & Sons, Chichester, pp. 90–111.

Tredici, M.R., 1999. Bioreactors, photo. In: Flickinger, M.C., Drew, S.W. (Eds.), Encyclopedia of Bioprocess Technology: Fermentation, Biocatalysis, and Bioseparation. John Wiley & Sons, Inc., pp. 364–419

Tredici, M.R., 2004. Mass production of microalgae: photobioreactors. In: Richmond, A. (Ed.), Handbook of Microalgal Culture Biotechnology and Applied Phycology, 2004. Blackwell Publishing, Iowa, pp. 179–214.

Ugwu, C.U., Ogbonna, J., Tanaka, H., 2002. Improvement of mass transfer characteristics and productivities of inclined tubular photobioreactors by installation of internal static mixers. Appl. Microbiol. Biotechnol. 58 (5), 600–607.

Uday, S., Ahluwalia, A., 2013. Microalgae: a promising tool for carbon sequestration. In: Mitigation and Adaptation Strategies for Global Change, 18 (1). Springer, pp. 73–95.

Ugwu, C.U., Aoyagi, H., Uchiyama, H., 2008. Photobioreactors for mass cultivation of algae. Bioresour. Technol. 99 (10), 4021–4028.

Venkata Mohan, S., Devi, M.P., 2014. Salinity stress induced lipid synthesis to harness biodiesel during dual mode cultivation of mixotrophic microalgae. Bioresour. Technol. 165, 288–294.

Watanabe, Y., Saiki, H., 1997. Development of a photobioreactor incorporating *Chlorella* sp. for removal of CO_2 in stack gas. Energ. Convers. Manag. 38, S499–S503.

Weissman, J.C., Goebel, R.P., Benemann, J.R., 1988. Photobioreactor design: mixing, carbon utilization, and oxygen accumulation. Biotechnol. Bioeng. 31, 336–344.

Wen, X., Du, K., Wang, Z., Peng, X., Luo, L., Tao, H., Xu, Y., Zhang, D., Geng, Y., Li, Y., 2016. Effective cultivation of microalgae for biofuel production: a pilot-scale evaluation of a novel oleaginous microalga *Graesiella* sp. WBG1. Biotechnol. Biofuel. 9, 123.

Xu, X.-Q., Beardall, J., 1997. Effect of salinity on fatty acid composition of a green microalga from an antarctic hypersaline lake. Phytochemistry 45 (4), 655–658.

Further reading

Bennett, M.C., Turn, S.Q., Chan, W.Y., 2014. A methodology to assess open pond, phototrophic, algae production potential: a Hawaii case study. Biomass Bioenergy 66, 168–175.

Bitoga, J.P., Lee, I.-B., Lee, C.-G., Kim, K.-S., Hwang, H.-S., Hong, S.-W., Seo, I.-H., Kwon, K.-S., Mostafa, E., 2011. Application of computational fluid dynamics for modeling and de- signing photobioreactors for microalgae production. Comput. Electron. Agric. 76, 131–147.

Chen, C.Y., Durbin, E.G., 1994. Effects of pH on the growth and carbon uptake of marine phytoplankton. Mar. Ecol.-Prog. Ser. 109, 83–89.

Chen, Y., Wang, J., Zhang, W., Chen, L., Gao, L., Liu, T., 2013. Forced light/dark circulation operation of open pond for microalgae cultivation. Biomass Bioenergy 56, 464–470.

Freitas, B.C.B., Cassuriaga, A.P.A., Morais, M.G., Costa, J.A.V., 2017. Pentoses and light intensity increase the growth and carbohydrate production and alter the protein profile of *Chlorella minutissima*. Bioresour. Technol. 238, 248–253.

Harun, I., Yahya, L., NazryChik, M., Abdul Kadir, N.N., Azmir Pang, M.A.M., 2014. Effects of natural light dilution on microalgae growth. Int. J. Chem. Engi. App. 5 (2), 112–116, 2014.

Irina, H., Liyana, Y., Chik, M.N., Abdul Kadir, N.N., Pang, M.A.M.A., 2014. Effects of natural light dilution on microalgae growth. Inte. J. Chem. Engi. App. 5, 2.

Jeong, H., Lee, J., Cha, M., 2013. Energy efficient growth control of microalgae using photobiological methods. Renew. Energy 54, 161–165.

Ketheesan, B., Nirmalakhandan, N., 2011. Development of a new airlift-driven raceway reactor for algal cultivation. Appl. Energy 88, 3370–3376.

Kitaya, Y., Azuma, H., Kiyota, M., 2005. Effects of temperature, CO_2/O_2 concentrations and light intensity on cellular multiplication of microalgae, Euglena gracilis. Adv. Space Res. 35 (9), 1584–1588.

Kumar, K., Dasgupta, C.N., Das, D., 2014. Cell growth kinetics of *Chlorella sorokiniana* and nutritional values of its biomass. Bioresour. Technol. 167, 358–366.

Lopes, E.J., Scoparo, C.H.G., Lacerda, L.M.C.F., Franco, T.T., 2009. Effect of light cycles (night/day) on CO_2 fixation and biomass production by microalgae in photobioreactors. Chem. Eng. Process. Process Intensif. 48, 306–310.

Roselet, F., Maica, P., Martins, T., Abreu, P.C., 2013. Comparison of open-air and semi- enclosed cultivation system for massive microalgae production in sub-tropical and temperate latitudes. Biomass Bioenergy 59, 418–424.

Rotatore, C., Colman, B., 1991. The acquisition and accumulation of inorganic carbon by the unicellular green alga *Chlorella ellipsoidea*. Plant Cell Environ. 14, 377–382.

Sato, T., Usui, S., Tsuchiya, Y., Kondo, Y., 2006. Invention of outdoor closed type photobioreactor for microalgae. Energy Convers. Manag. 47, 791–799.

Schenk, P., Thomas-Hall, S., Stephens, E., Marx, U., Mussgnug, J., Posten, C., Kruse, O., Hankamer, B., 2008. Second generation biofuels: high-efficiency microalgae for biodiesel production. Bioenergy Res. 1, 20–43.

Wahidin, S., Idris, A., Shaleh, S.R.M., 2013. The influence of light intensity and photoperiod on the growth and lipid content of microalgae *Nannochloropsis* sp. Bioresour. Technol. 129, 7–11.

CHAPTER

22

Fostering single cell oil synthesis by de novo and ex novo pathway in oleaginous microorganisms for biodiesel production

Radha Panjanathan, Sanjana Narayanan, Angana Chaudhuri, Sameena Anjum, Ramani Kandasamy

Biomolecules and Biocatalysis Laboratory, Department of Biotechnology, School of Bioengineering, SRM Institute of Science and Technology, Kattankulathur, Kanchipuram, Tamil Nadu, India

22.1 Introduction

Biodiesel has become a promising and alternative source of renewable energy for depleting petroleum supplies associated with growing global energy demand and concerns of pollutant and greenhouse gas emissions (Louhasakul et al., 2020). Biodegradability and decreased biodiesel toxicity render it a supercilious substitute for fossil fuels (Nguyen et al., 2018). For commercial biodiesel production, primary feedstocks derived from food crops, such as palm oil, rapeseed oil, soybean oil, and so on, have been employed. The high-cost feedstock of vegetable oils and their consumption in large amounts is a major challenge for biodiesel production (Zhang et al., 2020). Also, the entire demand for biodiesel production could not be met with animal fats and waste cooking oils as feedstock (Muniraj et al., 2015). To achieve higher yields in small areas, it is inevitable to find other oil sources that can be extracted rapidly and easily modified. Recently, biodiesel production using single cell oil (SCO) derived from various oleaginous microorganisms was conspicuous because of its high productivity, less workforce, season/climate flexibility, scalability, and short life cycle (Arous et al., 2019). SCOs are synthesized by oleaginous microorganisms such as yeasts, fungi, algae, and bacteria that, in the stationary growth process, can accumulate between 20% and up to 80% lipid per dry biomass under nutrient limitations, e.g., nitrogen or phosphorous, and by providing the surplus amount of carbon source (Ochsenreither et al., 2016).

The costly raw material, which accounts for up to 70% of the overall production cost, is a bottleneck in SCO production (Muniraj et al., 2015). The use of low-cost industrial waste could be a promising solution to SCO's sustainable and cost-effective development. It is therefore

Biofuels and Bioenergy
https://doi.org/10.1016/B978-0-323-85269-2.00018-6

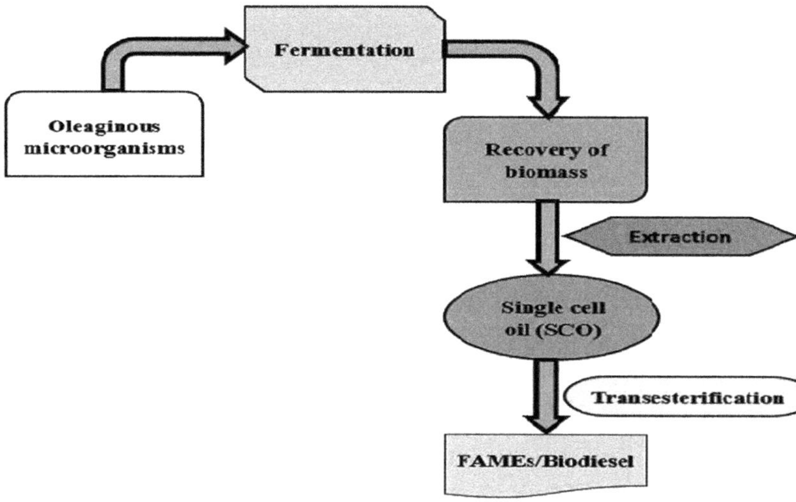

FIGURE 22.1 General process flow sheet for SCO production from oleaginous microorganisms. *SCO*, single cell oil.

essential to study different waste substrates for the aggregation of SCO by oleaginous microorganisms. The fermentation of SCO entails two different biopathways: the "de novo" pathway specifies the utilization of hydrophilic substrates, while the "ex novo" pathway pronounces the utilization of hydrophobic substrates. A number of high-value by-products such as biosurfactants, lipases, and organic acids are produced during lipid formation, and the SCO accumulation provides an extra economic value for SCO production processes (Kumar et al., 2019).

These oleaginous microorganisms convert the carbon present in different substrates into fatty acids stored in lipid bodies as SCOs comprising triacylglycerol (TAG) under suitable culture conditions. TAG biosynthesis is accomplished using different carbon sources, such as oils, sugars, alcohol, organic acids, and so on. In growth media, the surplus carbon and the limiting nitrogen condition activate TAG's biosynthesis within the microorganisms by restricting its growth after nitrogen depletion in the media and sequentially consuming the surplus carbon source for building up neutral lipids/SCO (Kumar et al., 2020). However, based on the strain of the oleaginous microorganism used and the applied cultivation conditions, the accumulated SCO may present diverse fatty acid compositions attributed to varying properties of the produced lipid. The properties grounded on the SCO's fatty acid composition predict its suitability for biodiesel production (Fig. 22.1).

This chapter explores the microbial synthesis of SCO and the simultaneous synthesis of biomolecules such as lipase, biosurfactants, organic acids, and so on, by employing diverse carbon sources. This chapter also encompasses the suitability of the synthesized SCO for the production of biodiesel. Lastly, this chapter also addresses the possibilities and challenges of using SCO as a suitable feedstock for biodiesel production.

22.2 Single cell oil and its unique properties

Each biological organism, whether micro- or macroorganisms, produces lipids for its vital functions and structural roles such as for formation of the lipid bilayer of cells and organelles (Ochsenreither et al., 2016). However, merely a few microorganisms can accumulate abundant cellular lipids, i.e., 20%–80% of the biomass as a lipid storage reserve. These microbial lipids or SCO and the microorganisms accumulating the SCOs are termed as oleaginous microorganisms, which comprise yeasts, fungi, algae, and bacteria. Carbon present in diverse substrates, under optimum culture conditions, is converted into TAG by the microorganisms. The main components of SCO are long-chain unsaturated fatty acids, which ideally suit as a feedstock in biodiesel production (Radha et al., 2020a,b). The fatty acid profiles help in understanding the varied applications in the industrial sector. An increased carbon–heteroatom ratio gives way for SCOs to be preferred in producing renewable fuels (Prabhu et al., 2019). SCOs are superior to vegetable oils due to their faster production process, lesser labor requirement, and easier scalability and deficit of climatic requirements. However, some algae and fungi appear to have the potential to synthesize SCO containing a high amount of (>20%) polyunsaturated fatty acids (PUFAs), which makes it commercially engaging for the production of arachidonic acid (20:4, ω-6) in nutritional supplementation (Ochsenreither et al., 2016).

22.2.1 Various carbon sources for the single cell oil production

The conversion and use of carbon substrates constitute a significant challenge in microbial SCO production. Exceptional attempts have thus been made to discover good sources of carbon for the SCO synthesis. Oleaginous microorganisms can employ diverse carbon substrates such as acetic acid, carbon dioxide, ethanol, and glucose to produce SCO (Table 22.1).

22.2.1.1 Conventional saccharides and other organic compounds

Different available saccharides have been explored as energy sources in the heterotrophic production of oleaginous microorganisms for SCO accumulation. Many oleaginous microorganisms have shown the ability to use typical conventional monosaccharides, including glucose, fructose, xylose, sucrose, and other disaccharides, and also polysaccharides such as starch for SCO aggregation (Table 22.1). Ghanavati et al. (2014) demonstrated comprehensive research on the competence of oleaginous yeast, *Cryptococcus aerius*, to synthesize SCO on xylose and glucose cultures. Xylose affected the production of SCOs, as the SCO content was pertinently higher (82.62%). Nevertheless, glucose cultures revealed higher SCO productivity and total dry weight of 6.21 and 7.92 g/L, respectively. Thus, glucose regulates the production of lipids by the cells in a different volume of culture. The authors also enumerated that the strain *C. aerius* produced SCO by utilizing either xylose or glucose as the carbon source (Ghanavati et al., 2014). Besides, a biphasic fermentation strategy was conducted by supplying glucose as well as xylose for growth and SCO production by *Lipomyces starkeyi*, which exhibited high biomass production and achieved an SCO content of at a range of 60%–85% (w/w) (Probst et al., 2017; Juanssilfero et al., 2018). *L. starkeyi* was studied for lipid production by potato starch, which reported a higher biomass yield of 12 g/L and SCO content of 40% (Wild et al., 2010). Other sources such as cassava (Gen et al., 2014; Chaturvedi et al., 2019) and sweet sorghum, containing a high amount of saccharides predominantly in the form of polysaccharides, have been extensively explored as carbon sources for SCO production (Antonopoulou et al., 2020). Papanikolaou et al. (2010) revealed high quantities of SCO accumulation in the mycelia ranging from 61.0% to 70.4%, irrespective of the carbon source used. Conventional saccharides are widely used as carbon sources for their easy availability, high production efficiency, and cheap maintenance requirement. However, the economic feasibility of utilizing these sources for large-scale production of SCO needs to be further evaluated.

TABLE 22.1 Single cell oil production by oleaginous microorganisms on various carbon sources.

Microorganism	Carbon source	Biomass yield (g/L)	Lipid yield (g/L)	Lipid content (%, w/w)	References
Cryptococcus aerius	Glucose	7.92	6.21	78.41	Ghanavati et al. (2014)
C. aerius	Xylose	5.41	4.47	82.62	Ghanavati et al. (2014)
Lipomyces starkeyi	Potato starch	12	4.8	40	Wild et al. (2010)
Rhodosporidium toruloides	Cassava starch	20.1	13.04	64.9	Gen et al. (2014)
Trichosporon fermentans	Sweet sorghum	16.70	3.66	21.91	Antonopoulou et al. (2020)
Rhodosporidium toruloides	Acetic acid	4.35	2.099	48.2	Huang et al. (2016)
C. curvatus	Acetic acid	10.75	5.43	50.5	Huang et al. (2018)
C. curvatus	Acetate	8.1	4.04	49.9	Gong et al. (2015)
Yarrowia lipolytica	Sugarcane bagasse	57.51	45.15	78.5	Unrean et al. (2017)
Trichosporon mycotoxinivorans	Rice straw	16.32	7.32	44.86	Sagia et al. (2020)
Trichosporon cutaneum CH002	Corncob	22.1	7.9	36	Chen et al. (2012)
Trichosporon cutaneum ACCC20271	Corncob	38.4	12.3	32	Gao et al. (2014)
C. curvatus	Wheat straw	17.2	5.76	33.5	Yu et al. (2011)
Rhodotorula glutinis	Wheat straw	13.8	3.5	25	Yu et al. (2011)
Lipomyces starkeyi	Wheat straw	14.7	4.6	31.2	Yu et al. (2011)
Y. lipolytica	Wheat straw	7.8	0.36	4.6	Yu et al. (2011)
Chlorella sorokiniana	Carbon dioxide	0.82	0.11	12.8	Zheng et al. (2012)

Certain oleaginous microorganisms utilized some organic compounds, such as acetic acid and acetate (i.e., C_2 compounds), to accumulate SCO. Acetic acid and acetates were employed as sources for SCO production (Table 22.1). *Cryptococcus albidus* was exhibited as a promising examinee for converting these organic compounds into lipid (Fei et al., 2011). Huang et al. (2016) evidenced that *Rhodosporidium toruloides* could conglomerate SCOs utilizing acetic acid (4–20 g/L), which yielded biomass of 4.3 g/L and 48.2% of lipid content. Moreover, an oleaginous yeast, *Cryptococcus curvatus*, was provided with a high concentration (40 g/L) of acetic acid as a carbon source for sequencing batch cultivation and improved extracellular lipid production with an SCO content of up to 50.5% and a high SCO yield of 5.43 g/L, respectively (Huang et al., 2018). The utilization of acetic acid as a carbon source has resulted in much greater SCO productivities and SCO accumulation by *C. curvatus* compared with glucose (Béligon et al., 2016). An analogous study using *C. curvatus* on acetate showed that the rate of SCO accumulation, SCO content, and SCO yield were all relatively higher under nitrogen-rich conditions. Competent SCO synthesis by *C. curvatus* was observed in a medium comprising 15.9 g/L of acetate (Gong et al., 2015). These findings have provided valuable information for the design and production of a more effective bioprocess.

22.2.1.2 Lignocellulosic resources

Lignocellulosic resources are the most prolific renewable resources in nature and are exploited substantially worldwide, with high research interest for SCO production. Compared with conventional sugars, lignocelluloses require pretreatment to obtain the fermentable saccharides such as glucose and xylose in the lignocellulosic hydrolysate. As carbohydrates are an essential substrate for microorganisms, carbohydrate degradation should be avoided. To prevent the generation of inhibitors linked to the lignin fraction, pretreatment of lignocellulosic biomass is necessary (Zuccaro et al., 2020). Therefore, owing to their tremendous effect on all subsequent phases of SCO development, pretreatment processes are pivotal. However, these processes are typically costly in terms of resources and have a considerable effect on the cost of the operation. Physical, chemical, physicochemical, and biological pretreatments are investigated (Moreno et al., 2019), concerning the pretreatment of the lignocellulosic resources, aiming to improve the polysaccharide hydrolysis rate, to decrease the formation of toxic by-products arising from the degradation of lignocellulose, and, ultimately, to reduce the cost of viability. Several lignocellulosic residues have been deployed as feedstock by oleaginous microorganisms to manufacture SCO (Table 22.1). Unrean et al. (2017) illustrated a study on providing sugarcane bagasse hydrolysate as a substrate for *Yarrowia lipolytica*, which reported 45.15 g/L of SCO and 78.5% SCO content, which represented the highest yield in lignocellulose-based bioprocess. Also, the sugarcane bagasse hydrolysates have been exploited as a fermentation medium for the cultivation of *Rhodotorula mucilaginosa* (Bandhu et al., 2019), *Y. lipolytica* (Tsigie et al., 2011), and *L. starkeyi* (Anschau et al., 2014). Recently, a study was conducted on the valorization of rice straw for SCO production by *Trichosporon mycotoxinivorans* under optimized conditions, which exhibited a lipid yield of 7.32 g/L and lipid content of 44.86% (Sagia et al., 2020). More specifically, for the cultivation of *Trichosporon cutaneum* CH002 (Chen et al., 2012), *Trichosporon cutaneum* ACCC 20271 (Gao et al., 2014), and *Trichosporon coremiiforme* (Huang et al., 2013), corncob hydrolysate was used. Moreover, wheat straw hydrolysates have been analyzed to cultivate *Rhodotorula glutinis*, *C. curvatus*, *R. toruloides*, *Y. lipolytica*, and *L. starkeyi* (Yu et al., 2011) for SCO production.

22.2.1.3 Carbon dioxide

Few microalgae species may utilize carbon dioxide as a natural source of carbon for SCO production, which can further be used as a feedstock to synthesize biodiesel (Table 22.1). Microalgae receive substantial recognition for their ability to directly fix carbon dioxide from the atmosphere or soluble carbonates and provide a high lipid yield. These species are cultured using carbon dioxide as the feedstock, and the harvested biomass-containing lipids were further employed for biodiesel production (Bhatia et al., 2019). In general, the concentration of carbon dioxide from 2% to 15% is seen to sustain a favorable growth rate for SCO production (Xu et al., 2013). Simultaneously, as low as 0.03% (v/v), the ambient carbon dioxide concentration is inadequate to sustain a high degree of growth and SCO accumulation (Ismail and Al-Zuhair, 2020). Kobayashi et al. (2015) studied the production of biomass and SCO in the *Chlorella* species: *Chlorella vulgaris*, *Chlorella kessleri*, *Chlorella protothecoides*, *Chlorella sorokiniana*, *Chlorella zofingiensis*, *Chlorella pyrenoidosa*, *Chlorella minutissima*, and *Chlorella fusca* var. *vacuolata* under different levels of carbon dioxide, thus resulting that additional carbon dioxide is crucial for growth and synthesis of SCO. Similarly, various studies were conducted on *Chlorella* species and reported that the dry weight (DW) and SCO content in *C. sorokiniana* were approximately 0.8 g/L and 13% DW, respectively (Zheng et al., 2012). Moreover, it was stated that for *C. minutissima*, the biomass and SCO content were 61–726 mg/L and 4%–4.8% DW, respectively (Tang et al., 2012). Other algal species such as *Botryococcus braunii* (Ashokkumar et al., 2014; Jin et al., 2016), *Scenedesmus obliquus* (Bagchi et al., 2019), and *Nannochloropsis gaditana* (Taher et al., 2020) were exemplified as potential microalgal oil producers, as the lipid accumulation magnified due to the utilization of carbon dioxide.

22.2.2 Exploration of industrial wastes as carbon sources for the single cell oil production

About 70% of the total production cost is elevated by the expensive raw material in SCO production (Prabhu et al., 2019). Low-cost carbon sources for SCO production decrease the cost of production and make it a viable alternative to conventional energy crops (Arous et al., 2019). Various industrial wastes utilized as carbon substrates by the oleaginous microorganisms are discussed in the following section (Table 22.2).

TABLE 22.2 Industrial wastes and by-products used by oleaginous microorganisms for single cell oil production.

Microorganism	Substrate	Biomass yield (g/L)	Lipid yield (g/L)	Lipid content (%, w/w)	References
Candida viswanathii Y-E4	Crude glycerol	17.8	7.3	41	Guerfali et al. (2020)
Yarrowia lipolytica Y-RC7	Crude glycerol	14.5	4.03	27.8	Guerfali et al. (2020)
Rhodotorula fluvial DMKU-RK253	Crude glycerol	6.0	3.9	65.2	Polburae et al. (2015)
Rhodotorula glutinis	Crude glycerol	9.76	4.0	41	Spier et al. (2015)
Sporidiobolus pararoseus	Crude glycerol	10.6	3.25	30.7	Manowattana et al. (2018)
Yarrowia lipolytica	Crude glycerol	25.8	13.6	52.7	Sara et al. (2016)
Rhodosporidium toruloides	Crude glycerol	21.1	11.2	53.2	Uprety et al. (2017)
Rhodosporidium toruloides	Pure glycerol	43.0	19.7	45.8	Kiran et al. (2013)
Rhodococcus opacus	Sugarcane molasses	12.0	3.6	30.0	Saisriyoot et al. (2019)
Rhodosporidium toruloides ATCC204091	Sugarcane molasses	50	39.2	78.4	Singh et al. (2020)
Cryptococcus curvatus	Cheese whey	7.2	4.7	65.0	Seo et al. (2014)
Yarrowia lipolytica	Cheese whey	7.4	4.3	58.0	Taskin et al. (2015)
Yarrowia lipolytica MTCC 9520	Chicken tallow	7.6	4.2	55.0	Radha et al., 2020b

TABLE 22.2 Industrial wastes and by-products used by oleaginous microorganisms for single cell oil production.—cont'd

Microorganism	Substrate	Biomass yield (g/L)	Lipid yield (g/L)	Lipid content (%, w/w)	References
Yarrowia lipolytica MTCC 9520	Goat tallow	3.8	2.6	69.0	Radha et al. (2020)
Debaryomyces etchellsii	Olive mill wastewater	4.2	0.7	17.9	Arous et al. (2016)
Wickerhamomyces anomalus	Olive mill wastewater	3.3	0.8	23.0	Arous et al. (2017)
Yarrowia lipolytica TISTR 5151	Palm oil mill wastewater	5.68	1.64	NA	Louhasakul et al. (2016)
Aspergillus oryzae	Potato processing wastewater	NA	3.5	NA	Muniraj et al. (2013)
Lipomyces starkeyi HL	Fishmeal wastewater + glucose	17.6	2.7	15.3	Huang et al. (2011)
Pichia kudriavzevii NCIM3653	Paper mill sludge extract	3.9	2.35	60.2	Prabhu et al. (2019
Cryptococcus vishniaccii	Papermill sludge extract	14.6	7.8	53.4	Deeba et al. (2017)
Chlorella pyrenoidosa	Municipal wastewater	0.35	NA	39.0	Zhou et al. (2020)

NA, Not available.

22.2.2.1 Biodiesel industry by-products

The major by-product of the biodiesel industry is crude glycerol, which is considered as a suitable feedstock for various industrial fermentation applications. Nearly 10% (w/w) of crude glycerol is engendered as a by-product during the biodiesel synthesis (Garlapati et al., 2016). As a result of its cheap and increasing availability, this by-product stream could be valorized through an integrated bioprocess. It is a practical approach to solve the environmental issues associated with its disposal (Tsouko et al., 2016; Uprety et al., 2017). Moreover, biodiesel produced from oilseeds (rapeseed, sunflower, soybean, etc.) generates significant by-products (excluding crude glycerol), namely oilseed meals, which could be recycled as a nutrient-rich supplement for SCO production. Particularly, crude glycerol has been extensively employed as feedstock for the SCO synthesis by oleaginous microorganisms, as shown in Table 22.2 (Arous et al., 2019; Chebbi et al., 2019). Guerfali et al. (2020) conducted studies for 13 different oleaginous yeast strains by utilizing crude glycerol obtained from the biodiesel production process as a carbon source for SCO production. It was reported that the yeast strain Y-E4 of *Candida viswanathii* displayed maximum lipid production (41% of SCO content). Similarly, *Rhodotorula fluvial* DMKU-RK253 achieved peak lipid production equal to 65.2% of its biomass among 23 newly isolated oleaginous yeast strains (3.9 g/L lipid and 6.0 g/L biomass) in crude glycerol (Polburee et al., 2015). Spier et al. (2015) have reported the maximal SCO production in *R. glutinis* (4 g/L of SCO yield and 41% SCO content). Manowattana et al. (2018) optimized the lipid production process of *Sporidiobolus pararoseus* utilizing crude glycerol, which reported 10.6 g/L of biomass yield and 30.7% (w/w) of SCO content. Sara et al. (2016) conducted similar studies on SCO production by *Y. lipolytica* in a 5-L fermenter fed with 89 g/L of glycerol, which showed a significant SCO content of 52.7%. Besides, Uprety et al. (2017) reported around 21.1 g/L of total dry weight and 53.2% (w/w) of lipid content in shake flask cultures of the oleaginous yeast *R. toruloides* using crude glycerol as a carbon source. Another study involved the cultivation of *R. toruloides* and *C. curvatus* on glycerol medium supplemented with sunflower meal, where both the yeasts achieved the total biomass yield in the range of 36.6–47.9 g/L and SCO yield of 18–19 g/L (Leiva-Candia et al., 2015). Similar to the previous study, Kiran et al. (2013) validated the coalescence of crude glycerol and rapeseed meal (nitrogen-rich stream) that resulted in higher lipid concentration compared with conventional nutrient sources (43 g/L of biomass and SCO content of 45.8% w/w), hence indicating a more robust balance of nutrients for lipid accumulation (Kiran et al., 2013). However, some impurities were impactful on microbial oil production by impeding the growth and SCO production (Chen et al., 2018; Kumar et al., 2019). Above all, lucrative measures must be conceived to discard these impurities before employing crude glycerol as a carbon source for SCO production.

22.2.2.2 *Food industry by-products*

The food industry by-products and waste streams such as molasses from a sugar refinery, and cheese whey from the dairy industry, have been used as substrates for SCO synthesis (Table 22.2). The by-product of the sugar industry, molasses, arises during refining sugarcane or sugar beets into sugar. Vieira et al. (2016) established the technical suitability of SCO production from *R. glutinis* on molasses. Bento et al. (2020) exemplified the potentiality of sugarcane molasses as a useful source for growth and SCO production by *Mucor circinelloides*. Recently, a study conducted by Singh et al. (2020) illustrated that *R. toruloides* produced maximum SCO and biomass among seven oleaginous yeasts. The FAME profile of SCO showed higher levels of unsaturated fatty acids, attributed to its eminence in biodiesel synthesis (Singh et al., 2020). Potentially, sugarcane molasses was utilized by *R. opacus* to maximize the lipid production, at high molasses concentration of 80 g/L, under batch culture conditions (Saisriyoot et al., 2019). These studies epitomized that sugarcane molasses could be employed as an excellent feedstock for biomass and SCO production, owing to their reduced price, increased supply, and high carbon content.

Cheese whey, one of the dairy industry's major by-products, is produced during the cheese-making process. Despite its high organic load and critical environmental pollution caused by cheese whey, several studies outlined that the high content of carbon in cheese whey makes it a putative candidate for microbial oil production (Carota et al., 2017). Seo et al. (2014) reported that the yeast strain *C. curvatus* could accumulate a lipid content of 65% (w/w) of the total cell dry weight and 7.2 g/L/d growth rate during fermentation on cheese whey medium. Taskin et al. (2015) revealed SCO production of *Y. lipolytica* on lactose-enriched ultrafiltered whey media, which yielded 58% (w/w) of lipid content (Table 22.2). Carota et al. (2017) focused on utilizing Ricotta cheese whey for lipid production by 18 strains of oleaginous yeasts, among which *C. curvatus* produced 6.8 g/L of SCO and *C. laurentii* produced 5.1 g/L of SCO. Nevertheless, cheese whey has proven to be a very promising feedstock; however, significant product development should be considered to design a sustainable approach to biorefinery.

Slaughterhouse and meat processing industries generate many wastes that include animal waste, fats, lards, and tallow derivatives, which can be used as substrates for SCO synthesis (Lopes et al., 2018; Radha et al., 2020a,b). Liang et al. (2010) demonstrated the utilization of animal fats as hydrophobic substrates that showed SCO accumulation in *Y. lipolytica* with an SCO content of 50% (w/w). Radha et al., 2020b symbolized the utilization of chicken tallow as a sole carbon source for SCO synthesis by *Y. lipolytica*, which resulted in maximum biomass of 7.6 g/L and SCO content of 55% (w/w). Similarly, Radha et al., 2020a conducted another study employing a slaughterhouse lipid waste, goat tallow, for SCO synthesis by *Y. lipolytica*, which yielded 69% (w/w) of lipid content (Table 22.2). Many studies currently focus on employing lipid-containing industrial wastes as feedstock for SCO production, followed by biodiesel production. Furthermore, the utilization of animal fat wastes and tallow derivatives makes SCO production affordable and develops a sustainable biodiesel production process (Donot et al., 2014).

22.2.2.3 Industrial wastewater streams and sludge

Studies have explored the application of wastewater and sewage sludge as substrates in the production of high-value products (Tsouko et al., 2016; Arous et al., 2019). Olive mill wastewater (OMW) is considered as one of the appropriate substrates for SCO production (Table 22.2). Arous et al. (2016) and Arous et al. (2017) demonstrated the utilization of OMW for SCO production by the strains *Debaryomyces etchellsii* and *Wickerhamomyces anomalus*. Evidently, Bellou et al. (2014) aimed at producing high-value-added SCOs that resulted from OMW conversion by various strains of zygomycetes, including *Thamnidium elegans* and *Zygorhynchus moelleri*. Interestingly, *T. elegans* and *Z. moelleri* exhibited maximum growth on the medium comprising of OMW, thus paving the way to the possible bioconversion of OMW into lipid.

Palm oil mill effluent (POME), a waste stream produced from the palm oil industry with high COD (chemical oxygen demand), has been explored for microbial lipid synthesis (Kamyab et al., 2017; Saenge et al., 2011). Moreover, the SCO accumulation of several strains by *Y. lipolytica* cultivated on POME substrate was studied, and it was observed that all the strains could accumulate SCOs more than 33% (w/w) (Louhasakul et al., 2016). POME contains high levels of carbohydrates and proteins, which promote cell growth and SCO accumulation (Marjakangas et al., 2015).

The applications of other wastewater streams, including starch-rich industrial wastewater, fishmeal wastewater (FW), and municipal wastewater, were investigated for SCO synthesis. Potato processing wastewater was exploited as a substrate for SCO production by *Aspergillus oryzae*, which showed an SCO accumulation of 3.5 g/L (Muniraj et al., 2013). FW was given as a medium to the oleaginous yeast *L. starkeyi* for producing SCO (Table 22.2). FW and glucose were supplemented for microbial oil production, resulting in enhanced biomass of 17.6 g/L and a lipid yield of 2.7 g/L (Huang et al., 2011). Municipal wastewater is an economically desirable culture medium to cultivate microalgae due to the presence of organic carbon (Kialashaki et al., 2019). Interestingly, the cultivation of *C. pyrenoidosa* in the presence of ammonia-oxidizing strain in municipal wastewater facilitated accelerated microalgal growth and lipid production by 0.35 g/L and 39%, respectively (Zhou et al., 2020).

Nowadays, industrial and municipal sludge is striking as an attractive renewable energy source owing to its massive carbon content (Kumar et al., 2019). Angerbauer et al. (2008) utilized pretreated sewage sludge as a raw material to synthesize SCO, which has a lipid content of 68%. Various sludge types such as primary municipal sludge, pulp, and paper industry sludge were used as feedstock for SCO synthesis by *T. oleaginous*, which showed maximum SCO yield (Zhang et al., 2014). Paper mill sludge is an extensively available resource to be harnessed as a feedstock used to recycle organic wastes, thus truncating major environmental issues. The sludge containing carbon could be transformed into lipids by the oleaginous microorganisms, thereby serving as a raw material for biodiesel production (Deeba et al., 2017;

Prabhu et al., 2019). Prabhu et al. (2019) studied the SCO production by *Pichia kudriavzevii*, cultivated on pretreated paper mill sludge extract (PMSE) with a C/N ratio of 200 for 120 h that gave an SCO yield of 2.35 g/L (Table 22.2). Similarly, the sonicated PMSE depicted enhanced lipid yield and lipid content 7.8 g/L, 53.40% (Deeba et al., 2017). Hence, using industrial wastewater or sludge as a substrate has a dual purpose benefitting both the economy and environment and synthesizing SCO to produce biodiesel further.

22.3 Oleaginous microorganisms and their potential for single cell oil accumulation

Oleaginous microorganisms such as fungi, yeast, and microalgae can accumulate lipids under suitable culture cultivations. This specific property has fetched attention toward its utilization in the biodiesel industry.

22.3.1 Oleaginous yeast and fungi

SCO synthesized by yeasts and fungi species are known to produce the fatty acid methyl esters (FAMEs) for biodiesel production, which also adds up an advantage of being a cheap substrate requirement (Raut et al., 2019). Yeast is considered as the most suitable microorganism to have its application in the biotechnology industry. Oleaginous yeasts comprised 30 of the 600 identified yeast species. These oleaginous yeasts have the potential to store lipids in correspondence to about 40% of their biomass. Nevertheless, the level might increase up to 70% of the total biomass. The oleaginous yeasts are of industrial importance, as they assimilate carbohydrates into biolipids under limited nitrogen conditions. These biolipids derived from oleaginous yeast cells mostly consist of long-chain fatty acids such as palmitic acid (C16:0), stearic acid (C18:0), palmitoleic acid (C16:1), oleic acid (C18:1, $n-9$), and linoleic acid (C18:2, $n-6$), which can be utilized for biodiesel production (Zuccaro et al., 2020). The best known oleaginous yeasts such as *R. toruloides*, *R. glutinis*, *L. starkeyi*, *T. fermentans*, and *Yarrowia* exhibit the potential to synthesize microbial oil (Beopoulos and Nicaud, 2012). Beopoulos and Nicaud (2012) observed microbial lipid accumulation by oleaginous yeasts due to deficiency of zinc, iron, or magnesium.

In the stationary growth phase, TAGs are accumulated as SCO in the cells. The acyl group distribution of TAGs is very similar to the acyl group distribution of plant oils. Mainly, the unsaturated group occupies the central position. Despite these molecules possessing any positive or negative charge, they are sequestered as hydrophobic particles and are further called microdroplets. These microdroplets are lipid particles or the lipid bodies, which are accumulated by the organism. Nevertheless, the lipid fraction of the cells is entirely made up of TAGs, and the latter can be extracted along with other lipid fractions associated with the cell membrane (phospholipids, sterols, steryl esters, and others) as well as free fatty acids associated with unregulated lipolysis that can occur throughout the extraction phase.

Thus, an increasing interest in applying oleaginous organisms for lipid synthesis and an opportunity to utilize them for biodiesel production has created a scientific approach to acquire an alternate method like those offered by oleaginous yeasts. Fungi and bacteria are of less application than microalgae and yeasts prominently due to the differences in the lipid production capacity. Some fungal species like *Mortierella alpine*, *Mortierella isabellina*, and *Aspergillus terreus* are also reported to produce special lipids such as arachidonic acid (ARA), gamma-linolenic acid (GLA), eicosapentanoic acid (EPA), and docosahexaenoic acid (DHA) (Kumar et al., 2019). Thus, fungal species particularly oleaginous Zygomycetes such as *M. isabellina* and *Cunninghamella echinulata* exhibiting large amounts of lipid (SCO) accumulation have also been considered to be a vital source for biodiesel production, as they may amass up to 86% and 57% of lipids in dry biomass, respectively (Bautista et al., 2012).

22.3.2 Oleaginous algae

Increasing energy consumption from fossil fuels and their rising concerns of depletion have paved the way for researchers to introduce a sustainable alternative. Considering these aspects, microalgae have been recognized as a potential feedstock for biofuels and other bioproducts (Feng et al., 2016). The cultivation of oleaginous algae offers more advantages than the yeasts in various aspects. Also, some oleaginous microalgae can produce microbial lipids and show great response to remove pollutants from various water bodies (Perez-Garcia et al., 2011). They also present a relatively high efficiency to utilize solar energy and could be used for carbon dioxide capture and recycling. High amounts of carbohydrates, lipids, and other trace elements help in easier cultivation to produce higher lipid content and cellular growth (Feng et al., 2016). For example, 42% w/w of lipids was accumulated by *C. vulgaris* while cultured on a synthetic wastewater medium (Feng et al., 2011). Olguín et al. (2015) suggested a dual-purpose system, which would be used in treating anaerobic effluents obtained from pig waste. This dual-purpose treatment utilizes *Neochloris oleoabundans*, oleaginous microalgae used to produce lipid-rich biomass. The strain was found to be very efficient at nutrient removal and a nonnegligible amount of lipid production (up to 22.4%, w/w) (Arous et al., 2019). Lately, four different algal species *Scenedesmus*, i.e., *Scenedesmus quadricauda, Scenedesmus dimorphus, Scenedesmus* sp. LX1, and *S. obliquus*, were tested to produce extensive lipid quantities in the range of 12.75%–66%, w/w (Zhan et al., 2016).

22.3.3 Oleaginous bacteria

Owing to its high organic degradation rate, the bacterial culture is primarily suitable for wastewater treatment. However, simultaneous lipid accumulation has also been studied in a few of the oleaginous bacterial species along with wastewater treatment capacity. For example, the oleaginous bacterium *R. opacus* has shown considerable promise in the conversion of raw dairy wastewater to lipids with the accumulation capacity of up to 14.28% w/w while supplying additional mineral salts along with dairy wastewater caused a tremendous increase in lipid content of about 33% (Arous et al., 2019). Oleaginous bacteria can be an excellent source for biodiesel production for yet another reason being easily reproducible, as they possess a comparatively high growth rate than any other microorganism. Although their tendency to produce SCO is limited, they can often accumulate complex lipids (for example, polyhydroxyalkanoates, glycolipids, phospholipids, and fatty alcohols) mainly in their outer membrane (Bautista et al., 2012). They tend to accumulate lipids under some distinct environmental conditions, and the composition of lipid accumulated thus varies from other microbial oils. However, some species of the *Actinomycetes* group of bacteria corresponding to the genera *Rhodococcus, Nocardia, Gordonia*, and *Mycobacterium* have been reported to accumulate lipids comparable to other oleaginous microorganisms, which can be directly used for biodiesel production. Reports say that *Gordonia* and *Rhodococcus* have the ability to accumulate 80% lipid content under specific culture conditions comprising higher carbon and lower nitrogen (Kumar et al., 2019).

22.4 Fermentation strategies to produce single cell oil by oleaginous organisms

The microbial oil production at the industrial level causes certain impediments related to the downstream processing and separation of carbon, lipids found intracellularly, and expenditure on the operation of the fermentation process. The establishment of an economically attainable and sustainable process requires specific criteria, which involves maximal productivity, the evolution of biorefinery concepts that lead to the production of different high-value-added coproducts. Studies conducted by Kopsahelis et al. (2018) represented a simultaneous production of certain coproducts, including tartrates, polyphenols, ethanol, and microbial oil production utilizing cheese whey and wine lees. The establishment of microbial lipids synthesis involves a crucial point, which involves profitable production and purification of fuels or other valuable products. This production can be ensured via two methods, i.e., submerged fermentation (SmF) and solid-state fermentation (SSF). The fermentation strategy influences the yield and composition of the lipid. They are also dependent on particular conditions such as nutrient media, pH value, temperature, nitrogen source, and aeration.

In SmF, the quantity of lipid accumulated intracellularly is dependent on the cultivation process opted along with the nitrogen source, pH, and temperature. Studies conducted by Ochsenreither et al. (2016) have reported a cellular content of 60% when xylose, glucose, and fructose have been used as substrates. It has been found out that carbon sources are necessary to trigger the lipid accumulation process, thus, utilizing low-cost materials such as glycerol is beneficial. Apart from carbon sources, nitrogen sources also affect SCO accumulation. Principally, the cultivation of oleaginous microorganisms can be ideal in cultivation based on batch, fed-batch, or continuous cultures. Baffled and unbaffled flasks cultured for *Schizochytrium* sp. represent the dissimilarities in biomass production and lipid accumulation. It showed that baffled flasks were responsible for incrementing biomass production (Ling et al., 2015). Stirred tank bioreactors provided a significant increase in lipid accumulation in *M. alpina* cells when compared with shaking flasks (Stressler et al., 2013). The observation obtained from this study was compared with the accumulation of lipid by *M. isabellina* in stirred tank reactors. An integrated process, recycling CO_2 in a phototrophic process, was conducted involving an oleaginous microalgae *Phaeodactylum tricornutum* and oleaginous yeast, *C. curvatus*. The processes reveal a lipid content of about 40%–45% (w/w). In SSF, higher productivity is obtained with cheap media besides comparatively low energy and wastewater costs. Despite these advantages, there is a difficulty in scaling up, managing the culture conditions, and increased cost during retrieval of the product (Couto and Sanroman, 2006; Asadi et al., 2015). The SSF process uses low-cost substrates such as agroindustrial wastes such as wheat straw, bran straw, and fruit wastes (Conti et al., 2001; Gema et al., 2002; Fakas et al., 2009; Jacobs et al., 2010; Lin et al., 2010; Stressler et al., 2013). These substrates require a different strategy, i.e., they need to be pretreated enzymatically to break insoluble materials down into simple monomers (Peng and Chen, 2008).

Studies have led to a process that contributes to the valorization of humongous waste streams originating from the confectionery or bakery industry, which comprises damaged, discarded, or out-of-date products. This strategy comprised a two-stage bioprocessing pathway, resulting in high lipid concentration and conversion yields (Tsakona et al., 2014, 2016). This particular process involved suitable conditions for the production. The bioreactor fermentations have been conducted in a 3 L bioreactor with a working volume of 1.5 L. The conditions maintained were the temperature of 27°C, aeration of 1.5 vvm, agitation rate of 150–500 rpm, and pH adjusted to 6. Furthermore, a 10% (v/v) inoculum was added in a 24 h preculture in a fed-batch process. Thus, every organism and substrate require different conditions and strategies to have better and maximal yield, which states the importance of suitable fermentation strategies, which further produce SCO (Tsakona et al., 2019).

22.5 Mechanism of single cell oil production by oleaginous microbes

The further development of the SCO processes requires understanding the mechanism of the SCO production by microbes. There are two different pathways for the synthesis of SCO—one being the de novo synthesis and the other being ex novo synthesis. The biochemistry of de novo synthesis can be subdivided into two distinct pathways—the intermediate cellular metabolism and biosynthesis of TAGs (Fig. 22.2). Pyruvic acid, the end product of glycolysis, further reaches the mitochondrial matrix via the mitochondrial membrane. Pyruvate dehydrogenase catalyzes the conversion of pyruvic acid to acetyl-CoA. This acetyl-CoA is further transported into the cytoplasm to augment the biosynthesis of cellular fatty acids. In certain microorganisms such as the oleaginous organism, the acetyl-CoA acts as a precursor for the intracellular biosynthesis of fatty acids (Ratledge et al., 2004). These fatty acids had been derived from the breakdown of preaccumulated mitochondrial citric acid. One of the essential criteria for fatty acid accumulation is excess carbon and nitrogen limitation. The exhaustion of nitrogen initiates a rapid decrease of the intracellular adenosine monophosphate (AMP) concentration. AMP-deaminase aids in the cleavage of AMP into inosine monophosphate (IMP) in the microorganism. Acetyl-CoA acts as the predecessor of intracellular biosynthesis of fatty acids, obtained from the breakdown of citric acid that might have been hoarded inside the mitochondria previously and later transported into the cytosol. Nitrogen limitation triggers a fast decrease in the concentration of intracellular AMP.

The hydrophobic substrates, such as FFA, n-alkanes, TAG, and sterol esters, are utilized via the ex novo pathway by oleaginous microorganisms for energy drives or accrue them as storage lipids (Fig. 22.3). The free fatty acids will be generated from the hydrolysis of TAG, which is catalyzed by lipases. These newly

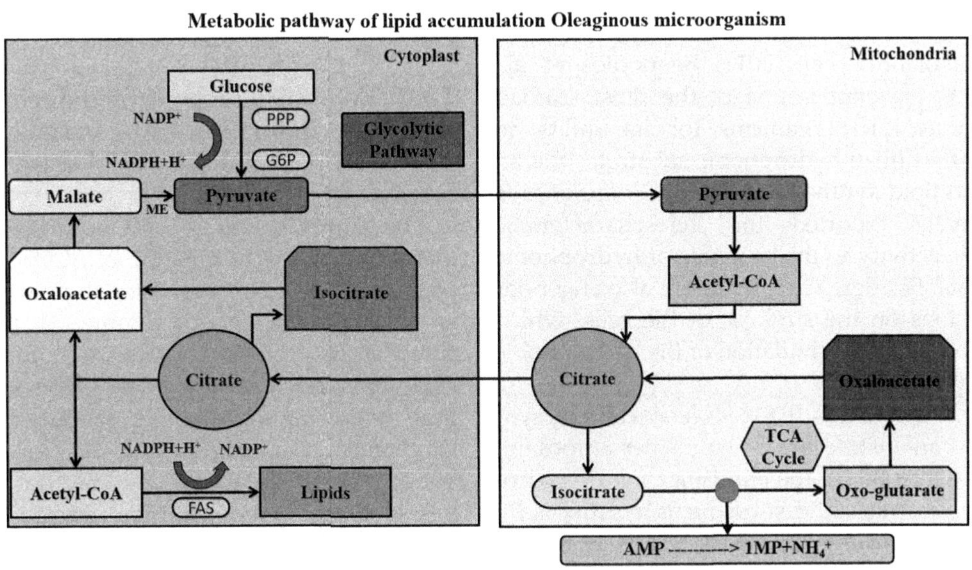

FIGURE 22.2 De novo metabolic pathway of lipid accumulation in oleaginous microorganisms.

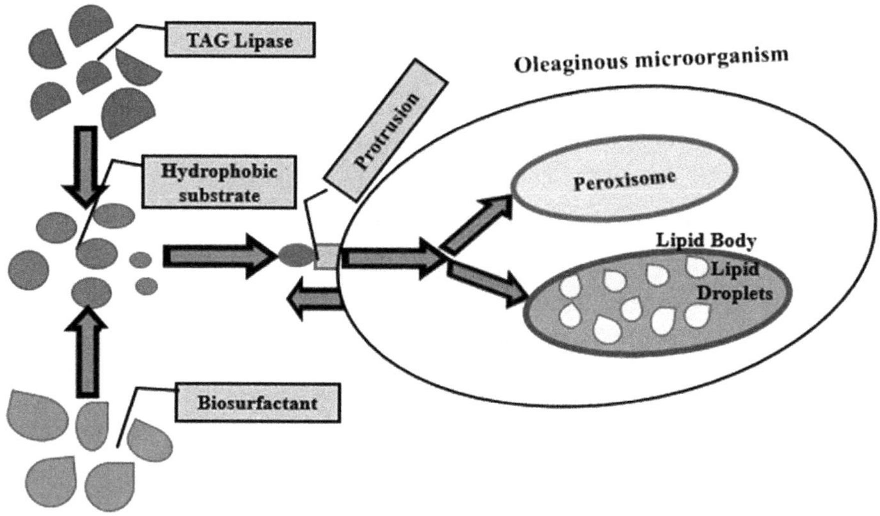

FIGURE 22.3 Ex novo metabolic pathway of lipid accumulation in oleaginous microorganisms.

altered fatty acids are converted through β-oxidation into short-chain acyl-CoAs and acetyl-CoAs by various acyl-CoA oxidases. They are used to provide the necessary energy for growth and cell refurbishment or bioformation of specific metabolites that enable new cellular materials to be synthesized (Papanikolaou et al., 2011; Beopoulos et al., 2009). *Y. lipolytica* is one of the most studied oleaginous microorganisms for its ability to efficiently utilize hydrophobic substrates via an ex novo lipid synthesis pathway. *Y. lipolytica* is extensively reported for the characteristic lipolytic activity in the presence of hydrophobic substrate. Besides, *Y. lipolytica* can also engender protrusions on the surfaces of the cells, which can promote the assimilation of the hydrophobic substrate.

In addition to this, the *Y. lipolytica* can also synthesize an extracellular emulsifier (liposan), which will enhance the emulsification of hydrophobic substrates and subsequent hydrolysis by lipase (Mlickova et al., 2004). The synergistic effect of emulsifier and lipase promotes the accumulation of small-sized hydrophobic lipid droplets inside the yeast cell. Besides, accumulated lipid droplets (LDs) can be degraded by different mechanisms, such as alkane oxidation, β-oxidation, and a mixture between the cycles of citrate and glyoxylate that exists in the endoplasmic reticulum, peroxisomes, and mitochondria. The final degradation process of the substrate takes place through β-oxidation in the peroxisome.

22.6 Co/by-products formed during single cell oil accumulation using various carbon sources and industrial wastes

The coproduction of certain metabolites is inexorable in the production of SCO. These coproducts may be essential for the environment as well as the market. Some of the most abundantly produced coproducts are organic acids, biosurfactants, and lipases. This coproduction of the main metabolic products, such as simultaneous production of SCO and value-added chemicals, makes use of the nutrients provided in the culture media, which further maintains the technical and economic balance of microbial lipid technology (Zhang et al., 2016). Recent research has led to SCO cogeneration processes and other chemicals such as carotenoids, fucoxanthin, bioethanol, and other pigments. However, not much importance has been given to the simultaneous production of other bulk chemicals and SCO (Qian et al., 2019). A report by Kumar et al. (2020) states that during the lipid production and fermentation process, other cogenerated products are formed such as citric acid and extracellular polymeric substances (EPS) (Kumar et al., 2019). This EPS is responsible for biopolymer formation around the cells themselves to protect them from the external environment. EPS has the additional role of serving as extra energy and nutrition to the cells, which have been deprived of food. EPS provides a considerable role in managing and controlling flocculation, sludge removal, and the removal of heavy metals. In particular, *Y. lipolytica*, a widely explored oleaginous microorganism for effective utilization of hydrophobic substrate through ex novo lipid accumulation pathway, has been reported for the synchronized synthesis of SCO, lipase, and biosurfactant. The amalgamation of the hydrophobic substrate was facilitated by the coproduced biosurfactant that reduces the surface tension and enhances the utilization of the oil substrate. Besides, *Y. lipolytica* has the proclivity to secrete lipase in the presence of the hydrophobic substrate (Radha et al., 2020b).

22.6.1 Lipases

Lipase is an essential enzyme that catalyzes the hydrolysis of fats. It can also be used as an enzymatic catalyst in transesterification. It has been found that lipase can be coproduced in SCO production as a cometabolite having a high-added value. In recent times, lipolytic enzymes have gained popularity due to their potential for transesterification and their hydrolytic activities and are therefore considered as a replacement for conventional catalysts, which are polluting in nature. They act as biological catalysts for the synthesis of biodiesel. The lipase production was observed to be critically affected by the medium composition. The increasing level of fat and extracellular nitrogen concentration was a crucial factor for lipid production. In several studies involving *Y. lipolytica* strains, it has been observed that extracellular lipase synthesis is enhanced by the existence of TAGs, organic nitrogen, and long-chain fatty acids (Dominguez et al., 2003; Fickers et al., 2004, 2005a, 2005b). The production of lipase is observed to be affected by the presence of surfactants, agitation, and aeration. It was also observed that extracellular lipase was produced notwithstanding substrate fat or nitrogen in medium. Ammonium sulfate and sulfur influenced the higher enzyme production. Coproduction of lipid and lipase has been reported rarely (Szczesna-Antczak et al., 2006). A large number of microorganisms have the ability to synthesize lipases with varied enzymological properties and their varied specificities. Growth parameters optimization in fed-batch cultivation produced high lipase concentration and biomass (Nicaud et al., 2002). Further to increase lipase production, genetic engineering and mutagenesis may be employed.

22.6.2 Biosurfactants

Biosurfactant, occurring in the different structures, is distinguished by its surface-active and amphiphilic nature. They are generally synthesized by microbes extracellularly and provide more benefits than synthetic surfactants. Biosurfactants have certain advantages over synthetic surfactants, including higher biodegradability, lower specificity, and selectivity (Rufino et al., 2014). In the cellular component of oleaginous yeasts, the internalization of hydrophobic substrates occurs by creating specific changes or protrusions on the surface, which increases the contact area between the yeast and the hydrophobic substrate (Carsanba et al., 2018). Surfactants generally form either bilayer micelles (for hydrophilic substances) or spherical micelles (for hydrophobic substances), which then function as follows: roll-up, emulsification, and solubilization. It has been found that yeasts such as *Candida* and *Yarrowia* that have a GRAS status are better for the production of biosurfactants.

Biosurfactants are secondary metabolites and structurally made with hydrophilic and hydrophobic groups. The inherent emulsification activity on hydrophobic material is exploited to assimilate hydrophobic materials and ascribing the enhanced utilization by oleaginous microorganisms. Among most of the oleaginous yeasts, *Y. lipolytica* has been linked with several different industrial and commercial pathways for biosurfactant production. Despite biosurfactants showing a promise, it has several disadvantages due to expensive production, inefficient recovery methods for production, and the usage of expensive substrates. A strategic way to reduce the production cost of biosurfactants is to use alternative sources of substrates, which are low cost and readily available, and providing high concentrations of biosurfactants. Nonetheless, suitable optimal culture conditions ensure better productivity of the biosurfactants. This biosurfactant has applications in various sectors such as agriculture, medicine, and the petroleum industry. Biosurfactants provide an aide to the microbes to adsorb soil particles trapped between pollutants, thus reducing the diffusion length between the uptake by microbes and the site of absorption (Makkar et al., 2003). The medical industry can utilize antimicrobial, anticancer, antiviral, and anti-adhesive properties. They have a wide range of applications in the food processing industries as a formulation agent and an antiadhesive agent. In the bioremediation field, biosurfactants have been used in metal removal (Juwarkar et al., 2007) and for the remediation of oil spillage (Liu et al., 2015).

22.6.3 Organic acids

Certain oleaginous yeasts can produce organic acids as a coproduct of SCO formation. The glycolytic pathway of yeast synthesizes intermediary compounds from glycerol by phosphorylation pathway or the oxidation pathway (Lazar et al., 2014). The most popular organic acid that has been produced alongside SCO is citric acid. The dihydroxyacetone in the glycolytic pathway is subsequently converted to stored lipids, citric acid, or other products. In an environment where the C/N ratio is higher, citric acid production follows the growth of the microbes. Furthermore, this results in decreasing lipid accumulation. Reports by Souza et al. (2014) described lipids and citric acid production by *Y. lipolytica* UFLA CM-Y9.4 by utilizing crude glycerol as the substrate. When the pH was controlled, the citric acid production increased and reached 11.10 g/L of production at 120 h. In contrast, in an uncontrolled pH reactor, the citric acid productivity decreased and reached 1.05 g/L at 120 h. The lipid production and accumulation decreased with the decreasing citric acid concentration. Hence, a pH-controlled environment is beneficial for citric acid and SCO production (Kumar et al., 2019).

Another coproduct that is produced along with the SCO is gluconic acid. The oleaginous yeast *Cryptococcus podzolicus* DSM 27192 could produce SCO and gluconic acid intracellularly and extracellularly, respectively. In certain research works, it has been seen that when glucose is used as a substrate, the gluconic acid could accumulate in high amounts (Schulze et al., 2014). Gluconic acid is an oxidative product of glucose with a wide range of applications in various sectors such as food, medicine, and cement industries (Ramachandran et al., 2006; Qi et al., 2015). The demand for gluconic acid in various sectors has been increasing globally due to its multiple applications. Prior studies inferred that the coproduction of SCO and gluconic acid is generally affected by three parameters: an excess of carbon, nitrogen limitation, and continuous aeration. In microbial fermentation, the nitrogen sources strongly affect the cell growth and distribution of metabolites. Synthesis of gluconic acid is suppressed by organic nitrogen sources, whereas high glucose greatly stimulates gluconic acid production (Kim et al., 2015).

Fumaric acid is a naturally occurring organic acid involved in the tricarboxylic acid (TCA) cycle. This fumaric acid has applications in the food, pharmaceutical, and chemical industries (Chen et al., 2015; Xu et al., 2016). Many different studies have focused on the production of fumaric acid or SCO but significantly less on the integrated production of both. The SCO can accumulate in the microbial cells under stress conditions such as higher carbon sources and nitrogen limitation (Papanikolou et al., 2011; Wang et al., 2018; Han et al., 2016). SCO and fumaric acid induction are generally under identical conditions. The SCO accumulates in the microbial cells under stressful conditions such as higher carbon concentration and limited nitrogen, which further triggers fumaric acid and malic acid accumulation. These identical conditions help in the simultaneous production of the fumaric acid and the SCO (Chen et al., 2015; Chi et al., 2016; Zelle et al., 2010; Romano et al., 1967). This report inferred that increasing the carbon/nitrogen ratio might increase the production of the coproducts along with the lipid yield. Wang et al. (2018) further had observed that the fumaric acid production was the highest at 27.8 g/L when the medium had a concentration of 120.0 g/L glucose.

22.7 Single cell oil for biodiesel production

Transesterification reaction primarily comprises converting oils and fatty acids into alkyl esters utilizing a surplus of methanol and acids or alkalis as a catalyst for biodiesel production (Moazeni et al., 2016).

22.7.1 Acid/alkali-based transesterification process

Acid-based transesterification practices the use of methanolic sulfuric acid (H_2SO_4), ferric sulfate ($Fe_2(SO4)_3$), sulfonic acid (H_2O_3S), methanolic hydrogen chloride (HCl), and methanolic boron trifluoride (BF_3). Due to their acidic properties and low-cost sulfuric acid, sulfonic acid and hydrochloric acid are used as the catalysts. Additionally, acidic catalysts are preferred more due to their insensitivity toward the free fatty acids (FFAs), thus eliminating undesired side reactions such as saponification (Lam and Lee, 2011). The transesterification step is carried out initially by mixing the acid-based catalyst properly with methanol in a small vessel, which is then further transferred into a large vessel accompanied by the lipids or oil (raw materials) for the transesterification process. The advantage of using an acid-based catalyst is that it provides a higher FAMEs yield, but a slow rate of reaction as a disadvantage is also followed. This involves a primary mechanism of oxonium ion formation by acid protonation, producing an intermediate due to an exchange reaction with alcohol, thus forming an ester after a proton loss. This complete process is rescindable at each phase; therefore, the excess usage of methanol allows the equilibrium to shift toward the product side, and the reaction is finished (Kumar et al., 2020).

Sodium hydroxide (NaOH) and potassium hydroxide (KOH) are frequently used as base catalysts. They are relatively low cost compared to heterogeneous and enzymatic catalysts and are effortlessly available and accelerate transesterification very effectively at very mild reaction conditions. Furthermore, base-catalyzed transesterification is 4000× faster than acidic catalysts. The transesterification is initiated by premixing the catalyst into methanol (alcohol) by vigorous shaking and further transesterifying accompanied with oils. The higher sensitivity of homogeneous base catalysts toward FFA content is a disadvantage to the biodiesel industry (Kumar et al., 2020; Lam et al., 2019).

22.7.2 Enzyme-catalyzed transesterification process

Enzymes are protein catalysts that can accelerate a chemical reaction and thus favor any required process. Enzyme catalysts are different from chemical catalysts and offer several advantages, as an intensive requirement of energy is reduced as the reactions usually proceed at room temperature. Moreover, they have an insensitive nature toward the FFA content and easy recovery of catalyst and glycerol after transesterification and minimize the added water wash step that leads to cost reduction in wastewater treatment. In this reaction process, an enzyme serves as a biocatalyst to carry out the transesterification. Like the noncatalyzed reaction method, the catalyzed reaction requires biodiesel purification techniques with considerably lower operating conditions. The widely used enzyme used for biodiesel production is lipases, which further convert TAGs to fatty acids and glycerol. Three various lipases were selected for transesterification of oil in a solvent-free environment from *Candida rugosa*, *Chromobacterium viscosum*, and *Porcine pancreas* for the process of biodiesel production. It was reported that only the lipase derived from *Chromobacterium viscosum* produced a substantial yield, thus enhancing the biodiesel production from 62% to 71%; this *C. viscosum* lipase was immobilized on Celite-545. Thus, it can be observed that combinational optimization of the process parameters involved in transesterification and lipase immobilization upsurges the biodiesel yield at certain instants (Kumar et al., 2020). Lipases can also be procured through various sources, including plants, fungal, animal, and bacteria. However, microbial lipases are the most common sources preferred for biodiesel production, as they are cheap and have fast production ability along with easiness in abundant cultivation as well as its independency on season aids the rapid growth of the microorganism in an inexpensive media (Moazeni et al., 2019).

22.7.3 Suitability of single cell oil for biodiesel production

The fatty acid methyl ester or FAME profile of transesterified SCO derived from oleaginous microorganisms ascertains its compatibility as a raw material for its production of biodiesel. It has been observed that their fatty acid profile immensely influences the physicochemical properties of biofuels. The enhanced engine behavior in a cold environment and having no oxidative degradation loss is due to the presence of monounsaturated fatty acids (MUFAs). The most frequently obtained unsaturated fatty acids are stearic (18:0), oleic (18:1), palmitic (16:0), and linoleic (18:2) acids. The fatty acid composition prophesied several principal characteristics of the SCO produced from microorganisms, i.e., viscosity, high heating value (HHV), cetane number (CN), oxidation stability (OS), iodine value (IV), and cold filter plugging point (CFPP). CN, HHV, and OS tend to enhance with the length of the chain and diminish with the unsaturation degree. A higher degree of unsaturation decreases the viscosity and resists frost (Radha et al., 2020a,b). CN can be correlated to fuel ignition and combustion quality and thus used as a biodiesel parameter. CN with a higher value (above 50) ensures better cold start properties and diminishes white smoke formation. The number and position of the double bonds in FAMEs influences the IV, which in turn affects the stability of the biodiesel. The higher the number of double bonds, the lower the stability of the fuel. CFPP is one of the critical parameters, which is used to predict the biodiesel properties at lower temperatures (Arous et al., 2019). The properties of the SCO-derived biodiesels complied with the standards stated by the American Society for Testing Materials (ASTM) and European Nations 14214 (EN 14214), as shown in Table 22.3. Several SCO-derived biodiesels from oleaginous microorganisms depicted remarkably higher CN values than those found in rapeseed oil biodiesel and palm oil biodiesel and fulfilled the ASTM and EN 14214 standards (Table 22.3). One of the most critical biodiesel parameters that affect fuel quality is density. This is of high significance for the accurate formulation of a fine blend of raw materials, which can further help in optimizing the cost of biodiesel production.

TABLE 22.3 Predicted properties for biodiesel from various sources.

Biodiesel source	Substrate	Density (g/cm^3)	Cetane number	Iodine value ($gI_2/100\ g$)	CFPP (°C)	References
Single cell oils						
Candida viswanathii Y-E4	Crude glycerol	0.86	47	120	−0.5	Guerfali et al. (2020)
Rhodotorula glutinis TISTR-5159	Palm oil mill effluent	NA	61.11	58.70	19.69	Saenge et al. (2011)
Aspergillus oryzae	Potato processing wastewater	NA	59.00	68.36	28.32	Muniraj et al. (2013)
Aspergillus mucilaginosa	Sugarcane bagasse hydrolysate	0.88	60.76	53.13	−5.93	Bandhu et al. (2019)
Pichia kudriavzevii NCIM 3653	Paper mill sludge extract	0.86	57.80	7.59	−12.34	Prabhu et al. (2019)
Yarrowia lipolytica MTCC 9520	Chicken tallow	0.86	60.68	54.89	10.84	Radha et al., 2020b
Yarrowia lipolytica MTCC 9520	Goat tallow	0.86	64.2	29.9	30.6	Radha et al. (2020)
Neochloris oleoabundans	Anaerobic effluents from pig waste	NA	52.12	90.37	0.35	Olguin et al. (2015)
Rhodosporidium toruloides Y2	Bioethanol wastewater	NA	54.92	79.51	17.42	Zhou et al. (2013)
Trichosporon cutaneum	Ethanol fermentation wastewater	NA	58.87	62.14	15.71	Wang et al. (2017)
Plant oils						
Rapeseed oil	NA	NA	54.33	NA	−18.93	Leiva-candia et al. (2015)
Palm oil	NA	NA	59.14	57.17	10.00	Ramos et al. (2009)
Sunflower oil	NA	NA	51.04	NA	−8.98	Leiva-candia et al. (2015)
Standards						
American society for testing material (ASTM)		NA	≥47	NA	NA	
European Nations 14214 (EN14214)		0.86–0.9	≥51	≤120	NA	

NA, not available.

This allows the fuel to meet the needed fuel quality standards. The density of microbial-derived biodiesels listed in Table 22.3 was within the range of EN 14214 and ASTM standards. These samples were observed to achieve an adequate degree of unsaturation since the estimated IV satisfied the limits imposed by EN 14214 standards. Comparing the traditional biodiesel concerning CFPP, the predicted models for microbial SCO biodiesel exhibited substantial differences (Table 22.3). Hence, the SCO-based biodiesels obtained from some of the oleaginous microorganisms could be used as a fuel blend for diesel engines.

22.8 Opportunities and challenges in employing single cell oil as a feedstock for biodiesel production

At present, the petroleum industries are encountering an exceptional need for environmentally sustainable fuel resources. Moreover, with an escalation in worldwide energy demand, the availability of petroleum-based resources has dropped significantly (Guerfali et al., 2020; Mohan et al., 2016). Biofuels such as biodiesel and bioethanol are commonly endorsed as a cost-effective and environmentally benign substitute to petroleum and other fossil fuels. However, using plant-based oils such as rapeseed oil, palm oil, soybean oil, and so on, as a substrate for biodiesel production is not economical, leading to exorbitant price hikes, food scarcity, and instability. Contemplating the growing demands and international regulations, prodigious attempts have been undertaken to extrapolate a novel biochemical pathway for producing fatty acids and biodiesel. Encouraging waste materials as feedstock for biodiesel production are evolving, viable, and ecofriendly resources (Kumar et al., 2020).

Three major challenges must be overcome for a lower value, higher volume biodiesel, and coproducts. One of the major challenges in the synthesis of SCO-derived biodiesel is the carbon source, as it adds up to 70% of the overall production cost. Therefore, the use of inexpensive carbon substrates from industrial, agricultural, and municipal wastes, including crude glycerol from biodiesel industry, cheese whey from the dairy industry, sludge from the wastewater treatment plant, hydrolyzed plant biomass, and molasses, would moderate the involved raw material expenses. Secondly, to develop a production strain with an optimal production rate, titer, and yield, metabolic engineering research must be performed in a much faster and cheaper way possible. Further analysis is required to understand the physiology, biochemistry, and genetics of wild types and engineered strains. The strain construction process would be accelerated by enhanced analytical modeling, metabolic flux analysis, and computational and programmed gene cloning. To reduce the time and expense of metabolic engineering, emerging technologies such as site-specific genetic modification techniques and sensor-based system biology methods would be crucial (Zhu et al., 2015). Lastly, the capital investment for manufacturing assets must be reduced. Industrial applications for biodiesel production using oleaginous microorganisms will be needed for advanced technology development in fermentation engineering and downstream processing. However, it is essential to optimize process parameters such as C/N ratio, pH, temperature, cultivation mode, and DO for improved SCO productivity. Research should therefore be aimed at isolating potential oleaginous microorganisms that may be able to use complex substrates. Low-cost, eco-sustainable solvents for SCO separation and conversion to biodiesel need to be explored to enhance downstream production efficiency (Kumar et al., 2019). Recently, blended feedstock through the biorefinery model is divulged as a promising alternative to the conventional process for biodiesel production. Nevertheless, these strategies are not reconnoitered profusely and strive for intense research and development before exploiting these methodologies in commercial processes.

22.9 Conclusion

The use of different waste substrates by oleaginous microbes to produce SCO has remarkable competence for industrial application owed to its ability to produce renewable oil for the production of biodiesel. The prophesied properties of biodiesel concerning fatty acid methyl esters composition of the SCOs synthesized by oleaginous microorganisms cultivated on various waste substrates were within the range defined by worldwide standard requirements of biodiesel, which may advocate that wastes-based medium could be used without sacrificing the quality of SCOs. Future research should emphasize improving the cost efficiency, accompanied by the attempt to scale up at the industrial level. In summary, processes with feedstock flexibility and biodiesel production from waste sources are outlined as the basis of sustainable biorefinery in the near future.

References

Angerbauer, C., Siebenhofer, M., Mittelbach, M., Guebitz, G.M., 2008. Conversion of sewage sludge into lipids by *Lipomyces starkeyi* for biodiesel production. Bioresour. Technol. 99 (8), 3051–3056.

Anschau, A., Xavier, M.C.A., Hernalsteens, S., Franco, T.T., 2014. Effect of feeding strategies on lipid production by *Lipomyces starkeyi*. Bioresour. Technol. 157, 214–222.

Antonopoulou, I., Spanopoulos, A., Matsakas, L., 2020. Single cell oil and ethanol production by the oleaginous yeast *Trichosporon fermentans* utilizing dried sweet sorghum stalks. Renew. Energy 146, 1609–1617.

Arous, F., Frikha, F., Triantaphyllidou, I.-E., Aggelis, G., Nasri, M., Mechichi, T., 2016. Potential utilization of agro-industrial wastewaters for lipid production by the oleaginous yeast *Debaryomyces etchellsii*. J. Clean. Prod. 133, 899–909.

Arous, F., Ben Atitallah, I., Nasri, M., Mechichi, T., 2017. A sustainable use of low-cost raw substrates for biodiesel production by the oleaginous yeast *Wickerhamomyces anomalus*. 3 Biotech 7, 268–277.

Arous, F., Jaouani, A., Mechichi, T., 2019. Oleaginous microorganisms for simultaneous biodiesel production and wastewater treatment: a review. In: Microbial Wastewater Treatment. Elsevier, pp. 153–174.

Asadi, S.Z., Khosravi-Darani, K., Nikoopour, H., Bakhoda, H., 2015. Evaluation of the effect of process variables on the fatty acid profile of single cell oil produced by *Mortierella* using solid-state fermentation. Crit. Rev. Biotechnol. 35, 94–102.

Ashokkumar, V., Agila, E., Sivakumar, P., Salam, Z., Rengasamy, R., Ani, F.N., 2014. Optimization and characterization of biodiesel production from microalgae *Botryococcus* grown at semi-continuous system. Energy Convers. Manage. 88, 936–946.

Bagchi, S.K., Patnaik, R., Sonkar, S., Koley, S., Rao, P.S., Mallick, N., 2019. Qualitative biodiesel production from a locally isolated chlorophycean microalga *Scenedesmus obliquus* (Turpin) Kützing GA 45 under closed raceway pond cultivation. Renew. Energy 139, 976–987.

Bandhu, S., Bansal, N., Dasgupta, D., Junghare, V., Sidana, A., Kalyan, G., Hazra, S., Ghosh, D., 2019. Overproduction of single cell oil from xylose rich sugarcane bagasse hydrolysate by an engineered oleaginous yeast *Rhodotorula mucilaginosa* IIPL32. Fuel 254, 115653.

Bautista, L.F., Vicente, G., Garre, V., 2012. Biodiesel from microbial oil. In: Advances in Biodiesel Production. Woodhead Publishing, pp. 179–203.

Béligon, V., Poughon, L., Christophe, G., Lebert, A., Larroche, C., Fontanille, P., 2016. Validation of a predictive model for fed-batch and continuous lipids production processes from acetic acid using the oleaginous yeast *Cryptococcus curvatus*. Biochem. Eng. J. 111, 117–128.

Bellou, S., Makri, A., Sarris, D., Michos, K., Rentoumi, P., Celik, A., Papanikolaou, S., Aggelis, G., 2014. The olive mill wastewater as substrate for single cell oil production by Zygomycetes. J. Biotechnol. 170, 50–59.

Bento, H.B., Carvalho, A.K., Reis, C.E., De Castro, H.F., 2020. Single cell oil production and modification for fuel and food applications: assessing the potential of sugarcane molasses as culture medium for filamentous fungus. Ind. Crop. Prod. 145, 112141.

Beopoulos, A., Nicaud, J.M., 2012. Yeast: a new oil producer? Oléagineux Corps gras Lipides 19 (1), 22–28.

Beopoulos, A., Chardot, T., Nicaud, J.-M., 2009. *Yarrowia lipolytica*: a model and a tool to understand the mechanisms implicated in lipid accumulation. Biochimie 91, 692–696.

Bhatia, S.K., Bhatia, R.K., Jeon, J.M., Kumar, G., Yang, Y.H., 2019. Carbon dioxide capture and bioenergy production using biological system—a review. Renew. Sustain. Energy Rev. 110, 143–158.

Carota, E., Crognale, S., D'Annibale, A., Gallo, A.M., Stazi, S.R., Petruccioli, M., 2017. A sustainable use of Ricotta Cheese Whey for microbial biodiesel production. Sci. Total. Environ. 584, 554–560.

Carsanba, E., Papanikolaou, S., Erten, H., 2018. Production of oils and fats by oleaginous microorganisms with an emphasis given to the potential of the nonconventional yeast *Yarrowia lipolytica*. Crit. Rev. Biotechnol. 38, 1230–1243.

Chaturvedi, S., Tiwari, R., Bhattacharya, A., Nain, L., Khare, S.K., 2019. Production of single cell oil by using cassava peel substrate from oleaginous yeast *Rhodotorula glutinis*. Biocatal. Agr. Biotechnol. 21, 101308.

Chebbi, H., Leiva-Candia, D., Carmona-Cabello, M., Jaouani, A., Dorado, M.P., 2019. Biodiesel production from microbial oil provided by oleaginous yeasts from olive oil mill wastewater growing on industrial glycerol. Ind. Crop. Prod. 139, 111535.

Chen, X.F., Huang, C., Xiong, L., Ma, L.L., 2012. Microbial oil production from corncob acid hydrolysate by *Trichosporon cutaneum*. Biotechnol. Lett. 34 (6), 1025–1028.

Chen, X., Wu, J., Song, W., Zhang, L., Wang, H., Liu, L., 2015. Fumaric acid production by *Torulopsis glabrata*: engineering the urea cycle and the purine nucleotide cycle. Biotechnol. Bioeng. 112 (1), 156–167.

Chen, J., Zhang, X., Tyagi, R.D., Drogui, P., 2018. Utilization of methanol in crude glycerol to assist lipid production in non-sterilized fermentation from *Trichosporon oleaginosus*. Bioresour. Technol. 253, 8–15.

Chi, Z., Liu, G.L., Liu, C.G., Chi, Z.M., 2016. Poly (β-l-malic acid) (PMLA) from *Aureobasidium* spp. and its current proceedings. Appl. Microbiol. Biotechnol. 100 (9), 3841–3851.

Conti, E., Stredansky, M., Stredanska, S., Zanetti, F., 2001. γ-Linolenic acid production by solid-state fermentation of *Mucorales* strains on cereals. Bioresour. Technol. 76, 283–286.

Couto, S.R., Sanromán, M.Á., 2006. Application of solid-state fermentation to food industry—a review. J. Food Eng. 76, 291–302.

Deeba, F., Pruthi, V., Negi, Y.S., 2017. Fostering triacylglycerol accumulation in novel oleaginous yeast *Cryptococcus psychrotolerans* IITRFD utilizing groundnut shell for improved biodiesel production. Bioresour. Technol. 242, 113–120.

Dominguez, A., Costas, M., Longo, M.A., Sanromán, A., 2003. A novel application of solid state culture: production of lipases by *Yarrowia lipolytica*. Biotechnol. Lett. 25 (15), 1225–1229.

Donot, F., Fontana, A., Baccou, J.C., Strub, C., Schorr-Galindo, S., 2014. Single cell oils (SCOs) from oleaginous yeasts and moulds: production and genetics. Biomass Bioenergy 68, 135–150.

Fakas, S., Papanikolaou, S., Batsos, A., Galiotou-Panayotou, M., Mallouchos, A., Aggelis, G., 2009. Evaluating renewable carbon sources as substrates for single cell oil production by *Cunninghamella echinulata* and *Mortierella isabellina*. Biomass Bioenergy 33, 573–580.

Fei, Q., Chang, H.N., Shang, L., 2011. Exploring low-cost carbon sources for microbial lipids production by fed-batch cultivation of *Cryptococcus albidus*. Biotechnol. Bioproc. E 16 (3), 482–487.

Feng, Y., Li, C., Zhang, D., 2011. Lipid production of *Chlorella vulgaris* cultured in artificial wastewater medium. Bioresour. Technol. 102 (1), 101–105.

Feng, Q., Chen, M., Wang, W., Chang, S., Zhang, L., Li, J., Li, S., 2016. Study on the harvest of oleaginous microalgae Chlorella sp. by photosynthetic hydrogen mediated autoflotation for biodiesel production. Int. J. Hydrogen Energy 41 (38), 16772–16777.

Fickers, P., Nicaud, J.M., Gaillardin, C., Destain, J., Thonart, P., 2004. Carbon and nitrogen sources modulate lipase production in the yeast *Yarrowia lipolytica*. J. Appl. Microbiol. 96 (4), 742–749.

Fickers, P., Benetti, P.H., Waché, Y., Marty, A., Mauersberger, S., Smit, M.S., Nicaud, J.M., 2005a. Hydrophobic substrate utilisation by the yeast *Yarrowia lipolytica*, and its potential applications. FEMS Yeast Res. 5 (6–7), 527–543.

Fickers, P., Fudalej, F., Nicaud, J.M., Destain, J., Thonart, P., 2005b. Selection of new over-producing derivatives for the improvement of extracellular lipase production by the non-conventional yeast *Yarrowia lipolytica*. J. Biotechnol. 115 (4), 379–386.

Gao, Q., Cui, Z., Zhang, J., Bao, J., 2014. Lipid fermentation of corncob residues hydrolysate by oleaginous yeast *Trichosporon cutaneum*. Bioresour. Technol. 152, 552–556.

Garlapati, V.K., Shankar, U., Budhiraja, A., 2016. Bioconversion technologies of crude glycerol to value added industrial products. Biotechnol. Rep. 9, 9–14.

Gema, H., Kavadia, A., Dimou, D., Tsagou, V., Komaitis, M., Aggelis, G., 2002. Production of gamma-linolenic acid by *Cunninghamella echinulata* cultivated on glucose and orange peel. Appl. Microbiol. Biotechnol. 58, 303–307.

Gen, Q., Wang, Q., Chi, Z.M., 2014. Direct conversion of cassava starch into single cell oil by co-cultures of the oleaginous yeast *Rhodosporidium toruloides* and immobilized amylases-producing yeast *Saccharomycopsis fibuligera*. Renew. Energy 62, 522–526.

Ghanavati, H., Nahvi, I., Roghanian, R., 2014. Monitoring growth and lipid production of new isolated oleaginous yeast *Cryptococcus aerius* UIMC65 on glucose and xylose cultures. Biotechnol. Bioprocess. Eng. 19 (3), 468–477.

Gong, Z., Shen, H., Zhou, W., Wang, Y., Yang, X., Zhao, Z.K., 2015. Efficient conversion of acetate into lipids by the oleaginous yeast *Cryptococcus curvatus*. Biotechnol. Biofuels. 8 (1), 1–9.

Guerfali, M., Ayadi, I., Sassi, H.E., Belhassen, A., Gargouri, A., Belghith, H., 2020. Biodiesel-derived crude glycerol as alternative feedstock for single cell oil production by the oleaginous yeast *Candida viswanathii* Y-E4. Ind. Crop. Prod. 145, 112103.

Han, M., Xu, Z.Y., Du, C., Qian, H., Zhang, W.G., 2016. Effects of nitrogen on the lipid and carotenoid accumulation of oleaginous yeast *Sporidiobolus pararoseus*. Bioprocess Biosyst. Eng. 39 (9), 1425–1433.

Huang, L., Zhang, B., Gao, B., Sun, G., 2011. Application of fishmeal wastewater as a potential low-cost medium for lipid production by *Lipomyces starkeyi* HL. Environ. Technol. 32, 1975–1981.

Huang, C., Chen, X.F., Xiong, L., Yang, X.Y., Ma, L.L., Chen, Y., 2013. Microbial oil production from corncob acid hydrolysate by oleaginous yeast *Trichosporon coremiiforme*. Biomass Bioenerg. 49, 273–278.

Huang, X.F., Liu, J.N., Lu, L.J., Peng, K.M., Yang, G.X., Liu, J., 2016. Culture strategies for lipid production using acetic acid as sole carbon source by *Rhodosporidium toruloides*. Bioresour. Technol. 206, 141–149.

Huang, X.F., Shen, Y., Luo, H.J., Liu, J.N., Liu, J., 2018. Enhancement of extracellular lipid production by oleaginous yeast through preculture and sequencing batch culture strategy with acetic acid. Bioresour. Technol. 247, 395–401.

Ismail, M., Al-Zuhair, S., 2020. Thermo-responsive switchable solvents for simultaneous microalgae cell disruption, oil extraction-reaction, and product separation for biodiesel production. Biocatal. Agr. Biotechnol. 26, 101667.

Jacobs, A., Botha, A., Van Zyl, W.H., 2010. Sunflower press cake as a substrate for eicosapentaenoic acid production by representatives of the genus *Mortierella*. Bioresources 5 (2), 1232–1243.

Jin, J., Dupré, C., Yoneda, K., Watanabe, M.M., Legrand, J., Grizeau, D., 2016. Characteristics of extracellular hydrocarbon-rich microalga *Botryococcus braunii* for biofuels production: recent advances and opportunities. Process. Biochem. 51 (11), 1866–1875.

Juanssilfero, A.B., Kahar, P., Amza, R.L., Miyamoto, N., Otsuka, H., Matsumoto, H., Kihira, C., Thontowi, A., Ogino, C., Prasetya, B., Kondo, A., 2018. Effect of inoculum size on single-cell oil production from glucose and xylose using oleaginous yeast *Lipomyces starkeyi*. J. Biosci. Bioeng. 125 (6), 695–702.

Juwarkar, A.A., Nair, A., Dubey, K.V., Singh, S.K., Devotta, S., 2007. Biosurfactant technology for remediation of cadmium and lead contaminated soils. Chemosphere 68 (10), 1996–2002.

Kamyab, H., Chelliapan, S., Shahbazian-Yassar, R., Din, M.F.M., Khademi, T., Kumar, A., Rezania, S., 2017. Evaluation of lipid content in microalgae biomass using palm oil mill effluent (Pome). JOM 69 (8), 1361–1367.

Kialashaki, M., Mahdavi, M.A., Gheshlaghi, R., 2019. Improved transesterification conditions for production of clean fuel from municipal wastewater microalgae feedstock. J. Clean. Prod. 241, 118388.

Kim, I., Seo, Y.H., Kim, G.Y., Han, J.I., 2015. Co-production of bioethanol and biodiesel from corn stover pretreated with nitric acid. Fuel 143, 285–289.

Kiran, E.U., Trzcinski, A., Webb, C., 2013. Microbial oil produced from biodiesel by-products could enhance overall production. Bioresour. Technol. 129, 650–654.

Kobayashi, N., Barnes, A., Jensen, T., Noel, E., Andlay, G., Rosenberg, J.N., Betenbaugh, M.J., Guarnieri, M.T., Oyler, G.A., 2015. Comparison of biomass and lipid production under ambient carbon dioxide vigorous aeration and 3% carbon dioxide condition among the lead candidate *Chlorella* strains screened by various photobioreactor scales. Bioresour. Technol. 198, 246–255.

Kopsahelis, N., Dimou, C., Papadaki, A., Xenopoulos, E., Kyraleou, M., Kallithraka, S., Kotseridis, Y., Papanikolaou, S., Koutinas, A.A., 2018. Refining of wine lees and cheese whey for the production of microbial oil, polyphenol-rich extracts and value-added co-products. J. Chem. Technol. Biotechnol. 93 (1), 257–268.

Kumar, L.R., Kaur, R., Yellapu, S.K., Zhang, X., Tyagi, R.D., 2019. Biodiesel production from oleaginous microorganisms with wastes as raw materials. In: Biofuels: Alternative Feedstocks and Conversion Processes for the Production of Liquid and Gaseous Biofuels. Academic Press, pp. 661–690.

Kumar, M., Rathour, R., Gupta, J., Pandey, A., Gnansounou, E., Thakur, I.S., 2020. Bacterial production of fatty acid and biodiesel: opportunity and challenges. In: Refining Biomass Residues for Sustainable Energy and Bioproducts. Academic Press, pp. 21–49.

Lam, M.K., Jamalluddin, N.A., Lee, K.T., 2019. Production of biodiesel using palm oil. In: Biofuels: Alternative Feedstocks and Conversion Processes for the Production of Liquid and Gaseous Biofuels. Academic Press, pp. 539–574.

Lam, M.K., Lee, K.T., 2011. Mixed methanol–ethanol technology to produce greener biodiesel from waste cooking oil: A breakthrough for $SO_4^{2-}/SnO_2–SiO_2$ catalyst. Fuel Process. Technol. 92 (8), 1639–1645.

Lazar, Z., Dulermo, T., Neuvéglise, C., Crutz-Le Coq, A.M., Nicaud, J.M., 2014. Hexokinase – a limiting factor in lipid production from fructose in *Yarrowia lipolytica*. Metab. Eng. 26, 89–99.

Leiva-Candia, D.E., Tsakona, S., Kopsahelis, N., García, I.L., Papanikolaou, S., Dorado, M.P., Koutinas, A.A., 2015. Biorefining of by-product streams from sunflower-based biodiesel production plants for integrated synthesis of microbial oil and value-added co-products. Bioresour. Technol. 190, 57–65.

Liang, Y., Sarkany, N., Cui, Y., Blackburn, J.W., 2010. Batch stage study of lipid production from crude glycerol derived from yellow grease or animal fats through microalgal fermentation. Bioresour. Technol. 101 (17), 6745–6750.

Lin, H., Cheng, W., Ding, H., Chen, X., Zhou, Q., Zhao, Y., 2010. Direct microbial conversion of wheat straw into lipid by a cellulolytic fungus of *Aspergillus oryzae* A-4 in solid-state fermentation. Bioresour. Technol 101, 7556–7562.

Ling, X., Guo, J., Liu, X., Zhang, X., Wang, N., Lu, Y., et al., 2015. Impact of carbon and nitrogen feeding strategy on high production of biomass and docosahexaenoic acid (DHA) by *Schizochytrium* sp. LU310. Bioresour. Technol. 184, 139–147.

Liu, J.F., Mbadinga, S.M., Yang, S.Z., Gu, J.D., Mu, B.Z., 2015. Chemical structure, property and potential applications of biosurfactants produced by *Bacillus subtilis* in petroleum recovery and spill mitigation. Int. J. Mol. Sci. 16 (3), 4814–4837.

Lopes, M., Gomes, A.S., Silva, C.M., Belo, I., 2018. Microbial lipids and added value metabolites production by *Yarrowia lipolytica* from pork lard. J. Biotechnol. 265, 76–85.

Louhasakul, Y., Cheirsilp, B., Prasertsan, P., 2016. Valorization of palm oil mill effluent into lipid and cell-bound lipase by marine yeast *Yarrowia lipolytica* and their application in biodiesel production. Waste. Biomass. Valori. 7 (3), 417–426.

Louhasakul, Y., Cheirsilp, B., Intasit, R., Maneerat, S., Saimmai, A., 2020. Enhanced valorization of industrial wastes for biodiesel feedstocks and biocatalyst by lipolytic oleaginous yeast and biosurfactant-producing bacteria. Int. Biodeteri. Biodegr. 148, 104911.

Makkar, R.S., Rockne, K.J., 2003. Comparison of synthetic surfactants and biosurfactants in enhancing biodegradation of polycyclic aromatic hydrocarbons. Environ. Toxicol. Chem. Int. J. 22 (10), 2280–2292.

Manowattana, A., Techapun, C., Watanabe, M., Chaiyaso, T., 2018. Bioconversion of biodiesel-derived crude glycerol into lipids and carotenoids by an oleaginous red yeast *Sporidiobolus pararoseus* KM281507 in an airlift bioreactor. J. Biosci. Bioeng. 125 (1), 59–66.

Marjakangas, J.M., Lakaniemi, A.M., Koskinen, P.E., Chang, J.S., Puhakka, J.A., 2015. Lipid production by eukaryotic microorganisms isolated from palm oil mill effluent. Biochem. Eng. J. 99, 48–54.

Mlickova, K., Roux, E., Athenstaedt, K., et al., 2004. Lipid accumulation, lipid body formation, and acyl coenzyme A oxidases of the yeast *Yarrowia lipolytica*. Appl. Environ. Microb. 70, 3918–3924.

Moazeni, F., Chen, Y.C., Zhang, G., 2016. Enzymatic transesterification for biodiesel production from used cooking oil, a review. J. Clean. Prod. 216, 117–128.

Moazeni, F., Chen, Y.C., Zhang, G., 2019. Enzymatic transesterification for biodiesel production from used cooking oil. J. Clean. Prod. 216, 117–128.

Mohan, S.V., Nikhil, G.N., Chiranjeevi, P., Reddy, C.N., Rohit, M.V., Kumar, A.N., Sarkar, O., 2016. Waste biorefinery models towards sustainable circular bioeconomy: critical review and future perspectives. Bioresour. Technol. 215, 2–12.

Moreno, A.D., Tomás-Pejó, E., Ballesteros, M., Negro, M.J., 2019. Pretreatment technologies for lignocellulosic biomass deconstruction within a biorefinery perspective. In: Biofuels: Alternative Feedstocks and Conversion Processes for the Production of Liquid and Gaseous Biofuels. Academic Press, pp. 379–399.

Muniraj, I.K., Xiao, L., Hu, Z., Zhan, X., Shi, J., 2013. Microbial lipid production from potato processing wastewater using oleaginous filamentous fungi *Aspergillus oryzae*. Water. Res. 47 (10), 3477–3483.

Muniraj, I.K., Uthandi, S.K., Hu, Z., Xiao, L., Zhan, X., 2015. Microbial lipid production from renewable and waste materials for second-generation biodiesel feedstock. Environ. Technol. Rev. 4 (1), 1–16.

Nguyen, V.P., Nguyen, H.H.M., Nguyen, D.T., Nguyen, H.L., Huynh, T.M., 2018. Optimization of biodiesel production from waste cooking oil using static mixer technology in Vietnam. Biofuels 9 (5), 567–574.

Nicaud, J.M., Madzak, C., van den Broek, P., Gysler, C., Duboc, P., Niederberger, P., Gaillardin, C., 2002. Protein expression and secretion in the yeast *Yarrowia lipolytica*. FEMS Yeast Res. 2 (3), 371–379.

Ochsenreither, K., Glück, C., Stressler, T., Fischer, L., Syldatk, C., 2016. Production strategies and applications of microbial single cell oils. Front. Microbiol. 7, 1539.

Olguín, E.J., Castillo, O.S., Mendoza, A., Tapia, K., González-Portela, R.E., Hernández-Landa, V.J., 2015. Dual purpose system that treats anaerobic effluents from pig waste and produce *Neochloris oleoabundans* as lipid rich biomass. New Biotechnol. 32 (3), 387–395.

Papanikolaou, S., Aggelis, G., 2011. Lipids of oleaginous yeasts. Part I: biochemistry of single cell oil production. Eur. J. Lipid Sci. Technol. 113, 1031–1051.

Papanikolaou, S., Diamantopoulou, P., Chatzifragkou, A., Philippoussis, A., Aggelis, G., 2010. Suitability of low-cost sugars as substrates for lipid production by the fungus *Thamnidium elegans*. Energy Fuel. 24 (7), 4078–4086.

Peng, X., Chen, H., 2008. Single cell oil production in solid-state fermentation by *Microsphaeropsis sp.* from steam-exploded wheat straw mixed with wheat bran. Bioresour. Technol. 99, 3885–3889.

Perez-Garcia, O., Escalante, F.,M., De-Bashan, L.,E., Bashan, Y., 2011. Heterotrophic cultures of microalgae: metabolism and potential products. Water Res. 45 (1), 11–36.

Polburee, P., Yongmanitchai, W., Lertwattanasakul, N., Ohashi, T., Fujiyama, K., Limtong, S., 2015. Characterization of oleaginous yeasts accumulating high levels of lipid when cultivated in glycerol and their potential for lipid production from biodiesel-derived crude glycerol. Fungal Biol. 119, 1194–1204.

Prabhu, K., Jayakumar, A., Sreelakshmi, K.P., Raha, A., Maitra, M., Radha, P., 2019. Utilization of microbial oil produced from *Pichia kudriavzevii* NCIM 3653 using paper mill sludge as an alternative substrate for biodiesel synthesis. Biofuels 1–8.

Probst, K.V., Vadlani, P.V., 2017. Single cell oil production by *Lipomyces starkeyi*: biphasic fed-batch fermentation strategy providing glucose for growth and xylose for oil production. Biochem. Eng. J. 121, 49–58.

Qi, P., Chen, S., Chen, J., Zheng, J., Zheng, X., Yuan, Y., 2015. Catalysis and reactivation of ordered mesoporous carbon-supported gold nanoparticles for the base-free oxidation of glucose to gluconic acid. ACS Catalysis 5 (4), 2659–2670.

Qian, X., Gorte, O., Chen, L., Zhang, W., Dong, W., Ma, J., Jiang, M., Xin, F., Ochsenreither, K., 2019. Co-production of single cell oil and gluconic acid using oleaginous *Cryptococcus podzolicus* DSM 27192. Biotechnol. Biofuels 12 (1), 1–9.

Radha, P., Narayanan, S., Chaudhuri, A., Anjum, S., Thomas, D.L., Pandey, R., Ramani, K., 2020a. Synthesis of single-cell oil by *Yarrowia lipolytica* MTCC 9520 utilizing slaughterhouse lipid waste for biodiesel production. Biomass Convers. Biorefin. 1–12.

Radha, P., Prabhu, K., Jayakumar, A., AbilashKarthik, S., Ramani, K., 2020b. Biochemical and kinetic evaluation of lipase and biosurfactant assisted ex novo synthesis of microbial oil for biodiesel production by *Yarrowia lipolytica* utilizing chicken tallow. Process Biochem. 95, 17–29.

Ramachandran, S., Fontanille, P., Pandey, A., Larroche, C., 2006. Gluconic acid: properties, applications and microbial production. Food Technol. Biotechnol. 44 (2).

Ramos, M.J., Fernández, C.M., Casas, A., Rodríguez, L., Pérez, Á., 2009. Influence of fatty acid composition of raw materials on biodiesel properties. Bioresour. Technol. 100 (1), 261–268.

Ratledge, C., 2004. Fatty acid biosynthesis in microorganisms being used for single cell oil production. Biochimie 86 (11), 807–815.

Raut, G., Kamat, S., RaviKumar, A., 2019. Trends in production and fuel properties of biodiesel from heterotrophic microbes. In: Advances in Biological Science Research. Academic Press, pp. 247–273.

Romano, A.H., Bright, M.M., Scott, W.E., 1967. Mechanism of fumaric acid accumulation in *Rhizopus nigricans*. J. Bacteriol. 93 (2), 600–604.

Rufino, R.D., Moura de Luna, J., de Campos Takaki, G.M., Asfora Sarubbo, L., 2014. Characterization and properties of the biosurfactant produced by *Candida lipolytica* UCP 0988. Electron. J. Biotechnol. 17 (1), 6–6.

Saenge, C., Cheirsilp, B., Suksaroge, T.T., Bourtoom, T., 2011. Efficient concomitant production of lipids and carotenoids by oleaginous red yeast *Rhodotorula glutinis* cultured in palm oil mill effluent and application of lipids for biodiesel production. Biotechnol. Bioproc. E. 16 (1), 23–33.

Sagia, S., Sharma, A., Singh, S., Chaturvedi, S., Nain, P.K.S., Nain, L., 2020. Single cell oil production by a novel yeast *Trichosporon mycotoxinivorans* for complete and eco-friendly valorization of paddy straw. Electron. J. Biotechn. 44, 60–68.

Saisriyoot, M., Thanapimmetha, A., Suwaleerat, T., Chisti, Y., Srinophakun, P., 2019. Biomass and lipid production by *Rhodococcus opacus* PD630 in molasses-based media with and without osmotic-stress. J. Biotechnol. 297, 1–8.

Sara, M., Brar, S.K., Blais, J.F., 2016. Lipid production by *Yarrowia lipolytica* grown on biodiesel-derived crude glycerol: optimization of growth parameters and their effects on the fermentation efficiency. RSC Adv. 6 (93), 90547–90558.

Schulze, I., Hansen, S., Großhans, S., Rudszuck, T., Ochsenreither, K., Syldatk, C., Neumann, A., 2014. Characterization of newly isolated oleaginous yeasts-Cryptococcus podzolicus, *Trichosporon porosum* and Pichia *segobiensis*. AMB Express 4 (1), 1–11.

Seo, Y.H., Lee, I., Jeon, S.H., Han, J.I., 2014. Efficient conversion from cheese whey to lipid using *Cryptococcus curvatus*. Biochem. Eng. J. 90, 149–153.

Singh, G., Sinha, S., Kumar, K.K., Gaur, N.A., Bandyopadhyay, K.K., Paul, D., 2020. High density cultivation of oleaginous yeast isolates in 'mandi' waste for enhanced lipid production using sugarcane molasses as feed. Fuel 276, 118073.

Souza, K.S.T., Schwan, R.F., Dias, D.R., 2014. Lipid and citric acid production by wild yeasts grown in glycerol. J. Microbiol. Biotechnol. 24 (4), 497–506.

Spier, F., Buffon, J.G., Burkert, C.A.V., 2015. Bioconversion of raw glycerol generated from the synthesis of biodiesel by different oleaginous yeasts: lipid content and fatty acid profile of biomass. Indian. J. Microbiol. 55 (4), 415–422.

Stressler, T., Eisele, T., Rost, J., Haunschild, E.M., Kuhn, A., Fischer, L., 2013. Production of polyunsaturated fatty acids by *Mortierella alpina* using submerse and solid state fermentation. Chem. Ingenieur. Tech. 85, 318–322.

Szczesna-Antczak, M., Antczak, T., Piotrowicz-Wasiak, M., Rzyska, M., Binkowska, N., Bielecki, S., 2006. Relationships between lipases and lipids in mycelia of two Mucor strains. Enzyme Microbial Technol. 39 (6), 1214–1222.

Taher, H., Giwa, A., Abusabiekeh, H., Al-Zuhair, S., 2020. Biodiesel production from *Nannochloropsis gaditana* using supercritical CO_2 for lipid extraction and immobilized lipase transesterification: economic and environmental impact assessments. Fuel. Process Technol. 198, 106249.

Tang, H., Chen, M., Simon Ng, K.Y., Salley, S.O., 2012. Continuous microalgae cultivation in a photobioreactor. Biotechnol. Bioeng. 109 (10), 2468–2474.

Taskin, M., Saghafian, A., Aydogan, M.N., Arslan, N.P., 2015. Microbial lipid production by cold-adapted oleaginous yeast *Yarrowia lipolytica* B9 in non-sterile whey medium. Biofuel. Bioprod. Biorefin. 9 (5), 595–605.

Tsakona, S., Kopsahelis, N., Chatzifragkou, A., Papanikolaou, S., Kookos, I.K., Koutinas, A.A., 2014. Formulation of fermentation media from flour-rich waste streams for microbial lipid production by *Lipomyces starkeyi*. J. Biotechnol. 189, 36–45.

Tsakona, S., Skiadaresis, A.G., Kopsahelis, N., Chatzifragkou, A., Papanikolaou, S., Kookos, I.K., Koutinas, A.A., 2016. Valorisation of side streams from wheat milling and confectionery industries for consolidated production and extraction of microbial lipids. Food Chem. 198, 85–92.

Tsakona, S., Papadaki, A., Kopsahelis, N., Kachrimanidou, V., Papanikolaou, S., Koutinas, A., 2019. Development of a circular oriented bioprocess for microbial oil production using diversified mixed confectionery side-streams. Foods 8 (8), 300.

Tsigie, Y.A., Wang, C.Y., Truong, C.T., Ju, Y.H., 2011. Lipid production from *Yarrowia lipolytica* Po1g grown in sugarcane bagasse hydrolysate. Bioresour. Technol. 102 (19), 9216–9222.

Tsouko, E., Papanikolaou, S., Koutinas, A.A., 2016. Production of fuels from microbial oil using oleaginous microorganisms. In: Handbook of Biofuels Production. Woodhead Publishing, pp. 201–236.

Unrean, P., Khajeeram, S., Champreda, V., 2017. Combining metabolic evolution and systematic fed-batch optimization for efficient single-cell oil production from sugarcane bagasse. Renew. Energy 111, 295–306.

Uprety, B.K., Dalli, S.S., Rakshit, S.K., 2017. Bioconversion of crude glycerol to microbial lipid using a robust oleaginous yeast *Rhodosporidium toruloides* ATCC 10788 capable of growing in the presence of impurities. Energy Convers. Manage. 135, 117–128.

Vieira, J.P.F., Ienczak, J.L., Costa, P.S., Rossell, C.E.V., Franco, T.T., Pradella, J.G.C., 2016. Single cell oil production integrated to a sugarcane-mill: conceptual design, process specifications and economic analysis using molasses as raw material. Ind. Crop. Prod. 89, 478–485.

Wang, J., Hu, M., Zhang, H., Bao, J., 2017. Converting chemical oxygen demand (COD) of cellulosic ethanol fermentation wastewater into microbial lipid by oleaginous yeast *Trichosporon cutaneum*. Appl. Biochem. Biotech. 182 (3), 1121–1130.

Wang, G., Bai, T., Miao, Z., Ning, W., Liang, W., 2018. Simultaneous production of single cell oil and fumaric acid by a newly isolated yeast *Aureobasidium pullulans* var. *aubasidani* DH177. Bioprocess Biosyst. Eng. 41 (11), 1707–1716.

Wild, R., Patil, S., Popović, M., Zappi, M., Dufreche, S., Bajpai, R., 2010. Lipids from *Lipomyces starkeyi*. Food. Technol. Biotech. 48 (3), 329–335.

Xu, J., Du, W., Zhao, X., Zhang, G., Liu, D., 2013. Microbial oil production from various carbon sources and its use for biodiesel preparation. Biofuel. Bioprod. Biorefin. 7 (1), 65–77.

Xu, Q., Liu, Y., Li, S., Jiang, L., Huang, H., Wen, J., 2016. Transcriptome analysis of *Rhizopus oryzae* in response to xylose during fumaric acid production. Bioprocess Biosyst. Eng. 39 (8), 1267–1280.

Yu, X., Zheng, Y., Dorgan, K.M., Chen, S., 2011. Oil production by oleaginous yeasts using the hydrolysate from pretreatment of wheat straw with dilute sulfuric acid. Bioresour. Technol. 102, 6134–6140.

Zelle, R.M., De Hulster, E., Kloezen, W., Pronk, J.T., Van Maris, A.J., 2010. Key process conditions for production of C4 dicarboxylic acids in bioreactor batch cultures of an engineered *Saccharomyces cerevisiae* strain. Appl. Environ. Microbiol. 76 (3), 744–750.

Zhan, J., Zhang, Q., Qin, M., Hong, Y., 2016. Selection and characterization of eight freshwater green algae strains for synchronous water purification and lipid production. Front. Environ. Sci. Eng. 10 (3), 548–558.

Zhang, X., Yan, S., Tyagi, R.D., Surampalli, R., Valéro, J.R., 2014. Wastewater sludge as raw material for microbial oils production. Appl. Energy 135, 192–201.

Zhang, C., Shen, H., Zhang, X., Yu, X., Wang, H., Xiao, S., Wang, J., Zhao, Z.K., 2016. Combined mutagenesis of *Rhodosporidium toruloides* for improved production of carotenoids and lipids. Biotechnol. Lett. 38 (10), 1733–1738.

Zhang, Q., Zhang, Y., Deng, T., Wei, F., Jin, J., Ma, P., 2020. Sustainable production of biodiesel over heterogeneous acid catalysts. In: Biomass, Biofuels, Biochemicals. Elsevier, pp. 407–432.

Zheng, Y., Chi, Z., Lucker, B., Chen, S., 2012. Two-stage heterotrophic and phototrophic culture strategy for algal biomass and lipid production. Bioresour. Technol. 103 (1), 484–488.

Zhou, W., Wang, W., Li, Y., Zhang, Y., 2013. Lipid production by *Rhodosporidium toruloides* Y2 in bioethanol wastewater and evaluation of biomass energetic yield. Bioresour. Technol. 127, 435–440.

Zhou, X., Jin, W., Wang, Q., Guo, S., Tu, R., Han, S.F., Chen, C., Xie, G., Qu, F., Wang, Q., 2020. Enhancement of productivity of *Chlorella pyrenoidosa* lipids for biodiesel using co-culture with ammonia-oxidizing bacteria in municipal wastewater. Renew. Energy 151, 598–603.

Zhu, Q., Jackson, E.N., 2015. Metabolic engineering of *Yarrowia lipolytica* for industrial applications. Curr. Opin. Biotechnol. 36, 65–72.

Zuccaro, G., Pirozzi, D., Yousuf, A., 2020. Lignocellulosic biomass to biodiesel. In: Lignocellulosic Biomass to Liquid Biofuels. Academic Press, pp. 127–167.

CHAPTER

23

Production of biofuel from genetically modified microalgal biomass and its effects on environment and public health

Pratyush Kumar Das[1], *Bidyut Prava Das*[2], *Patitapaban Dash*[1], *Baskar Gurunathan*[3]

[1]Centre for Biotechnology, Siksha 'O' Anusandhan (Deemed to be University), Bhubaneswar, Odisha, India; [2]Department of Botany, Sailabala Women's Autonomous College, Cuttack, Odisha, India; [3]Department of Biotechnology, St. Joseph's College of Engineering, Chennai, Tamil Nadu, India

23.1 Introduction

The growing economy, industrial development, and accompanying changes in the public lifestyle widen the gap between underlying fuel demands and available pool of fuel resource. The energy-intensive developmental programs create the possibility for exhaustion of fossil fuel in the near future. To bridge the gap between demand and supply of energy resources and to protect the environment from the combustion of conventional fuel, novel alternative sources of energy shall be added to the energy pool. The energy forms required in the present context should be ecofriendly, cost-effective, and sustainable. Non-conventional renewable energy option is preferable over non-renewable energy forms. Fuel from biomass emerged as an alternative fuel option to meet the growing energy demands. It is a promising opportunity to generate biofuel from renewable energy—rich sources such as microalgal biomass. Autotrophic microalgal populations represent diversified groups of living organisms with high energy macromolecules such as lipids, enriched in its biomass.

Rapidly growing microalgal biomass can be considered as feedstock for production of biofuel. High photosynthetic yield with energy-rich biomass of microalgae retains the features for its use as a feedstock for biofuel production. Improvement in energy yield from microalgal biomass is essential to make it competitive with other available sources of renewable energy. The genome editing of microalgae is a viable option for increasing the synthesis of biofuel or its precursors. The efficiency of accumulation of lipids from biomass is possible through genetic engineering of microalgae. It is the time to manipulate microalgae at the genetic level for

Biofuels and Bioenergy
https://doi.org/10.1016/B978-0-323-85269-2.00004-6

carbon metabolism. It is supported by the findings of high terpenoids synthesis by engineered strain of *Chlamydomonas reinhardtii* (Wichmann, 2018). Manipulation of regulator gene is a possible approach for upstream regulation of lipid production in microalgae, qualitatively or quantitatively. The enzymatic regulation of the targeted metabolic pathways in microalgae can open possibility for efficient production of feedstock for biofuel production (Miao, 2018).

The beneficial aspects of biofuel production from microalgae rest upon minimization in emission of greenhouse gases (GHGs). The current energy-demanding scenario to meet the requirements is searching for an ecofriendly, affordable, and non-conventional energy. The energy-rich lipid contents with quick growth of microalgal biomass show direction for their possible utility as a non-conventional renewable energy source. The low cost of production and high biomass yield with ability to grow on wastewater are value addition features for consideration of microalgal biomass as resource for biofuel production.

The use of genetically engineered microalgae for production of biofuel at pilot-scale condition is not the ultimatum. Success of the production biofuel at field-scale level from genetically engineered microalgae is essential for its commercialization.

23.2 Biofuel as a source to supplement global fuel demands

The ever-rising population coupled with large-scale industrial development has led to an increase in fuel demand year after year (Das et al., 2020a). The conventional resources, mostly the fossil fuels that include oil, coal, and natural gas, are the most relied-upon sources of energy to meet the global fuel needs. Oil and coal still remain the most preferred sources of energy and are widely used in several industries and other sectors as the base for any energy-driven process (IEA, 2019). Petroleum fuels are one of the important energy sources in this era of rapid industrial growth, which is quite visible from its utilization (more than 70%) in the transportation sector. The current utilization pace of these petroleum fuels is predicted to result in a short fall of these resources by 2080 (Malode et al., 2020). Fossil fuels as already known are a major contributor of global carbon dioxide concentrations in the atmosphere. Large-scale combustion of fossil fuels keeps up piling the CO_2 load in the atmosphere (Table 23.1) (IEA, 2020), thereby having serious impact on the environment. These impacts are quite evident from the catastrophic changes occurring in the atmosphere, which includes increase in global temperature, melting of polar ice caps, rise in sea level, flooding of the coastal areas, and many more.There is an immediate need to develop alternative sources of energy that are renewable in nature, easily accessible, environmentally friendly, and cost-effective so as to lessen the burden on the existing conventional fuel sources (Li et al., 2017).

TABLE 23.1 Annual changes in global energy-driven carbon dioxide emissions (1980–2020).

Year	CO_2 emission (in Gt)
1980	−0.2
1990	0.1
2000	0.7
2010	1.7
2020	−2.6

Based on: IEA (2020). Annual change in global energy-related CO_2 emissions, 1900–2020. https://www.iea.org/data-and-statistics/charts/annual-change-in-global-energy-related-co2-emissions-1900-2020.

23.3 Current status of biofuel production

Advancement in the fossil fuel production technology has reduced the extraction cost to a larger extent, thereby affecting the growth of the biofuel industry (Gul, 2016). The environmental issues related to the combustion of fossil fuels, however, cannot be ignored. Moreover, the fossil fuel reserves are mostly concentrated in few parts of the world leading to its biased allocation. This raises the risk of some countries running out of the reserves and simultaneously some others becoming wealthier and thereby controlling the foreign exchange market (Suganya et al., 2016). Production of biofuels to meet global energy demands in this context is quite important. Biofuels in comparison with other renewable sources such as wind, solar, and geothermal are more feasible economically and therefore can provide energy at cheaper rates. Advanced nations are also showing increased interest in the biofuel sector. By the year 2022, the United States has decided to replace 20% of its fuel used in road transport services with the biofuels (Alberici and Toop, 2013). The European Union (EU) to reduce the CO_2 emissions has made it mandatory for car manufacturers to control CO_2 emission levels from newly sold car maximum within 95 gm/km. This can only be achieved by using biofuels that have a very low carbon footprint as compared with the conventional fuel sources (Hombach et al., 2016).

The biofuel industry is steadily growing and thus signifies the growing knowledge of people toward environmental changes. The global mindset has also now started to realize the consequences behind reckless use of fossil fuels and the advantages posed by biofuels. This growing acceptance of biofuels is quite evident from the gradual increase in global production of biofuels during the past decade. A rise of almost 79% in the biofuel production was observed in 2019 as compared with the year 2009. A 78% increase in the biofuel consumption was also evident in the year 2019 as compared with 2009 (Fig. 23.1).

23.4 Prospects of production of biofuel from microalgal biomass

The biofuels generated from the algal sources are not only effective but also easy to obtain and can be produced on a large scale, thereby absorbing large amount of CO_2 from the atmosphere (Collet et al., 2013). Algae have an inherent ability of converting energy obtained from the sun into chemical energy, hence making it a better option as compared with the earlier

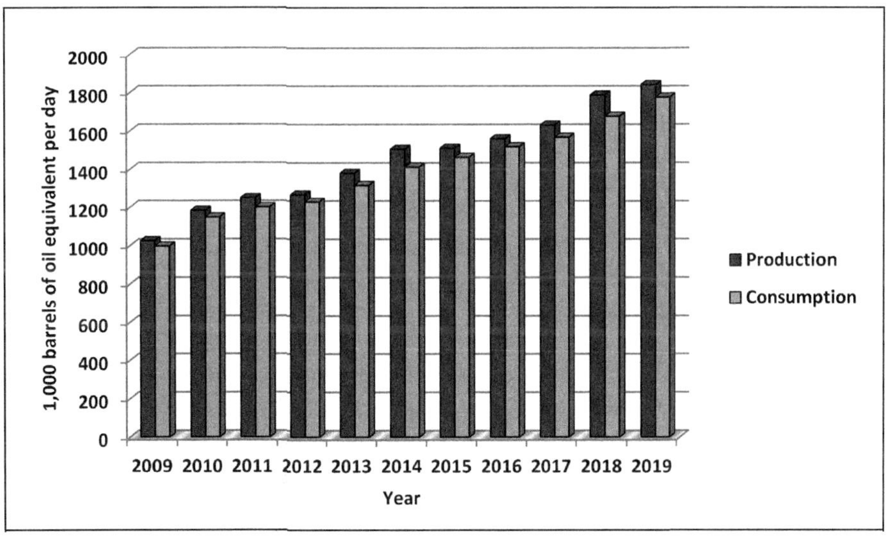

FIGURE 23.1 Global biofuel production and consumption values (2009–2019).

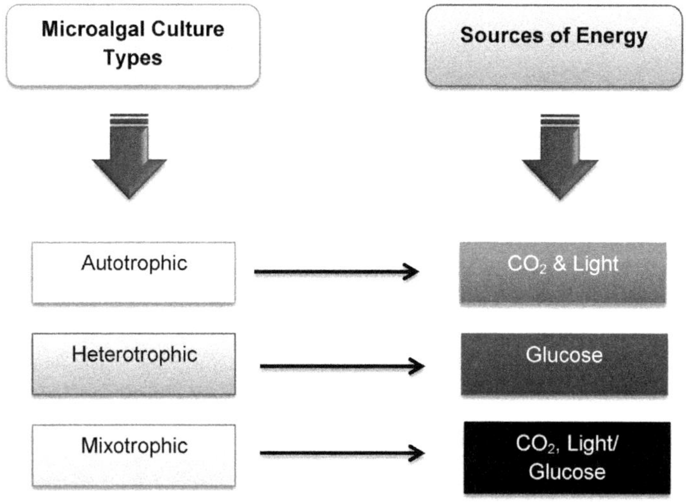

FIGURE 23.2 Types of microalgal cultures and their corresponding energy sources.

two generations (Liew et al., 2014). The algal oil is mostly converted into biodiesel by a process known as transesterification.

Microalgal biofuel production has been found to be quite efficient and easy method for the production of biofuels. Cultivation of microalgal feedstock is quite easy and can be carried out using any sources of water. Wastewater can be utilized for the purpose, thereby minimizing pollution and also utilizing atmospheric CO_2 from the atmosphere. Microalgal cultures make use of the photosynthetic process, thereby converting solar energy into chemical form of energy (Brennan and Owende, 2010). The basic nutrients required for microalgal cultivation can be obtained from water and thus eliminate the input of additional nutrients. The CO_2 thus formed is converted into lipids, proteins, and carbohydrates, as carbohydrate is the end product of photosynthetic process (IEA, 2016).

Microalgae are mostly rich in lipid content accounting for around 20%–40% of the biomass. Besides lipids, the microalgal biomass is also composed of proteins, carbohydrates, and nucleic acids (Prajapati et al., 2013). Microalgal cultures can be classified into three different types based upon their source of energy (Fig. 23.2). The mixotrophic cultures, which form the combination of both autotrophic and heterotrophic cultures, are characterized by high growth rate and lipid production and are the most preferred source for production of algal biofuels (Tan et al., 2018).

23.5 Biodiesel production from genetically modified microalgal biomass as a fourth-generation biofuel

Biofuels are energy sources that are mostly derived from biomass of plants and other organisms and are non-toxic, less polluting, and renewable in nature. Commonly used biofuels can be generally categorized into bioethanol, biodiesel, and biogas. These alternative fuels obtained can be further categorized into different generations (first, second, third, and fourth) based on their sources of production.

23.5.1 Advantages of genetically modified microalgal biomass

The first, second, and third generation of biofuels had several shortcomings (Fig. 23.3), thereby paving the path for the more efficient genetically engineered microalgal strains. The first-generation biofuels are obtained from edible crops (Baskar et al., 2012). The edible crops, although a rich source of raw materials required for biofuel production, gave rise to a food-fuel debate thereby discouraging their use. The second-generation biofuels are mostly obtained from lignocellulosic plant biomass (Andrée et al., 2017; Das et al., 2020b), which are rich in lignin, hemicellulose, and cellulose (Malode et al., 2020). However, the complexity in the biomass structure requires several pretreatment steps, thereby increasing the cost of production. The third-generation biofuels makes use of algal biomass as the source. The microalgal biomasses are effective and easy to obtain and can be easily upscaled. Algae have an inherent ability to absorb large amount of CO_2 from the atmosphere (Collet et al., 2013) and simultaneously convert solar energy into chemical form, making it a better option as compared with the earlier two generations (Liew et al., 2014). However, algal cultures are more vulnerable to contaminations and environmental fluctuations. Genetically modified microalgae mitigate the problems faced by previous generations and thus provide a sustainable and more effective option as compared with the earlier ones. Moreover, the genetically modified organisms have an enhanced ability of producing lipids that can lead to higher biofuel generation.

23.5.2 Genetically modified microalgal strains for improved biodiesel production

The cell size of microalgae and their cell constituents, which include mainly lipids and starch, vary from species to species. This implies that all the available species are not suited for efficient production of biofuels. Certain species can synthesize and store high concentration of starch and lipids, thus making them a better option for synthesis of biofuels. Such type of high yielding species can be further improvised by means of modifications at the gene level (Urtubia et al., 2016). Algal genomes are simple in their organization and can be genetically manipulated so as a means to increase their carbon capturing

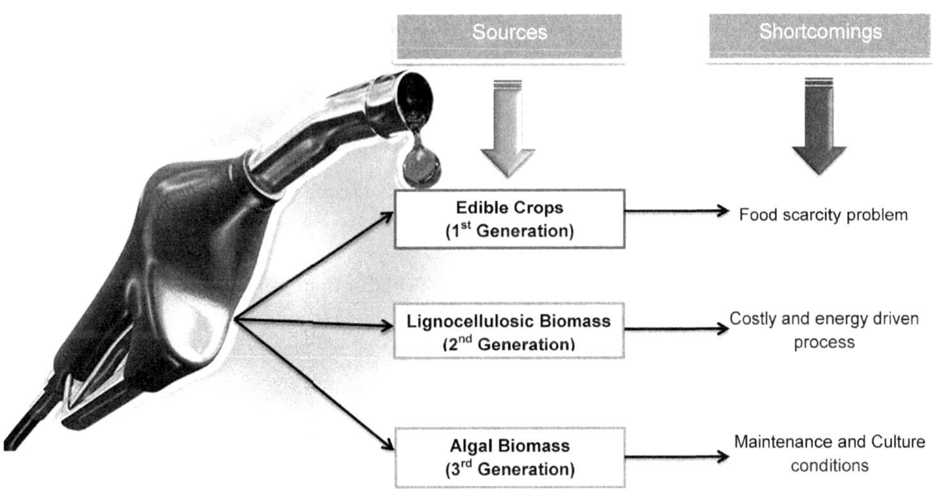

FIGURE 23.3 Shortcomings possessed by the first, second, and third generation of biofuels.

and lipid synthesis ability for the production of important products such as biodiesel.

Genetic engineering of selected algal species can help in higher production of biodiesel. Increased biodiesel production can be achieved genetically by either enhancing the synthesis of algal lipids and fatty acids or by enhancing the carbon capturing and lipid accumulation ability in the species (Khan and Fu, 2021). Enhanced microalgal fatty acid and lipid synthesis can be carried out by overexpressing related enzymes (Yunus and Jones, 2018; Takemura et al., 2019; Kao and Ng, 2017). Enhanced biosynthesis of triacylglycerol (TAG) (Fukuda et al., 2018) and fatty alcohols (Kaczmarzyk et al., 2018) can lead to increased biodiesel production. Genetic modification of DGAT has been found to enhance the production of TAG in algal endoplasmic reticulum (Banerjee et al., 2020). Another approach through which enhancement in biodiesel production can be achieved is by knocking down of genes involved in biosynthesis of starch and/or lipid catabolism (Shin et al., 2019). Genetic recombination has also been used to produce *C. reinhardtii* starchless mutants for improved lipid production. Higher levels of lipids have been successfully produced in algae by blocking the flow of carbon utilized toward the synthesis of starch. Over expression of genes responsible for capturing CO_2 from the atmosphere has been used as a strategy for higher production of lipids (Oh et al., 2018). Table 23.2 provides an account of genetic modifications of microalgal strains undertaken by several research groups for improved biodiesel production.

23.6 Criteria for involving genetically modified microalgal species in biofuel production system

Cultivation of microalgal species is a major problem during biofuel production process, thereby leading to decreased production of microalgal metabolites. It is quite essential to maximize the microalgal growth and production irrespective of the cultivation type being followed. The role of genetically engineered microalgae is of high importance in this regard. Microalgae can be genetically modified to have

TABLE 23.2 Improvement of microalgal strains for biodiesel production through genetic engineering.

Microalgae	Type of genetic modification	Improvement	References
Phaeodactylum tricornutum	DGAT2 overexpression	35% higher lipid content	Niu et al. (2013)
Chlamydomonas Reinhardtii	Nuclear alteration of acetyl-CoA synthetase (ACS2)	2.4 times increased production of triacylglycerol	Rengel et al. (2018)
Chlorella vulgaris	Transformation of carbonic anhydrase through electroporation	2.2-fold increased lipid accumulation	Lin et al. (2018)
C. reinhardtii (CC-4349)	CRISPR-Cas9-based knockout of PLA2 gene	64.25% higher lipid content	Shin et al. (2019)
P. tricornutum	Genetic transformation of glucose-6-phosphate dehydrogenase	Lipid content increased by 2.7 times	Xue et al. (2017)
Schizochytrium sp.	Acetyl-CoA synthetase (ACS) overexpression	11.3% increase in fatty acid	Yan et al. (2013)
Cyanidioschyzon merolae	Overexpression of CmGPAT in endoplasmic reticulum	56.1 times increase in triacylglycerol production	Fukuda et al. (2018)

enhanced growth rates and higher yield (Snow and Smith, 2012). Microalgae can be modified genetically in several ways. Modification of DNA in microalgal antennae would enable the microalgal cultures to make more amount of light available for the species, thereby enhancing the photosynthesis process and leading to higher yield. Restructuring of microalgal DNA could help the species to strive under drastic conditions (Dana et al., 2012). Modification of specific sites within the DNA of microalgae could also reflect upon the microalgal metabolism, thereby enhancing yield. Some microalgae could also be genetically modified so as to develop certain new qualities that would prove to be beneficial in the production of biofuels.

It is highly essential to explore for new genes that could help enhance growth rates and yield quality in microalgal species. However, genetic engineering is a very complicated process and requires a huge capital, which can only be met by the financial support from government as well as private funding.

23.7 Mechanism of biofuel production from genetically modified microalgal biomass

The use of microalgal source is one of the best options for the production of biofuels. Microalgae are genetically modified with prior objectives of supporting the growth conditions and thus increasing the biomass yield qualitatively and quantitatively.

23.7.1 Priority for production of biofuel from genetically modified microalgal biomass

These microflora have been given priority over other as the sources of biomass for biofuel production on the basis of certain specific features. One can obtain their fast biomass yield from aquatic bodies rich in freshwater, marine water, or wastewater. Most important is the fact that biofuel generation from microalgal biomass is an energy-efficient and cost-effective process as compared with other sources of biofuel production.

The genetically modified microalgae may show higher growth during the cultivation phase, higher resistance toward fluctuations in environmental parameters, and higher yield of lipids. Higher growth rate can be achieved by making these species able to utilize light efficiently, even at low intensity. Acclimatization to fluctuations in environmental parameters such as temperature, pH, CO_2 concentrations, nutrient, and water availability can also be achieved by genetically modified microalgal species. It will lead to the development of environmental-resistant species with growth promotion. High growth rate is linked with high biomass yield and thus higher biofuel yield. Modifications at the genetic level can enhance lipid production, thereby making them more suitable for production of biodiesel. The microalgae species can be made tolerant toward invasive organisms such as fungus, bacteria, and other algal species and promote stable growth rate by genetic engineering. Fig. 23.4 provides an overview of the biofuel production from genetically engineered microalgal species.

23.7.2 Growth of microalgal species in natural and artificially constructed environment

Microalgal species can be cultivated in both natural and artificial constructed environments. Natural processes include cultivation in open ponds or lakes (Suganya et al., 2016) and are provided with paddle wheels or air pumps for aeration (Trivedi et al., 2015). The open cultivation system is characterized by simplicity in construction and operation, and cost-effectiveness. The major setback for such a type of setup is the

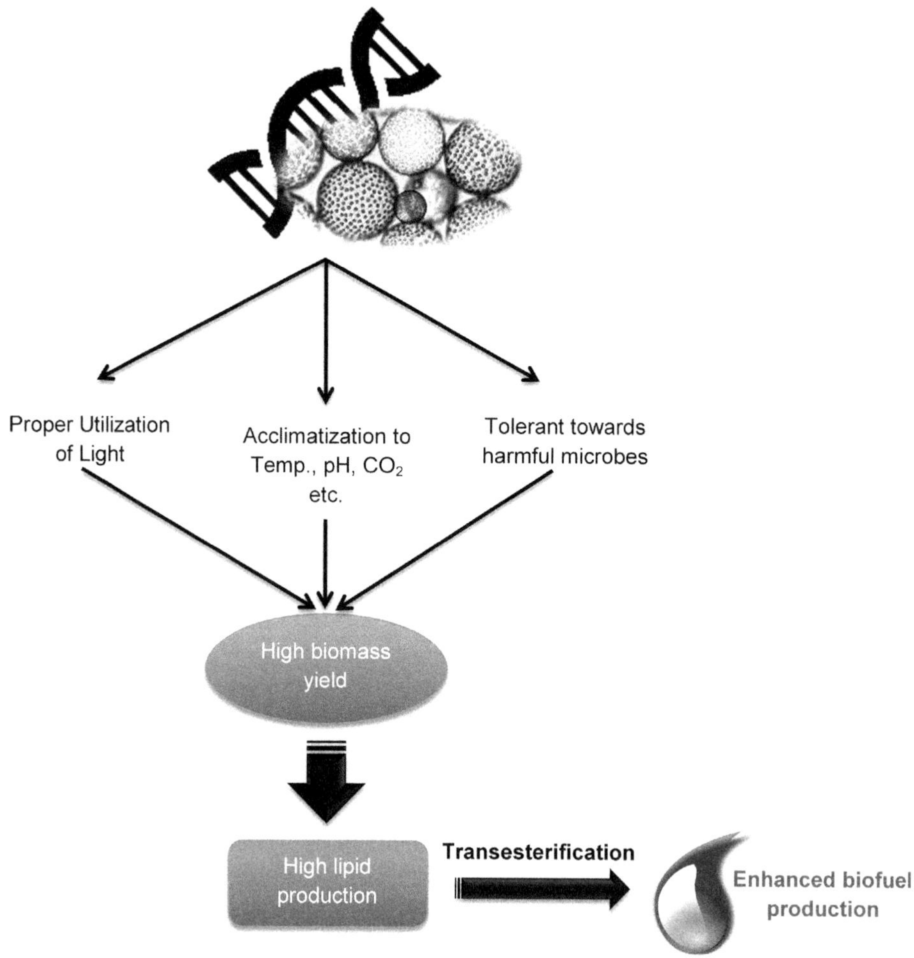

FIGURE 23.4 Overview of biofuel production from genetically engineered microalgae.

difficulty in maintaining the culture conditions. Proper utilization of light, contamination during growth, temperature fluctuation, evaporation of water, and CO_2 diffusion into the surrounding atmosphere cannot be regulated (Lee and Lee, 2016).

Artificial cultivation systems are mostly carried out in photobioreactors that cultivate microalgae under a closed and controlled environmental condition. All the required nutrients along with CO_2 and algal biomass are fed into the reactor where factors such as pH, temperature, cell density, and intensity of light are governed and controlled as per the need (Kandiyoti et al., 2017). The closed or artificial cultivation system eliminates all the drawbacks posed by the natural systems; however, the major barrier is the high cost of setup and maintenance. The drawbacks posed by natural and artificial microalgal cultivation systems are different on certain aspects (Fig. 23.5).

The cultivation of microalgae is followed by harvesting, wherein the cell biomass is isolated from the culture medium. The harvesting

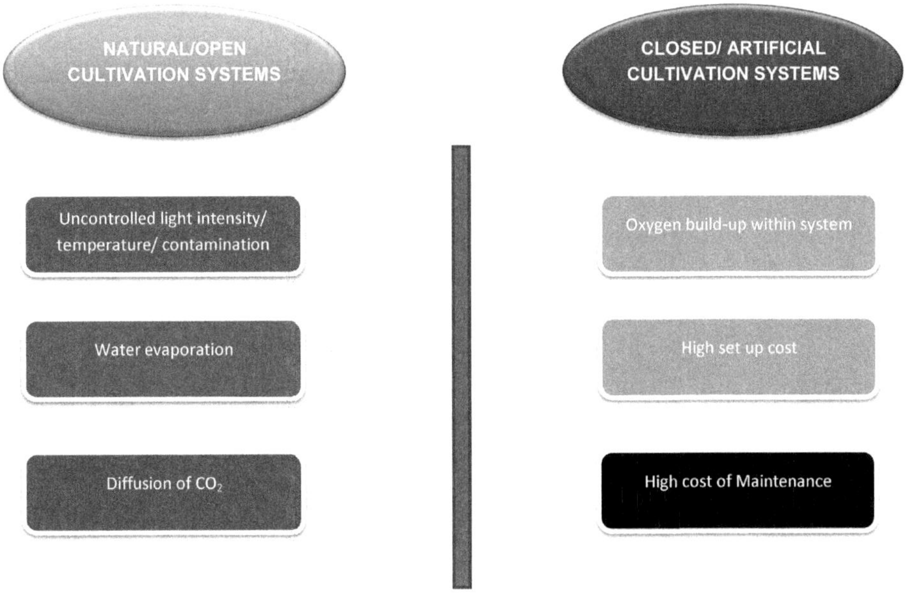

FIGURE 23.5 Comparative details of drawbacks posed by natural and artificial microalgal cultivation systems.

method purely depends on the property of microalgae such as cell density, physiology, and nature of the final intended product. Most commonly employed harvesting techniques are mechanical, chemical, electrical, and biological. Mechanical methods are the most preferred mode of harvesting microalgal cells, which can be attributed to their higher efficiency.

Microalgae being rich in lipids are treated as vital source for the production of biodiesel. Lipids obtained from microalgae can be converted into biodiesel following transesterification process. It is necessary to extract lipid content from microalgal biomass prior to transesterification. The extraction of lipid could be carried out following mechanical methods or solvent extraction method or by using suitable enzymes (biological method). The solvent extraction methods make use of solvent or a mixture of solvent to separate lipids (Jones et al., 2012). The mechanical methods make use of sheer pressure or force to rupture microalgal cells, thereby extracting the lipids. The latest development toward extraction principle is the use of biological enzymes that are non-toxic, sensitive, rapid, and compatible. Enzymatic extraction is more energy efficient and result oriented when compared with other techniques (Ranjith et al., 2015). Triglycerides and triglycerol form the major part of microalgal lipids. These ester compounds undergo transesterification process to produce fatty acid methyl esters, commonly known as FAME. The advantages of FAME are its non-toxic, low viscosity, and biodegradable nature (Kiran et al., 2014). The process is reversible in nature and includes the addition of excess of methanol along with an appropriate catalyst resulting in the formation of biodiesel and glycerol as the end product (Suganya et al., 2016). The catalysts include the use of alkaline or acidic substances. Alkaline catalysts are prepared by mixing sodium or potassium hydroxide in methanol, while acid catalysts are prepared by mixing hydrochloric or sulfuric acid with methanol in a particular proportion (Lam and Lee, 2012). Large-scale industrial production processes frequently use alkaline catalyst, which may be attributed to their higher catalytic activity (more than 4000 times) than the acid catalysts. Enzyme catalysts are also used as an alternative; however, these catalysts carry major drawbacks, which include lower operating temperatures and high cost of production (Kiran et al., 2014). Microalgae are also an efficient source for production of ethanol via the fermentation process and yield at least two times more ethanol as compared with second-generation sources (Veillette et al., 2018). The residues left behind after the process of fermentation can be reutilized for producing biomethane (Fig. 23.6) by the anaerobic digestion process (Suganya et al., 2016).

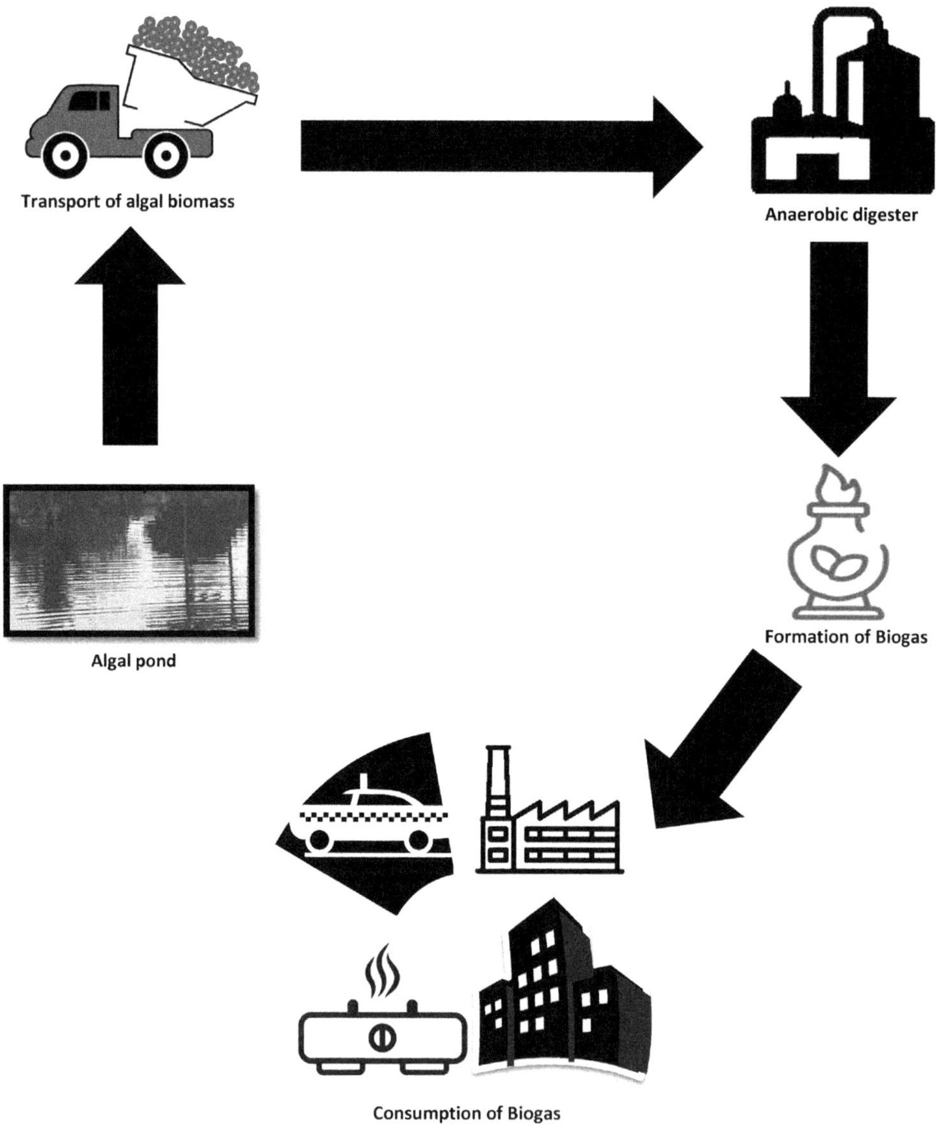

FIGURE 23.6 Utilization of algal biomass in production of biogas.

23.8 Evaluation of safety linked with the use of microalgae as a biofuel feedstock

23.8.1 Effects on public health

The microalgae used for biofuel production represent the group of plants involved with trapping of solar light and conversion of solar energy to bioenergy. The genome editing and expression of altered genome may be able to express adverse effects on public health.

23.8.2 Effects on environment

Use of biomass from genetically modified microalgal species has the possibility to affect the energy flow and cycling of matters in the environment (Rastogi et al., 2018). Being the primary producer, it can express responses on the distribution of organisms at subsequent trophic levels in ecosystems. Natural selection of organisms at trophic levels maintains the balance of nature and biodiversity in ecosystems. Genetically modified microalgae can disturb this balance of nature to cause damage to the environment (Singh et al., 2016).

23.8.3 Effects of introduction of genetically modified algal species on wild cultivars

The introduction of genetically modified algal species is required for qualitative and quantitative enrichment of biofuel. The introduction of genetically modified algal species can bring about changes in the genomic construction of wild strains through breeding. It has the possibility of destabilizing the gene pool. The mixed cultivation of genetically modified algal species and their wild counterparts may lead to intercrossing and generation of new genetic combination.

23.8.4 Effects of obeying safety provisions on public health and environment

The formulation and obeying of biosafety regulations and risk assessment techniques are essential for safe use of genetically modified microalgal biomass. The adverse effects of its use for biofuel production can be observed on public health and exposed environmental components. It can be minimized with strict adherence and timely monitoring of these regulations. Evaluation of the impacts of these genetically modified microalgal species on the exposed environmental components is essential to take corrective measures (Glass, 2015; Randhawa et al., 2017).

23.9 Socio economic aspects of microalgal use as a feedstock for biofuel production

23.9.1 Social acceptance

Society needs cheap, eco-friendly, and sustainable sources of energy to meet its gap in demand and availability of energy. The renewable sources of energy are better option as compared to conventional non-renewable energy. The bioenergy from microalgae is promising in this context. The biomass of microalgae with high lipid content, up to 80% of the biomass, is a feasible choice for biofuel production under emerging situation (Chisti, 2008). It is possible to use biomass of microalgae as feedstock for biofuel production in all the seasons, sustainably. It produces fuel without affecting the landmasses earmarked for agriculture and forestry. Society likes green technology. Cultivation of microalgae and their use for biofuel production channelize the sequestration of carbon dioxide (CO_2) from environmental components to the growing biomass and in return release oxygen (O_2) to the environment.

23.9.2 Protection of environment

The greenhouse effects and the outcome of the burning of fossil fuels as source of energy are the burning environmental problems. To protect the environment, it is essential to curtail the release of GHGs from anthropogenic activities. The consumption of GHGs such as carbon dioxide from environment during assimilation process of a microalga is a positive trend to reduce the clouds of GHGs from the environment. The biofuel production from the biomass of microalgae is eco-friendly as it emits few GHGs as compared with conventional fossil fuels. The microalgae balance the generation of carbon dioxide during burning of biofuel by their consumption during assimilation for biomass yield. The efficiency of the carbon cycling makes the biofuel residues almost carbon free.

23.9.3 Energy security

The global energy sector is passing through a scenario of acute energy shortage. With the approach of energy insecure condition in the coming decades, it seems to be a boon for extraction of renewable energy from genetically modified microalgal biomass. It is an energy form to support daily requirements with minimal release of CO_2. This form of energy is available for use in multiple energy consuming sectors following the standards fixed by the Kyoto protocol (Maamoun, 2019).

23.9.4 Reduced dependence on fossil fuel

The blending of biofuels with fossil fuels such as gasoline and diesel can reduce the quantity of these fuels demanded from energy consumption sector. The extraction of energy from algal biomass can be effective if the biomass used as feedstock is rich in energy-rich molecules such as lipids. The editing of algal genomic resources can lead to alternation in the conventional synthesis of molecules. It can possibly be channelized for the production of excessive amount of energy-rich lipids in intracellular conditions.

23.9.5 Success under field-scale use

The generation of biofuel from microalgae has already achieved success at laboratory scale. The transfer of this technology to field scale is challenging. Use of microalgae for biofuel production is not yet established under field-scale condition. The successful trial of microalgal biofuel under field-scale condition needs microalgal strain improvement.

23.9.6 Economic feasibility

The cost of production of biofuel using microalgal biomass is a challenging factor behind its success as a source of renewable energy and its acceptability at the community level. Biotechnology offers to give relief to this challenge to some extent. The cost of production of microalgal biomass can be reduced up to 20% with its application (Chung et al., 2017). The production of biofuel from microalgal biomass is cost-effective under growth of microalgae under open aquatic ecosystems (Kumar et al., 2018). The yield of microalgal biomass is even possible from unused or underused water bodies (Greenwell et al., 2010). The separation of energy-rich lipids from microalgal biomass is an efficient process. Further improvement in efficiency is possible with biotechnological applications. The modification of microalgae genetically is a viable step for increasing the efficiency of biofuel production.

23.9.7 Livelihood generation

Algal farming, harvesting of biomass, processing of biomass, and production of biofuel are not only required for finding solution to the energy scarcity, but it can also engage unemployed people in taking biofuel production from microalgal biomass as a profession. Provision for seed money at the initial stages for beginners may inspire them to accept this profession for generating green energy. The short span of production of biofuel from voluminous algal biomass is an encouraging step and attracts the rural households to take it up as a profession. It can be used for livelihood generation without much difficulty, as its requirements are easily available such as aquatic bodies, solar light, nutrients for growth, and carbon dioxide for assimilation (Sasaki et al., 2020; Rafiee et al., 2020).

23.10 Future prospects

The current trend of price hike in fossil fuel sector and shortage of its availability in reservoirs puts severe strain on the global socio-economic conditions. It is the time to harness energy from microalgal thalloid biomass having enormous growth potential in tropical and subtropical aquatic conditions. The group of microalgal species such as Cyanophyceae, Chlorophyceae, and Bacillariophyceae may be proved promising for this purpose. The modulation of environmental parameters relating to the growth of microalgal species can improve the calorific value of biomass yield and efficiency of bioenergy generation. Besides, mapping and editing at the genomic level is useful for alteration in the chemical composition of algal biomass and to explore the pathway for efficient bioenergy production.

The deficiencies in this area of study include complete characterization of microalgal species and optimization of process parameters for efficient biofuel production. Lack of awareness on its use among people at the grass root level of the society is another bottleneck in this area of study. The future prospects of this work would be based on removal of these constraints and to make it a cheap source of renewable energy available readily for the people present in energy-starved condition.

23.11 Conclusion

Biofuel from microalgae offers the production of environmental friendly, renewable energy without compromising with the yield of food crops. Efficient production of energy with bio-economy from microalgae is an added advantage. The use of biomass of genetically engineered microalgae is an encouraging approach for efficiency of biofuel production. Despite limited constraints, it is useful as an emerging alternative energy source. It can be able to replace the use of fossil fuels to a certain extent in the near future. Further work in this area is required to remove the limitations linked with the microalgae-based biofuel production, to reduce its production cost, and to improve its production efficiency.

References

Alberici, S., Toop, G., 2013. UK Biofuel Industry Overview: Input to DRAFT PIR. ECOFYS UK Ltd, UK.

Andrée, B.P.J., Diogo, V., Koomen, E., 2017. Efficiency of second-generation biofuel crop subsidy schemes: spatial heterogeneity and policy design. Renew. Sustain. Energy Rev. 67, 848–862.

Banerjee, S., Banerjee, S., Ghosh, A.K., Das, D., 2020. Maneuvering the genetic and metabolic pathway for improving biofuel production in algae: present status and future prospective. Renew. Sustain. Energy Rev. 133, 110155. https://doi.org/10.1016/j.rser.2020.110155.

Baskar, C., Baskar, S., Dhillon, R.S. (Eds.), 2012. Biomass Conversion: The Interface of Biotechnology, Chemistry and Materials Science. Springer, Berlin.

Brennan, L., Owende, P., 2010. Biofuels from microalgae—a review of technologies for production, processing, and extractions of biofuels and co-products. Renew. Sustain. Energy Rev. 14 (2), 557–577.

Chisti, Y., 2008. Biodiesel from microalgae beats bioethanol. Trends Biotechnol. 26, 126–131.

Chung, Y.S., Lee, J.W., Chung, C.H., 2017. Molecular challenges in microalgae towards cost-effective production of quality biodiesel. Renew. Sustain. Energy Rev. 74, 139–144. https://doi.org/10.1016/j.rser.2017.02.048.

Collet, P., Spinelli, D., Lardon, L., Helias, A., Steyer, J.-P., et al., 2013. Life-cycle assessment of microalgal-based biofuels. In: Pandey, A., Lee, D.J., Chisti, Y., Soccol, C.R. (Eds.), Biofuels from Algae. Elsevier, Amsterdam, p. 287e312.

Dana, G.V., Kuiken, T., Rejeski, D., Snow, A.A., 2012. Synthetic biology: four steps to avoid a synthetic-biology disaster. Nature 483, 29.

Das, P.K., Das, B.P., Dash, P., 2020a. Role of energy crops to meet the rural energy needs: an overview. Biomass Valoriz. Bioenergy 11–30.

Das, P.K., Das, B.P., Dash, P., 2020b. Potentials of Postharvest Rice Crop Residues as a Source of Biofuel. Refining Biomass Residues for Sustainable Energy and Bioproducts. Academic Press, pp. 275–301.

Fukuda, S., Hirasawa, E., Takemura, T., Takahashi, S., Chokshi, K., Pancha, I., Tanaka, K., Imamura, S., 2018. Accelerated triacylglycerol production without growth inhibition by overexpression of a glycerol-3-phosphate acyltransferase in the unicellular red alga *Cyanidioschyzon merolae*. Sci. Rep. 8, 12410.

Glass, D.J., 2015. Pathways to obtain regulatory approvals for the use of genetically modified algae in biofuel or bio-based chemical production. Ind. Biotechnol. 11, 71–83. https://doi.org/10.1089/ind.2015.1503.

Greenwell, H.C., Laurens, L.M., Shields, R.J., Lovitt, R.W., Flynn, K.J., 2010. Placing microalgae on the biofuels priority list: a review of the technological challenges. J. R. Soc. Interface 7 (46), 703–726.

Gul, T., 2016. Renewable transport fuel obligation statistics. Period 9, 1–6.

Hombach, L.E., Cambero, C., Sowlati, T., Walther, G., 2016. Optimal design of supply chains for second generation biofuels incorporating European biofuel regulations. J. Clean. Prod. 133, 565–575.

IEA, 2016. IEA Data Shows Global Energy Production and Consumption Continue to Rise. International Energy Agency, Paris, France. www.iea.org/newsroom/news/2016/august/iea-data-shows-globalenergy-production-and-consumption-continue-to-rise.html. (Accessed 25 January 2017).

IEA, 2019. World Energy Outlook 2019. https://www.iea.org/reports/world-energy-outlook-2019.

Jones, J., Manning, S., Montoya, M., Keller, K., Poenie, M., 2012. Extraction of algal lipids and their analysis by HPLC and mass spectrometry. J. Am. Oil Chem. Soc. 89 (8), 1371–1381. https://doi.org/10.1007/s00216-011-5376-6.

Kaczmarzyk, D., Cengic, I., Yao, L., Hudson, E.P., 2018. Diversion of the long-chain acyl-ACP pool in Synechocystis to fatty alcohols through CRISPRi repression of the essential phosphate acyltransferase PlsX. Metab. Eng. 45, 59–66.

Khan, S., Fu, P., 2021. Biotechnological perspectives on algae: a viable option for next generation biofuels. Curr. Opin. Biotechnol. 62, 146–152.

Kandiyoti, R., Herod, A., Bartle, K., Morgan, T., 2017. 1 - Fossil Fuels and renewables.Solid Fuels and Heavy Hydrocarbon Liquids, second ed. Elsevier, Oxford, United Kingdom, pp. 1–9.

Kao, P.H., Ng, I.S., 2017. CRISPRi mediated phosphoenolpyruvate carboxylase regulation to enhance the production of lipid in *Chlamydomonas reinhardtii*. Bioresour. Technol. 245, 1527–1537.

Kiran, B., Kumar, R., &Deshmukh, D., 2014. Perspectives of microalgal biofuels as a renewable source of energy. Energy Conver. Manag. 88, 1228–1244.

Kumar, P., Kumar, D., Nehra, P., Sharma, P.K., 2018. Green algae biomass cultivation, harvesting and genetic modifications for enhanced cellular lipids. In: Patra, J.K., Das, G., Shin, H.S. (Eds.), Microbial Biotechnology. Springer Nature), Singapore, pp. 119–140.

Lam, M.K., Lee, K.T., 2012. Microalgae biofuels: a critical review of issues, problems and the way forward. Biotechnol. Adv. 30 (3), 673–690.

Li, M., Luo, N., Lu, Y., 2017. Biomass energy technological paradigm (BETP): trends in this sector. Sustainability 9 (4), 567. https://doi.org/10.3390/su9040567.

Lee, O.K., Lee, E.Y., 2016. Sustainable production of bioethanol from renewable brown algae biomass. Biomass Bioenergy 92, 70–75.

Liew, W.H., Hassim, M.H., Ng, D.K.S., 2014. Review of evolution, technology and sustainability assessments of biofuel production. J. Clean. Prod. 71, 11e29. https://doi.org/10.1016/j.jclepro.2014.01.006.

Lin, W.R., Lai, Y.C., Sung, P.K., Tan, S.I., Chang, C.H., Chen, C.Y., Chang, J.S., Ng, I.S., 2018. Enhancing carbon capture and lipid accumulation by genetic carbonic anhydrase in microalgae. J. Taiwan Inst. Chem. Eng. 93, 131–141.

Maamoun, N., 2019. The Kyoto protocol: empirical evidence of a hidden success. J. Environ. Econ. Manag. 95, 227–256.

Malode, S.J., Prabhu, K.K., Mascarenhas, R.J., Shetti, N.P., Aminabhavi, T.M., 2020. Recent advances and viability in biofuel production. Energy Conver. Manag. 10, 100070. https://doi.org/10.1016/j.ecmx.2020.100070.

Miao, R., Xie, H., Ho, F.M., Lindblad, P., 2018. Protein engineering of α-ketoisovalerate decarboxylase for improved isobutanol production in Synechocystis PCC 6803. Metab. Eng. 47, 42–48.

Niu, Y.F., Zhang, M.H., Li, D.W., Yang, W.D., Liu, J.S., Bai, W.B., Li, H.Y., 2013. Improvement of neutral lipid and polyunsaturated fatty acid biosynthesis by overexpressing a type 2 diacylglycerol acyltransferase in marine diatom Phaeodactylum tricornutum. Mar. Drugs 11, 4558e4569.

Oh, Y.K., Hwang, K.R., Kim, C., Kim, J.R., Lee, J.S., 2018. Recent developments and key barriers to advanced biofuels: a short review. Bioresour. Technol. 257, 320–333.

Prajapati, S.K., Kaushik, P., Malik, A., Vijay, V.K., 2013. Phycoremediation coupled production of algal biomass, harvesting and anaerobic digestion:Possibilities and challenges. Biotechnol. Adv. 31 (8), 1408–1425.

Rafiee, P., Ebrahimi, S., Hosseini, M., WahTong, Y., 2020. Characterization of soluble algal products (SAPs) after electrocoagulation of a mixed algal culture. Biotechnol. Rep. 25, e00433.

Randhawa, K.S., Relph, L.E., Armstrong, M.C., Rahman, P.K.S.M., 2017. Biofuel production: tapping into microalgae despite challenges. Biofuels 8, 261–271. https://doi.org/10.1080/17597269.2016.1224290.

Ranjith Kumar, R., HanumanthaRao, P., Arumugam, M., 2015. Lipid extraction methods from microalgae: a comprehensive review. Front. Energy Res. 2 (612), 61. https://doi.org/10.3389/fenrg.2014.00061.

Rastogi, R.P., Pandey, A., Larroche, C., Madamwar, D., 2018. Algal green energy – R & D and technological perspectives for biodiesel production. Renew. Sustain. Energy Rev. 82, 2946–2969. https://doi.org/10.1016/j.rser.2017.10.038.

Rengel, R., Smith, R.T., Haslam, R.P., Sayanova, O., Vila, M., Leon, R., 2018. Overexpression of acetyl-CoA synthetase (ACS) enhances the biosynthesis of neutral lipids and starch in the green microalga *Chlamydomonas reinhardtii*. Algal Res. 31, 183–193.

Sasaki, M., Takagi, A., Ota, S., Kawano, S., Sasaki, D., Asayama, M., 2020. Coproduction of lipids and extracellular polysaccharides from the novel green alga *Parachlorella* sp. BX1.5 depending on cultivation conditions. Biotechnol. Rep. 25, e00392. https://doi.org/10.1016/j.btre.2019.e00392.

Shin, Y.S., Jeong, J., Nguyen, T.H.T., Kim, J.Y.H., Jin, E., Sim, S.J., 2019. Targeted knockout of phospholipase A2 to increase lipid productivity in *Chlamydomonas reinhardtii* for biodiesel production. Bioresour. Technol. 271, 368–374.

Singh, S.K., Sundaram, S., Sinha, S., Rahman, M.A., Kapur, S., 2016. Recent advances in CO_2 uptake and fixation mechanism of cyanobacteria and microalgae. J. Crit. Rev. Environ. Sci. Technol. 46, 1297–1323. https://doi.org/10.1080/10643389.2016.1217911.

Snow, A.A., Smith, V.H., 2012. Genetically engineered algae for biofuels: a key role for ecologists. Bioscience 62 (8), 765–768.

Suganya, T., Varman, M., Masjuki, H.H., Renganathan, S., 2016. Macroalgae and microalgae as a potential source for commercial applications along with biofuels production: a biorefinery approach. Renew. Sustain. Energy Rev. 55, 909–941.

Takemura, T., Imamura, S., Tanaka, K., 2019. Identification of a chloroplast fatty acid exporter protein, CmFAX1, and triacylglycerol accumulation by its overexpression in the unicellular red alga *Cyanidioschyzon merolae*. Algal Res. 38, 101396.

Tan, X.B., Lam, M.K., Uemura, Y., Lim, J.W., Wong, C.Y., Lee, K.T., 2018. Cultivation of microalgae for biodiesel production: a review on upstream and downstream processing. Chin. J. Chem. Eng. 26 (1), 17–30.

Trivedi, J., Aila, M., Bangwal, D.P., Kaul, S., Garg, M.O., 2015. Algae based biorefinery-how to make sense? Renew. Sustain. Energy Rev. 47, 295–307.

Urtubia, H.O., Betanzo, L.B., Vasquez, M., 2016. Microalgae and cyanobacteria as green molecular factories: tools and perspectives. Algae Organ. Imminent Biotechnol. 1–27. https://doi.org/10.5772/100261. IntechOpen.

Veillette, M., Giroir-Fendler, A., Faucheux, N., Heitz, M., 2018. Biodiesel from microalgae lipids: from inorganic carbon to energy production. Biofuels 9 (2), 175–202.

Wichmann, J., Baier, T., Wentnagel, E., Lauersen, K.J.J., Kruse, O., 2018. Tailored carbon partitioning for phototrophic production of (E)-α-bisabolene from the green microalga *Chlamydomonas reinhardtii*. Metab. Eng. 45, 211–222.

Xue, J., Balamurugan, S., Li, D.W., Liu, Y.H., Zeng, H., Wang, L., Yang, W.D., Liu, J.S., Li, H.Y.Y., 2017. Glucose-6-phosphate dehydrogenase as a target for highly efficient fatty acid biosynthesis in microalgae by enhancing NADPH supply. Metab. Eng. 41, 212–221.

Yunus, I.S., Jones, P.R., 2018. Photosynthesis-dependent biosynthesis of medium chain-length fatty acids and alcohols. Metab. Eng. 49, 59–68.

Yan, J., Cheng, R., Lin, X., You, S., Li, K., Rong, H., Ma, Y., 2013. Overexpression of acetyl-CoA synthetase increased the biomass and fatty acid proportion in microalga *Schizochytrium*. Appl. Microbiol. Biotechnol. 97, 1933–1939.

SECTION VIII

Techno-economic analysis and life cycle assessment

CHAPTER

24

Sustainability assessment of third-generation biofuels: a life cycle perspective

B. Senthil Rathi[1], P. Senthil Kumar[2,3], V. Vinoth Kumar[4]

[1]Department of Chemical Engineering, St. Joseph's College of Engineering, Chennai, Tamil Nadu, India; [2]Department of Chemical Engineering, Sri Sivasubramaniya Nadar College of Engineering, Chennai, Tamil Nadu, India; [3]Centre of Excellence in Water Research (CEWAR), Sri Sivasubramaniya Nadar College of Engineering, Chennai, Tamil Nadu, India; [4]Department of Biotechnology, SRM Institute of Science and Technology, Chennai, Tamil Nadu, India

24.1 Introduction

The rising appetite for fuel, together with the exhaustion of conventional fuels as well as the adverse environmental effects related to the use of fossil energy in the transportation sector, indicates that alternate solution, ecofriendly fuels should be adopted on a massive scale (Aboelazayem et al., 2018; Foteinis et al., 2020). There is also a necessity reduce the reliance on conventional fuels and to shift toward low-emission sustainability (Pardo et al., 2010). The oil shortage as well as environmental problems has been a central area of research in recent times. Industrial development has driven the human race to replenish coal and oil, and as a result, gas emissions have risen worldwide. A variety of studies were conducted to find renewable fuels to satisfy growing energy needs. Biofuel has newly been launched as an economic green and sustainable fuel, referred to as an ecologically responsible fuel (Jeswani et al., 2020). Approximately 350 petroleum crops were studied, and a few of them may be seen as possible substitute fuels such as biodiesel. This involves pure vegetable oils and vegetable oil residue. *Brassica*, Euphorbiaceae, *Glycine max*, and Arecaceae oil are perhaps the most major sources of biofuel (Murugesan et al., 2009; Hosseini and Wahid, 2012; Othman et al., 2017).

Bioenergy sources of the very first generation are plentiful since the technological advancements are quite well known. Even so, the development of raw resources is in dispute with food and nutrition security, such that biodiesel of the very first generation is not quite as convincing. The second generation of biodiesel would not communicate directly toward food but will need many electricity methods to generate it, which would also raise property transition, which also limits its ecological and financial viability. The biofuel production of third generation eliminates the problems encountered by biodiesel of first and second generation—food

Biofuels and Bioenergy
https://doi.org/10.1016/B978-0-323-85269-2.00002-2

competitiveness, property transition, and so forth—and is therefore deemed a feasible renewable energy option (Manikandan et al., 2014). A life cycle evaluation strategy is more applicable to all aspects of sustainable development to prevent challenges in the moving of challenges. The need for agricultural residue and carbon dioxide in exhaust gases for bioprocessing further improves the sustainable development of third-generation biofuel production, as it minimizes greenhouse effect pollution and waste management issues (Singh et al., 2011) The biofuel production from algae is promising for the aforementioned purposes. Algae could grow easily, and all products provided are nontoxic as well as environmentally friendly, but during that development, there seems to be a chance to induce greenhouse gases trapping (Saravanan et al., 2021). In addition, since producing algae does not really require agricultural land, this can be cultivated without even interfering with food or feeding commodities, and the oil produced by it is "third-generation" renewable energy (Alaswad et al., 2015; Sadatshojaei et al., 2020). Biodiesel derived from plants, livestock, or seaweed commodities may provide a potential to cut our reliance on fossil fuels and help to sustain a healthier climate system. Microalgae are now a common production of biofuels nominee, resulting in high lipid, simplicity of propagation, and high reproduction (Alam et al., 2015; Aravind et al., 2020; Callegari et al., 2020). Seaweeds propose different green, clean energy sources since they have a high capacity for generating large quantities of compost and, in addition, could be used for the manufacturing of specific third-generation bioenergy on a massive scale. Microalgae process solar power into carbon storage commodities, which contribute to lipid aggregation, such as triacylglycerol that can then be turned into bioethanol biodiesel and bio methanol (Maity et al., 2014; Bhatia et al., 2020; Liu et al., 2020).

The cost of biofuel changes is based mostly on raw material, the geographical location, the rates of methanol, and the changes in demand of agricultural production. Most biofuels are actually manufactured from *G. max*, canola, and Areca oils. Even so, there are many vast quantities of reduced fats and oils (for example, catering residue, animal fats, pork lard, and yellow fat) that can be turned to biodiesel. Crop varieties, farming activities, land and labor rates, field sizes, technology advancements, and regulatory frameworks in various areas vary greatly in the rate of fuel inflation rates by area (Demirbas, 2011; Bai et al., 2012; Millinger et al., 2017). Two different types of biofuels energy conversion are transesterification, hydrothermal liquefaction, and pyrolysis for sustainable diesel generation and anaerobic digestion without/with hydrothermal pretreatment for biogas processing. The life cycle effects of such processes have been measured in percentage of overall energy proportions and greenhouse emissions. The findings demonstrate that perhaps the anaerobic digestion method including hydrothermal pretreatment is much more commercially viable and environmentally sustainable on a huge application owing to its poor total energy ratios and greenhouse gases (Sun et al., 2019).

This chapter summarizes the third-generation biofuels, feedstock cost and biofuel competition and technical aspect of biofuel production. It also discusses the life cycle assessment of biofuel production.

24.2 Third-generation biofuels

Biodiesel of the very first generation from farm productivity is presently not known to be a clean energy source. Comparably, second-generation biodiesel generated via nonfood raw material really is not commercially feasible due to its high service charges. By comparison, microalgae biodiesel of the third generation is seen as a realistic and renewable alternative to potential energy consumption by addressing the limitations of previous generations of

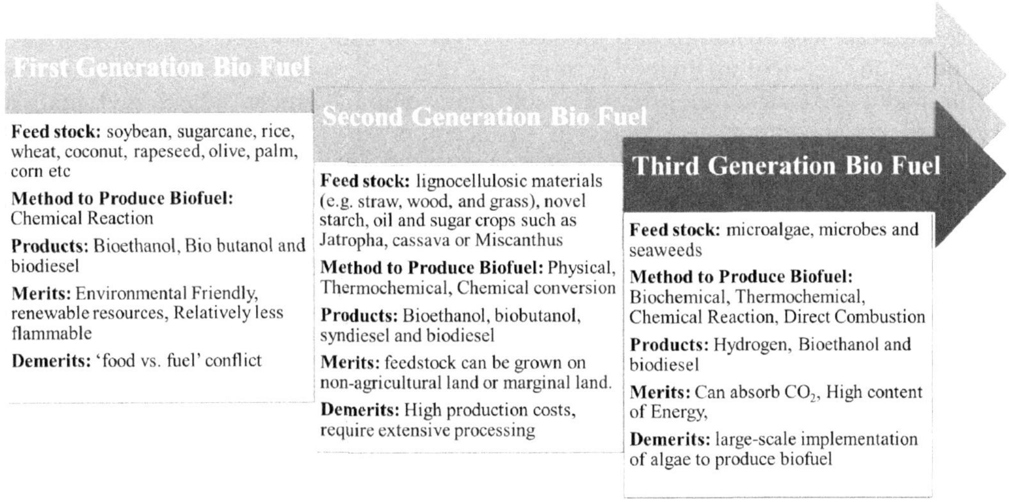

FIGURE 24.1 Overview of biofuels in three generations.

bioenergy (Chowdhury and Loganathan, 2019). Fig. 24.1 depicts the overview of the biofuels in all three generations.

Irrespective of vegetable oils, several other materials, such as algae, surplus oil animal products, fatty fish, pyrolysis oil, and so on, are a third-generation origin of bioenergy (Rajamohan et al., 2021). This feasible biofuel supply resolves the challenges established by different generation fuel sources such as supply, financial viability, effects on the food web, and flexibility to environment situations. Seaweeds may be a possible raw material for the processing of biodiesel. Because certain microalgae have the potential to survive in brutal environments, they are ideally suited to regional areas with low cost of cultivation (Behera et al., 2015; Sakthivel et al., 2018). Regardless of the source of the algae, leftover frying oil is a budget and extremely heterogeneous raw material for the production of bioethanol (Kumar et al., 2015; Gunasundari, 2017). The use of waste oil further decreases the cost of sewer systems and decreases water pollution. In recent times, animal fats, such as pork, beef, livestock, and lamb, have emerged as a possible and effective origin of biodiesel production. This pattern is attributable to the fact which improvements in the feeding diets of person and wash industries are unable to absorb all the fats created (Jambo et al., 2016). Unrefined fish oil generated from scrapped fish sections is considered to become a simple and affordable origin for the extraction of biofuel. Because the lipid quality of meat products sea fish is higher, it is usually favored for the extraction of the raw fuel. Pyrolysis is among the simplest techniques used, which turns organic matter into an effective oil for specific application. The fluid produced from the pyrolysis system is measured as biooil, which could be used as a diesel substitute just after required processing. In the meantime, some experiments have shown that gas produced from biomass pyrolysis could also be used as oil in IC engines (Gajraj et al., 2018; Sakthivel et al., 2018).

24.3 Feedstock and biofuel competition

The massive rise of the biofuel production deflects a vast number of agricultural production as fuel feedstocks, which, in consequence, impacts the distribution of agricultural land, the balance of the feedstock demand, and the economic production of agriculture in local communities. The major feedstocks for production of biofuel based on generations are depicted in Fig. 24.2. The bioenergy sector is expected to develop dramatically in the upcoming years as well as the effects of the development of the sector mostly on farming sector, and the local economy needs to be studied. That "food versus energy" rivalry was being fiercely observed in recent years (Bai et al., 2012). The research reveals that although second-generation bioenergy could make a substantial contribution to the potential total energy mix, price is a significant obstacle to development projects throughout the close to immediate future. Based on the type of bioenergy, raw material prices, and operational expenses, the price of cellulose ethanol is estimated to also be two to three times greater than the actual fuel equivalent cost of gas. The average income of biodiesel derived from algae, a possible raw material, is seven times greater than the current rate of fuel, because even higher budget plans were published. Compared with the situation of the first-generation biofuels, where raw material may compensate for more than two-thirds of their overall cost, a proportion of raw material in the overall cost of second-generation biofuels is comparatively smaller (30%–50%). Although substantial cost decreases are required for all forms of second-generation biofuels, the crucial obstacles will be at specific phases of development. The prices of converting biomass to ethanol production have to be decreased. The biggest problem for biofuel, on the other side, is the price of raw material (Carriquiry et al., 2011).

Algae-based bioenergy are physically and commercially feasible and cost-effective, need no additional ground, need limited water usage, and minimize CO_2 Emissions. Even so, owing to the low amount of residues and expensive downstream processes, industrial processing of microalgae biodiesel is not yet viable (Suganya et al., 2017). The feasibility of the manufacture

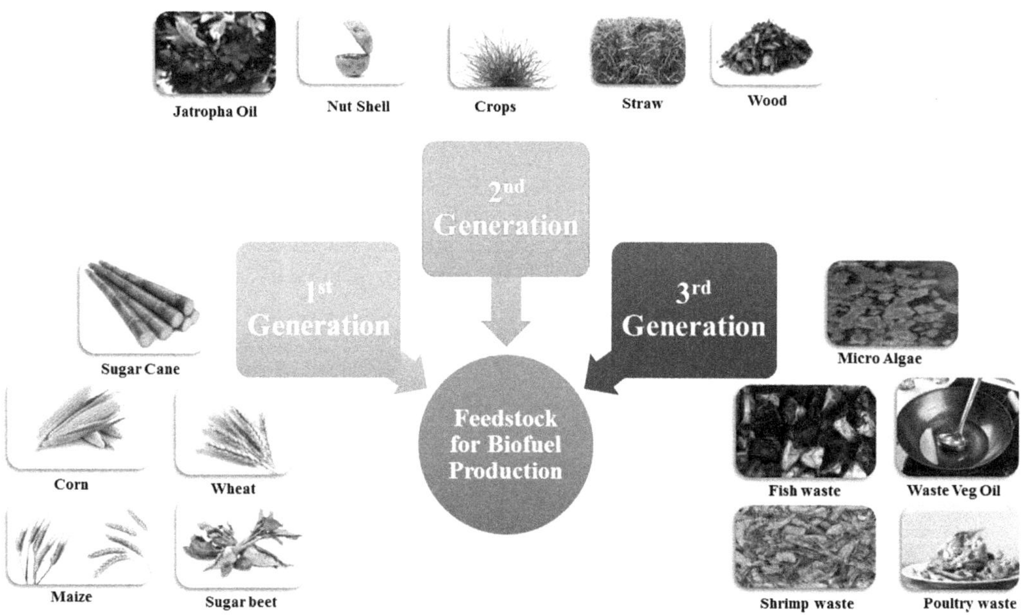

FIGURE 24.2 Various feedstocks for biofuel based on different generations.

of microalgae biodiesel could be accomplished through the construction of sophisticated photobioreactors and the implementation of relatively low processing, drying, and oil extraction techniques. Market development can also be achieved by improving genetic engineering to regulate chronic pollution and designing pathways for high lipids production (Medipally et al., 2015; Miranda et al., 2016).

Due to the fragile price of oil production and biodiesel refining and the uncertainty of algal biomass production, long-term price measures for algal manufacturing must concentrate mostly on manufacturing process of an oil-rich biomass itself. It needs to be addressed by optimizing microbial biology and cultural identity technology. In comparison, using all elements of algae to generate real worth goods in addition to algal fuel such as advanced biorefinery is an attractive means of reducing the cost of production of microbial bioenergy. In fact, microbes constitute a significant fraction of oil, with both the remaining pieces comprising great numbers of carbohydrates, proteins, and other nutrients. This allows residues after oil extraction desirable to be used as livestock feed and in other real worth goods (Wen and Johnson, 2009; Ratnavathi et al., 2011; Chen et al., 2015).

World energy consumption is rising each day, with a large rise in the danger to the atmosphere. Both fuel supply and demand are forecast to expand exponentially next to each other, but the use of fossil fuels is projected to have negative environmental consequences. Each one of these reasons has significantly drawn the interest of scientists to seeking certain alternate solution green energy options. Biodiesel is an excellent example of renewable energy that can be generated using biological life, potentially contributing to a decrease in reliance on fossil fuels (Kour et al., 2019). Bioenergy is also a desirable and viable form of clean energy, in contrast to the global volatility, limited existence, and negative world consequences of coal and oil. Renewable energies are primarily energy-enriched substances, which are produced whether directly by biological systems or via the chemical conversion of organic matter from previous life forms. Biofuels are primarily provided by photosynthesizing species, such as photosynthetic microbes, micro- and macroalgae, and vascular soil plants. Seaweeds are known to become the most desirable origin of biofuel production among all of these species used (Mueller et al., 2011). Bioenergy development is still difficult on a commercial level, and further exploration of new varieties with revenue value is still required (Hoogeveen et al., 2009; Joshi et al., 2017). It would definitely be beneficial to incorporate several genetic engineering techniques to maximize the production of biodiesel. Thus, to resolve the energy shortage, global collaboration activities are quite critical to turn bioenergy into our existing energy infrastructure, which will further promote the advancement of cultivation technique and also technical developments in the manufacture of bioenergy (Van der Horst and Vermeylen, 2011; Kour et al., 2019).

24.4 Technical aspect of biofuel production

Biodiesel is a combination of monoalkyl esters of fatty acids that can be derived from various natural lipid fuel sources and biomass. It could be found specifically in various engines. Researches to investigate microbes as a raw material for the processing of renewable resources started in the 1980s. Microbes are single-cell microscopic species that occur naturally in marine- and freshwater habitats. There are even more over 300,000 algal organisms, many of which are far more diverse than crops. Microbes are typically more effective converters of solar energy relative to plants (Alam et al., 2015). In particular, growing microalgae is much more costly and technically difficult than crop production. Photosynthesizing algal development includes light, carbon dioxide, water, and salts.

The temperature policy must be tightly regulated (Gunasundari and Kumar, 2016; Karthik et al., 2020; Gunasundari et al., 2020). For many of these microalgae cultivations, the temperature is usually between 20 and 30°C. In terms of costs, biofuel should depend on readily available sun, considering frequent and annual fluctuations in sunlight intensity (Behera et al., 2015). The production of bioenergy from microalgae biomass continues in the subsequent steps: (1) algae cultivation, (2) algae processing (or dewatering), (3) the extraction of oil from biomass, (4) the restoration of algae oil, and (5) the conversion of oil into bioenergy (Ho et al., 2014; Ganesan et al., 2020). Fig. 24.3 depicts the typical production of biofuel from algae.

24.4.1 Algae cultivation and processing

Numerous approaches are typically used in the processing of seaweed, including chemically based, mechanically based, biologically based, and, to a smaller degree, electrically driven growth although separate variations or sequences of such approaches are still frequently utilized. The algae average diameter is quite tiny (Pavithra et al., 2020). Chemical flocculation is also mostly done as a pretreatment to maximize the size distribution of algae before using other techniques, such as floating, to extract algae. In a mechanically based procedure, centrifugation, which is the most efficient and fastest method, is being used for the recovery of suspending seaweed (Kröger and Müller-Langer, 2012; Alam et al., 2015; Suparmaniam et al., 2019; Adeniyi et al., 2018).

24.4.2 The extraction of oil from biomass and conversion of bioenergy

Unicellular microbes create a cell wall comprising lipids and fatty acids that vary among larger plants and animals. Alternative techniques of oil exploration from seaweed, such as mechanical and solvent extraction, are available in the literature. Even so, the removal of lipids from microbes is expensive and energy consuming (Gunasundari et al., 2017). Transesterification is a method for transforming microbial oil to biofuel, which requires several reactions

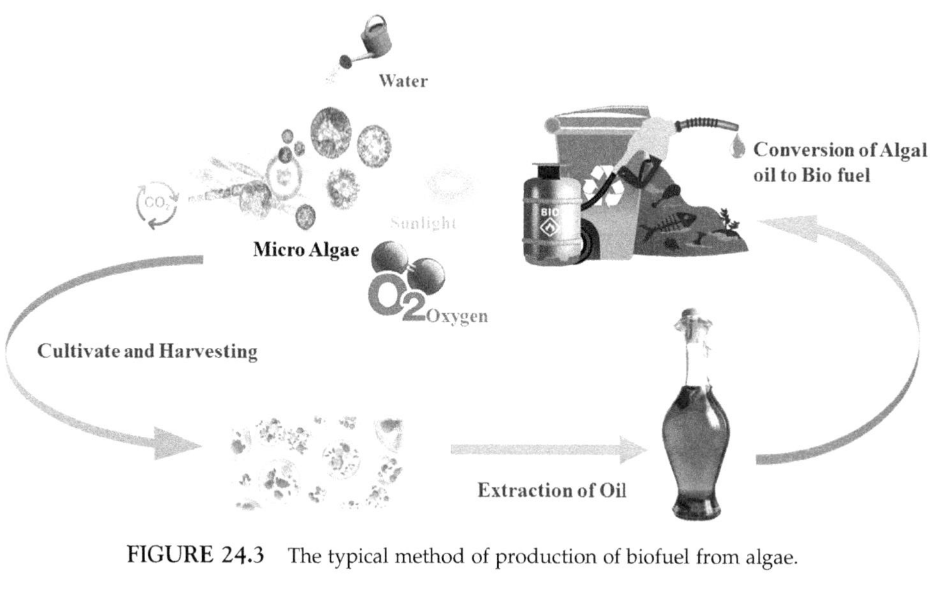

FIGURE 24.3 The typical method of production of biofuel from algae.

among triglycerides or fatty acids and alcohol (Lee et al., 2020). With this reaction, various alcohols such as butanol, ethanol, methanol, propanol, and amyl alcohol could be used. Even so, because of their cheap price and chemical and physical benefits, ethanol and methanol are often used for mass production (Jalilian et al., 2020; Ganesan et al., 2020).

A variety of ways, wherein microbial residue can be transformed into sources of energy, include (1) chemical conversion, (2) biochemical reactions, (3) direct combustion, and (4) thermochemical conversion. Based on the variety of microbes, several substances can also be obtained, with useful applications in various industrial sectors, in terms of production and consumption of biofuels. Bioenergy processing is a dynamic process. Even so, such methods are complicated and technically demanding in terms of payment (Um and Kim, 2009; Han et al., 2015; Rodionova et al., 2017). Table 24.1 summarizes the production of biodiesel from different feedstock.

24.5 Life cycle assessment of biofuel production

Life cycle assessment is among the most appropriate methods for uniform methodology for the evaluation of the viability of bioenergy.

TABLE 24.1 Production of biodiesel from different feedstocks.

Feedstock	Methanol-to-oil ratio	Reaction time	Yield of fatty acid methyl esters	Method	Flash point	Acid value	References
Fish waste	9:1	90 min	95.39%	Alkaline transesterification	–	–	García-Moreno et al. (2014)
Waste cooking oil	20:1	4 h	97.22%	Two-step catalysis process: (1) Ferric sulfate–catalyzed methanolysis (2) Alkaline transesterification	–	–	Wang et al. (2006, 2007)
Waste cooking oils	6:1	40 min	–	Sonochemical reactors	182°C	2.805 mg KOH/g oil	Hingu et al. (2010)
Waste cooking oils	–	–	–	Supercritical transesterification	485 K	1.32 mg KOH/g oil	Demirbas (2009)
Waste cooking oils	1:1	–	91.08%	Immobilized lipase in fixed-bed reactor	195°C	0.12 mg KOH/g oil	Chen et al. (2009)
Spirulina,	–	1 h	41.21%	Oil extraction followed by transesterification	130°C	0.45 mg KOH/g oil	Nautiyal et al. (2014)
Pond water algae	–	1 h	38.39%	Oil extraction followed by transesterification	–	0.40 mg KOH/g oil	Nautiyal et al. (2014)
Spirogyra	3:1	24 h	>95%	Oil extraction followed by transesterification	140°C	–	Ahmed et al. (2015)
Schizochytrium limacinum	–	2.25 h.	66.37%	Oil extraction followed by transesterification	204°C	0.11 mg KOH/g oil	Johnson and Wen (2009)

Even so, there have been no specific criteria for the life cycle evaluation of bioenergy applications. The reported life cycle evaluation uses a variety of methods, various network boundaries, effect definitions and organizational departments, various distribution methods, conclusions about by-products and coproducts, and also varying proportions structures to which bioenergy system is contrasted (Czyrnek-Delêtre et al., 2017). This life cycle appraisal addressed the energy requirement and ecological consequences of the production and refining of macroalgae into bioethanol and biogas, with a special emphasis on various species and cultivation (Singh et al., 2010; Sills et al., 2013; Aitken et al., 2014).

Use of seaweed for bioenergy processing is projected to play a significant role in ensuring power generation in the upcoming years. Follow-up life cycle assessments as well as a power study of the development of marine algae have been conducted out under Northern European conditions to record and enhance the efficiency of the operation. Different alternatives were studied for the development of brown algae, respectively, biogas (case 1) and bioethanol and biogas (Case 2). Extreme pollution, acidification, and terrestrial eutrophication were the possible environmental effects classes under review. The development of algae was described as the most environmentally damaging stage. Case 1 gives promising performance relative to case 2 with all effect groups, partially due to power bioethanol separation process and thus reduced average device reliability. To enhance sustainability impact, the emphasis must be on maximizing the manufacturing of seaweed, distilling bioethanol, and controlling digestion on property (Alvarado-Morales et al., 2013; Peters et al., 2015).

The extension of farmland used during food crops generates directly and indirectly complex greenhouse gases as well as other fiscal, environmental, and social impacts. We examine the use of life cycle analysis to quantify the carbon emissions of biofuel output from indirect land transition. There are two critical methods: straightforward, subjective life cycle analysis, and consequential life cycle analysis (Kauffman et al., 2011). The suggested composite "coupled paradigm" of the two approaches to indirect land use change study is based on the first description of the device border of the subsequent complete life cycle analysis. Decisions then are taken with regard to the simulation approach (economic balance or cause–effect), statistical procedures, total land assessment, ecosystem carbon reporting, and complexity assessment to be used. We concluded that a follow-up life cycle study is appropriate to the calculation of historical pollution from indirect land use transition, while changes to the based scheme solution, combined with continuous adjustments, are needed, and unknown values need to be appropriately defined; furthermore, the degree and scope of the extension of the control limits are needed for the consequential life of the system (Sanchez et al., 2012; Li et al., 2020).

The life cycle analysis for biodiesel fuel was performed to examine and verify the common belief that palm biodiesel is a renewable and sustainable fuel. The life cycle assessment analysis was classified into three main categories, including farming operations, oil puttering, and the method of transesterification of producing biodiesel. The energy balance and greenhouse gas evaluations were introduced and addressed at each point. These are essential data for the socioeconomic and environmental viability assessment of palm biodiesel. The use of biodiesel fuel was observed to produce a power density ratio of 3.53 (output energy/input energy) showing the positive net energy produced and maintaining its sustainable development. The power proportion for palm biodiesel was observed to become more than twice it for rapeseed biodiesel, which were calculated to be just 1.44, suggesting that oil palm would have been a more sustainable feedstock relative to rapeseed oil. In comparison, palm biodiesel combustion has been found to be more environmentally sustainable than petroleum-derived diesel, as a substantial 38% drop in emissions of CO_2 per liter of fuel can be obtained (Yee et al., 2009).

24.6 Future prospects

In view of the existing state-of-the-art technology, biofuel cannot be a complete substitute of fossil fuels at least in the near run. Clearly, it has been shown that microalgae bioenergy is pure, sustainable, and green, and it is also costly relative to coal and oil. However, to compete with, or even replace, coal fuel, large-scale production must be planned at an equal cost. In this context, revolutionary strategies are required to establish an integrated multiple cultivation of microbes at a fair cost. Diverse problems exist in the processing of biofuel through algal biomass feedstock. Present statistics show that the cost of processing bioenergy from microalgae biomass is still greater than most other suppliers. Microalgae have not yet been seen to be a financially feasible origin of bioenergy. The harvesting methods such as biological, chemical, mechanical, or electrical techniques are still costly. Further scientific research is required to make such harvesting processes more reliable and to reduce the price costs in the future.

24.7 Conclusion

The promise of third-generation raw material for biofuel production was assessed in this chapter, along with its important technological aspects of production, biofuel life cycle assessment, and possible commercialization. As far as third-age biofuels are worried, they are only in the initial process of development. After such a target, the monetary safety of its employees of such phases would be pursued. Commercial exploitation involves launching experiment cycles to improve efficiency and economic productivity. It is ecologically responsible and far less contaminating, making it an energy resource, which is more desirable. These results are focused on studies of bioenergy growth, multiple sources, new processing methods, and a wide range of feedstocks used in the production of biofuels. In particular, biofuel derived from algae is perhaps the most productive choice in terms of farmland utilization, since it does not need farmland for the production of biomass and prevents conflict with crop production. Especially in comparison with many other bioenergy sources, certain biodiesel production routes may also work satisfactorily in respect of greenhouse gases, but others could be much worse than conventional diesel. Nonetheless, algae biofuels do not contend with the other bioenergy or coal and gas in terms of energy consumption.

Acknowledgment

The authors express their gratitude to Management of Sri Sivasubramaniya Nadar College of Engineering, Chennai, and St Joseph's College of Engineering, Chennai.

References

Aboelazayem, O., Gadalla, M., Saha, B., 2018. Valorisation of high acid value waste cooking oil into biodiesel using supercritical methanolysis: experimental assessment and statistical optimisation on typical Egyptian feedstock. Energy 162, 408–420.

Adeniyi, O.M., Azimov, U., Burluka, A., 2018. Algae biofuel: current status and future applications. Renew. Sustain. Energy Rev. 90, 316–335.

Ahmed, I., Ali, M., Ahmad, N., Ahmad, I., 2015. Production of biodiesel from algae. J. Pure Appl. Microbiol. 9 (1), 79–85.

Aitken, D., Bulboa, C., Godoy-Faundez, A., Turrion-Gomez, J.L., Antizar-Ladislao, B., 2014. Life cycle assessment of macroalgae cultivation and processing for biofuel production. J. Clean. Prod. 75, 45–56.

Alam, F., Mobin, S., Chowdhury, H., 2015. Third generation biofuel from algae. Proc. Eng. 105, 763–768.

Alvarado-Morales, M., Boldrin, A., Karakashev, D.B., Holdt, S.L., Angelidaki, I., Astrup, T., 2013. Life cycle assessment of biofuel production from brown seaweed in Nordic conditions. Bioresour. Technol. 129, 92–99.

Aravind, S., Kumar, P.S., Kumar, N.S., Siddarth, N., 2020. Conversion of green algal biomass into bioenergy by pyrolysis. A review. Environ. Chem. Lett. 18 (3), 829–849.

Bai, Y., Ouyang, Y., Pang, J.S., 2012. Biofuel supply chain design under competitive agricultural land use and feedstock market equilibrium. Energy Econ. 34 (5), 1623–1633.

Behera, S., Singh, R., Arora, R., Sharma, N.K., Shukla, M., Kumar, S., 2015. Scope of algae as third generation biofuels. Front. Bioeng. Biotechnol. 2, 90.

Bhatia, L., Bachheti, R.K., Garlapati, V.K., Chandel, A.K., 2020. Third-generation biorefineries: a sustainable platform for food, clean energy, and nutraceuticals production. Biomass Convers. Biorefin. 1–16.

Callegari, A., Bolognesi, S., Cecconet, D., Capodaglio, A.G., 2020. Production technologies, current role, and future prospects of biofuels feedstocks: a state-of-the-art review. Crit. Rev. Environ. Sci. Technol. 50 (4), 384–436.

Carriquiry, M.A., Du, X., Timilsina, G.R., 2011. Second generation biofuels: economics and policies. Energy Policy 39 (7), 4222–4234.

Chen, G., Zhao, L., Qi, Y., 2015. Enhancing the productivity of microalgae cultivated in wastewater toward biofuel production: a critical review. Appl. Energy 137, 282–291.

Chen, Y., Xiao, B., Chang, J., Fu, Y., Lv, P., Wang, X., 2009. Synthesis of biodiesel from waste cooking oil using immobilized lipase in fixed bed reactor. Energy Convers. Manage. 50 (3), 668–673.

Chowdhury, H., Loganathan, B., 2019. Third-generation biofuels from microalgae: a review. Curr. Opin. Green Sustain. Chem. 20, 39–44.

Czyrnek-Delêtre, M.M., Smyth, B.M., Murphy, J.D., 2017. Beyond carbon and energy: the challenge in setting guidelines for life cycle assessment of biofuel systems. Renew. Energy 105, 436–448.

Demirbas, A., 2009. Biodiesel from waste cooking oil via base-catalytic and supercritical methanol transesterification. Energy Convers. Manage. 50 (4), 923–927.

Demirbas, A., 2011. Competitive liquid biofuels from biomass. Appl. Energy 88 (1), 17–28.

Foteinis, S., Chatzisymeon, E., Litinas, A., Tsoutsos, T., 2020. Used-cooking-oil biodiesel: life cycle assessment and comparison with first-and third-generation biofuel. Renew. Energy 153, 588–600.

Gajraj, R.S., Singh, G.P., Kumar, A., 2018. Third-generation biofuel: algal biofuels as a sustainable energy source. Biofuels: Greenh. Gas Mitig. Global Warm. 307–325.

Ganesan, R., Manigandan, S., Samuel, M.S., Shanmuganathan, R., Brindhadevi, K., Chi, N.T.L., Duc, P.A., Pugazhendhi, A., 2020. A review on prospective production of biofuel from microalgae. Biotechnol. Rep. e00509.

García-Moreno, P.J., Khanum, M., Guadix, A., Guadix, E.M., 2014. Optimization of biodiesel production from waste fish oil. Renew. Energy 68, 618–624.

Gunasundari, E., Kumar, P.S., 2016. Higher adsorption capacity of Spirulina platensis alga for Cr (VI) ions removal: parameter optimisation, equilibrium, kinetic and thermodynamic predictions. IET Nanobiotechnol. 11 (3), 317–328.

Gunasundari, E., 2017. Adsorption isotherm, kinetics and thermodynamic analysis of Cu (II) ions onto the dried algal biomass (*Spirulina platensis*). J. Ind. Eng. Chem. 56, 129–144.

Gunasundari, E., Kumar, P.S., Christopher, F.C., Arumugam, T., Saravanan, A., 2017. Green synthesis of metal nanoparticles loaded ultrasonic-assisted *Spirulina platensis* using algal extract and their antimicrobial activity. IET Nanobiotechnol. 11 (6), 754–758.

Gunasundari, E., Kumar, P.S., Rajamohan, N., Vellaichamy, P., 2020. Feasibility of naphthol green-B dye adsorption using microalgae: thermodynamic and kinetic analysis. Desalin. Water Treat. 192, 358–370.

Han, S.F., Jin, W.B., Tu, R.J., Wu, W.M., 2015. Biofuel production from microalgae as feedstock: current status and potential. Crit. Rev. Biotechnol. 35 (2), 255–268.

Hingu, S.M., Gogate, P.R., Rathod, V.K., 2010. Synthesis of biodiesel from waste cooking oil using sonochemical reactors. Ultrason. Sonochem. 17 (5), 827–832.

Ho, S.H., Ye, X., Hasunuma, T., Chang, J.S., Kondo, A., 2014. Perspectives on engineering strategies for improving biofuel production from microalgae-a critical review. Biotechnol. Adv. 32 (8), 1448–1459.

Hoogeveen, J., Faurès, J.M., Van de Giessen, N., 2009. Increased biofuel production in the coming decade: to what extent will it affect global freshwater resources? Irrig. Drain. 58 (S1), S148–S160.

Hosseini, S.E., Wahid, M.A., 2012. Necessity of biodiesel utilization as a source of renewable energy in Malaysia. Renew. Sustain. Energy Rev. 16 (8), 5732–5740.

Jalilian, N., Najafpour, G.D., Khajouei, M., 2020. Macro and micro algae in pollution control and biofuel production—a review. ChemBioEng. Rev. 7 (1), 18–33.

Jambo, S.A., Abdulla, R., Azhar, S.H.M., Marbawi, H., Gansau, J.A., Ravindra, P., 2016. A review on third generation bioethanol feedstock. Renew. Sustain. Energy Rev. 65, 756–769.

Jeswani, H.K., Chilvers, A., Azapagic, A., 2020. Environmental sustainability of biofuels: a review. Proc. R. Soc. A. 476 (2243), 20200351.

Johnson, M.B., Wen, Z., 2009. Production of biodiesel fuel from the microalga *Schizochytrium limacinum* by direct transesterification of algal biomass. Energy Fuel 23 (10), 5179–5183.

Joshi, G., Pandey, J.K., Rana, S., Rawat, D.S., 2017. Challenges and opportunities for the application of biofuel. Renew. Sustain. Energy Rev. 79, 850–866.

Karthik, V., Kumar, P.S., Vo, D.V.N., Sindhu, J., Sneka, D., Subhashini, B., Saravanan, K., Jeyanthi, J., 2020. Hydrothermal production of algal biochar for environmental and fertilizer applications: a review. Environ. Chem. Lett. 1–18.

Kauffman, N., Hayes, D., Brown, R., 2011. A life cycle assessment of advanced biofuel production from a hectare of corn. Fuel 90 (11), 3306–3314.

Kour, D., Rana, K.L., Yadav, N., Yadav, A.N., Rastegari, A.A., Singh, C., Negi, P., Singh, K., Saxena, A.K., 2019. Technologies for biofuel production: current development, challenges, and future prospects. In: Prospects of Renewable Bioprocessing in Future Energy Systems, pp. 1–50.

Kröger, M., Müller-Langer, F., 2012. Review on possible algal-biofuel production processes. Biofuels 3 (3), 333–349.

Kumar, P.S., Pavithra, J., Suriya, S., Ramesh, M., Kumar, K.A., 2015. *Sargassum wightii*, a marine alga is the source for the production of algal oil, bio-oil, and application in the dye wastewater treatment. Desalin. Water Treat. 55 (5), 1342–1358.

Lee, S.Y., Khoiroh, I., Vo, D.V.N., Kumar, P.S., Show, P.L., 2020. Techniques of lipid extraction from microalgae for biofuel production: a review. Environ. Chem. Lett. 1–21.

Li, G., Lu, Z., Zhang, J., Li, H., Zhou, Y., Zayan, A.M.I., Huang, Z., 2020. Life cycle assessment of biofuel production from microalgae cultivated in anaerobic digested wastewater. Int. J. Agric. Biol. Eng. 13 (1), 241–246.

Liu, Z., Wang, K., Chen, Y., Tan, T., Nielsen, J., 2020. Third-generation bio refineries as the means to produce fuels and chemicals from CO_2. Nat. Catal. 3 (3), 274–288.

Maity, J.P., Bundschuh, J., Chen, C.Y., Bhattacharya, P., 2014. Microalgae for third generation biofuel production, mitigation of greenhouse gas emissions and wastewater treatment: present and future perspectives—a mini review. Energy 78, 104–113.

Manikandan, G., Kumar, P.S., Prakalathan, R., 2014. Comparison of biodiesel production from macro and micro algae. Int. J. ChemTech Res. 6 (9), 4143–4147.

Medipally, S.R., Yusoff, F.M., Banerjee, S., Shariff, M., 2015. Microalgae as sustainable renewable energy feedstock for biofuel production. Biomed. Res. Int. 2015, 519513 https://doi.org/10.1155/2015/519513.

Millinger, M., Ponitka, J., Arendt, O., Thrän, D., 2017. Competitiveness of advanced and conventional biofuels: results from least-cost modelling of biofuel competition in Germany. Energy Policy 107, 394–402.

Miranda, A.F., Biswas, B., Ramkumar, N., Singh, R., Kumar, J., James, A., Roddick, F., Lal, B., Subudhi, S., Bhaskar, T., Mouradov, A., 2016. Aquatic plant Azolla as the universal feedstock for biofuel production. Biotechnol. Biofuels 9 (1), 1–17.

Mueller, S.A., Anderson, J.E., Wallington, T.J., 2011. Impact of biofuel production and other supply and demand factors on food price increases in 2008. Biomass Bioenergy 35 (5), 1623–1632.

Murugesan, A., Umarani, C., Subramanian, R., Nedunchezhian, N., 2009. Bio-diesel as an alternative fuel for diesel engines-a review. Renew. Sustain. Energy Rev. 13 (3), 653–662.

Nautiyal, P., Subramanian, K.A., Dastidar, M.G., 2014. Production and characterization of biodiesel from algae. Fuel Process. Technol. 120, 79–88.

Othman, M.F., Adam, A., Najafi, G., Mamat, R., 2017. Green fuel as alternative fuel for diesel engine: a review. Renew. Sustain. Energy Rev. 80, 694–709.

Pardo, Y., Kafarov, V., Acevedo, P., 2010. Life cycle assessment of third generation biofuels production based on "cradle to cradle" methodology. Chem. Eng. Trans. 21, 1177–1182.

Pavithra, K.G., Kumar, P.S., Jaikumar, V., Vardhan, K.H., SundarRajan, P., 2020. Microalgae for biofuel production and removal of heavy metals: a review. Environ. Chem. Lett. 1–19.

Peters, J.F., Iribarren, D., Dufour, J., 2015. Simulation and life cycle assessment of biofuel production via fast pyrolysis and hydroupgrading. Fuel 139, 441–456.

Rajamohan, N., Kumar, P.S., Rajasimman, M., Al Qasmi, F., 2021. Treatment of methanol industry effluent using algal biomass, *Gelidium omanense*-kinetic modeling. Chem. Eng. J. Adv. 5, 100068.

Ratnavathi, C.V., Chakravarthy, S.K., Komala, V.V., Chavan, U.D., Patil, J.V., 2011. Sweet sorghum as feedstock for biofuel production: a review. Sugar Technol. 13 (4), 399–407.

Rodionova, M.V., Poudyal, R.S., Tiwari, I., Voloshin, R.A., Zharmukhamedov, S.K., Nam, H.G., Zayadan, B.K., Bruce, B.D., Hou, H.J.M., Allakhverdiev, S.I., 2017. Biofuel production: challenges and opportunities. Int. J. Hydrog. Energy 42 (12), 8450–8461.

Sadatshojaei, E., Wood, D.A., Mowla, D., 2020. Third generation of biofuels exploiting microalgae. In: Sustainable Green Chemical Processes and their Allied Applications, pp. 575–588.

Sakthivel, R., Ramesh, K., Purnachandran, R., Shameer, P.M., 2018. A review on the properties, performance and emission aspects of the third generation biodiesels. Renew. Sustainan. Energy Rev. 82, 2970–2992.

Sanchez, S.T., Woods, J., Akhurst, M., Brander, M., O'Hare, M., Dawson, T.P., Edwards, R., Liska, A.J.,

Malpas, R., 2012. Accounting for indirect land-use change in the life cycle assessment of biofuel supply chains. J. R. Soc. Interface 9 (71), 1105–1119.

Saravanan, A., Kumar, P.S., Varjani, S., Jeevanantham, S., Yaashikaa, P.R., Thamarai, P., Abirami, B., George, C.S., 2021. A review on algal-bacterial symbiotic system for effective treatment of wastewater. Chemosphere 129540.

Sills, D.L., Paramita, V., Franke, M.J., Johnson, M.C., Akabas, T.M., Greene, C.H., Tester, J.W., 2013. Quantitative uncertainty analysis of life cycle assessment for algal biofuel production. Environ. Sci. Technol. 47 (2), 687–694.

Singh, A., Olsen, S.I., Nigam, P.S., 2011. A viable technology to generate third-generation biofuel. J. Chem. Technol. Biotechnol. 86 (11), 1349–1353.

Singh, A., Pant, D., Korres, N.E., Nizami, A.S., Prasad, S., Murphy, J.D., 2010. Key issues in life cycle assessment of ethanol production from lignocellulosic biomass: challenges and perspectives. Bioresour. Technol. 101 (13), 5003–5012.

Suganya, S., Saravanan, A., Senthil Kumar, P., Yashwanthraj, M., Sundar Rajan, P., Kayalvizhi, K., 2017. Sequestration of Pb (II) and Ni (II) ions from aqueous solution using microalga *Rhizoclonium hookeri*: adsorption thermodynamics, kinetics, and equilibrium studies. J. Water Reuse Desalin. 7 (2), 214–227.

Sun, C.H., Fu, Q., Liao, Q., Xia, A., Huang, Y., Zhu, X., Reungsang, A., Chang, H.X., 2019. Life-cycle assessment of biofuel production from microalgae via various bioenergy conversion systems. Energy 171, 1033–1045.

Suparmaniam, U., Lam, M.K., Uemura, Y., Lim, J.W., Lee, K.T., Shuit, S.H., 2019. Insights into the microalgae cultivation technology and harvesting process for biofuel production: a review. Renew. Sustain. Energy Rev. 115, 109361.

Um, B.H., Kim, Y.S., 2009. A chance for Korea to advance algal-biodiesel technology. J. Ind. Eng. Chem. 15 (1), 1–7.

Van der Horst, D., Vermeylen, S., 2011. Spatial scale and social impacts of biofuel production. Biomass Bioenergy 35 (6), 2435–2443.

Wang, Y., Ou, S., Liu, P., Zhang, Z., 2007. Preparation of biodiesel from waste cooking oil via two-step catalyzed process. Energy Convers. Manage. 48 (1), 184–188.

Wang, Y., Ou, S., Liu, P., Xue, F., Tang, S., 2006. Comparison of two different processes to synthesize biodiesel by waste cooking oil. J. Mol. Catal. A: Chem. 252 (1–2), 107–112.

Wen, Z., Johnson, M.B., 2009. Microalgae as a Feedstock for Biofuel Production.

Yee, K.F., Tan, K.T., Abdullah, A.Z., Lee, K.T., 2009. Life cycle assessment of palm biodiesel: revealing facts and benefits for sustainability. Appl. Energy 86, S189–S196.

Index

Note: 'Page numbers followed by "f" indicate figures and "t" indicate tables.'

C

N

9780323852692